INTEGRATED CIRCUITS APPLICATIONS HANDBOOK

THE WILEY ELECTRICAL AND ELECTRONICS TECHNOLOGY HANDBOOK SERIES

The Wiley Electrical and Electronics Technology Handbook Series provides technicians and engineers with up-to-date information on a wide range of topics in a rapidly evolving field. Every chapter in each volume is written by an authority on the subject, and takes a practical, how-to-do-it approach with numerous worked-out examples and step-by-step calculations. These procedures provide productive approaches to many problems encountered in electronics and electrical technology.

TOPICS OF FORTHCOMING HANDBOOKS

Op Amps
Active Filters
Digital Circuits
Microprocessors and Microcomputers
Communications
Electrical Machines
Automatic Control Systems
System Troubleshooting
Electrical and Process Measurements
Semiconductor Memories
Microwave Measurements

Arthur H. Seidman
Series Editor

INTEGRATED CIRCUITS APPLICATIONS HANDBOOK

Arthur H. Seidman
Pratt Institute

JOHN WILEY & SONS
New York • Chichester • Brisbane • Toronto • Singapore

Library of Congress Cataloging in Publication Data:

Integrated circuits applications handbook.

(Wiley electrical and electronic technology handbook series)
Includes bibliographies and index.
1. Integrated circuits–Handbooks, manuals, etc.
I. Seidman, Arthur H. II. Series.

TK7874.I546 1983 621.381′73 82-10903
ISBN 0-471-07765-8

Printed in the United States of America

10 9 8 7 6 5 4 3 2 1

In Memory of Our
Dear Son, Ben

PREFACE

The *IC applications handbook* is designed for electronic technicians, technologists, engineers, scientists, and students. Theory is minimized, applications are emphasized, and the math used is essentially limited to arithmetic and elementary algebra.

Because the IC field has mushroomed into many specialties, no single individual can be an expert on all the different types of integrated circuitry available today and their applications. A distinctive feature of this handbook is that each chapter is written by an expert. The contributors are affiliated with such organizations as Motorola, Fairchild, RCA, Signetics, and Texas Instruments.

Another distinctive feature of the work is its practical orientation. Over 200 worked-out examples, accompanied by step-by-step solution procedures and numerous reference tables, are provided. For example, this approach enables the reader to calculate the fan out of a T^2L gate or the output of a summing amplifier, to select a suitable IC for an FM IF strip, or to design an active low pass filter.

The first section of the handbook deals with digital ICs. Topics covered include T^2L, I^2L, ECL, MOS/CMOS logic families, charge coupled devices, semiconductor and bubble memories, and the microprocessor.

Linear ICs are the concern in the second section of the book. Topics considered are op amps, active filters, waveform generators, analog-to-digital and digital-to-analog converters, communications ICs, voltage regulators, interfacing circuits, and phase locked loops.

The concluding section treats thin film, thick film, and IC fabrication technologies.

I gratefully acknowledge the fine efforts and cooperation of the various contributors who helped to make this handbook a reality. Thanks also to the reviewers whose suggestions helped improve the presentation of material. Because of the broad coverage of the book, it is inevitable that some errors went undetected. I will be appreciative if any remaining errors are called to my attention.

It has been a pleasure working with a splendid staff at Wiley. Two individuals in the Engineering Technology Group deserve special acknowledgment. I am grateful to Alan Lesure, Executive Editor, for his faith and support in launching a series of handbooks in electrical and electronics technology, of which this is the first published volume. The assistance, cooperation, and wisdom furnished by my editor, Judy Green, helped to make the editing of this book bearable and, at times, even fun.

Arthur H. Seidman

CONTENTS

INTEGRATED CIRCUITS APPLICATIONS HANDBOOK

PART ONE

DIGITAL INTEGRATED CIRCUITS

CHAPTER 1

Transistor Transistor Logic

Howard R. Cohen
Digital Design Engineer
Norden Systems

1.1 INTRODUCTION

This chapter covers transistor transistor logic (TTL, T^2L) circuits: their function, electrical performance, and applications. Throughout, many examples are presented to emphasize the practical implementation of this logic family. TTL produces the highest performance-to-cost ratio of all logic types. This makes it the most widely used logic family. It provides the user with a large variety of logic functions for different operating (temperature, speed, and power) requirements. TTL was introduced by Texas Instruments as a standard product line in 1964.

A partial listing of available TTL functions is provided in Table 1.1. The prefix 54 designates the operating temperature range from −55 to +125°C; prefix 74 devices cover the temperature range of 0 to +70°C. For example, a 5400 quad 2-input NAND gate operates over the temperature range of −55 to +125°C; a 7400 device will operate only from 0 to +70°C.

There are different members in the TTL family. These include standard, high-speed (HTTL), Schottky (STTL), and low-power Schottky (LSTTL). The differences between these types will be discussed later in this chapter.

Table 1.1 **A partial list of TTL functions**

Circuit	*Part number*
Quad 2-input NAND gate	54/7400
Quad 2-input NOR gate	54/7402
Quad 2-input AND gate	54/7408
Triple 3-input NOR gate	54/7427
Quad 2-input OR gate	54/7432
BCD/decimal decoder/driver	54/7445
JK flip flop	54/7470
4-bit binary full adder	54/7483
Decade counter	54/7490
4-bit shift register	54/7494

1.2 BASIC LOGIC FUNCTIONS

The basic logic functions and operations are the AND, OR, and the NOT (Invert) functions. Positive logic will be assumed; that is, the voltage representing a logic 1 is greater than the voltage representing a logic 0. The opposite is true for negative logic.

In Fig. 1.1*a*, the AND function is a logic 1 if all logic input variables are 1s. If we consider two logic variables A and B, $Y = AB$ can be represented by the truth table in Fig. 1.1*a*. The truth table shows that output Y is a logic 1 only when input variables *A and B* are logic 1s.

The OR function is a logic 1 if at least one of the inputs is a logic 1. For two inputs, A and B, $Y = A + B$ can be represented by the truth table in Fig. 1.1*b*. The truth table shows that output Y is a logic 1 when one *or* both inputs are a logic 1.

For the NOT function (Fig. 1.1*c*), if the input is a logic 1, the output is a logic 0, and vice versa.

Logic functions also can be implemented using NAND and NOR gates. The truth tables and logic symbols for these gates are provided in Figs. 1.1*d* and *e*, respectively. In the NAND gate of Fig. 1.1*d*, for two inputs A and B, output Y is a 1 only if one or both inputs are a logic 0. If $A = B = 1$, output $Y = 0$.

Referring to the truth table of the NOR gate in Fig. 1.1*e*, we find that output $Y = 1$ only when both inputs are logic 0s. If one or both inputs are at a logic 1, output $Y = 0$.

Another gate that is used is the exclusive-OR of Fig. 1.1*f*. Referring to its truth table, $Y = 1$ only if either A or B is a 1. If both inputs are the same (0 or 1), $Y = 0$.

1.3 BASIC TTL OPERATION

Operation of the standard TTL NAND gate of Fig. 1.2 will be examined. Referring to Fig. 1.1*d*, all inputs must be a logic 1 in order that the output be a

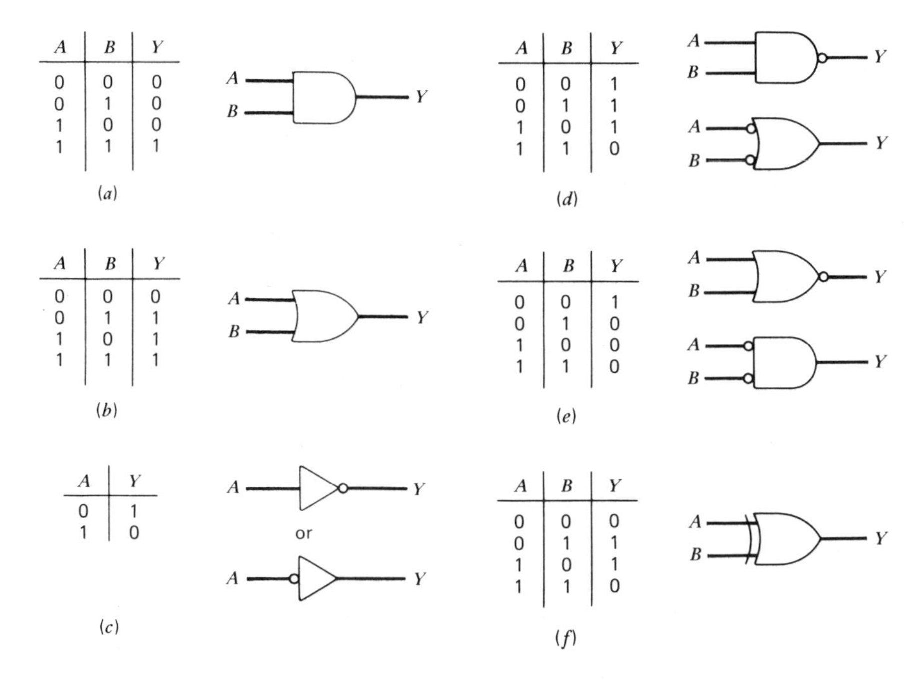

(a)

A	B	Y
0	0	0
0	1	0
1	0	0
1	1	1

(b)

A	B	Y
0	0	0
0	1	1
1	0	1
1	1	1

(c)

A	Y
0	1
1	0

(d)

A	B	Y
0	0	1
0	1	1
1	0	1
1	1	0

(e)

A	B	Y
0	0	1
0	1	0
1	0	0
1	1	0

(f)

A	B	Y
0	0	0
0	1	1
1	0	1
1	1	0

Figure 1.1 Truth tables and logic symbols for basic functions. (*a*) AND. (*b*) OR. (*c*) NOT (inverter). (*d*) NAND. (*e*) NOR. (*f*) Exclusive-OR.

logic 0. Any input at a logic 0 produces a logic 1 at the output. The TTL voltage threshold for a logic 0 is defined as 0.6 V maximum, and for a logic 1 it is 1.3 V minimum.

If the voltage at input emitter A or B of the multiple-emitter transistor Q_1 is less than 0.6 V (logic 0), transistors Q_1 and Q_3 are turned off. This causes transistor Q_4 to conduct, resulting in a logic 1 output. The logic 1 output voltage level is related to the forward voltage drop across diode D, the base-emitter voltage of Q_4, and the current supplied to the load, which results in a voltage drop across R_4. Subtracting these items from V_{CC}, which is typically 5 V, defines the logic 1 output voltage. Transistors Q_3 and Q_4 constitute a *totem-pole output stage*.

When the voltages at inputs A and B are greater than 1.3 V (logic 1), transistors Q_2 and Q_3 conduct, Q_4 is off, and the output is at logic 0. The logic 0 output voltage is determined by the amount of current Q_3 must sink from the load and the saturation resistance of Q_3. In practice, a guaranteed maximum voltage of 0.4 V is specified while sinking 16 mA of current.

The most common application for a logic gate is to drive another similar logic gate. The input and output logic levels, therefore, must be compatible. Listed in Table 1.2 are typical input/output logic levels for Series 54/74 TTL. The circuit designer must distinguish the different load characteristics of various circuits. Before finalizing a design, the *fan out* and *fan in* loading must be considered.

Table 1.2 **Input/output logic levels for TTL**

V_{IL}	The maximum voltage level for a logic 0 at an input. Its guaranteed maximum is 0.8 V.
V_{IH}	The minimum voltage level for a logic 1 at an input. Its guaranteed minimum is 2.0 V.
V_{OL}	The maximum voltage level for a logic 0 at an output. Its guaranteed maximum is 0.4 V.
V_{OH}	The minimum voltage level for a logic 1 at an output. Its guaranteed minimum is 2.4 V.

Fan in and Fan out

Fan in is defined as the amount of input current a driven gate loads a driving gate. Knowledge of the input characteristics of a TTL gate is necessary to utilize these devices. An input parameter that must be specified is the input current for the logic 1 state. When both inputs of a TTL gate are at a logic 1, the emitter-base junctions of Q_1 (Fig. 1.2) are reverse biased and the collector-base junction is forward biased.

Input current is a function of the leakage current of Q_1. When reverse biased, the emitter acts like the collector and the collector becomes the emitter. Current flow from the emitter to collector, therefore, determines the logic 1 load current in an input terminal. Each input of a multiple-emitter TTL gate represents a unit load (U.L.) of 1 at an input current of 40 μA, or less, at a voltage of 2.4 V for a logic 1. For a logic 0, the emitter current is 1.6 mA.

Fan out is defined as the amount of output current a gate can deliver in the logic 0 or 1 states. In the logic 1 state, the driving gate must source current

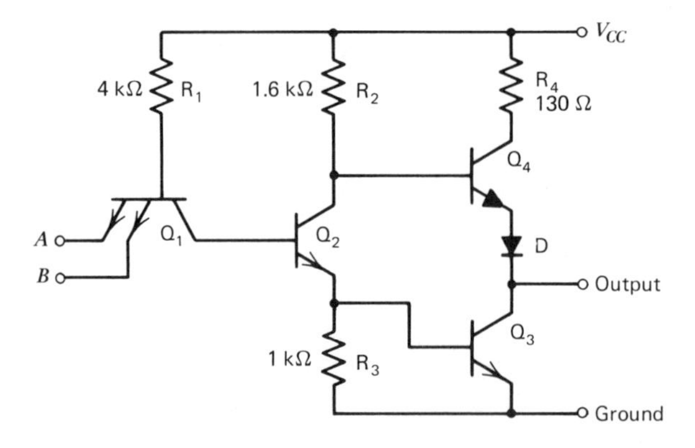

Figure 1.2 Circuit diagram of a standard TTL 2-input NAND gate.

to the load; in the logic 0 state, the driving gate must sink current from the load. If a gate input has a unit load of 1, the driving gate must source 40 μA in the logic 1 state and 1.6 mA in the logic 0 state.

For example, the loading imposed by the input of a gate is specified as 2 unit loads. The device that can drive 10 unit loads could then be loaded with up to five of these 2 unit load inputs, with ten of 1 unit load inputs, or with any other combination that keeps the total loading at 10 unit loads or less.

EXAMPLE 1.1 Fig. 1.3 shows a $\overline{\text{RESET}}$ line distributed to two clear (CLR) inputs of a 5474 (dual D-type positive-edge triggered flip flop), one clear input of a 54109 (dual JK positive-edge triggered flip flop), and the master reset (MR) of a 54161 (synchronous 4-bit counter). Table 1.3 lists the logic 1 and 0 input currents for each of the chips. Calculate the fan in and fan out, and select a suitable driving gate.

PROCEDURE

1 By listing the number of inputs tied to the $\overline{\text{RESET}}$ line, the fan in and fan out loading may be calculated.
2 The calculation for logic 1 loading is:

$$\begin{aligned}
\text{2 CLRS (5474)} &= 2 \times 120\ \mu\text{A} = 240\ \mu\text{A} \\
\text{1 MR (54161)} &= 1 \times 40\ \mu\text{A} = 40\ \mu\text{A} \\
\text{1 CLR (54109)} &= 1 \times 160\ \mu\text{A} = \underline{160\ \mu\text{A}} \\
&\qquad\text{Total} \quad 440\ \mu\text{A}
\end{aligned}$$

Table 1.3 **Fan in and fan out loads for Ex. 1.1**

	5474 (CLR input)	*54161 (MR input)*	*54109 (CLR input)*
High-level input current (logic 1)	120 μA (3 U.L.)	40 μA (1 U.L.)	160 μA (4 U.L.)
Low-level input current (logic 0)	3.2 mA (2 U.L.)	1.6 mA (1 U.L.)	4.8 mA (3 U.L.)

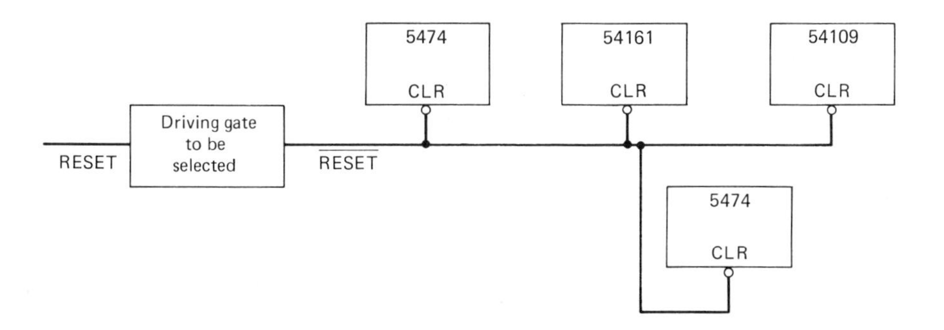

Figure 1.3 Selecting a suitable driving gate in Ex. 1.1.

Table 1.4 **Logic gate driving capabilities**

	5400, 04	*54H00, H04*	*5440*
Low-level output current	16 mA	20 mA	48 mA
High-level output current	400 μA	500 μA	1.2 mA

3 The calculation for logic 0 loading is:

$$\begin{aligned} &2 \text{ CLRS } (5474) = 2 \times 3.2 \text{ mA} = 6.4 \text{ mA} \\ &1 \text{ MR } (54161) = 1 \times 1.6 \text{ mA} = 1.6 \text{ mA} \\ &1 \text{ CLR } (54109) = 1 \times 4.8 \text{ mA} = \underline{4.8 \text{ mA}} \\ &\qquad\qquad \text{Total} \qquad 12.8 \text{ mA} \end{aligned}$$

4 A current of 440 μA for the logic 1 state is equivalent to 11 unit loads (fan in = 11), and a current of 12.8 mA in the logic 0 state is equivalent to 8 unit loads (fan out = 8).

5 Table 1.4 lists representative gates with their respective logic-level output currents. A reasonable selection for the driving gate is the 54H00 quad 2-input NAND gate. Letter H indicates that this is a high-speed version of a NAND gate (see Section 1.4).

Propagation Delay

Propagation delay is defined as the time it takes for an input pulse to travel through a medium, such as a transistor, wire, or cable, and appear at the output of the medium. The propagation delay determines the speed at which a circuit may be operated. In digital processing, requirements for speed can be a major factor.

The switching time parameters that are specified for TTL gates are propagation delay time from a logic 1 to a logic 0 level at the output (t_{pHL}) and propagation delay time from a logic 0 to a logic 1 level at the output (t_{pLH}). These parameters are measured with respect to the input pulse. Typical TTL delays from high to low are less than 7 ns and low to high are less than 11 ns.

EXAMPLE 1.2 The output of a register is to be added to a fixed value (1001) in an ALU and then stored in an accumulator (Fig. 1.4). Calculate the maximum frequency of operation for the circuit.

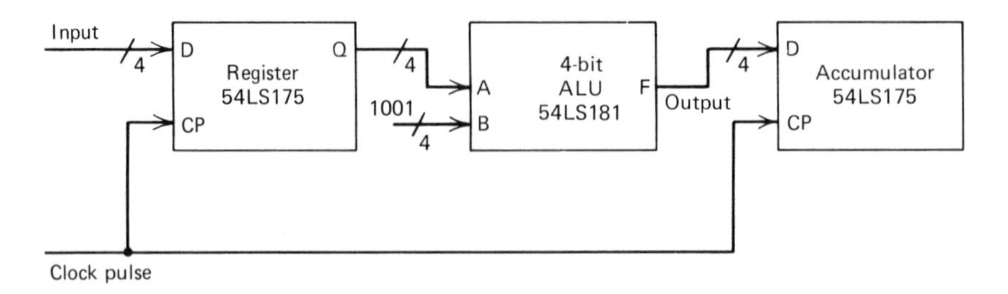

Figure 1.4 Logic diagram for Ex. 1.2.

PROCEDURE

1 The delays for the chips in the circuit, from manufacturers' data sheets, are:

54LS175: from CP to Q = 25 ns maximum
54LS181: from A and B to output = 32 ns maximum
54LS175: data setup time = 10 ns maximum

2 The total delay from the input register through the ALU to the accumulator is equal to the sum of the delays: 25 + 32 + 10 = 67 ns maximum. A value of 67 ns is equivalent to a clock rate of 1/(67 ns) = 14.9 MHz. The circuit of Fig. 1.4, therefore, can operate at frequencies up to 14.9 MHz.

Noise Margin

The *dc noise margin* is defined as the difference between the guaranteed logic state voltage limits of a driving gate and the input voltage thresholds of a driven gate. The threshold voltage for a gate is the voltage that, when applied to all inputs, produces a level change at the output. Figure 1.5*a* illustrates a 2-input NAND gate with both inputs tied together. If input voltage V_{in} slowly changes as indicated in Fig. 1.5*b* and is less than V_1, the output voltage is unaffected. As V_{in} increases above V_1, the output voltage, V_o, decreases. When V_{in} is greater than V_2, the output voltage no longer changes. The threshold voltage, V_T, is the voltage at which $V_o = V_{in}$.

Referring to Table 1.2, the guaranteed gate input voltage for a logic 0 is less than 0.8 V, and for a logic 1 input is greater than 2.0 V. For a logic 0 output, the output is less than 0.4 V, and for a logic 1 output is greater than 2.4 V. The guaranteed dc noise immunity is calculated by taking the guaranteed minimum

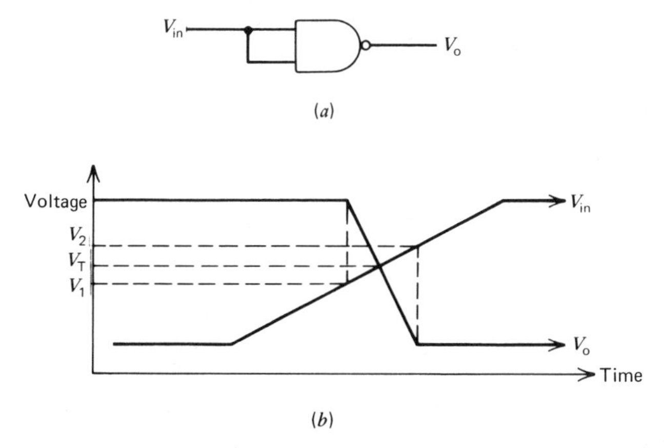

Figure 1.5 Definition of threshold voltage, V_T. (*a*) NAND gate with both inputs tied together. (*b*) Variation in the input and output voltages with time.

or maximum output voltage of the driving gate for the logic level and subtracting it from the receiving gate's input threshold level.

For example, in the logic 0 state a gate has an output voltage of 0.4 V and the receiving gate has a threshold voltage of 0.8 V. The dc noise immunity, therefore, is $0.8 - 0.4 = 0.4$ V. In the logic 1 state, the driving gate output is a minimum of 2.4 V, while the receiving gate's minimum logic 1 threshold is 2 V. The noise immunity is $2.4 - 2 = 0.4$ V. TTL has a guaranteed noise margin of 0.4 V; however, it often exhibits a noise margin in excess of 1.0 V.

The *ac noise immunity* refers to a circuit's ability to be unaffected by the coupling of pulses onto a logic line. Pulses with short widths in the low-nanosecond region require a significant pulse amplitude to propagate through a device. High-amplitude pulses are needed to effect a change at the output. Typical ac noise immunity for the 54/74 TTL family exhibits a sharp knee in the curve at about 7.0 ns pulse widths. The noise immunity goes from 3.5 to 2.0 V at 9 ns. Immunity to a 1 to 0 transition of a NAND gate is better than a 0 to 1 transition. This stems from the longer propagation time for the output to go to the 1 state.

1.4 TYPES OF TTL GATES

The gates in this section are variations of the basic TTL structure of Fig. 1.2 and are part of the TTL family.

Open Collector

An open-collector gate is very similar to a standard TTL gate at its input side. The basic difference between the two gates is that the open-collector gate has no pullup transistor at its output. That is, the collector of the pulldown transistor goes directly to the output. This configuration allows wiring of outputs together to perform logic operations such as wired-OR, dot-OR, and wired-AND. This is not possible with the totem-pole output stage in Fig. 1.2. A simplified schematic for an open-collector gate is provided in Fig. 1.6.

EXAMPLE 1.3 Design a circuit to realize the logic function $Y = \overline{AB} \cdot \overline{CD} \cdot \overline{EFG}$ using totem-pole and open-collector output gates.

PROCEDURE

1. A circuit using totem-pole gates is shown in Fig. 1.7*a*. Four NAND gates and one NOT are required.
2. A circuit using open-collector gates (wire-ANDing) is given in Fig. 1.7*b*. Only three NAND gates are required.

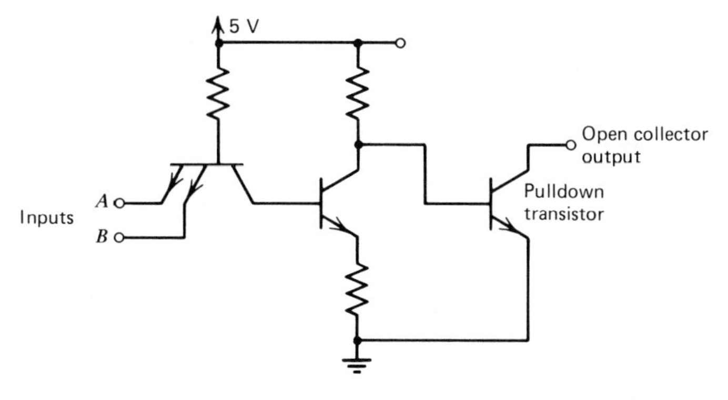

Figure 1.6 Simplified schematic of an open collector gate.

As seen from the solution of Ex. 1.3, a savings of two logic gates was realized by employing open-collector gates.

The value of the pullup resistor, R_L, in Fig. 1.7*b* must be carefully selected. A large value will affect the output impedance and tend to increase propagation delay and noise susceptibility. A maximum value for R_L must be found to ensure that the output is equal to, or greater than, 2.4 V in the logic 1 state. In the

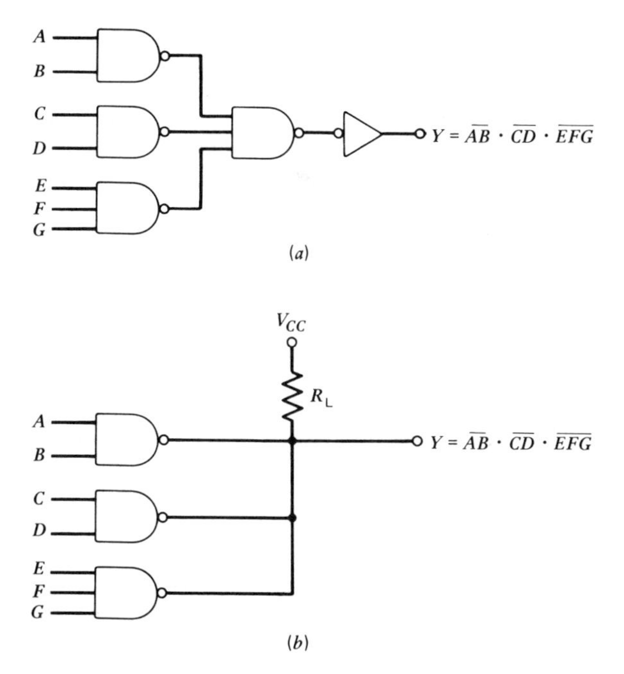

Figure 1.7 Realization of logic function $Y = \overline{AB} \cdot \overline{CD} \cdot \overline{EFG}$ using (*a*) conventional and (*b*) open collector (wire-ANDing) NAND gates.

logic 0 state, a minimum value for R_L must be selected to ensure that the voltage output will not exceed 0.4 V with the combined load of R_L and the fan in of other gate inputs.

EXAMPLE 1.4 Figure 1.8 illustrates a wired-AND gate in the logic 1 state. The voltage at the node must be at least $V_{OH} = 2.4$ V. Calculate the maximum value of R_L, $R_{L(max)}$.

PROCEDURE

1 The voltage across R_L is the difference between V_{CC} (typically 5 V) and $V_{OH} = 2.4$ V (the voltage at the node). The current through R_L is the sum of the currents into the driven loads and the leakage currents into the open-collector output transistors.

2 The maximum value of R_L is given by:

$$R_{L(max)} = (V_{CC} - V_{OH})/(NI_{OH} + MI_{IH}) \tag{1.1}$$

where V_{CC} is the power supply voltage, V_{OH} is the voltage required at the node, N is the number of collectors tied together, I_{OH} is the leakage current of an open-collector transistor, M is the number of gate loads, and I_{IH} is the load current of an input gate.

3 Typical value for I_{IH} is 40 μA maximum, and for leakage current I_{OH} it is 250 μA maximum.

4 From Fig. 1.8, $N = 4$ and $M = 3$. Substitution of the given values in Eq. 1.1 yields:

$$\begin{aligned} R_{L(max)} &= (5 - 2.4)/(4 \times 250 \times 10^{-6} + 3 \times 40 \times 10^{-6}) \\ &= 2.3 \text{ k}\Omega \end{aligned}$$

5 The maximum value of 2.3 kΩ will produce an output voltage at the node of 2.4 V. A value greater than 2.3 kΩ will result in a lower output voltage.

EXAMPLE 1.5 Calculate the minimum value of R_L for the circuit of Fig. 1.9. Voltage at the node represents a logic 0 ($V_{OL} = 0.4$ V). The sink current, I_S, is 16 mA, and the current for a gate load I_{IL} is 1.6 mA.

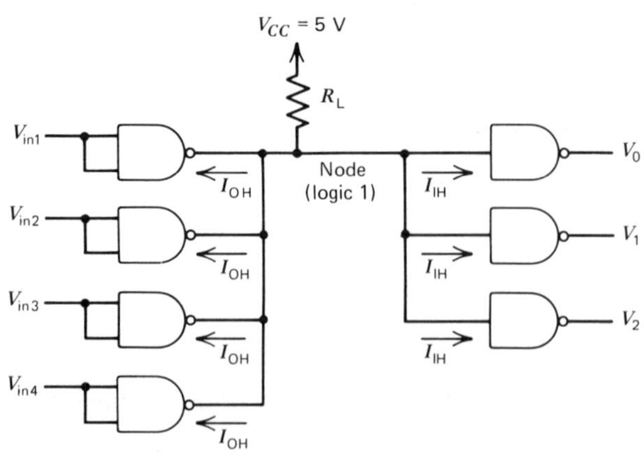

Figure 1.8 Calculation of $R_{L(max)}$ in Ex. 1.4.

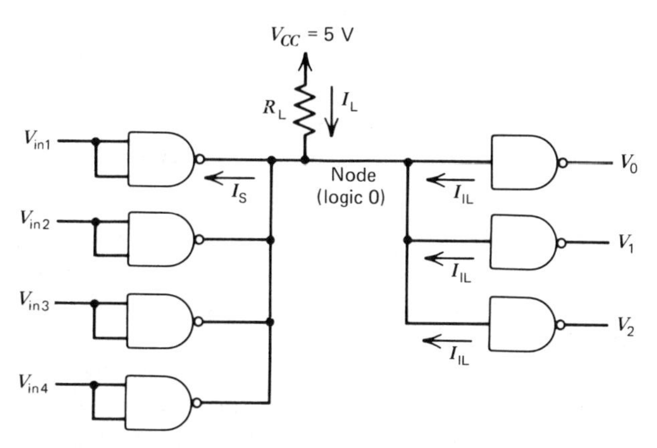

Figure 1.9 Calculation of $R_{L(min)}$ in Ex. 1.5.

PROCEDURE

1 The voltage at the wired-AND node must be no greater than $V_{OL} = 0.4$ V. Current through R_L and the gate loads must be limited to the maximum sink current capability of one output open-collector stage.
2 The minimum value of R_L, $R_{L(min)}$, is given by:

$$R_{L(min)} = (V_{CC} - V_{OL})/(I_S - MI_{IL}) \tag{1.2}$$

3 Substituting the given values in Eq. 1.2 and solving,

$$R_{L(min)} = (5 - 0.4)/(16 \times 10^{-3} - 3 \times 1.6 \times 10^{-3})$$
$$= 410\ \Omega$$

As the number of open-collector outputs tied to a wired-AND node increases, the value of $R_{L(max)}$ decreases. The maximum allowable number of open-collector outputs tied together is reached when $R_{L(max)} = R_{L(min)}$.

AND gates

An AND gate may be implemented by following one NAND gate with another whose inputs are tied together, as illustrated in Fig. 1.10. To minimize the number of components and propagation delays, a TTL AND gate may be used. The AND function is available as 2-, 3-, or 4-input gates, 5408, 5411, and 5421, respectively.

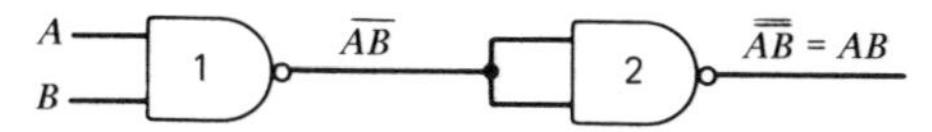

Figure 1.10 Implementing the AND function using NAND gates.

NOR gates

The basic NOR function may be implemented from NAND gates; however, significant additional logic levels are required. The NOR function is available in TTL as a quad 2-input (5402) and as a triple 3-input (5427) package.

Schmitt Input NAND Gates

When the inputs to devices change state slowly, various problems can arise. False triggering may result as the input slowly crosses the threshold region of the input device. Propagation delays may become difficult to predict, causing timing problems. Devices may not operate owing to rise time sensitivity. For these reasons, many companies have developed a Schmitt trigger gate that provides pulse shaping by introducing positive feedback to obtain high gain and hysteresis (see Chap. 16).

Hysteresis is a built-in safeguard that prevents the input from switching at the same levels for positive and negative waveforms. A typical application for a Schmitt NAND gate is an interface to a switch. The Schmitt gate shapes the switching waveform, making it a practical and inexpensive means for interfacing with front panel controls. A popular TTL Schmitt trigger is the 5413.

High-speed TTL (HTTL)

High-speed NAND gate circuitry (Fig. 1.11) is almost the same as that of a standard TTL NAND gate (Fig. 1.2). The basic differences include lower resistance values and input clamping diodes to reduce transmission line effects, which become more apparent with fast rise and fall times.

The output section of the high-speed gate uses a Darlington pair (Q_3 and Q_4) that provides a somewhat higher gate speed than the standard TTL gate and low output impedance. The output impedance for a logic 1 is typically 10 ohms (Ω) and for a logic 0 it is 100 Ω. High-speed TTL has the disadvantage of using more power than standard TTL.

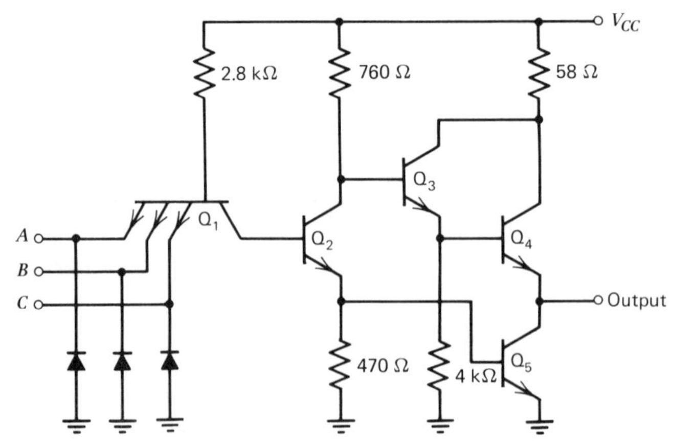

Figure 1.11 High-speed TTL (HTTL) NAND gate.

High-speed TTL circuits are available in a full complement of gates, flip flops, and arithmetic elements.

Schottky TTL (STTL)

Schottky TTL (STTL) is a super high-speed TTL series that features Schottky-barrier diode clamping of transistors (Fig. 1.12). Utilization of a Schottky-barrier diode as a clamp from base to collector diverts excess base current from flowing to the base and prevents the transistor from saturating. A great reduction in storage time is thereby achieved, and the switching speed is improved. Schottky also offers an improvement in propagation delays over temperature. STTL gates, flip flops, and medium-scale integration (MSI) functions can be used to realize higher system speeds and still maintain TTL noise immunity.

The advantage of STTL is the significant reduction of throughput delay over high-speed and standard TTL. This reduction means that critical path delay problems can be eliminated by replacing a standard TTL gate, pin for pin, with a Schottky device. TTL memory speeds can be improved by using Schottky devices for data registers, address registers, address buffer/drivers, and output data bus drivers. CPU data processing time can be reduced for ALUs, multiplexers, registers, accumulators, and data bus drivers by using Schottky devices.

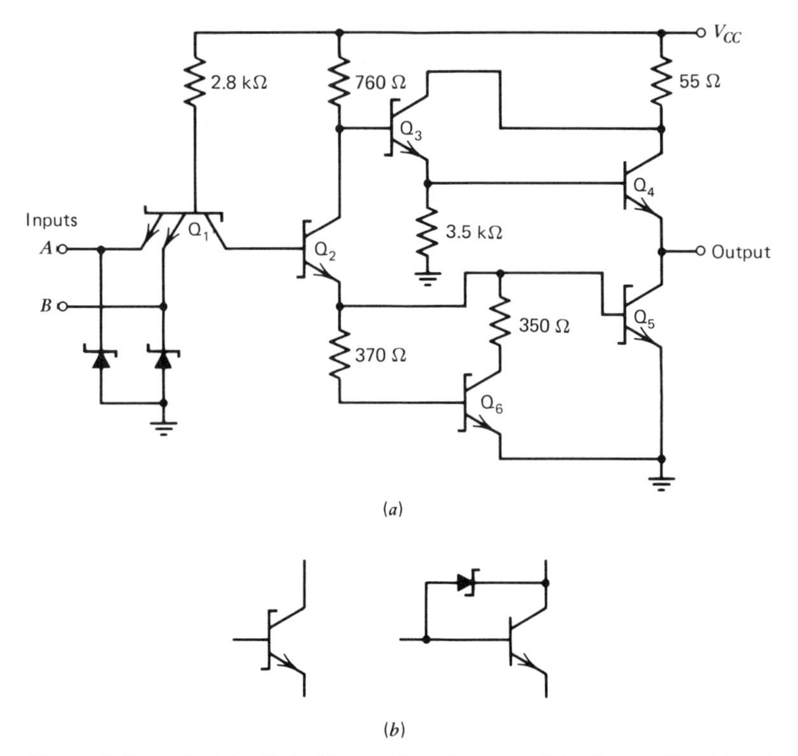

Figure 1.12 Schottky TTL (STTL) NAND gate. (*a*) Circuit. (*b*) Schottky transistor is a junction transistor with a Schottky diode connected across the collector and base.

The power consumption of Schottky circuits is generally higher than for standard TTL; however, power consumption does not increase as rapidly when the circuit is operating at high clock rates. The actual power consumption of a Schottky device may be less than an equivalent standard TTL device in a high-speed system.

The rise time of Schottky TTL outputs is typically 3 ns, or two to three times faster than standard TTL. A transition time of 3 ns can cause ringing and other transmission-line effects. If signal-line lengths are kept under 4 inches, transmission-line effects are minimized.

A Schottky gate in the logic 0 state is guaranteed to sink 20 mA at 0.5 V maximum. In the logic 1 state, it can source 1 mA at a minimum output voltage of 2.5 V.

Low-power Schottky (LSTTL)

As the complexity of TTL circuits increases, the cost and size of the power supply and heat dissipation of the circuit become increasingly important factors. Improvements in semiconductor technology have made it possible to reduce power consumption and improve speed.

The low-power Schottky TTL (Fig. 1.13) family has a power reduction of a factor of 5 over standard TTL. Low-power Schottky is available in a full complement of TTL functions. Its advantages include the following:

1 Less supply current allows the use of smaller power supplies, thus reducing cost.

2 Lower power consumption simplifies thermal design because less heat is generated.

3 Increased reliability is achieved because lower dissipation causes less chip temperature rise above ambient.

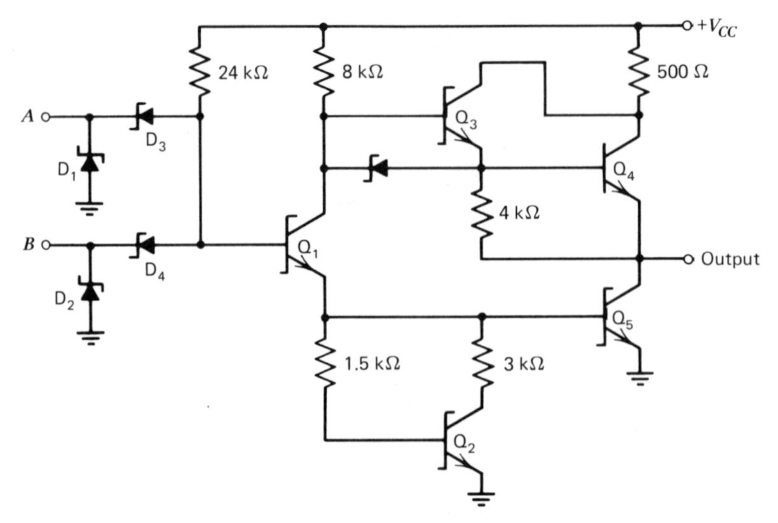

Figure 1.13 Low-power Schottky TTL (LSTTL) NAND gate.

4 Lower currents are used to switch transistors. Smaller current spikes than in standard TTL occur and, therefore, less noise is generated. The need for fewer decoupling capacitors increases board space, allowing higher packing densities.

5 The LSTTL can interface directly to other TTL types.

Standard TTL

Standard TTL offers a combination of speed and power dissipation, which is suitable for many applications. The basic standard NAND gate of Fig. 1.2 has an output stage with low impedance that results in improved noise immunity and fast switching times. Many circuits are offered in the standard TTL version. These include shift registers, counters, decoders, memories, arithmetic elements, and a wide variety of gates. A comparison of the various TTL families is provided in Table 1.5.

Tri-State Logic

The newer TTL families have an additional control input whereby the output of a gate is disabled. This allows the output to remain in a high-impedance (high-Z) state, permitting different circuits to be connected to a common bus. A typical tri-state circuit is provided in Fig. 1.14.

The output enable function is connected through diodes to the bases of Q_2 and Q_3. A low signal at the input turns off both Q_2 and Q_3, thereby disabling the pullup and pulldown transistors, Q_4 and Q_5, respectively. This represents the high-Z state. When the circuit is enabled, the output high and low states are the same as TTL circuits without the tri-state feature.

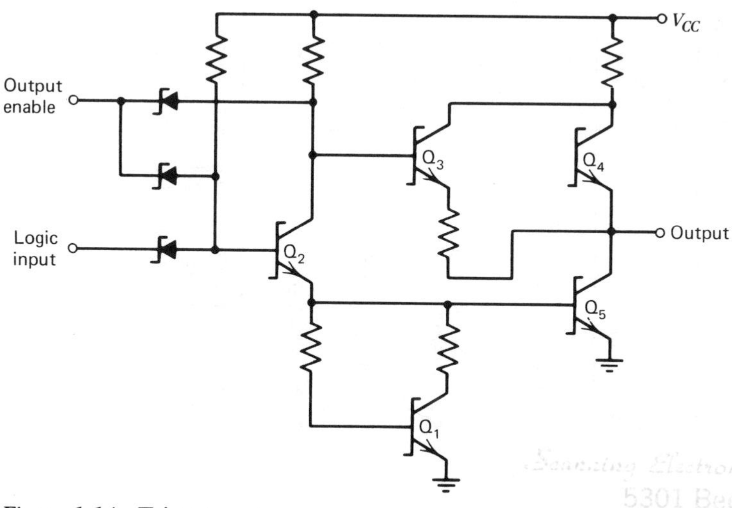

Figure 1.14 Tri-state gate.

Table 1.5 TTL logic family comparisons

Logic family gates	Propagation delay (ns)	Power dissipation (mW)	Fan in	Fan out	Low-level noise margins (mV)	High-level noise margins (mV)
TTL 54/74 Standard series	t_{pLH}[b] 22 t_{pHL}[c] 15	110	40 μA/1.6 mA	800 μA/16 mA	400	400
HTTL, 54/74H	t_{pLH} 10 t_{pHL} 10	200	50 μA/2 mA	500 μA/20 mA	400	400
STTL, 54/74S	t_{pLH} 4.5 t_{pHL} 5.0	180	50 μA/2 mA	1 mA/20 mA	300	500
LSTTL, 54/74LS	t_{pLH} 10 t_{pHL} 10	22	20 μA/0.4 mA	400 μA/8 mA	300	500
FTTL,[a] 54/74F	t_{pLH} 4.0 t_{pHL} 3.5	51	20 μA/0.6 mA	1.0 mA/20 mA	300	500

[a] FTTL — Fairchild FAST TTL family.
[b] t_{pLH} — Propagation delay from low to high output transition.
[c] t_{pHL} — Propagation delay from high to low output transition.

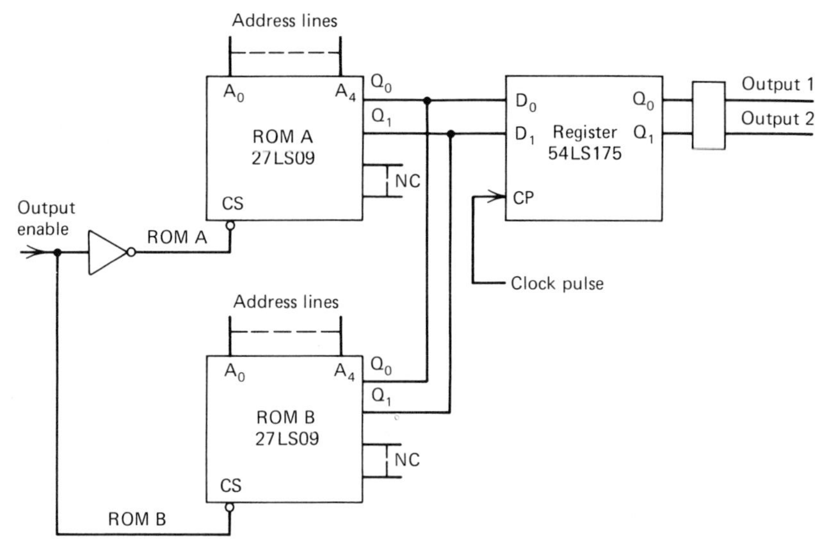

Figure 1.15 Outputs of two ROMs stored in register (Ex. 1.6).

EXAMPLE 1.6 The outputs of two ROMs (see Chap. 6) are to be stored in a register and outputted to an external circuit. Implement the circuit with tri-state logic.

PROCEDURE

1 The resulting circuit is given in Fig. 1.15.

2 The output enable controls the selection of either ROM A or B. When the enable is high, ROM A is selected and ROM B is disabled (its output is in a high-impedance state).

3 The output lines of ROM A are clocked into the 54LS175 and outputted as output 1 and 2 lines.

4 When the enable is low, ROM A is disabled and ROM B enabled. The output lines of ROM B are now clocked into the 54LS175 and outputted as output 1 and 2 lines.

1.5 IMPLEMENTATION OF LOGIC FUNCTIONS

A logic function can be implemented in many ways. The type of implementation depends on consideration of propagation delay, kinds and number of gates, and available signals.

EXAMPLE 1.7 Implement the following logic function: $Y = AB + AC + BC$.

PROCEDURE

1 Using a single chip, the 54/7454, which contains three 2-input AND gates, one 3-input AND, and one 4-input OR, the logic function is implemented as shown in Fig. 1.16*a*.

2 Another solution using three of the NAND gates of the quad 54/7400 and a 54/7410 for the 3-input NAND gate, the logic function may be implemented as illustrated in Fig. 1.16*b*.

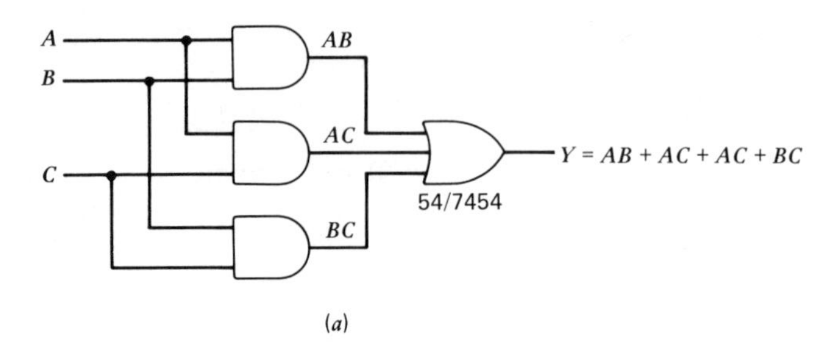

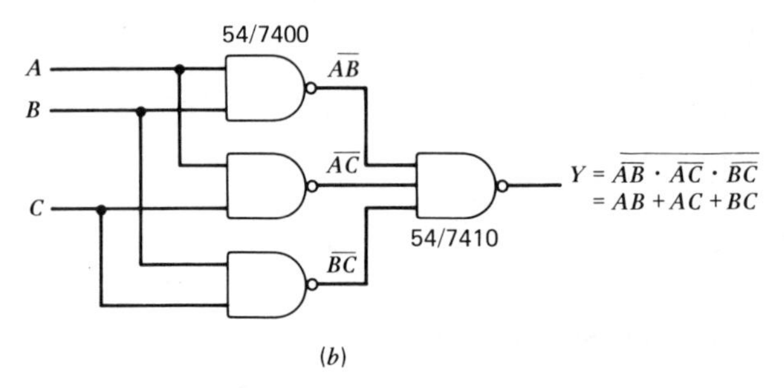

Figure 1.16 Implementing $Y = AB + AC + BC$ in Ex. 1.7 using (*a*) AND and OR gates and (*b*) NAND gates only.

EXAMPLE 1.8 Implement the following logic function with NAND gates only: $Y = \overline{A}C + A\overline{C} + \overline{C}\overline{D} + A\overline{B}$

PROCEDURE

1 To implement any function with NAND gates, one must first convert the given function to a NAND expression. Rewrite the expression as:

$$Y = \overline{\overline{\overline{A}C + A\overline{C} + \overline{C}\overline{D} + A\overline{B}}}$$

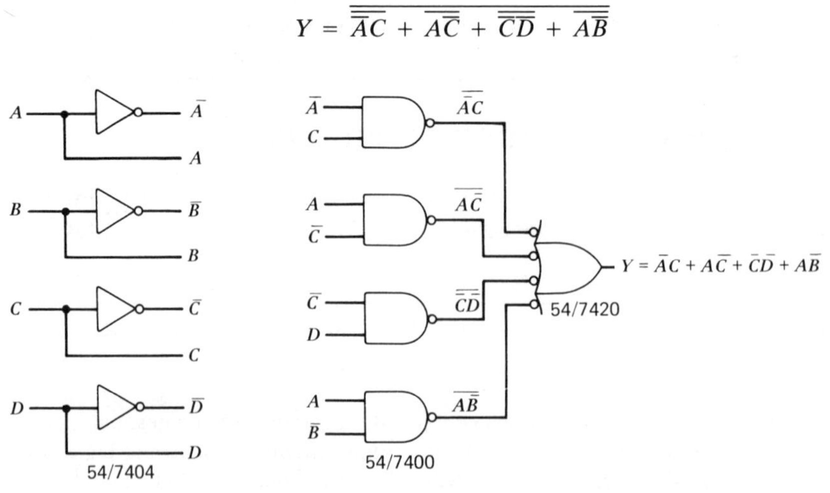

Figure 1.17 Implementing $Y = \overline{A}C + A\overline{C} + \overline{C}\overline{D} + A\overline{B}$ with NAND gates only in Ex. 1.8.

2 By de Morgan's theorem,

$$Y = \overline{(A + \overline{C})(\overline{A} + C)(C + D)(\overline{A} + B)} = \overline{A}C + A\overline{C} + \overline{C}\overline{D} + A\overline{B}$$

3 The implementation of the function is given in Fig. 1.17.

1.6 FLIP FLOPS

The fundamental and most important characteristic of a flip flop is that it acts like a one-bit memory. The logic level at the output of a flip flop can store a logic 1 or 0 indefinitely.

SR Flip Flop

A set-reset, SR, flip flop using NAND gates and its truth table are illustrated in Fig. 1.18. The flip flop changes state when either one of the inputs is at a logic 0. The output will not change when both inputs are at logic 0; the Q and $\overline{\text{Q}}$ outputs would be forced to a logic 1 and, hence, this state is not used.

In addition to its use in shift registers, the SR flip flop is employed to *debounce* a switch. The contacts of a switch, when thrown from one position to another, may bounce many times before coming to rest.The SR flip flop will change state as soon as contact closure is made and, in effect, debounce the switch contacts.

JK Flip Flop

The JK flip flop is a clocked circuit that is similar to the SR flip flop. The JK has many desirable features, making it a very versatile element. The logic symbol for the JK flip flop is shown in Fig. 1.19*a*. Examples of TTL JK flip flops include the 54/7470, the 7472, the 7476, the 74H101, and the 74109.

Its two input signals, J and K, provide a range of conditions that is described in the truth table of Fig. 1.19*b*. When $J = K = 0$, the flip flop output remains unchanged. When either J or K is a 1, but not both, a logic 1 and logic 0 will appear at the output, respectively. If $J = K = 1$, then the output will *toggle* each time the clock pulse, C_p, occurs. That is, if the state before the clock pulse is a logic 0, it will change to a logic 1 after the clock, and vice versa.

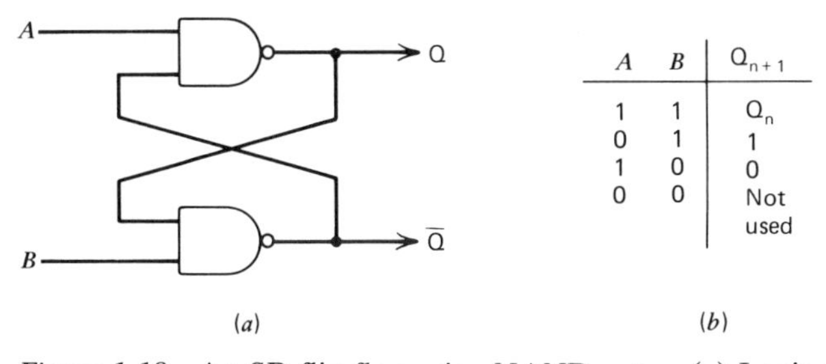

A	B	Q_{n+1}
1	1	Q_n
0	1	1
1	0	0
0	0	Not used

Figure 1.18 An SR flip flop using NAND gates. (*a*) Logic diagram. (*b*) Truth table.

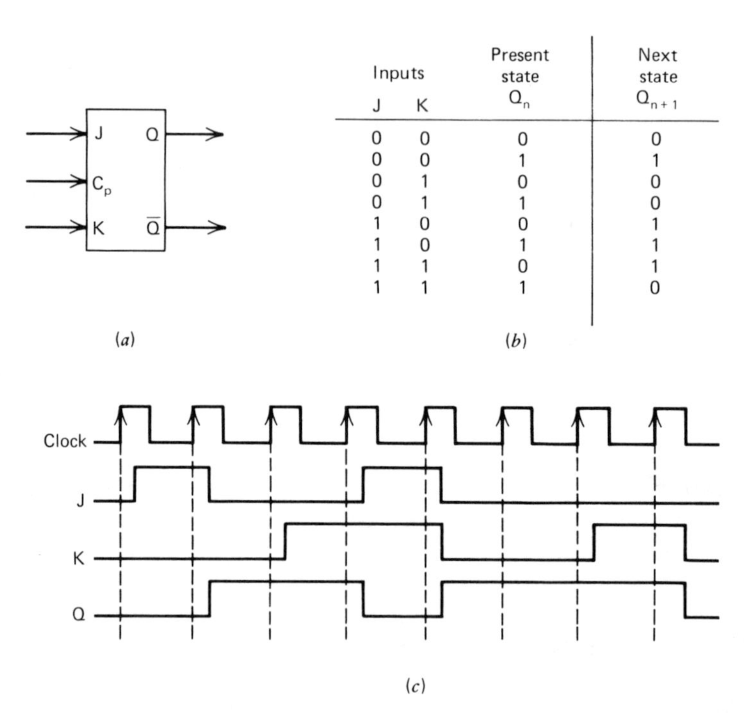

Inputs J	Inputs K	Present state Q_n	Next state Q_{n+1}
0	0	0	0
0	0	1	1
0	1	0	0
0	1	1	0
1	0	0	1
1	0	1	1
1	1	0	1
1	1	1	0

Figure 1.19 A JK flip flop. (*a*) Logic symbol. (*b*) Truth table. (*c*) Timing diagram.

The timing diagram of Fig. 1.19*c* illustrates the operating characteristics of a JK flip flop. Output Q changes state on the positive-going clock edge. This is referred to as a *positive edge-triggered* flip flop. There also exist *negative edge-triggered* flip flops. An example of a positive edge-triggered JK flip flop is the 54/74109 and that of a negative edge-triggered flip flop is the 54H/74H101.

D Flip Flop

It is often required to delay a digital signal in time by one clock pulse. The delay, or D, flip flop accomplishes this. The D flip flop has two input signals, C_p and D, as shown in Fig. 1.20*a*; its truth table is provided in Fig. 1.20*b*. It is seen that the logic output state following a clock pulse, Q_{n+1}, is equal to the logic state of the D input before the clock pulse. An example of a TTL D flip flop is the 54/7474. The D flip flop is primarily used in data storage and register applications.

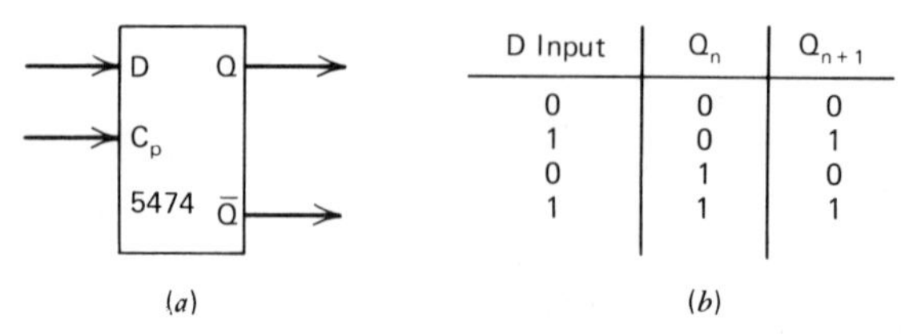

D Input	Q_n	Q_{n+1}
0	0	0
1	0	1
0	1	0
1	1	1

Figure 1.20 D flip flop. (*a*) Logic symbol. (*b*) Truth table.

1.7 COUNTERS

The simplest counter to design and implement is the *ripple counter*. Because the flip flops in a ripple counter are not under command of a single clock pulse, it is an example of an *asynchronous* counter. Figure 1.21*a* shows a 3-bit binary ripple counter. Initially all flip flops are reset to the logic 0 state. A clock pulse applied to the clock (C_p) input of flip flop A causes Q_A to change from a logic 0 to a logic 1. Flip flop B does not change its state until Q_A changes from a logic 1 to 0.

When clock pulse 7 arrives, all flip flops will be in their logic 1 states. Clock pulse 8 will cause Q_A, Q_B, and Q_C to reset to logic 0. The timing diagram of Fig. 1.21*b* demonstrates the operation of the counter. As shown, all flip flop output transitions occur on the negative-going edge of the clock. The state table for the counter is provided in Fig. 1.21*c*.

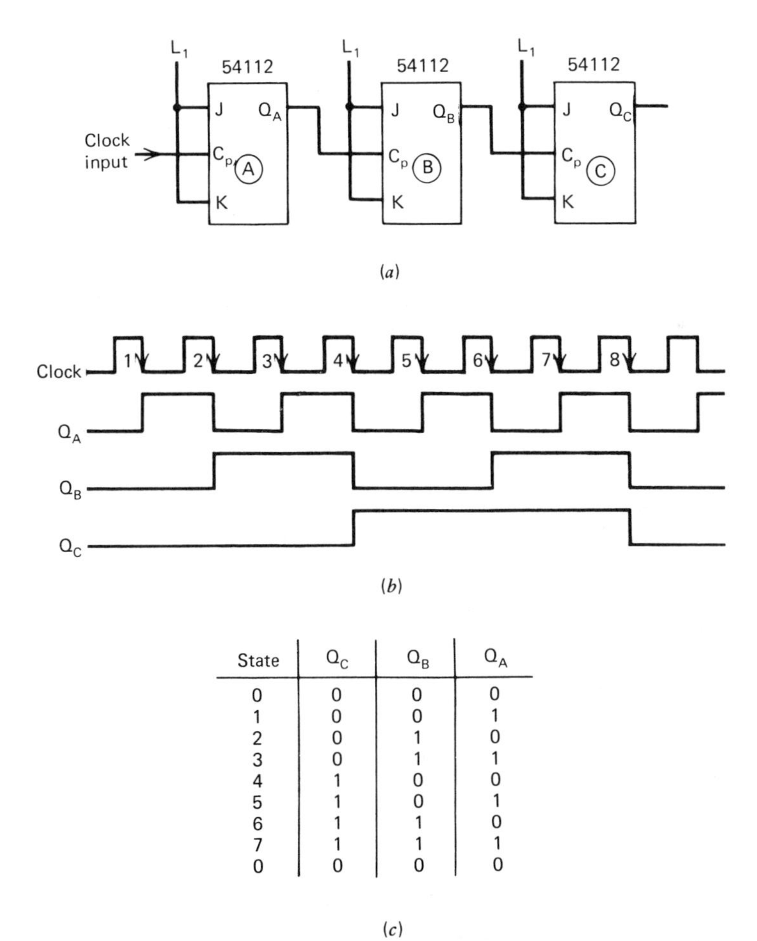

State	Q_C	Q_B	Q_A
0	0	0	0
1	0	0	1
2	0	1	0
3	0	1	1
4	1	0	0
5	1	0	1
6	1	1	0
7	1	1	1
0	0	0	0

Figure 1.21 A 3-bit ripple counter. (*a*) Logic diagram. (*b*) Timing diagram. (*c*) State table.

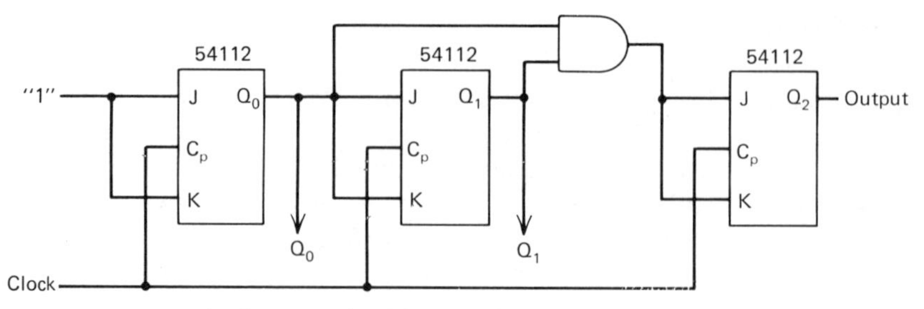

Figure 1.22 Logic diagram of a 3-bit synchronous counter.

Synchronous Counter

If all flip flops are clocked by the same waveform, they are said to operate *synchronously*. In such a case, all state changes of a flip flop occur simultaneously. The speed of counters can be increased by operating them synchronously. An example of a synchronous 3-bit parallel counter is shown in Fig. 1.22.

The sequence of states assumed by a flip flop and the timing diagram are the same as for the ripple counter. The synchronous parallel counter is the fastest counter; it is not, however, without disadvantages. As stages are added to this circuit, each succeeding AND gate requires one additional input. Also, as each additional gate increases by one, the fan out required of each flip flop also increases.

EXAMPLE 1.9 Design a mod-5 counter (reset to zero on every fifth input pulse) to operate as specified in Fig. 1.23.

PROCEDURE

1 For each counter output one can construct a next-state truth table according to the following rules:

- α indicates that the flip flop is to change from a 0 to a 1.
- β indicates that the flip flop is to change from a 1 to a 0.
- 1 indicates that the flip flop remains at a 1.
- 0 indicates that the flip flop remains at a 0.
- X indicates a don't care condition. (A don't care is a logic variable whose value does not matter in a given application.)

Counter state	Q_2	Q_1	Q_0
0	0	0	0
1	0	0	1
2	0	1	0
3	0	1	1
4	1	0	0

Figure 1.23 Truth table for a 5-mod counter (Ex. 1.9).

2 Employing JK flip flops to implement a mod-5 counter, we use the following generalized input equations: $J = \alpha$ and don't care's $= 1$, X, β, and $K = \beta$ and don't care's $= 0$, α, X.

3 *Transition maps* are drawn as illustrated in Fig. 1.24*a*. The transition map plots the changes in the state of one flip flop as a function of everything which affects it. This technique allows the circuit designer to solve numerous problems.

It is noted that at counter states 0 through 3, Q_2 does not change its state (Fig. 1.23). At counter state 4, Q_2 changes to a 1. Therefore, an α is inserted on the transition map for stage 3 at state 011. At state 4, a β is inserted because at counter state 0, Q_2 returns to a 0 state. Transition maps for stages 1 and 2 are obtained in the same manner.

4 By utilizing the transition maps and grouping the proper variables, one can obtain the equations listed in Fig. 1.24*a*. The implementation of these equations results in the mod-5 synchronous counter shown in Fig. 1.24*b*.

1.8 SHIFT REGISTERS

An array of flip flops that permit data to be transferred from one point to another is called a shift register. A 4-bit shift register implemented with D flip

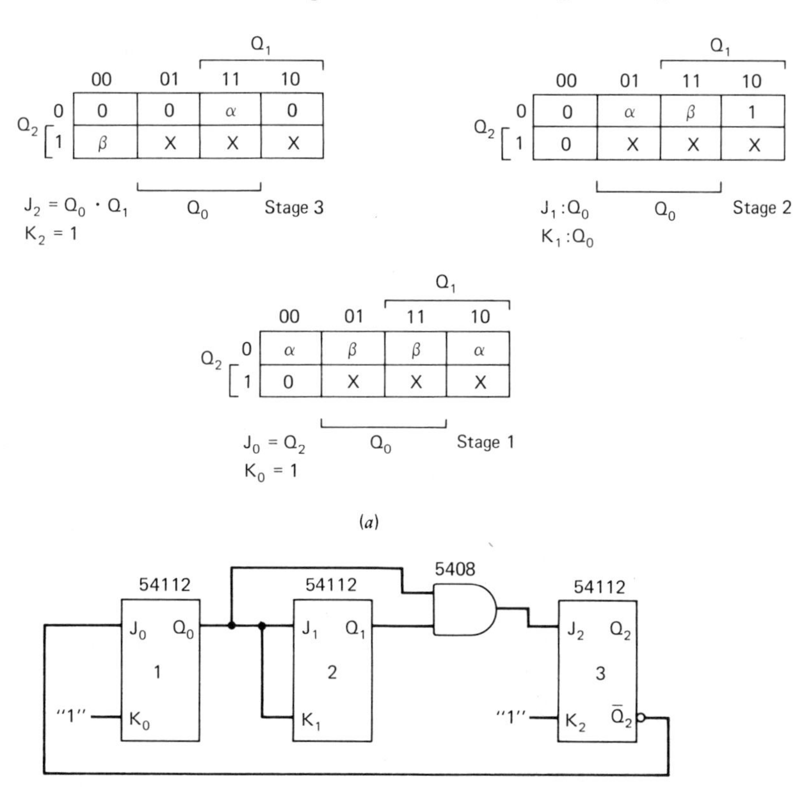

Figure 1.24 A mod-5 counter (*a*) Transition maps. (*b*) Logic diagram.

flops is illustrated in Fig. 1.25*a*. As data appear at the D_0 input, it is shifted into the D_0 flip flop after the clock transition. On the next clock pulse, the D_1 flip flop stores the data, and subsequently the data are transferred to the D_3 stages after two additional clock pulses.

The timing diagram of Fig. 1.25*b* describes the operation of a shift register. Data are shifted one clock pulse at the output of each register stage. The data at output Q_3 are the same as the input data at D_0; however, it appears 4 clock pulses later in time.

There are several types of 4-bit shift registers available. These include serial in/serial out, parallel in/serial out (54/7495), parallel in/parallel out (54/74195), and bidirectional (54/74194) shift registers.

EXAMPLE 1.10 A 3-clock wide pulse is required to be generated. Use a single chip with parallel load capability to generate the pulse.

PROCEDURE

1 The circuit, using a 54/74191 synchronous binary up/down counter, is given in Fig. 1.26*a*.

2 The S_0 and S_1 command lines control the mode of operation. When S_0 is low and S_1 is high, a shift right (SR) function is performed. When S_0 and S_1 are both high, a parallel load of inputs A, B, C, and D is performed. Input S_1 is always maintained at a logic 1.

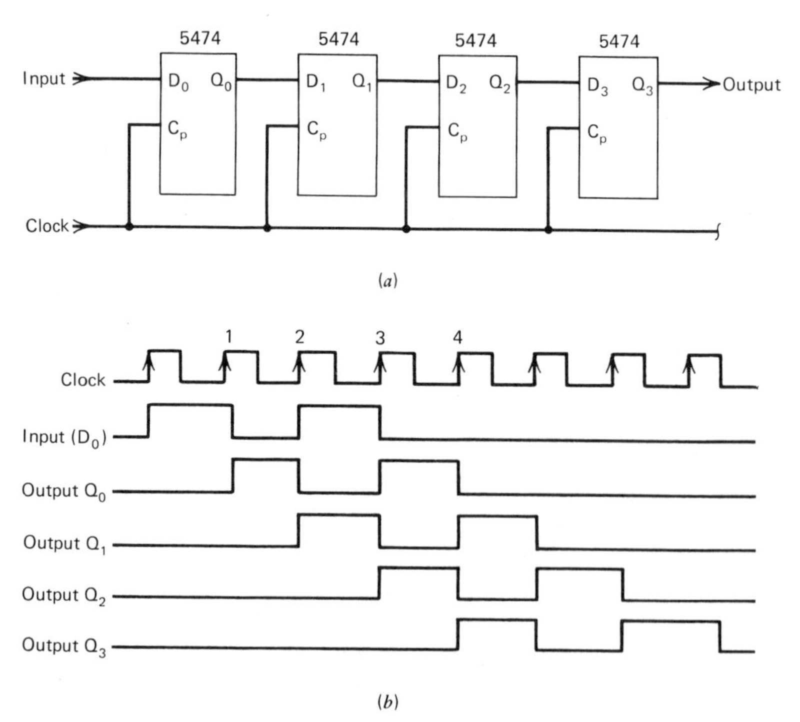

Figure 1.25 A 4-bit shift register. (*a*) Logic diagram. (*b*) Timing diagram.

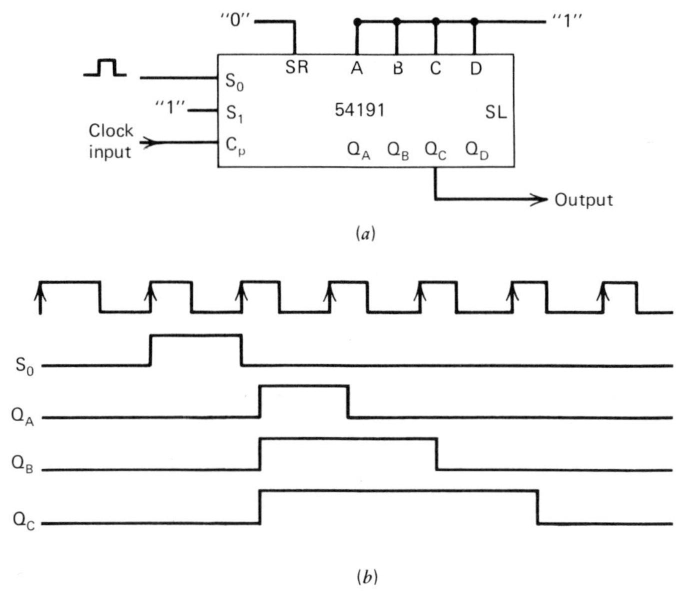

Figure 1.26 Generating a 3-clock wide pulse. (*a*) Circuit. (*b*) Timing diagram. (Ex. 1.10.)

3 Referring to the timing diagram of Fig. 1.26*b*, when S_0 goes to 1, all flip flop outputs go to a logic 1 state on the next clock pulse. Because the SR input is a logic 0, the Q_A output remains at logic 1 for one clock duration. The Q_A output shifts to the Q_B output with one more clock pulse and to Q_C on the next clock pulse. As shown, the Q_C output remains at logic 1 for the duration of three clock pulses.

1.9 OTHER APPLICATIONS

Examples of converting one code to another include Gray to binary, binary to Gray, excess-3 to BCD, decimal to BCD, BCD to binary, and so on. In designing for any conversion, one must consider the number of ICs required and power consumption before choosing the best technique.

EXAMPLE 1.11 Design the required logic to convert the code from a set of three input switches as indicated in Fig. 1.27. The switches are located on a front panel and are control mode (S_1), speed of processing (S_2), and disabling (S_3).

PROCEDURE

Approach I

1 Using combinational logic, we construct Karnaugh maps (Fig. 1.28) to minimize the gate count.
2 Implementation of the equations results in the logic diagram of Fig. 1.29.

Inputs			*Outputs*								
			Speed			*Mode*			*Control*		
S_1	S_2	S_3	*lo*	*hi*	*med*	*add*	*sub*	*mult*	X_1	X_2	X_3
0	0	0	1	0	0	1	0	0	1	0	0
0	0	1	1	0	0	1	0	0	1	0	0
0	1	0	0	1	0	1	0	0	0	1	0
0	1	1	0	1	0	0	1	0	0	1	1
1	0	0	0	0	1	0	1	0	1	0	0
1	0	1	0	0	1	0	0	1	1	1	1
1	1	0	1	0	0	0	0	1	0	0	0
1	1	1	0	1	0	0	0	1	1	0	1

Figure 1.27 Truth table for a code conversion in Ex. 1.11.

3 Total number of ICs required is:

eight 2-input AND gates (5408):		2 packages
four 3-input AND gates (5411):		2 packages
eight 2-input OR gates (5432):		2 packages
3 inverters (5404):		1 package
	Total	7 packages

Approach II

1 A ROM (see Chap. 6) is a memory device that is addressed by external inputs and may be coded to output a logic 1 or 0 for a particular number of inputs. Most ROMs

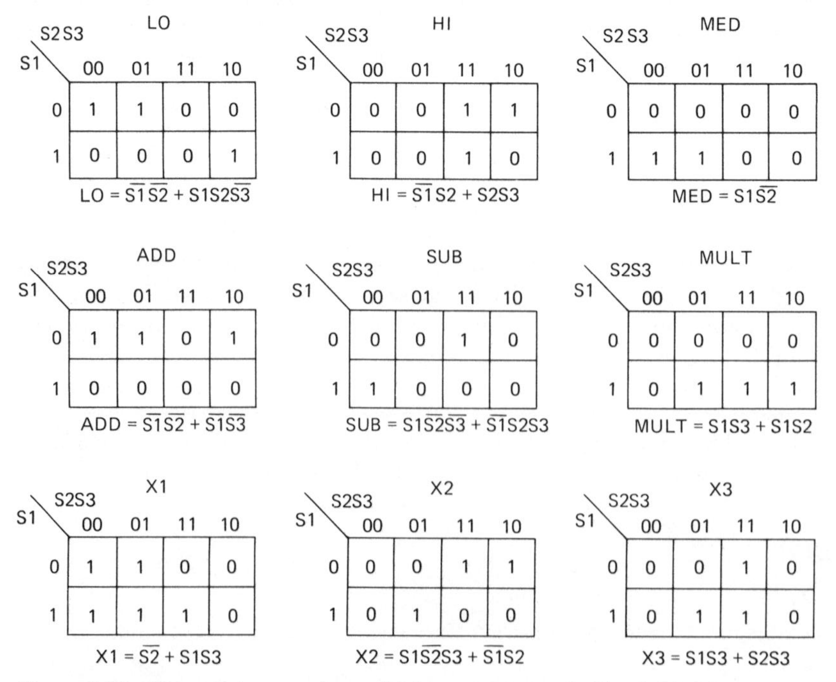

Figure 1.28 Karnaugh maps for reducing gate count in Ex. 1.11.

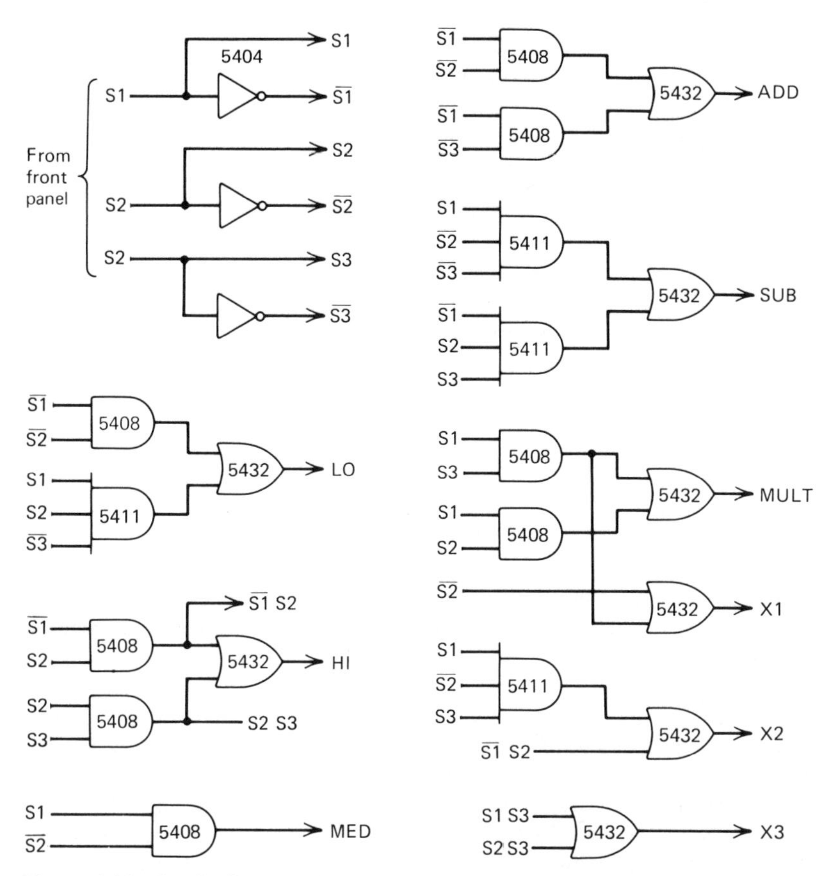

Figure 1.29 Logic diagram for implementing code conversion in Ex. 1.11 using combinational logic.

are configured in some binary interval of bits; the smallest ROM is 32 words by 8 bits wide. If we use input switches S_1, S_2, and S_3 to address the ROM and code its outputs to select or not to select, according to Fig. 1.27, the circuitry required is given in Fig. 1.30.

2 Comparison of the two methods shows that Approach II is a better choice. Two ROMs, as compared to 7 IC packages results in less power, fewer interconnections, and less space. Cost, however, must be considered before choosing each approach. The ROM approach could be the more expensive one. If board space, however, is critical, then it would be an acceptable choice.

Many system problems contain the requirement of generating pulses that occur at different time intervals. Let us consider the problem of generating four distinct pulses that occur for different time durations.

EXAMPLE 1.12 Timing required for a data processing system to perform its operations is given in Fig. 1.31. The signals instruct the processor to add, store results, route data

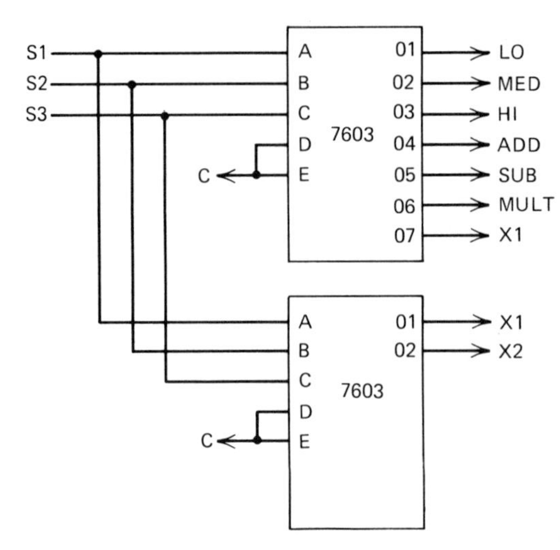

Figure 1.30 Implementing code conversion with two ROMs.

through a data selector, and compute a new value. The system is timed off the positive edge of clock pulses. Design, using standard TTL, circuitry to generate the timing signals.

PROCEDURE

1 First, we consider the number of clock cycles between repetitive signal intervals. Referring to Fig. 1.31, the number of clock cycles is 7; therefore, a 3-bit binary counter is needed. A standard 4-bit binary counter is selected (Fig. 1.32*a*).

2 The basic approach in solving this problem is first to select a way of cycling the counter to add every 7 pulses. If we use the Q_3 output as the ADD pulse and use it to preset the counter on the next clock pulse, Q_3 is forced to last one clock interval and cause a new timing cycle to begin. This is illustrated in Fig. 1.32*a*, but is not yet the final solution.

3 The complete circuit is given in Fig. 1.33. If we preset a binary number 2 into the counter, it will take 7 clock pulses to complete a full cycle. To generate the pulse we simply decode the state of the counter that follows the state assigned to the ADD pulse. That is, we decode the preset stage 2 to generate store results. If we use store results to reset a flip flop and state 4 as defined in Fig. 1.32*b*, we generate a pulse that has a duration of 2 clock pulses, referred to as Data SEL. The fourth pulse, Compute V_a, occurs 1 clock pulse between when Data SEL ends and it begins. By decoding

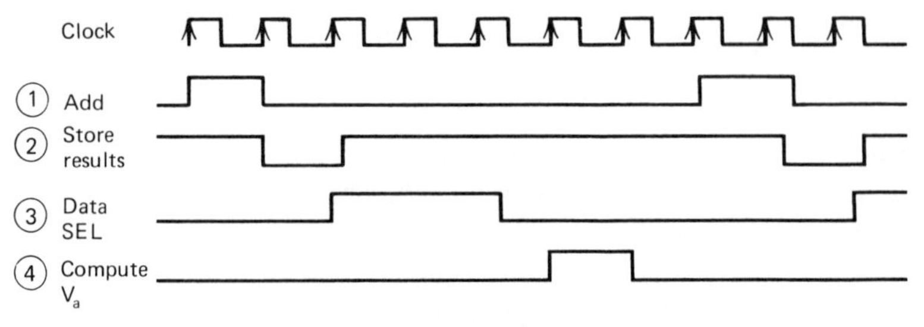

Figure 1.31 Timing required for a data processor in Ex. 1.12.

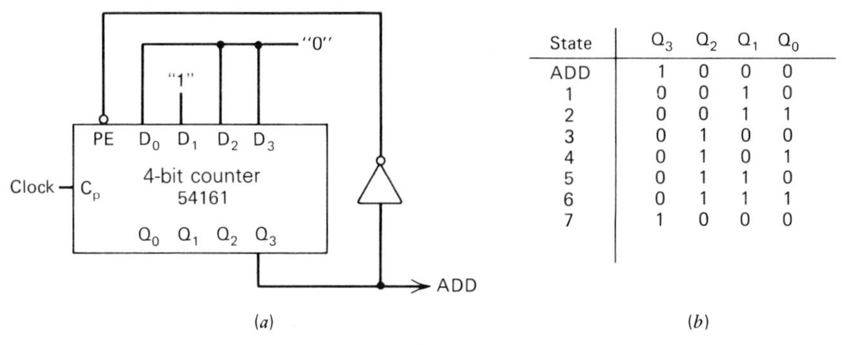

State	Q_3	Q_2	Q_1	Q_0
ADD	1	0	0	0
1	0	0	1	0
2	0	0	1	1
3	0	1	0	0
4	0	1	0	1
5	0	1	1	0
6	0	1	1	1
7	1	0	0	0

Figure 1.32 Developing the necessary logic in Ex. 1.12. (*a*) Using a 4-bit counter. (*b*) State table.

state 6, we can generate Compute V_a by again using simple logic gates. The interval continually repeats every time an ADD pulse is generated.

4 The total number of IC packages needed to generate the timing pulses is four: a binary counter, a JK flip flop, a triple 3-input NAND gate, and a hex inverter.

1.10 ADVANCED TTL

Recently, a new advanced Schottky TTL technology was developed and introduced by Fairchild. Fairchild Advanced Schottky Technology (referred to as

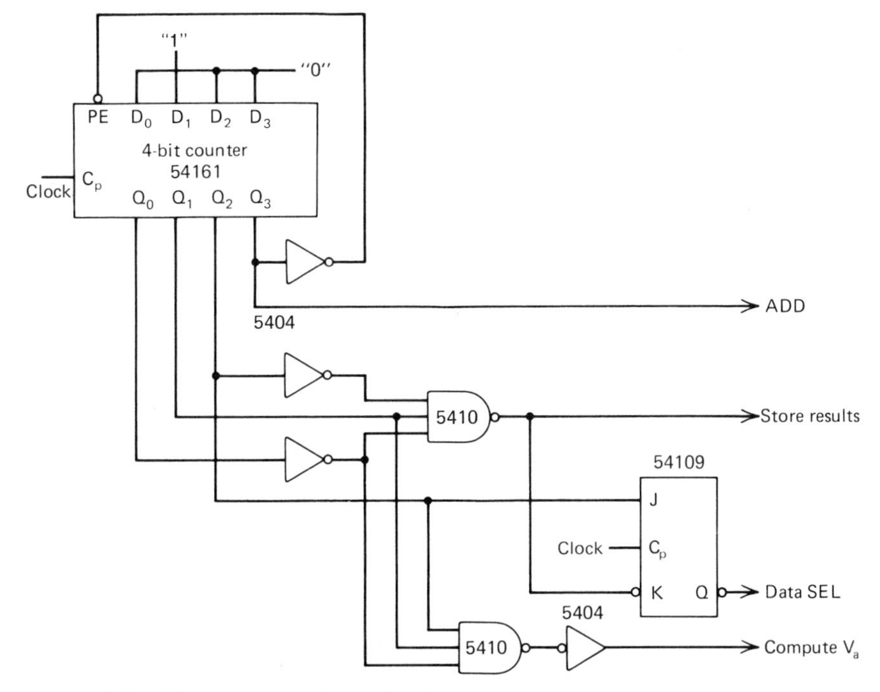

Figure 1.33 The complete circuit for Ex. 1.12.

FAST) exhibits a combination of speed and power which are superior to TTL. FAST circuits are fabricated with an advanced process which produces devices exhibiting very high frequency of operation and switching speeds.

The circuit of a basic FAST 2-input NAND gate is illustrated in Fig. 1.34. Unlike TTL, the gate has three stages of gain provided by Q_1, Q_2, and Q_3, raising the input-threshold voltage and increasing output drive. A higher input threshold allows the use of pn junction diodes, D_1 and D_2, to perform the AND function. The capacitance of a junction diode is low and the ac noise immunity is improved.

Diodes D_3 and D_4 act as low-resistance paths to discharge the parasitic capacitance at the base of Q_2. When Q_2 turns on and its collector voltage falls, D_7 provides a discharge path for the capacitance. Diode D_8 rapidly discharges the load capacitance by supplying additional base current to Q_3 through Q_2. Schottky-clamping diodes built into the transistors prevent saturation, thereby minimizing storage time turn-on and turn-off problems.

Table 1.6 summarizes typical family characteristics of the 54F series of FAST logic. With respect to TTL, the input load of a typical FAST gate is 20 μA, or 0.5 U.L., in the high state and 0.6 mA, or 0.375 U.L., in the low state. The output drive capability in the logic 1 state is 25 U.L., or 1 mA, and in the logic 0 state is 12.5 U.L., or 20 mA. Therefore, FAST logic can drive many more gates than TTL. The propagation delay of FAST logic is typically 3.9 ns from a low to a high transition and 3.5 ns from a high to low.

Presently, many FAST devices are available with pin-for-pin replacements for most TTL gates.

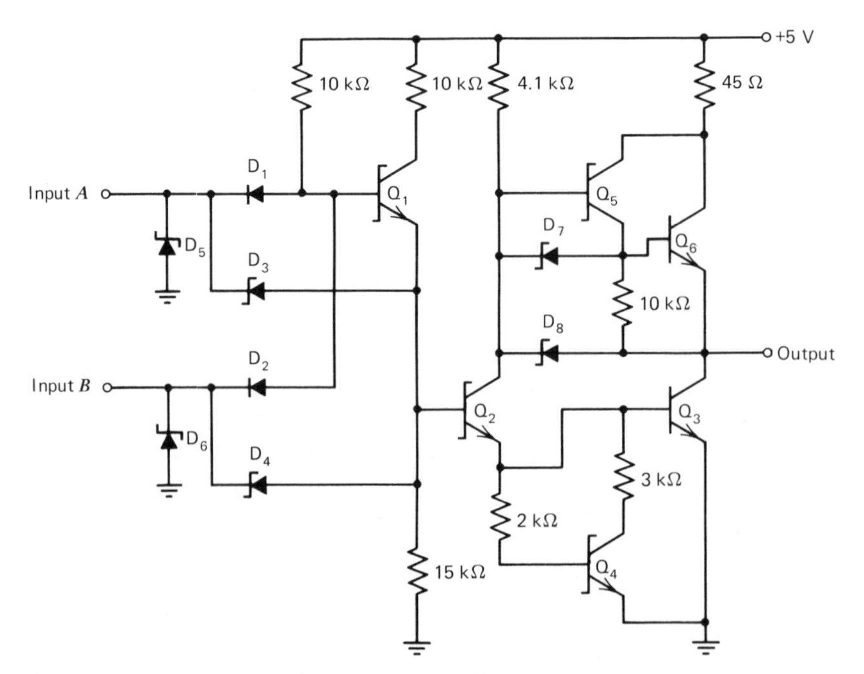

Figure 1.34 Circuit of a FAST 2-input NAND gate.

Table 1.6 **Characteristics of the 54F FAST logic family**

Input loading/fan out	1 U.L. = 40 μA in the 1 state 1 U.L. = 1.6 mA in the 0 state		
High/low inputs (U.L.)	0.5/0.375		
High/low outputs (U.L.)	25/12.5		
Propagation delay (ns)	min.	typical	max.
t_{pLH}	2.5	3.9	5.5
t_{pHL}	2.5	3.5	5.0

1.11 REFERENCES

Fairchild, *FAST Applications Manual*, latest edition.

Greenfield, J. D., *Practical Digital Design Using IC's*, second edition, Wiley, New York, 1982.

Hill, F. J. and G. R. Peterson, *Digital Systems: Hardware Organization and Design*, second edition, Wiley, New York, 1978.

Lancaster, D., *TTL Cookbook*, Sams, Indianapolis, 1974.

Peatman, J. P., *The Design of Digital Systems*, McGraw-Hill, New York, 1972.

Texas Instruments, *Designing with TTL Integrated Circuits*, McGraw-Hill, New York, 1971.

Texas Instruments, *TTL Applications Manual*, latest edition.

CHAPTER 2

Integrated Injection Logic

H. K. Hingarh
Manager Circuit Design Advanced Logic
Fairchild Camera and Instrument Corporation

2.1 INTRODUCTION

Integrated injection logic (I^2L) represents an innovation in bipolar integrated circuits that achieves high packing density and low power-delay product. The basic logic cell is realized through the integration of a lateral pnp and a vertical multicollector npn transistor. In a sense, I^2L is not a new technology, but an extension of bipolar technology into LSI by new circuit design techniques. The logic offers the following features:

1 High packing density.
2 Lowest power–delay product of major logic families.
3 Compatible linear and digital functions on the same chip.
4 Higher radiation resistance than MOS.
5 Low-power and low-voltage operation.

Its disadvantages includes:

1 Low current gains.
2 Limited fan out capability.

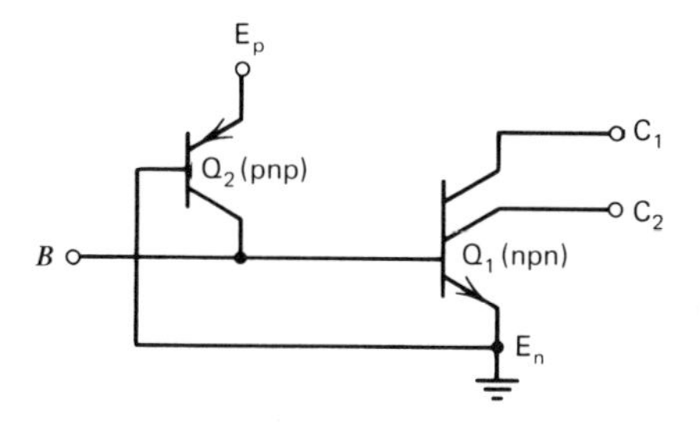

Figure 2.1 Basic I²L structure. Vertical npn transistor Q_1 with multiple collectors C_1 and C_2 operates as an inverter. Lateral pnp transistor Q_2 serves as a current source and load.

2.2 BASIC I²L STRUCTURE

In Fig. 2.1, the basic I²L structure has isolated multicollector outputs. Resistors are replaced by pnp transistor current sources that have their collectors connected to the bases of the npn device. The emitters of the npn transistors are common with the base of the pnp transistor, which injects charge directly into the npn bases. The npn transistors are operated in the *inverse* mode. In this mode, conventional npn emitters operate as collectors.

In all cases, the npn transistor serves as an inverter, and logic is implemented by wiring their collectors. Figure 2.2 shows the implementation of a NAND gate. Other logic functions, such as NOR/OR and AND gates, can be easily obtained. These and other gates are discussed in Sec. 2.4.

The operation of the basic I²L gate can be explained with reference to Fig. 2.3, which shows the output of one gate being cascaded with the inputs of two following gates. When input *A* to the first gate is at a high state (logic 1), the injector current (pnp collector current) is applied directly to the base of Q_0. This causes Q_0 to saturate and all its outputs go to a low state (logic 0). Each output is at approximately 50 to 100 mV above ground. This results in the collector currents, I_1 and I_2 of Q_4 and Q_5, respectively, being diverted from the bases of Q_1 and Q_2. Consequently, Q_1 and Q_2 are turned OFF and go to a high state (clamped to the base-emitter voltage of succeeding stages).

It should be noted that, for proper operation of an inverter stage, each output should be able to sink collector current equal to, or greater than, the injector current of a succeeding stage. Hence, each of the I²L multicollector outputs is generally restricted to drive one gate only.

Structure Details

Figure 2.4 shows the cross section and topology of a typical three-collector I²L gate realized with standard bipolar IC technology. The logic gate is formed in an n-epitaxial region over an n+ buried layer. Selective n+ diffusion is used to define the boundaries of both the individual gates and the lateral pnp transistor. This is followed by a selective p-type diffusion to form the npn base and the emitter of the lateral pnp devices. The next process is an n+ diffusion to define the multicollector outputs of the logic gate. The process is then completed by formation of contacts and interconnections. (These processes are explained in Chap. 20.)

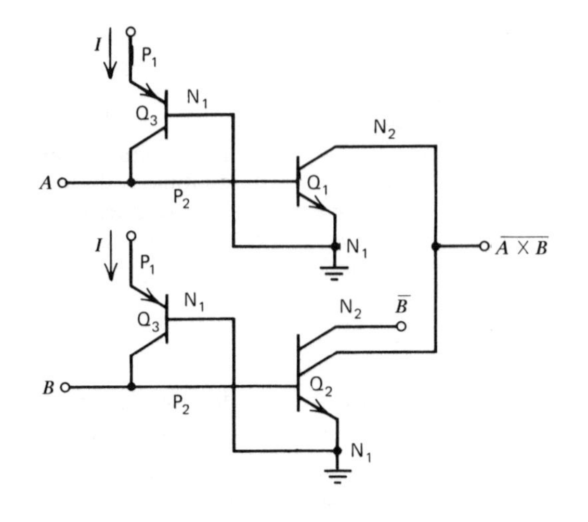

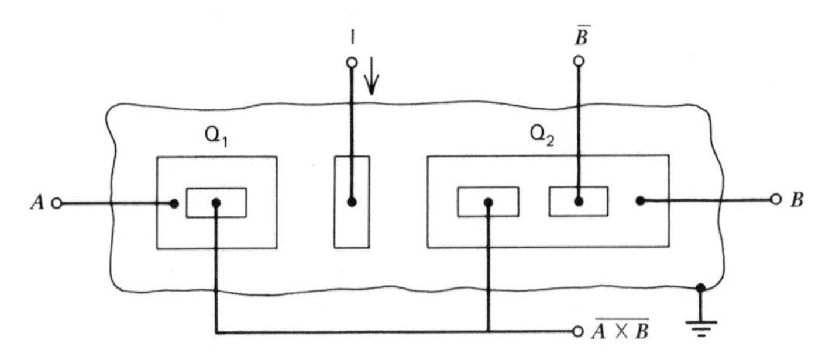

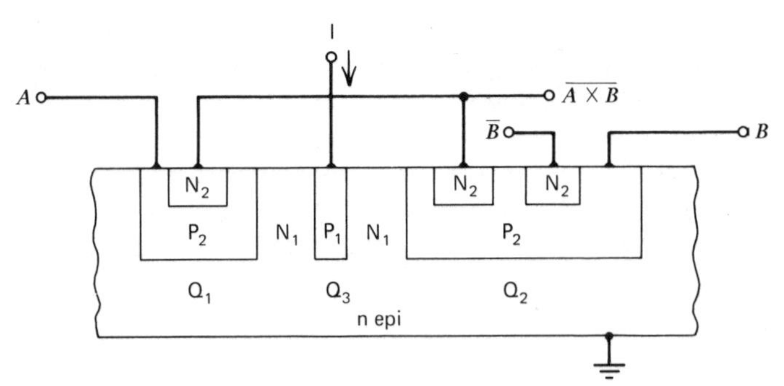

Figure 2.2 Implementation of NAND gate using I^2L. (*a*) Circuit schematic. (*b*) Chip layout.

A simplified version of I^2L that omits the buried n+ layer and utilizes n+ substrates has been used for specific logic applications. The full approach described above, however, is more desirable because selective n+ diffusion has two important effects:

1 It improves the current gain of an I^2L transistor.

2 It reduces parasitic carrier injection from injectors adjacent to gates.

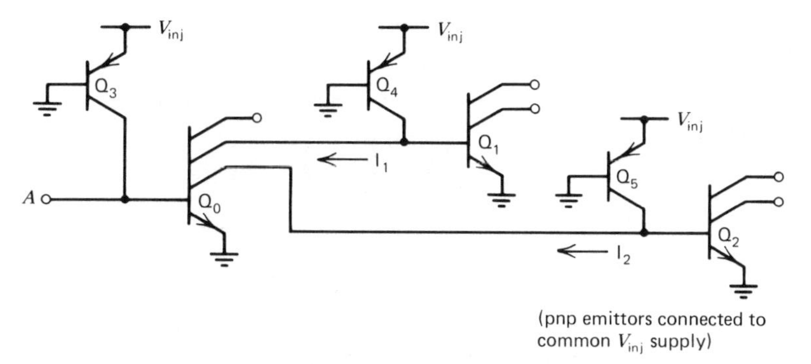

Figure 2.3 Operation of basic I²L gate.

Figure 2.5 illustrates a cross section of a conventional npn transistor and an I²L cell in p-substrate material. This process makes it possible to combine analog, ECL, and TTL circuits with I²L.

Packing Density

The high packing density of I²L stems from:

1 The absence of area-consuming diffused resistors.

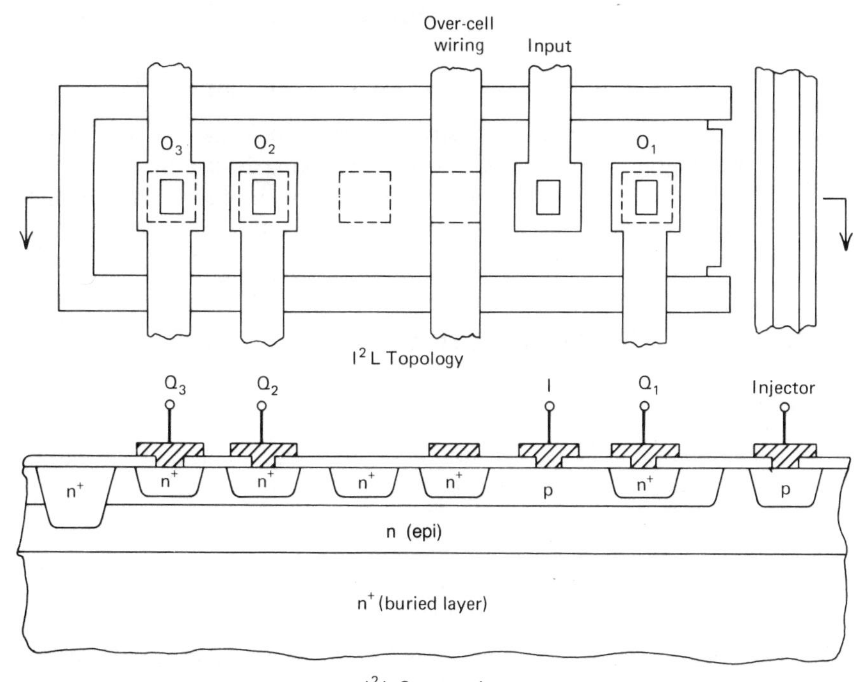

Figure 2.4 Topology and cross section of an I²L cell. Intercell wiring can be routed over unused outputs. This contributes to the high packing density of I²L. (Courtesy *IEEE Spectrum,* June 1977.)

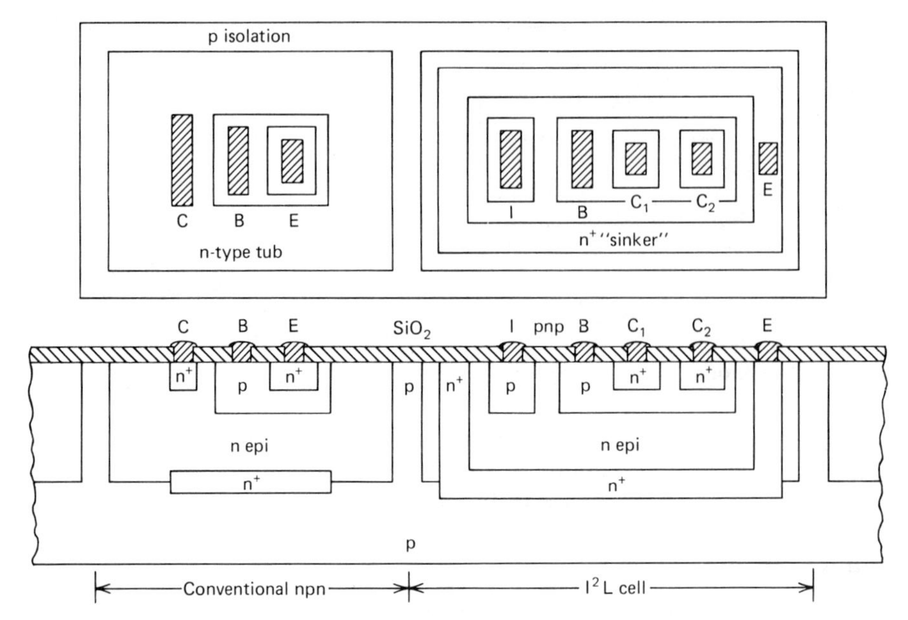

Figure 2.5 Bipolar compatible I^2L structure.

2 The small number of contact windows required per gate (fan out plus gate contacts).

3 Intercell wiring that can be routed over unused base or output areas.

These factors, plus the repetitive structure of a logic gate, result in extremely dense and highly optimized IC layouts. Applying 10 μm minimum tolerances, a packing density of 40 to 80 gates/mm^2 has been obtained. Reducing the minimum dimension to 5 μm, a packing density of 160 to 300 gates/mm^2 has been realized.

2.3 ELECTRICAL CHARACTERISTICS

The dc Characteristics

The important dc parameters are the npn beta, the pnp alpha, and the fan in and fan out capabilities. The npn beta, β, is:

$$\beta = \frac{I_C}{I_B} \tag{2.1}$$

where I_C is the collector current and I_B is the base current. Figure 2.6 provides a plot of β versus I_C for a four-collector device. Beta in the inverse mode is quite low, 3 to 10, compared to beta of 50 to 100 for a conventional transistor. For dc operation, a minimum beta of 1 to 2 is required.

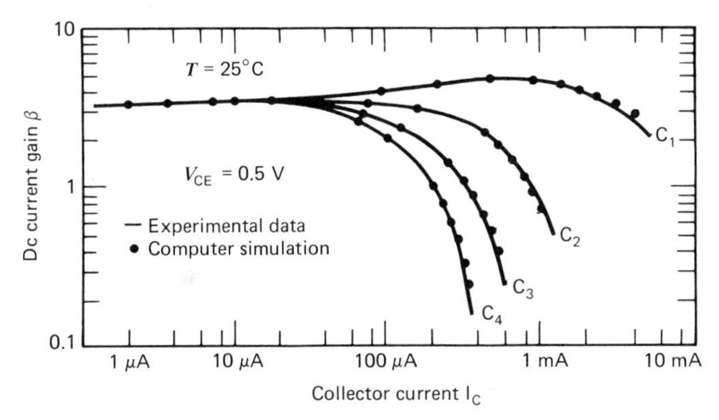

Figure 2.6 Npn dc current gain versus collector current for a fan out of four gate.

For the pnp lateral transistor, the common base current gain, α_p, is:

$$\alpha_p = \frac{I_{Cp}}{I_{inj}} \tag{2.2}$$

where I_{Cp} is the collector current of the pnp transistor and I_{inj} is the base current. Figure 2.7 shows a plot of alpha versus collector current in the forward (α_F) and reverse (α_R) modes of operation. There is a strong dependence of the npn beta on the reverse pnp alpha. The decay at low currents is directly proportional to the forward pnp alpha.

Cell Gain

In typical circuit layouts, many I²L gates have their injectors tied together. In such an array, the gate output must be able to sink the injector current of an

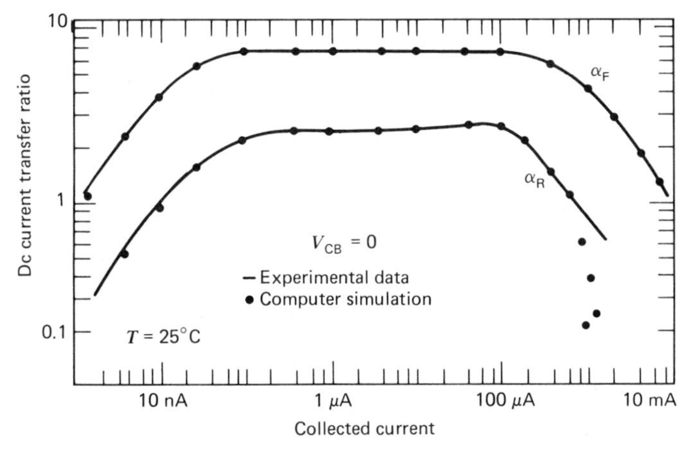

Figure 2.7 Forward (α_F) and reverse (α_R) dc current transfer ratios versus collector current for a lateral pnp transistor.

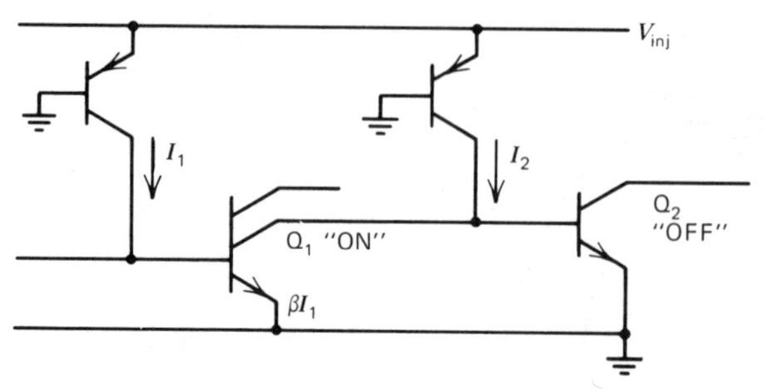

Figure 2.8 Significance of cell gain.

adjacent cell. An important parameter relating to the fan out capability is the collector cell gain:

$$\text{cell gain} = \frac{I_C}{I_{inj}} \approx \beta\alpha_p \tag{2.3}$$

If more than one collector is used, the cell gain is defined with respect to each output, as indicated in Fig. 2.6. In Fig. 2.8, for proper operation, $\beta I_1 > I_2$ assuming that $I_1 = I_2$. Thus, the cell gain of each collector must be greater than, or equal to, one. This requirement applies over supply current range, temperature range, and collector fan out. For this reason, it is desirable to have a minimum cell gain of 2 to 5.

Fan Out

In I²L, the fan out of a gate is limited for the following reasons:

1 As the number of collectors increases, the gain of each individual collector decreases.

2 Propagation delay increases with fan out owing to an increase in capacitance.

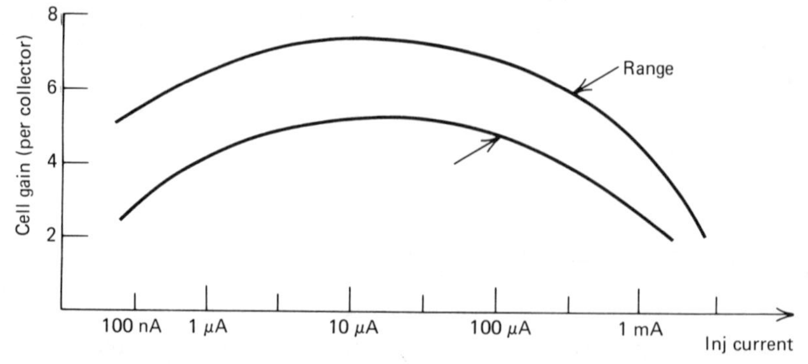

Figure 2.9 Cell gain versus injector current.

Figure 2.9 is a plot of cell gain versus injector current. With a range of injector current from nanoamperes to milliamperes, the cell gain varies with cell design and the manufacturing process. In practice, a high value of cell gain is desirable. Experimental and theoretical work indicates, for example, that both the gate noise margin and propagation delay improve with increasing cell gain. It is possible to raise the cell gain by paralleling additional outputs.

In regard to fan in, no limit exists for most designs except that ac performance degrades with increasing fan in.

Noise Margin

The I^2L gate has a low voltage noise margin but high current noise margin. The voltage noise margin, NM, is:

$$\mathrm{NM} = (kT/q) \ln (\beta_{\mathrm{eff}}) \tag{2.4}$$

where β_{eff} is the npn beta with the pnp injector grounded, k is Boltzmann's constant (1.38×10^{-23} J/°K), q is charge of electron (1.6×10^{-19} C), and T is temperature (°K).

If, for example, $\beta_{\mathrm{eff}} = 10$, $kT/q = 26$ mV at room temperature, the voltage noise margin NM = 60 mV. This translates into a 10 to 1 change in current. For this reason, an I^2L cell can tolerate large noise currents.

The ac Characteristics

Figure 2.10 shows a typical plot of propagation delay, t_{D}, as a function of gate power, P_{D}, measured in a ring oscillator. For high power dissipation, 20 μW or more, the power–delay product begins to deviate from linearity and approaches an intrinsic delay limit of 15 to 20 ns.

Because the delay is a function of the layout, the propagation delay of a gate increases as more outputs are used. This stems from the fact that each output adds to the parasitic capacitance that must be charged and discharged during switching.

From Fig. 2.11, the region of constant power–delay product, called the *extrinsic region of operation*, extends over a wide range of operating current. In this region, the power–delay product is capacitance limited. The propagation delay time, t_{D}, depends on C_{o}—the total capacitance including depletion layer, stray, and load capacitance. For an output voltage swing, ΔV:

$$t_{\mathrm{D}} = \frac{C_{\mathrm{o}} \Delta V}{2 I_{\mathrm{Cp}}} \tag{2.5a}$$

$$= \frac{C_{\mathrm{o}} \Delta V}{2 \alpha_{\mathrm{p}} I_{\mathrm{inj}}} \tag{2.5b}$$

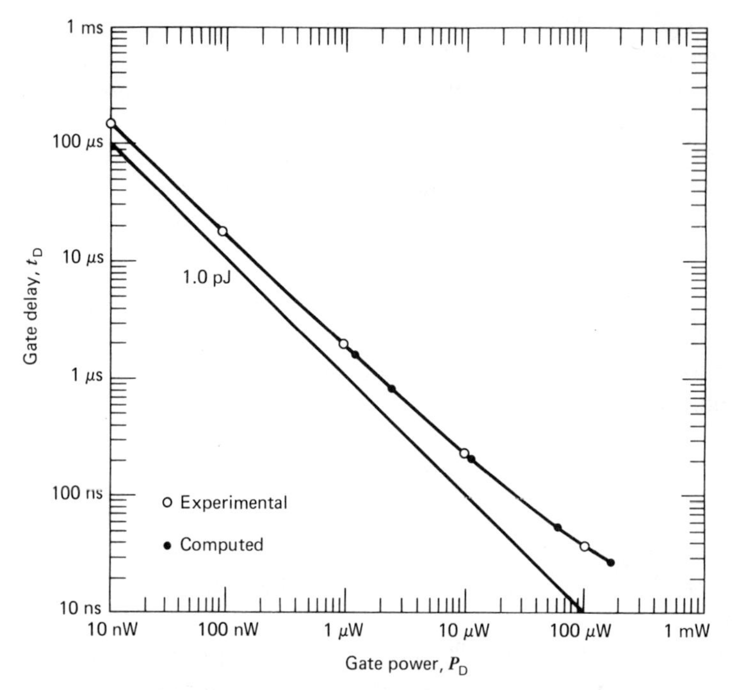

Figure 2.10 The power-delay product of I^2L exhibits an inverse linear relationship over a wide range of delay and dissipation. (Courtesy *IEEE Spectrum,* June 1977.)

The power–delay product can be expressed in terms of the injector voltage, V_{inj}, by:

$$\text{power–delay product} = \frac{C_o \Delta V}{2\alpha_p} V_{inj} \text{ joules} \tag{2.6}$$

EXAMPLE 2.1 Given that $C_o = 1$ pF, $\Delta V = 0.6$ V, $V_{inj} = 0.8$ V, $I_{inj} = 0.05\ \mu$A, and $\alpha_p = 0.6$. Calculate the gate delay and power–delay product.

PROCEDURE

1 By Eq. 2.5*b*,

$$t_D = (10^{-12} \times 0.6)/(2 \times 0.6 \times 0.05 \times 10^{-6}) = 10\ \mu\text{s}$$

2 By Eq. 2.6,

$$\begin{aligned}\text{power–delay product} &= (10^{-12} \times 0.6 \times 0.8)/(2 \times 0.6) \\ &= 0.4 \times 10^{-12}\ \text{J} = 0.4\ \text{pJ}\end{aligned}$$

Because V_{inj} is constant, the power–delay product also may be considered to be constant. It is possible to improve the power–delay product (see Eq. 2.6) by reducing the signal swing and the device capacitance.

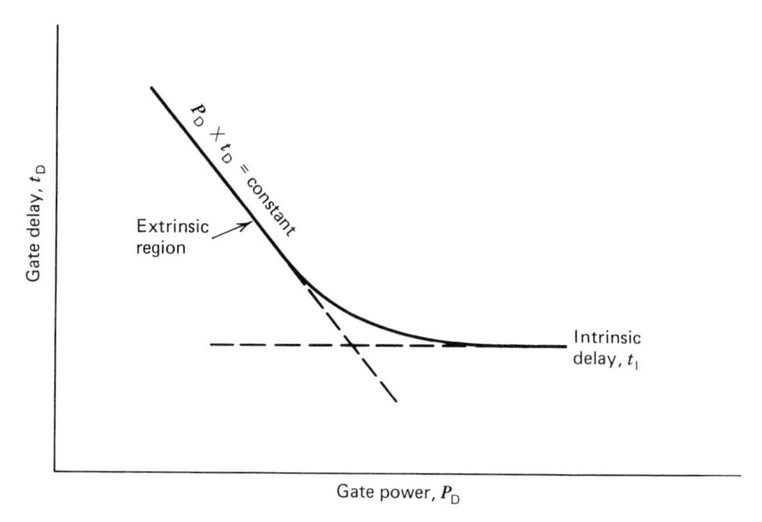

Figure 2.11 The limiting intrinsic delay stems from minority carrier charge storage effects.

An *intrinsic delay*, t_I, is ultimately reached as the gate power increases (Fig. 2.11). This limiting occurs predominantly because of minority-carrier charge storage effects. By minimizing the minority-carrier injection into the n-epitaxial region and making modification in the device structure and doping levels, we can move the onset of the intrinsic delay to a higher current level.

The power–delay product derived from a ring oscillator is optimistic, because a ring oscillator layout does not include parasitic capacitances. A more realistic measurement for design purposes is the maximum toggle frequency of a D flip flop in a divide-by-two mode.

2.4 LOGIC DESIGN WITH I²L GATES

Whereas other logic families are built around multiple-input, single-output gates, I²L is designed around single-input, multiple-output gates. One of the important steps in designing with I²L, therefore, is to convert conventional logic diagrams into I²L logic diagrams.

EXAMPLE 2.2 Develop a schematic of an I²L inverter and show its logic symbol.

PROCEDURE

1 The basic circuit is given in Fig. 2.12*a* and its logic symbol in Fig. 2.12*b*.
2 When $A = 1$, the injector current flows into the base of Q_1 and the transistor saturates ($F = 0$). If $A = 0$, the injector current is diverted from the base of Q_1 and $F = 1$.
3 Note that at any one of the three outputs, $F = \overline{A}$.

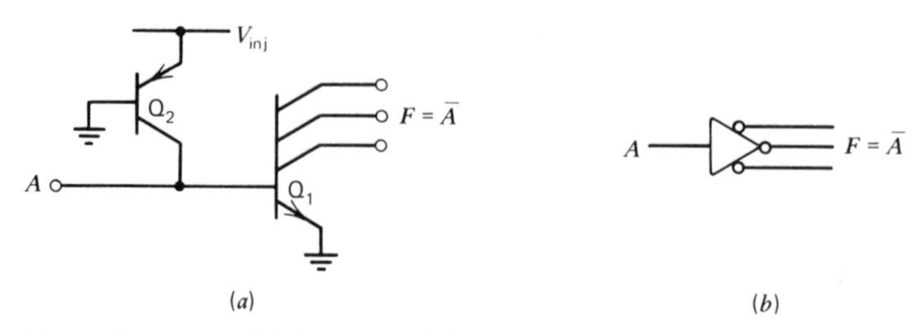

Figure 2.12 An I²L inverter. (*a*) Circuit schematic. (*b*) Logic symbol.

EXAMPLE 2.3 Show how AND and NAND gates may be implemented using I²L.

PROCEDURE

1 An AND function is formed by wire-ANDing the separate output collectors of the preceding gates.
2 In Fig. 2.13, $X = AB$ exists at the wired-AND connection. If X is applied to the inverter of Fig. 2.12, the NAND function $F = \overline{AB}$ is obtained.
3 Logic symbols are provided in Fig. 2.14.

EXAMPLE 2.4 Show how a NOR and OR gate may be implemented using I²L.

PROCEDURE

1 Two I²L inverters are wired as illustrated in Fig. 2.15 to realize the NOR gate.
2 By connecting an inverter to the output of the NOR gate of Fig. 2.15, an OR gate is obtained (Fig. 2.16).

EXAMPLE 2.5 Using I²L, realize an AND-OR-INVERT (AOI) gate.

PROCEDURE

The realized circuit is given in Fig. 2.17.

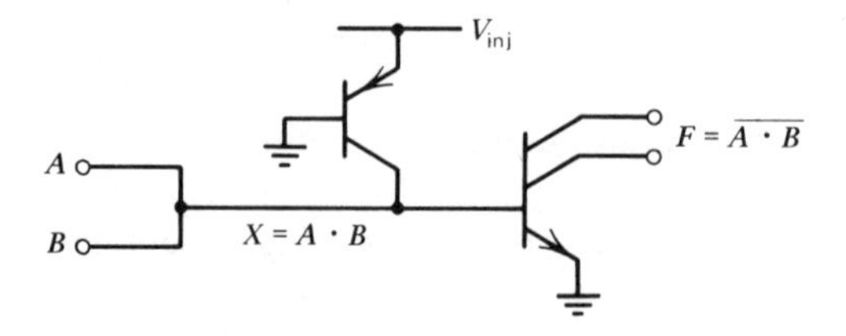

Figure 2.13 Implementing a 2-input NAND gate.

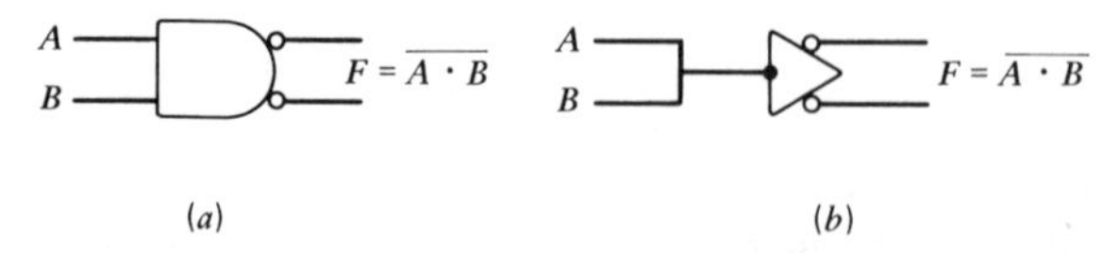

Figure 2.14 Logic symbols for (*a*) NAND gate and (*b*) wired-AND and INVERT.

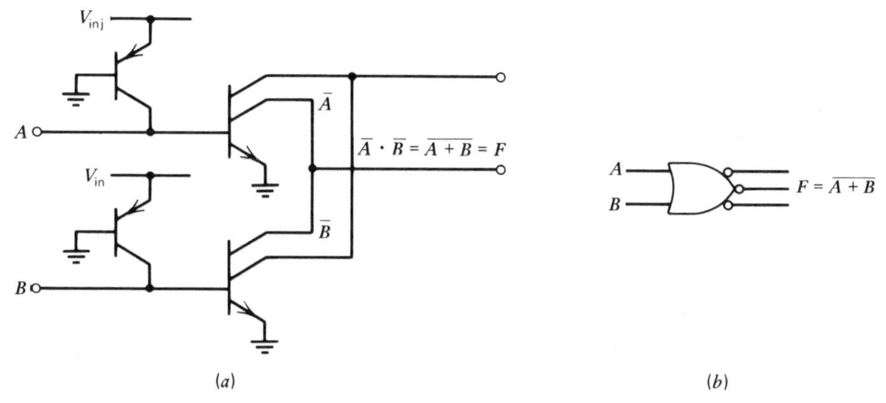

Figure 2.15 An I²L 2-input NOR gate. (*a*) Circuit schematic. (*b*) Logic symbol.

EXAMPLE 2.6 Using I²L, implement an Exclusive-NOR gate.

PROCEDURE

The realized circuit is given in Fig. 2.18.

EXAMPLE 2.7 Reduce the logic diagram of the full adder of Fig. 2.19*a* to a minimum gate I²L circuit. Calculate the delay through one long path in the layout and the total required power.

PROCEDURE

1 Redraw the logic diagram using multiple-output collectors as shown in Fig. 2.19*b*.
2 Signals A, B, and C in Fig. 2.19*a* have more than one fan out. In Fig. 2.19*b*, if the preceding gates for signals A, B, and C have enough fan out capability, then extra buffering gates 2, 4, and 6 are not required.

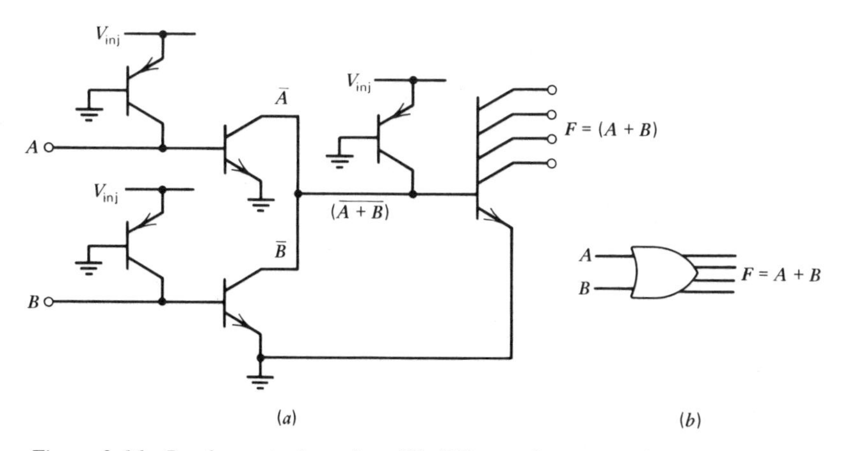

Figure 2.16 Implementation of an I²L OR gate by connecting an inverter to output of a NOR gate. (*a*) Circuit schematic. (*b*) Logic symbol.

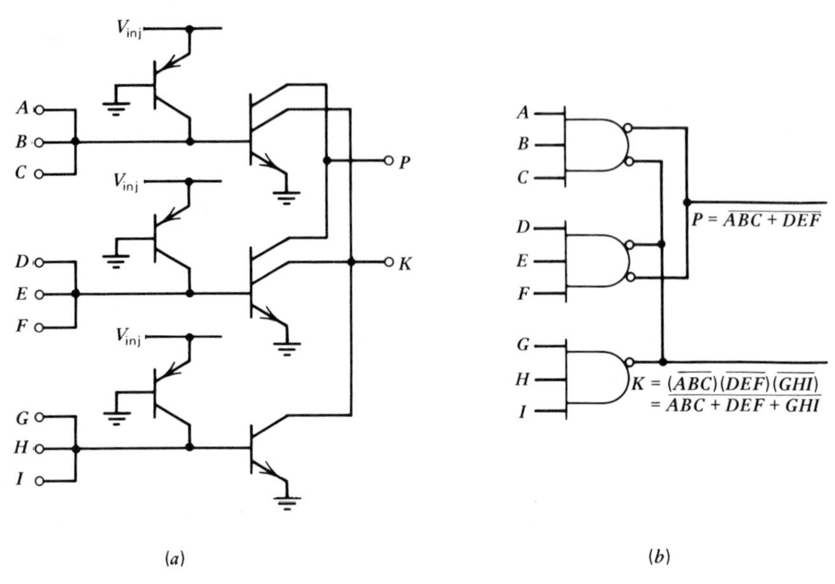

Figure 2.17 An AOI gate. (*a*) Circuit Schematic. (*b*) Logic symbol.

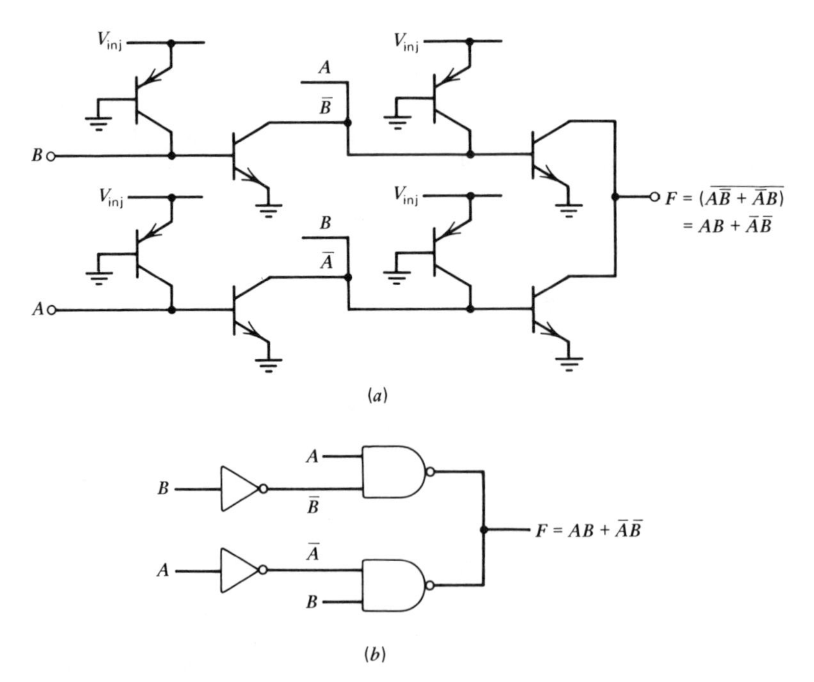

Figure 2.18 An exclusive-NOR gate. (*a*) Circuit schematic. (*b*) Logic symbol.

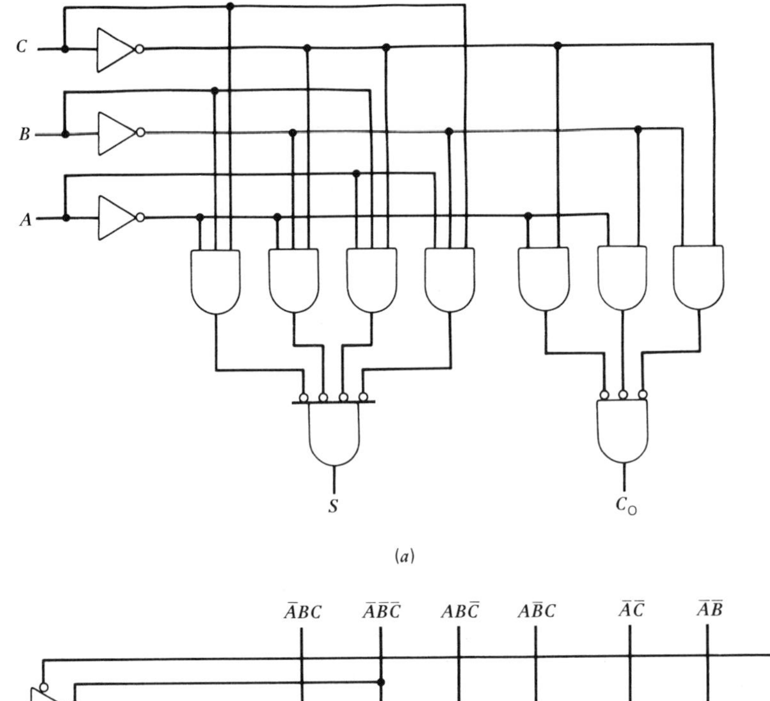

(a)

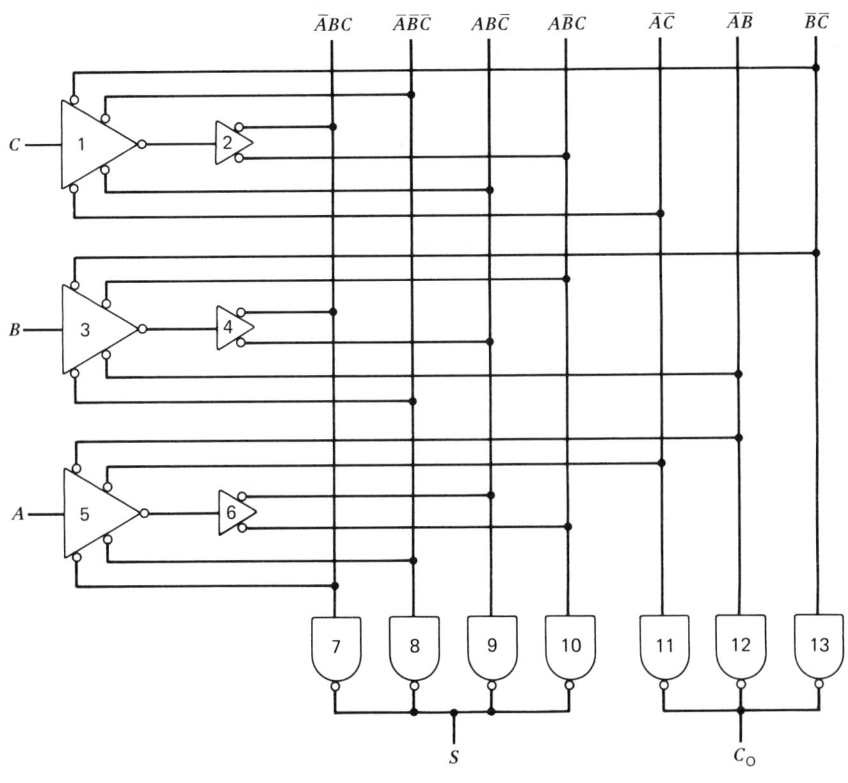

(b)

Figure 2.19

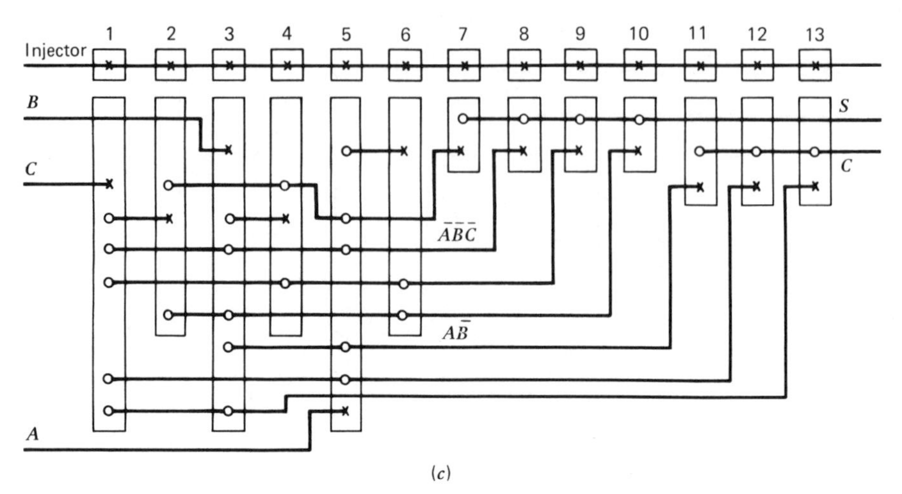

Figure 2.19 Implementing a full adder using I²L. (*a*) Given logic diagram. (*b*) I²L logic diagram. (*c*) Cell layout.

3 The logic generated in steps 1 and 2 above yields a minimum I²L gate layout, shown in Fig. 2.19*c*. The layout interconnections have the following notations:
  a. A small circle represents a collector output.
  b. A small x is an input to a gate.
  c. A large rectangle defines the gate area.

4 AND logic is implemented by wiring together the collectors of the appropriate stage.

5 The long path in the layout is from input A to output S. Assume that $B = C = 0$ and A goes from high to low. Then gate 5 will switch from high to low and gate 8 will go high. Assuming the power/cell = 100 μW, from Fig. 2.10, the delay per gate is 30 ns. The total delay is, therefore, $2 \times 30 = 60$ ns.

6 For 13 gates, the total power required is 13×100 μW = 1.3 mW.

EXAMPLE 2.8 Draw the layout for the I²L latching flip flop of Fig. 2.20*a*.

PROCEDURE

The resultant layout is provided in Fig. 2.20*b*.

EXAMPLE 2.9 For an edge-triggered D flip flop whose logic diagram is given in Fig. 2.21*a*, develop the layout for the circuit using I²L.

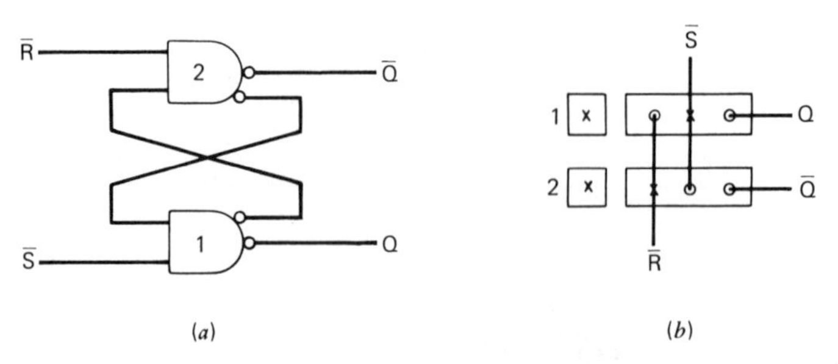

Figure 2.20 Latching flip flop. (*a*) Logic diagram. (*b*) Cell layout.

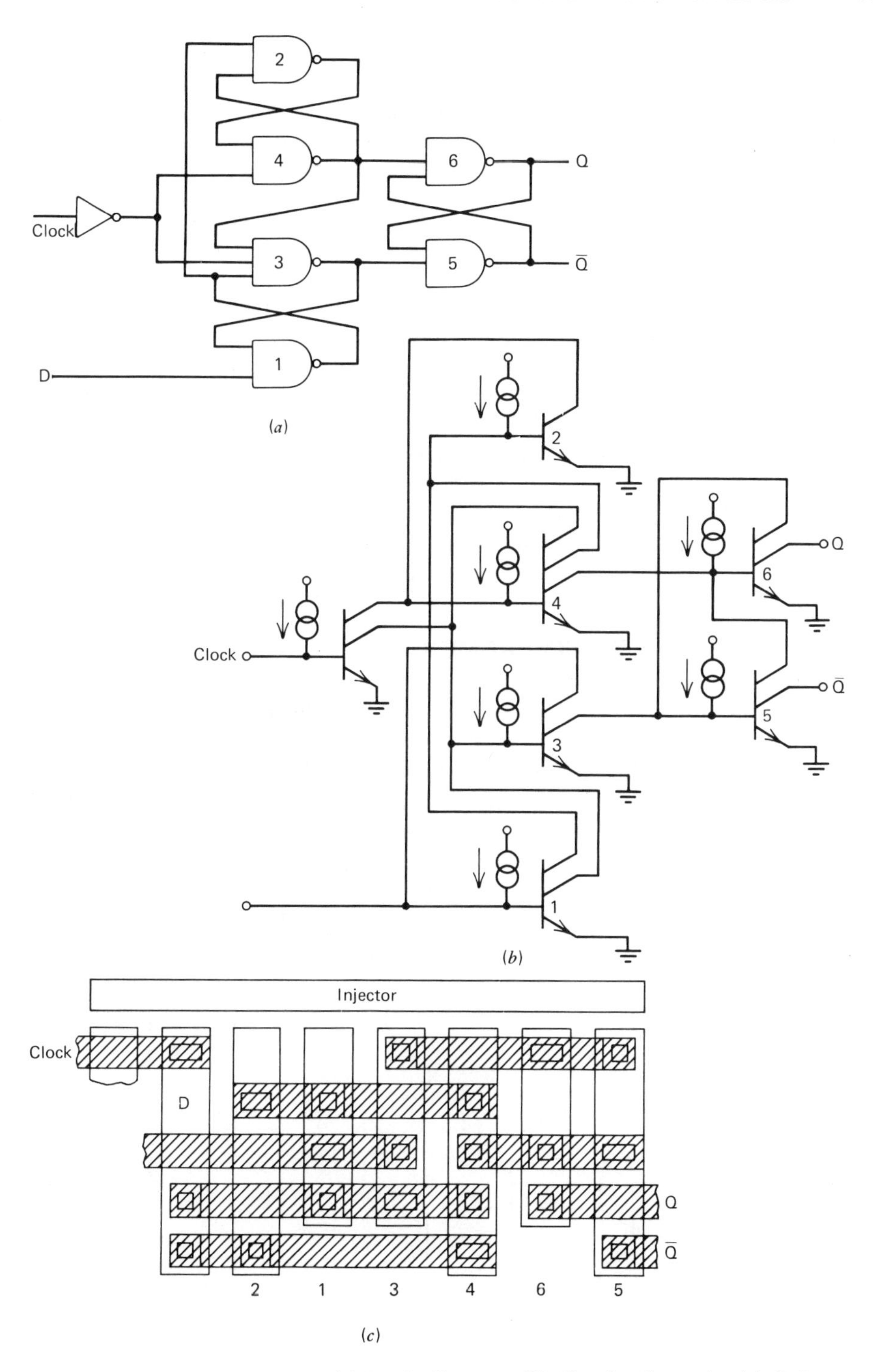

Figure 2.21 D-type flip flop. (*a*) Logic diagram. (*b*) Circuit schematic. (*c*) Cell layout which occupies 19 mm².

PROCEDURE

1 The schematic, based on I^2L, is first drawn (Fig. 2.21*b*).
2 From the schematic obtained in step 1, the layout is developed as shown in Fig. 2.21*c*.

2.5 INTERFACING

The input and output levels of a typical I^2L gate are not compatible with other logic families. Circuit modifications are, therefore, required to provide the desired compatibility. Figure 2.22 shows an example of TTL input-output interfacing with I^2L. By using conventional Schottky TTL, the components can be integrated on the same chip with I^2L.

Other examples of input interfacing are shown in Fig. 2.23. The circuit of Fig. 2.23*a* uses a simple resistor divider to provide the TTL input threshold. Figure 2.23*b* shows an input stage having a threshold equal to $2V_{BE}$ V. Resistor R_2 must be greater than R_1. Figure 2.24 illustrates a simple inverter transistor used as an open-collector TTL output. Such a stage requires a pullup resistor.

2.6 POWER SUPPLY REQUIREMENTS

One of the important features of I^2L is that circuit power and performance can be controlled externally for a given design. An I^2L circuit looks like a forward-biased diode to ground at the power supply pin. Thus, the power supply voltage is fixed by the circuit at approximately 0.8 V and the supply current is controlled by an external current-limiting resistor. In large I^2L chips, the power dissipation and the performance of the circuit can be selected to any value along the power–delay curve (Fig. 2.10) by varying the value of the resistor.

If a 5 V TTL power supply is used, then a series dropping resistor can be connected between the 5 V bus and the injector pin of the chip (Fig. 2.25). In

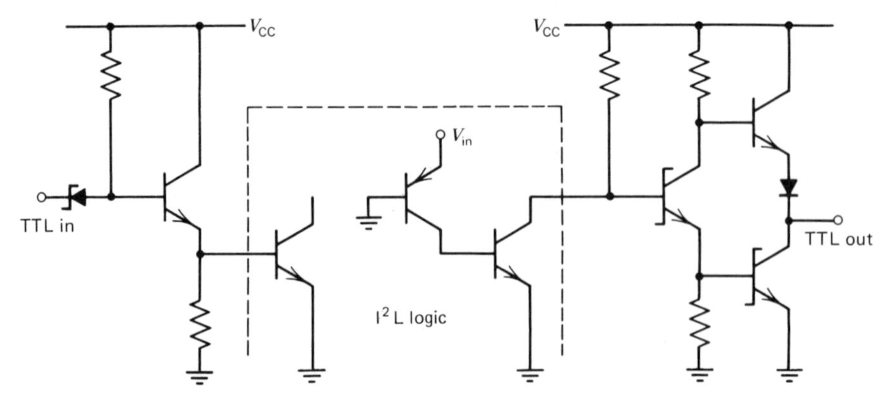

Figure 2.22 An example of interfacing I^2L with TTL. (Courtesy Wescon 1975.)

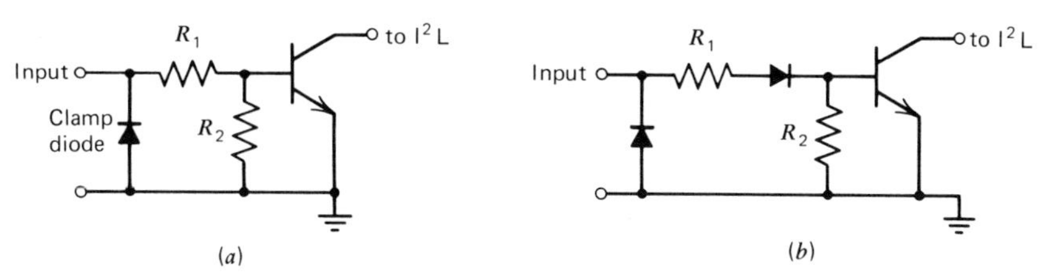

Figure 2.23 Input interfacing to I^2L. (*a*) Using a resistor divider. (*b*) Using two diodes.

general, the value of resistance R is:

$$R = (V_{CC} - 0.8)/I_{inj} \tag{2.7}$$

In some designs, it is possible to power the entire circuit from a single 5 V supply. To reduce power dissipation in a package, stacked I^2L logic cards may be used.

2.7 MAKING A SELECTION

When considering IC designs at the LSI level, several factors have to be examined. While power and speed are prime performance factors, packing density, noise immunity, and other design factors need to be considered. A comparison of I^2L with other logic families is provided in Table 2.1.

Cell packing density, a major LSI parameter, is a critical determinant in the overall chip area for a given complex function. Maximum cell density and array complexity are realized by I^2L and NMOS technologies. Gate densities of 600 gates/mm^2 have been reported for advanced I^2L. Isoplanar Integrated Injection Logic, I^3L^{TM}*, is an example of an advanced I^2L technology. It utilizes:

1 Oxide isolation, which reduces device size and parasitic capacitances.

2 Ion implantation (see Chap. 20).

3 Dual-layer metal for interconnects.

* I^3L is a trademark of Fairchild Camera & Instrument Corp.

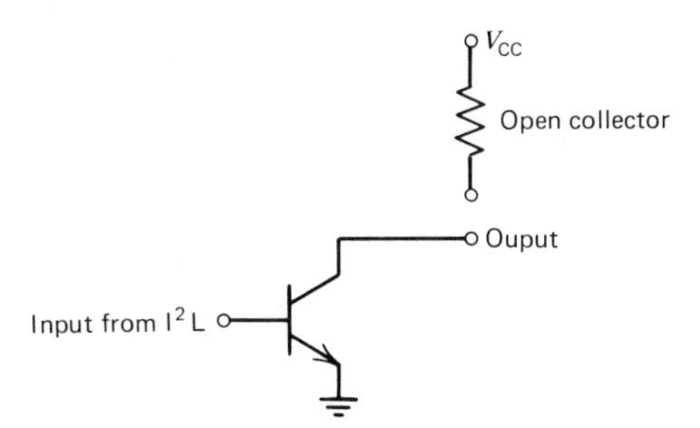

Figure 2.24 Inverter used as an open collector TTL output.

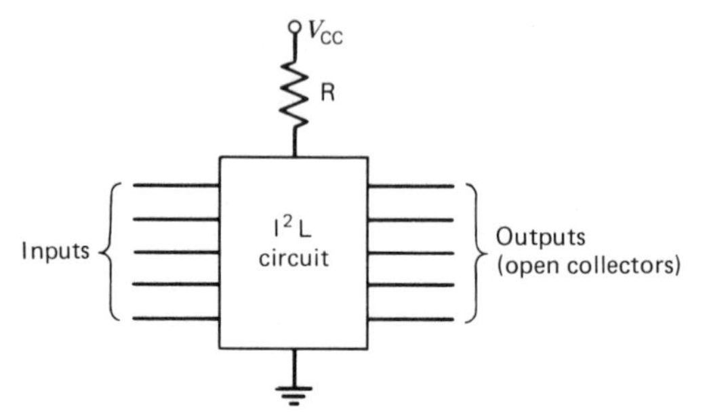

Figure 2.25 Connecting a power supply to an I²L circuit.

All these features produce a technology with high packing density and faster speed. The device structure and its gate equivalent are shown in Fig. 2.26.

One of the key performance criteria for a logic family is the power–delay product. Increasing data processing and control requirements seem to demand continuously greater array speeds. Power dissipation is another critical parameter in LSI designs owing to the limitations of the chip heat-sinking capabilities and power requirements. Figure 2.27 provides a comparison of speed and power requirements for various logic families. It is seen that I²L is superior to other logic families in power–delay product. Clearly, I²L cannot obtain the delay times of Schottky TTL and ECL. Using advanced I²L, however, D flip flops operating between 30 and 50 MHz have been realized.

A detailed geometric area comparison for the various logic families is shown in Fig. 2.28. Using a 6 μm design rule, even though considerable progress has been made in all technologies, the layouts in Fig. 2.28 are still valid. Table 2.2 provides a summary of features for commonly used logic families.

Table 2.1 **Subjective comparison of LSI technologies**

	S/C TTL	*ECL*	*I²L*	*NMOS*	*CMOS*	*CMOS/ SOS*
Cell density	0	−	+ +	+ +	0	+
Switching speed	+	+ +	0	0	−	+ +
Static power dissipation	−	− −	+	−	+ +	+ +
Dynamic power dissipation	+	+	+ +	+	0	+
Speed-power product	0	0	+ +	+	0	+ +
Output drive capability	+	+	0	0	−	− −
Noise immunity	+	0	− −	0	+ +	+ +
Temperature range	+	+	+	−	0	−
Neutron damage	0	+ +	−	+	+	+ +
Long-term ionization damage	+	+ +	+	− −	−	− −
Transient logic upset level	0	0	+	0	+	+ +

+ + Superior, + Good, 0 Average, − Below average, − − Weak
Source: IEEE Transactions on Nuclear Science, vol. NS-24, December 1977.

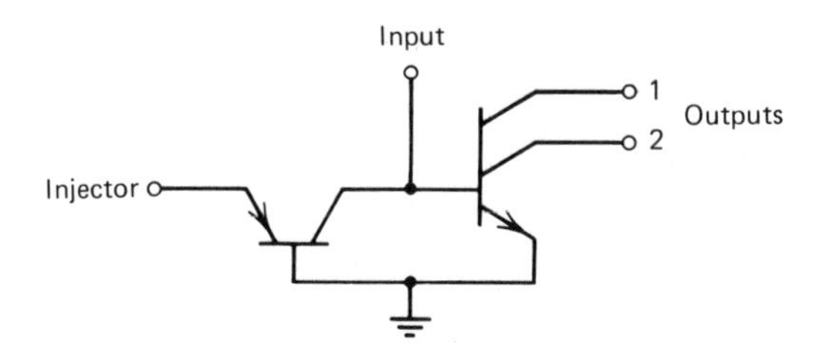

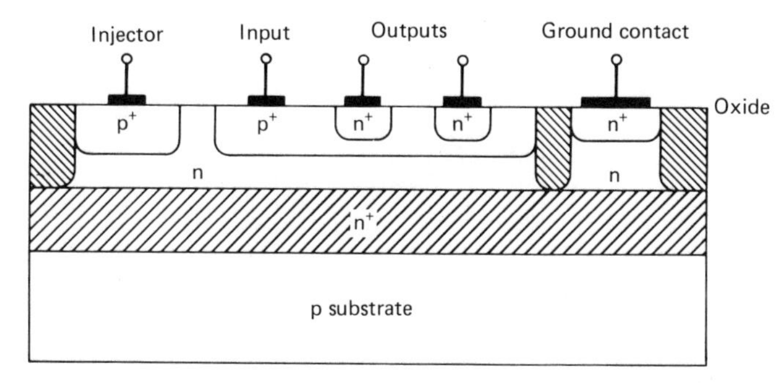

Figure 2.26 Basic I³L(TM) gate structure. (Courtesy Fairchild Camera and Instrument Corp.)

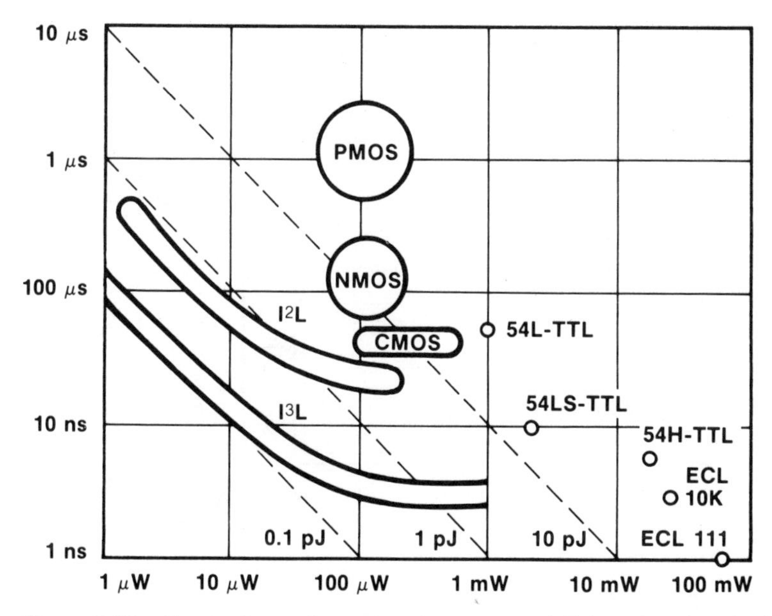

Figure 2.27 Comparison of speed and power capabilities of various logic families. (Courtesy Fairchild Camera and Instrument Corp.)

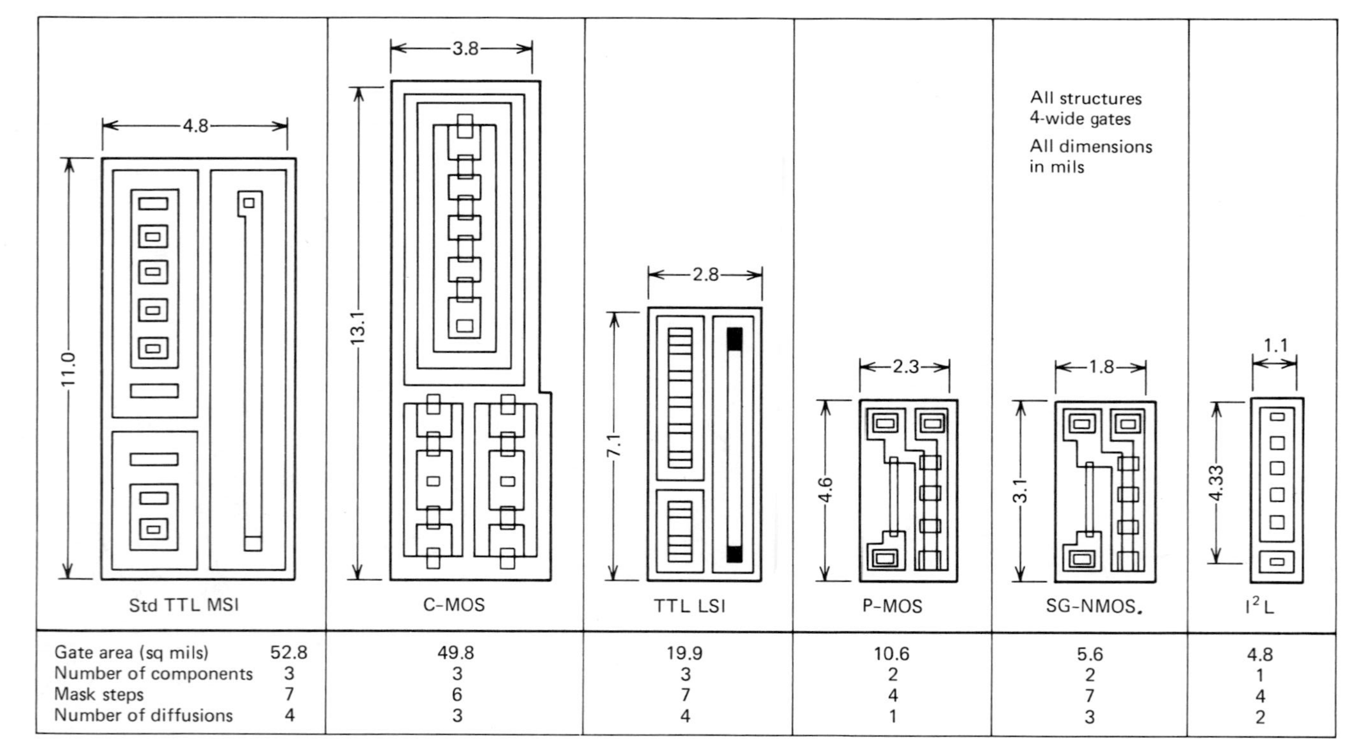

Figure 2.28 Comparison of gate areas used for various logic families. (Courtesy Texas Instruments, Inc.)

Table 2.2 **Summary of features of logic families**

Schottky TTL Summary	*CMOS*
• Played dominant role in digital IC technology • Consumes only $1/5$ of the power of saturated TTL • Fast circuit operation • High silicon area — uses multichip organization for microprocessor applications	• Utilizes both p-channel and n-channel on same substrate • More complex processing • Improved speed-power product over p- and n-channel MOS
p-Channel MOS	*SOS*
• Enhancement mode • Excellent packing density • Slow speed	• Similar to CMOS • Forms devices on insulating substrate of sapphire • Reduced device capacitance • Improved speed • More costly
n-Channel MOS	*I^2L*
• Faster than p-channel MOS • Normally depletion mode • Needs channel stoppers for isolation • Self-aligned silicon gate improves performance • High packing density	• Eliminates load resistors and current sources of TTL • Reduced power consumption over bipolar • Greater packing density than bipolar • Mixes speed of bipolar with packing density of MOS

Courtesy of Texas Instruments.

2.8 APPLICATIONS OF I^2L

The applications of I^2L to date fully utilize both the high functional density and on-chip circuit versatility. Initial applications include a 1500-gate 4-bit microprocessor element, developed by Texas Instruments, and I^2L and CMOS competitive circuits for applications in electronic watches and timers. In addition, TTL compatible I^2L devices have been developed by various manufacturers.

More recently, advanced forms of I^2L have been used in new LSI products. Texas Instruments has produced a family of components that constitute an I^2L microcomputer chip set. The family includes a 16-bit single-chip microprocessor (SBP 9900). Fairchild Camera and Instrument Corporation has applied their isoplanar techniques in I^3L. They have produced several LSI devices using this technology, including microprocessors.

The marriage of I^2L with on-chip linear and digital functions has been applied effectively by several manufacturers, notably Analog Devices and Exar. Linear applications are possible because I^2L processing can be made compatible with that of linear bipolar devices. Typical commercial circuits consist of timer/counters, phase-locked loops, frequency synthesizers, and A/D and D/A converters. Other I^2L/linear developments are focused on designs intended for the telephone industry.

The highly regular nature of I^2L gate layout is well suited for uncommitted gate arrays for semicustom LSI. Several manufacturers have developed chips ranging from 300 to 5000 gates in complexity.

Finally, LSI circuits using I^2L can be built to operate over the full military temperature range and exhibit good radiation resistance characteristics.

2.9 REFERENCES

Berger, H. H. and S. K. Wiedman, "Merged Transistor Logic—a Low Cost Bipolar Logic Concept," *IEEE Journal of Solid State Circuits*, vol. SC-7, pp. 340–346, October 1972.

Brown Jr., P. M., "Complex LSI Design Using I^2L Gate Arrays," Wescon, 1980.

Hart, K. and A. Slob, "Integrated Injection Logic—A New Approach to LSI," *IEEE Journal of Solid State Circuits*, vol. SC-7, pp. 346–351, October 1972.

Hingarh, H. K., "A High Performance 4000-Gate I^3L Array," Wescon, 1980.

Horton, R. L. et al., "I^2L Takes Bipolar Integration a Significant Step Forward," *Electronics*, pp. 83–90, February 6, 1975.

Millman, J., *Microelectronics*, McGraw-Hill, New York, 1979, Chapter 9.

Raymond, J. P. et al., "A Comparative Evaluation of Integrated Injection Logic," *IEEE Transactions on Nuclear Science*, vol. NS-24, pp. 2327–35, December 1977.

Taub, H. and D. Schilling, *Digital Integrated Electronics*, McGraw-Hill, New York, 1977, Chapter 4.

CHAPTER 3

Emitter Coupled Logic

Walter C. Seelbach

Engineer, Research and Development
Motorola Inc.
Semiconductor Products Sector

3.1 INTRODUCTION

Integrated circuit versions of Emitter Coupled Logic (ECL) became commercially available in 1962. After a slow start, ECL has become a significant digital logic circuit technique for the vast high-speed data processing, digital communication, and test equipment industries. The principal reasons for its use in these industries are its ultra-high-speed switching properties and transmission line drive capability.

Basic transmission line loaded gate delays have improved from 8 ns in 1962 to less than 1 ns with today's modern IC manufacturing technologies (see Chap. 20). Owing to its extreme speed, careful attention and understanding of high frequency and transmission-line properties of interconnectors are paramount. Because controlled characteristic impedance interconnect systems are usually more costly than point to point, or harness-wired systems, ECL should be chosen when the system's speed requirement dictates. Table 3.1 summarizes the characteristics of ECL.

In cases where high-speed switching is not important and ECL is chosen for reasons other than speed, for instance, hardness to atomic radiation (see Table 3.2), care must still be exercised in interconnect grounding, lead dress, and so on. If an ECL system were constructed, for example, using some of the

Table 3.1 **Characteristics of ECL**

	Gate delay (ns)		
Features (typical)	*0.75*	*1*	*2*
Bias regulator parameters			
voltage compensation	(100K, 11C)[a]		(F95K)[a]
temperature compensation	(100K)[a]		(F95K)[a]
tracking, power supply and/or temperature	temp. only (11C)[a]	(MECLIII)[a]	(F10K, MECL10K)[a]
Built-in output load resistors	no	no	no
Built-in input pulldown resistors	yes	yes	yes
Input loading current	350 μA	350 μA	240 μA
Output current	>20 mA	>20 mA	>20 mA
Transmission line application	yes	yes	yes
Dc loading, fan out	>50	>50	>75
Input capacitance	2.0 pF	3.3 pF	2.9 pF
Output impedance	7 Ω	5 Ω	7 Ω
Gate edge speed (t_r, t_f)	0.7 ns	1 ns	3.5 ns
Flip flop toggle speed	1 GHz	500 MHz	125 MHz
Gate power dissipation	40 mW	60 mW	25 mW
Power speed product	30 pJ	60 pJ	50 pJ
Package availability			
flat package	yes	no	yes
dual in-line	yes	yes	yes

[a] Motorola Inc. is the prime source for MECL, 10K, and MECL III.
Fairchild is the prime source for F10K, F95K, F100K, and F11C.

interconnect techniques that work well with CMOS and T^2L, it is quite possible the builder would spend considerable time correcting system malfunctions caused by ringing and oscillation problems.

3.2 THE BASIC EMITTER COUPLED SWITCH

A basic emitter coupled switch is shown in Fig. 3.1. While this circuit is the basic building block for digital ECL logic circuits, it is also the basic differential amplifier configuration used widely in modern linear integrated circuits (Chap. 9). As a matter of interest, the original vacuum-tube differential amplifier was invented and built by J. F. Toennies in 1938.

Figure 3.1 shows a simplified version of the basic ECL switch so that we can more easily understand its operation in an actual integrated circuit. Note that transistors Q_1 and Q_2 control the path the current source I_0 can take between the power supply rials V_{CC} and V_{EE}. Because the base of Q_2 is connected to a

Table 3.2 **Radiation susceptibility of various semiconductors (Courtesy Fairchild Camera and Instrument Corp.)**

Radiation environment	*Bipolar trans. & JFET discretes*	*SCR*	*TTL*	*LSTTL*	*Analog IC*	*CMOS*	*NMOS*	*LED*	*ISO II ECL*
Neutrons n/cm^2	$10^{10}-10^{12}$	$10^{10}-10^{12}$	10^{14}	10^{14}	10^{13}	10^{15}	10^{15}	10^{13}	$>10^{15}$
Ionizing total dose Rads[a] (Si)	$>10^4$	10^4	10^6	10^6	5×10^4 10^5	10^3 10^4	10^3	$>10^5$	10^7
Transient dose rate rads (Si)/s (upset or saturation)		10^3	10^7	5×10^7	10^6	10^7	10^5		$>10^8$
Transient dose rate rads (Si)/s (survival)	10^{10}	10^{10}	$>10^{10}$	$>10^{10}$	$>10^{10}$	10^9	10^{10}	$>10^{10}$	$>10^{11}$
Dormant total dose (zero bias)	$>10^4$	10^4	10^6	10^6	10^5	10^6	10^4	$>10^5$	$>10^7$

[a] The unit of absorbed radiation dose that superseded the roentgen

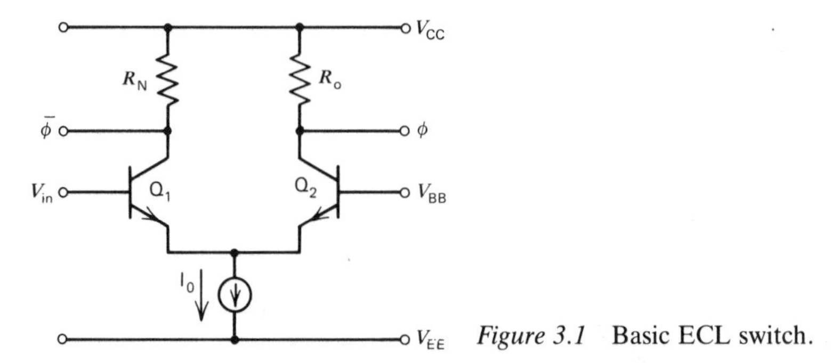

Figure 3.1 Basic ECL switch.

fixed dc potential (V_{BB}), current will flow through Q_1 when V_{in} is more positive than V_{BB} and conversely through Q_2 when V_{in} is negative with respect to V_{BB}. Current flow in Q_1 causes an out of phase ($\bar{\phi}$) voltage drop across R_N and a corresponding in phase (ϕ) voltage drop across R_o when Q_2 is conducting. The $\bar{\phi}$ response is shown in Fig. 3.2. These voltage excursions across R_N and R_o set the voltage logic swing (V_L) of the basic switch as indicated in Fig. 3.2.

A very important property of a logic gate is its *transition width*. The transition width, ΔT_W, is measured in volts and is that amount of voltage change in the input (V_{in}) that will cause the output to change from a high (V_{OH}) to a low (V_{OL}) state or vice versa. The ratio of logic swing V_L to ΔT_W is a measure of the circuit's noise immunity and/or ability to recognize reliably the difference between a one and zero input. If we define ΔT_W as the input difference between

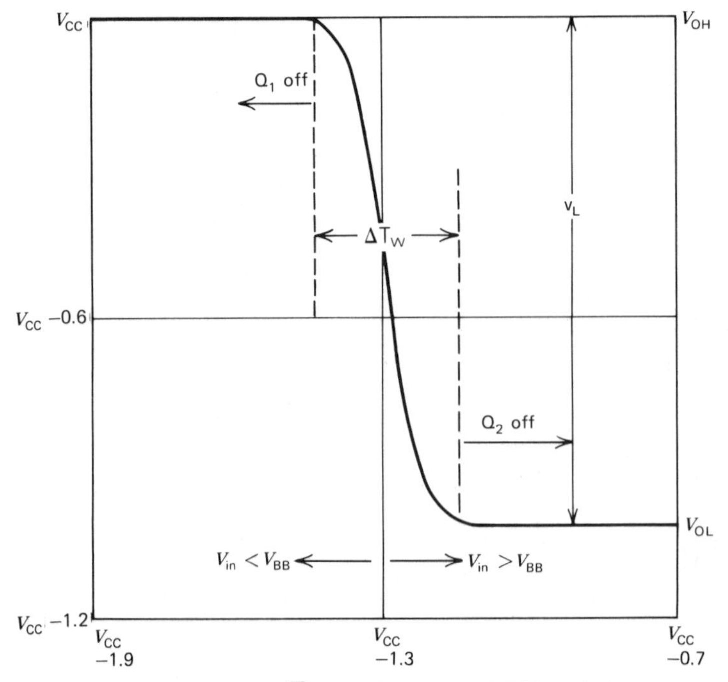

Figure 3.2 Out of phase ($\bar{\phi}$) transfer curve of ECL switch.

the two points of the transfer curve where $\Delta V_o/\Delta V_{in} = \pm 1$, then the transition width, neglecting transistor parasitics, is:

$$\Delta T_W = \frac{2kT}{q} \ln \alpha \tag{3.1}$$

where $\alpha = 0.8q/kT$.

It should be noted that ΔT_W is determined by the electronic charge (q), Boltzman's constant k (1.38×10^{-23}J/°K), and the absolute temperature T in degrees Kelvin (°K = 273° + °C).

EXAMPLE 3.1 Calculate the values of the transition width, ΔT_W, for chip temperatures of −50°C, 27°C (room temperature), and 125°C.

PROCEDURE

1 At −50°C,

$$\alpha = \frac{0.8 \times 1.6 \times 10^{-19}}{1.38 \times 10^{-23} \times (273 - 50)} = 41.6$$

By Eq. 3.1,

$$\Delta T_W \text{ (at } -50°\text{C)} = \frac{2 \times 1.38 \times 10^{-23} \times 223}{1.6 \times 10^{-19}} \times \ln(41.6) = 143 \text{ mV}$$

2 Similarly, at 27°C, ΔT_W = 178 mV and at 125°C, ΔT_W = 216 mV

From Ex. 3.1, we see that the worst case ΔT_W, in terms of our previous discussions on logic swing V_L to ΔT_W ratio, occurs at high temperature (assuming the logic swing is constant). This is important for the user to understand, because it now follows that packaged ECL systems will operate more reliably in a given noise environment as chip temperatures are reduced by proper cooling. Modern ECL, in particular the Fairchild F100K series, have by clever circuit design desensitized parameters other than ΔT_W in the basic ECL gate to temperature and power supply changes. This guarantees a more noise-immune product than older ECL product families. The transition width ΔT_W as we now understand it is set by absolute temperature (T) and can, therefore, not be effectively compensated for by using circuit design techniques. It behooves the user to design effective thermal management into an ECL system, realizing the noise-immunity benefits along with the more generally known long-term reliability advantages of reduced chip operating temperatures.

3.3 THE BASIC ECL GATE

The current switch of Fig. 3.1 shows only a portion of a complete ECL gate. Additional devices, resistors and transistors, have been added by ECL chip designers to make the basic switch capable of driving like chips over a wide range of temperatures—0 to 75°C (commercial grade) and −55 to +125°C (military grade)—with power supply voltage ranges of typically ±10% about the 5.2 V nominal utilized for most ECL product families. In addition, series gating was developed by the chip designers to provide additional logic function.

Figure 3.3*a* shows the addition of emitter followers (Q_3 and Q_4) to make the basic switch input/output compatible. Figure 3.3*b* shows the addition of Q_5 and Q_6 to make the basic current switch a logic NOR/OR gate and, finally, Fig. 3.3*c* illustrates the technique of series gating, which provides an additional level

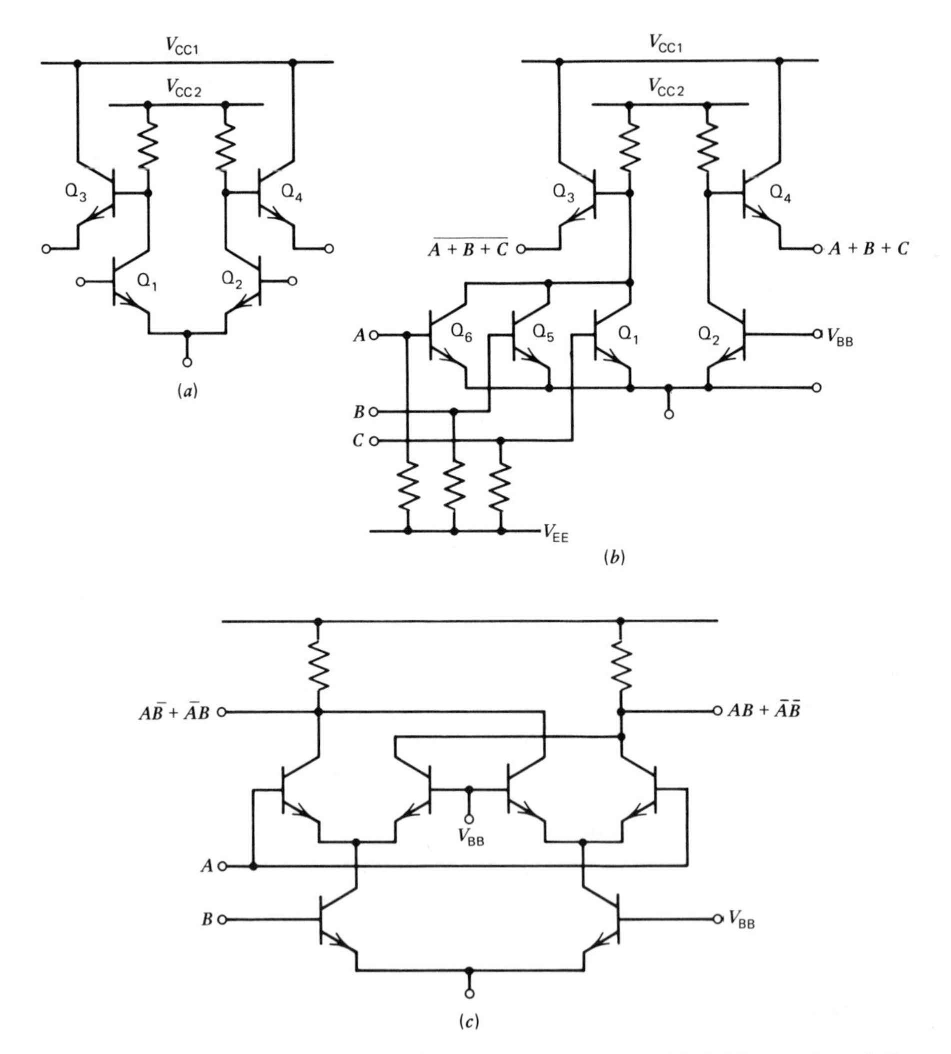

Figure 3.3 Expanding the basic ECL switch of Figure 3.1. (*a*) Adding emitter followers for I/O compatability. (*b*) Adding input transitors to increase fan in. (*c*) Exclusive OR/NOR—an example of series gating.

of logic without increasing power dissipation. The example of series gating (Fig. 3.3*c*) is the exclusive OR/NOR and is not limited to that function. As a matter of fact, some ECL MSI products use three levels of series gating.

In both the basic-gate and series-gate ECL configurations, the logic function (ϕ) and its complement ($\overline{\phi}$) are simultaneously available. This is important to the user because both functions are generally needed to implement logic. In other product lines, for example, CMOS and T^2L, ϕ and $\overline{\phi}$ are obtained by inversion using an additional gate. This approach is less desirable in high-speed systems because the additional gate would cause a time skew between ϕ and $\overline{\phi}$ that stems from the delay of the inverter.

3.4 THE ON-CHIP V_{BB} REFERENCE GENERATOR

We have up to this point discussed the basic current switch, the addition of emitter follower buffers for I/O compatibility, multiple inputs Q_5 and Q_6 to perform logic, and series gating. The last step is the generation of the internal reference supplies V_{BB} shown in the preceding figures. The purpose of the V_{BB} reference generator is to provide a temperature and power supply tracking reference potential that is somewhere midway between a V_{OH} (maximum available gate voltage for a logic 1) and V_{OL} (minimum gate voltage for a logic 0)

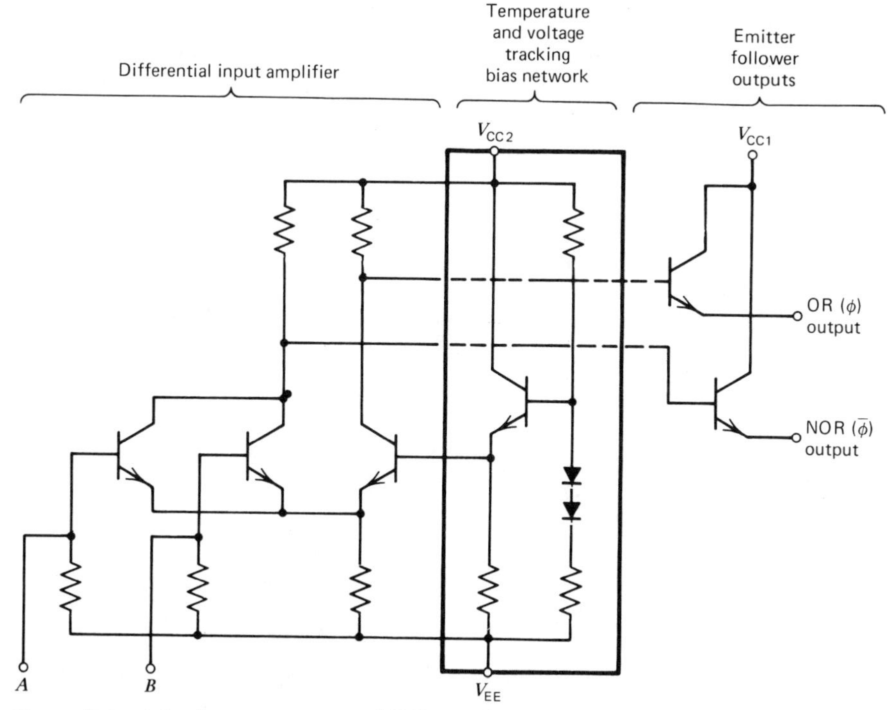

Figure 3.4 A basic uncompensated ECL gate of the MECL III, 10K variety.

signal input level and is set by the transistor whose base is connected to the V_{BB} supply. Voltages V_{OH}, V_{OL}, and V_{BB} are referenced to V_{CC}, which is nominally connected to ground (0 V). This is done to optimize noise immunity so that a gate can reliably detect the difference between a logic 1 high state (~ -0.8 V) and a logic 0 low state (~ -1.6 V). In addition, a centered V_{BB} minimizes switching time skews that would occur if the reference supply were not centered.

The original uncompensated bias network is illustrated in Fig. 3.4. This network was developed originally by Motorola and is currently used in some MECL III and 10K products. With this approach, the internal reference voltages generated track well with the external variations in temperature and power supply voltages. Figure 3.5*a* shows the transfer characteristics of a MECL 10K uncompensated gate. Note the high V_{OH}, low V_{OL}, and the transition region moves with both the external effects of temperature and power supply voltage. Figure 3.5*b* shows the transfer curve of a Fairchild fully compensated F100K

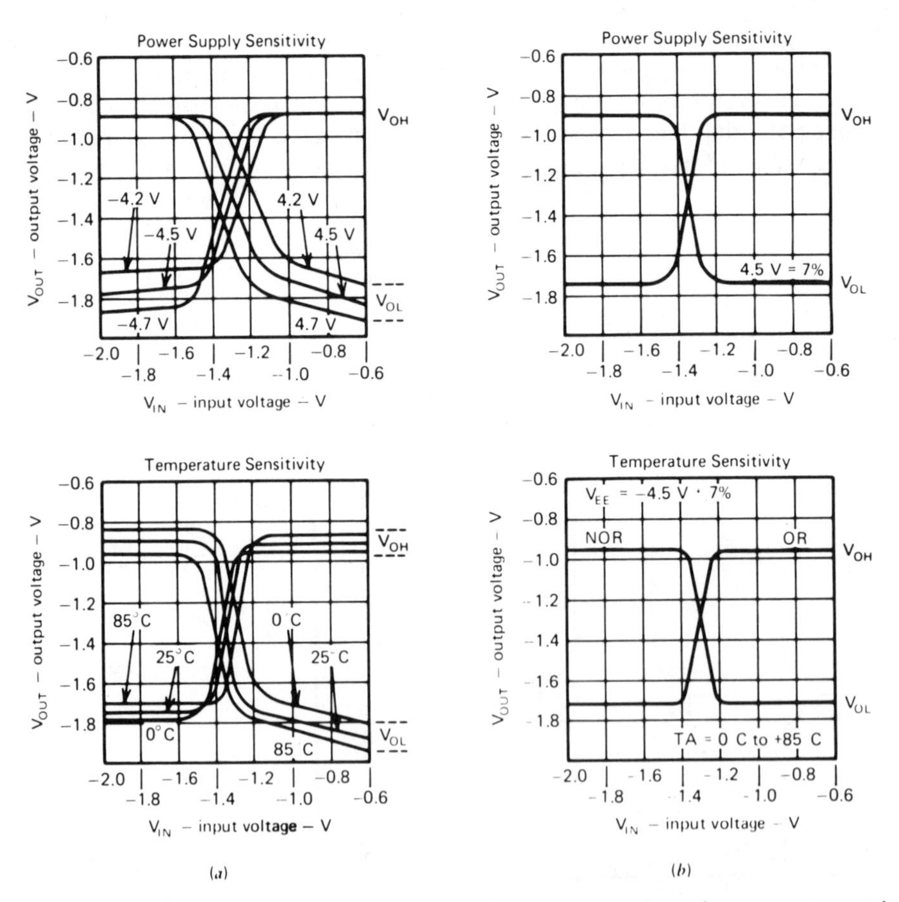

Figure 3.5 Changes in the transfer curves with power supply and temperature variations for (*a*) uncompensated ECL gates (MECL III, 10K) and (*b*) fully compensated ECL gates (F100K). Note the insensitivity to power supply and temperature variations for the fully compensated gate.

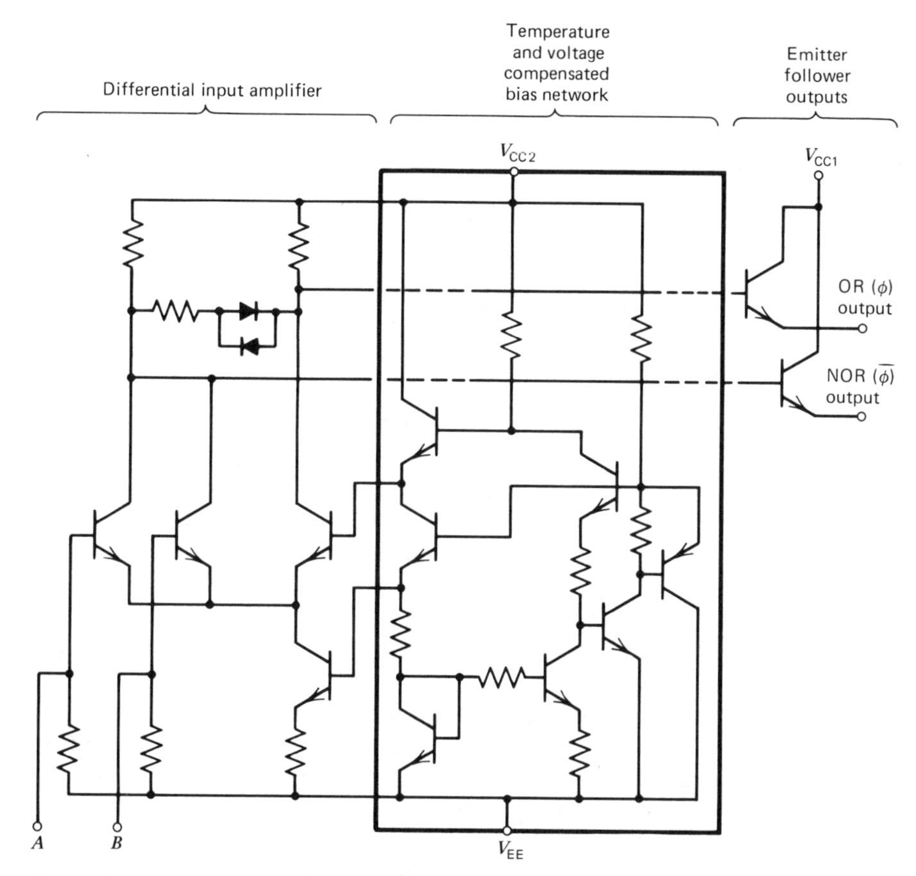

Figure 3.6 The basic fully compensated ECL gate of the F100K variety.

ECL gate. Now V_{OH}, V_{OL}, and the transition regions are virtually unaffected by the external effects of temperature and power supply voltage. Figure 3.6 shows the modified gate circuit and accompanying on-chip V_{BB} regulator that was employed in the F100K series to achieve fully compensated ECL products.

The system designer will find the F100K series ECL with fully compensated circuitry easier to use because these circuits are less critical of power supply regulation, line drops, chip temperature, and temperature differences between chips. A forerunner to the fully temperature and power supply compensated F100K series was the F9500 series of ECL products introduced by Fairchild in 1971. This product is temperature compensated, that is, transfer properties vary with power supply variations but are unaffected by chip temperature. Readers interested in more information on the differences of these 10K, 100K, and 9500 ECL lines are encouraged to read the Fairchild and Motorola Handbooks (listed in Sec. 3.9, References, at end of chapter).

At present, a third type of regulator scheme, not identified as a product family per se, is found in some 10K and ECL memory parts. This third scheme is a compromise between the uncompensated and compensated circuit approach.

By modification of the circuit in Fig. 3.6, this class of products is voltage compensated like the F100K series, but varies with temperature as with the MECL III and 10K approach. This technique has become more popular with some chip designers because it provides a better way for the user to build a system from a mix of 10K and 100K parts, thereby taking advantage of the merits of each product line.

3.5 APPLICATION CONSIDERATIONS

The dc Loading Effects

One should understand the input and output loading and drive capability of ECL circuits before building an ECL system. In Fig. 3.7 ECL input and output current, voltage, and polarities are defined. Notice that input current is defined as flowing into the unit while the output current is defined as flowing out of the unit. Because the output of one logic circuit must provide the input current required to the driven circuits (magnitude and polarity), it is more convenient to define input and output in this manner than using the standard IEEE convention. (Standard IEEE conventions define all I/O port currents and voltages as having the same polarity.)

Figure 3.8 shows a composite of input and output ECL current–voltage relationships. The intersection of the input current characteristics with the output V_{OH} and V_{OL} characteristics defines an operating point for that particular case (output driving one input with no pull down, or external, load resistor). It is seen from Fig. 3.8 that the difference between input and output impedance is quite large. If it were not for ac considerations, the fan out drive capability of ECL would be very large, much greater than any practical application would require.

Figure 3.9 shows the same V_{OH} and V_{OL} output characteristics as in Fig. 3.8, but with various resistor loads connected between the output and minus two volts (-2 V). Note that even with this substantially heavier loading, the changes in V_{OH} and V_{OL} are relatively small when compared to the logic swing ($V_{OH} - V_{OL}$). It should be evident that the dc loading effects of inputs (Fig. 3.8) are negligible when compared to the loading effects caused by resistors in the range of 35 to 150 Ω (Fig. 3.9). Because external resistors are normally used to load the outputs of ECL circuits for ac considerations, one can usually neglect the dc loading effects caused by inputs. They are relatively small when compared to the loading effects of external resistors.

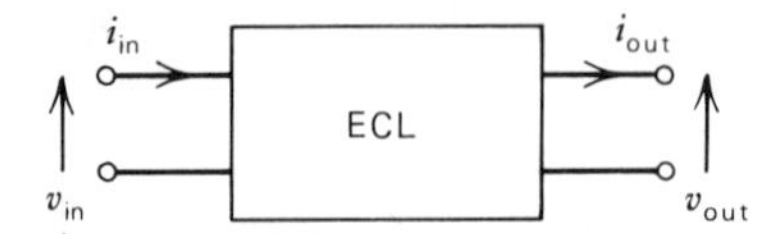

Figure 3.7 Definition of I/O port currents and voltages.

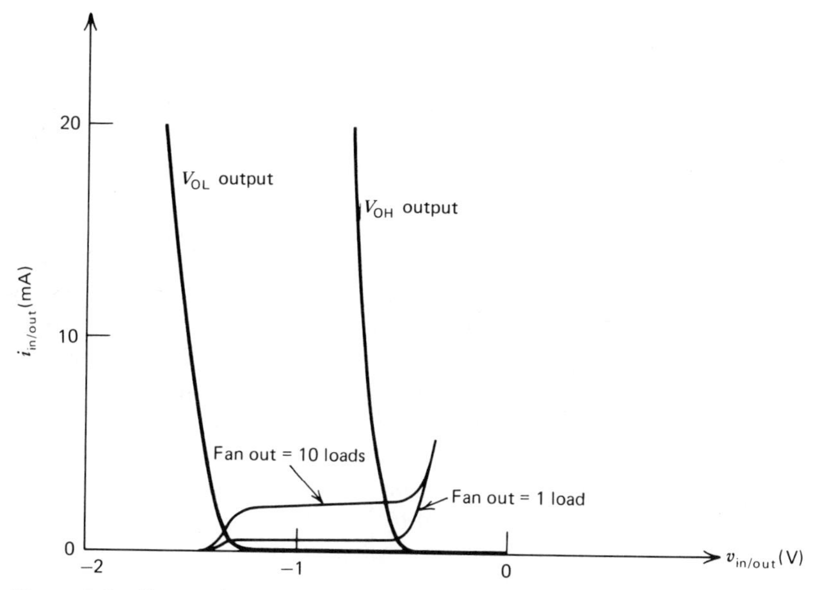

Figure 3.8 Composite of ECL I/O voltages and currents.

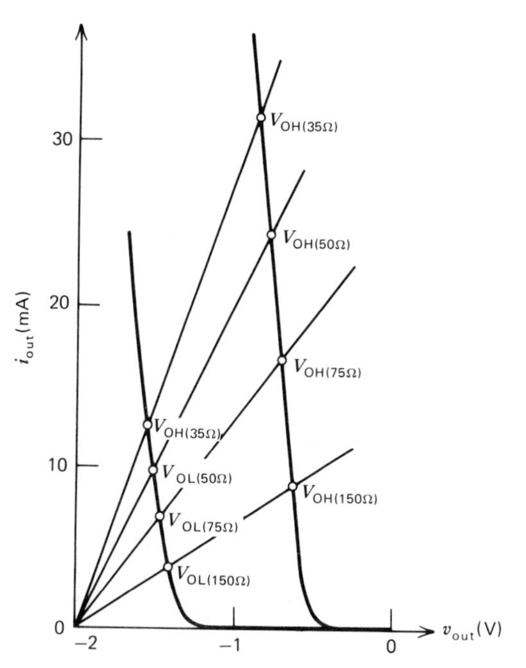

Figure 3.9 Composite of ECL output and resistor load voltage and current.

The Minus 2 Volt Termination Supply

In Fig. 3.9, a -2 V is indicated as the terminating supply for external output load resistors. The -2 V supply is generally used in ECL systems to provide the load supply for transmission line terminators. Use of the -5.2 V rail for output load termination would draw excessive current for a 50 Ω system. For this reason, only nontransmission line output loads, ~ 500 Ω or greater, are connected to -5.2 V. In large systems, where transmission lines are used quite extensively, a second -2 V supply is usually provided. If, however, transmission lines are used infrequently, for example, in a small system, one would like to eliminate the need for a second power (-2 V) supply. The problem now is how to provide a -2 V that can be derived from the main V_{EE} (-5.2 V) supply.

In Fig. 3.10, a resistor network, comprised of R_1 and R_2, is shown connected between ground (V_{CC}) and -5.2 V (V_{EE}) with a divider tap voltage set at -2 V (V_{TT}). Because we want this network to be 50 Ω in order to terminate a transmission line of alike characteristic impedance (Z_o), the parallel combination of R_1 and R_2 are required then to be equal to Z_o:

$$\frac{R_1 R_2}{R_1 + R_2} = Z_o \tag{3.2}$$

Because V_{TT} is obtained by the divider ratio of R_1 and R_2 then,

$$\frac{R_1 V_{EE}}{R_1 + R_2} = V_{TT} \tag{3.3}$$

or

$$\frac{R_1}{R_1 + R_2} = \frac{V_{TT}}{V_{EE}} \tag{3.4}$$

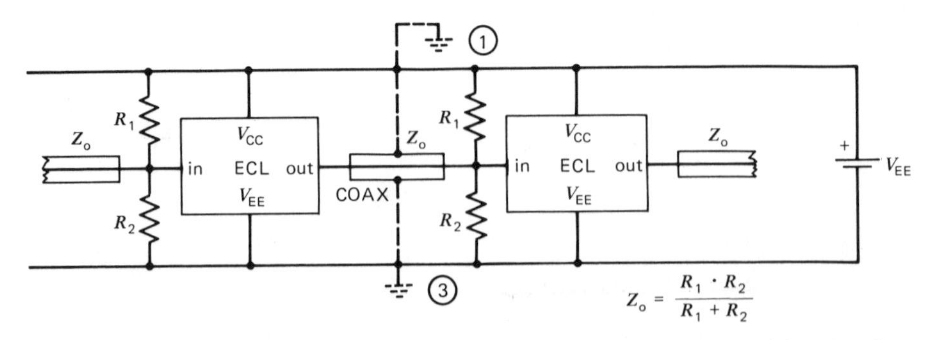

Figure 3.10 Thevenin equivalent network for obtaining V_{TT}. Two possible ground reference points are (1) and (3).

Substituting Eq. 3.4 in 3.2, gives us:

$$\frac{R_2 V_{TT}}{V_{EE}} = Z_o \tag{3.5}$$

Solving for R_1 and R_2, yields

$$R_1 = \frac{Z_o V_{EE}}{V_{EE} - V_{TT}} \tag{3.6a}$$

$$R_2 = \frac{Z_o V_{EE}}{V_{TT}} \tag{3.6b}$$

EXAMPLE 3.2 For a transmission line having a characteristic impedance $Z_o = 50\ \Omega$ and $V_{EE} = 5.2$ V, calculate the values of R_1 and R_2 for $V_{TT} = 2$ V.

PROCEDURE

1 By Eq. 3.6*a*,

$$R_1 = \frac{50 \times 5.2}{5.2 - 2} = 81.25\ \Omega$$

2 By Eq. 3.6*b*,

$$R_2 = \frac{50 \times 5.2}{2} = 130\ \Omega$$

Depending on the degree of matching desired, 1% or 5% resistors may be used.

From the preceding analysis we see that a 50 Ω resistor can be obtained by the Thevenin equivalent of two resistors. Other values of termination resistance and termination voltage can be calculated using Eqs. 3.6*a* and 3.6*b*.

The power dissipation requirements of these resistors can be obtained as follows:

$$P_{R_1} \geqslant \frac{V_{OL}^2}{R_1} \tag{3.7}$$

$$P_{R_2} \geqslant \frac{(V_{EE} - V_{OH})^2}{R_2} \tag{3.8}$$

EXAMPLE 3.3 Calculate the power dissipated in resistors R_1 and R_2 found in Ex. 3.2. Assume that $V_{OH} = -0.8$ V and $V_{OL} = -1.6$ V.

PROCEDURE

1 Power dissipated in R_1 is:

$$P_{R_1} = V_{OL}^2/R_1 \tag{3.7}$$

2 Power dissipated in R_2 is:

$$P_{R_2} = (V_{EE} - V_{OH})^2/R_2 \tag{3.8}$$

3 Substitution of given values in Eqs. 3.7 and 3.8 yields

$$P_{R_1} = (-1.6)^2/81.25 = 31.2 \text{ mW}$$

$$P_{R_2} = (-5.2 + 0.8)^2/130 = 149 \text{ mW}$$

A 50 Ω resistor connected to −2 V, if the −2 V termination voltage is provided, would require a single 50 Ω resistor that dissipates a maximum of only 28.8 mW, or a ⅛ W resistor in a calculation similar to Ex. 3.3. A single resistor system to −2 V would require one-half as many resistors and requires less power and less board space in packaging; an additional voltage bus, however, is required in the power distribution system. In systems that are packaged with multilayer printed circuit boards having ground (V_{CC}), termination (V_{TT}), and power planes (V_{EE}), the 50 Ω system to −2 V would be the most optimum. One should, however, carefully weigh the cost-performance tradeoffs, because the cost of multilayer boards increases with additional layers. Continuous ac ground planes are used in these multilayer boards to provide low impedance power distribution connections to ICs owing to the low inductance and large capacitance between the ac ground planes in a multilayer board system.

Ground System Considerations

ECL circuits require a nominal supply voltage (V_{EE}) anywhere from −4.5 V to −5.2 V to operate, depending on whether the F100K or MECL 10K type parts are used. In addition, a termination supply of −2 V with respect to V_{CC} is required to serve as a transmission line (V_{TT}) termination supply. The −2 V supply can, as we saw in the previous discussion on dc loading effects, be obtained directly from a second supply of −2 V or by two resistors connected across the main V_{EE} supply.

In Fig. 3.11, we have three choices for a ground reference connection in the dual-supply system. These are (1) a connection of V_{CC} to ground, (2) V_{TT} to ground, or (3) V_{EE} to the ground reference point. At first glance the ground reference may seem irrelevant, but as we shall see, the choice of V_{CC} as the ground point is the best choice with V_{TT} as second and V_{EE} a distant last choice. In determining the best ground reference point we shall assume the following:

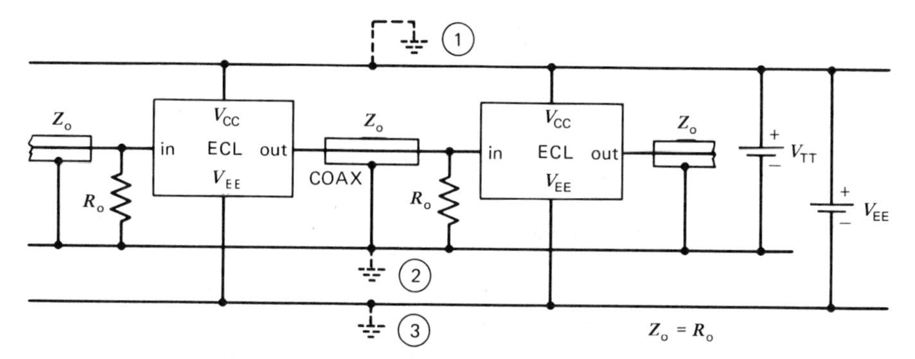

Figure 3.11 Separate termination supply V_{TT}. Three possible ground reference points are (1), (2), and (3).

1 During the life of the system, numerous accidental shorting of the outputs to ground will occur, usually during system test.

2 If only one low-impedance rail is available it also will be the ground reference point. Any ac systems measurements will be made between a signal point and this ground reference point.

3 Because the dc component of ac measurements, for instance, V_{OH} and V_{OL}, also are important, in addition to ac time measurements between system signals, the signal dc offset with respect to ground should be kept to a minimum to facilitate ease in application of oscilloscope measurements.

4 ECL subsystems should be able to interface directly with one another.

Clearly, grounding V_{CC} meets all four requirements. Grounding V_{TT} violates (4) since portions of a system that uses the Thevenin equivalent for termination cannot be ground referenced to V_{TT} like the other parts of the system that used a separate V_{TT} supply. Grounding V_{EE} violates (1) and (3). Because ECL chips are speed optimized by the designers, no output current-limiting resistors, as used in T^2L, are found in ECL products. An accidental short of an output to V_{EE}, therefore, would destroy the unit. For this reason, V_{EE} wiring should be designed to minimize the possibility of accidental shorting to an output when testing the system.

3.6 LOGIC FUNCTIONS: POSITIVE NOR/OR, NEGATIVE AND/NAND

It is important for the new ECL user to understand the fundamental logic operation performed by the basic ECL gate. Gates may be viewed as the glue parts that tie together the whole system of complex functions, for example, counters, multipliers, arithmetic logic units, and memories, and make them work as defined by system requirements.

The input/output level and timing requirements for complex ECL functions required to perform these respective functions are adequately defined in the ECL product specifications. Achieving the user's required system function is usually the job of the basic ECL gate. ECL gates come in a variety of configurations, such as dual four, quad two, and so on. Figure 3.12 shows a simplified 3-input gate which we shall use to study the basic logic operation performed.

Let us define a logic 1 as the most positive voltage level of a signal and a logic 0 as the most negative. The opposite convention (most positive = logic 0 and most negative = logic 1) could also be used. Practice, however, has favored the positive notation, which will be used in our analysis.

In Sec. 3.2 (The Basic Emitter Coupled Switch), we noted that ECL chip designers chose a logic swing (V_L) of 0.8 V for the basic gate. For a swing of 0.8 V with V_{CC} chosen as ground, along with the V_{BE} offset, the output emitter followers set the gate I/O levels to approximately -0.8 V and -1.6 V with respect to ground at room temperature, for both compensated and uncompensated products.

Here is where some new users find difficulty with ECL. Most users of MOS and T^2L have little difficulty with voltage polarity and the logic state definition because a positive voltage (V_{OH}) greater than 3 V corresponds easily with the logic 1 designation along with a near-zero signal voltage (V_{OL}) designated as a logic 0. In ECL, on the other hand, the logic 1 is the -0.8 level because it is the *most positive* signal level and -1.6 is the *most negative* signal level.

The new user should become comfortable with the ECL level/logic notation before the logic design of gates and MSI complex function proceeds. Experience has shown a tendency for the inexperienced to identify the -0.8 V signal level as a logic 0 because it is closest to zero volts as prejudiced by prior T^2L MOS experience. In Fig. 3.12, any input A or B or C when high (more positive than V_{BB}) will cause the current to flow on the input side and a voltage drop (V_L) across R_{NOR}. Conversely, when inputs A, B, and C are all low (more negative

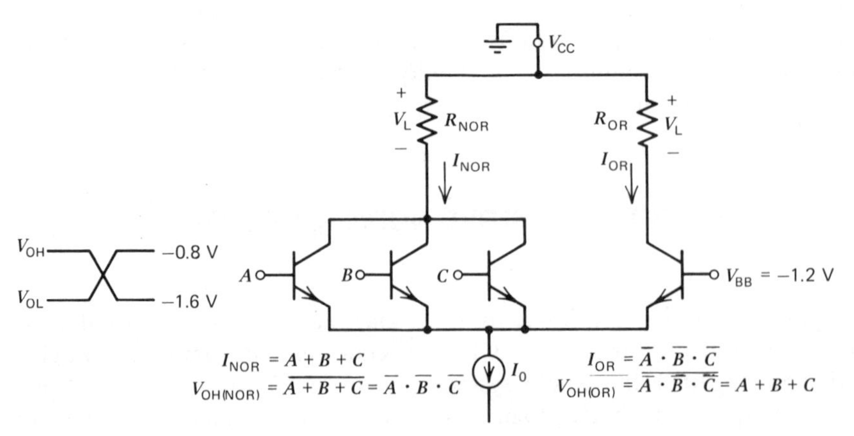

Figure 3.12 Simplified ECL gate illustrating the basic logic operations performed.

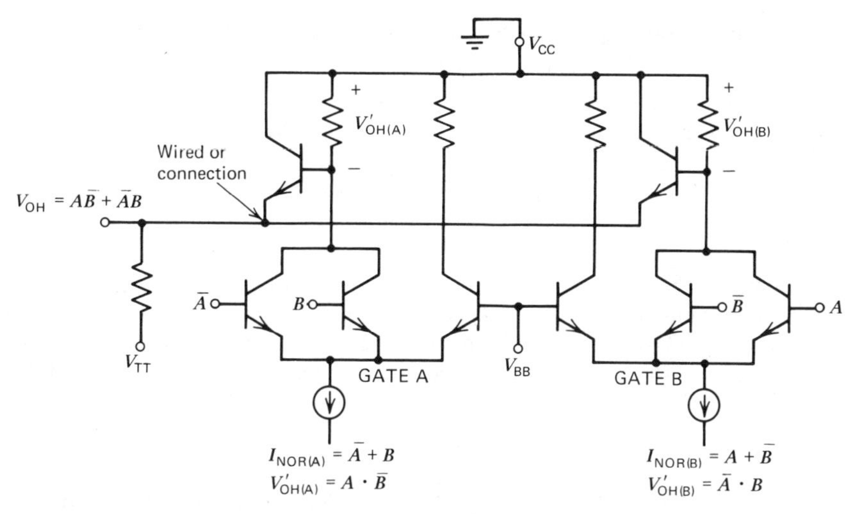

Figure 3.13 The basic *wired OR* operation.

than V_{BB}), current will flow on the V_{BB} side causing a voltage drop V_L across R_{OR}.

Writing the current flow equation in Boolean notation, we get:

$$I_{NOR} = A + B + C \tag{3.9}$$

$$I_{OR} = \overline{A} \cdot \overline{B} \cdot \overline{C} \tag{3.10}$$

Current I_{NOR} will flow when A or B or C is a logic 1 (-0.8 V). Current I_{OR} will flow when A and B and C are *all* logic 0 (-1.6 V).

Because the ECL gates respond to voltage signals, we convert the current Boolean equations to voltage equations. The logic 1 output, V_{OH}, occurs when the current does not flow in the corresponding side; hence:

$$V_{OH(NOR)} = \overline{I}_{NOR} = \overline{A + B + C} \qquad \text{The NOR function of input variables } A, B, C \tag{3.11}$$

Also,

$$V_{OH(OR)} = \overline{I}_{OR} = \overline{\overline{A} \cdot \overline{B} \cdot \overline{C}} = A + B + C \qquad \text{The OR function of input variables } A, B, C \tag{3.12}$$

The preceding is a rather simplified Boolean analysis of the basic ECL gate and serves only as a refresher and to point out the differences between ECL and T²L or MOS. (See Chaps. 1 and 4.)

Wired OR

In addition to the basic NOR/OR logic function, another logic operation can be achieved by connecting the outputs of ECL parts to a common signal point, as in Fig. 3.13. Because any one of the outputs connected to the common point can cause it to go to the V_{OH}, or logic 1 state, this connection, called *wired OR*, performs the logic OR of the two outputs.

EXAMPLE 3.4 Generate the following logic function using 2-input gates:

$$V_{OH} = A \cdot \overline{B} + \overline{A} \cdot B \tag{3.13}$$

PROCEDURE

1 In Fig. 3.13 we see that by applying $\overline{A}$ and B to the inputs of gate A, $A \cdot \overline{B}$ is obtained on its NOR side. Similarly, A and $\overline{B}$ to gate B yields $\overline{A}B$.

2 By connecting the NOR outputs of both gates together we obtain the logic OR of these two functions, which happens to be the exclusive OR of the variables A and B.

Collector Dotting

Collector dotting, which performs the logic AND function in positive notation, is usually available only to the chip designer. As we shall see later, however, it is available to the user in the custom design of gate-array products. The technique of collector dotting is illustrated in Fig. 3.14, where again we see two, 2-input gates connected as in Fig. 3.13 with the use of the emitter wired OR to perform the exclusive OR function. If we connect the collectors of both V_{BB} transistors to a common load R_{OR}, and provide a means for limiting the logic swing because twice the amount of current can flow through R_{OR}, the V_{BB} transistors are driven into saturation. Saturation (forward bias of collector to base junctions) is always avoided by ECL chip designers because substantial degradation of switching performance occurs as a result of stored charge effects that cause increased

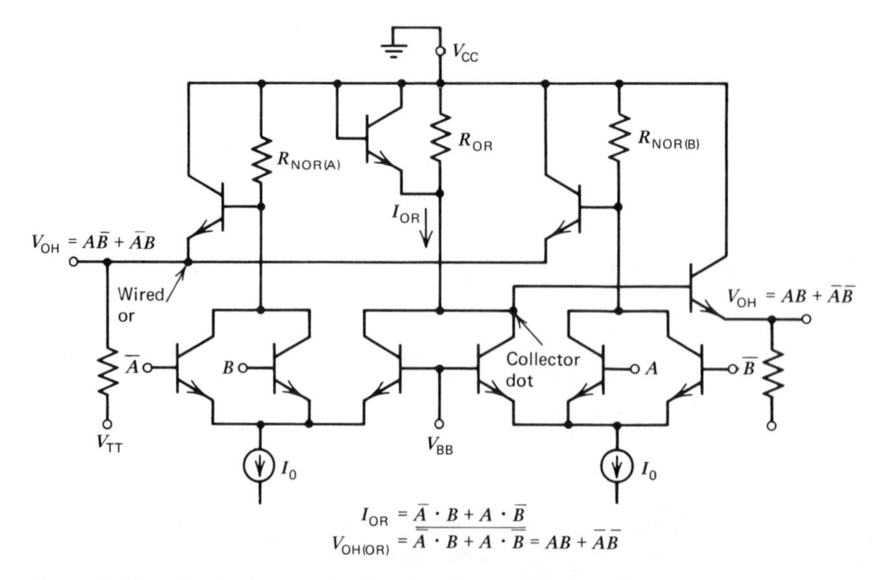

Figure 3.14 The basic *wired OR* and *collector dot* operations.

switching delay. By making this connection and providing a diode clamp to limit the swing to $\sim V_L$, we see the current Boolean expression as:

$$I_{OR} = \overline{A} \cdot B + A\overline{B} \quad (3.14)$$

The voltage signal equation is:

$$V_{OH} = \overline{\overline{A}B + A\overline{B}} = (A + \overline{B})(\overline{A} + B)$$
$$= A\overline{A} + AB + B\overline{B} + \overline{A}\overline{B} = AB + \overline{A}\overline{B} \quad (3.15)$$

Examination of Eqs. 3.13 and 3.15 shows that the emitter wired OR and collector dotting have generated the exclusive OR and its complement, respectively. As mentioned previously, collector dotting is usually available only to the chip designer. This node would be extremely sensitive to capacitance and, under certain conditions, could lead to instability and oscillation if it were generally available as a package pin. The exclusive OR function generated by wired OR of the outputs and collector dotting of two 2-input gates is offered solely as an example of the power of this technique. Exclusive-OR MSI functions are available, as well as many other more complex popular logic functions, that could be implemented less effectively and at more cost with individual basic gates. As we shall see later, the ECL gate-array products have made this mode and other ECL circuitry available to the user through custom design of some of the lithographic masks used to construct today's integrated circuits.

Gate Arrays

Standard off-the-shelf ECL and custom ECL designs are the mainstay of the ECL components industry. Custom ECL parts are similar to standard products in function, electrical, and mechanical package properties. They differ in that the custom product is usually available only to the customer who has developed the product with the IC vendor. Gate arrays are Large Scale Integration (LSI) in nature. Hundreds to thousands of equivalent logic gate functions are available by customizing usually three of the lithographic mask (see Chap. 20) levels used in fabrication. The number of gate functions per gate array chip has steadily increased as the IC manufacturer's yield and complexity (number of gates per chip) has improved.

ECL gate array products were pioneered in the 1960s by Motorola and have been primarily used by the computer industry since that time. Numerous domestic and foreign ECL vendors are currently supplying gate array products. Figure 3.15 shows the functional arrangement of a Motorola ECL Macrocell gate array. There are three basic circuit cell types available on this gate array. Of the 106 total cells available 48 are major, 32 interface, and 26 are output cells each customizable by the customer. In addition to the customizable feature of the individual cells, interconnection of the cells is provided to the customer via the vertical and horizontal metal interconnect channels.

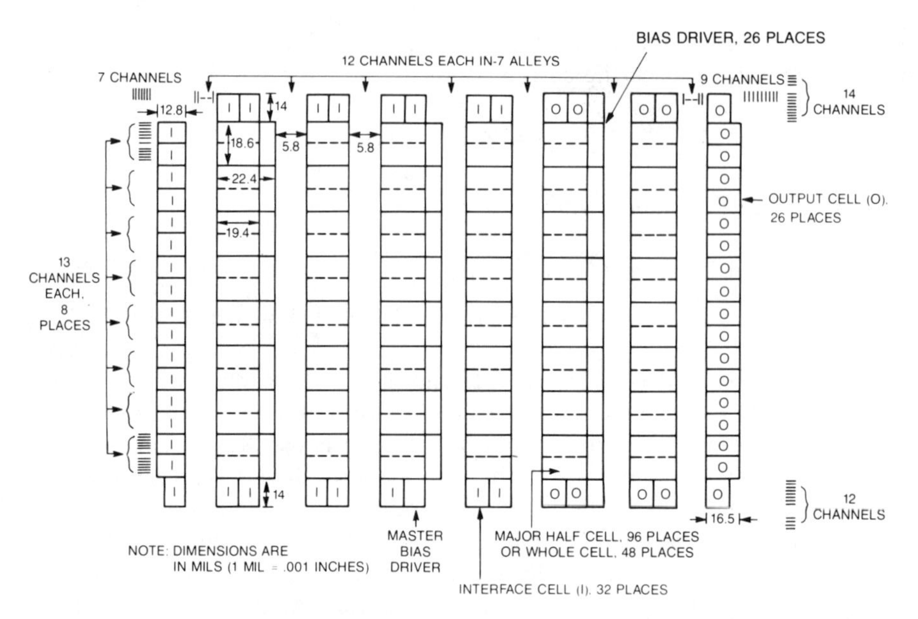

Figure 3.15 A MECL Macrocell array showing the metal interconnect channels.

Table 3.3 lists the basic Macrocell features for this 106-cell gate array product. Figures 3.16, 3.17, and 3.18 are schematics showing the components available for customizing the major interface and output cells. For this particular type of gate array, circuit design flexibility is limited only to the number and

Table 3.3 **Basic Macrocell array features**

106 total cells—48 major, 32 interface, and 26 output cells.
Up to 1192 equivalent gates if full adders and latches are used in all the cells.
Up to 904 equivalent gates if flip flops and latches are used in all the cells.
Die size—221 × 249 mils.
Power dissipation—4 W typical.
4.4 mW per equivalent gate (for 904 gates and 4 W).
Interface cell delay—0.9 ns typ (1.3 ns max).
Major cell delay—0.9 to 1.3 ns typ (1.3 to 1.8 ns max).
Output cell delay—1.5 ns typ (2.2 ns max).
Any output cell (up to a total of 8) can drive a 25 Ω load.
All output cells can drive 50 Ω loads.
Edge speed—1.5 ns typ 20 to 80% (1 ns min).
85 Macros in cell library—54 macros for major cells, 14 macros for interface cells, 17 macros for output cells.
Ambient temperature range with heat sink and 1000 lfpm air flow = 0 to 70°C.
Θ_{JA} = 15°C/W with heatsink and 1000 lfpm air flow.
Absolute maximum function temperature, T_J = 165°C.
Voltage compensated, $V_{EE} = -5.2$ V ±10%.
MECL 10K compatible.

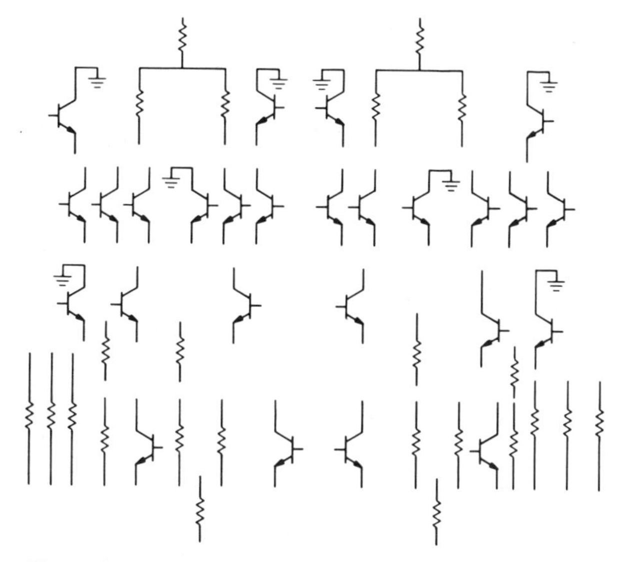

Figure 3.16 Gate array half-cell components (one-half of major cell).

types of components available. Considerable thought went into these choices by the chip designers. The cell components were optimized for ECL series gating. Continuing with our prior example on how to generate the exclusive OR and its complement with two 2-input gates, let us see what advantages a gate array approach might offer.

EXAMPLE 3.5 Figure 3.19 shows a partially wired major cell. Logic variables A and B are inputs to Q_3, Q_6, and Q_{11} as shown. What logic functions are realized?

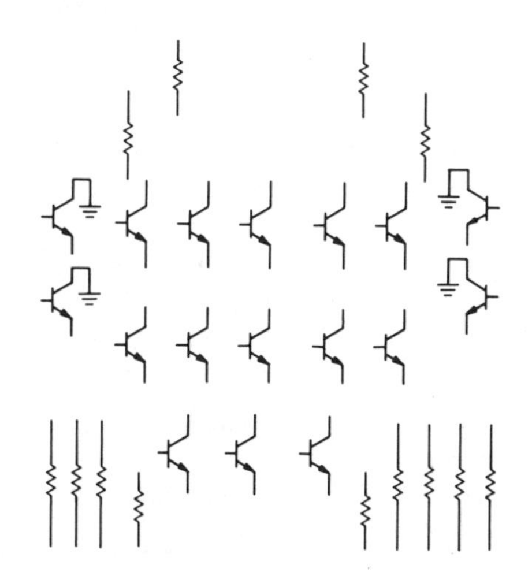

Figure 3.17 Gate array interface cell components.

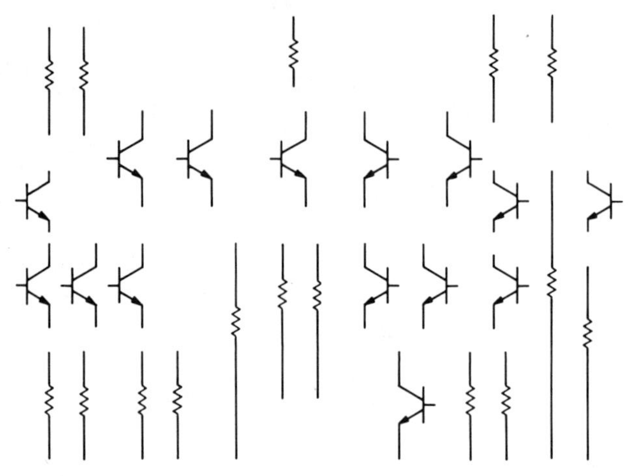

Figure 3.18 Gate array output cell components.

PROCEDURE

1 Write the Boolean equations for the collector currents I_C for Q_3, Q_4, Q_5, and Q_6.

$$I_{C(Q3)} = A \cdot B \tag{3.16}$$

$$I_{C(Q4)} = \overline{A} \cdot B \tag{3.17}$$

$$I_{C(Q5)} = \overline{A} \cdot \overline{B} \tag{3.18}$$

$$I_{C(Q6)} = A \cdot \overline{B} \tag{3.19}$$

2 Write the desired output voltage responses for the exclusive-OR and complement of logic variables A and B:

$$V_{(F)} = A\overline{B} + \overline{A}B \tag{3.20}$$

$$V_{(\overline{F})} = AB + \overline{A}\,\overline{B} \tag{3.21}$$

3 From our previous example of the two 2-input gates, we remember that the current Boolean expression is obtained by taking the complement of the desired voltage expressions:

$$I_{(F)} = \overline{A\overline{B} + \overline{A}B} = (\overline{A} + B)(A + \overline{B}) = AB + \overline{A}\,\overline{B} \tag{3.22}$$

$$I_{(\overline{F})} = \overline{AB + \overline{A}\,\overline{B}} = (\overline{A} + \overline{B})(A + B) = A\overline{B} + \overline{A}B \tag{3.23}$$

4 From the above current equations we see that connecting collector outputs of Q_3 and Q_5 generates the voltage exclusive-OR and connecting Q_4 to Q_6 generates the complement. Figure 3.20 shows the completed wiring for ¼ of a Macrocell required to implement the exclusive-OR and complement functions.

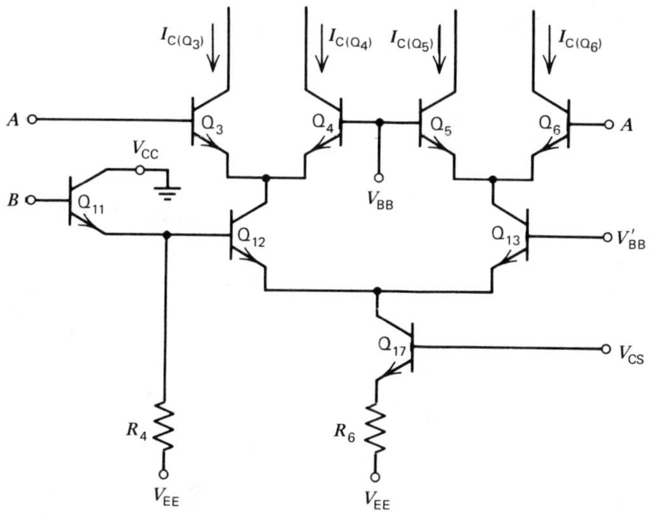

Figure 3.19 A partially wired major cell.

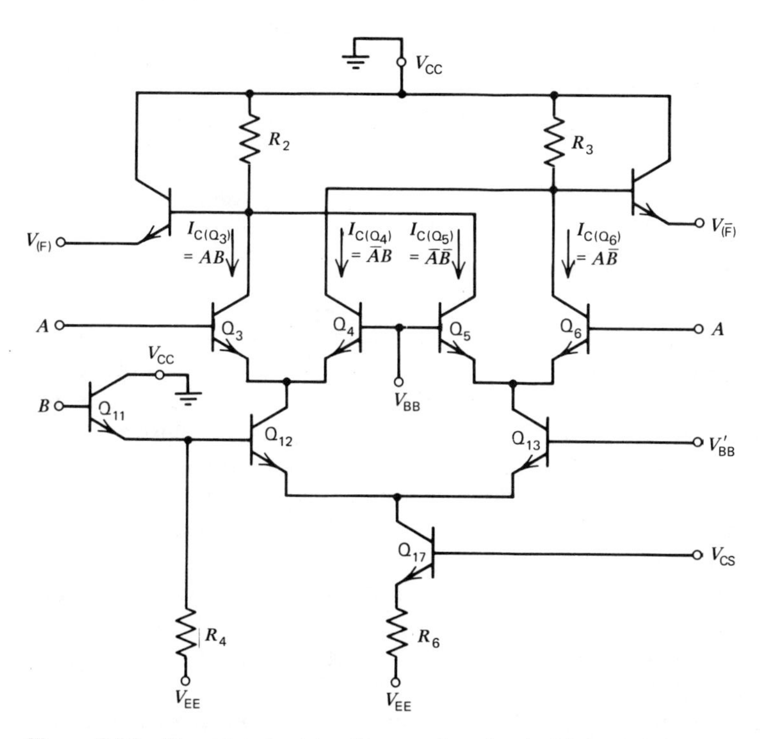

Figure 3.20 Completed wiring for one-fourth of a Macrocell implementing the exclusive OR function and its complement.

The series gating implementation requires only one current source (less power and components) and does not require collector clamping as in the two-gate example. In addition, it can be implemented with only the A and B inputs, whereas the two-gate example requires 4 inputs, A, $\overline{A}$, B, and $\overline{B}$.

Figures 3.21, 3.22, and 3.23 show other possible logic implementation examples for ½ of a major cell, an interface, and an output cell. Figure 3.24 is a representation of a Macrocell chip mounted in a 68-lead package to give the reader some idea of the power and flexibility offered by the gate array approach over the standard/custom IC product approach. Of course, all of this does not come without cost to the user. Gate array development costs are in the thousands of dollars, but should be considered as a viable alternative to PC board interconnected gates and complex standard and custom IC products.

3.7 PACKAGED INTERCONNECTION OF ECL CONSIDERATIONS

Successful interconnection of ECL in a user's system requires a good working knowledge of transmission line techniques. Ample information is available to the new user of ECL in the Motorola and Fairchild ECL handbooks. Basic transmission line theory and application will not be repeated here, but it is strongly recommended that the new user acquire, read, and understand the wealth of practical information available in these handbooks. A few considerations, however, will be discussed which the author feels will add to what is already available.

Ringing and Oscillation

There are tendencies for ECL to ring or, even worse, to become unstable and break into sustained oscillations owing to improper grounding, lead dress, decoupling, and the like. Figure 3.25 shows the real and imaginary parts of the small signal input impedance of a typical ECL gate. In this illustration the imaginary part is always negative (capacitance) and the real part (resistance) varies in magnitude with frequency. As long as the real part of the input impedance is positive, any inductance introduced to the R-C network due to lead dress would result only in ringing.

Ringing itself is not usually a problem if it is limited to the rise or fall time of the incident signal used in the setting of the system's flip flops because they are clocked after the steady-state (V_{OH} or V_{OL}) levels have occurred. If we accept the criterion that ringing be confined to the rise or fall transition times of an ECL gate, we can calculate the maximum length of direct wire interconnection that can be made between two ICs beyond which a terminated transmission line should be used. In free space, an electromagnetic wave travels at

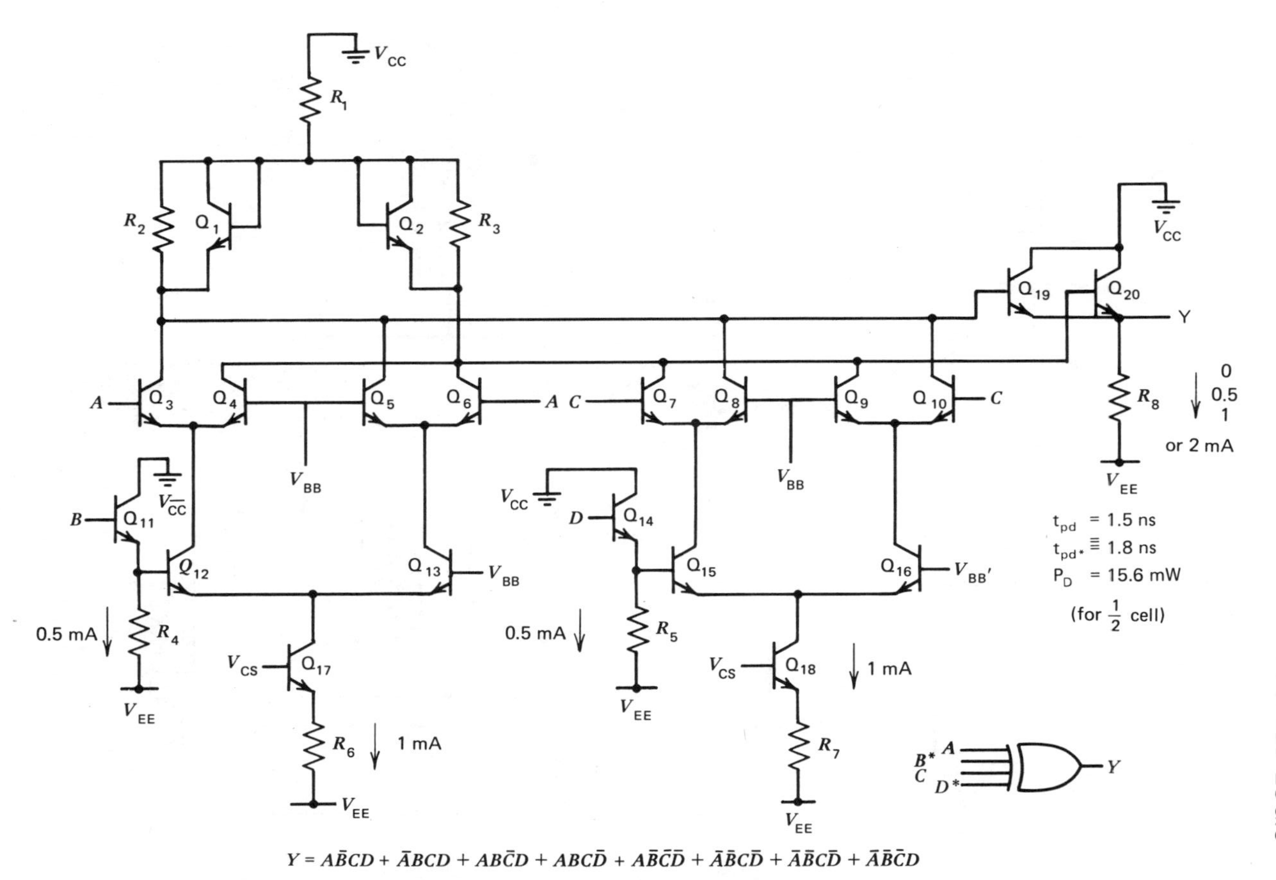

$Y = A\bar{B}CD + \bar{A}BCD + AB\bar{C}D + ABC\bar{D} + A\bar{B}\bar{C}\bar{D} + \bar{A}\bar{B}C\bar{D} + \bar{A}\bar{B}C\bar{D} + \bar{A}\bar{B}\bar{C}D$

Figure 3.21 Schematic of a 4-input exclusive OR gate (one-half of a major cell).

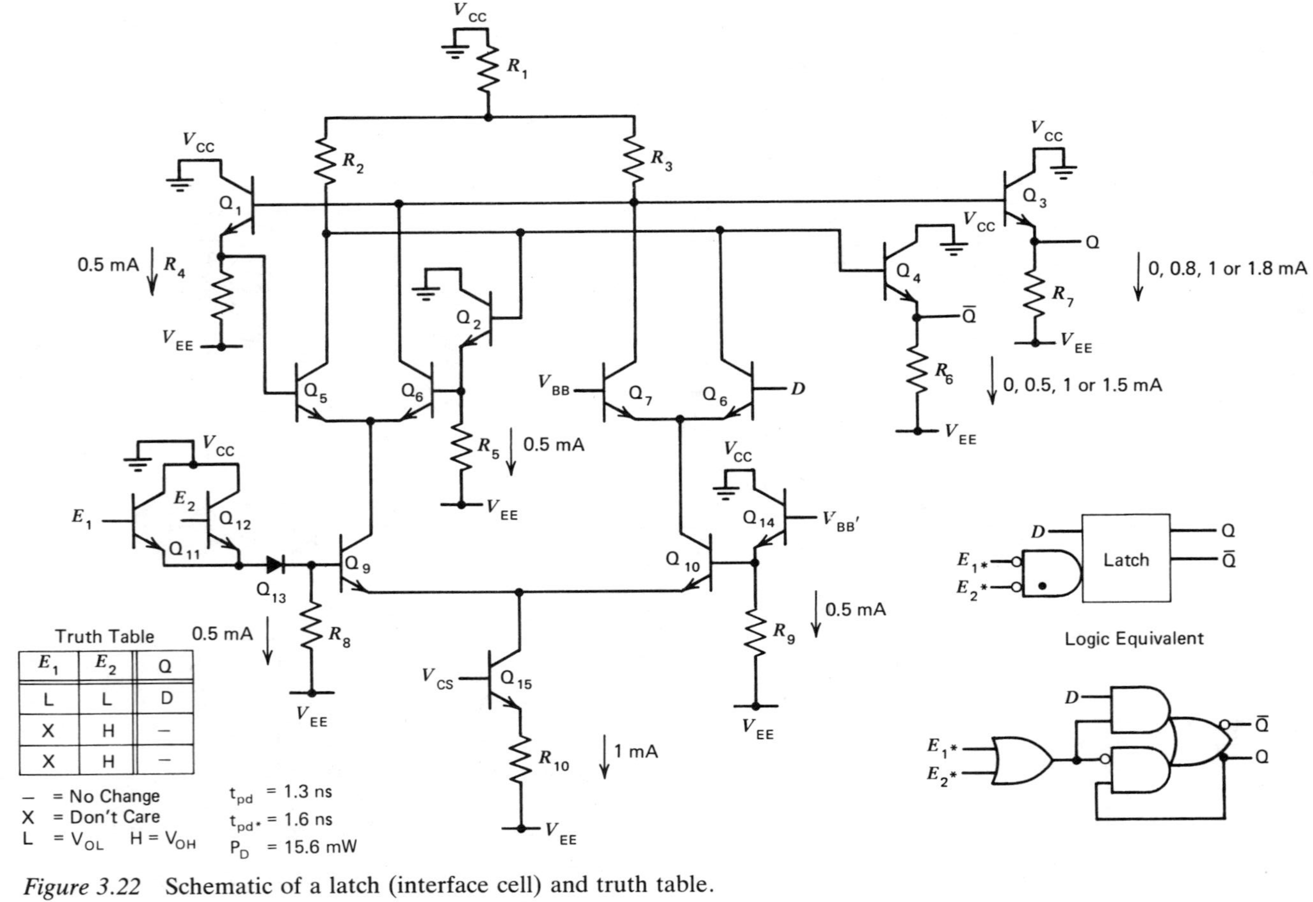

Truth Table

E_1	E_2	Q
L	L	D
X	H	–
X	H	–

– = No Change
X = Don't Care
L = V_{OL} H = V_{OH}

t_{pd} = 1.3 ns
t_{pd*} = 1.6 ns
P_D = 15.6 mW

Figure 3.22 Schematic of a latch (interface cell) and truth table.

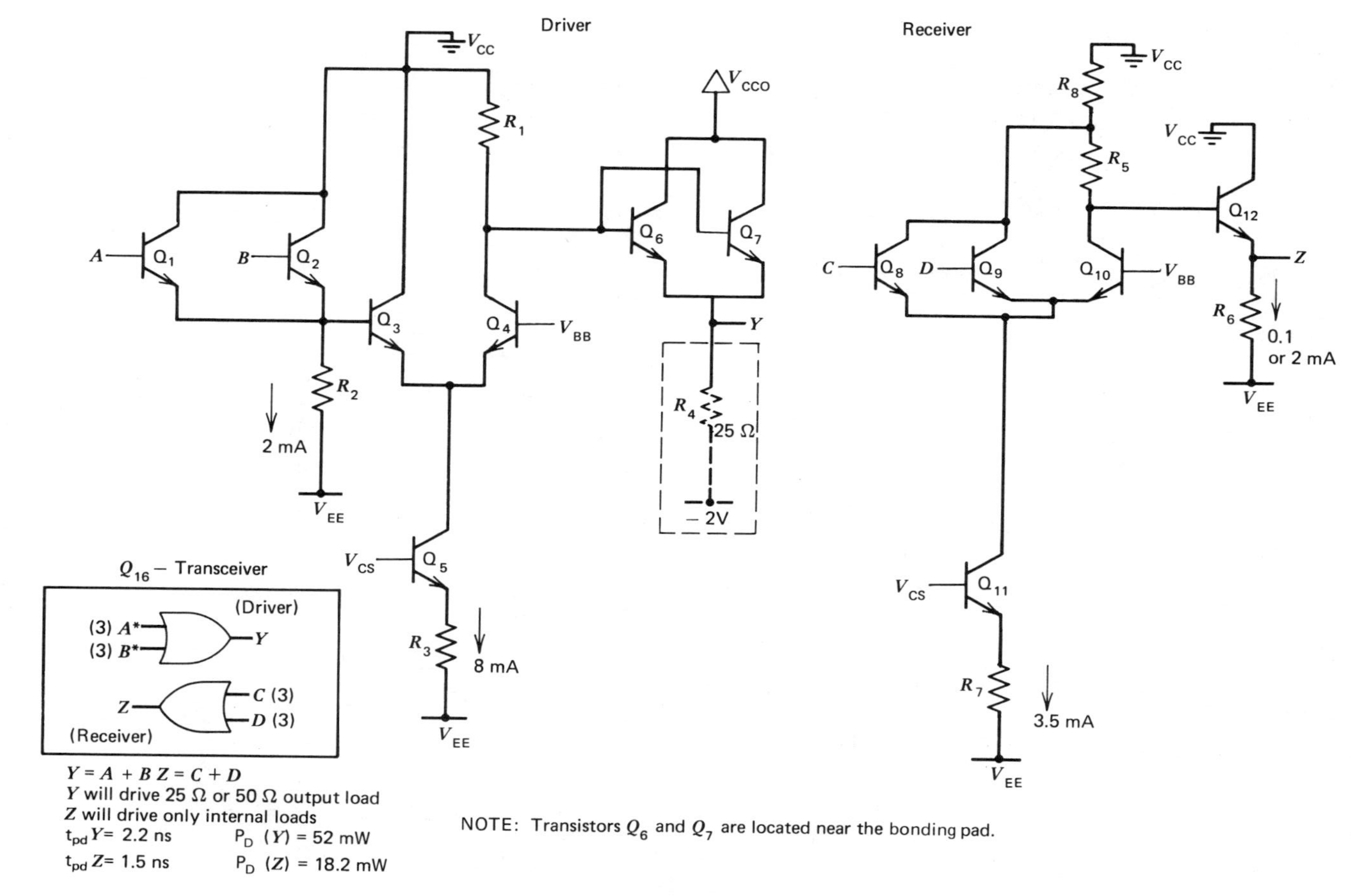

NOTE: Transistors Q_6 and Q_7 are located near the bonding pad.

Figure 3.23 Schematic of a line driver/receiver (output cell).

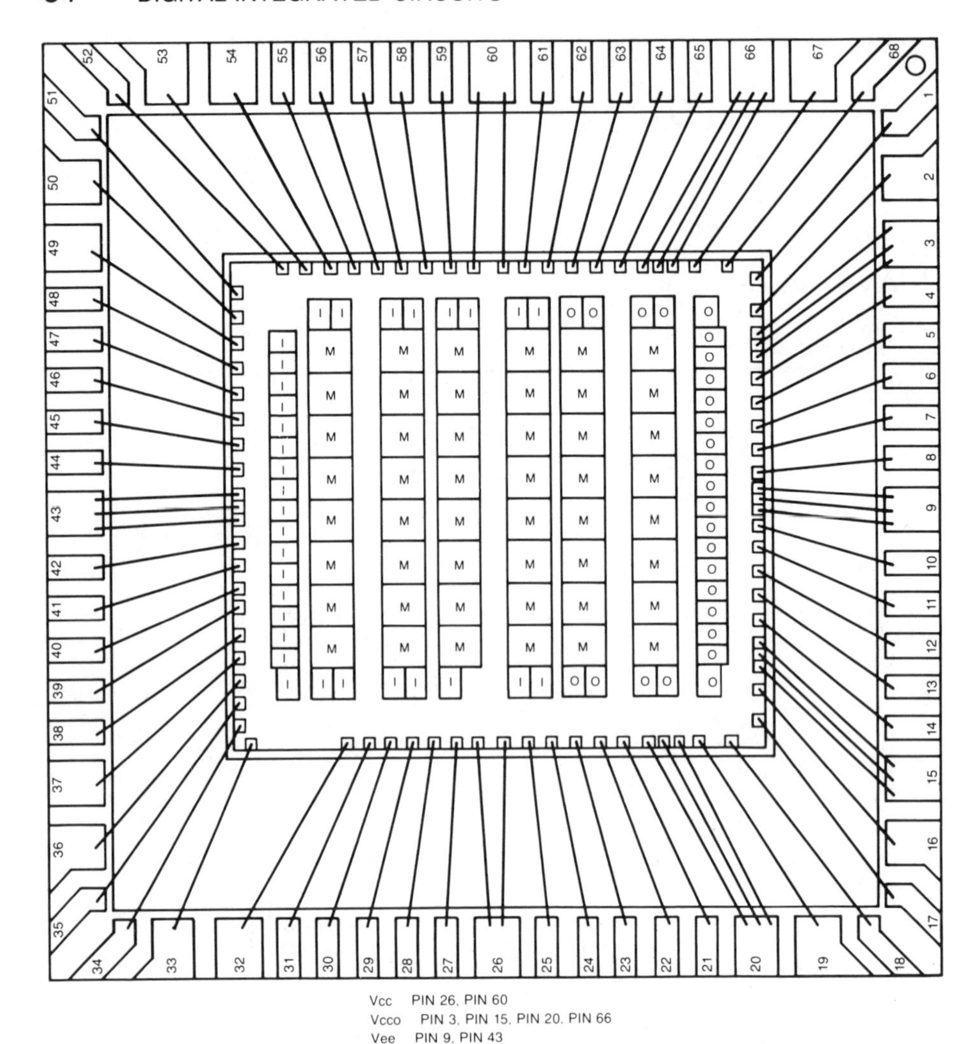

Figure 3.24 Macrocell array mounted in a 68-lead chip carrier.

the speed $v_o = 3 \times 10^8$ m/s. In a lossless dielectric constant media (K), its speed v_d is reduced by the $\sqrt{K}$.

$$v_o = 3 \times 10^8 \text{ m/s} \tag{3.24}$$

$$v_d = \frac{3 \times 10^8 \text{ m/s}}{\sqrt{K}} \tag{3.25}$$

The distance, X_D, a wave travels during the rise time (t_r) or fall time (t_f) transition of an ECL gate is:

$$X_D = t_{r,f} \times v_d = \frac{3 \times 10^8}{\sqrt{K}} t_{r,f} \tag{3.26}$$

For the ringing to be confined to the switching transition, the interconnection length (ℓ_D) must be then kept to ½ X_D because of the down and back time interval of the reflection:

$$\ell_D = \frac{X_D}{2} = \frac{3 \times 10^8}{2\sqrt{K}} t_{r,f} \tag{3.27}$$

EXAMPLE 3.6 Compute ℓ_D, the maximum direct (nonterminated) wire interconnection distance between two ECL gates, if $t_r = t_f = 1$ ns and $K = 4$.

PROCEDURE

1 Substituting the given values in Eq. 3.27,

$$\ell_D = \frac{3 \times 10^8 \times 10^{-9}}{4} = 0.075 \text{ m} = 7.5 \text{ cm}$$

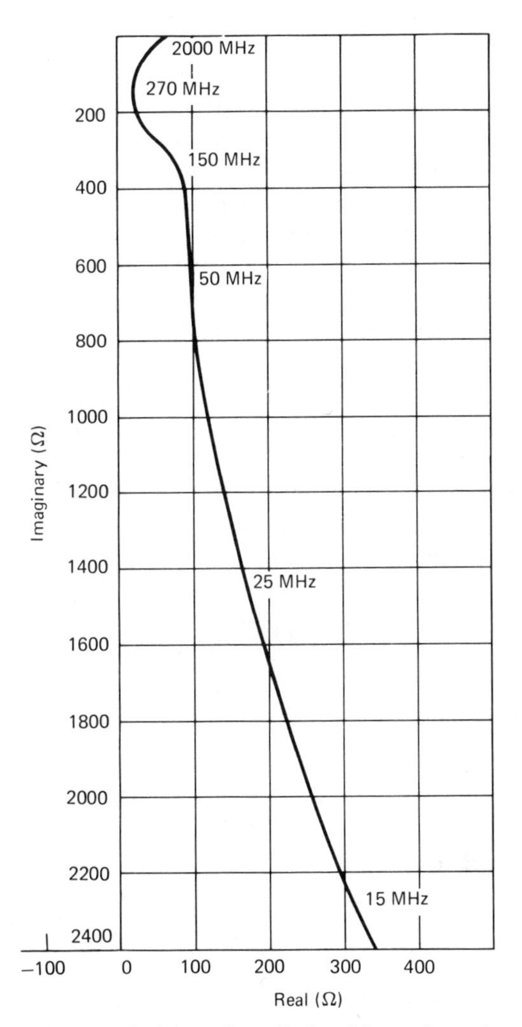

Figure 3.25 Plot of small signal input impedance of a typical ECL gate as a function of frequency.

2 We see that about 3 inches (7.5 cm) is the maximum interconnection length allowable. This distance decreases as t_r and t_f decrease and/or K increases. Teflon dielectric ($K \sim 2$) is, therefore, more desirable than, for example, PC material ($K \sim 5$).

Oscillation in an improperly packaged ECL system is perhaps the most difficult to cure, short of repackaging. For this reason a bit of qualitative understanding of the problem could be important to the new ECL user. Referring to Fig. 3.25 again, we note that the real part (resistance) decreases with signal frequency and then increases after reaching a minimum value at 270 MHz. This characteristic was experimentally obtained in a carefully constructed test fixture where extraneous lead inductances at the V_{CC}, $V_{CC(aux)}$, and V_{EE} connections (Fig. 3.26) were kept low relative to what might occur in a poorly constructed ECL system. The net effects of these stray V_{CC} and V_{EE} inductances can shift the real part of the input impedance into the negative real region.

The requirement for sustained oscillation is some interconnection inductance along with the capacitive imaginary part of input Z. Early ECL products of the 1960s to early 1970s were more prone to self-oscillation because no input compensation was used by the chip designers to guarantee a positive real input Z. The incorporation of these input compensators in current products has vastly reduced, but not eliminated, the problem. The inputs that are most sensitive are the lower levels of gating signals that must be on-chip level shifted by an emitter follower in series gated products.

Input B (Q_{11} of Fig. 3.21) is less forgiving (prone to oscillation) than input A. Qualitatively, input B is always more unstable because Q_{11} and Q_{12} are a Darlington pair when compared to Q_3, which is representative of an emitter follower for stability consideration. As mentioned previously, oscillations are difficult to cure; however, if reasonably low Z connections to V_{TT}, V_{CC}, and $V_{CC(aux)}$ are available, an R_o (50 to 150 Ω) connection between a suspected nonterminated oscillating input and V_{TT} can cure the problem. Applying a resistor in parallel to a suspected input will usually be more effective and less degrading on switching times than applying a series resistor to the input (another but less desirable possibility). If we had, for example, a negative input real part (R_1) of -100 Ω, it would require a 100 Ω or greater series resistor (R_X) to

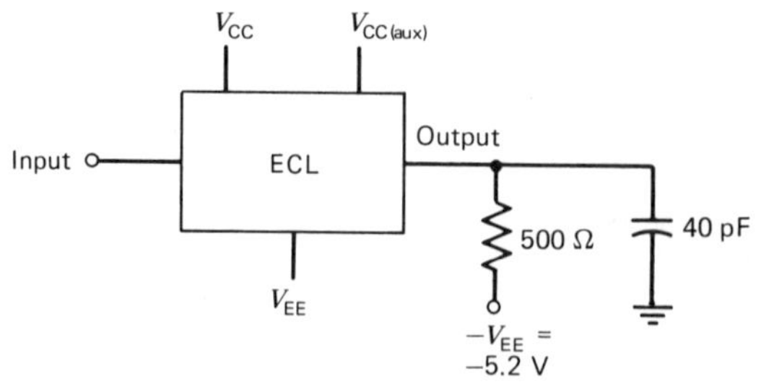

Figure 3.26 An ECL gate driving a lumped capacitance (40 pF) load. $V_{CC(aux)}$ is output emitter follower supply.

stabilize the circuit since:

$$R_1 + R_X \geqslant 0 \qquad R_X > |R_1| \tag{3.28}$$

Placing a resistor in parallel requires that the parallel combination be positive, or

$$\frac{R_1 R_X}{R_1 + R_X} \geqslant 0 \qquad R_X < |R_1| \tag{3.29}$$

Series resistance stabilization is a less desirable solution because an additional delay occurs stemming from $R_1 C_{in}$ effects. Applying a parallel R_X to V_{TT} could result in a speed improvement because the time constants are reduced.

3.8 SYSTEM SWITCHING PERFORMANCE

Two primary factors determine the switching performance of a system. These are the individual IC delays and the associated wiring delays. Let us consider a system where a critical timing path can be defined between n parts each having a delay of $t_{D_{IC}}$, and a wire connection delay characterized by t_{DW}. In simplistic terms, the total path delay t_P then is:

$$t_P = \sum_1^n t_{D_{IC}} + \sum_1^n t_{DW} \tag{3.30}$$

If we assume the critical path is controlled by a clock, it is then reasonable to say that the clock period will be related to t_P. If we assume a 50% duty cycle (t_P occurs during ½ of the clock period t_C) then:

$$t_C = 2\sum_1^n t_{D_{IC}} + \sum_1^n t_{DW} \tag{3.31}$$

It is currently fashionable to quote a megaherz or gigaherz property for a digital system. Care must be used, however, because by Eq. 3.31 it is seen that n, the number of logic elements delay in a critical path chain, and the total wire delay, which varies proportional to n, determine the system clock frequency.

How Can the User Affect System Performance? (Secondary Effects)

The user in the construction of an ECL system controls the following parameters, which can add to the fundamental delay as shown by Eq. 3.31. Some of these parameters are:

1 a. Lengths of interconnections (IC package density).

b. Signal wiring capacitance (choice of dielectric constant K).
c. Fan out (number of IC loads driven by an output).

2 Resistor load value connected between output(s) and V_{EE} or V_{TT}.

3 The magnitude of V_{TT}.

4 Decoupling of V_{CC}, $V_{CC(aux)}$, and V_{TT}.

5 Rise and fall time of clock drivers.

Wiring delays where the interconnections between ICs are in a homogeneous media, for example, terminated coaxial cable or strip lines sandwiched between a ground plane, can be easily estimated. The delay t_{DO} of an electromagnetic wave traveling in free space is:

$$t_{DO} \cong 32.8 \text{ ps/cm} \tag{3.32}$$

For a terminated lossless transmission-line system of dielectric constant K, the delay is:

$$t_D \cong \sqrt{K}\, 32.8 \text{ ps/cm} \tag{3.33}$$

For cases where the down and back delays are equal to, or less than, the rise time, we can make the following calculation for delay. For a nontransmission line analysis, all the capacitances can be lumped to calculate the delay. Because the output of an ECL IC is an npn emitter follower, it can supply a large range of transient charge current when the output is rising without appreciable change in rise delay, especially when compared to the increase in fall delay for the same amount of capacitance. The reason the fall delay is more sensitive to capacitance stems from the fact that the discharge of this capacitor is through the output load resistor. Its value, and to which power-supply connection is made (V_{TT} or V_{EE}), affect the fall delay response.

EXAMPLE 3.7 Figure 3.26 shows an ECL gate driving a lumped capacitance load of 40 pF with a 500 Ω output load resistor connected to -5.2 V. Delay $t_{DR} = 1.0$ ns and $t_{DF} = 1.2$ ns into 500 ohms, but with negligible capacitance loading ($C_L \leqslant 1$ pF). What would the loaded (40 pF) delay be?

PROCEDURE

1 For this case, the output rise delay would essentially remain unchanged. The resulting fall delay can be calculated as follows:

$$V_{BB} = \frac{|V_{OH}| + |V_{OL}|}{2} \tag{3.34}$$

$$t_{DF} = t_{DO} + \frac{R_o C_o V_L}{2(V_{EE} - V_{BB})} \tag{3.35}$$

where $t_{DO} = 1$ ns
$R_o = 0.5$ kΩ
$-V_{EE} = -5.2$ V
$C_o = 40 \times 10^{-12}$ F
$V_L = 0.8$ V
$V_{OH} = -0.8$ V
$V_{OL} = -1.6$ V

2 By Eq. 3.34,

$$V_{BB} = \frac{0.8 + 1.6}{2} = 1.2 \text{ V}.$$

3 By Eq. 3.35,

$$t_{DF} = 1 \text{ ns} + \frac{0.5 \times 10^3 \times 40 \times 10^{-12} \times 0.8}{2(5.2 - 1.2)} = 3 \text{ ns}$$

4 Terminating the load resistor to V_{EE} gives us

$$t_{DF} = 1 \text{ ns} + 0.1 \text{ ns/k}\Omega\text{-pF} \tag{3.36}$$

5 If we connected our 500 Ω resistor to V_{TT} (−2 V) rather than V_{EE} (−5.2 V), the delay is:

$$t_{DF} = t_{DO} + 0.7RC \tag{3.37}$$

or

$$t_{DF} = 1 \text{ ns} + 0.7 \times 0.5 \times 10^3 \times 40 \times 10^{-12} = 15 \text{ ns}$$

6 For a load terminated to V_{TT},

$$t_{DF} = 1 \text{ ns} + 0.7 \text{ ns/k}\Omega\text{-pF} \tag{3.38}$$

7 From the above analysis we see that 500 Ω to V_{TT} results in too large a delay, considering the larger increase in fall delay when loaded with 40 pF.
8 Substituting 50 Ω in Eq. 3.38, for R_o we get:

$$t_{DF} = 1 \text{ ns} + 0.7 \text{ ns} \times \frac{50}{1000} \times 40 = 2.4 \text{ ns}$$

This is a much better solution if we wish to use the V_{TT} supply to terminate R_o.

Power dissipation and maximum IC output current limits must be considered before the final choice of resistance and to which supply it is connected. If it were not for power and current limits, termination of the load resistor to

V_{EE} would seem to be the best choice. Equations 3.36 and 3.38 can be used to calculate fairly accurately the fall delay for lumped capacitance load cases.

EXAMPLE 3.8 Calculate the total delay of a 30.5-cm coaxial cable if the fan out $m = 5$, $R_o = 50\ \Omega$, $C_{in} = 0.5$ pF, and $K = 2$ (Teflon).

PROCEDURE

1 A Thevenin equivalent circuit of the terminated transmission line loaded with m inputs is given in Fig. 3.27. The delay caused by the ICs input capacitance is:

$$t_{Di} = 0.7 \frac{R_o}{2} mC_{in} \tag{3.39}$$

2 Adding Eq. 3.33 to 3.39, we have

$$t_D = \sqrt{K}\ 32.8 \frac{\text{ps}}{\text{cm}} \times \text{length of line (cm)} + t_{Di} \tag{3.40}$$

3 Substitution of given values in Eq. 3.40 yields:

$$\begin{aligned} t_D &= \sqrt{2} \times 32.8 \text{ ps} \times 30.5 + \frac{0.7 \times 50}{2} \times 0.5 \times 10^{-12} \times 5 \\ &= 1.414 \text{ ns} + 0.04375 \text{ ns} \simeq 1.46 \text{ ns} \end{aligned}$$

If we compare this result with the lumped case where R_o is 50 Ω to V_{TT} (−2 V), we note that a lumped capacitance of 40 pF (Ex. 3.7) increases the fall delay the same amount a 30.5 cm (1 foot) length of Teflon dielectric 50 Ω coaxial cable would. The transmission line delays both the rise and fall responses equally, whereas the lumped equivalent 40 pF affects primarily the fall delay. From this simple analysis we see the equivalence of capacitance on switching delay whether it be lumped or distributed, as in a transmission line. Knowing a cable's char-

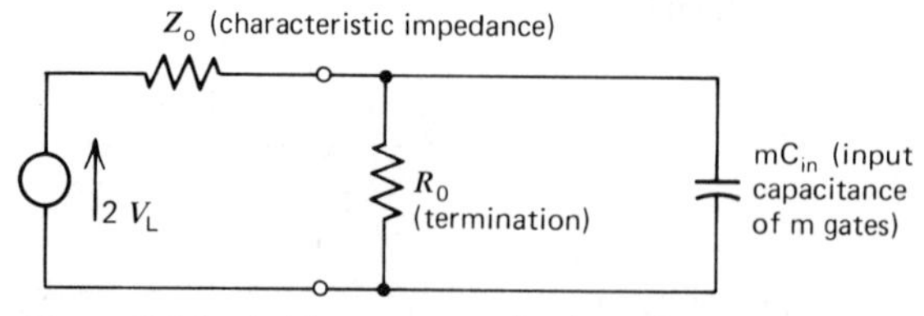

Figure 3.27 A Thevenin equivalent circuit for a terminated transmission line driving m ECL inputs.

acteristic impedance (Z_o) and its delay, we can calculate the lumped capacity equivalence by:

$$C_{\text{coaxial}} = \frac{t_D}{0.7\,Z_o} = \frac{1.414 \times 10^{-9}}{0.7 \times 50} = 40.41 \text{ pF} \tag{3.41}$$

The magnitude of V_{TT} also can affect switching performance. As V_{TT} approaches V_{OL}, the rise delay will start to increase rapidly owing to changing effects of the emitter follower transistor's diffusion capacitance as the bias current decreases toward zero, which would be the case when $V_{TT} = V_{OL}$. The problem is further intensified when output wire-ORing is used. Any small bias current is further reduced because it is shared by the multiple emitter followers when the output is in the low state.

Figure 3.28 shows a representative emitter follower rise response delay as a function of bias current in the low state. In general, a 0.1 V minimum difference between V_{OL} and V_{TT} will eliminate this problem.

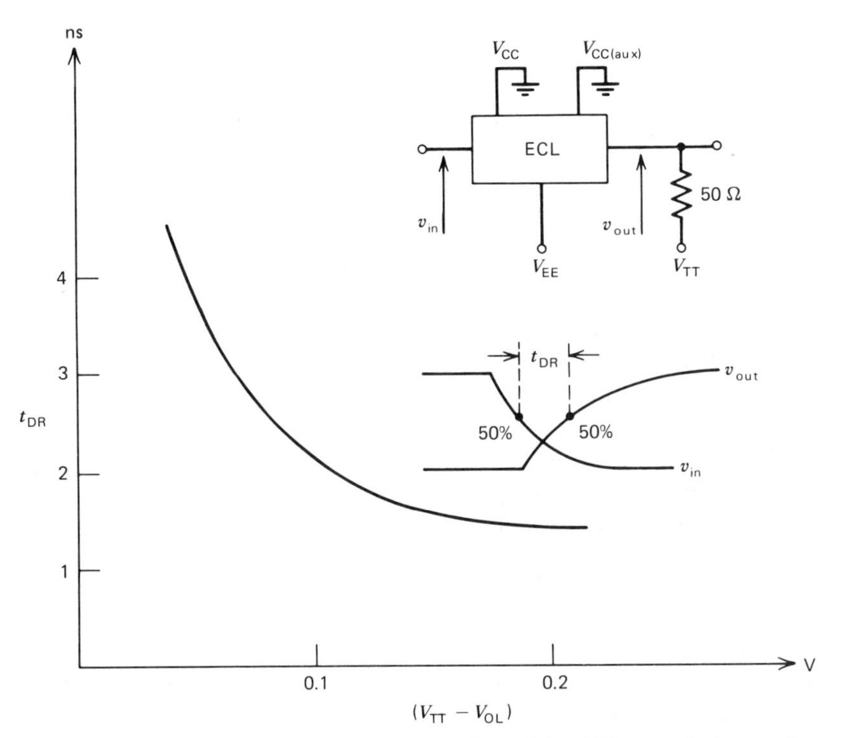

Figure 3.28 Rise delay response as a function of the difference between the termination supply (V_{TT}) and the logic 0 output of the gate (V_{OL}).

Most ECL gates, especially the higher speed variety, for instance, Motorola's MECL III and Fairchild IIC, have three power supply connections: V_{CC}, $V_{CC(aux)}$, and V_{EE}. Voltage V_{CC} is the most sensitive to power supply noise and should always be connected to the lowest impedance and quietest power distribution line. Voltage $V_{CC(aux)}$ is the output emitter follower supply for ECL chips. It is quite common for current transients of 25 mA/ns to occur on these lines when transmission lines are driven. While this line is not as critical to noise as the V_{CC} line, care must be taken to ensure that the distribution inductance is kept reasonably low; otherwise the emitter follower could saturate owing to excessive $L(\Delta I/\Delta t)$ voltage drops in the collector. At 25°C, a forward voltage of 0.1 V is reasonable without incurring saturation effects.

EXAMPLE 3.9 Calculate the maximum output inductance for the $V_{CC(aux)}$ line.

PROCEDURE

1 From the previous discussion,

$$\frac{\Delta I}{\Delta t} \cong \frac{25 \times 10^{-3}}{10^{-9}} = 25 \times 10^{6} \text{ A/s}$$

2 For no saturation,

$$L \frac{\Delta I}{\Delta T} \leqslant 0.1 \text{ V}$$

3 Therefore,

$$L \leqslant \frac{0.1}{25 \times 10^{6}} \leqslant 4 \times 10^{-9} \text{ H} = 4 \text{ nH}$$

For a gate that has two independent outputs terminated in 50 Ω, L_{max} would be 2 nH, for four outputs 1 nH, and so on. Some compensation takes place when ϕ and $\overline{\phi}$ are loaded simultaneously into 50 Ω because one output, which is rising, is canceled by the other output, which is falling. Sometimes system ac performance can be improved when unused outputs of the opposite phase are loaded with an output load resistor because of the noise cancelation effect. Balancing the output in this manner should not be used if the nominal output is wire-ORed with other ICs. In this case, loading the unused output could generate additional noise.

An obvious question at this point is how much wire length corresponds to a few nanohenries? Inductance of a wire, unlike its capacitance, is usually more difficult to calculate. A simple way to estimate a wire's inductance can be arrived at by extending our previous transmission-line analysis to determining a line's equivalent inductance and using that result to estimate a wire's inductance of the type found on PC board wiring. For a lossless transmission line, the relationship between inductance, capacitance, and characteristic impedance is:

$$Z_o \cong \sqrt{\frac{L}{C}}$$

Solving for L:

$$L = Z_o^2 C \tag{3.42}$$

EXAMPLE 3.10 In our previous analysis (Ex. 3.8) of a 30.5 cm line and $Z_o = 50\ \Omega$ having a Teflon dielectric ($K = 2$), we found its equivalent lumped capacitance was 40.41 pF. Calculate the total and unit inductance of the line.

PROCEDURE

1 By Eq. 3.42,

$$L_{(30.5\ \text{cm})} = 50^2 \times 40.41 \times 10^{-12} = 101\ \text{nH}$$

2 The unit length inductance is:

$$L/\text{cm} = 101\ \text{nH}/30.5\ \text{cm} = 3.3\ \text{nH/cm}$$

While this approach is based on a lossless transmission-line analysis, it does provide a reasonably accurate estimate for short wire lengths typically found on printed circuit boards.

The V_{EE} distribution system is the least sensitive to distribution inductance effects. Generally, V_{EE} wiring can be printed board traces with V_{CC} and $V_{CC(aux)}$ connected to ground planes within the multilayer board. Considerable detailed information on this subject is available in the Motorola and Fairchild ECL handbooks, as well as detailed theory and practical application of transmission line theory.

Rise and fall times of the various system clock drives that drive ECL MSI/LSI gates will affect the overall performance of the system. Care should be used in providing these pulses to ensure maximum performance from the MSI/LSI circuits used in the system. In general, these drivers are SSI ECL gates driving well terminated, strip line, or coaxial clock distribution systems.

It is hoped that in reading this chapter the reader has gained some additional insight into the nature and idiosyncrasies of ECL. Some personal experience and observations by the author not previously documented in any known publication were presented in this chapter.

3.9 REFERENCES

Fairchild, *ECL Data Book*, Fairchild Camera and Instrument Corporation. Latest edition.

Marley, R. R., "On-chip Temperature Compensation for ECL," *Electronics Products*, pp. 36–39, March 1, 1971.

Motorola, *MECL System Design Handbook*, Motorola Semiconductor Products, Inc. Latest edition.

Meyer, C. S., D. K. Lynn, and D. J. Hamilton, *Analysis and Design of Integrated Circuits*, McGraw-Hill, New York, 1968.

Rao, N. N., *Elements of Engineering Electromagnetics*, Prentice-Hall, Englewood Cliffs, NJ, 1977.

Seelbach, W. C., "The Unfolding of a Technology," *Motorola Monitor*, pp. 17–23, March 1974.

CHAPTER 4

MOS and CMOS Logic

John D. Virzi
Member of the Technical Staff
RCA Solid State Division

4.1 INTRODUCTION

Since the late 1960s, the metal oxide semiconductor field effect transistor (MOSFET) has supplanted the saturated bipolar transistor in many digital circuit designs. Its small geometry allows high-density circuits and makes possible large-scale integration (LSI).

In this chapter, we shall examine the structure of the MOSFET, how it operates electrically, and concentrate on its several possible configurations in the most basic element of digital logic—the inverter. The operational constraints of the inverter will be discussed and finally examples of how it may be used to perform more complex logic functions will be given. The chapter concludes with an examination of the power MOSFET structure.

4.2 MOS DEVICE STRUCTURE AND THEORY

The MOSFET is a *unipolar* device, in contrast to conventional npn and pnp transistors, which are *bipolar* devices. While the bipolar transistor relies on the flow of minority and majority carriers for their operation, the MOSFET's operation involves the flow of one type of carrier only. In addition, it may be thought of as a voltage-controlled device in much the same way as a vacuum-tube triode is controlled by a grid voltage.

MOSFETs with two characteristics have been developed. These are *en-*

hancement-mode and *depletion-mode* devices. The presence of a gate-control voltage is necessary to permit carrier conduction in the enhancement-mode device; no such voltage, however, is necessary to permit conduction in the depletion-mode device. The enhancement-mode device is normally "off" and will remain so until its gate is properly biased; the depletion-mode device is normally "on" with zero gate bias. The enhancement-mode device is used as a switching element while the depletion-mode device is often used as a load element in integrated logic designs.

The two types of enhancement-mode devices are shown in Fig. 4.1. Two-dimensional, cross-sectional views illustrate the n-channel MOSFET (NMOS) and the p-channel MOSFET (PMOS).

When a sufficiently positive voltage is applied to the gate electrode with respect to the source of the n-channel device, electron current flows between the source and drain regions, which are highly doped n-type silicon diffused areas (see Chap. 20). When a negative gate-to-source voltage is applied to the p-channel device, hole current flows between source and drain. These regions are diffused p-type silicon (Si).

The metal electrodes are most often aluminum. In some cases the gate electrode is made of polycrystalline silicon. The insulating material is silicon dioxide (SiO_2) and the semiconductor material is doped silicon.

When a small voltage is impressed across the gate and source electrodes,

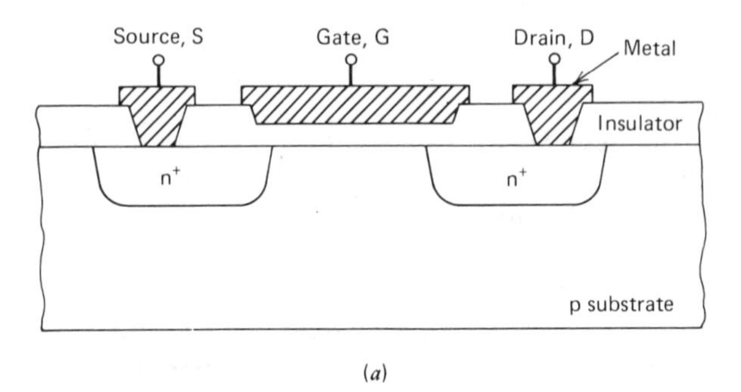

(*a*)

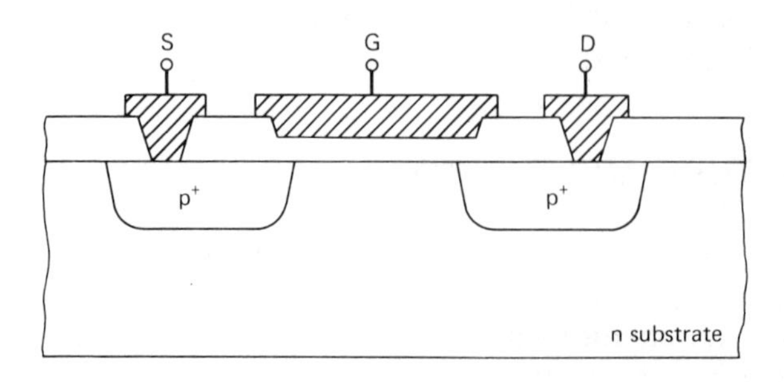

(*b*)

Figure 4.1 Two types of enchancement transistor IC structures. (*a*) NMOS. (*b*) PMOS.

a depletion region forms beneath the gate, just below the oxide layer between source and drain regions. When this voltage is increased to a certain threshold voltage (V_{th}), the region just below the oxide is inverted as to its charge. That is, for the NMOS transistor the p region will appear to be n-type Si. At this inversion voltage, minority carriers are attracted into the depletion region, a channel is formed between source and drain, and current can begin to flow.

A characteristic of MOSFETs is that they have a very high input impedance. The contact to the channel region is the SiO_2 layer. Silicon dioxide is an excellent insulator. This is an advantage as the resulting input leakage current is very low, typically less than 10^{-14} A. The oxide layer serves as a natural dielectric for the capacitance associated with the gate. Commonly used terms for characterizing a MOSFET are summarized in Table 4.1.

EXAMPLE 4.1 Calculate the oxide capacitance of an MOS transistor having a channel length of 0.24 mil, a channel width of 0.36 mil, and a gate oxide thickness of 1000 Å.

PROCEDURE

1 As we shall later see, the numerical value of input capacitance for the MOSFET is of importance in determining its maximum operating frequency. It is, therefore, pertinent to calculate the oxide capacitance, C_G. In general, capacitance is proportional to the area (width times length) of the conducting elements divided by their separation; hence,

$$C_G = \frac{\varepsilon WL}{d_o} \tag{4.1}$$

for an integrated transistor of channel width W and length L. The thickness of the oxide layer under the gate electrode is d_o and ε is the relative permittivity of the oxide.

2 Because $\varepsilon = k_o\varepsilon_0$, where k_o is the relative dielectric constant of the oxide and ε_0 is the permittivity of free space, the oxide capacitance per unit surface area may be written as:

$$\begin{aligned} C_o &= k_o\varepsilon_0/d_o \\ &= (8.86 \times 10^{-14}\ \text{F/cm})(3.9)/d_o \\ &= (3.46 \times 10^{-13})/d_o\ \text{F/cm} \end{aligned} \tag{4.2}$$

for SiO_2.

3 The oxide layer below the gate has a thickness of approximately 1000 Å for a standard process. Therefore, a minimum size NMOS device having a width $W = 0.36$ mil and length $L = 0.24$ mil has an oxide capacitance of $C_G = C_oWL$, or:

$$\begin{aligned} C_G &= \frac{(3.46 \times 10^{-13}\ \text{F/cm})}{1000 \times 10^{-8}\ \text{cm}}(0.24\ \text{mil})(0.36\ \text{mil})(6.45 \times 10^{-6}\ \text{cm}^2/\text{mil}^2) \\ &= 0.0193\ \text{pF} \end{aligned}$$

While the MOSFET input capacitance may be closely approximated by C_G, the total input capacitance, C_{input} *per unit area*, is the *series combination* of

Table 4.1 **Commonly used terms for characterizing a MOSFET**

Term	Definition
Inversion layer	A layer of doped silicon beneath the gate oxide which has been inverted as to its charge. It serves as a conduction channel between source and drain regions.
Threshold voltage (V_{th})	The gate-source potential necessary to permit conduction in the MOSFET by forming an inversion layer beneath the gate oxide.
Enhancement-mode device	A MOSFET requiring the application of sufficient gate-source bias to permit conduction by forming an inversion.
Depletion-mode device	A MOSFET in which a channel is formed even in the absence of gate bias. Channel conduction is enhanced by the proper application of gate bias.
Pinch OFF voltage (V_P)	The gate-source voltage which causes the mobile charge concentration at a particular point in the channel to become zero. For a depletion-mode device, the channel "pinches off" when $V_D = V_G - V_P$.
Gate oxide capacitance (C_G)	$C_G = \varepsilon WL/d_o$ approximates the MOSFET input capacitance.
Maximum operating frequency (f_o)	$f_o = g_m/C_G$
Saturation region	The region of the MOSFET I–V characteristics were $V_D \geqslant V_G - V_{th}$ and the channel current assumes a fairly constant value.
Channel "ON" resistance (R_{ON})	$R_{ON} = \left[(V_{GS} - V_{th})(W/L)\ \mu\varepsilon/d_o\right]^{-1}$ for a MOSFET which is turned "on."
Noise immunity	The amount the input voltage of the MOSFET may vary without significantly affecting the output voltage; expressed as a percentage of supply voltage.
Noise margin	The noise voltage that can be impressed upon v_{in} of a logic gate without upsetting the logic or causing an output to exceed given v_{out} ratings.
Fanout	The number of identical gates which a logic gate is able to drive from its output.

the oxide capacitance C_o and C_d, the capacitance associated with the depletion region below the gate oxide. This depletion layer capacitance per unit area is given by:

$$C_d = \frac{k_d \varepsilon_0}{d_d} \tag{4.3}$$

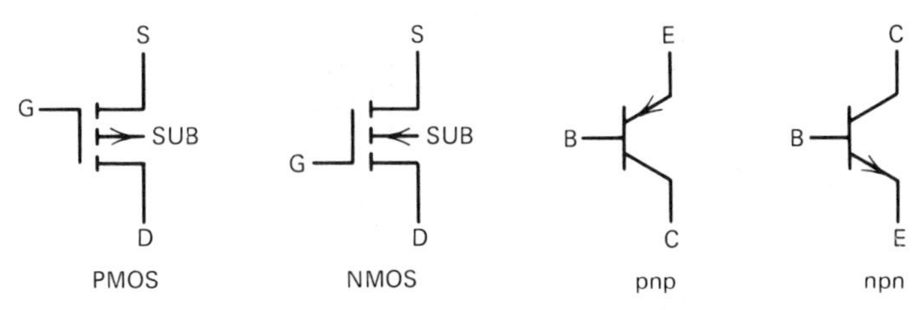

Figure 4.2 Electrical symbols for MOSFETs and bipolar junction transistors (BJTs).

where k_d, the dielectric constant for Si, is 11.7 as opposed to 3.9 for SiO_2. The depth of the depletion region, d_d, is not a constant from source-to-drain regions. It is a function of the applied gate voltage. Therefore, the total input capacitance per unit surface area, given by $C_{input} = C_o C_d/(C_o + C_d)$ is also a function of gate voltage because d_d is a function of the gate voltage. Typical design practice is to approximate C_{input} as C_o for all gate voltages, as this is the maximum value encountered.

Figure 4.2 shows the schematic designations of the NMOS and PMOS enhancement-type transistors with terminal assignments given. They are compared with their npn and pnp counterparts. MOS devices may be used to perform functions analogous to those performed by bipolar transistors.

For example, two methods of implementing a current mirror are shown in Fig. 4.3. The circuits are called current mirrors because, in these configurations, whatever current flows through Q_1 is reflected (or *mirrored*) almost perfectly in device Q_2. If the two devices are in a 1:1 matching geometry ratio, these two currents are equal. Therefore, a reference current, I_{REF}, may be reflected, as I_2, to form a current source.

4.3 MOS DEVICE CHARACTERISTICS

Typical current–voltage curves for n-channel and p-channel transistors are shown in Figs. 4.4 and 4.5, respectively. These I–V characteristics are for the large geometry, enhancement NMOS and PMOS transistors found, for example, in the RCA CD4007UB package.

Below a certain gate-source threshold voltage (V_{th}) current will not flow. For voltages $>V_{th}$, the channel conductivity may be modulated by varying the

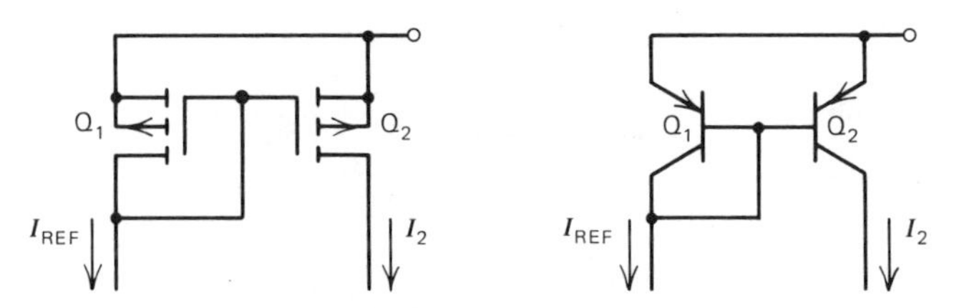

Figure 4.3 MOS and bipolar current mirrors.

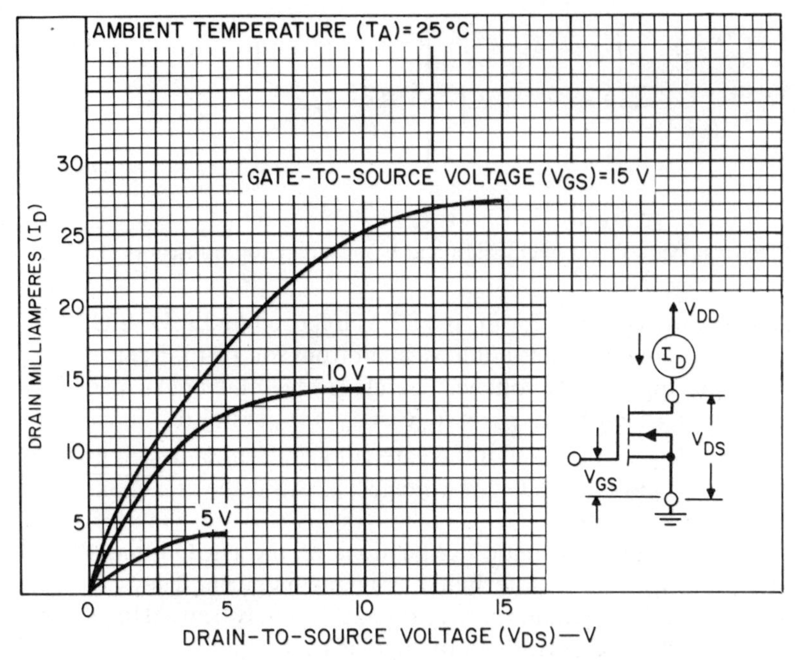

Figure 4.4 Typical n-channel drain characteristics. (Courtesy RCA Corp.)

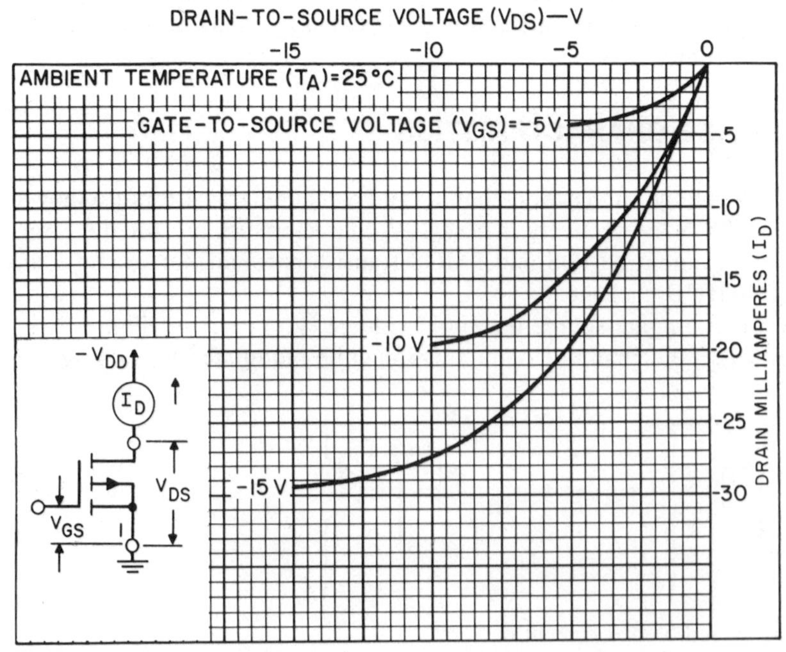

Figure 4.5 Typical p-channel drain characteristics. (Courtesy RCA Corp.)

gate-to-source potential. Examination of Fig. 4.6 shows the generic regions in which the MOSFET may operate. These are the linear and saturated regions.

In the following discussion, all voltages are measured relative to the source. For small drain voltages, the MOSFET operates in the linear region where the channel acts like a linear resistor. Channel conductance g is essentially constant and is given by:

$$g \simeq \left.\frac{\Delta I_D}{\Delta V_D}\right|_{V_G = \text{const}} = \frac{W}{L}\mu C_o(V_G - V_{th}) \tag{4.4}$$

where μ is the carrier mobility in the inversion layer. The conductance will increase linearly if V_G is increased.

The transconductance g_m in the linear region is given by:

$$g_m \simeq \left.\frac{\Delta I_D}{\Delta V_G}\right|_{V_D = \text{const}} = \frac{W}{L}\mu C_o V_D \tag{4.5}$$

The drain current I_D in the linear region is:

$$I_D = \mu \frac{W}{L} C_o\left[(V_G - V_{th})V_D - \frac{V_D^2}{2}\right] \tag{4.6}$$

While the drain voltage is slowly increased, the channel resistance begins to increase. When the drain voltage is increased such that $V_D = V_G - V_{th}$, as shown by the dashed line of Fig. 4.6, the MOSFET enters its saturated region. The curves flatten out and the channel current "saturates" to a fairly constant value. The transconductance in the saturated region increases linearly as V_G increases.

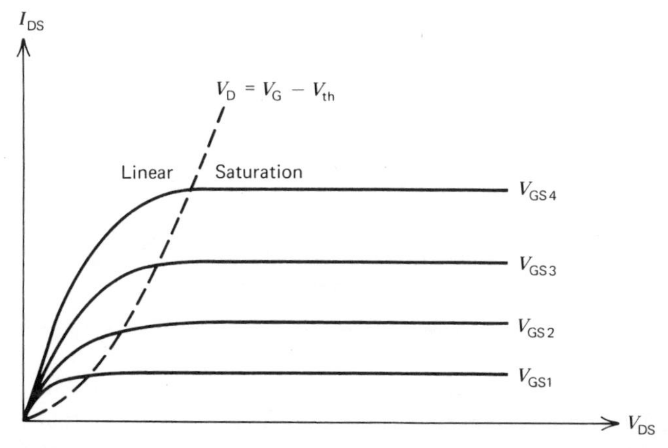

Figure 4.6 Current-voltage (I–V) characteristics showing regions of linear and saturation operation.

The drain current in the saturated region, I_{DSAT}, is given by:

$$I_{DSAT} = \frac{\mu W}{2L} C_o(V_G - V_{th})^2 \tag{4.7}$$

for $V_{DSAT} \geqslant V_G - V_{th}$.

EXAMPLE 4.2 Calculate the transconductance and maximum operating frequency for the MOSFET of Ex. 4.1.

PROCEDURE

1 The maximum frequency of operation, f_0, for the MOSFET is given by:

$$f_0 = \frac{g_m}{C_G} \tag{4.8}$$

where C_G is the gate capacitance of Ex. 4.1 and g_m is the transconductance given by Eq. 4.5 for $V_D \leqslant V_{DSAT}$.

2 Thus,

$$\begin{aligned} g_m &= \frac{W}{L} \mu C_o V_D \\ &= \frac{0.36 \text{ mil}}{0.24 \text{ mil}} \times \frac{450 \text{ cm}^2}{\text{Vs}} \times \frac{0.0346 \ \mu\text{F}}{\text{cm}^2} \times 1.5 \text{ V} \\ &= 34.98 \ \mu\text{s} \end{aligned}$$

given a drain-source voltage of 1.5 V and recaling that capacitance may be expressed as Amperes-second/volts. The electron mobility, μ_n, is given as 450 cm²/Vs and is a function of concentration and the type of silicon.

3 By Eq. 4.8,

$$\begin{aligned} f_0 &= \frac{g_m}{C_G} = \frac{\mu_n V_D}{L^2} \\ &= \frac{450 \text{ cm}^2}{\text{Vs}} \times 1.5 \text{ V} \times (0.0576 \text{ mil}^2 \times 6.45 \times 10^{-6} \text{ cm}^2/\text{mil}^2)^{-1} \\ &= 1.817 \text{ GHz} \end{aligned}$$

Because of the higher mobility of electrons, an NMOS transistor's maximum operating frequency will be greater than that of a PMOS device of the same geometry.

It can be seen from this example that the intrinsic MOSFET is a fairly fast device. However, actual switching speeds for the MOSFET, as used in a circuit, are considerably slower. This stems from the parasitic and loading capacitances present in the circuit.

4.4 THE MOS INVERTER

The logic inverter is a fundamental element in any digital circuit. It may take several forms, four of which are discussed in this section.

NMOS with Saturated Load

One type of inverter used for NMOS logic has a saturated NMOS transistor as its load and is shown in Fig. 4.7. Both devices are of the enhancement type. It will be helpful to recall that the MOSFET may act as a voltage variable, nonlinear resistor. When connected in its saturated mode, with gate tied to drain, this resistance is particularly high. The use of diffused resistors instead of NMOS transistors as load elements for inverters in an integrated circuit consumes too much area on the chip and is not practical for the layout of a complex digital integrated circuit.

An expression governing the inverter transfer function follows:

$$v_{out}^2(1 + \beta) - v_{out}[2V_{HI} + 2\beta_R(v_{in} + V_{th})] + V_{HI}^2 = 0 \qquad (4.9)$$

where

$$\beta_R = \frac{k_D}{k_L} = \frac{\mu_L(W_L/L_L)C_o/2}{\mu_D(W_D/L_D)C_o/2} = \frac{W_D L_L}{W_L L_D}$$

is the ratio of the conduction factors of the driver and load devices. Because the gate is tied to the drain for the load device, $V_{HI} \triangleq V_{DD} - V_{th}$ and no current may flow until $V_{DD} = V_D > V_{th}$. Therefore, the output voltage for this design always falls short of the upper rail (V_{DD}) by the transistor's threshold voltage. Equation 4.9 may be solved for v_{out} as a function of v_{in}. A typical inverter transfer function is shown in Fig. 4.8.

In the transition region, both devices are saturated, the curve is linear, and its slope:

$$\frac{\Delta v_{out}}{\Delta v_{in}} = -(\beta_R)^{1/2} \qquad (4.10)$$

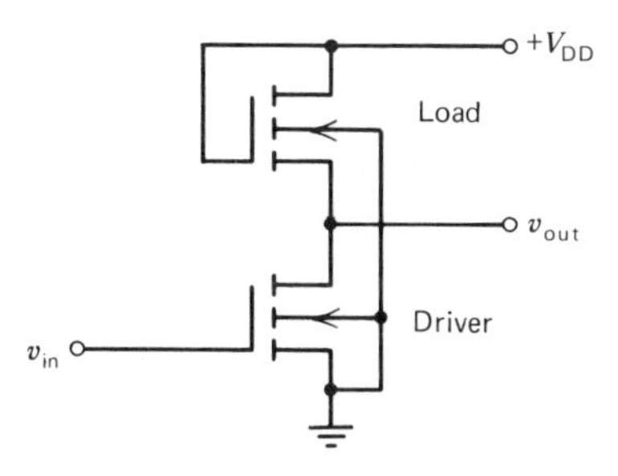

Figure 4.7 NMOS inverter with saturated load.

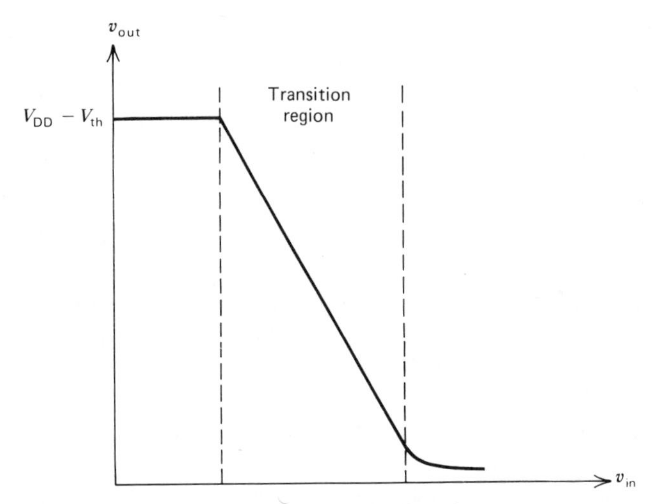

Figure 4.8 Typical inverter transfer function.

defines the voltage gain of the inverter. For small values of v_{in}, both devices are saturated, because $V_D >> V_G$ for the driver, and v_{out} is V_{HI} (logic 1). For large values of v_{in}, the load remains saturated, but the driver enters its linear region because v_{out}, which is V_D for the driver, is at a low voltage V_{LO} (logic 0).

EXAMPLE 4.3 The following parameters describe an NMOS inverter with a saturated enhancement-mode load transistor: $W_D = 550\ \mu m$, $W_L = 22\ \mu m$, $L_L = L_D = 10\ \mu m$, $V_{DD} = 10$ V, and $V_{th} = 2$ V.
Calculate the

a. Logic levels V_{HI}, V_{LO}.
b. Voltage gain A in the transition region.
c. Driver linear and saturated voltage regions.

PROCEDURE

a.
1 $V_{HI} = V_{DD} - V_{th} = (10 - 2)\text{V} = 8$ V
2 Solving for V_{LO} from Eq. 4.9, we obtain:

$$v_{out}^2(1 + \beta_R) - v_{out}[2V_{HI} + 2\beta_R(v_{in} - V_{th})] + V_{HI}^2 = 0$$

where

$$\beta_R = \frac{W_D L_L}{W_L L_D} = \frac{(550)(10)}{(22)(10)} = 25$$

and

$$v_{in} = 8 \text{ V}$$

3 Then

$$\begin{aligned} 0 &= v_{out}^2(26) - v_{out}[2 \times 8 + 50(8 - 2)] + 8^2 \\ &= 26v_{out}^2 - 316v_{out} + 64 \end{aligned}$$

and

$$V_{LO} = V_{out}|_{vin=V_{HI}} = \frac{316 - [(316)^2 - 4(26)(64)]^{1/2}}{52}$$

$$= 0.206 \text{ V}$$

b.

1 $A = \dfrac{\Delta v_{out}}{\Delta v_{in}} = -(\beta_R)^{1/2} = -(25)^{1/2} = -5$

c.

1 For small v_{in}, the driver operates in the saturation region, for example, for v_{in} less than,

$$v_{in} = V_{th} + \frac{V_{DD} - V_{th}}{1 + (\beta_R)^{1/2}} = 3.33 \text{ V}$$

and greater than V_{th}.

2 For all v_{in} greater than 3.33 V, the driver is nonsaturated.

3 If $v_{in} = 3.33$ V, then,

$$v_{out} = \frac{V_{DD} - V_{th}}{1 + (\beta_R)^{1/2}} = 1.33 \text{ V}$$

Figure 4.9 illustrates these regions.

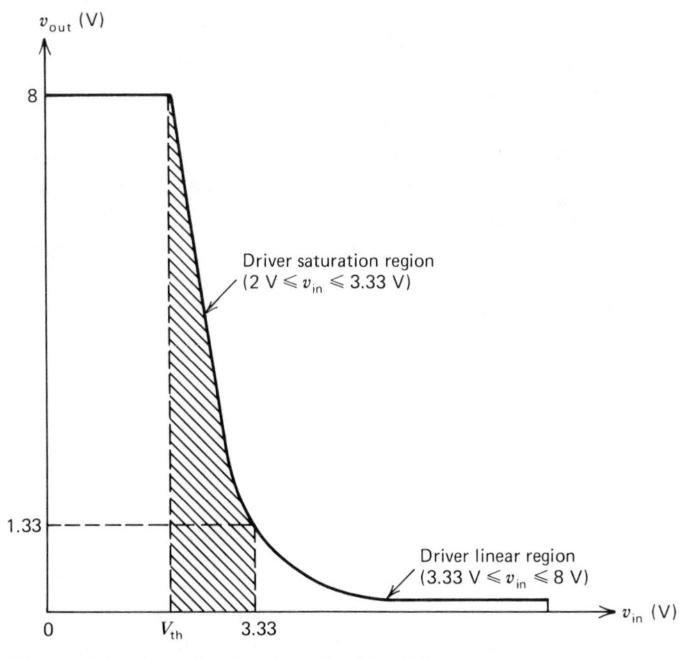

Figure 4.9 Transfer function for Ex. 4.3.

NMOS with Nonsaturated Load

The load for the NMOS driver may be reconnected with two power supplies to operate in its linear (nonsaturated) region, as illustrated in Fig. 4.10. To operate in its linear region, $V_{GG} - V_{DD}$ must be greater than V_{th} of the load device.

The equation governing the transfer characteristic for this inverter configuration is given by:

$$v'^2_{out} + 2V_{th}v'_{out} = \beta_R[2v'_{in}(V_{DD} - v'_{out}) - (V_{DD} - v'_{out})^2] \quad (4.11)$$

where $v'_{out} \triangleq V_{DD} - v_{out}$ and $v'_{in} \triangleq v_{in} - V_{th}$

The advantage of an inverter having a nonsaturated load over one having a saturated load is that the former has a greater logic swing. Output voltage v_{out} may reach the upper power supply rail since the device's threshold voltage does not create an offset. The disadvantage of this inverter is that it requires two power supplies as opposed to just one for the inverter with a saturated load transistor.

NMOS with Depletion-Mode Load

If the enhancement-mode load device (discussed until now) is replaced by an NMOS depletion-mode load device and is operated in its nonsaturated mode, large logic swings may be obtained while using only a single power supply. The onset of saturation occurs in the depletion-mode device when a voltage called the pinch-off voltage, V_P, is impressed across the gate oxide so that the mobile charge concentration at a particular point in the inversion channel is minimum. Hence, the depletion layer extends to this point and "pinches off" the channel.

The depletion layer will not block the flow of channel current because any carriers entering it will be swept across it by the presence of a high electric field. The pinch-off voltage for an NMOS depletion device is negative. If the gate and drain of the load device are tied together, then the inversion channel of the load device cannot be pinched off because the inversion channel will only pinch off when $V_D = V_G - V_P$. Thus, the load remains in its nonsaturated region when the gate and drain are connected to V_{DD}.

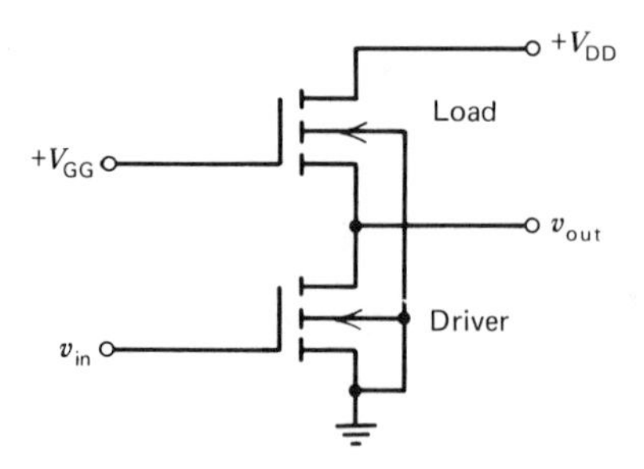

Figure 4.10 NMOS inverter with nonsaturated load.

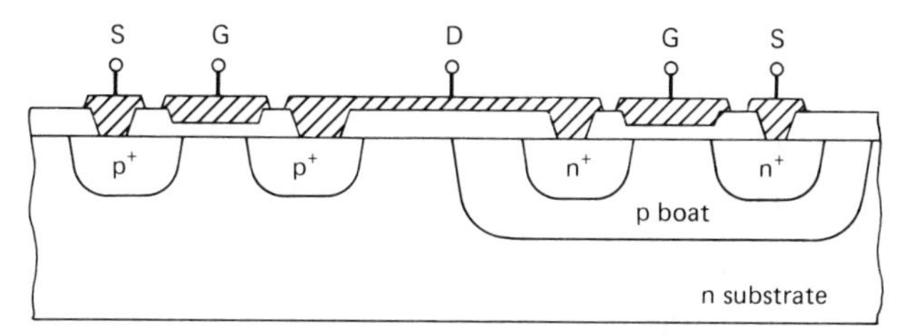

Figure 4.11 Integrated CMOS inverter structure.

CMOS Inverter

An early advance in the wafer fabrication of MOS integrated circuits was the ability to produce both PMOS and NMOS devices on the same chip. When a PMOS and NMOS enhancement device are fabricated side by side, as shown in Fig. 4.11, they are connected together in a complementary symmetry (CMOS) circuit to form an integrated inverter.

The CMOS inverter is shown in Fig. 4.12. In this configuration, the PMOS source is connected to $+V_{DD}$ and the NMOS source to V_{SS}. Their gates are tied together to form the input, and their drains form the common output. In addition, one device is isolated from the other by biasing the n substrate to V_{DD} and the p boat to V_{SS}, commonly grounded. This provides a reverse-biased pn junction between devices.

Figure 4.13 presents the I–V characteristics of both the PMOS and NMOS devices (superimposed on one another) when connected in the manner described above. It illustrates the interaction of both devices in forming the inverter. The threshold voltage for the p-channel transistor (V_{thp}) is assumed here to equal the threshold voltage of the n-channel transistor (V_{thn}), namely, $V_{thp} = V_{thn} = V_{th}$.

For small values of v_{in} and less than V_{th}, the NMOS device is completely off and no current flows through it. Output voltage v_{out} equals V_{DD} under this condition, because the PMOS device is fully turned on and is in its linear region of operation. As v_{in} increases above V_{th}, both devices enter their saturation regions. This corresponds to the inverter transition region. Voltage V_m, the high-to-low switching point of the transition region, occurs at approximately 50% of the magnitude of V_{DD} (see Ex. 4.4). As v_{in} is further increased, v_{out} drops to V_{SS} and the NMOS device becomes fully on, conducting in its linear region; the PMOS device is completely off.

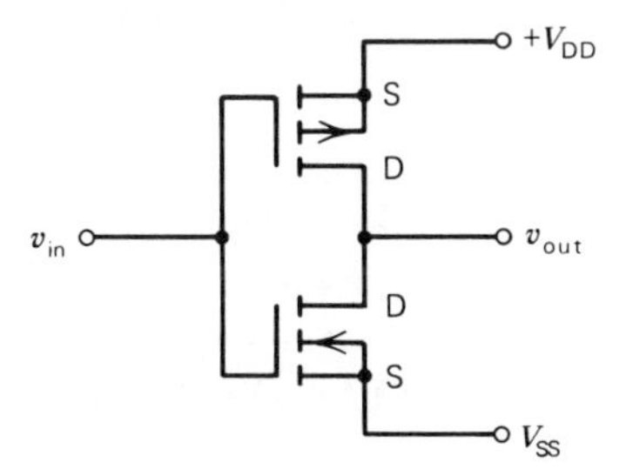

Figure 4.12 A CMOS inverter.

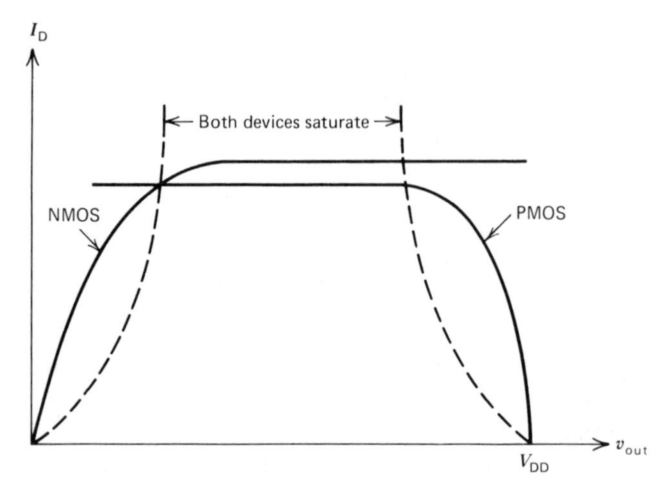

Figure 4.13 I–V characteristics for PMOS and NMOS devices corresponding to their arrangement in the CMOS inverter (simplified for only one V_{GS} value per device).

A representation of the CMOS inverter transfer function, showing regions of active device operation, is given in Fig. 4.14 for $V_{DD} = 5$ V, $V_{SS} = 0$ V, and a typical value of $V_{th} = 1.5$ V.

CMOS inverters have sharper transfer characteristics than pure NMOS inverters; hence, their noise immunity is superior. (Noise immunity will be discussed in a later section.)

The equation governing the transfer characteristic of the CMOS inverter is given by:

$$v_{out} = v_{in} + V_{th} + [(v_{in} + V_{th})^2 + V_{DD}^2 - 2v_{in}V_{DD} - 2V_{DD}V_{th} - (v_{in} - V_{th})^2]^{1/2} \qquad (4.12)$$

In the layout of a balanced CMOS inverter, the W/L ratio of the PMOS transistor is sized to be approximately 2.5 times greater than that of the NMOS transistor. The mobility of electrons is typically greater than the mobility of holes, the effective carrier in the PMOS transistor, by this amount. This is done to achieve a more balanced switching speed by making the inverter's output rise and fall times essentially the same.

EXAMPLE 4.4 Describe the effect on the dc switching point, V_m, of a CMOS inverter whose transfer characteristic is illustrated in Fig. 4.14, by adjusting the ratio $(W/L)_P$:$(W/L)_N = K'_P$:K'_N to the following values: **a** 1:1, **b** 2:1, **c** 2.5:1. Let the nominal value of $K'_P = 1.5$ and that of $K'_N = 1.5$ for the 1:1 ratio and let $\mu_N = 1000$ cm²/Vs and $\mu_P = 400$ cm²/Vs, or $\mu_N = 2.5\ \mu_P$.

PROCEDURE

1 When $v_{in} = V_m$, both the PMOS and the NMOS transistors will be saturated and, from Eq. 4.7,

$$I_D = K_P(V_G - V_{th})^2 = K_N(V_G - V_{th})^2$$
$$= K_P(V_{DD} - v_{in} - V_{th})^2 = K_N(v_{in} - V_{th})^2$$

Therefore,

$$V_l^2 K_N = V_h^2 K_P \quad \text{or} \quad V_l/V_h = (K_P/K_N)^{1/2}$$

2 Solving for v_{in}, from the preceding equation governing I_D, we obtain:

$$v_{in} = V_m = \frac{V_{DD}}{1 + (\beta_R)^{1/2}} - \frac{V_{th}(1 + (\beta_R)^{1/2}}{1 + (\beta_R)^{1/2}}$$

$$\beta_R = \frac{K_N}{K_P} \quad \text{reduces to the form}$$

$$\beta_R = \frac{K'_N \mu_N C_o/2}{K'_P \mu_P C_o/2} = \frac{K'_N \mu_N}{K'_P \mu_P}$$

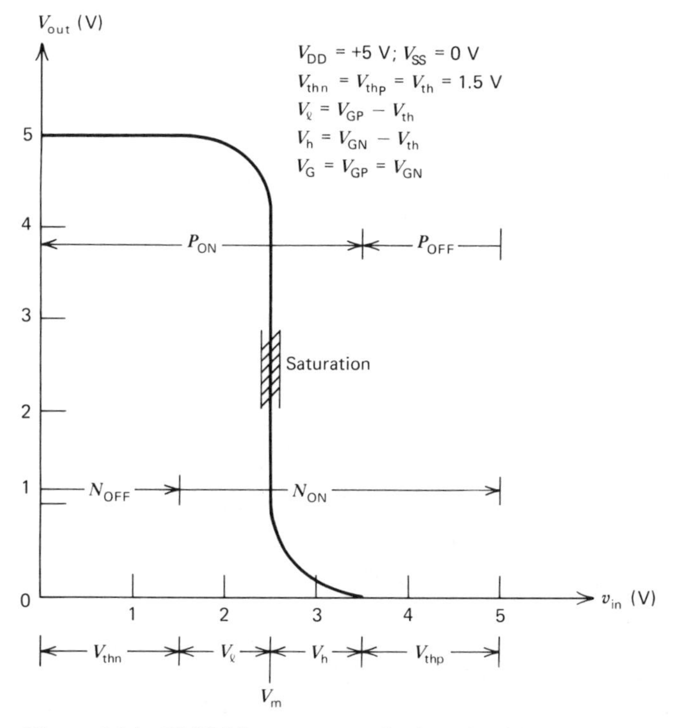

Figure 4.14 CMOS inverter transfer function illustrating regions of active device operation.

3 a. $K'_P = 1.5, \quad K'_N = 1.5$

$$\beta_R = 2.5$$

$$V_m = \frac{5}{1 + 1.58} - \frac{1.5(1 - 1.58)}{1 + 1.58} = 2.275 \text{ V}$$

$$V_l = 0.63V_h$$

b. $K'_P = 3, \quad K'_N = 1.5$

$$\beta_R = 1.25$$

$$V_m = \frac{5}{1 + 1.12} - \frac{1.5(1 - 1.12)}{1 + 1.12} = 2.443 \text{ V}$$

$$V_l = 0.80V_h$$

c. $K'_P = 3.75, \quad K'_N = 1.5$

$$\beta_R = 1.00$$

$$V_m = \frac{5}{1 + 1} - \frac{1.5(1 - 1)}{1 + 1} = 2.50 \text{ V}$$

$$V_l = V_h$$

It may be concluded from the above example that the switching point, V_m, will be significantly affected only if the ratio of device sizes is altered to a large extent.

Channel Resistance

When the transistor is turned "on," the channel resistance, R_{ON}, is given by:

$$R_{ON} = \frac{1}{(W/L)K''(V_{GS} - V_{th})} \tag{4.13}$$

where $K'' = \mu\varepsilon/d_o$.

With increasing reverse bias on the gate, an increasing penetration of the depletion region reduces the cross-sectional area of the channel and hence the channel resistance increases. The channel resistance for both the PMOS and NMOS devices varies with gate voltage and, when either transistor is fully turned off, its channel resistance, R_{OFF}, exceeds 10^4 MΩ.

EXAMPLE 4.5 Calculate R_{ON} for both PMOS and NMOS transistors in a typical minimum-size inverter where $V_{GS} = 5$ V, $V_{th} = 1.75$ V, $W_N = 0.36$ mil, $L_N = 0.24$ mil, $W_P = 0.84$ mil, and $L_P = 0.26$ mil. $K''_N = 5\times10^{-6}$ and $K''_P = 10\times10^{-6}$.

PROCEDURE

1 $(W/L)_N = 1.5$; $(W/L)_P = 3.23$
2 By Eq. 4.13,

$$(R_{ON})_N = \frac{1}{(1.5)(10\times10^{-6})(5 - 1.75)} = 20.51\ \text{k}\Omega$$

$$(R_{ON})_P = \frac{1}{(3.23)(5\times10^{-6})(5 - 1.75)} = 19.05\ \text{k}\Omega$$

4.5 PERFORMANCE CHARACTERISTICS

The user and designer of an MOS system usually have several performance objectives in mind which influence the choice of a particular MOS logic family, such as NMOS or CMOS. These performance objectives include noise performance, power dissipation, fan out, and speed requirements.

Device Current

Maximum device current flows in single-channel MOSFET inverters when the output voltage is in its logic 0 state. For the following equations describing the current of the load transistor, I_L, let $v_{out} \simeq 0$ V.

For the MOSFET inverter with a saturated load, I_L is:

$$I_L \simeq K_L(V_{DD} - V_{th})^2 \tag{4.14}$$

The load current for the inverter with a nonsaturated load is:

$$I_L \simeq K_L V_{DD}[2(V_{GG} - V_{th}) - V_{DD}] \tag{4.15}$$

Lastly, the load current for the inverter with a depletion-mode load is:

$$\begin{aligned} I_L &\simeq K_L[2(V_{DD} - V_p)V_{DD} - V_{DD}^2] \\ &\simeq K_L V_{DD}[2(V_{DD} - V_p) - V_{DD}] \end{aligned} \tag{4.16}$$

The current drawn by the CMOS inverter in either of the steady-state conditions is negligible and is defined by the device leakage currents only.

Power Dissipation

Recalling that maximum device current flows when $v_{out} \simeq 0$ V, maximum power, P_{max} is dissipated for this condition because,

$$P_{max} = V_{DS}I_L = V_{DD}I_L \tag{4.17}$$

where $V_{DS} = V_{DD} - v_{out} = V_{DD}$

EXAMPLE 4.6 Calculate the maximum power dissipated by the three single-channel MOSFET inverters discussed in the previous section, given $V_{DD} = 5$ V, $V_{GG} = 8$ V, $V_P = -1.5$ V, $V_{th} = 1.5$ V, and $K_L = 11.67 \times 10^{-6}$.

PROCEDURE

Using Eqns. 4.14 through 4.17 we get:

1 NMOS inverter with saturated load:

$$I_L \simeq (11.67 \times 10^{-6})(5 - 1.5)^2 = 0.143 \text{ mA}$$
$$P_{max} \simeq I_L V_{DD} = 0.715 \text{ mW}$$

2 NMOS inverter with nonsaturated load:

$$I_L \simeq (11.67 \times 10^{-6})[2(5 + 1.5)5 - 25] = 0.467 \text{ mA}$$
$$P_{max} \simeq I_L V_{DD} = 2.33 \text{ mW}$$

3 NMOS inverter with depletion-mode load:

$$I_L \simeq (11.67 \times 10^{-6})[2(5 + 1.5)5 - 25] = 0.467 \text{ mA}$$
$$P_{max} \simeq I_L V_{DD} = 2.33 \text{ mW}$$

The static power dissipated by a CMOS inverter can be approximated by considering the equivalent resistive network representing the inverter in its output logic 0 and logic 1 states. Let the channel resistances, $R_{ON} = 20$ kΩ, $R_{OFF} = 10^4$ MΩ and $V_{DD} = 5$ V. The inverter may then be considered as two resistors connected in series. With $v_{out} = 0$ V, the equivalent resistive network is as shown in Fig. 4.15*a*. With $v_{out} = 5$ V, the resistive network is as shown in Fig. 4.15*b*. In both cases,

$$P_{max} = \frac{V_{DD}^2}{R_{ON} + R_{OFF}} \cong \frac{25}{10^4 \text{ M}\Omega} = 2.5 \text{ nW}$$

It may, therefore, be concluded that the steady-state power dissipation of the CMOS inverter is negligible.

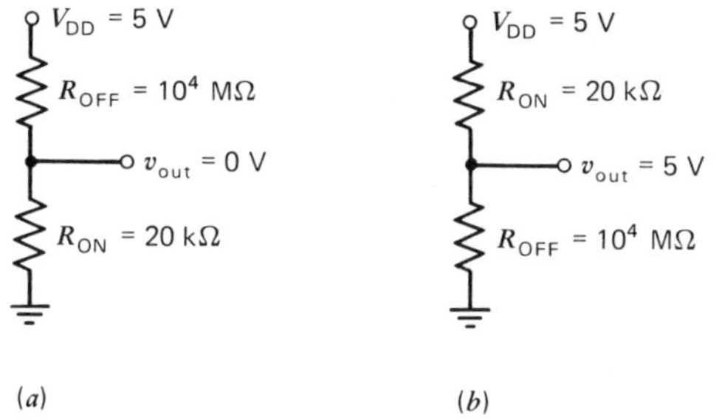

V_{DD} = 5 V
R_{OFF} = 10^4 MΩ
v_{out} = 0 V
R_{ON} = 20 kΩ

V_{DD} = 5 V
R_{ON} = 20 kΩ
v_{out} = 5 V
R_{OFF} = 10^4 MΩ

(*a*) (*b*)

Figure 4.15 Equivalent resistive networks for a CMOS inverter when (*a*) $v_{out} = 0$ V and (*b*) $v_{out} = 5$ V.

The most important consideration concerning the power dissipated by the CMOS inverter is its ac power dissipation. This is the power dissipated when switching between logic 0 and logic 1 states or the power dissipated during the time the inverter operates in its transition region. If the CMOS inverter drives another MOS logic gate, this is the time required to charge and discharge the input capacitance presented by this load. The ac power dissipated is thus:

$$P_{ac} = C_L V_{DD}^2 f \tag{4.18}$$

where C_L is the equivalent load capacitance and f is the switching rate of the inverter. Dissipated power P_{ac} increases with both increasing frequency and increasing power-supply voltage.

EXAMPLE 4.7 Calculate the ac power dissipated by a CMOS inverter which drives a 20 pF load given $f = 1$ MHz and $V_{DD} = 10$ V.

PROCEDURE

By Eq. 4.18,

$$\begin{aligned} P_{ac} &= C_L V_{DD}^2 f \\ &= (20\ \text{pF})(100\ \text{V})^2(1\ \text{MHz}) \\ &= 2\ \text{mW} \end{aligned}$$

Noise Considerations

The presence of electrical noise in an integrated circuit or logic system may cause erroneous logic operation. Noise, for example, may enter the circuit through the power supply lines. Input noise picked up along interconnections may cause an inverter or other logic gate to change logic state in an unpredictable fashion.

Noise immunity may be defined as the amount the input voltage of the device may vary without significantly affecting the output voltage. It is expressed as a percentage of supply voltage.

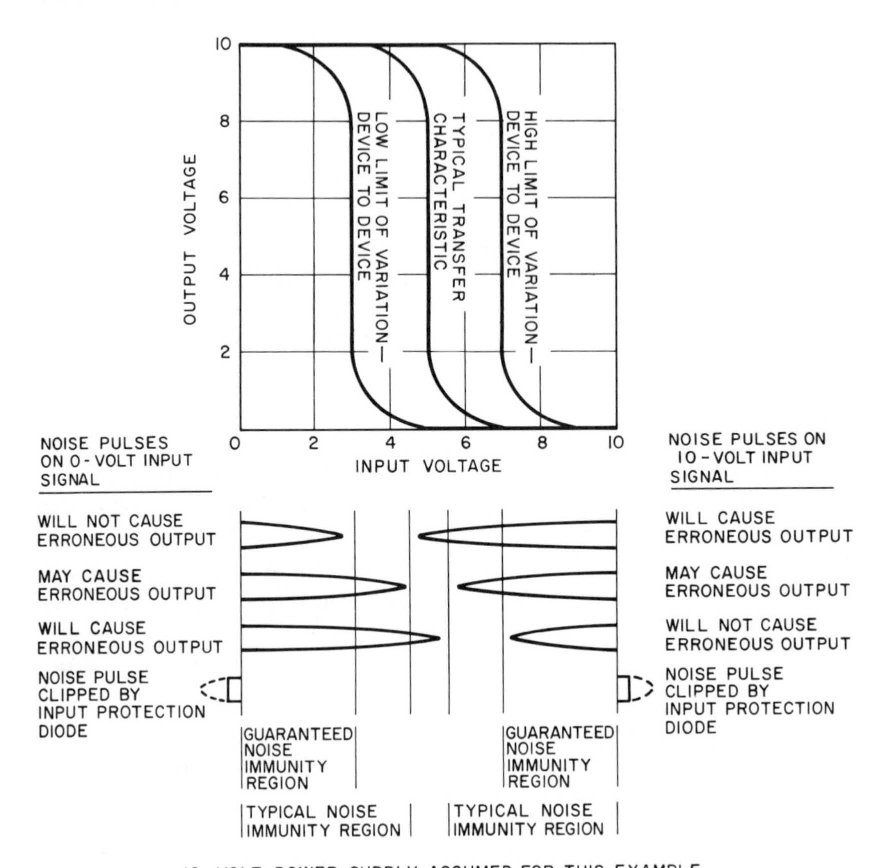

Figure 4.16 Illustrating noise immunity. (Courtesy RCA Corp.)

Figure 4.16 shows the effect that noise pulses have on the output logic states of an inverter whose supply voltage is 10 V. For example, 3 V is the maximum amplitude of a noise pulse which definitely will not cause erroneous operation. Therefore, 30% is the guaranteed noise immunity for this inverter. The maximum voltage of a noise pulse that may cause erroneous operation is 4.5 V. Therefore, the typical noise immunity of this inverter is 45%.

A high noise immunity is desirable. A sharp transfer characteristic, as is the case in CMOS, contributes to high noise-immunity characteristics.

Noise margin is defined as that noise voltage that can be impressed upon v_{in} of a logic gate without upsetting the logic or causing an output to exceed given v_{out} ratings. Figure 4.17 shows one inverter driving another to form a noninverting pair. The following definitions, as used in Fig. 4.17, apply to the calculation of noise margin:

$V_{IL(max)}$: The maximum allowed input voltage for a logic 0 input.

$V_{OH(min)}$: The minimum allowed output voltage for a logic 1 output.

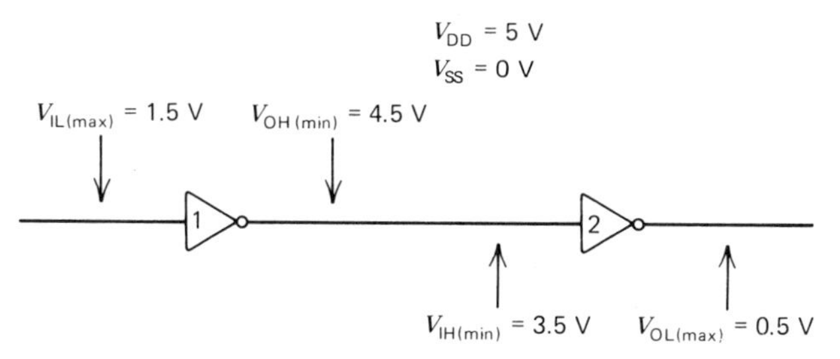

Figure 4.17 Noise margin for noninverting pair.

$V_{IH(min)}$: The minimum allowed input voltage for a logic 1 input.
$V_{OL(max)}$: The maximum allowed output voltage for a logic 0 output.

Values for these parameters, as given in Fig. 4.17, allow 0.5 V to exist between the actual gate output voltage and the supply rail to which it is nearest. Actual ratings are determined by each manufacturer.

The noise margin in Fig. 4.17 is $V_{OH(min)} - V_{IH(min)} = 4.5\text{ V} - 3.5\text{ V} = 1\text{ V}$. Therefore, 1 V of noise may exist between gates 1 and 2 without upsetting the correct logic operation of the noninverting pair.

Output Drive Capability

Fan out may be defined as the number of identical gates that a logic gate is able to drive at its output. Thus an inverter driving one identical inverter has a fan out of 1. The number of gates may be expressed in terms of the total input capacitance of all gates being driven. It is theoretically possible for an MOS logic gate to drive an infinite number of gates, and hence, charge an infinite capacitance. Practically, however, this is not the case because it would take an infinite amount of time to do so. Therefore, the amount of capacitance to be charged or discharged, and hence the number of logic gates, must be chosen so that the circuit may switch within certain limits of a specified operating frequency.

Switching Speed

The factors that limit the speed of an MOS circuit are the device size, the power supply voltage, and the amount of load capacitance that one or more logic gates present to another gate's output.

We consider Fig. 4.18 in analyzing the output rise time, t_r, and output fall time, t_f, for the NMOS inverter having a saturated load transistor. All capacitances are lumped into the equivalent capacitance C_L. The time required for v_{out} to rise from 0 V to V_H is:

$$t_r = \frac{3C_L}{K_L V_H} \tag{4.19}$$

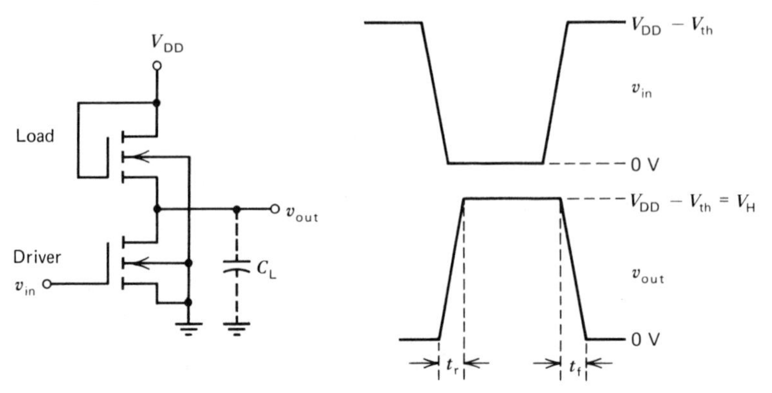

Figure 4.18 NMOS transient response.

The time required for v_{out} to fall from V_H to 0 V is:

$$t_f = \frac{3C_L V_H}{K_D}\left[\left(2 + \frac{V_{th}}{V_H}\right)(V_H - V_{th})^2 - \frac{V_H^2}{\beta_R}\right]^{-1} \tag{4.20}$$

For the CMOS inverter of Fig. 4.19, t_r is defined as the time required for v_{out} to rise from 10 to 90% of V_{DD}, and t_f is defined as the time required for v_{out} to fall from 90 to 10% of V_{DD}.

Output fall and rise times may be approximated by:

$$t_f \simeq \frac{3C_L V_{DD}^2}{K_N(V_{DD} - V_{th})^2(2V_{DD} + V_{th})} \tag{4.21a}$$

$$t_r \simeq t_f \tag{4.21b}$$

for a balanced inverter.

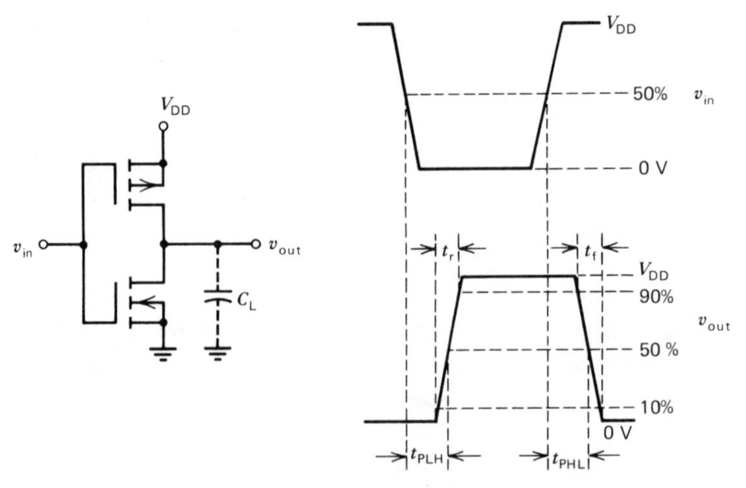

Figure 4.19 CMOS transient response.

Another parameter indicated in Fig. 4.19 is the propagation delay times, from input to output for the inverter, labeled as t_{PLH} and t_{PHL}. Propagation delay time is inherent in logic gates of all types and is measured between the 50% points of the input and output waveforms.

By increasing the supply voltage, we may increase the speed of a CMOS logic gate. In a like manner, increasing the load capacitance decreases the logic gate speed.

EXAMPLE 4.8 Calculate t_r and t_f for the CMOS inverter of Fig. 4.19, given $C_L = 4$ pF, $V_{DD} = 10$ V, $V_{th} = 2$ V, and $K_N = 1.167 \times 10^{-5}$.

PROCEDURE

By Eqns. 4.21*a* and 4.21*b*,

$$t_f \simeq \frac{3(4 \times 10^{-12})(100)}{(1.167 \times 10^{-5})(64)(22)} = 73 \text{ ns}$$

and $t_r \simeq t_f = 73$ ns

The speed of the CMOS inverter also may be increased by increasing the size of the inverter structure. As W/L increases, the speed also increases. A limiting speed is eventually reached; that is, where increasing this ratio will not contribute to an additional increase in speed.

The ring oscillator of Fig. 4.20 may be constructed to examine the maximum operating frequency of an MOS circuit. This circuit consists of a NAND gate connected to two inverters (see Chap. 1). Assuming the initial value of v_{out} is undetermined as to its logic state with the control voltage, V_c, at logic 0, the ring oscillator output, v_{out}, will become stable and remain in the logic 1 state. When V_c is set to a logic 1, the NAND gate acts as an inverter and the circuit begins to oscillate. A series of pulses is generated at the output until V_c is again returned to its logic 0 state. The frequency of operation is only limited by the propagation delays of each gate.

Tying the output of a single MOS inverter back to its own input will not achieve the same result. In this configuration, the inverter will bias itself into its transition region and just remain there drawing current. Tying together the input and output of an MOS inverter is also not a method of "shorting out" the inverter.

State of the art technology has achieved gate delays of 1 and 2 ns for NMOS and CMOS gates, respectively. The slight speed advantage obtainable with NMOS logic stems from the greater device density currently possible in the layout of an NMOS IC.

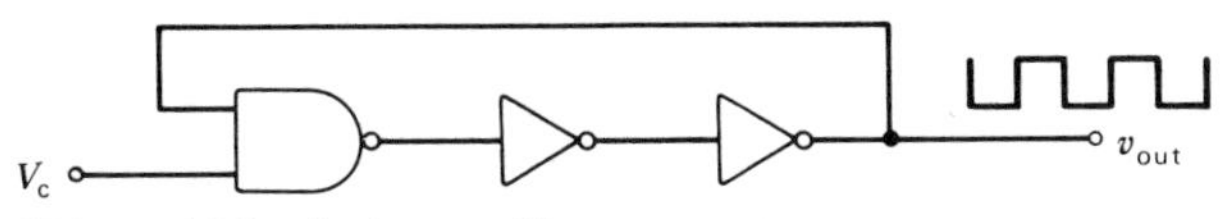

Figure 4.20 A ring oscillator.

4.6 OPERATING CONDITIONS

Unused Inputs

Leaving an MOS input unterminated is not desirable, because the logic state at the input of a gate remains undefined. For example, an unterminated MOS gate input is particularly susceptible to the buildup of electrostatic charge. This charge may develop a sufficient voltage to exceed the switching threshold voltage of the device. Several results are then possible. First, the device, such as an inverter, may unintentionally switch logic states and cause erroneous results in system operation. Second, even if the voltage generated by the static charge is not sufficient to change a logic state, it may be enough to exceed the MOSFET threshold and thus cause excessive current to flow, thereby increasing circuit power dissipation.

All unused input pins must, therefore, be connected either to V_{SS} or V_{DD}, which ever is appropriate to the logic function involved. This applies even to MOSFETs that play no role in the circuit. For example, when breadboarding a circuit with MOSFET inverter arrays, an unused inverter's gate input must be terminated to prevent spurious current flow.

Input Protection

A third result of an unterminated input may prove to be a catastrophic one. The accumulation of static charge on the gate structure may be sufficient to cause permanent damage to the device. Because the gate input is equivalent to a very small capacitance in parallel with a very large resistance, the accumulation of even a small amount of charge may be sufficient to develop a voltage capable of rupturing the gate oxide and causing a gate-channel short. The necessary amount of charge to cause such damage could be obtained from energy injected through stray circuit wiring capacitances or even by leaving the input pin unterminated. Thus, leaving an unterminated MOS circuit on a shelf could permanently damage it.

In order to avoid this, all MOS gate inputs include some form of protective network on the chip. Figure 4.21 shows a typical diode-resistor network used to protect a CMOS inverter circuit. In this figure, diode D_2 is a distributed

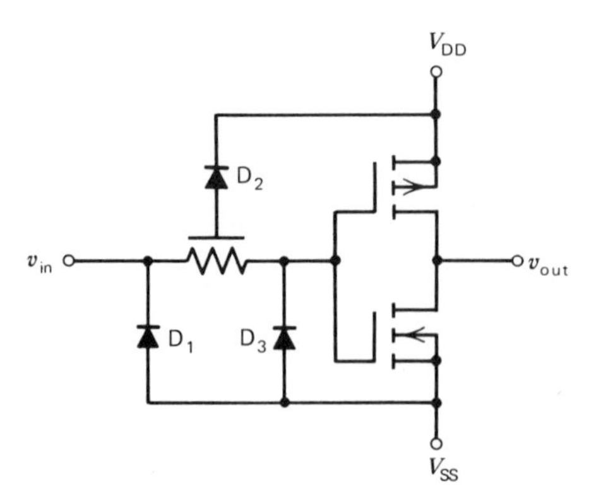

Figure 4.21 Input protection network.

resistor-diode network. If the input voltage rises owing to some type of transient voltage spike, and exceeds V_{DD}, diode D_2 becomes forward biased and clamps the gate voltage, preventing it from further increase and damage to the device. Likewise, diodes D_1 and D_3 protect the inverter from damage owing to a negative voltage spike.

Handling and Operating MOS Devices

Even though protective networks are used in MOS devices, precautions must be exercised in their use and handling to prevent damage. For example, MOS devices should be handled and stored in conductive carriers to guard against exposing them to excessive static charges. In addition, signals should not be applied to the input terminals while the device power supply is off, unless the input current is limited to a steady-state value of typically less than 10 mA. Otherwise, as can be seen by studying Fig. 4.21, input diode D_2 becomes conductive and could be damaged by excessive current flow.

Lastly, care must be taken never to exceed the recommended maximum rating for power supply voltage, as determined by the manufacturer.

4.7 APPLICATION TO LOGIC DESIGN

Static Logic

The 2-input NOR function, illustrated in the truth table of Fig. 4.22*a*, can be implemented in the NMOS saturated-load arrangement, as shown in Fig. 4.22*b*; its CMOS implementation is given in Fig. 4.22*c*.

A	B	Out
0	0	1
0	1	0
1	0	0
1	1	0

(*a*)

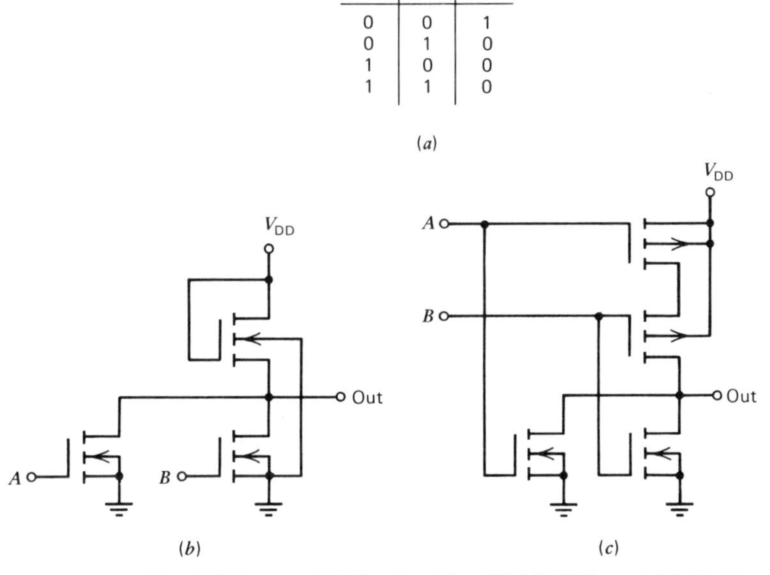

Figure 4.22 The NOR gate. (*a*) Truth table. (*b*) NMOS, and (*c*) CMOS implementations.

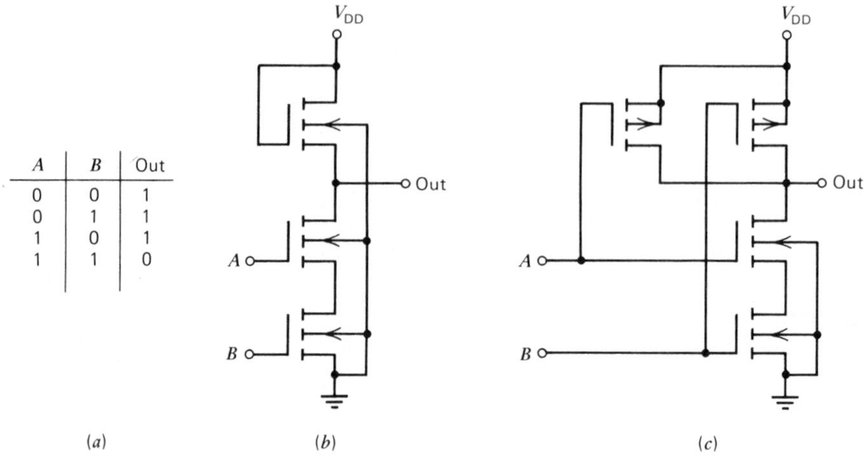

A	B	Out
0	0	1
0	1	1
1	0	1
1	1	0

Figure 4.23 The NAND gate. (*a*) Truth table. (*b*) NMOS, and (*c*) CMOS implementations.

The 2-input NAND truth table, together with NMOS and CMOS implementations used to perform the NAND function, appear in Fig. 4.23.

Three-input NOR- and NAND-gate arrangements also can be formed by the appropriate addition of MOS transistors to the circuits shown in Figs. 4.22 and 4.23, respectively.

EXAMPLE 4.9 Write the truth table for a CMOS 3-input NAND gate and draw the circuit schematic for this configuration.

PROCEDURE

The truth table and circuit schematic for the 3-input NAND gate appear in Fig. 4.24.

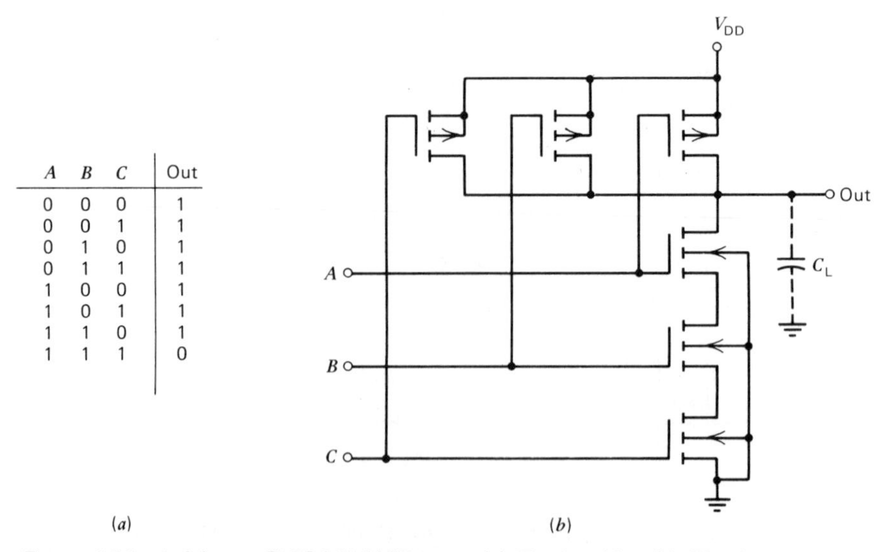

A	B	C	Out
0	0	0	1
0	0	1	1
0	1	0	1
0	1	1	1
1	0	0	1
1	0	1	1
1	1	0	1
1	1	1	0

Figure 4.24 A 3-input CMOS NAND gate. (*a*) Truth table. (*b*) Circuit.

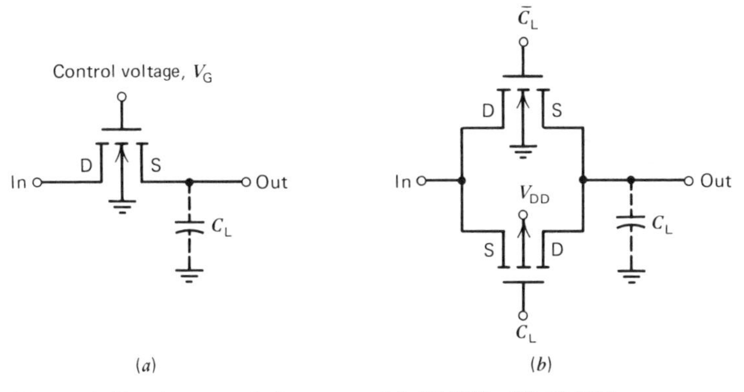

Figure 4.25 A transmission gate. (*a*) NMOS. (*b*) PMOS.

When series connecting NMOS transistors, as in Fig. 4.24, it may be desirable to use NMOS transistors having a greater W/L ratio than that of the PMOS transistors. Without this precaution, if the inputs of the NAND gate are clocked at a particularly high rate, the load capacitance, C_L, may not completely discharge through the high resistance of the three series-connected NMOS devices. If the W/L ratios of these devices are deliberately increased, then the total R_{ON} seen by C_L will be much smaller and the output voltage will be able to swing between its upper and lower rails. This type of discharge problem is most prone to occur when lowering the supply voltage in a CMOS circuit containing the NAND gate, because this will result in a lowering of v_{in} to the NAND gate. Hence, the V_{GS} of each NMOS transistor will decrease. From Eq. 4.13, R_{ON} will increase as V_{GS} decreases.

The transmission gate is useful as a switch in MOSFET designs. In Fig. 4.25, both NMOS and CMOS versions are shown driving a load capacitance, C_L. When the control voltage (V_G) applied to the circuit is made sufficiently positive to turn on the NMOS transistor, C_L charges to a voltage ($V_G - V_{th}$). The CMOS transmission gate, shown in Fig. 4.25*b*, is easily controlled from a single-terminal input driving an inverter, as in Fig. 4.26.

Flip Flops

The structure of all flip flops is based on two cross-coupled inverters (see Chap. 1). Variations of this structure can accommodate set and reset functions, as illustrated in Fig. 4.27 for the static-type flip flop, known as the reset-set (R-S) flip flop. Two NOR gates are cross coupled.

As the truth table indicates, when the set and reset inputs are low, the circuit can assume either of two stable states; that is, one NOR gate output is low and the other is high. If the set input is raised to a logic 1 level, the Q output will assume a logic 1 level and the flip flop is said to be in the SET state. Raising the reset input to the logic 1 level forces the Q output into the logic 0 state; the flip flop is then said to be in the RESET state. Forcing both set and reset inputs to a logic 1 state, drives both Q and $\overline{\text{Q}}$ into logic 0 states. This is a

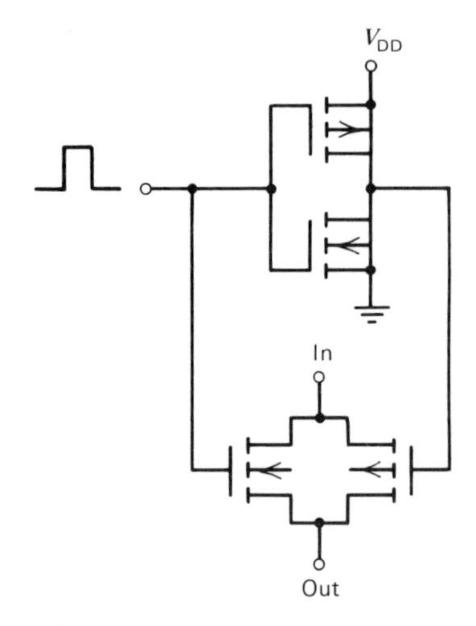

Figure 4.26 Combination of a CMOS transmission gate and inverter to form a switch.

logical inconsistency and, therefore, the condition $S = 1$, $R = 1$ is said to lead to an undefined state; it is forbidden and must not be allowed to occur.

EXAMPLE 4.10 Draw the circuit diagram for the R-S flip flop using NMOS transistors only.

PROCEDURE

The NMOS configuration of the R-S flip flop is shown in Fig. 4.28. It uses two less transistors than an equivalent CMOS version.

A somewhat more sophisticated flip flop is the data (D-type) flip flop. The logic diagram and a CMOS circuit schematic for the D flip flop, illustrating the use of transmission gates, are shown in Fig. 4.29. The D flip flop is clocked, whereas the R-S flip flop shown previously is of the unclocked type.

The D flip flop consists of a master flip flop that drives a slave flip flop having a similar configuration. When the input signal is at a low level, the TG1 transmission gates are closed and the TG2 gates are open. In this configuration, the master flip flop samples the input data while the slave flip flop retains the data from the previous input and feeds it to the output. When the clock input

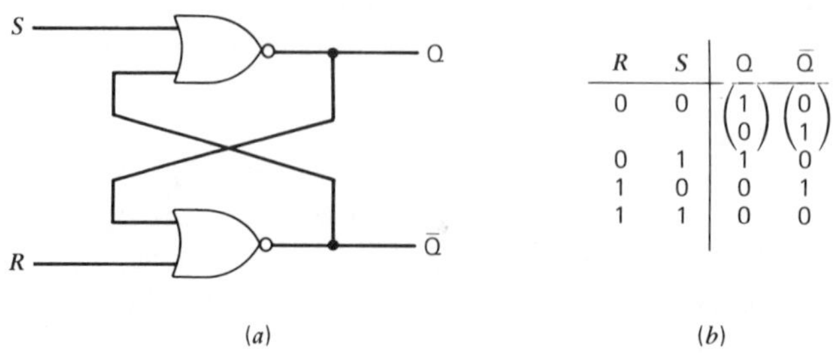

R	S	Q	$\bar{Q}$
0	0	(1 / 0)	(0 / 1)
0	1	1	0
1	0	0	1
1	1	0	0

Figure 4.27 The R-S flip flop. (*a*) Logic diagram. (*b*) Truth table.

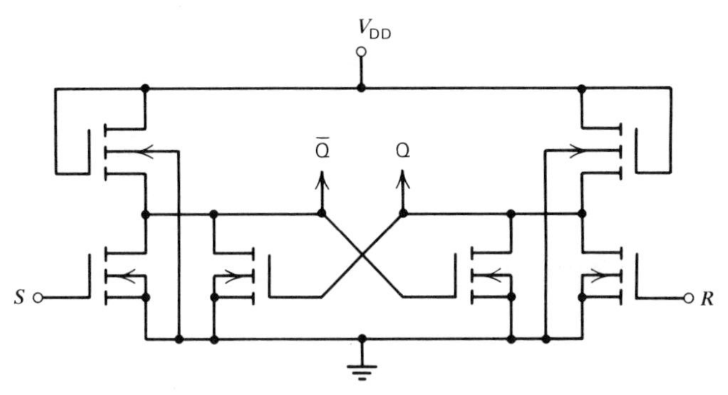

Figure 4.28 NMOS realization of an R-S flip flop.

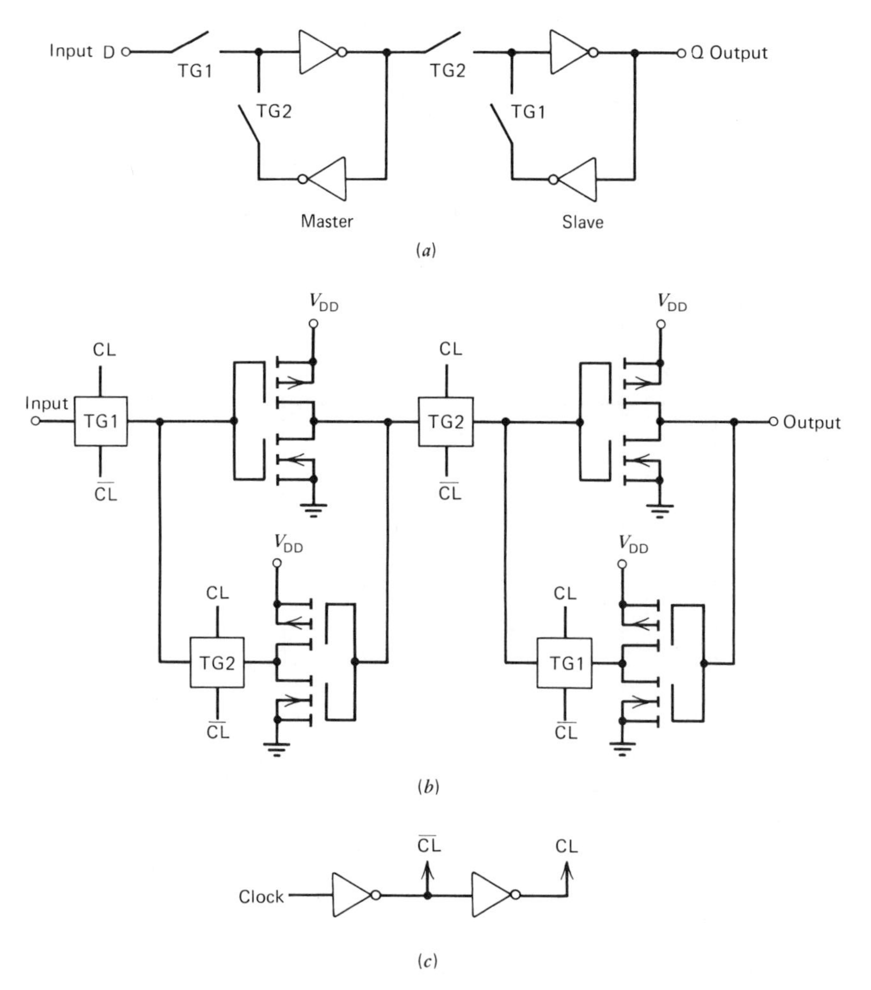

Figure 4.29 CMOS D flip flop. (*a*) Logic diagram. (*b*) Circuit. (*c*) Clock inversion.

is high, the TG1 transmission gates act as an open switch and the TG2 transmission gates act as a closed switch. In this way, the master holds the input data, transferring it to the output during the positive-going transition of each clock pulse. The clock signals, CL and $\overline{CL}$, are provided by an inverter within the integrated circuit (see Fig. 4.29*c*).

EXAMPLE 4.11 Draw a circuit schematic for the D flip flop using a CMOS implementation. Include a reset function.

PROCEDURE

See Fig. 4.30. Note the use of the MOS transistors embedded in the NOR structure to implement the switching function of the transmission gate.

EXAMPLE 4.12 Use the D flip flop to implement a four-stage shift register. The D flip flop need not have a reset function.

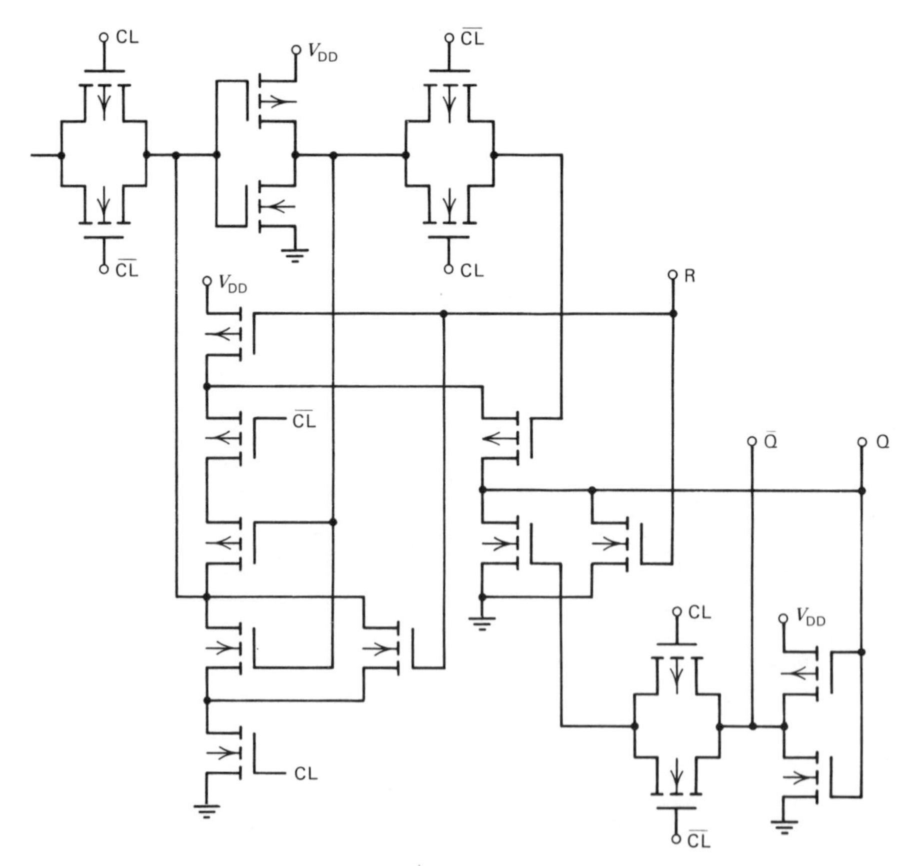

Figure 4.30 D flip flop with reset. (Substrate connections are not shown.)

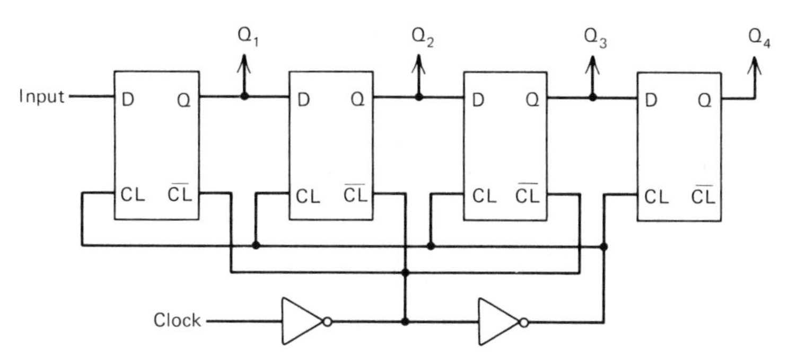

Figure 4.31 A shift register.

PROCEDURE

1 The logic implementation of the shift register is shown in Fig. 4.31. The D flip flop is shown as a simple block, with the terminal notation used previously.

2 Note that the clock input signal is buffered before it is applied to the flip flop CL and $\overline{CL}$ inputs. These buffer inverters are important if the shift register is increased to n stages and the drive capability of the clock may need to be enhanced. Larger geometry transistors may then be used to provide this capability.

Dynamic Logic

The random access memory (RAM) is another important application of MOSFET technology (see Chap. 6). Highly ordered memory cells are arrayed with the maximum density permitted by the state of the art. At this writing, RAMs capable of storing 256K bytes of data have been developed.

An example of a four-transistor dynamic RAM cell used in such memories is shown in Fig. 4.32. Once again, the cross-coupled transistor arrangement is used. This type of logic is called "dynamic" rather than "static" because charge is stored on the parasitic capacitance associated with the transistor's gate. The charge must be periodically refreshed (recharged) to compensate for the small leakage current at the storage nodes. To read data, the ROW SELECT line is enabled, turning on Q_2 and Q_4. If C_1 is holding a charge, Q_1 conducts and the capacitance associated with the data line discharges through Q_1. Capacitance C_1 is simultaneously refreshed by Q_4. The current in the data lines is detected by sense amplifiers.

To write data into the memory cell, information is placed on the DATA and $\overline{\text{DATA}}$ lines while the ROW SELECT line is raised to a logic 1 state. The bistable circuit is then set to the desired state by Q_2 and Q_4.

Dynamic memory cells using less than four transistors have been developed. The dynamic RAM cell has evolved from the four-transistor cell to the one-transistor cell. This has enabled the development of very dense memory circuits. Refer to Chap. 6 for additional details.

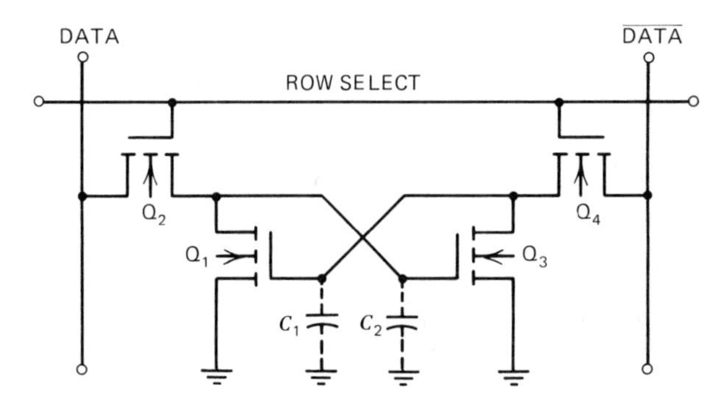

Figure 4.32 A four-transistor dynamic RAM cell.

Recently, CMOS has begun to emerge as the major technology in which to design VLSI circuits. CMOS seems well suited to VLSI design by virtue of its performance features and its versatility in digital, linear, and mixed circuitry design. As circuit dimensions shrink and as chips accommodate more and more functions, the low power dissipation of CMOS circuits make them attractive for keeping the total chip power dissipation at a manageable level.

4.8 THE POWER MOSFET

The lateral MOSFET structure, which has been discussed, could not compete against the bipolar junction transistor in power applications. The lateral MOSFET has a large channel resistance that is determined by the length of the channel. The length of the channel is controlled by the photomask spacing of source and drain regions. This spacing tends to be relatively wide and contributes to long channel lengths that increase the value of channel resistance.

The power MOSFET is formed by placing thousands of individual MOSFET structures in parallel on the same chip. Because source, drain, and gate conductors lie on the same surface level of this lateral structure, problems are encountered in connecting common lines and in busing each of these lines to separate pads. This is a layout task that consumes precious area on the chip.

The introduction of the V-groove structure in 1976 brought the MOSFET back into serious consideration as a power device. A cross-sectional view of the V-groove MOSFET is shown in Fig. 4.33. An n^- epitaxial layer is grown on an n^+ substrate. A p region is then diffused into this epi region followed by the diffusion of an n^+ source layer into the p region. The V-groove is then etched into the surface of the device. An oxide layer is grown over the entire surface and finally gate, source, and drain metal electrodes are deposited. Note that the drain is at the bottom of the MOSFET structure. This helps to alleviate some of the metallization problems encountered in the layout of the lateral power MOSFET.

The p region close to the gate becomes inverted when an appropriate gate-source potential is applied. The length of this channel region is controlled by a

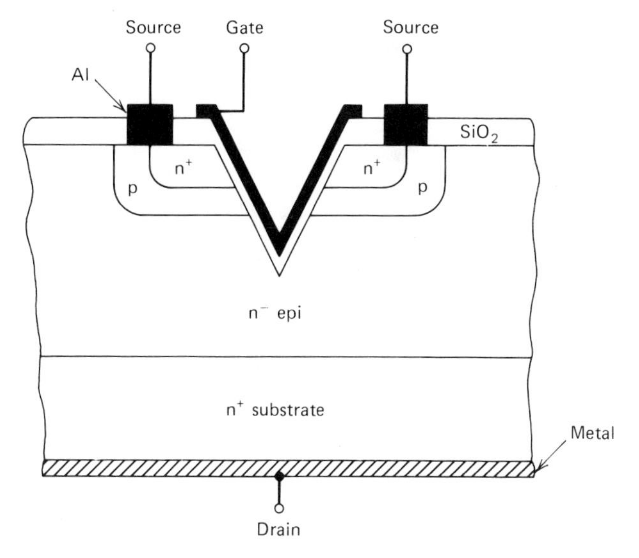

Figure 4.33 Cross section view of a V-groove power MOSFET.

diffusion process rather than a photomask process. It is because of this that smaller channel lengths and channel resistances may be achieved. Note also that this structure allows the current between drain and source to flow in a vertical direction. Devices of this type are called vertical MOS (VMOS) devices.

The V-groove structure has found common use in devices with less than 100 V drain-source breakdown voltage ($V_{(BR)DSS}$). However, it has its own drawbacks. The sharp point at the bottom of the groove makes it susceptible to gate oxide breakdown. This results in limited high-voltage capability.

The V-groove structure was soon replaced by the U-groove structure shown in Fig. 4.34. Breakdown problems are alleviated by flattening the bottom of the groove. The U-groove devices are widely used in applications requiring device breakdown voltages less than 150 V.

MOSFETs did not begin to compete with bipolar transistors requiring voltages greater than 150 V until the introduction of a modified VMOS process called DMOS (double diffused MOS). A cross-sectional view of the DMOS structure is shown in Fig. 4.35. A polycrystalline silicon gate is imbedded in a layer of silicon dioxide that insulates the gate from the source metal above it. The source metal covers the surface of the chip, connecting each device. A metal gate pad is placed at the edge of the chip and connects to the layer of polycrystalline silicon that forms the gate. Current flow from drain to source is initially vertical and then horizontal. The geometry of the source contact is sometimes hexagonal rather than square. In such cases, the structure is sometimes called a HEXFET.*

* HEXFET is a trademark of International Rectifier.

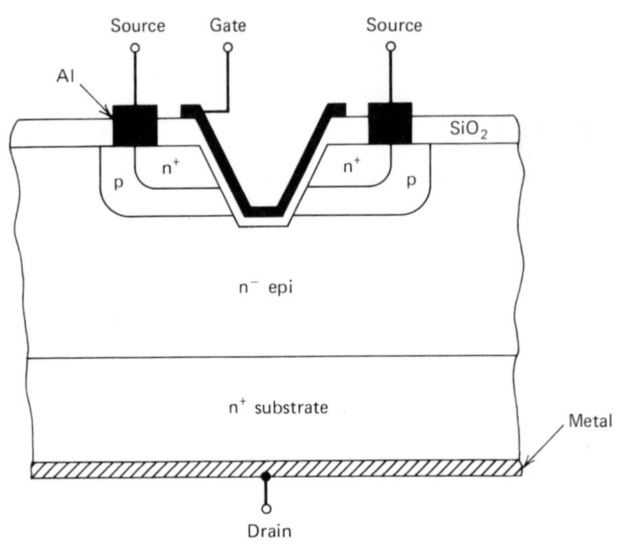

Figure 4.34 The U-groove structure.

All of the above power MOSFETs rely on an interdigitated structure that places thousands of transistors in parallel for the purpose of increasing current handling capabilities and lowering drain-source on-resistance.

A basic limitation of all MOSFETs is that as breakdown voltage is increased, the on-resistance for a given chip area increases exponentially with a power of 2.3 to 2.7. In order to increase the breakdown voltage capability of a MOSFET while maintaining a constant on-resistance, the die area must be increased.

Table 4.2 lists parameters for some discrete n channel, enhancement-mode power MOSFETs using the DMOS configuration.

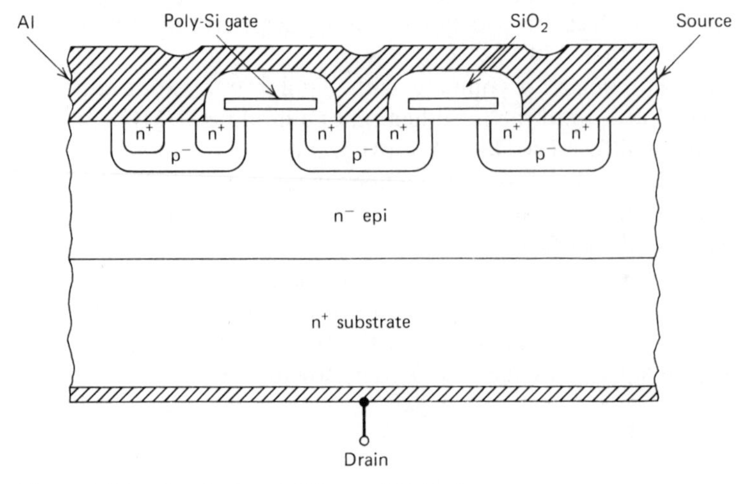

Figure 4.35 The doubled diffused (DMOS) structure.

Table 4.2 **Parameters for some discrete n-channel enhancement-mode MOSFETs using the DMOS structure**

Continuous current I_D *(A)*	*Breakdown voltage* $V_{(BR)DSS}$ *(V)*	*Drain-source on-resistance* $r_{DS(on)}$ *max* *(Ω)*	@ I_D *(A)*	*Resistive switching max @* $T_J = 100°C$ t_{on} *(ns)*	t_{off} *(ns)*	@ I_D *(nA)*
1.5	120	4	1	20	20	1
2	450	2.5	1	100	150	1
6	150	1	3	100	200	3
18	150	0.2	12	200	250	12

Because MOSFETs trade off breakdown voltage, on-resistance, power-handling capability, and die area much variation is possible from the parameter values listed in the table. For example, a MOSFET with a 1000 V rating is commercially available that can handle 4.7 A continuously with an on-resistance of 2 Ω and a turn-off time of 500 ns. MOSFETs with on-resistance of 0.18 Ω, $V_{(BR)DSS}$ of 100 V and current-handling ability of 12 A are also available.

Power MOSFETs are suited for application in switching power supplies, motor speed controls, voltage regulators, and in numerous applications where fast switching speeds are required.

4.9 REFERENCES

Antoniuk, R., "Special Report: Power Semiconductors," *Electronic Engineering Times*, pp. 39–57, August 17, 1981.

Barna, A. and D. Porat, *Integrated Circuits in Digital Electronics*, McGraw-Hill, New York, 1975.

Glaser, A. and G. Subak-Sharpe, *Integrated Circuit Engineering—Design, Fabrication and Application*, Addison-Wesley, Reading, PA, 1977.

Grove, A. S., *Physics and Technology of Semiconductor Devices*, John Wiley, New York, 1967.

Hamilton, D and W. Howard, *Basic Integrated Circuit Engineering*, McGraw-Hill, New York, 1975.

Krausse, J., Tihanyi, J., and P. Tillmans, "Power MOSFETs run directly off TTL," *Electronics*, pp. 145–147, August 28, 1980.

Ohr, S., "Power components meld the strengths of MOS, bipolar," *Electronic Design*, pp. 65–71, July 5, 1980.

Penney, W. and L. Laud, eds., *MOS Integrated Circuits—Theory, Design and Systems Applications of MOS LSI*, VNR, New York, 1972.

RCA, *COS/MOS Integrated Circuits Manual*, RCA Corporation, 1971.

Richman, P., *Characteristics and Operation of MOS Field-Effect Devices*, McGraw-Hill, New York, 1967.

Severns, R., "MOSFETs rise to new levels of power," *Electronics*, pp. 143–152, May 22, 1980.

CHAPTER 5

Charge Coupled Devices

Frank H. Bower
Manager, Technology Planning
Fairchild Advanced Research and Development

5.1 INTRODUCTION

The charge coupled device (CCD) is one of a family of *charge transfer* devices that also includes charge injection devices (CIDs). The CCD is a semiconductor device that operates on the principle of *charge coupling*. Charge coupling is the collective transfer of the mobile electric charge stored within a semiconductor storage element to a similar adjacent storage element by the external manipulation of voltages. Transfer is accomplished by application of clock voltages to gate electrodes located above the silicon dioxide layer (see Chap. 20) on the surface of the silicon. The quantity of charge stored in each element is an analog value anywhere from essentially zero to complete saturation of the element, which could consist of 100,000 or 1,000,000 or more electrons. At the low end, a thermally generated electron charge might be in the order of 50 to 100 electrons.

The quantity of charge stored in each element is called a *charge packet*. Each packet is a finite amount of charge that is transferred from one storage element to the next by manipulation of the gate voltages. Because the amount of charge in each packet can vary widely, a sequence of packets can be used to represent analog information over a very wide dynamic range.

Figure 5.1 pictures some typical CCDs manufactured by Fairchild. The two devices at the top of the picture are area array image sensors. The six devices

Figure 5.1 Typical examples of charge coupled devices (CCDs). (Courtesy Fairchild Camera and Instrument Corp.)

in the center are a family of line-scan image sensors, and the two at the bottom are analog video delay lines. The operation and characteristics of these devices are examined in Sec. 5.5.

5.2 CCD OPERATION

As mentioned in Sec. 5.1, finite amounts of electrical charge are created in specific locations in the semiconductor material. Each specific location, called a *storage element,* is created by the field of a pair of electrodes very close to the surface of the silicon at that location. If we place the storage elements adjacent to each other, in a line for instance, voltages on the adjacent gate electrodes can be alternately raised and lowered to move the individual charge packets beneath them from one storage element to the next (Fig. 5.2). Because each charge packet may be a different size, the line of elements becomes a very simple analog shift register.

Figure 5.2 is an example of a two-phase CCD shift register. The two complementary clock-voltage waveforms, ϕ_1 and ϕ_2, are connected to alternate closely spaced gate electrodes on the surface of the thin insulating layer on the silicon. A deep potential well, which attracts electrons, is created under the electrodes at clock voltage HIGH and disappears under the electrodes at clock voltage LOW. At $t = 0$, the ϕ_1 voltage is HIGH and the finite charge packet of seven electrons is in the potential well under gate electrode #2 in storage element A. At $t = \frac{1}{2}$ cycle later, the potential well under gate #2 has collapsed owing to ϕ_1 having gone LOW.

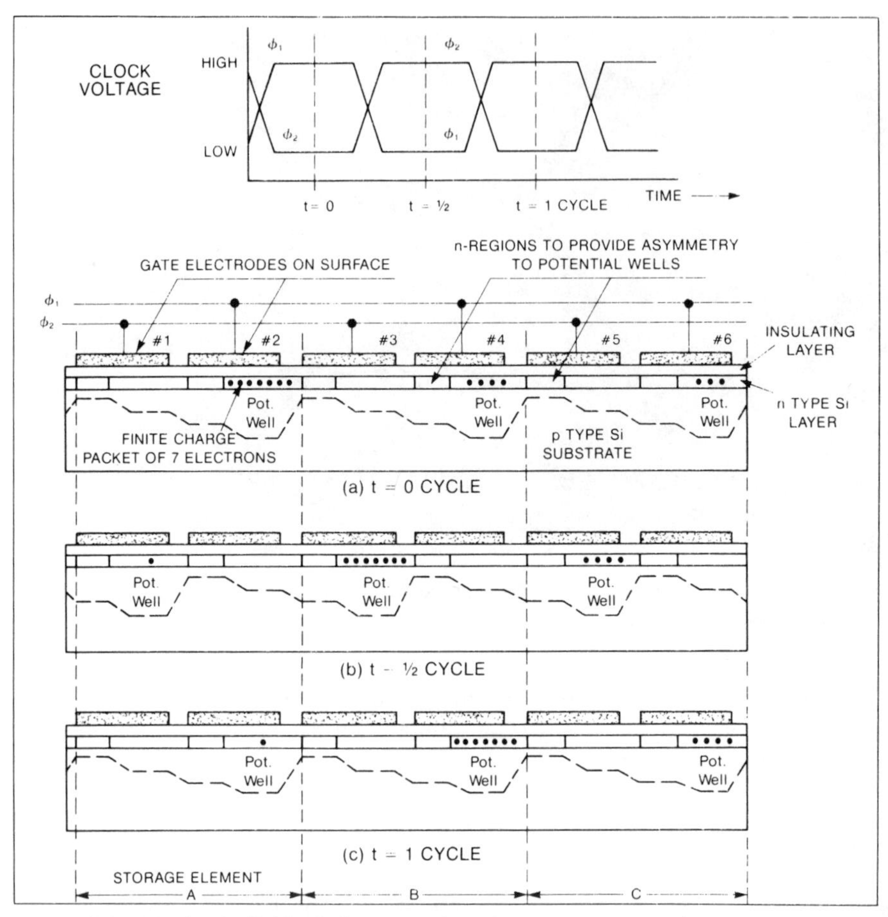

Figure 5.2 Moving individual charge packets from one storage element to the next in a two-phase CCD. (Courtesy Fairchild Camera and Instrument Corp.)

At the same time, because the adjacent electrode #3 connected to ϕ_2 has gone HIGH, the seven-electron charge packet has been attracted to the new potential well under electrode #3. Another half-cycle later, at $t = 1$ cycle, the potential well under electrode #3 has collapsed with ϕ_2 going LOW and the electron packet moves to the new well under electrode #4, which has gone HIGH with clock voltage ϕ_1.

All CCDs are basically shift registers and, because the transfer of charge from each storage element to the next is very efficient, the amount of charge in each packet stays essentially the same, even after it has been passed from one element to as many as 1000 sequentially adjacent elements. Because the amount of charge in each packet is unique, the string of charge packets can represent analog information. The device is, in a sense, storing that information until it is delivered as an electrical signal from the charge detector built into the device at the end of the charge coupled register.

This shift register performance is the basic characteristic of a CCD used in analog signal processing and memory devices. Figure 5.3 shows a diode gate structure for entering or removing information into and from the CCD register so that the device can operate in an electronic system with currents and voltages, rather than with the charge packets manipulated with a CCD.

For digital operation, if the charge detector at the end of the register is set at a given threshold level, then a packet of electrons reaching the detector, which is larger than the threshold level, can indicate a logic 1; if the packet is less than the threshold level, it indicates a logic 0. Consequently, the shift register can be used for binary memory storage. Without the preset detector threshold, the output voltage from the detector is linearly related to the number of electrons in the delivered packet. The result is an analog shift register memory.

The transfer of electrons through 100 or 1000 stages is very efficient, but not perfect. Each time a transfer occurs, a very small portion of the charge packet is left behind. This is referred to as *charge transfer inefficiency* (CTI); the remaining fraction is the *transfer efficiency*. The transfer efficiency of a charge coupled device is a function of the material from which it was made, the fabrication process and device structure, the clocking voltages, and the length of the transfer periods.

The maximum operating speed, or data rate (bits/second) of a CCD may be limited by a direct increase in CTI, by the RC time constant of the driving electrodes and their intraconnection wiring, or by the frequency response of the on-chip amplifier. On many CCDs, it is the speed of operation of the detector and the amplifier that is the limiting factor in the maximum data delivery rate, rather than the transfer speed of electron charge packets from well to well. Buried-channel CCDs have higher operating speeds owing to the fringing fields under the clocking gates that accelerate electron transfer from well to well.

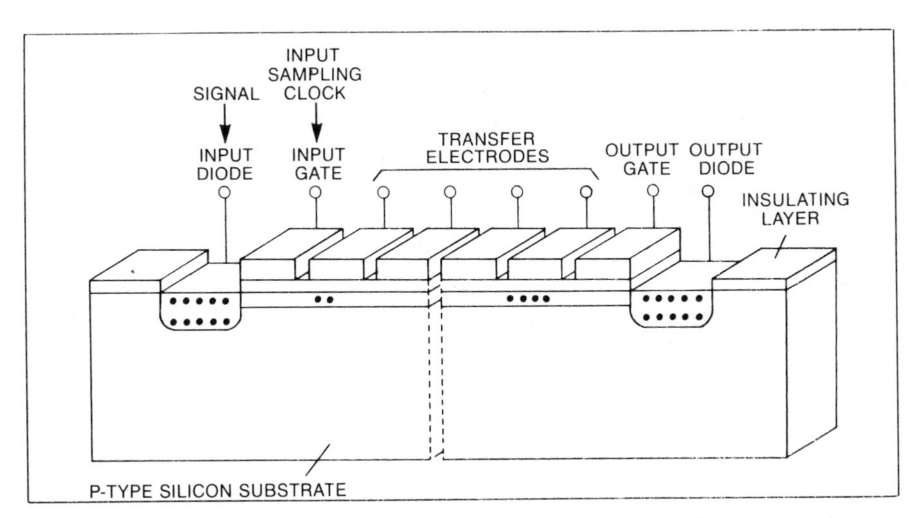

Figure 5.3 Input and output diode gate structures for electrical input and output to a CCD shift register. (Courtesy Fairchild Camera and Instrument Corp.)

EXAMPLE 5.1 Calculate the output data rate of a sensor if 1500 bits of data per line is transmitted at a scan rate of 30 ms per line.

PROCEDURE

1 The output data rate, *DR,* is defined by:

$$DR = D/S \text{ bits/s} \tag{5.1}$$

where D = data in bits/line,
S = scan rate in seconds/line

2 Substitution of the given values in Eq. 5.1 yields:

$$DR = 1500/0.030 = 50{,}000 \text{ bits/s} = 50 \text{ kb/s}$$

The on-chip amplifiers and the sense-node detectors have an exceedingly low noise level because the charge coupled device has no junctions. In addition, there is no conversion of electrical voltage or current to charge with its inherent noise generation. The device can, therefore, have a very low noise figure and a very high dynamic range. Consequently, the output detector and amplifier must be very carefully designed and built to avoid adding significant noise to the signal coming through the device. Modern CCDs use very sensitive low-noise-figure amplifiers.

As stated earlier, the number of electrons that can be stored in a well is a finite maximum. That maximum is determined by the area of the well, the height of the surrounding potential barriers, and the capacitance per unit area. The larger the well volume, the larger is its capacity. When a well is completely filled, it is considered to be saturated. If one attempts to put more electrons in a saturated well, they will spill over into adjacent wells and cause malfunction of a memory or smearing of the output image, known as *blooming*.

Oversaturation occurs in an image-sensing device when the exposure exceeds a critical level stemming from the incidence of stronger illumination or extended integration time, thereby generating more electrons than the well can handle. The storage well has walls which are barriers built either into the silicon by processes such as diffusion and ion implanation (see Chap.20), or resulting from the clock voltages applied to the gates. The barriers have a characteristic potential height and, for example, are referred to as 3 V or 4 V barriers.

EXAMPLE 5.2 A CCD structure has a built-in 3 V barrier. Calculate the maximum number of electrons that can be stored in a storage well without oversaturating. The capacity of the well is 100,000 electrons/V.

PROCEDURE

1 The number of electrons, *e,* stored in a well is:

$$e = Cv \tag{5.2}$$

where C = capacity of well, electrons/V,
v = barrier voltage, V

2 Substitution of the given values in Eq. 5.2 yields:

$$e = 100{,}000 \text{ electrons/V} \times 3 \text{ V} = 300{,}000 \text{ electrons}$$

Dark Current

Dark current is a term applied to the rate at which electrons enter a well because of thermal generation within the silicon itself. The term "dark" means "unilluminated." Precisely as reverse leakage current in a semiconductor diode increases with temperature from the thermally generated carriers, dark current in a CCD increases with temperature. At very low temperatures (−50°C), the dark current is extremely small and may be unmeasurable. At room temperature, however, thermally generated carriers accumulate faster so that the longer a potential well exists without being "pumped out" by clock voltage operation, the greater the number of thermally generated carriers accumulated.

A very small charge packet being transferred down a shift register and accumulating significant dark current charge over a period of time becomes, therefore, a large charge packet and no longer accurately represents the original information. To alleviate this problem, devices can either be cooled to lower the dark current levels or can be operated at higher clock rates so that the potential wells gather dark current over a shorter period of time.

5.3 CCD STRUCTURES

There are two basic CCD structures: *surface-channel* CCD called SCCD and *buried-channel* CCD called BCCD. Each type has significant performance differences. The channel is the region in the silicon in which a line of potential wells is formed to create a CCD shift register. Each well is created by a positive potential on a gate electrode very slightly above the silicon surface and separated from the silicon surface by a thin oxide layer. In other words, this is basically an MOS structure.

In a surface-channel device, the electrons in the packet are attracted by the positive gate potential to the silicon surface at the interface between the silicon and silicon dioxide. In this region at the surface, there are a large number of irregularities called *surface traps*. Electrons in the well at the surface can be trapped and when the potentials on the gates are changed to move a packet of electrons down the shift register, some electrons remain behind in the traps; they cannot be moved, at least for some period of time. For this reason, surface-channel devices have greater charge transfer inefficiency than buried-channel devices.

Trapped electrons may be released at a later time and can cause smearing of the image or distortion of the signal. This particular difficulty in performance can be overcome by using what is called a *fat zero,* that is, providing a continuous low level of charge into the shift register wells to fill up all, or almost all, of the surface traps. The signal charge passing through the register can then be transferred more efficiently, unaffected by the surface traps.

In a buried-channel device, the wells are below the surface because of the implant which creates the buried channel. Surface traps, therefore, have no effect on transfer efficiency.

5.4 IMAGE SENSORS AND ANALOG DELAY LINES

We now consider two types of CCD devices: the *image sensor* and the *analog delay line*. Commercial examples of the line imaging device (LID), the area imaging device (AID), and the analog delay line (ADL) are described in Sec. 5.5.

A CCD image sensor can be configured as a line-scan device or as a X–Y TV-type device. The line-scan device has a single line of sense elements and scans itself electronically in one axis only along the center line of the sense elements. The X–Y device is an area matrix of sense elements capable of being electronically scanned in both the X and Y axes to produce an area TV picture.

Most CCD image sensors have a wide spectral range and are nominally useful from 450 to 1000 nm, that is, visible through the near-infrared regions. Standard commercial charge coupled image sensors will operate well up to a wavelength of about 800 to 900 nm. Beyond that, they lose resolution rapidly. Loss stems from the infrared image photons generating electrons much deeper in the silicon and, therefore, beyond the effect of the field created by the gate electrodes on the silicon surface.

The generated electrons diffuse in the bulk of the silicon until they are either lost by recombination or move randomly near to the surface where they are captured by the field of one of the gate electrodes. Because of the time delay, however, they may arrive too late or appear in a sense element other than the one through which their exiting photons entered the silicon. The practical result is a loss of resolution or smearing of the image. In some laboratories work is being done to develop special CCDs for long-wavelength IR image sensing.

All CCD image sensors consume low power and operate at low voltages. They do not exhibit lag or memory and are not damaged by intense incident light. Some current devices oversaturate and bloom under intense illumination, but are not permanently damaged. Antiblooming structures have been built and are now appearing in commercial CCD image sensors.

Line Imaging Devices (LIDs)

LIDs are configured as a single line of sensor elements on a long, narrow chip. These devices are commercially available with 256, 512, 1024, 1728, and 2048 sense elements. LIDs are used in facsimile machines and in spectrometers where there is a line image pattern. When there is relative motion of the object scene with respect to the sensor, the line array can present a high resolution TV-type picture. For example, a continuous real-time picture can be obtained from a LID sensor in an aircraft or satellite passing over the surface of the earth at a constant altitude and velocity. The use of a scanning mirror in the optical system can accomplish a similar result.

LID applications include high-speed, high-resolution facsimile for text, maps, fingerprints, and photographs. They are also used for aerial mapping with high measuring accuracy, real-time reconnaissance and surveillance, bar-code reading, sorting of conveyorized parts, mail, currency and food, noncontact inspection, and automatic warehouse routing and palletizing control. Combined with an analog delay line, a LID can be used as a sensor for moving target indication (MTI).

EXAMPLE 5.3 Application of a CCD line-scan image sensor in a high-speed facsimile machine requires careful consideration of the image resolution desired; the illumination type and intensity which can be provided; the lens speed, focal length, and flatness of field; the rate of travel of the document being scanned; and the bandwidth of the channel over which the data is to be transmitted.

Assume the system requires 200 lines per inch resolution, that is, 200 picture elements per inch in each direction. The document to be transmitted is $8\frac{1}{2}$ inches wide. The number of sense elements required in the line-scan sensor is therefore $200 \times 8\frac{1}{2} = 1700$. CCD line-scan image sensors with 1728 elements are commercially available. Calculate the output data rate of the sensor and consider the optics required.

PROCEDURE

1 To obtain 200 lines/inch resolution in the Y direction, the direction of document travel, the sensor must scan at a line rate determined by the rate of travel of the document being scanned. This, in turn, is determined by the desired page transmission speed of the facsimile machine.

If the machine is to transmit at a page rate of 1 minute/page, then the 11 inches of page length must be scanned in 60 s with a resolution of 200 lines/inch. The number of line-scans per page is, therefore, equal to 11 inches times 200 lines/inch, or 2200 lines per page.

2 The line-scan rate is 2200 lines per 60 s, or 36.7 lines/s. The reciprocal is the scan time per line, 27.3 ms.

3 If a full line of 1728 pieces of data is to be transmitted every 27.3 ms, then the output data rate for the sensor, by Eq. 5.1, is:

$$DR = 1728/0.0273 = 63.3 \text{ kb/s}$$

If simple black and white information is desired and if no data compression is used, the data transmission channel would be required to handle 63.3 kb/s.

4 *Optical path lens.* The image of the $8\frac{1}{2}$ inch line across the scanned page must have a length equal to the length of the sense element array of the line-scan sensor chosen. For example, the Fairchild CCD122 1728 element line-scan array has an element array length of 0.88 inch. The lens and optical path length would be chosen to provide a 0.9 inch long in-focus flat field image of the line across the $8\frac{1}{2}$ inch wide page.

5 *Lens speed/illumination intensity and spectrum.* The line rate of 36.7 lines/s allows 27.3 ms of image integration time per line. The lens speed and illumination intensity must be chosen so that signals from the "white" image areas will be well above the device dark-signal at this integration time. The illumination source must be chosen with careful consideration of spectrum, uniformity, availability, operating life, and power (heat) dissipation.

Area Imaging Devices (AIDs)

AIDs are made to produce a TV picture. They are built in an array capable of being self scanned in both the X and Y directions. These devices are available in a variety of resolutions such as 100 × 100 element, 244 × 190 element, 400 × 400 element, and 488 × 380 element. The 488 × 380 element device has a four to three aspect ratio, dissipates approximately 100 mW, and is on a chip measuring approximately $\frac{1}{2}$ inch along the diagonal.

AIDs are used for low-light level search and surveillance, missile and remotely piloted vehicle (RPV) guidance, star tracking, remote and projectile TV reconnaissance, cockpit TV cameras, space telescopes, and optical robot sensors. Both single and multisensor color TV cameras have been built experimentally by a number of laboratories and manufacturers. Such products should be commercially available for electronic news gathering within the next two to three years and for home color TV movies within the next three to five years.

Analog Signal Processors

The CCD has been shown to be a nearly ideal analog shift register. The simplest analog signal processor is a variable analog delay line. The delay obtained is a direct function of the clocking frequency and the number of storage elements in the register.

EXAMPLE 5.4 In a television application, a one-line delay is required which is equal to 63.5 μs in the NTSC standard. In the TV set, a color subcarrier frequency of 3.5795 MHz is available. Calculate the number of shift register stages needed to obtain a 63.5 μs delay.

PROCEDURE

1 One can obtain a 63.5 μs delay by using a multiple of the available 3.5795 MHz as the sampling frequency and the shift register clock frequency.
2 The total delay, D, for N shift register stages is:

$$D = N/(f_C M) \text{ s} \tag{5.3}$$

where f_C is the clock frequency and M is a multiple of the sampling frequency. Hence,

$$63.5 \times 10^{-6} = N/(3.5795 \times 10^6 \times M)$$

Solving for N,

$$N = 227.33M \text{ required stages}$$

3 Assuming a bandwidth of 6 MHz and a requirement of at least two samples/cycle, there must be a minimum of $2 \times 6 \times 63.5 = 762$ samples. Therefore, M must be at least 4 ($N = 227.33 \times 3 < 762$).

4 For $M = 4$, the value of N is $223.7 \times 4 = 909.32$; use $N = 910$ stages.

Differential phase of one degree and differential gain of one percent or less are available in commercial devices. Differential phase and differential gain are measures of the linearity of the device over a range of signal level (voltage) and signal frequency. Both should be low to avoid distortion of the delayed signal.

Tapped CCD delay lines are excellent analog filters that can be externally programmed to change filter characteristics, scan a frequency spectrum, or provide correlation of weak signals in a strong noise background. CCD analog signal processor applications include video and audio variable delay lines, moving target indicator filters, signal correlation and convolution, sonic imaging, voice compression and scrambling, video frame grabbing, communications filtering, scan-rate conversion, and spread-spectrum filtering.

5.5 TYPICAL COMMERCIAL CCD PRODUCTS

Line-Scan CCD Image Sensors

Important terms used in CCD line-scan image sensor specifications include:

Transfer Clock, ϕ_X The voltage waveform applied to the transfer gate to move the accumulated charge from the image sensor elements to the CCD transport shift registers.

Transport Clock, ϕ_T The clock applied to the gates of the CCD transport shift registers to move the charge packets received from the image sensor elements to the gated charge-detector/amplifier.

Gated-Charge Detector/Amplifier The output circuit which receives the charge packets from the CCD transport shift registers and provides a signal voltage proportional to the size of each charge packet received. Before each new charge packet is sensed, a reset clock returns the charge-detector voltage to a fixed base level.

Reset Clock, ϕ_R The voltage waveform required to reset the voltage on the charge-detector.

Sample-and-Hold Clock, ϕ_{SH} An internally supplied voltage waveform applied to the sample-and-hold gate in the amplifier to create a continuous sampled video signal at the output. The sample-and-hold feature can be defeated by connecting ϕ_{SH} to V_{DD}, the output amplifier drain voltage.

Dark Reference Video output level generated from sensing elements covered with opaque metalization providing a reference voltage equivalent to device operation in the dark. Permits use of external dc restoration circuitry.

White Reference Video output level generated by on-chip circuitry providing a reference voltage allowing external automatic gain control circuitry to be used. The reference voltage is produced by charge-injection under the control of the electrical input bias voltage, V_{EI}. The amplitude of the reference is typically 70% of the saturation output voltage.

Isolation Cell A site on-chip producing an element in the video output that serves as a buffer between valid video data and dark and white reference signals. The output from an isolation cell contains no valid video information and should be ignored.

Dynamic Range The saturation exposure divided by the peak-to-peak noise equivalent exposure. (This does not take into account any dark signal components.) Dynamic range is sometimes defined in terms of root mean square noise. To compare the two definitions, a factor of four to six is generally appropriate because peak-to-peak noise is approximately equal to four to six times the root mean square noise.

Peak-to-Peak Noise Equivalent Exposure The exposure level that gives an output signal equal to the peak-to-peak noise level at the output in the dark.

Saturation Exposure The minimum exposure level that will produce a saturated output signal. Exposure is equal to the light intensity multiplied by the photosite integration time.

Charge Transfer Efficiency Percentage of valid charge information that is transferred between each successive stage of the transport registers.

Spectral Response Range The spectral band in which the response per unit of radiant power is more than 10% of the peak response.

Responsivity The output signal voltage per unit exposure for a specified spectral type of radiation. Responsivity equals output voltage divided by exposure level.

Dark Signal The output signal in the dark caused by thermally generated electron-hole pairs which is a linear function of integration time and highly sensitive to temperature.

Total Photoresponse Nonuniformity The difference of the response levels between the most and least sensitive elements under uniform illumination.

Saturation Output Voltage The maximum usable signal output voltage, measured from the zero reference level (refer to Fig. 5.4). Any photoelement whose video output is less than the saturation output voltage has an in-spec charge transfer efficiency, CTE. The value of CTE will be below the specification if the video output is equal to, or greater than, the saturation output voltage.

Integration Time The time interval between the falling edges of any two successive transfer pulses ϕ_X, as shown in the timing diagram of Fig. 5.4. The integration time is the time allowed for the photosites to collect charge.

Pixel Picture element (photosite).

Structure of Line Image Sensors

The CCD122 and 142 are examples of monolithic 1728 and 2048 element line image sensors, respectively. The devices are designed for page scanning applications including facsimile, optical charcter recognition, and other imaging applications which require high resolution and sensitivity.

The 1728 sensing elements of a CCD122 provide a 200 lines/inch resolution across an $8\frac{1}{2}$ page adopted as an international facsimile standard. The 2048 sensing elements of a CCD142 provide an 8-line per millimeter resolution across a 256 mm page adopted as the Japanese facsimile standard. Both devices also incorporate on-chip clock drivers. The photoelement size is 13 μm (0.51 mil) by 13 μm on 13 μm centers, as illustrated in Fig. 5.5.

The CCD122/142 consists of the following functional elements as shown in the block diagram of Fig. 5.6:

1 Image Sensor Elements A line of 1728/2048 image sensor elements are separated by diffused channel stops and covered by a silicon dioxide surface passivation layer. Image photons pass through the transparent silicon dioxide layer and are absorbed in the single crystal silicon creating electron-hole pairs.

The photon generated electrons are accumulated in the photosites. The amount of charge accumulated in each photosite is a linear function of the incident illumination intensity and the integration period. The output signal will vary in an analog manner from a thermally generated noise background at zero illumination to a maximum at saturation under bright illumination.

2 Transfer Gate Gate structure adjacent to the line of image-sensor elements. The charge packets accumulated in the image sensor elements are transferred out through the transfer gate to the transport registers whenever the transfer gate voltage goes HIGH. Alternate charge packets are transferred to the analog transport shift registers. The transfer gate also controls the exposure time for the sensing elements and permits entry of charge to the end-of-scan (EOS) shift registers, creating the end-of-scan waveforms.

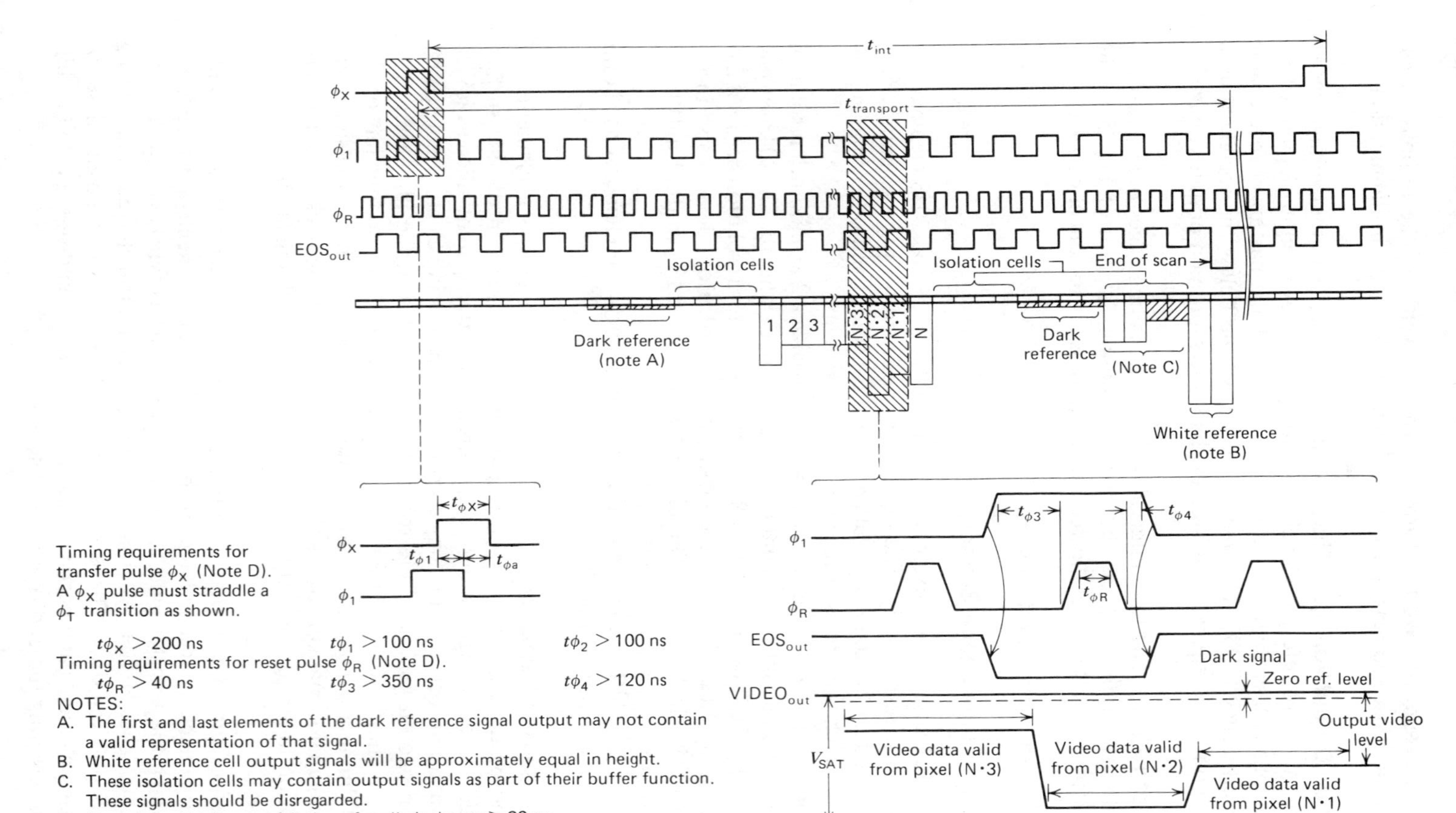

Figure 5.4 Timing diagram of drive signals for a CCD122/142 line image sensor. (Courtesy Fairchild Camera and Instrument Corp.)

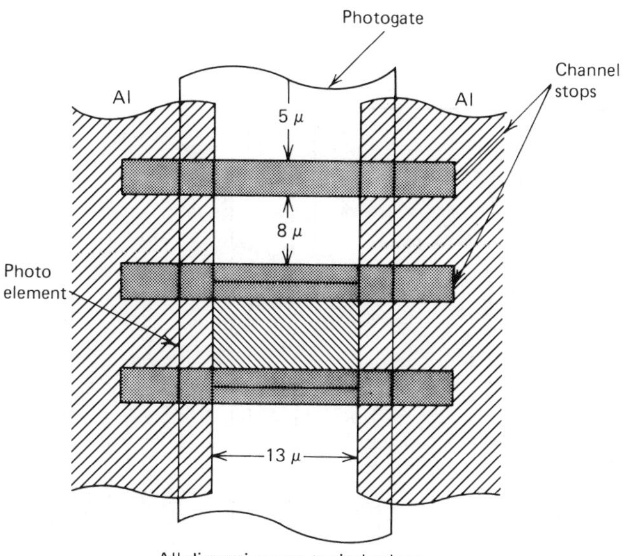

Figure 5.5 Photoelement dimensions of a CCD122/142 line image sensor. (Courtesy Fairchild Camera and Instrument Corp.)

3 Four 879/1039-Bit Analog Shift Registers Two shift registers on each side of the line of image sensor elements and separated from the line by the transfer gate. The two inside registers, called transport shift registers, are used to move the image-generated charge packets delivered by the transfer gate serially to the charge-detector/amplifier. The complementary phase relationship of the last elements of the two transport shift registers provides for alternate delivery of charge packets to establish the original serial sequence of the line of video in the output circuit. The outer two registers serve to deliver the end-of-scan waveform and reduce peripheral electron noise in the inner shift registers.

4 Gated Charge-Detector/Output Amplifier Charge packets are transported to a precharged diode whose potential changes linearly in response to the quantity of the signal charge delivered. This potential is applied to the gate of an n-channel MOS transistor producing a signal which passes through the sample-and-hold gate to the output at $VIDEO_{OUT}$. The sample-and-hold gate is a switching MOS transistor in the output amplifier that allows the output to be delivered as a sampled-and-held waveform. A reset transistor is driven by the reset clock, ϕ_R, and recharges the charge-detector diode capacitance before the arrival of each new signal charge packet from the transport registers.

5 Clock Drivers Allows the CCD122/142 to be operated using only three external clocks: (i) a reset clock signal that controls the integrated output signal amplifier, (ii) a square wave transport clock that operates at half the reset clock frequency and controls the readout rate of video data from the sensor, and (iii)

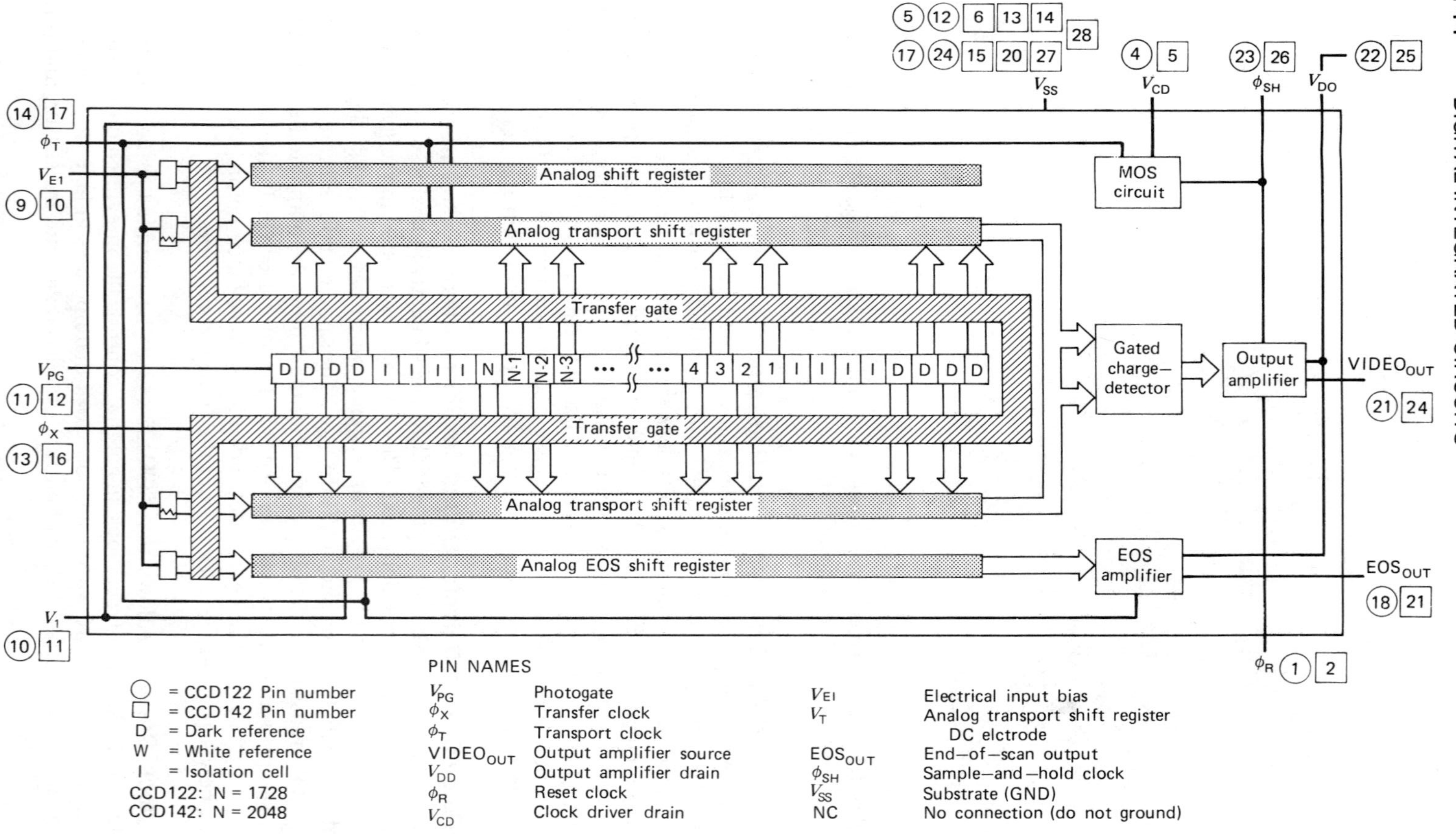

Figure 5.6 Functional block diagram of the CCD122/142. (Courtesy Fairchild Camera and Instrument Corp.)

a transfer clock pulse that controls exposure time of the sensor. The external clocks should be able to supply TTL level power (see Chap. 1).

6 Dark and White Reference Circuitry Four additional sensing elements at both ends of the 1728/2048 array are covered by opaque metalization. They provide a dark (no illumination) signal reference which is delivered at both ends of the line of video output representing the illuminated 1728/2048 sensor elements (labeled "D" in the block diagram). Also included at one end of the array is a white signal reference level generator which likewise provides a reference in the output signal (labeled "W" in the block diagram). These reference levels are useful as inputs to external dc restoration and/or automatic gain control (AGC) circuitry.

Table 5.1 Dc, ac, and clock characteristics for the CCD122/144 image line sensor. (Notes are provided in Table 5.2.)

DC CHARACTERISTICS: T_P = 25°C (Note 1)

SYMBOL	CHARACTERISTIC	RANGE			UNITS	CONDITIONS
		MIN	TYP	MAX		
V_{CD}	Clock Driver Drain Supply Voltage	12.0	13.0	14.0	V	
I_{CD}	Clock Driver Drain Supply Current		6.9	12.5	mA	
V_{DD}	Output Amplifier Drain Supply Voltage	12.0	13.0	14.0	V	
I_{DD}	Output Amplifier Drain Supply Current		6.9	12.5	mA	
V_{PG}	Photogate Bias Voltage	6.5	7.0	7.5	V	
V_T	DC Electrode Bias Boltage	4.5	5.0	5.5	V	Note 2
V_{EI}	Electrical Input Bias Voltage		11.4		V	Note 3
V_{SS}	Substrate (Ground)		0.0		V	

AC CHARACTERISTICS: (Note 1)
T_P = 25°C, $f_{\phi R}$ = 0.5 MHz, t_{int} = 10 ms, light source = 2854°K + 3.0 mm thick Corning 1-75 IR-absorbing filter. All operating voltages nominal specified values.

SYMBOL	CHARACTERISTIC	RANGE			UNITS	CONDITIONS
		MIN	TYP	MAX		
DR	Dynamic Range					Note 9
	(relative to peak-to-peak noise)	250:1	500:1			
	(relative to rms noise)	1250:1	2500:1			
NEE	RMS Noise Equivalent Exposure		0.0002		μj/cm²	Note 10
SE	Saturation Exposure		0.4		μj/cm²	Note 11
CTE	Charge Transfer Efficiency		0.999995			Note 12
V_O	Output DC Level	3.0	5.5	10.0	V	
Z	Output Impedance		1.4	3.0	kΩ	
P	On-Chip Power Dissipation					
	Clock Drivers		90	150	mW	
	Amplifiers		90	150	mW	
N	Peak-to-Peak Noise		2.0		mV	

CLOCK CHARACTERISTICS: T_P = 25°C (Note 1)

SYMBOL	CHARACTERISTIC	RANGE			UNITS	CONDITIONS
		MIN	TYP	MAX		
$V_{\phi TL}$	Transport Clock LOW	0.0	0.3	0.5	V	Notes 4, 5
$V_{\phi TH}$	Transport Clock HIGH	9.75	10.0	10.5	V	Note 5
$V_{\phi XL}$	Transfer Clock LOW	0.0	0.3	0.5	V	Notes 4, 6
$V_{\phi XH}$	Transfer Clock HIGH	9.75	10.0	10.5	V	Note 6
$V_{\phi RL}$	Reset Clock LOW	0.0	0.3	0.5	V	Note 7
$V_{\phi RH}$	Reset Clock HIGH	9.75	10.0	10.5	V	Note 7
$f_{\phi R}$	Maximum Reset Clock Frequency (Output Data Rate)	1.0	2.0		MHz	Note 8

(Courtesy Fairchild Camera and Instrument Corp.)

The dc, ac, and clock characteristics are provided in Table 5.1. Performance characteristics are given in Table 5.2. Typical performance curves are shown in Fig. 5.7. Note that DARK SIGNAL (dark current) increases rapidly with temperature. A timing diagram showing recommended clock drive signals is provided in Fig. 5.4. The drive signals may be obtained by the circuitry of Fig. 5.8.

The CCD133/143 family is similar to the CCD122/142, except for the use of two output amplifiers and two output terminals. These allow the device to perform at 10 times the data rate.

Area CCD Image Sensors

Important terms used in CCD area imaging device specifications include:

Photogate clock, ϕ_P The voltage waveform applied to the photogate to move the accumulated charge from the image sensor elements to the vertical transport registers.

Table 5.2 Performance Characteristics of the CCD122/144.

PERFORMANCE CHARACTERISTICS: (Note 1)

T_P = 25°C, $f_{\phi R}$ = 0.5 MHz, t_{int} = 10 ms, light source = 2854°K + 3.0 mm thick Corning 1-75 IR-absorbing filter. All operating voltages nominal specified values.

SYMBOL	CHARACTERISTIC	RANGE			UNITS	CONDITIONS
		MIN	TYP	MAX		
PRNU*	Photoresponse Non-uniformity					
	Peak-to-Peak		160	210	mV	Note 16
	Peak-to-Peak without Single-Pixel Positive and Negative Pulses		100		mV	Note 16
	Single-pixel Positive Pulses		85		mV	Note 16
	Single-pixel Negative Pulses		130		mV	Note 16
	Register Imbalance ("Odd"/"Even")		20		mV	Note 16
DS	Dark Signal					
	DC Component		5	15	mV	Notes 13, 14
	Low Frequency Component		5	10	mV	Notes 13, 14
SPDSNU	Single-pixel DS Non-uniformity		20	40	mV	Notes 13, 15
R	Responsivity	2.0	3.5	5.0	Volts per μj/cm²	Note 17
V_{SAT}	Saturation Output Voltage	800	1400	1600	mV	Note 18

*All PRNU Measurements taken at a 700 mV output level using an f/2.8 lens and excluded the outputs from the first and last elements of the array. The "f" number is defined as the distance from the lens to the array divided by the diameter of the lens aperture. As the f number increases, the resulting more highly columnated light causes the package window aberrations to dominate and increase PRNU. A lower f number results in less columnated light causing device photosite blemishes to dominate the PRNU.

NOTES:

1. T_P is defined as the package temperature.
2. V_T should be equal to (1/2) $V_{\phi TH}$.
3. V_{EI} is used to generate the end-of-scan output and the white reference output. These two signals can be eliminated by connecting V_{EI} to a voltage level equal to $V_{\phi XH}$ + 5 V.
4. Negative transients on any clock pin going below 0.0 V may cause charge-injection which results in an increase of apparent DS.
5. $C_{\phi T} \cong$ 700 pF
6. $C_{\phi X} \cong$ 300 pF
7. $C_{\phi R} \cong$ 5 pF
8. Minimum clock frequency is limited by increase in dark signal.
9. Dynamic range is defined as V_{SAT}/peak-to-peak (temporal) or V_{SAT}/rms noise.
10. 1 $\mu j/cm^2$ = 0.02 fcs at 2854°K, 1 fcs = 50 $\mu j/cm^2$ at 2854°K.
11. SE for 2854°K for light without 3.0 mm thick Corning 1-75 IR-absorbing filter is typically 0.8 $\mu j/cm^2$.
12. CTE is the measurement for a one-stage transfer.
13. See photographs for DS definitions.
14. Dark signal component approximately doubles for every 5°C increase in T_P.
15. Each SPDSNU is measured from the DS level adjacent to the base of the SPDSNU. The SPDSNU approximately doubles for every 8°C increase in T_P.
16. See photographs for PRNU definitions.
17. Responsivity for 2854°K light source without 3.0 mm thick Corning 1-75 IR-absorbing filter is typically 2 V per $\mu j/cm^2$.
18. See test load configurations.

(Courtesy Fairchild Camera and Instrument Corp.)

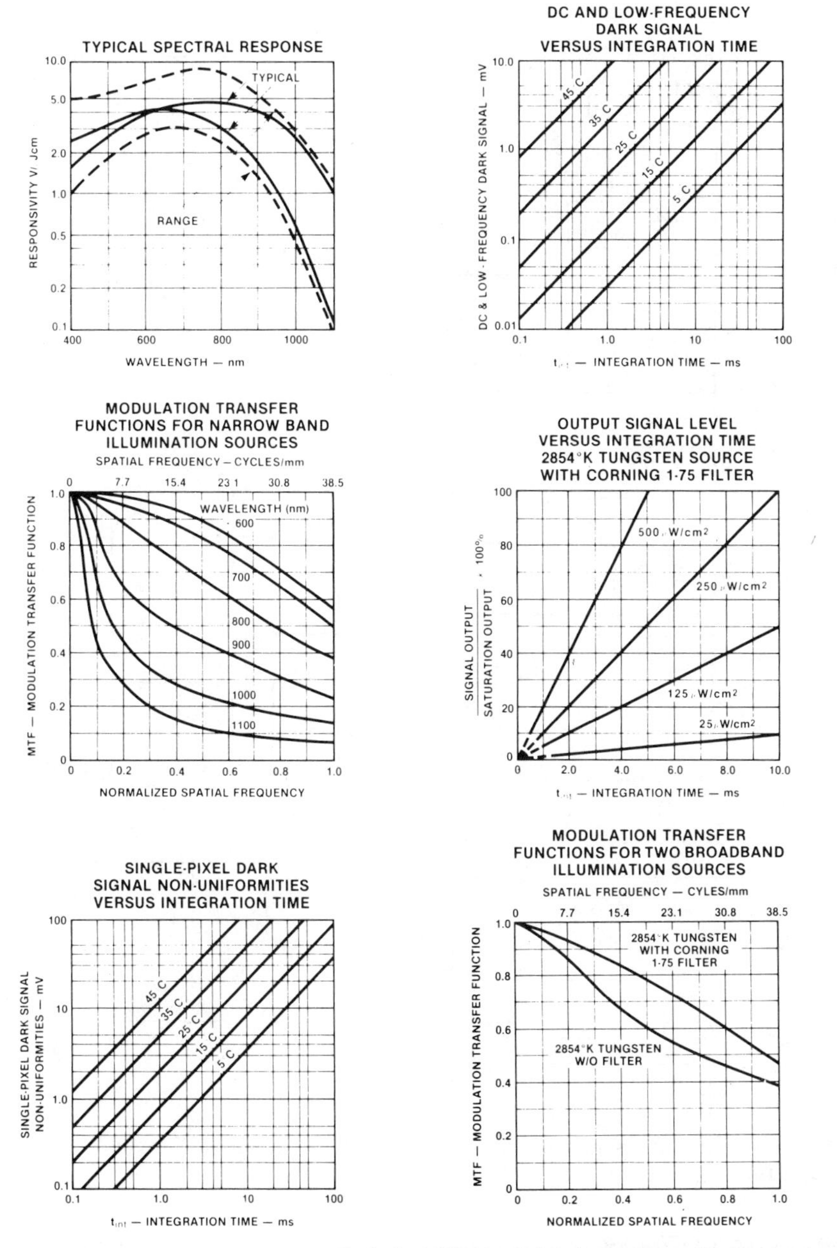

Figure 5.7 Typical performance curves of the CCD122/142. (Courtesy Fairchild Camera and Instrument Corp.)

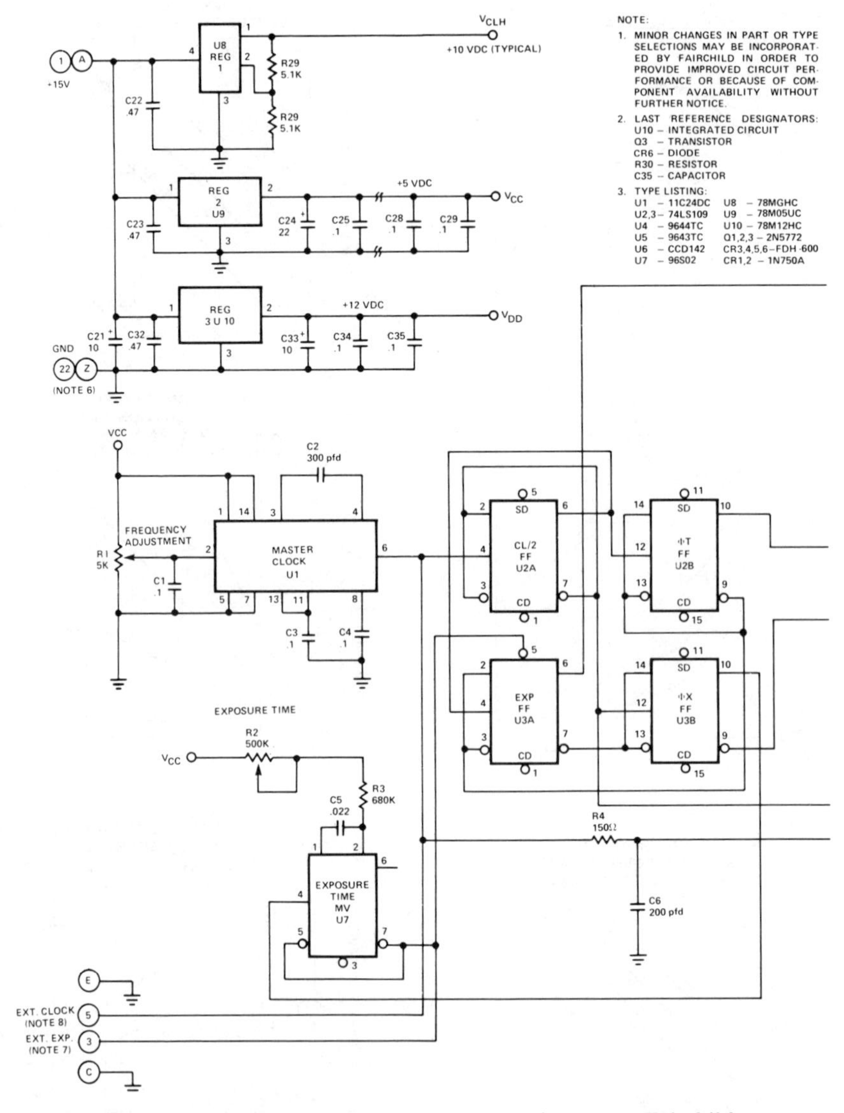

Figure 5.8 Drive circuitry schematic for the CCD122/142. (Courtesy Fairchild Camera and Instrument Corp.)

Vertical Transport Clocks, ϕ_{V1} ***and*** ϕ_{V2} The two clocks applied to the vertical transport registers to move the charge packets received from the image-sensor elements towards the CCD horizontal transport register.

Horizontal Transport Clocks, ϕ_{H1} ***and*** ϕ_{H2} The two clocks applied to the horizontal transport register to move the charge packets received from the vertical transport registers towards the floating-gate amplifier.

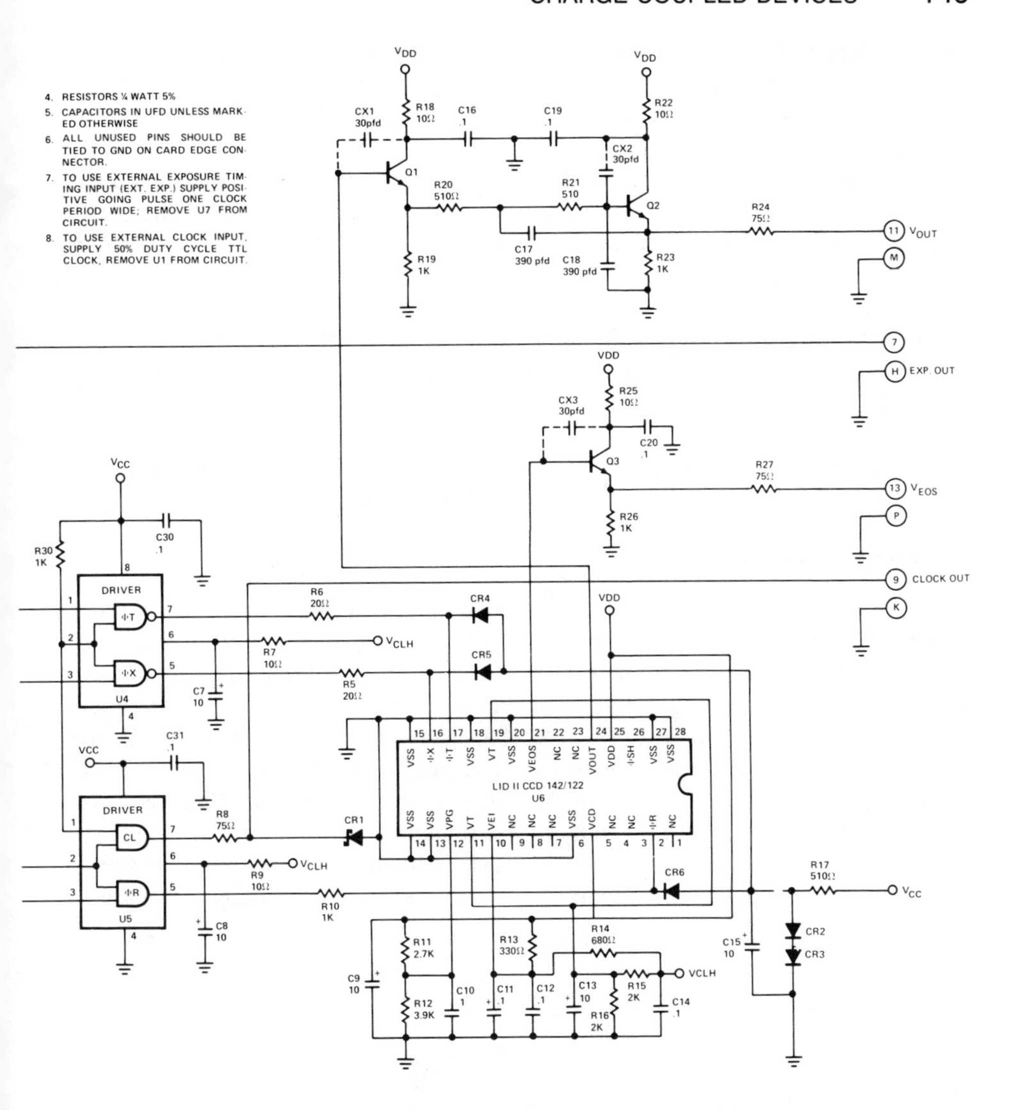

Floating-Gate Amplifier The first stage of the on-chip amplifier which develops a signal voltage linearly proportional to the number of electrons contained in each sensed charge packet. For low-noise signal detection, the floating-gate is coupled to the charge transport channel exclusively by electrostatic fields.

Sample-and-Hold Clock, ϕ_S The clock applied to the sample-and-hold gate of the amplifier. The sample-and-hold feature may be disabled by connecting ϕ_S to V_{DD}.

Dark Reference Video output level generated from photoelements covered with opaque metalization. The video output from these elements provides a reference voltage equivalent to sensor operation in the dark.

Dynamic Range The saturation level output signal voltage of the sensor divided by the root mean square noise output of the sensor in the dark. The peak-to-peak random noise output of the device is four to six times the root mean square noise output.

Saturation Exposure The minimum exposure level that will produce a saturated output signal. Exposure is equal to the light intensity multiplied by the photosite integration time.

Spectral Response Range The spectral band in which the response per unit of radiant power is more than 10% of the peak response.

Structure of Image Sensors

The CCD221 is an example of a 488 × 380 element charge coupled area image sensor intended for use as a high-resolution detector in a variety of scientific and industrial optical instrumentation systems. The device is organized as a matrix array of 488 horizontal lines by 380 vertical columns of charge coupled photoelements. The dimensions of the 185,440 photoelements of the CCD are 12 μm horizontally by 18 μm vertically. The photoelements are precisely positioned on 30 μm horizontal centers and 18 μm vertical centers. Dimensions of the active area are 8.8 by 11.4 mm, with a diagonal of 14.4 mm (Fig. 5.9).

The low-noise performance of the buried-channel CCD structure can provide excellent low-light level capabilities when cooled. The geometric accuracy

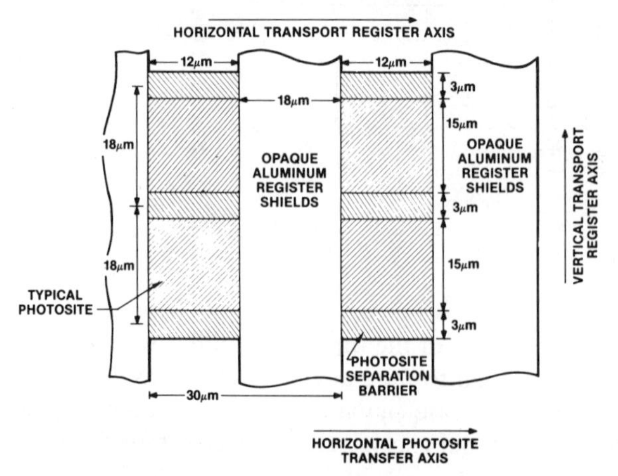

Figure 5.9 Photoelement dimensions of a CCD221 area image sensor. (Courtesy Fairchild Camera and Instrument Corp.)

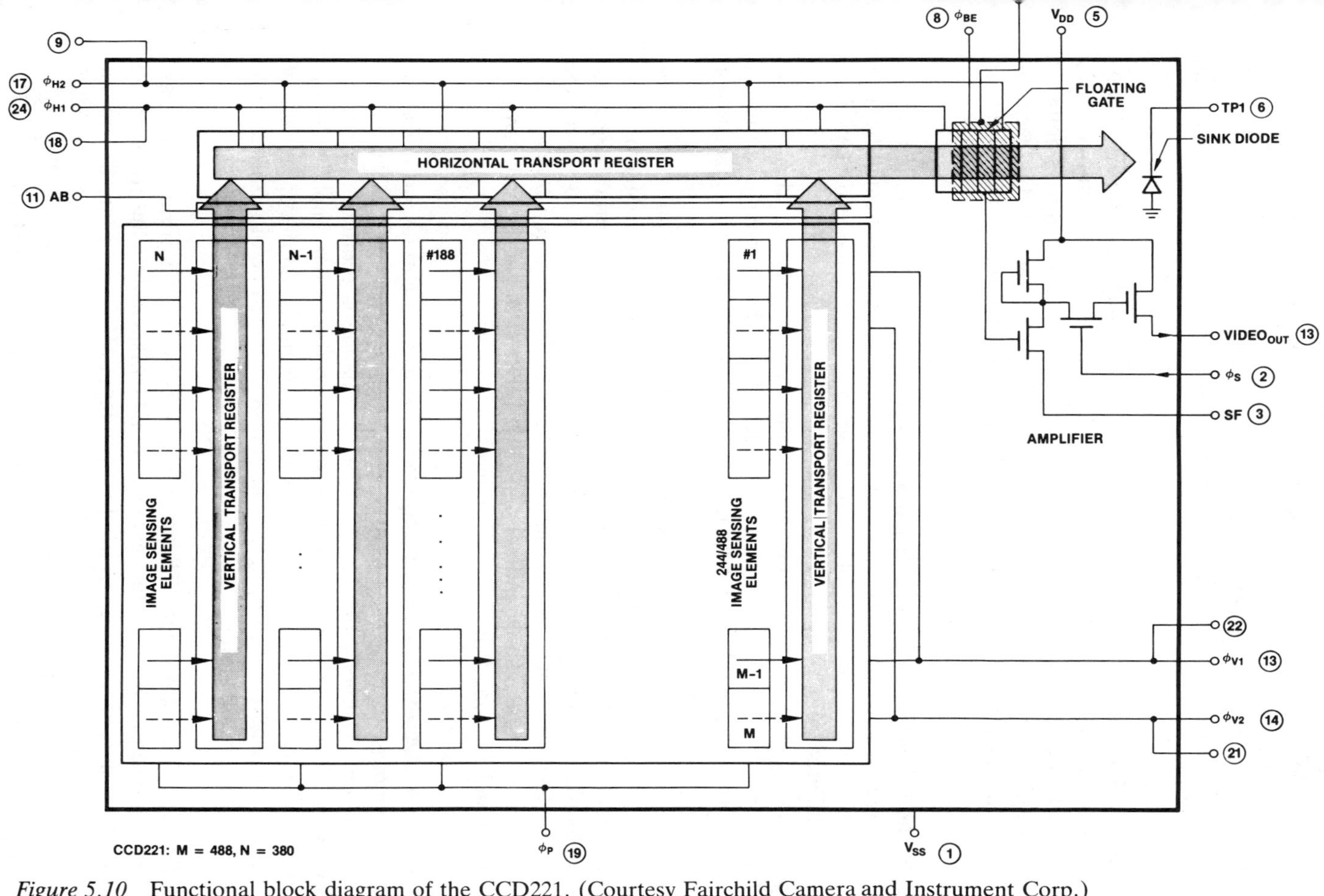

Figure 5.10 Functional block diagram of the CCD221. (Courtesy Fairchild Camera and Instrument Corp.)

of the device structure, combined with a video readout which is controlled by digital clock signals, allows the signal output from each photoelement to be precisely identified for easy realization of computer-based image processing systems. The devices can be used in video cameras that require low power, small size, high sensitivity, high reliability, and rugged construction.

The CCD221 contains the following elements, as shown in the block diagram of Fig. 5.10:

1 Image Sensor Elements Image photons pass through a transparent polycrystalline silicon gate and are absorbed in the silicon crystal structure creating electron-hole pairs. The resulting photoelectrons are collected in the photosites during the integration period. The amount of charge accumulated in each site is a linear function of the localized incident illumination intensity and the integration period.

2 Vertical Analog Transport Registers At the end of the integration period, the charge packets are transferred out of the array in two sequential fields of 244 lines each. When the photogate voltage is lowered, charge packets from odd-numbered photosites (1, 3, 5, . . ., 487) are transferred to the vertical transport registers at the beginning of readout of an odd field when the ϕ_{V1} clock is HIGH. Clocking ϕ_{V1} and ϕ_{V2} then transports the charge packets up the vertical transport registers, line by line, to the output horizontal transport register.

Before the readout of the next even field and when the photogate voltage is again lowered, the ϕ_{V2} clock is held HIGH causing the transfer of even-numbered photosite charge packets (2, 4, 6, . . ., 488) to the vertical registers. A minimum of 245 vertical clock pulses are required per field to deliver the entire field to the output. The additional clock cycle is required owing to the existence of a nonsensitive antiblooming line between the horizontal transport register and the top of the vertical columns.

3 Horizontal Analog Transport Register The horizontal transport register is a 380-element, 2-phase register that receives the charge packets from the vertical registers, line by line. After each line of information is transferred from the vertical transport registers, it is moved serially to the output amplifier by the complementary horizontal clocks ϕ_{H1} and ϕ_{H2}. A minimum of 385 horizontal clock pulses is required to complete the transfer of one line of information to the floating-gate amplifier.

4 Floating-Gate Amplifier The charge packets from the horizontal transport register are sensed by a floating gate whose potential changes linearly with the quantity of signal charge and which drives an input MOS transistor. The output signal from this transistor, in turn, drives the gate of an output n-channel MOS transistor which produces the video signal at terminal $\text{VIDEO}_{\text{OUT}}$. The signal is sampled under control of clock ϕ_S through an MOS transistor switch. The

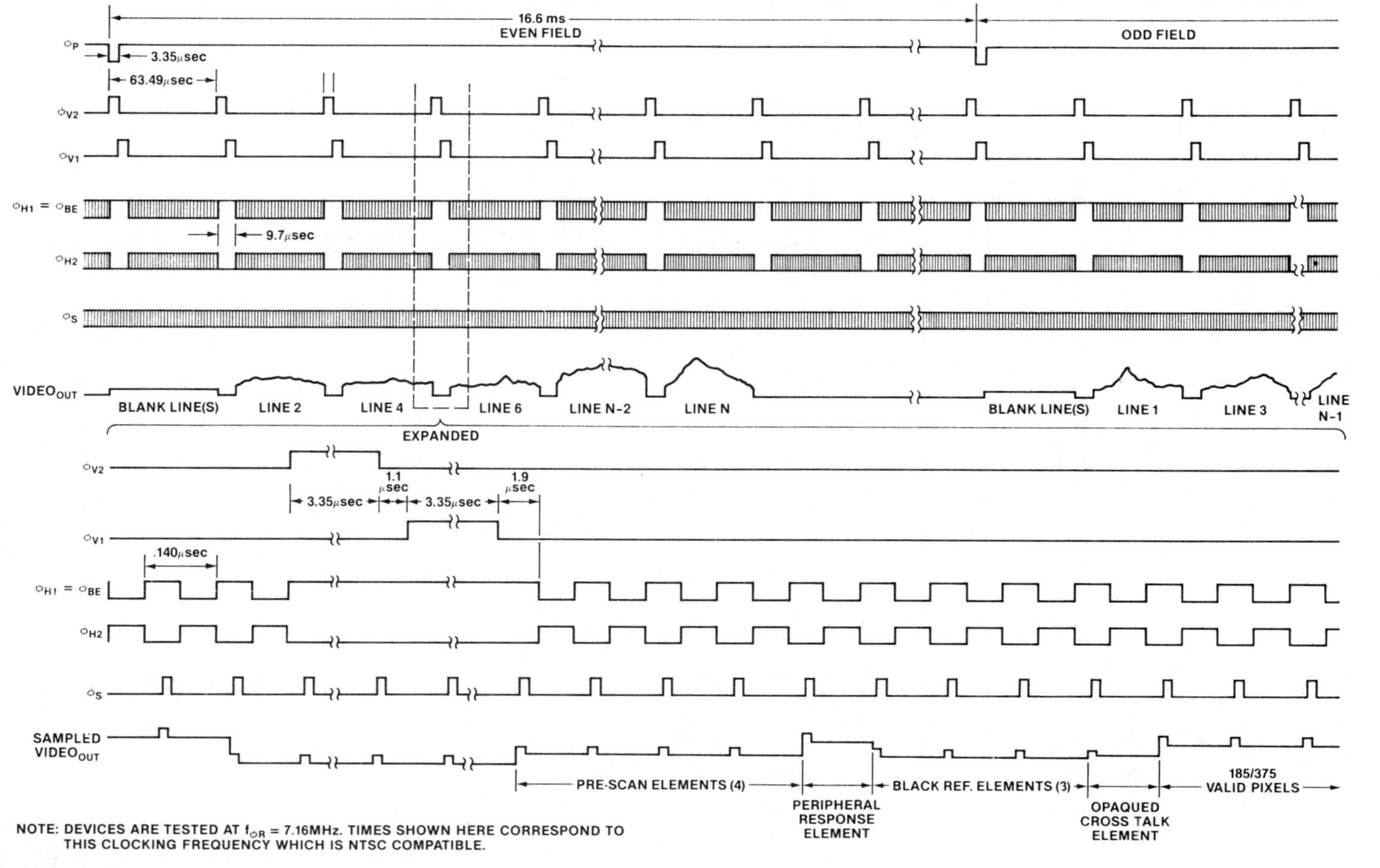

Figure 5.11 Timing diagram of drive signals for a CCD221. (Courtesy Fairchild Camera and Instrument Corp.)

resultant video output signal is a sampled-and-held, clock-controlled analog signal representing the spatial distribution of the sensor surface exposure.

5 Sampled Video Output The output waveform of the CCD221 is shown in detail in the timing diagram of Fig. 5.11. Each frame (488 horizontal lines) is delivered to the output in two sequential fields of 244 horizontal lines each. Each horizontal line is 380 elements long.

The sequence of data comprising each horizontal line is as follows:

1 At the beginning of each line are 4 prescan elements which contain no video information, but are representative of the dark current levels in the horizontal register.

2 The output then contains information from five elements that are covered with opaque aluminum including:

a. A peripheral response element containing information representative of

Table 5.3 Dc operating and clock conditions for the CCD221 area image sensor. (Notes are provided in Table 5.4.)

DC Operating Conditions and Characteristics: Devices are tested at nominal conditions except for V_{SF}, V_{BE}, and V_{AB} which are adjusted for individual sensors.

Symbol	Parameter	Range			Unit	Remarks
		Min.	Nom.	Max.		
V_{DD}	DC Supply Voltage	12.0	15.0	16.5	V	
V_{AB}	Anti-Blooming Bias Voltage	6.0	10.0	V_{DD}	V	Note 1
V_{SF}	Source of Floating-Gate Amplifier	4.0	7.0	10.0	V	Note 1
V_{BE}	Bias Electrode	-5.0	0.0		V	Note 1
TP_2, TP_4	Test Points		0.0		V	
TP_1, TP_3, TP_5	Test Points		V_{DD}		V	
I_{DD}	DC Supply (V_{DD}) Current		3.5		mA	$T_C = 0°C$
I_{SF}	Floating-Gate Amplifier Current		1		μA	$T_C = 0°C$

Clock Conditions

Symbol	Parameter	Range			Unit	Remarks
		Min.	Nom.	Max.		
$V_{\phi PL}$	Photogate Clock LOW	−6.0	0.0		V	Note 2, 10
$V_{\phi PH}$	Photogate Clock HIGH	3.0	5.0	7.0	V	Note 2
$V_{\phi BEL}$	Bias Electrode of FGA Clock LOW	−3.0	0.0	0.0	V	
$V_{\phi BEH}$	Bias Electrode of FGA Clock HIGH	0.0	5.0	7.0	V	Note 1
$V_{\phi H1L}$, $V_{\phi H2L}$	Horizontal Transport Clock LOW	−5.0	0.0	0.0	V	Note 3
$V_{\phi H1H}$, $V_{\phi H2H}$	Horizontal Transport Clock HIGH	5.0	9.0	12.0	V	Note 1, 3
$V_{\phi V1L}$, $V_{\phi V2L}$	Vertical Transport Clock Low	−6.0	0.0	0.0	V	Note 2, 10
$V_{\phi V1H}$, $V_{\phi V2H}$	Vertical Transport Clock HIGH	5.0	9.0	12.0	V	Note 4
$V_{\phi SL}$	Sample-and-Hold Clock LOW	−3.0	0.0	0.0		
$V_{\phi SH}$	Sample-and-Hold Clock HIGH	3.0	5.0	7.0	V	
$f_{\phi H1}$, $f_{\phi H2}$	Max Horizontal Transport Clock Frequency	7.2		20.0	MHz	Note 5

(Courtesy Fairchild Camera and Instrument Corp.)

Table 5.4 Performance characteristics of the CCD221.

Performance Specifications: Standard Test conditions are TV format data output at a 30 Hz frame rate, 60 Hz field rate, 15.75 kHz line rate, 7.16 MHz pixel rate, T_C = 0°C. Light source is 2854°K incandescent with 2.0 mm thick Schott BG-38 IR reject filter.

Symbol	Parameter	CCD211/221			Unit	Condition
		Min	Typ	Max		
V_{SAT}	Saturation Output Voltage	200	700		mVp-p	Note 8
DR	Dynamic Range		1000			See definition of terms
SE	Saturation Exposure		0.28		μJ/cm²	Note 6
R	Responsivity		2.5		V/μJcm⁻²	Note 6
Z	Output Impedance		1000		ohm	
CTF_H	Contrast Transfer Function, Horizontal		75		%	At 190/380 line pairs/ picture width
CTF_V	Contrast Transfer Function, Vertical		70		%	At 244/488 line pairs/ picture height
DSSNU	Dark Signal Shading		1	10	% V_{SAT}	Measured with a 1.5 kHz cutoff low pass filter. Note 8, 9
PRSNU	Photo Response Shading		1	10	% V_{OUT}	Measured at V_{OUT} = 50% V_{SAT} with a 1.5 kHz low pass filter. Note 8

Notes

1. Adjustment is required within the indicated range for optimum operation.
2. $C_{\phi P}$ = 4,000 pF for CCD211; $C_{\phi P}$ = 16,000 pF for CCD221
3. $C_{\phi H1} = C_{\phi H2}$ = 100 pF for CCD211; $C_{\phi H1} = C_{\phi H2}$ = 200 pF for CCD221.
4. $C_{\phi V1} = C_{\phi V2}$ = 3,000 pF for CCD211; $C_{\phi V1} = C_{\phi V2}$ = 12,000 pF for CCD221.
5. Devices are tested at a clock rate of 7.2 MHz. This gives a standard NTSC rate at 30 frames per second. Higher clock rates are possible. Operation of the device at lower or higher frequencies will not damage the device. Two factors contribute to the fundamental low frequency limit: dark current contributions from the photosites and associated dark current non-uniformities, and dark current contributions in the register which will result in increased average dark signal at the output. The longer the intergration time, the higher the spatial non-uniformities.
6. 1 μJ/cm² = (1 μW – S)/cm²
   1 μW/cm² = 3.5 lux with 2854°K + BG-38 filter
   1 lux = 0.03 μW/cm² with 2854°K + BG-38 filter
   Energy is measured *after* the filter.
7. Measured with a 100% contrast bar pattern as a test target. The saturation level is where the video peaks just start to flatten out as the incident illumination is increased
8. Measurement excludes single point blemishes, line and column defects and outer edge elements on a line or field basis
9. DSSNU reduces (increases) in magnitude by a factor of 2X for every 7-10° reduction (increment) in chip temperature.
10. Minimum increase DSNU for certain arrays results when the low level for these clock signals is between 0 and −6V with respect to V_{SS}

(Courtesy Fairchild Camera and Instrument Corp.)

the charge generated around the periphery of the device. This element output should be ignored.

b. Three dark reference cells that contain no video information, but correspond to the true dark current (the sum of register plus photosite currents) of the particular line. These elemental outputs may be used as dark reference levels in post-output dc restoration circuitry.

c. A peripheral response reduction element that is partially covered by aluminum

3 Following are the 375 elements which contain the true video information (valid pixels) showing the spatial distribution of incident brightness for that line.

The dc operation conditions and characteristics, and the clock conditions are provided in Table 5.3. Performance specifications are given in Table 5.4. Typical performance curves are shown in Fig. 5.12.

Analog Delay Lines

The CCD321A is an example of an electrically variable analog delay line intended to be used in analog signal processing systems which include delay and temporary

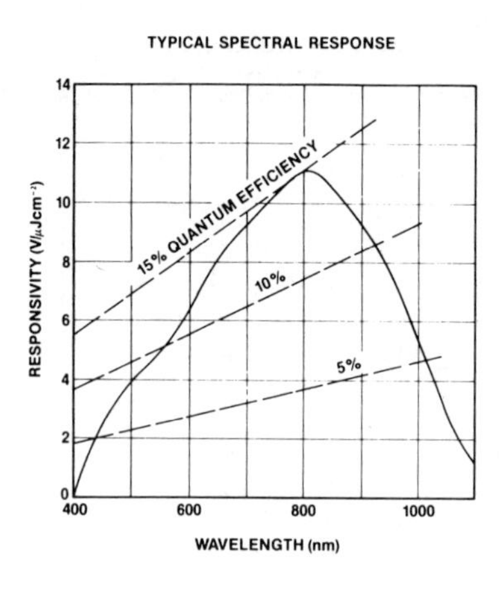

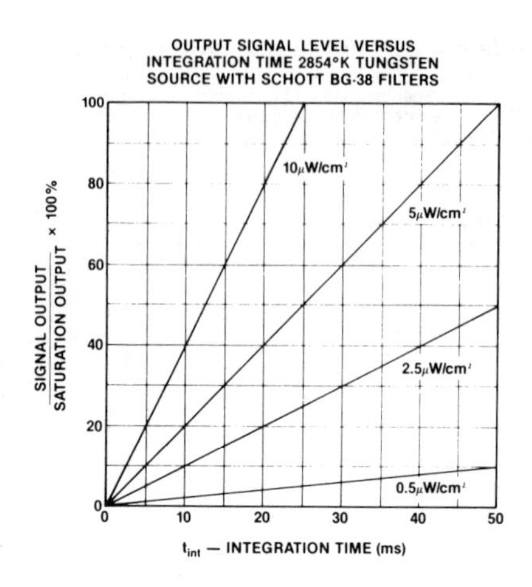

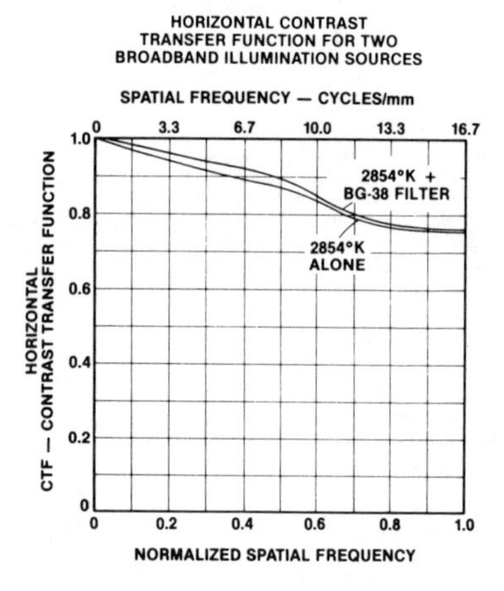

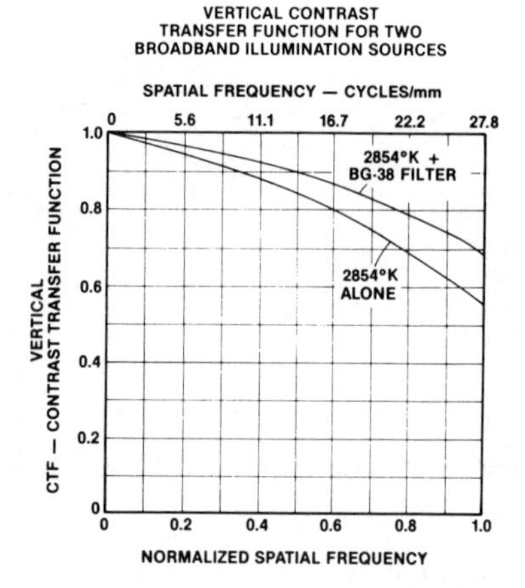

Figure 5.12 Typical performance curves of the CCD221. (Courtesy Fairchild Camera and Instrument Corp.)

storage of analog information. The CCD321A consists of two 455-bit analog shift registers, each with its own charge injection port, transport clock, and output port allowing the device to be used as two 455 or one 910 bit analog delay line.

The CCD321A can be used in applications ranging from video frequencies to audio frequencies. A complete TV line of 63.5 μs can be stored with a sampling frequency of 14.318 MHz (four times the color subcarrier frequency of 3.58 MHz). Applications in video systems include time base correction, signal-to-noise enhancing, and comb filtering. Comb filtering separates chrominance and luminance in the composite signal after detection and ahead of the video amplifiers in a TV receiver, while maintaining full bandwidth.

Audio applications include variable delay of audio signals, reverberation effects in stereo equipment, tone delay in organs and musical instruments, as well as voice scrambling applications. The device also finds application in time base compression and expansion where analog data can be fed at one rate to the device, (the clock can be temporarily stopped) and then data is clocked out at a different rate.

Structure of Analog Delay Lines

The CCD321A contains the following elements, illustrated in the block diagram of Fig. 5.13:

1 Two Charge Injection Ports The analog information in voltage form is applied to two input ports at V_{1A} (V_{1B}). Upon the activation of the analog sample clocks ϕ_{SA} (ϕ_{SB}), a charge packet linearly dependent on the voltage difference between V_{1A} and V_{RA} (V_{1B} and V_{RB}) is injected into analog shift register A (B).

2 Two 455-Bit Analog Shift Registers Each register transports the charge packets from the charge injection port to its corresponding output amplifier. Both registers are operated in the $1\frac{1}{2}$ phase mode where one phase (ϕ_{1A} or ϕ_{1B}) is a clock and the other phase (V_2) is an intermediate dc potential. Phases ϕ_{1A} and ϕ_{1B} are completely independent. Voltage V_2 is a dc voltage common to both registers.

3 Two Output Amplifiers Charge packets from each analog shift register are delivered to their corresponding output amplifier, as indicated in Fig. 5.13. Each output amplifier consists of three source follower stages with constant current bias sources. A sample-and-hold transistor is located between the second and

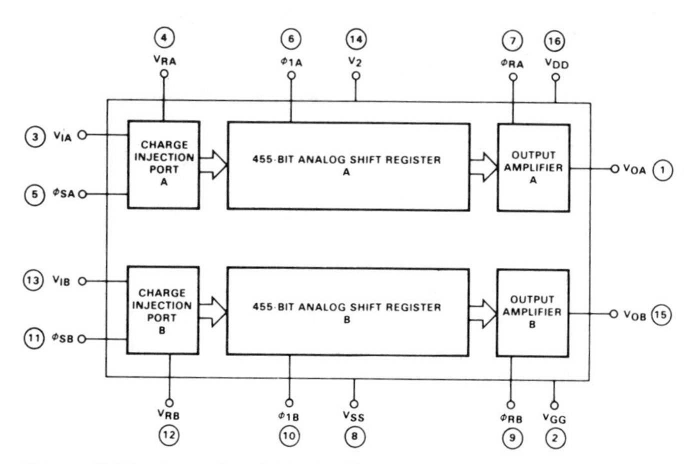

Figure 5.13 Functional block diagram of a CCD321A analog delay line. (Courtesy Fairchild Camera and Instrument Corp.)

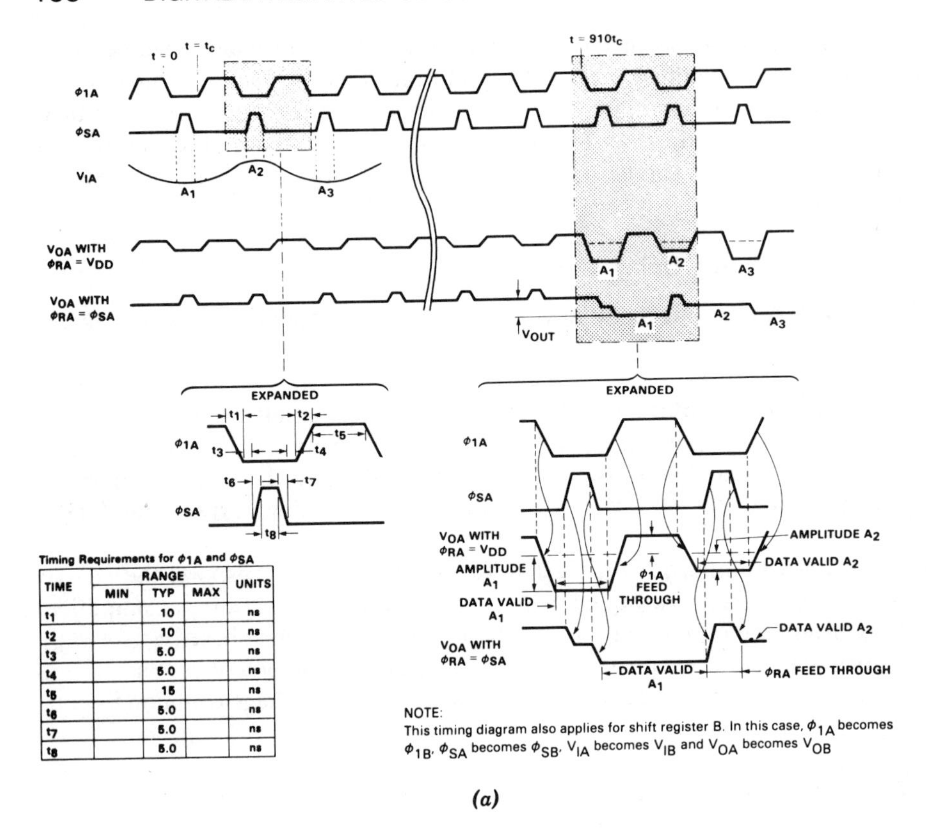

Timing Requirements for ϕ_{1A} and ϕ_{SA}

TIME	RANGE MIN	TYP	MAX	UNITS
t_1		10		ns
t_2		10		ns
t_3		5.0		ns
t_4		5.0		ns
t_5		15		ns
t_6		5.0		ns
t_7		5.0		ns
t_8		5.0		ns

(a)

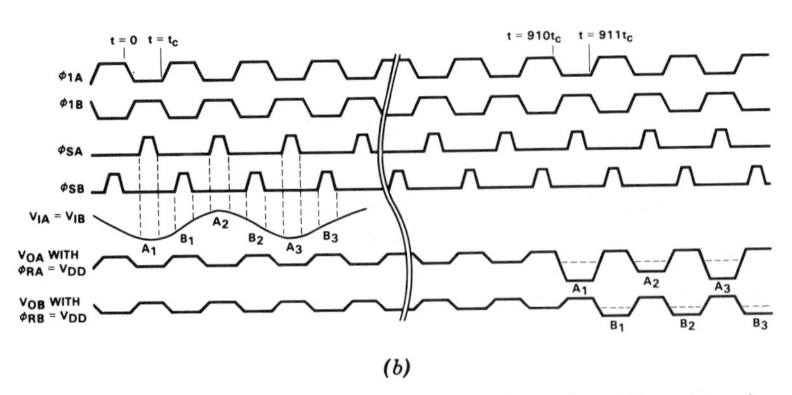

(b)

Figure 5.14 Timing diagrams for the CCD321A. (*a*) Analog shift register A or B operation. (*b*) Analog shift register A and B operation in the multiplexed mode. (Courtesy Fairchild Camera and Instrument Corp.)

third stage of the amplifier. When the gate of the sample-and-hold transistor is clocked (ϕ_{RA} or ϕ_{RB}), a continuous output waveform is obtained as shown in the timing diagram of 5.14*a*. The sample-and-hold transistor can be defeated by connecting ϕ_{RA} and/or ϕ_{RB} to V_{DD}. In this case, the output is a pulse modulated waveform, as indicated in the timing diagram.

Modes of Operation

The CCD321A can be operated in four different modes:

1 455-bit Analog Delay Either 455-bit analog shift register can be operated independently as a 455-bit delay line. The driving waveforms to operate shift register A are shown in Fig. 5.14*a*. The input voltage signal is applied directly to V_{IA}. The input sampling clock ϕ_{SA} samples this input voltage and injects a proportional amount of charge packet into the first bit of register A. Input voltage V_{IA}, which is sampled between $t = 0$ and $t = t_c$, appears at the output terminal V_{OA} at $t = 910t_c$.

If the sample-and-hold circuit is not used, then the output appears as a pulse amplitude modulated waveform, as shown in the timing diagram of Fig. 5.14*a*. In that case, ϕ_{RA} should be connected to V_{DD}. If the sample-and-hold

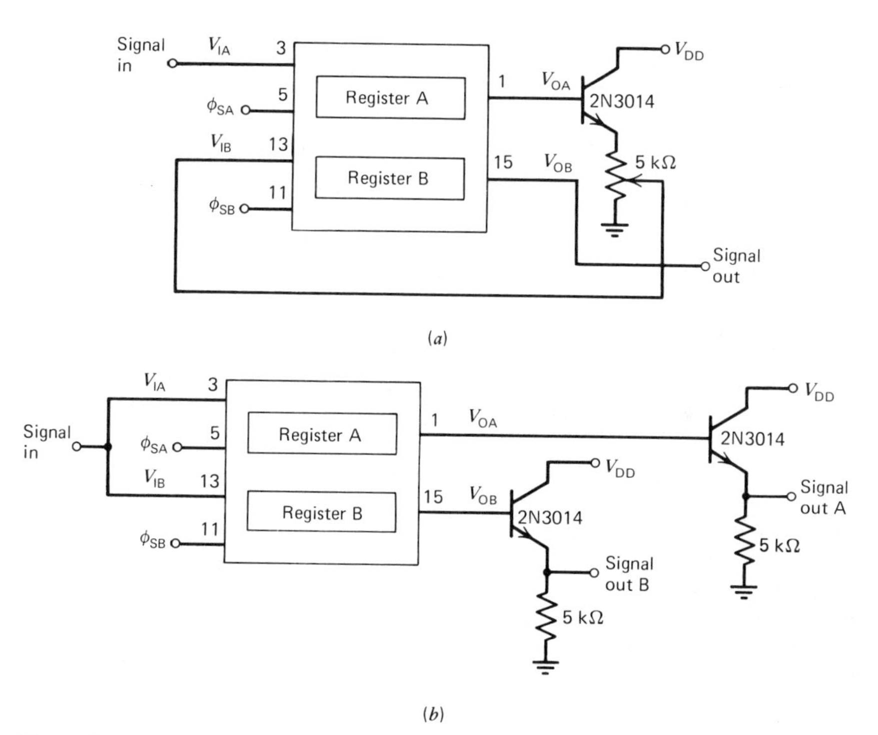

Figure 5.15 Realizing a 910-bit delay in (*a*) series and (*b*) multiplexed modes. (Courtesy Fairchild Camera and Instrument Corp.)

Table 5.5 Ac Characteristics of the CCD321A family of analog delay lines.

CCD321A-1 AC CHARACTERISTICS: T_A = 55°C. Both registers in the multiplexed mode, clock rate = 7.16 MHz. Sampling rate = 14.32 MHz. $V_{out} \cong$ 700 mV.

Symbol	Characteristic	Range			Units	Conditions
		Min	Typ	Max		
BW	Signal bandwidth (3 dB down)	5.0			MHz	Note 7
IG	Insertion gain	0	3.0	6.0	dB	Note 8
Δ G	Differential gain			1.0	%	Note 9
$\Delta\ \phi$	Differential phase			1.0	degrees	Note 9
S/N	Signal-to-noise ratio	58			dB	Note 10
$V_{I\ (max)}$	Maximum input signal voltage		1.0		$V_{pk\text{-}pk}$	

CCD321A-2 AC CHARACTERISTICS: T_A = 55°C. Both registers in the multiplexed mode, clock rate = 7.16 MHz, sampling rate = 14.32 MHz. $V_{out} \cong$ 700 mV.

Symbol	Characteristic	Range			Units	Conditions
		Min	Typ	Max		
BW	Signal bandwidth (3 dB down)	4.2	5.0		MHz	Note 7
IG	Insertion gain	0	3.0	6.0	dB	Note 8
Δ G	Differential gain			3.0	%	Note 9
$\Delta\ \phi$	Differential phase			3.0	degrees	Note 9
S/N	Signal-to-noise ratio	58			dB	Note 10
$V_{I\ (max)}$	Maximum input signal voltage		1.0		$V_{pk\text{-}pk}$	

CCD321A-3 AC CHARACTERISTICS: T_A = 55°C. Both registers in the multiplied mode, clock rate = 7.16 MHz, sampling rate = 14.32 MHz. Clocks are stopped for 300 μs. $V_{out} \cong$ 700 mV after 4.2 MHz low pass filter.

Symbol	Characteristic	Range			Units	Conditions
		Min	Typ	Max		
BW	Signal bandwidth (3 dB down)	4.2	5.0		MHz	Note 7
IG	Insertion gain	0	3.0	6.0	dB	Note 8
Δ G	Differential gain			3.0	%	Note 9
$\Delta\ \phi$	Differential phase			3.0	degrees	Note 9
S/N	Signal-to-noise ratio	55			dB	Note 10
SN	Spacial noise		10.0	20.0	mV	Notes 11, 12
$V_{I\ (max)}$	Maximum input signal voltage		1.0		$V_{pk\text{-}pk}$	

NOTES:

1. V_2 level should be 1/2 of the ϕ_{1A} or ϕ_{2A} HIGH level. Adjustment in the range of ±1 V may be necessary to maximize signal bandwidth.
2. Signal charge injection is proportional to the difference V_I and V_R. Adjustment of either V_I or V_R is necessary to assure proper operation.
3. Negative transients below ground of fast rise and fall times of the clocks may cause charge injection from substrate to the shift registers. A negative bias on V_{SS} of −2.0 to −5.0 Vdc will eliminate the injection phenomenon.
4. $C\phi_{1A} = C\phi_{1B}$ = 30 pF = capacitance with respect to V_{SS}.
5. $C\phi_{SA} = C\phi_{SB}$ = 10 pF = capacitance with respect to V_{SS}.
6. $C\phi_{RA} = C\phi_{RB}$ = 10 pF = capacitance with respect to V_{SS}.
7. Signal bandwidth is typically 1/3 to 1/2 of the sampling rate.
8. Insertion gain = 20 log V_{OUT}/V_{IN}.
9. Differential gain and differential phase are measured with Tektronix NTSC Signal Generator (147A) and Vector Scope (520A).
10. Video S/N is defined as the ratio of the peak-to-peak output signal to RMS random (temporal) noise. The peak-to-peak signal is the maximum output level that satisfies the ΔG and $\Delta\phi$ specs.
11. In the start/stop mode of operation it is recommended that the rise and fall times of ϕ_{1A} and ϕ_{1B} exceed 20 ns to eliminate charge injection.
12. Spacial noise is the peak-to-peak spacial variation (fixed pattern noise) in the device output after clocks have been stopped. It is usually caused by the variation of leakage current density in the shift registers. Spacial noise is a function of the clock stop period and temperature.
13. Input signal = 1 kHz sine wave.
14. Audio S/N is defined as the ratio of RMS signal to RMS noise at 23 kHz bandwidth. Both are measured with an HP3400A RMS Voltmeter.
15. Rate of average-signal offset is caused by leakage current in the registers. It is a function of temperature.
16. Devices are tested using the values shown in the typical columns.
17. Devices can be operated beyond 20 MHz without damage. The minimum clock rate can be lower than 10 kHz.

(Courtesy Fairchild Camera and Instrument Corp.)

Table 5.6 Dc and clock characteristics of the CCD321. (Notes are provided in Table 5.5.)

DC CHARACTERISTICS: T_A = 55° C, Note 16

SYMBOL	CHARACTERISTICS	RANGE			UNITS	CONDITIONS
		MIN	TYP	MAX		
V_{DD}	Output Drain Voltage	14.5	15.0	15.5	V	
V_2	Analog Shift Register DC Transport Phase Voltage		6.0		V	Note 1
V_{RA}, V_{RB}	Analog Reference Inputs Voltage		3-7		V	Note 2
V_{GG}	Signal Ground		0.0			
V_{SS}	Substrate Ground		0.0			Note 3
V_{IA}, V_{IB}	Input DC Level		3-7		V	Note 2
V_{OA}, V_{OB}	Output DC Level		6-11		V	V_{DD} = 15 V
R_{IN}	Small Signal Input Resistance		1.0		MΩ	Resistance from Pins 3, 4, 12 or 13 to V_{SS}. V_{IA} = V_{IB} = 3 V
C_{IN}	Small Signal Input Capacitance		10		pF	Capacitance from Pins 3, 4, 12 or 13 to V_{SS}. V_{IA} = V_{IB} = 3 V
R_{OUT}	Small Signal Output Resistance		250		Ω	V_{DD} = 15 V
ODM	Output DC Mismatch Between A & B Registers		±1		V	
OAM	Output AC Mismatch Between A & B Registers		±20		%	

CLOCK CHARACTERISTICS: T_A = 55° C, Note 16

SYMBOL	CHARACTERISTICS	RANGE			UNITS	CONDITIONS
		MIN	TYP	MAX		
$V\phi_{1AL}$, $V\phi_{1BL}$	Analog Shift Register Transport Clocks LOW	0	0.5	0.8	V	Note 4
$V\phi_{1AH}$, $V\phi_{1BH}$	Analog Shift Register Transport Clocks HIGH	12.0	13.0	15.0	V	Note 4
$V\phi_{SAL}$, $V\phi_{SBL}$	Input Sampling Clocks LOW	0	0.5	0.8	V	Note 5
$V\phi_{SAH}$, $V\phi_{SBH}$	Input Sampling Clocks HIGH	12.0	13.0	15.0	V	Note 5
$V\phi_{RAL}$, $V\phi_{RBL}$	Output Sample and Hold Clocks LOW	0	0.5	0.8	V	Note 6
$V\phi_{RAH}$, $V\phi_{RBH}$	Output Sample and Hold Clocks HIGH	12.0	13.0	15.0	V	Note 6
$f\phi_{1A}$, $f\phi_{1B}$	Analog Shift Register Transport Clock Frequency	0.02		20	MHz	See Note 17
$f\phi_{SA}$, $f\phi_{SB}$	Input Sampling Clocks Frequency	0.02		20	MHz	See Note 17
$f\phi_{RA}$, $f\phi_{RB}$	Output Sample and Hold Clocks Frequency	0.02		20	MHz	See Note 17

(Courtesy Fairchild Camera and Instrument Corp.)

circuit is used, then the output appears as a continuous waveform. Here, ϕ_{RA} should be clocked simultaneously with ϕ_{SA} and the two pins can be connected together.

2 910-Bit Analog Delay in Series Mode The two analog shift registers A and B can be connected in series to provide 910 bits of analog delay as in Fig. 5.15*a*. The analog signal input voltage is applied to V_{IA}. The output of register A is connected to the input of register B with a simple emitter follower buffer stage. In order to ensure proper charge injection of register B, V_{RB} should be adjusted. The timing diagram of Fig. 5.14*a* applies in this mode of operation. Here $\phi_{1A} = \phi_{1B}$, $\phi_{SA} = \phi_{SB}$, $\phi_{RA} = V_{DD}$, and ϕ_{RB} is clocked.

3 910-Bit Analog Delay in Multiplexed Mode The two analog shift registers can be connected in parallel to provide 910 bits of analog delay, as in Fig. 5.15*b*. The analog signal input voltage is applied to both V_{IA} and V_{IB}. The outputs as V_{OA} and V_{OB} can be combined to recover the analog input information.

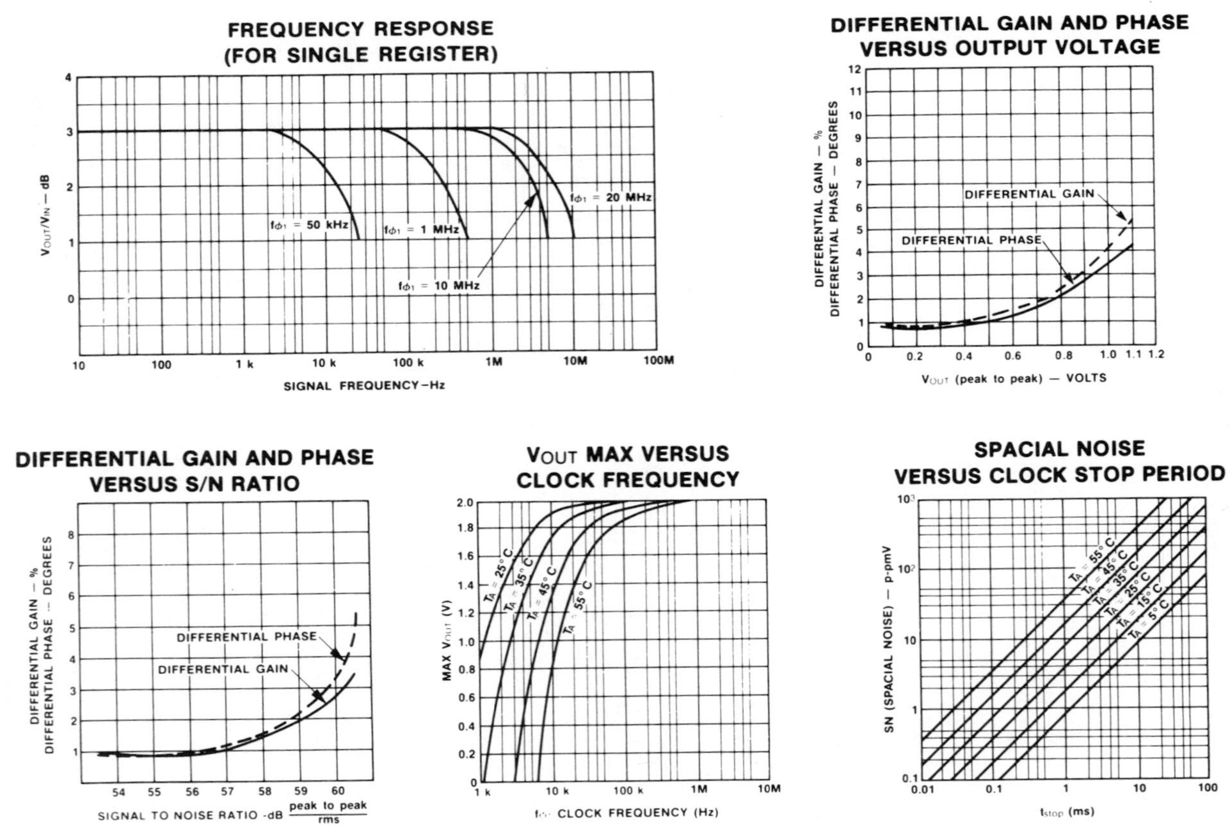

Figure 5.16 Typical video performance curves for the CCD321A. (Courtesy Fairchild Camera and Instrument Corp.)

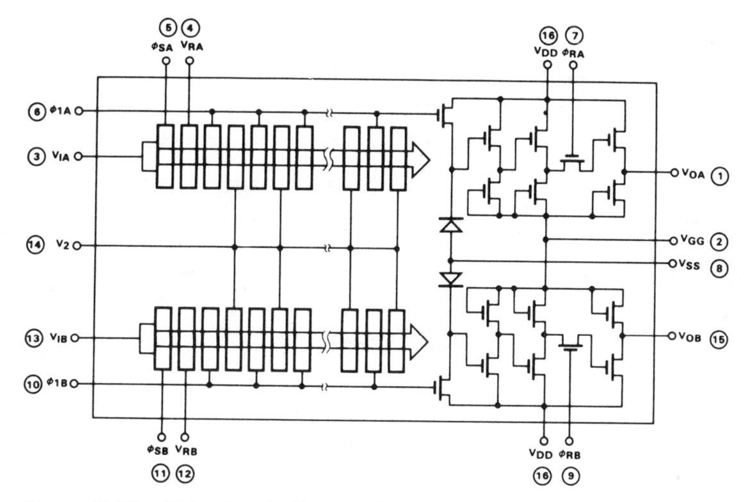

Figure 5.17 Chip circuit diagram for the CCD321A. (Courtesy Fairchild Camera and Instrument Corp.)

The necessary waveforms to operate the device in this mode are provided in Fig. 5.14*b*. In this case, ϕ_{SA} samples the analog input A_1 at V_{IA} between $t = 0$ and $t = t_c$. Clock ϕ_{SB} samples the analog input B, at V_{IB}, between $t = t_c$ and $t = 2t_c$. The output corresponding to B_1 appears at V_{OB} at $t = 911t_c$. This mode of operation results in an effective sampling rate of twice the rate of ϕ_{1A}, ϕ_B, ϕ_{SA}, and ϕ_{SB}.

4 Stop/Start Mode The charge packets in the two analog shift registers can be held stationary by stopping ϕ_{1A} and ϕ_{1B} in their LOW states. Clocks ϕ_{SA}, ϕ_{SB}, ϕ_{RA}, and ϕ_{RB} also can be stopped in the LOW state or kept clocking as usual. The two shift registers should not be connected in series in the stop-start mode of operation.

The CCD321A is available in three different types for different applications. The CCD321A-1 is a high-quality broadcast 1H delay line for video systems with 1% differential gain and 1° differential phase. The CCD321A-2 is a high-quality video delay line with 3% differential gain and 3° differential phase. The CCD321A-3 is tested in the stop/start mode of operation and parameters are guaranteed in this mode. The dc and clock characteristics of the three types are the same and the ac characteristics vary as shown in Table 5.5.

(***CAUTION:*** The CCD321 devices have limited built-in gate protection. Charge buildup should be minimized. Care should be taken to avoid shorting pins V_{OA} and V_{OB} to ground during operation.)

The dc and clock characteristics for the CCD321 are supplied in Table 5.6. Typical video performance curves are illustrated in Fig. 5.16. The chip circuit diagram is provided in Fig. 5.17. The schematic of a video delay module using the CCD321A is given in Fig. 5.18.

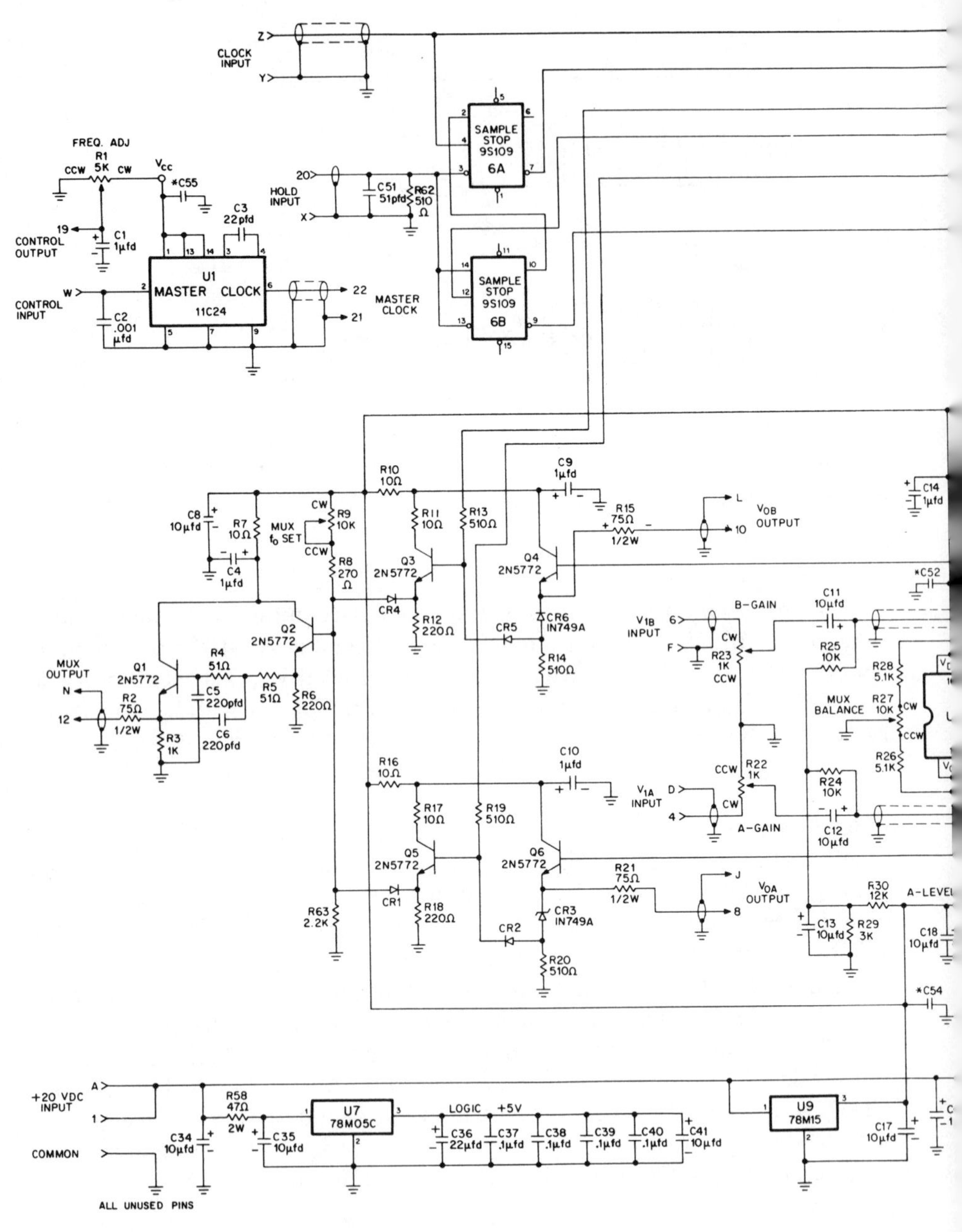

NOTES:

1. MINOR CHANGES IN PART VALUE OR TYPE SELECTIONS MAY BE INCORPORATED IN ORDER TO PROVIDE IMPROVED CIRCUIT PERFORMANCE OR BECAUSE OF COMPONENT AVALILABILITY WITHOUT FURTHER NOTICE.
2. ALL DIODES ARE FDH-600 UNLESS MARKED OTHERWISE.
3. 50% DUTY CYCLE CLOCK REQUIRED FOR PROPER OPERATI
4. RESISTORS 1/4 WATT UNLESS MARKED.
5. *OPTIONAL PARTS

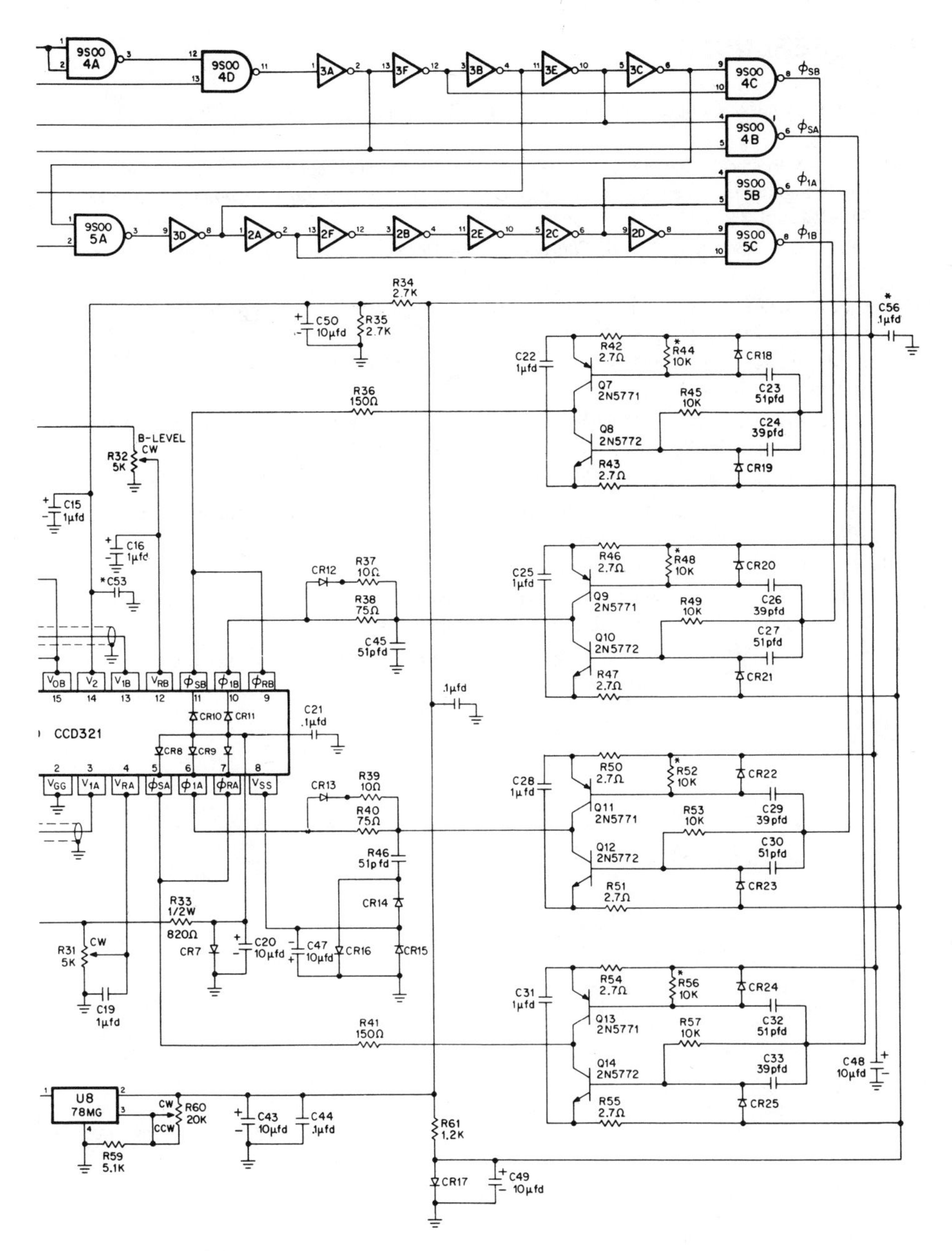

Figure 5.18 Schematic of a video delay module employing the CCD321A. (Courtesy Fairchild Camera and Instrument Corp.)

5.6 REFERENCES

Amelio, G., "Charge Coupled Devices," *Scientific American,* pp. 22–31, vol. 230, no. 2, February 1974.

Barton, S., "A Practical Charge-Coupled Device Filter for the Separation of Luminance and Chrominance Signals in a Television Receiver," *IEEE Transactions on Consumer Electronics,* pp. 342–354, vol. CE-23, no. 3, August 1977.

Beynon, J. D. E. and D. R. Lamb, *Charge-Coupled Devices and their Applications*, McGraw-Hill, New York, 1982.

Bower, F., "CCD Fundamentals," *Military Electronics/Countermeasures,* pp. 16–34, vol. 4, no. 2, February 1978.

Farrier, M. and R. Dyck, "A 128 × 1030 Element CCD Image Sensor for a Periscope Camera System," ISSCC, 1980.

Murphy, H. and J. Rothstein, "Fast Sorting with CCD," *PROGRESS, Fairchild Journal of Semiconductors,* pp. 14–19, vol. 5, no. 5, September/October 1977.

Sequin, C. and M. Tomsett, *Charge Transfer Devices,* Academic Press, New York, 1975.

Solomon, P., "Charge-Coupled Device (CCD) Trackers for High Accuracy Guidance Applications," *SPIE,* vol. 203, *Recent Advances in TV Sensors and Systems,* 1979.

Steffe, W., "CCD's in Industrial Camera Applications," ELECTRO, 1977.

Vicars–Harris, M., "Solid State CCD Cockpit Television System," Military Electronics Defense Expo, Rhein-Main Halle Wiesbaden, W. Germany, 1979.

Wen, D., "Advanced Charge-Coupled Device (CCD) Line Imaging Devices," *SPIE,* vol. 203, *Recent Advances in TV Sensors and Systems,* 1979.

CHAPTER 6

Semiconductor Memories

James D. Heiman, P.E.
Senior Electrical Engineer
IDR Incorporated

6.1 INTRODUCTION

Memory applications involve careful examination of various tradeoffs, especially cost and size evaluations. Increased chip size accompanied by reduced circuit cost have been the dominant trends in semiconductor memory development. Most new products reflect advances for the designer in terms of net system cost for a given memory size requirement. PC board space requirements may be reduced by replacing two or four small memory parts with one new larger part. Power supply design may be made simpler and cheaper by using "five volt only" parts instead of multiple voltage devices. Special microprocessor bus compatible memories may eliminate the need for additional support logic. In short, the economics of application is behind a great deal of the thrust of semiconductor manufacture, research, and development programs.

The designer is faced with a wide array of memory parts, when it comes to a particular application. The IC MASTER* lists fourteen categories of memory components. The actual memory component data listings and detailed product information is over five-hundred pages. Manufacturers' data books and application guides could easily fill an entire bookcase.

In this chapter, we shall discuss the major semiconductor memory devices that are currently being used. Dynamic and static RAM, ROM, PROM, and

* IC MASTER, United Technical Publications, Inc. Garden City, New York; published annually.

EAROM applications will be examined, with particular attention paid to device specifications, so that the user can pick the correct device for the job.

6.2 RANDOM ACCESS MEMORY (RAM)

RAM semiconductor memory components make up the largest segment of the memory IC market. The name RAM is actually a misnomer because RAM is an acronym for "random access memory." Mostly all memories are random access in the sense that any given address can be interrogated for a "read" or a "write" without the need to access the IC sequentially. The IC part that is called a RAM by manufacturers is actually a *read/write* memory. Data can be written into the IC and later read out from the IC.

RAM applications revolve around situations where data must be accessed and changed. Many computer applications involve calculations, data manipulation, or processing (see Chap. 8). Most programs work with new data or yield new results so that it is necessary to pull data from RAM memory, perform a calculation or manipulation, and then place the result into the same changeable RAM memory.

Most RAMs are *volatile*. Volatile means that data is lost or corrupted when power is removed. When power is supplied to a system with a volatile memory, the memory data are usually indeterminate. A typical procedure is to "zero-out" the RAM memory after power-up, and then RAM initialization can take place. Low-power RAMs, such as CMOS, have been used with battery back-up systems in order to form a *pseudo-nonvolatile* memory. This approach has been used successfully, but it is only nonvolatile for the life of the battery or until the RAM power voltage drops to the level where data becomes volatile again.

Magnetic core memories and the ROM (described later in this chapter) are examples of "nonvolatile" memories. System power can be shut off and data will be preserved. When power is reapplied, program execution can use the memory data without worrying about reloading programs or initialization routines.

Almost all logic families are represented when one examines the semiconductor RAM. TTL, ECL, I^2L, and many MOS-type processes are used to make RAMs of various speeds and sizes. Of the MOS type, CMOS, PMOS, NMOS, and VMOS are represented (see Chap. 4).

There are two basic types of RAM devices: *static* and *dynamic*. The static RAM uses a flip flop element to latch and store data. The dynamic RAM holds charge in individual storage cells in a manner that is similar to charging a bank of capacitors. The dynamic RAM must maintain this charge by dynamically "refreshing" the charge in each cell with periodic current pulses.

Selecting a RAM

Factors to consider in the selection of a RAM include:

1 Static RAMS are often used as a benchmark for system design since they are easy to use.

2 Dynamic RAMS are generally cheaper on a per-chip basis.

3 Dynamic RAM refresh logic adds a complication to the system design and timing.

4 Static RAMs are available in single-voltage, "five-volt only" configurations.

5 Dynamic RAMS may require three different dc power-supply voltages.

6 Overall power consumption for a dynamic RAM system may be much less than for a static RAM system.

7 The PC board layout for a dynamic RAM design is more critical than for a static RAM layout. Poor layout in a dynamic RAM design could contribute to data errors.

8 IC memories are broken down into "N" words of "M" bits wide. For example, a 16K static RAM with a byte-wide output is designated as a 2K × 8 RAM (1K = 1024 bits). Dynamic RAMs are usually organized in a "N × 1" architecture. A 16K dynamic RAM, for example, holds 16,384 data words of 1 bit each.

9 Static RAMs are becoming available in wider and wider architectures such as "N × 4" or "N × 8".

Selection guidelines for a RAM are summarized in Table 6.1.

6.3 STATIC RAM

Static RAMS are classified according to size and speed. Size is given as either the total cell count or the word organization of N words by M bits. Speed usually refers to the RAM access time. *Access time* is the generally accepted indication of "how fast" a particular chip is. It is the time required to access the RAM data cell after presenting a valid address. For example, you could read valid data, out of one particular (type 2114A) static RAM, 200 ns after a valid address. Therefore, this part has an access time of 200 ns.

The Mostek MK4802 static RAM, for example, is organized in a "2K × 8" structure which corresponds to 2,048 words of eight bits each, or a total of 16,384 bits. Mostek is aiming this part at the lucrative 8-bit microprocessor market so the speed is geared toward the access time requirements of most microprocessors. The access time specification is 90 ns.

Table 6.1 **Selection guidelines for RAMs**

Is read/write memory required?	Use RAM
Is fast speed required?	Use bipolar RAM
Is low power required?	Use CMOS RAM
What size memory?	Use "N" words of "M" bits
Is final assembly cost important?	Use single-chip memory part
Is memory size extremely large?	Evaluate dynamic RAM design
Is system power only +5 V?	Use single-voltage memory part
Is the system microprocessor controlled?	Use byte-wide (8 bits) memory part

The internal block diagram of the Mostek MK4802 RAM is shown in Fig. 6.1*a*. It is divided into seven sections. The heart of any RAM chip is the memory cell array which is organized in this part as 128 by 16 by 8. This means that there are 128 rows by 16 columns for 2048 words of 8 bits for a total of 16,384 bits, or individual cells. The address lines are buffered in order to obtain stable voltages on the silicon chip. The buffers also protect the chip from unsafe input levels or static discharges. The buffered address bits are used to drive two decoders that select the column and row of the desired data word. The "Y" or column decoder is fed four address lines at the input and generates a select pulse on one of 16 lines going to the column sense amplifiers or write drivers. The "X" or row decoder receives the other seven address lines and pulses one of the 128 row lines in the memory-cell matrix.

The remaining two sections are the I/O (input or output) buffer and the control logic. The I/O buffer is essentially an eight-bit transceiver that is used to output the eight bits of data from the memory-cell matrix via the sense amplifiers to the eight data lines or it is used to write into the memory cell matrix with the write drivers. Because the external chip interface is usually to a tri-state™* data bus, the output portion of the transceiver functions like a tri-state™ buffer. The control logic is used to activate and steer the other sections of the IC. The address buffer is enabled or disabled by the control section. The I/O buffer is steered to input or output mode by the control logic. Either the sense amps or the write drivers are selected by the control logic, depending upon whether the function to be performed is a "read" or a "write."

Pin Outs

The term "pin-outs" refers to the use and location of each pin of an IC package. The Mostek MK4802 is a 24 pin (Fig. 6.1*b*) IC package and is designed to be used with a single dc power supply voltage of 5 V. The power supply pin is labeled V_{CC} and ground is labeled V_{SS}. The address inputs are labeled A_0 thru A_{10} that is an eleven-bit address. Because the total word count is 2,048, it is possible to address all of the binary combinations of eleven bits. Note that the address lines and the word count are related by powers of two with 2,048 being equal to 2^{11}.

The data lines serve a dual function because data is "read" out from the memory cell array during a "read" operation or data is written into the array during a "write" operation. This two way path is called *bidirectional* or tri-state™ and it is usually connected to a system data bus. Because this IC contains words organized as eight bits each, there are eight data lines labeled DQ_0 thru DQ_7.

The remaining three pins are used for control purposes. The chip select or $\overline{CS}$ pin is used to enable internal logic in the memory IC when a logic 0 level is present. The output enable or "OE" pin is used to enable the tri-state™ output buffers on the data lines. When the $\overline{OE}$ pin is brought to a logic 0 level,

* Tri-state™ is a trademark of National Semiconductor.

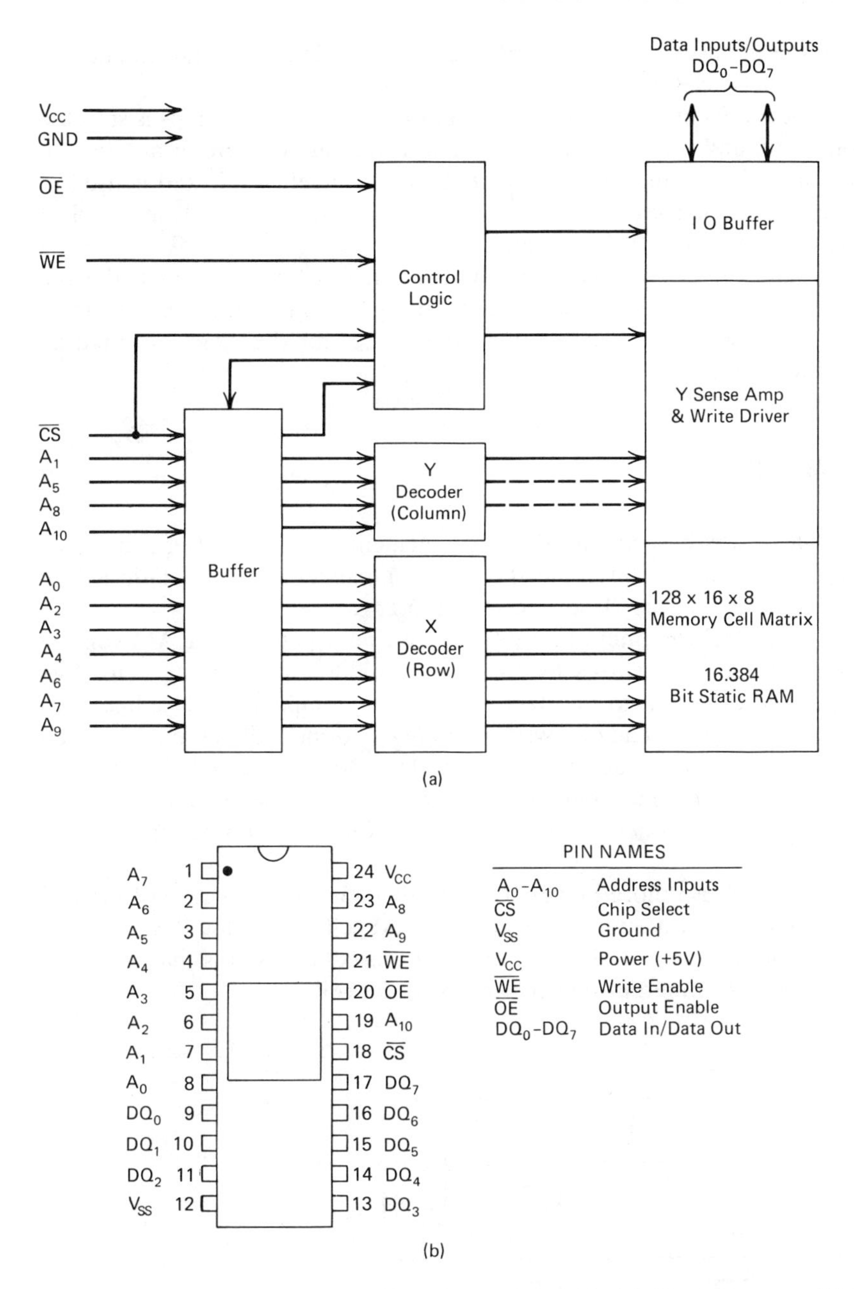

Figure 6.1 Example of organization of a static RAM. (*a*) Block diagram. (*b*) Pin connections (pin outs). (Copyright Mostek 1980. Specifications subject to change.)

the data in the memory cell array is put out onto the data bus. The $\overline{CS}$ and $\overline{OE}$ pins are especially useful in multiple chip memory configurations.

Logic decoders are used to ensure that only one group of chips is selected at any one time and only one chip is placed in the output state at any instant on a data bus. The write strobe or write enable is labeled $\overline{WE}$ and is used to signify whether the current operation is a read operation with $\overline{WE}$ at a logic 1 or a write operation with $\overline{WE}$ at a logic 0.

Most static RAM ICs will have similar pin functions, although the pin names and pin connection numbers may be markedly different. It is extremely important to consult the manufacturer's data sheet for the exact information needed for a design.

Static RAM Cell

A static RAM cell is basically a "flip flop" element with connections to a one row word line, a data bit line, and the data complement line. A static RAM cell, using MOS devices, is illustrated in Fig. 6.2.

The static RAM is operated by first selecting a particular RAM cell via the word select line. The word selection "turns on" the two coupling transistors and this ties the cell latch to the data line and the complement data line. Depending upon whether a read or a write is being performed, data is either placed on the data bit lines or is read into the latch. The four cross-coupled transistors form the cell latch. The input data, during a write, forces the latch into a "zero/one" or "one/zero" state. After the word select signal is removed, this state is maintained.

Static RAMS use more power than the dynamic type because of the cell latch loads that are not used in the dynamic RAM cell. The static RAM buffers also use considerable power. The net result of this comparison is that the static RAM design has an inherent power penalty at the cell level.

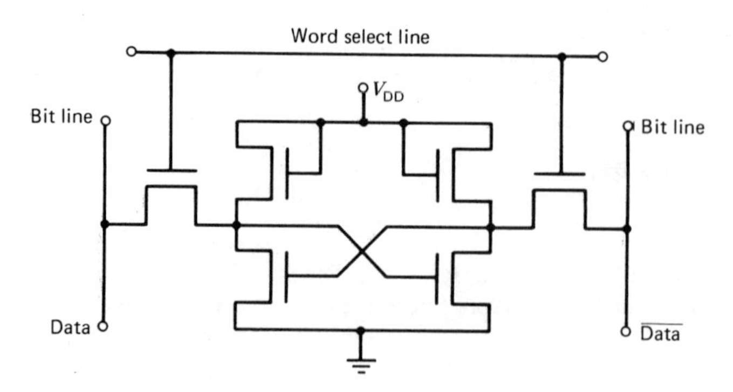

Figure 6.2 The static RAM cell is basically a flip flop. (Courtesy National Semiconductor Corp.)

EXAMPLE 6.1 Design an interface for a static RAM to be used with a microprocessor.

PROCEDURE

1 In this application, a static RAM memory is used with an eight-bit microprocessor for a program read/write memory. For purposes of illustration, a 6800 processor (see Chap.8) is used with a 2114A static RAM memory. The 2114A static RAM is "1K × 4" or 4K while the MK4802 is "2K × 8" or 16K. We want to examine the details of this interface with particular regard for the additional support circuitry needed, the timing required, and the analysis for selecting the appropriate 2114A part.
2 A schematic of the static RAM interface is shown in Fig. 6.3. The 6800 processor address bus is sent to the two 2114A static RAM chips and to the decoding logic (LS138) that is required. The bidirectional data bus links the processor and the static RAM I/O pins. The static RAM write enable pins are tied to the processor "read (high)/write (low)" line. The RAM chip select is generated from a 74LS138 decoder IC that sends a logic "low" level to the RAMs whenever the RAM address block is selected. Because the 2114A static RAM is a 1K × 4 device, we need two of them to form an eight-bit data word.
3 Because we are using address bits AD15, AD14, AD13, and AD12, our outputs are a function of the 32K words covered by this address range. This is divided into eight blocks of 4K words for simplicity. Note that we have allocated an address block of 4K words using the decoder IC, so we are "wasting" 3K out of this 4K block. This is wasted space in the sense that we only need a 1K block of memory, but we allocated a 4K block. We cannot use this extra 3K space for something else.

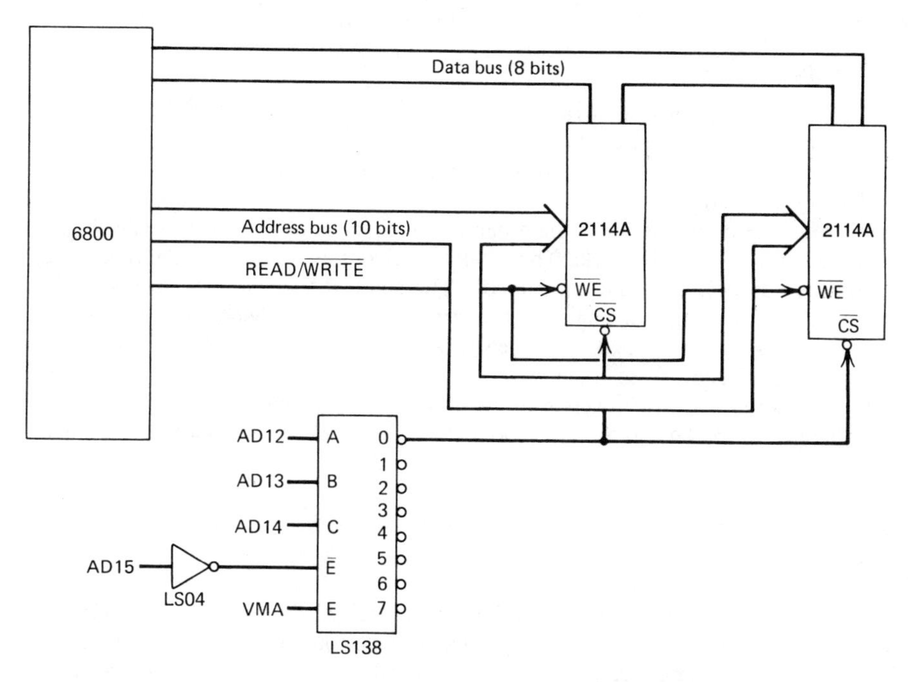

Figure 6.3 Example of a static RAM/6800 microprocessor interface.

4 The start of our address block is at address zero or 0000_{16} (hexadecimal). The additional support circuitry for the 6800 is not shown, but is assumed to be part of the overall design. This circuit can be used to store microprocessor data in memory and this data can also be read back by the microprocessor.

EXAMPLE 6.2 Describe the operation of a static RAM during a microprocessor "write" cycle.

PROCEDURE

1 A "write" cycle starts when the microprocessor program encounters the instruction that signifies that internal microprocessor data is to be stored in the address range of the static RAM. The microprocessor outputs the full static RAM address on the 16 address lines. Some of the upper address bits are used for the decoder function that selects the static RAM ICs. The lower 10 address bits are sent to the static RAMs in order to select the cell in the RAMs where data is to be stored.
2 The microprocessor also puts the read/write line in the write or "low" state. The tri-state™ data bus is placed in the write state with the microprocessor in an output mode and the rest of the logic is in the receive or input mode. The eight microprocessor data lines are set to levels that correspond to the internal microprocessor data word. This state is held for a certain time period to enable the static RAM chips to get setup for the end of the write cycle.
3 Data is stored as the read/write line returns "high" to the read state. Data must be held stable as the write pulse changes state. At this point, the new data is stored in the static RAM cell that corresponds to the address selected.

EXAMPLE 6.3 Describe the operation of a static RAM during a "read" cycle.

PROCEDURE

1 A "read" cycle starts like a "write" cycle, with the microprocessor placing the full address word on the 16 address lines. The high-order bits are used to select the static RAMs via the decoder. The 10 low-order bits are used to locate the particular RAM cell that will be read.
2 The microprocessor instruction in the program causes the read/write line to stay "high" in the read state during the cycle. The tri-state™ data bus is placed in the read state with the microprocessor in an input mode. The static RAM selection causes RAM data to be placed on the data bus after a certain time period delay. The rest of the data bus logic is in a high impedance state so it appears as if the static RAM is "talking" directly to the microprocessor.
3 Data is stored in the microprocessor at the end of the cycle when the microprocessor clock changes state. The static RAM holds the data stable as the address lines just start to change state.

EXAMPLE 6.4 Calculate the static RAM access time in a "read" cycle.

PROCEDURE

1 Because of some other system constraints, we are given that the processor clock is nominally 1.5 MHz, or a period of $t_{CLK} = 667$ ns. We could use the 68A00 processor. By referring to the timing diagrams of Figs. 6.4 and 6.5, we select a 2114A part to fit our application. We will treat a "read cycle" and "write cycle" separately. We will also ignore rise and fall time in order to simplify the analysis.

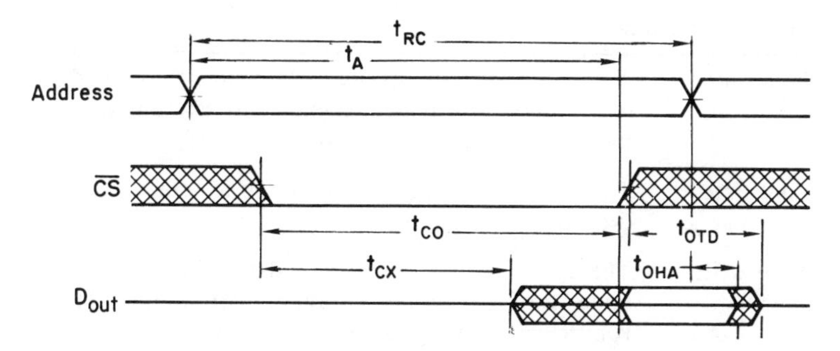

Figure 6.4 Static RAM read cycle timing. NOTE: $\overline{WE}$ is high for a read cycle. (Courtesy of and copyrighted by Synertek, Inc.)

2 In the "read cycle" case of Fig. 6.4, the processor is running a program and it reaches an instruction where it must read data from the "zero page" memory block. As the processor executes this cycle, a valid memory address (VMA) is presented on the address bus after a delay, t_{AD} (t_{AD} is the delay time measured from the start of a clock cycle to the beginning of a valid memory address; it is a specified value of the 68A00 processor), of possibly 180 ns. The decoder delay, t_{DD}, is a maximum of 41 ns. The 2114A RAMs will output valid data after the access time, t_A. This data will remain valid until either the address lines or the chip select change. The processor has a *data set-up time requirement,* t_{DSR}, of 60 ns, minimum. This means that data must be valid for at least the last 60 ns of the cycle when the address lines change.

3 We can calculate the required maximum access time for the 2114A as follows:

$$\begin{aligned} t_A &= t_{CLK} - t_{AD} - t_{DSR} - t_{DD} \qquad (6.1) \\ &= 667 - 180 - 60 - 41 \\ &= 386 \text{ ns maximum} \end{aligned}$$

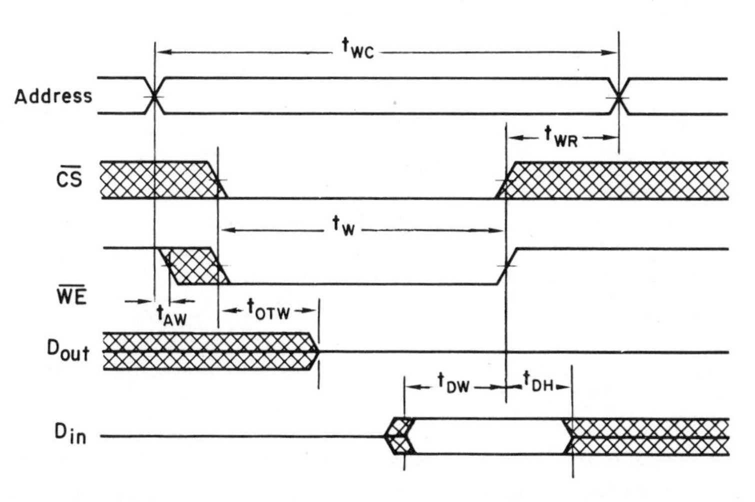

Figure 6.5 Static RAM write cycle timing. (Courtesy of and copyrighted by Synertek, Inc.)

Table 6.2 **Definitions of some time intervals**

t_H = processor hold time	The processor requires that data being sent to it, during a write, must remain stable 10 ns after the end of the clock cycle.
t_{OHA} = 2114A output time from address change	The 2114A keeps data stable for 50 ns after the address lines change.
t_{DW} = 2114A data-to-write overlap time	The 2114A requires that stable data values be maintained during the later portion of a write pulse. This is necessary so that the internal data paths in the chip are settled as the trailing edge of the write pulse ends the write cycle.

4 The processor hold time, t_H, of 10 ns is safely met by the RAM output hold time from the address change, t_{OHA}, of 50 ns. (See Table 6.2.)

EXAMPLE 6.5 Calculate the static RAM "data-to-write time overlap" for the "write" cycle.

PROCEDURE

1 In the "write cycle" of Fig. 6.5, the processor writes data to the RAM address block. The processor executes this cycle with a VMA delay of 180 ns. The processor outputs valid data after the data bus enable signal goes true. This data delay time is a t_{DDW} of 200 ns. Data remains valid with a hold time, t_H, of 10 ns after the cycle ends. The 2114A chip receives a valid address, a valid chip select, and a low signal on the write enable line during the beginning of the cycle. The 2114A RAM requires that data must be valid, with the write line low, for a minimum time at the end of the period where the write line changes.

2 The date-to-write overlap, t_{DW}, varies with each type of 2114A. The duration of valid processor data must match, or be greater than, the t_{DW} we choose for the 2114A:

$$t_{DW} = \frac{1}{2} t_{CLK} - t_{DDW} \tag{6.2}$$

$$= 333.5 - 200$$

$$= 133.5 \text{ ns maximum}$$

The 2114A hold time, t_{DH}, is zero so it does not affect our choice.

3 Examining our choices of 2114A chips we must use a 2114AL-2 part (for example, a Synertek 2114AL-2). A 2114AL-2 has a t_{DW} of 70 ns minimum and an access time, t_A, of 120 ns maximum. A data page for the part is shown in Fig. 6.6.

EXAMPLE 6.6 Design a CMOS static RAM to be used as a pseudo-nonvolatile memory.

PROCEDURE

1 In this application, a CMOS RAM memory is made to operate in a nonvolatile manner by adding a battery backup circuit, as shown in Fig. 6.7. A rechargeable nickel-cadium (Ni-CAD) battery is trickle charged when the main logic power supply is on. When

ABSOLUTE MAXIMUM RATINGS

Temperature Under Bias	$-10^{\circ}C$ to $80^{\circ}C$
Storage Temperature	$-65^{\circ}C$ to $150^{\circ}C$
Voltage on Any Pin with Respect to Ground	$-3.5V$ to $+7V$
Power Dissipation	1.0W

COMMENT

Stresses above those listed under "Absolute Maximum Ratings" may cause permanent damage to the device. This is a stress rating only and functional operation of the device at these or any other conditions above those indicated in the operational sections of this specification is not implied.

D.C. CHARACTERISTICS $T_A = 0^{\circ}C$ to $+70^{\circ}C$, $V_{CC} = 5V \pm 10\%$ (Unless otherwise Specified)

Symbol	Parameter	2114AL-1/ L-2/L-3/L-4		2114A-4 2114A-5		Unit	Conditions
		Min	Max	Min	Max		
I_{LI}	Input Load Current (All input pins)		10		10	μA	$V_{IN} = 0$ to $5.5V$
I_{LO}	I/O Leakage Current		10		10	μA	$\overline{CS} = 2.0V$, $V_{I/O} = 0.4V$ to V_{CC}
I_{CC1}	Power Supply Current		35		65	mA	$V_{CC} = 5.5V$, $I_{I/O} = 0$ mA, $T_A = 25^{\circ}C$
I_{CC2}	Power Supply Current		40		70	mA	$V_{CC} = 5.5V$, $I_{I/O} = 0$ mA $T_A = 0^{\circ}C$
V_{IL}	Input Low Voltage	-0.5	0.8	-0.5	0.8	V	
V_{IH}	Input High Voltage	2.0	6.0	2.0	6.0	V	
V_{OL}	Output Low Voltage		0.4		0.4	V	$I_{OL} = 3.2$ mA
V_{OH}	Output High Voltage	2.4	V_{CC}	2.4	V_{CC}	V	$I_{OH} = -1.0$ mA

CAPACITANCE $T_A = 25^{\circ}C$, f = 1.0 MHz

Symbol	Test	Typ	Max	Units
$C_{I/O}$	Input/Output Capacitance		5	pF
C_{IN}	Input Capacitance		5	pF

NOTE: This parameter is periodically sampled and not 100% tested.

A.C. CHARACTERISTICS $T_A = 0^{\circ}C$ to $70^{\circ}C$, $V_{CC} = 5V \pm 10\%$ (Unless otherwise Specified)

Symbol	Parameter	2114AL-1		2114AL-2		2114AL-3		2114A-4/L-4		2114A-5		Unit
		Min	Max	Min	Max	Min	Max	Min	Max	Min	Max	
Read Cycle												
t_{RC}	Read Cycle Time	100		120		150		200		250		nsec
t_A	Access Time		100		120		150		200		250	nsec
t_{CO}	Chip Select to Output Valid		70		70		70		70		85	nsec
t_{CX}	Chip Select to Output Enabled	10		10		10		10		10		nsec
t_{OTD}	Chip Deselect to Output Off		30		35		40		50		60	nsec
t_{OHA}	Output Hold From Address Change	15		15		15		15		15		nsec
Write Cycle												
t_{WC}	Write Cycle Time	100		120		150		200		250		nsec
t_{AW}	Address to Write Setup Time	0		0		0		0		0		nsec
t_W	Write Pulse Width	75		75		90		120		135		nsec
t_{WR}	Write Release Time	0		0		0		0		0		nsec
t_{OTW}	Write to Output Off		30		35		40		50		60	nsec
t_{DW}	Data to Write Overlap	70		70		90		120		135		nsec
t_{DH}	Data Hold	0		0		0		0		0		nsec

A.C. Test Conditions

Input Pulse Levels . . . . 0.8V to 2.0V
Input Rise and Fall Time . . . . 10 n sec
Timing Measurement Levels: Input 1.5V
Output . . . . 0.8 and 2.0V
Output Load . . . . 1TTL Gate and 100pf

Figure 6.6 A data page from the 2114A family of RAMs. (Courtesy of and copyrighted by Synertek, Inc.)

power fails or the system is shut off, the Ni-CAD battery maintains an acceptable CMOS V_{CC} supply level until the main power can be restored.

2 The battery charger circuit is used to keep the battery in a fully charged state. It is also used to generate a "power failure" indication if the main logic supply falls below a certain level. The "power failure" signal is used to inhibit spurious write pulses from occurring whenever the main logic supply is at a level that causes invalid circuit operation.

3 Circuit operation depends upon the "power fail" signal causing logic "high" levels at the output of the CMOS "OR" gate. This inhibits the enable line and write strobe inputs into the CMOS static RAM chips. Data held in the RAMs are protected and can later be read when normal circuit operation is resumed. Care must be taken to ensure that the address pin inputs and data lines do not assume an undefined logic level, or else the normally low supply current for the CMOS RAMs will rise drastically.

4 Typical CMOS RAMs will safely hold data without loss as long as the CMOS V_{CC} pin is 2 V or higher. Power consumption and supply current depend upon the type and amount of RAM used. If a small printed circuit mounted Ni-CAD of 65 mA-hour capacity (for example, the General Electric Data Sentry) is used, it is possible to backup a CMOS RAM array using 100 μA for many days without main logic power. When power is returned, the Ni-CAD battery will be recharged to its normal voltage. If power is not returned in time, the battery will be discharged below the required 2 V level and data will be lost.

5 The CMOS RAM control circuit is both a battery charging circuit and a "power fail" inhibit circuit. The transistors and zener diode are used to form a "power fail" detector. When the system power rises above the zener voltage of 3.6 V, the zener conducts and turns on transistor Q_1. Transistor Q_1 then turns on transistor Q_2 and this makes the "power fail" signal low. If the system power supply causes the voltage to fall below the zener voltage, the transistors turn off and the battery will pull-up the "power fail" line. The "power fail" line inhibits further CMOS RAM operation.

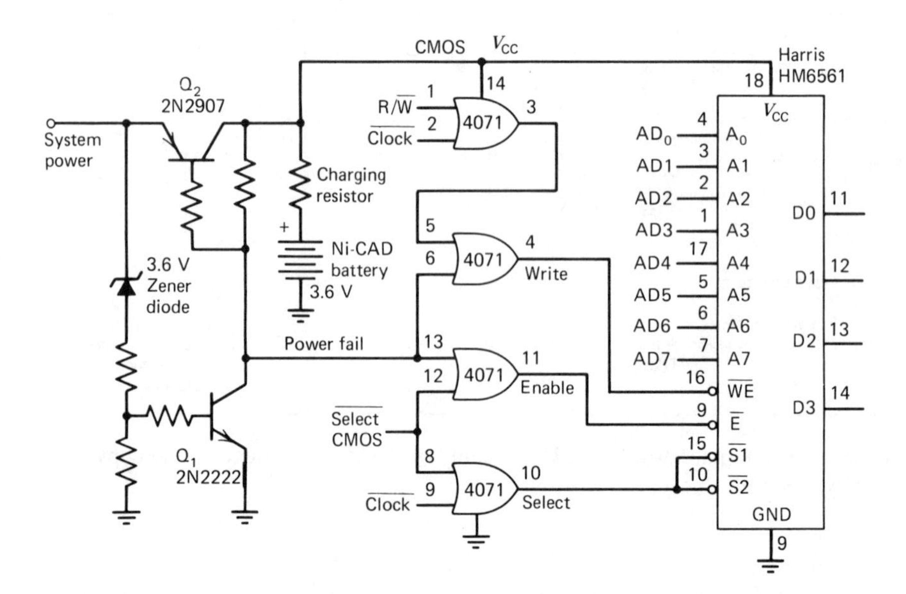

Figure 6.7 A CMOS RAM with a battery back-up circuit.

The battery is charged whenever the system power is "on" and the battery voltage is lower than the CMOS V_{CC} voltage. The charging resistor is used to limit the current flow into the battery when the battery is in a discharged state. Usually the manufacturer specifies the maximum charging rate for a Ni-CAD battery. Sometimes one-tenth of the total capacity is used as the specified rate. For example, a 100 mA-hour rated battery requires an appropriate resistor to limit the maximum possible current to 10 mA.

6.4 DYNAMIC RAM

Dynamic RAMs have critical timing requirements with select strobes being applied during distinct time phases. There are many different control states during a dynamic RAM operation cycle, whereas static RAMs are usually placed in one particular state during a whole operational cycle.

Modern dynamic RAMs employ a multiplexed address scheme in order to use the smallest possible package size for the best memory system bit-density. Package size decreased when manufacturers moved to 16K dynamic RAM designs from 4K designs. Static RAMs use a straightforward address technique where each internal address line is brought out to an individual pin.

Dynamic RAMs require a "refresh" cycle periodically or else data is lost. Static RAMs hold data indefinitely without further access or refresh requirements.

The 4116 dynamic RAM (Fig. 6.8*a*) is currently a very popular MOS memory and it is organized as 16,384 data words of one bit each. The 4116 has a 16-pin package (Fig. 6.8*b*). Pin count is minimized because the address pins are multiplexed between the row address bits and the column address bits. Only seven address lines are required to form a full 14-bit address. (*Note:* 2^{14} is 16,384.) This particular dynamic RAM uses multiple dc power supplies so pins are provided for +12 V, +5 V, −5 V, and the power supply ground. The data output and data input pins are separated for flexibility in application. The remaining three pins are used for control purposes. The "write enable" pin signals whether the current cycle is a "read" or a "write." The $\overline{CAS}$ pin is used to strobe the seven column address bits into the chip. The $\overline{RAS}$ pin is used to strobe the address pins for the seven row address bits.

A dynamic RAM memory IC has more internal subsections than a static RAM part, yet the typical dynamic RAM structure is geared toward a one-bit data word. The internal block diagram of the 4116 dynamic RAM in Fig. 6.8*a* shows the seven row address buffers and the seven column address buffers. The seven row bits are decoded into 128 row decodes, while the seven column bits are decoded into 128 column decodes. The row address decoder is used to drive the 128 sense-refresh amplifiers.

The I/O data is buffered at the memory cell array input and output. Input data is latched in a one-bit register before being stored in the memory cell. Output data from the memory cell is latched before being outputted from the IC.

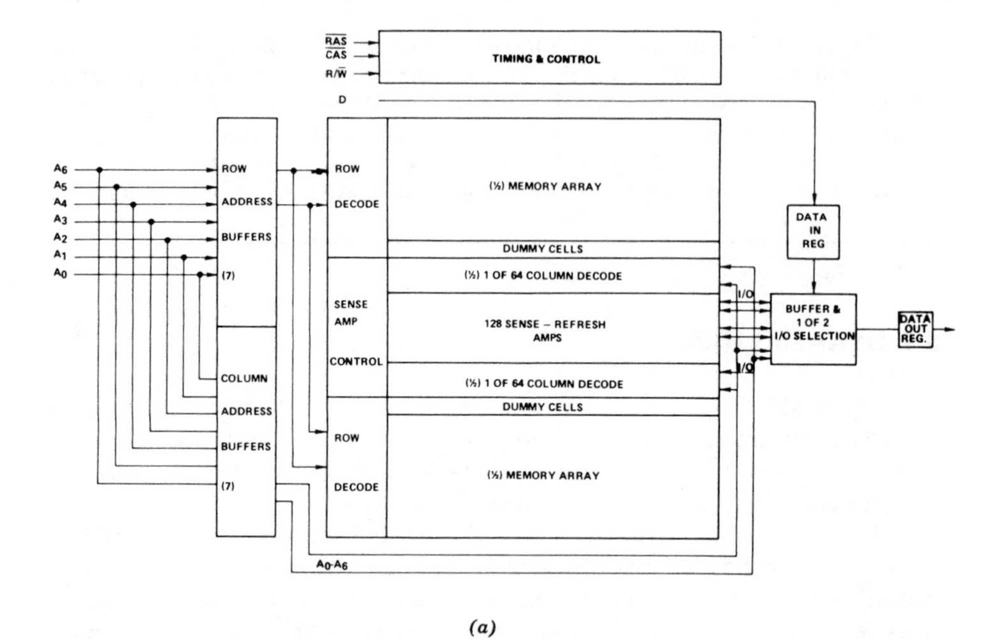

(a)

16-PIN CERAMIC AND PLASTIC
DUAL-IN-LINE PACKAGES
(TOP VIEW)

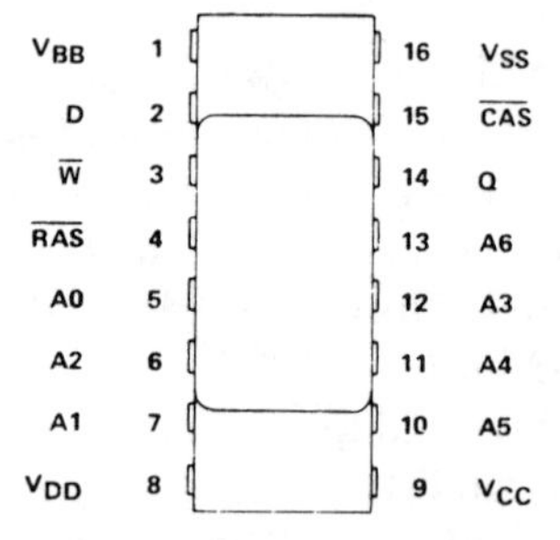

PIN NOMENCLATURE			
A0-A6	Address Inputs	$\overline{W}$	Write Enable
$\overline{CAS}$	Column address strobe	V_{BB}	−5-V power supply
D	Data input	V_{CC}	+5-V power supply
Q	Data output	V_{DD}	+12-V power supply
$\overline{RAS}$	Row address strobe	V_{SS}	0 V ground

(b)

Figure 6.8 Example of organization of a dynamic RAM (TI4116). (*a*) Block diagram. (*b*) Pin connections. (Courtesy Texas Instruments, Inc.)

The timing and control section determines whether a read or write operation is in progress by the state of the R/$\overline{\text{W}}$ pin. Pins $\overline{\text{RAS}}$ and $\overline{\text{CAS}}$ are used to control address decoding and refresh.

Refreshing Memory Cells

A dynamic RAM requires periodic current pulses that must reach all cell locations before the stored cell charge dissipates. Usually, refresh time is specified in terms of the maximum interval that can elapse before each chip row is replenished. The 4116 part requires that a full "refresh" sequence must be performed every 2 ms. All 128 row addresses must be strobed by pulses on the $\overline{\text{RAS}}$ line. Some of the 64K dynamic RAM parts have improved sense amplifier designs which allow the refresh cycle to be stretched to 4 ms.

Dynamic RAM refresh requires additional system logic in order to perform the refresh function. A counter is needed to generate the refresh row addresses. A multiplexer (MUX) may be required in order to select between computer use of memory and a refresh cycle. Some form of timing logic is required to ensure that the refresh sequence is repeated within the required cycle time. This system penalty is offset by the advantages of using a dynamic RAM in a large system, because only one control section is required for many RAM chips. Many manufacturers offer a single IC that may be used to implement a dynamic RAM controller. This approach can be cost effective even for dynamic RAM memory systems as small as 16K.

Dynamic RAM Cell

The dynamic RAM cell is much simpler than the static type. The basic cell element is formed in the integrated circuit from a capacitor and transistor, as illustrated in Fig. 6.9. The capacitor holds the cell charge that reflects whether a "one" or a "zero" is stored in the cell.

The transistor operates as a switch to select a particular capacitor and then gate the cell charge on or off. Because of leakage current, it is necessary to "refresh" or replenish the charge on the capacitor before the cell charge is dissipated. Row and column decoders are used to select a particular cell in the cell matrix while sense amplifiers are used to store and recover data from the cell.

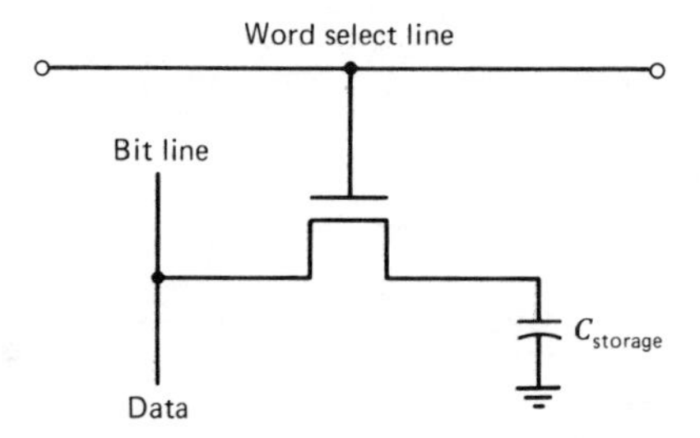

Figure 6.9 The basic dynamic RAM cell. (Courtesy National Semiconductor Corp.)

EXAMPLE 6.7 Describe the timing sequence for a dynamic RAM read operation.

PROCEDURE

1. The 4116 dynamic RAM is a 16K MOS chip. The read cycle timing is shown in Fig. 6.10. Essentially there are three steps that take place during a "read."
2. First, the seven row address bits are presented to the RAM and strobed into the chip on the falling edge of the RAS signal.
3. Second, the seven column address bits are strobed on the trailing edge of the $\overline{\text{CAS}}$ signal. During this time, the write line must stay in the inactive or high state.
4. Finally, as the $\overline{\text{CAS}}$ signal returns high, output data is valid and available to be read by the computer.

EXAMPLE 6.8 Describe the timing sequence for a dynamic RAM write operation.

PROCEDURE

1. The 4116 dynamic RAM write cycle is shown in Fig. 6.11. During the write operation, the $\overline{\text{RAS}}$ signal is used to strobe the row addresses and $\overline{\text{CAS}}$ is used to strobe the column addresses as in a read operation.

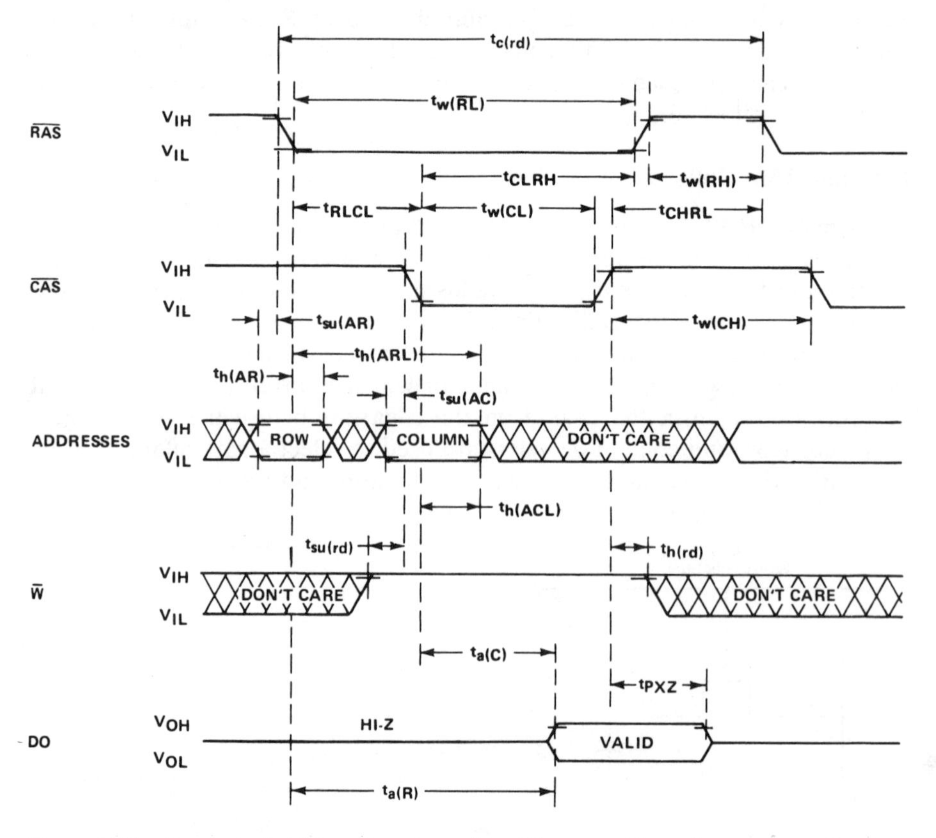

Figure 6.10 Dynamic RAM read cycle timing. (Courtesy Texas Instruments, Inc.)

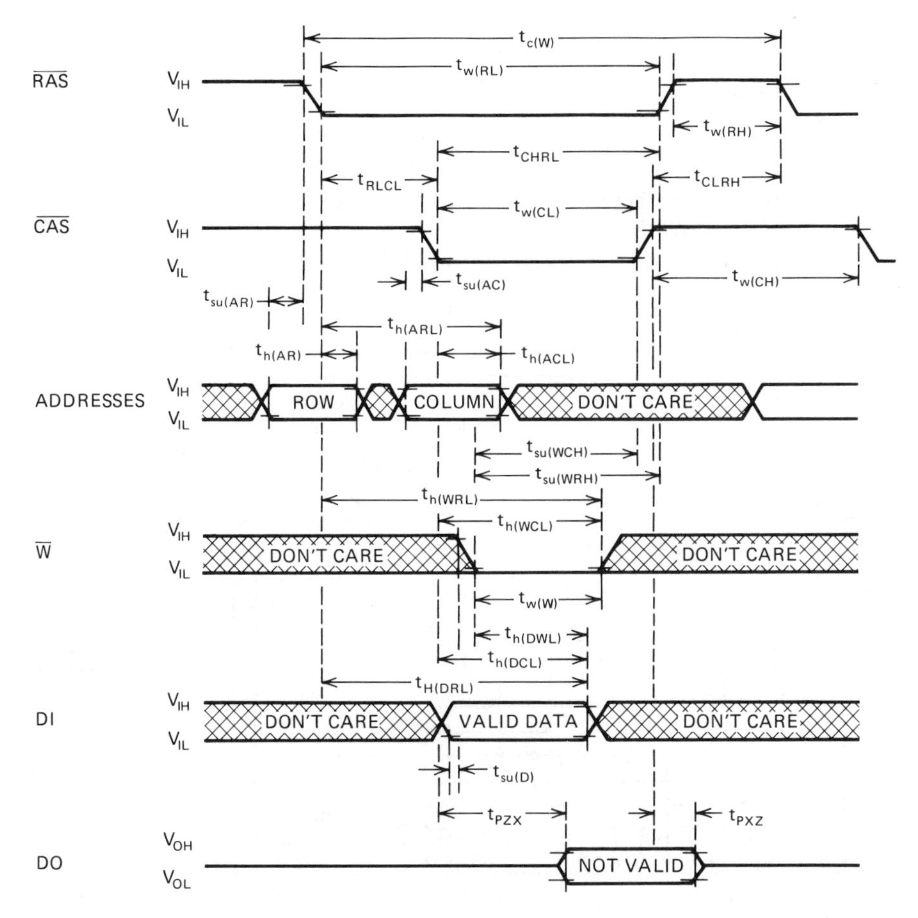

Figure 6.11 Dynamic RAM write cycle timing. (Courtesy Texas Instruments, Inc.)

2 The control logic in the IC recognizes that a read operation is not taking place because the write line is low immediately after the $\overline{CAS}$ signal goes low. The input data is also presented at this time.

3 Data must be valid and stable while the write line is held low for at least the minimum specified time period. Data is stored and the write line then returns high.

EXAMPLE 6.9 Design a dynamic RAM memory for a microprocessor-based system.

PROCEDURE

1 A block diagram of a possible design is provided in Fig. 6.12. The memory subsystem contains eight 4116 dynamic RAM ICs and a dynamic RAM address multiplexer. The microprocessor subsystem contains the microprocessor, the system clock generator, and the data transceiver. These elements form a microprocessor memory of 16K bytes. Additional microprocessor control circuitry may be required in particular system applications for address decode, reset, DMA and interrupts. A single ROM chip could hold the program memory (for example, 2K bytes).

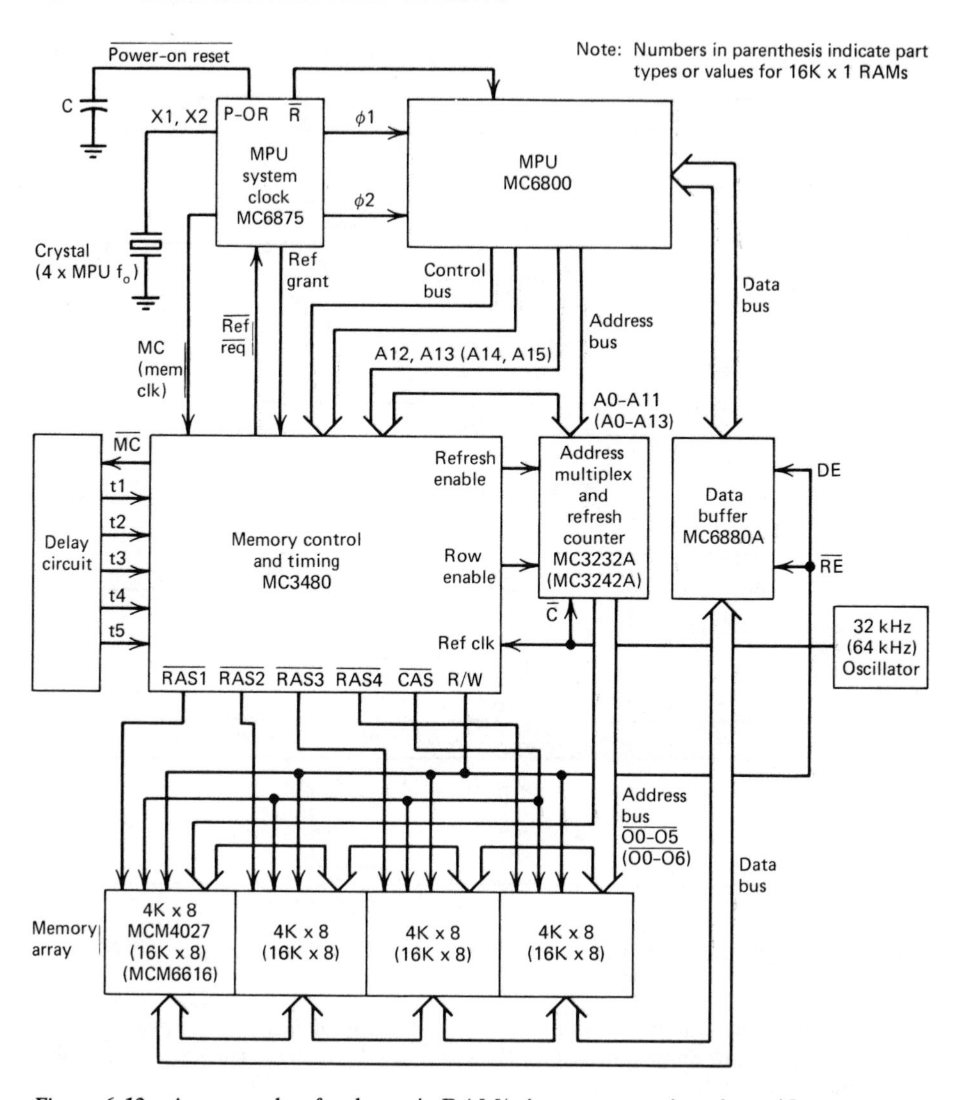

Figure 6.12 An example of a dynamic RAM/microprocessor interface. (Courtesy Motorola, Inc.)

If we assume that the microprocessor is executing a stored ROM program that uses a dynamic RAM for the read/write memory, it will be easy to follow what is happening at each system element.

2 The MPU is running the stored program by using the clocks being generated by the system oscillator. The MPU can access the dynamic RAM memory by presenting an address in the lower portion of available memory address space. For example, if the microprocessor had 16 address lines it could address 64K of total memory. We have arbitrarily chosen to place the dynamic RAM at the bottom of this memory space.

Our 16K dynamic RAM would occupy the addresses from 0000_{16} to $3FFF_{16}$ in hexadecimal notation. When a valid dynamic RAM address is presented, the address MUX sends to the RAMs the lower seven address lines as the row address and then sends the next seven address lines as the column address.

3 The clock generator is designed to ensure that timing specifications for the dynamic RAM and the microprocessor are both met. The $\overline{\text{RAS}}$ and $\overline{\text{CAS}}$ lines are used to strobe the row and column address. Pulse widths are designed to use a particular oscillator frequency in order to operate with dynamic RAMs of a certain speed, since many options are available. Refresh is scheduled by a timer in the clock generator circuit. When refresh is due, the clock generator raises the refresh enable line high, while "freezing" the processor clocks in a fixed state. The dynamic RAM MUX has a counter to generate the 128 refresh addresses from a refresh clock input. The dynamic RAMs get refreshed while the CPU is frozen and after refresh, normal operation continues.

EXAMPLE 6.10 Design a dynamic RAM CRT display memory.

PROCEDURE

1 A fairly large, five-page display memory can be formed by using 16K dynamic RAM chips. The latest CRT terminal designs are using display pages of 25 rows of 80 characters. If we use a 4116 dynamic RAM (Fig. 6.8) for each of the seven data bits we require, the 128 rows by 128 columns structure of the device allows us to form a display memory of 128 rows. We are wasting some of the display memory because we only use 80 characters out of the possible 128 column positions available. This allows us to tradeoff the wasted memory for simplicity in the design.

2 The schematic of the memory-CRT controller interface is shown in simplified form in Fig. 6.13. (The computer interface to the memory and the memory refresh logic are not shown.) It is assumed that the computer will write to the memory in a "burst mode" during the vertical blanking interval of the display (for example, the time period when we see the familar black bar on a home television set when the picture rolls). It will, therefore, appear to the eye that the display data has changed instantaneously when the computer writes new data.

3 Refresh is required at least every 2 ms because we are using dynamic RAMs. If we update or refresh five consecutive row addresses during each horizontal blanking interval (for example, the left and right borders of the display), we will refresh the whole memory after every 26 scan lines. The refresh cycle repeats after every 1.65 ms, which satisfies our requirement.

4 One of the modes of operation available with the 4116 dynamic RAM is called a "page-read" mode. This mode is ideally suited to our application because we only want to read one row in our RAM for each scan of our display row. Using the horizontal blanking signal for the $\overline{\text{RAS}}$ strobe, we load the seven bit row address as we leave the horizontal blanking interval. For each character in the row, we load, via the $\overline{\text{CAS}}$ strobe, the seven bit column address of the RAM data word. The 4116 RAM "remembers" internally the current row address, so we do not have to update this value until we go to the next display row. (Refer to the timing diagrams in Fig. 6.14.) This approach simplifies the logic requirements.

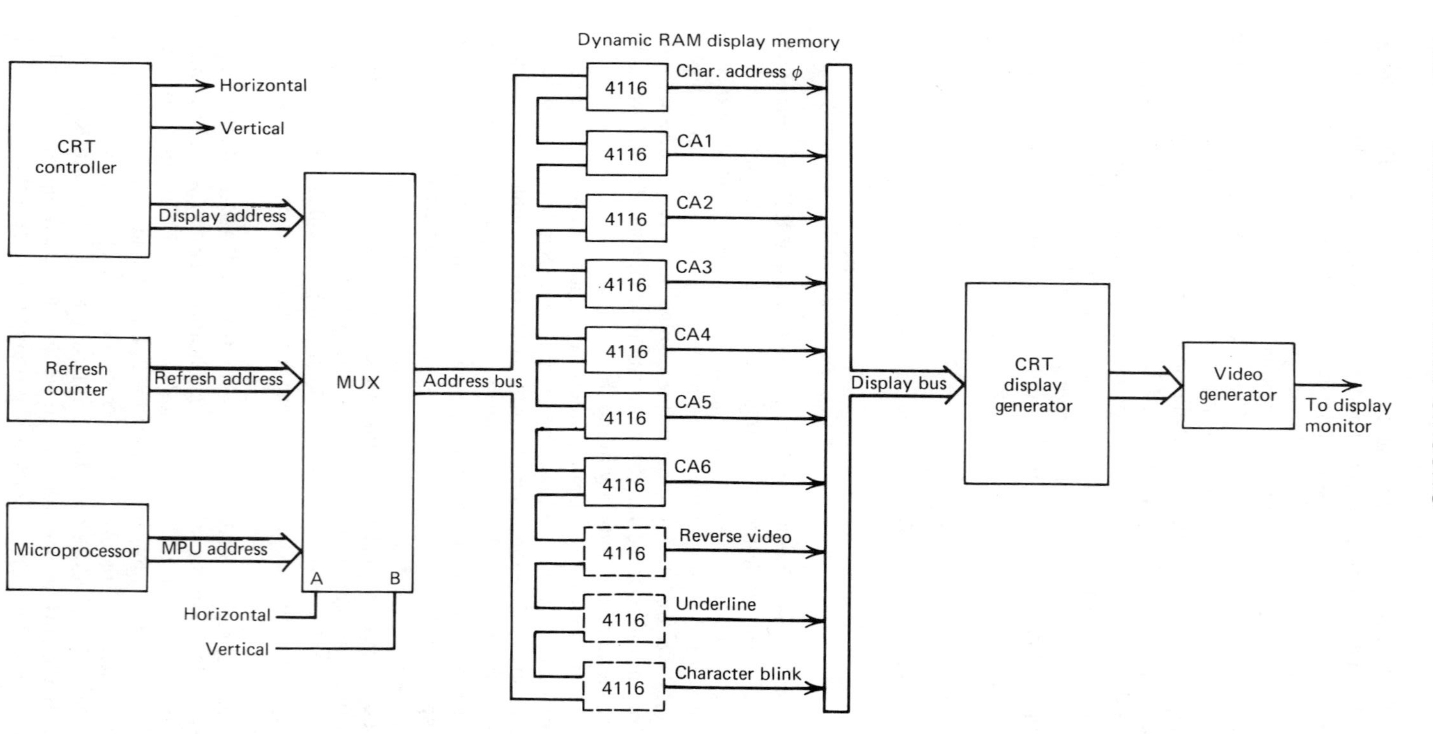

Figure 6.13 Dynamic memory CRT display.

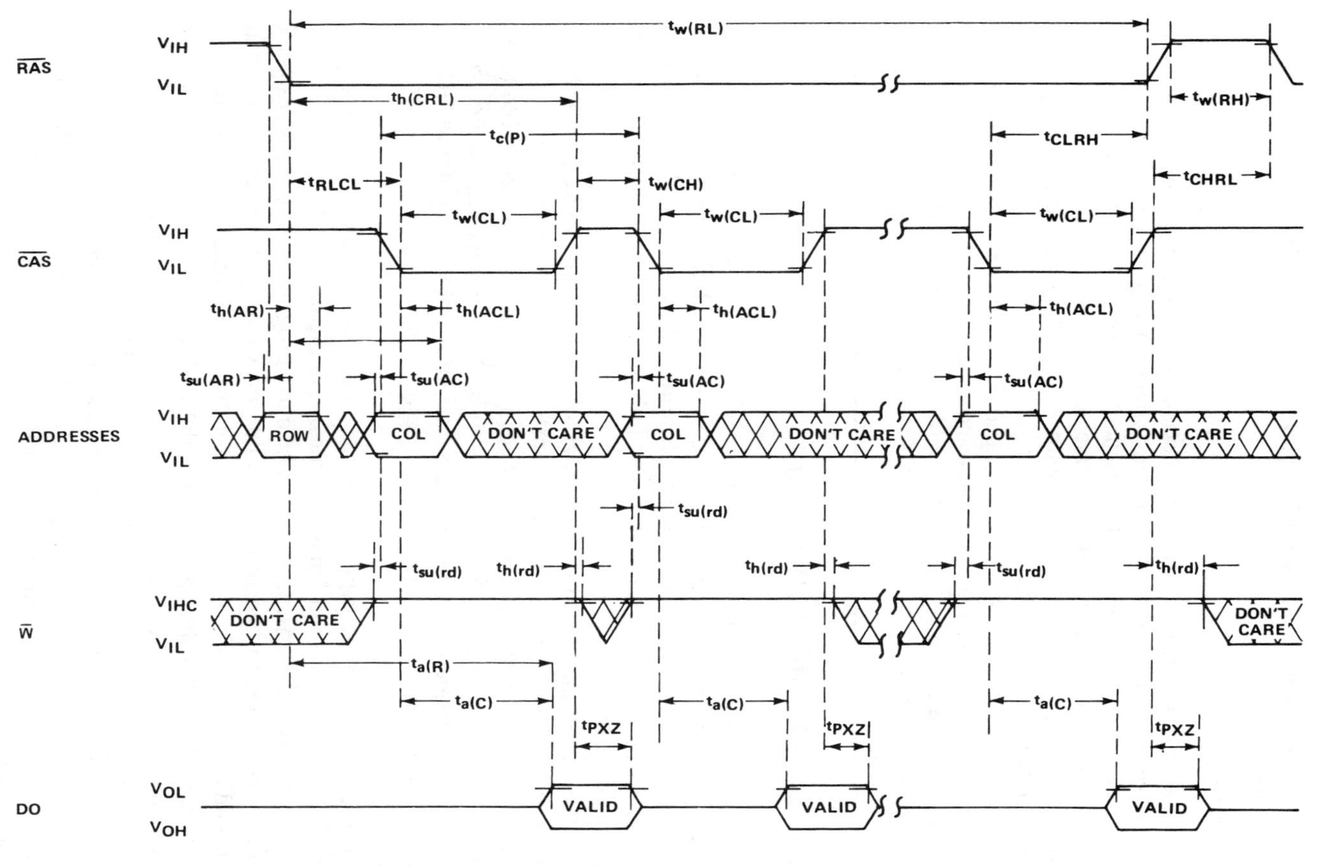

Figure 6.14 Dynamic RAM page-read cycle timing. (Courtesy Texas Instruments, Inc.)

EXAMPLE 6.11 Describe how the CRT display of Ex. 6.10 can be enhanced by adding more memory.

PROCEDURE

1 Since we are using a memory part that is organized as 16K words of one bit each, we can enhance our display features by adding a single 4116 RAM for each new display attribute we desire. Many new CRT controller ICs have pins for character underline, reverse video, and character blinking. Three additional 4116 RAMs are used to add these features to our design. We also must expand the interface to the computer controlling the display memory.
2 If we add these three 4116 RAMs, as indicated by dashed lines in the bottom of Fig. 6.13, we have an enhanced product. We could make one design and merely remove the three chips to return to a straightforward alphanumeric "dumb terminal." The "16K by one" organization of the 4116 RAM allows us to create two different products from one design.

6.5 READ ONLY MEMORY (ROM)

The term ROM is an acronym for "read-only memory." A ROM is used where data must remain fixed or constant. There is no write line because new data is not written into the IC in the intended application. There is no clock line because a ROM is a static device. Stored data can only be read from a ROM after it is programmed or masked.

ROMs are employed in a wide range of applications, some of which are not necessarily for computer-oriented systems. In microprocessor systems, ROMs are used for program storage, address decoders, program patches, table look-ups, and security passwords. ROMs can also be used as logic sequencers, computer microcode sequencers and decoders, and interrupt vector pointers. ROMs are utilized in many character generator designs and waveform generator applications.

There are two basic types of ROMs: mask programmed and user programmable. The designation "mask" comes from the name of the photographic artwork that is used to fabricate the IC. The mask programmed ROM is designed at the factory to correspond to a code table provided by the user. The actual table is laid out in the IC by a transistor pattern. The transistor pattern reflects the "one/zero" data which is contained in the customer's ROM table. When the transistor pattern is formed, a connection is made to form one state, while the lack of a connection is read as the opposite state. In many cases, the mask drawings for the IC will visually match the user table pattern.

The user programmable version of a ROM is usually called a PROM for *programmable* read-only memory. Most PROM users utilize a machine to program the PROM with their table of information. "PROM programmers," "prom burners," "prom blowers," and "programming cards" are available from a variety of manufacturers. In most applications, there is no performance difference between ROMs and PROMs of similar configuration. Selection guidelines for ROMs and PROMs are summarized in Table 6.3.

Table 6.3 **Selection guidelines for ROMs and PROMs**

Is fixed memory data required?	Use ROM
Are large production quantities envisioned?	Use factory masked ROM
Is the application small or is only a prototype required?	Use PROM
Is high speed required?	Use bipolar ROM or PROM
Is low power required?	Use CMOS ROM
Will the prototype be changed during development?	Use erasable UV PROM

The PROM is usually used for low-volume production and prototype applications. ROMs are cost effective in higher volume applications because the mask development costs must be amortized over the total quantity of ICs used. PROMs are usually programmed on a fairly inexpensive programming machine that many companies own from previous development work. Usually a new type of PROM requires that a new "personality module" be purchased for the basic PROM programmer. A "personality module" or "programming head" contains the appropriate electronics to program and verify a particular PROM or group of PROMs. There is no universal programming specification for all PROMs, and even PROMs from the same manufacturer might have different programming specifications. It is necessary to program a PROM according to its unique "personality" or specifications. One must evaluate the purchase prices of the ROM and PROM version of an IC, the programming equipment and operator cost must be calculated for the PROM version, and the masking charge for the ROM version must be examined before the total "real cost" can be calculated.

6.6 TYPES OF PROMS

There are two types of PROMS available: one-time programmable and various erasable or reprogrammable versions. Bipolar and fusible link PROMs are programmed once. In the fusible link version, actual metalization links are "blown" open in the programming process. Once the links are blown, the PROM cannot be modified except to "blow" other links. An open link is irreversible.

There are three kinds of EPROMs or *erasable* PROMs currently being offered. UV PROMs use ultraviolet light to erase all PROM locations back to the all "one" state. These EPROMs actually have "little windows" on the chip where the internal IC die can be exposed to high-intensity UV light for the required erase time.

The other two erasable PROMs are EAROM and EE PROM. EAROM stands for electrically alterable ROM and EE PROM means electrically erasable PROM. Each type uses different technologies, and present designs warrant slightly different applications. The EAROM uses MNOS or metal-nitride-oxide semiconductor to achieve a charge tunneling effect when a high gate voltage (25 to 30 V) is applied during programming. In many applications, data is guaranteed to last for years. The drawback, for some potential EAROM applications, is

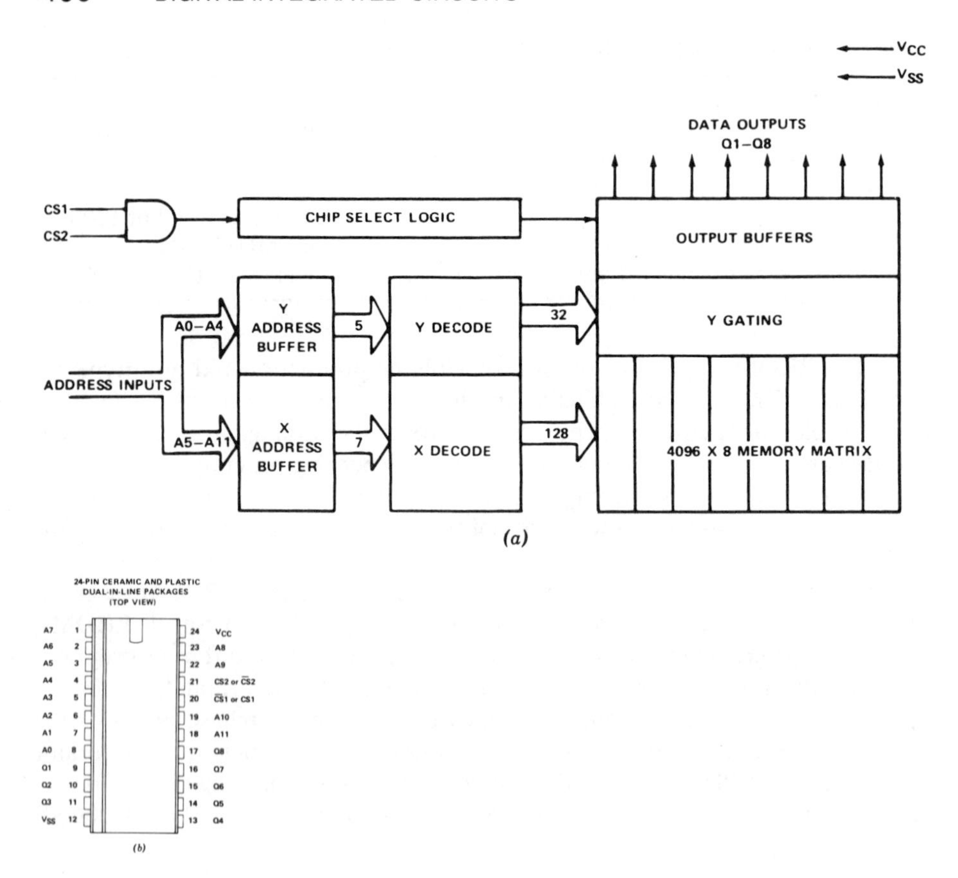

Figure 6.15 Example of organization of a ROM (TI4732). (*a*) Block diagram. (*b*) Pin connections. (Courtesy Texas Instruments, Inc.)

that current MNOS parts have a finite limit to the number of erase/write cycles which can be performed. These parts could wear out if misapplied.

EE PROMs are being developed by Intel, Hitachi, and others. The EE PROM would be an alternative to the UV EPROM with the main purpose of eliminating the use of UV and possibly allowing in-circuit erasure. Intel uses FLOTOX, or floating gate oxide semiconductor, while Hitachi uses MNOS. Both approaches rely on a tunneling effect in the semiconductor to retain stored charge. Owing to the stored charge, these parts can be considered nonvolatile because data is held when power is removed, as opposed to a read-write RAM. Table 6.4 lists some typically available PROMs.

6.7 APPLICATIONS OF ROMs and PROMs

The block diagram of a 4K × 8 ROM is shown in Fig. 6.15*a*. The TMS4732 is an NMOS chip. The chip is in a 24-pin package (Fig. 6.15*b*) and requires only

Table 6.4 **Representative PROMs**

Size	*Type*	*Typical manufacturer*
64 × 8	T^2L	Harris
256 × 4	CMOS	Harris
	ECL	Fairchild, Motorola
	T^2L	AMD, Intersil, Texas Instruments
256 × 8	PMOS	National Semiconductor
	UV erasable	
1024 × 8	NMOS	AMD, Fairchild, Fijutsu, Intel
	UV erasable	
2048 × 8	T^2L	Fairchild, Intel, National Semiconductor
	CMOS	Intersil
	UV erasable	
8192 × 8	NMOS	Intel, Motorola
	UV erasable	

a single dc power supply voltage of 5 V. It is a fully static part with an access time of 450 ns. It has three state outputs for microprocessor bus-type applications. There are seven logic sections that form the 4732.

The 12 address lines, A0 through A11, are sent through input buffers. Address lines A0 to A4 are fed to the "Y" decoder, while A5 through A11 are fed to the "X" decoder. One of the 128 "X" decoder lines is presented to the memory array. The memory array is organized as a matrix of 128 by 32 by 8 for a total of 32,768 cells. The X decoder input causes an output of 32 eight-bit words. One of these 32 words is selected by the Y decoder gating. Depending upon the CS1 and CS2 states, the chip select logic enables the eight output buffers. When the chip is selected the output buffers go from three state to the active state. The ROM output word is then available at the output pins.

There are two important parameters in using a ROM: access time and select enable time. The access time is the time required to read valid data at the output of the ROM after inputting a valid address, as shown in Fig. 6.16.

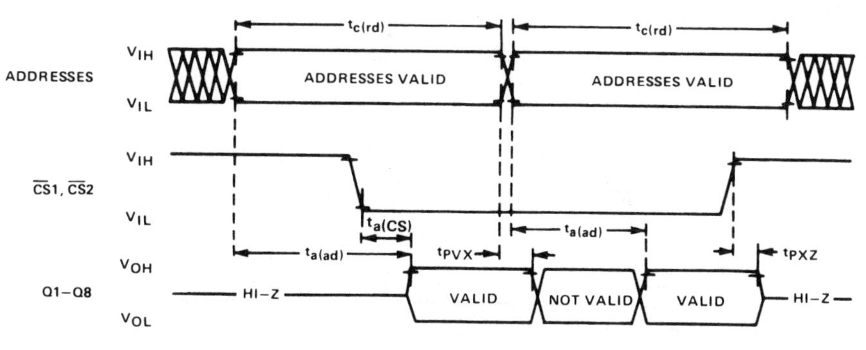

Figure 6.16 ROM read cycle timing. (Courtesy Texas Instruments, Inc.)

The maximum access time from valid address for the 4732 is 450 ns. The output enable is controlled by the chip select inputs. The access time from chip select is 200 ns for the 4732. These parameters usually overlap such that the minimum cycle time, to obtain data from two different addresses, becomes equal to the maximum access time from address, or 450 ns.

EXAMPLE 6.12 Design a ROM-based waveform generator.

PROCEDURE

1. A waveform generator can be formed from a counter, a ROM and a D/A converter. This circuit could be used, for example, as a test signal generator. A diagram of this application is shown in Fig. 6.17.
2. The counter is driven by a single clock input and the counter outputs are used to form the address sequence for the ROM. The ROM is programmed with a digitized version of the analog waveform we are trying to reproduce. A digital-to-analog converter is used to form an analog voltage from the digital output of the ROM.
3. The key design parameters in this application are clock speed, the number of sample points, and the resolution of the D/A converter (see Chap. 13).

Circuits like this one can be used to generate pseudo-sinewaves or cosine-squared functions, as well as staircases, ramps, and trapezoids. High-speed, high-resolution D/A converters have been used in similar circuits to form color television video signals and "computer-generated speech."

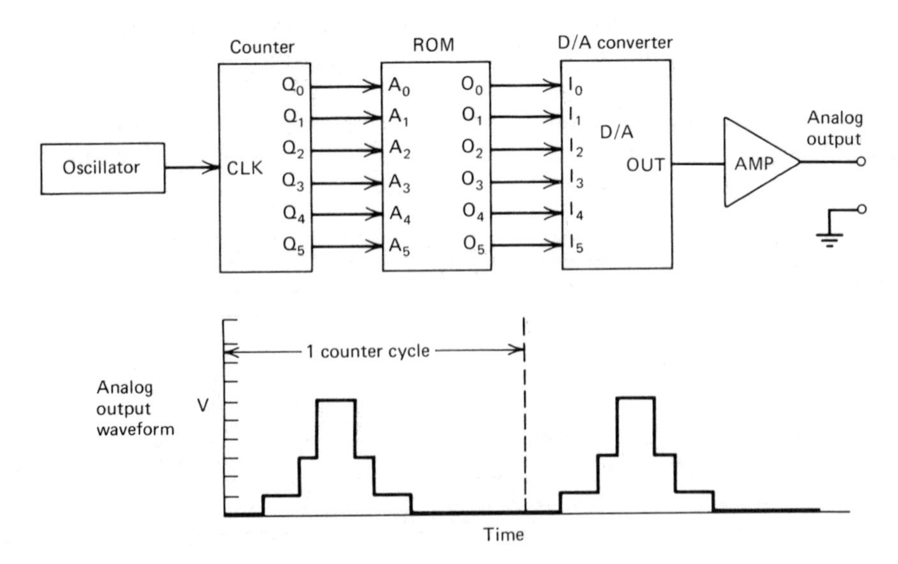

Figure 6.17 A ROM-based waveform generator.

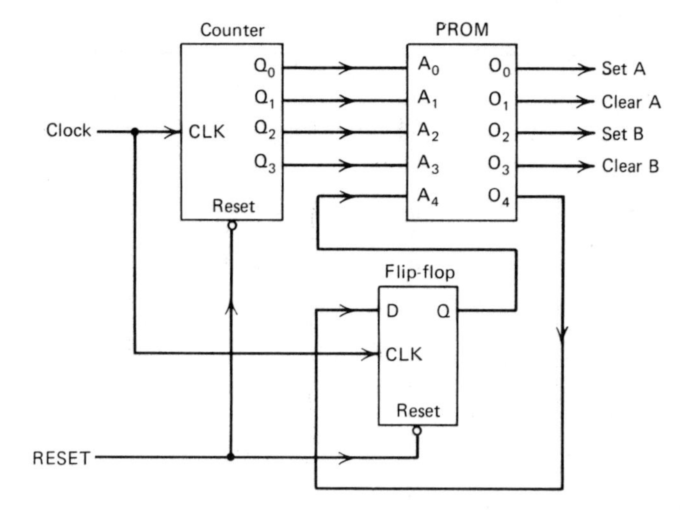

Figure 6.18 A PROM sequencer for a logic state controller.

EXAMPLE 6.13 Design a PROM-sequencer for a logic state controller.

PROCEDURE

1 The heart of many computer designs is the state controller that is used to implement a particular machine's operation code (op code). In many minicomputer designs, the state controller is formed from counters and ROMs instead of random logic, such as flip flops and gates. A simple sequencer is illustrated in Fig. 6.18.

2 The counter is driven from the system clock or oscillator. The counter outputs provide the address inputs to the PROM. The PROM outputs are used to generate strobes during certain timing states by programming a particular bit at the address corresponding to one state. In addition, "sequence jumps" can be made by using one or more of the outputs to help form the "next address." The state diagram is illustrated in Fig. 6.19, and the PROM code is provided in Table 6.5.

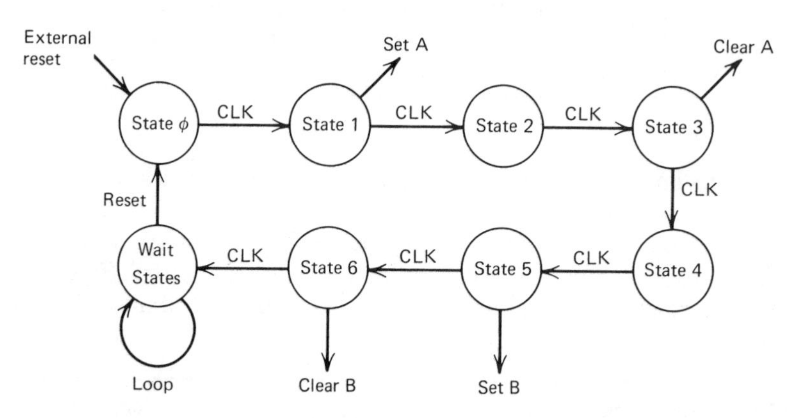

Figure 6.19 State diagram of PROM sequencer of Fig. 6.18.

Table 6.5 **PROM code**

	Input					*Output*					
State	A_4	A_3	A_2	A_1	A_0	O_4	O_3	O_2	O_1	O_0	*Rule*
0	0	0	0	0	0	0	0	0	0	0	Reset
1	0	0	0	0	1	0	0	0	0	1	Set A
2	0	0	0	1	0	0	0	0	0	0	
3	0	0	0	1	1	0	0	0	1	0	Clear A
4	0	0	1	0	0	0	0	0	0	0	
5	0	0	1	0	1	0	0	1	0	0	Set B
6	0	0	1	1	0	1	1	0	0	0	Clear B, go to WAIT
7	0	0	1	1	1	0	0	0	0	0	
8	0	1	0	0	0	0	0	0	0	0	
9	0	1	0	0	1	0	0	0	0	0	
10	0	1	0	1	0	0	0	0	0	0	
11	0	1	0	1	1	0	0	0	0	0	
12	0	1	1	0	0	0	0	0	0	0	
13	0	1	1	0	1	0	0	0	0	0	
14	0	1	1	1	0	0	0	0	0	0	
15	0	1	1	1	1	0	0	0	0	0	
16	1	0	0	0	0	1	0	0	0	0	WAIT
17	1	0	0	0	1	1	0	0	0	0	
18	1	0	0	1	0	1	0	0	0	0	
19	1	0	0	1	1	1	0	0	0	0	
20	1	0	1	0	0	1	0	0	0	0	
21	1	0	1	0	1	1	0	0	0	0	
22	1	0	1	1	0	1	0	0	0	0	
23	1	0	1	1	1	1	0	0	0	0	
24	1	1	0	0	0	1	0	0	0	0	
25	1	1	0	0	1	1	0	0	0	0	
26	1	1	0	1	0	1	0	0	0	0	
27	1	1	0	1	1	1	0	0	0	0	
28	1	1	1	0	0	1	0	0	0	0	
29	1	1	1	0	1	1	0	0	0	0	
30	1	1	1	1	0	1	0	0	0	0	
31	1	1	1	1	1	1	0	0	0	0	↓

Although the application illustrated is fairly simple, this PROM-based technique can be applied in fairly complex situations. It is not unusual to see this technique used with a minicomputer microcode of over 50-bits wide and with multiple jumps and loops. Equivalent performance is being offered in sequencer ICs, where the entire application circuit is incorporated into a single programmable chip.

EXAMPLE 6.14 Design a memory interface for program storage in a microprocessor system.

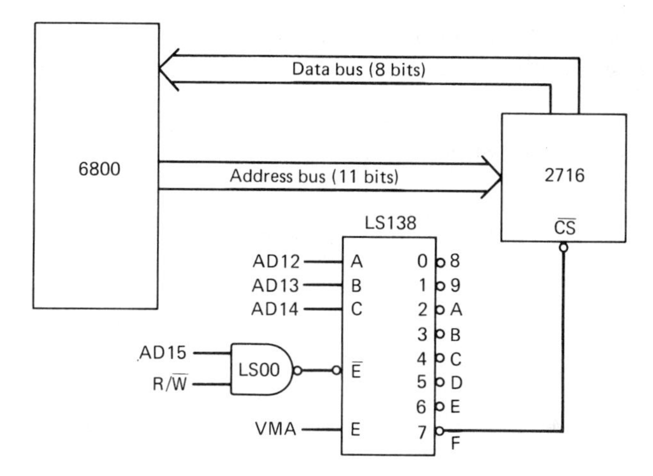

Figure 6.20 A UV PROM computer interface for program storage.

PROCEDURE

1 We have chosen to use an UV PROM for program storage so we may conveniently modify the program as the need arises. In the block diagram of Fig. 6.20, a 2716 UV erasable NMOS 2048 × 8 PROM is used to hold the program. We could store up to 2K words of eight bits each in this part.

2 This particular microprocessor jumps to the highest two address locations when power is applied. In our case, these addresses are $FFFE_{16}$ and $FFFF_{16}$. The processor uses the data in these two addresses to form a pointer address to the start of our program. Because the 2716 is used for total program storage, we will position it at the top of the available memory by external decode logic. Our PROM occupies the space from $F800_{16}$ to $FFFF_{16}$.

3 The sequence of operation is a two-step process. First, power is turned on and the microprocessor starts to function. The processor goes to the top two words in the 2716 and reads them. The processor manipulates this data to form the address of the start of the program at $F800_{16}$. The processor then jumps to $F800_{16}$ and begins executing the steps in the program. Each step involves a memory read of the next PROM location or locations, followed by an execution of the instruction that was read.

6.8 REFERENCES

Allen, J., "64K-bit Dynamic MOS RAM Slow to Take Hold," *Computer Business News,* p. 25, May 19, 1980.

Bursky, D., "Special Report: Memories pace systems growth," *Electronic Design,* p. 63, September 27, 1980.

E.E. Times, "Electrically Erasable PROMs: Floating Gate Challenges NMOS," *Electronic Engineering Times,* p.18, March 3, 1980.

Mhatre, G., "IC Memories: RAMS—An EET Special Report," *Electronic Engineering Times,* p. 41, February 4, 1980.

Olsen, S. C., "Semiconductor Memories," *Digital Design,* p. 72, January 1980.

Young, D. C., "Guidelines for Designing Battery Backup Circuits for CMOS RAMS," *Computer Design,* p. 117, August 1979.

Databooks from the following manufacturers and sources: Harris Semiconductor, IC Master, Mostek, Motorola, National Semiconductor, Synertek, and Texas Instruments.

CHAPTER 7

Magnetic Bubble Memories

Russell MacDonald

Program Manager, Telecommunications Integrated Circuits
Texas Instruments, Incorporated

7.1 INTRODUCTION

The concept of using magnetic bubbles for mass data storage was developed by Bell Laboratories in 1967. It was found that millions of bits of data could be stored on a single garnet chip. Also, making these chips could be very similar to the way silicon is used in the manufacture of semiconductor products. This discovery set off the electronic industry's wildest race since the invention of the integrated circuit. Currently, magnetic bubble memories (MBMs) are available that can store in excess of one million bits of data.

Designing a magnetic bubble memory system requires more understanding of the MBM device itself than of any other component within the system. This stems from the unique fundamental differences between an MBM device and a typical IC. It is important to understand from the start that an MBM device is not an integrated circuit; in fact, it is not even a semiconductor device. It may, however, be considered a solid-state device, because it has no moving parts.

Semiconductor manufacturers took on the task of making these devices for two basic reasons. First, MBMs enhance semiconductor systems immensely. Second, the manufacture of these devices involves many steps that are similar to the manufacture of semiconductors.

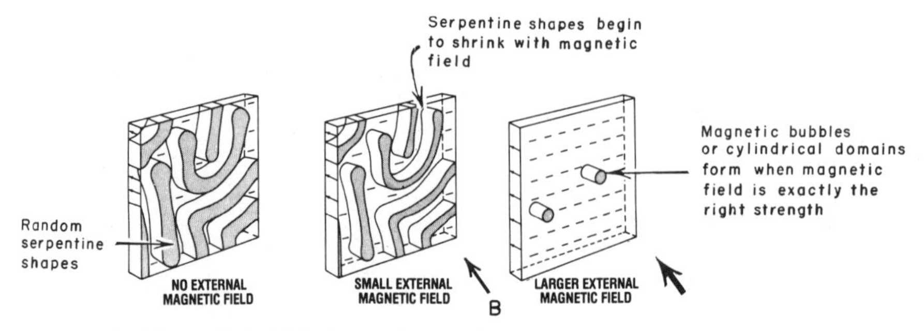

Figure 7.1 Magnetic bubble formation.

7.2 THE MAGNETIC BUBBLE MEMORY

The term *magnetic bubble* is somewhat misleading. A magnetic bubble is actually a *cylinder* of highly mobile magnetic charge, or domain, residing in a thin magnetic film of opposite charge. Referring to Fig. 7.1, an external magnetic bias field, B, is needed to maintain the cylindrical shape of the bubble. With no field, the patterns formed in the film are random serpentine shapes. Increasing the field causes these patterns to shrink and most, but not all, of the patterns to disappear (*collapse*). The ones that remain become stable cylinders, or bubbles. Decreasing the field will cause the bubbles to grow larger and finally elongate, or *stripe out*.

These two phenomena, *collapse* and *stripe out,* are extremely important. At temperature extremes, either can occur because, as temperature changes, the strength of the field of the biasing magnets and the magnetic film changes. Although the field and the film are made to track as closely as possible, they do not track together exactly. It is important that this phenomenon be understood by the MBM system designer. Failures of this nature can indeed be baffling.

Figure 7.2 shows a complete MBM device. The bubble memory chip consists of a nonmagnetic substrate made of GGG (gadolinium gallium garnet) with aluminum conductor lines. Permalloy (nickel-iron) patterns, and a thin magnetic epitaxial film is grown upon it in a precise manner. The materials are chosen to provide, as nearly as possible, the same coefficient of expansion over temperature. The aluminum conductor lines are used to introduce highly localized magnetic fields by running an electric current through them. The permalloy patterns are extremely soft magnetic material that respond very rapidly to changing external magnetic fields. These are used to provide positional control of bubbles in the epitaxial film.

The chip is mounted within two orthogonal field coils and two precision permanent magnets. The permanent magnets provide the correct bias field to maintain stable bubble existence. The orthogonal field coils, driven by triangular current waveforms, provide the required magnetic gradient to cause the bubbles to move within the epitaxial film. The outside covering of the MBM is a magnetic shield to ensure data integrity against external magnetic fields.

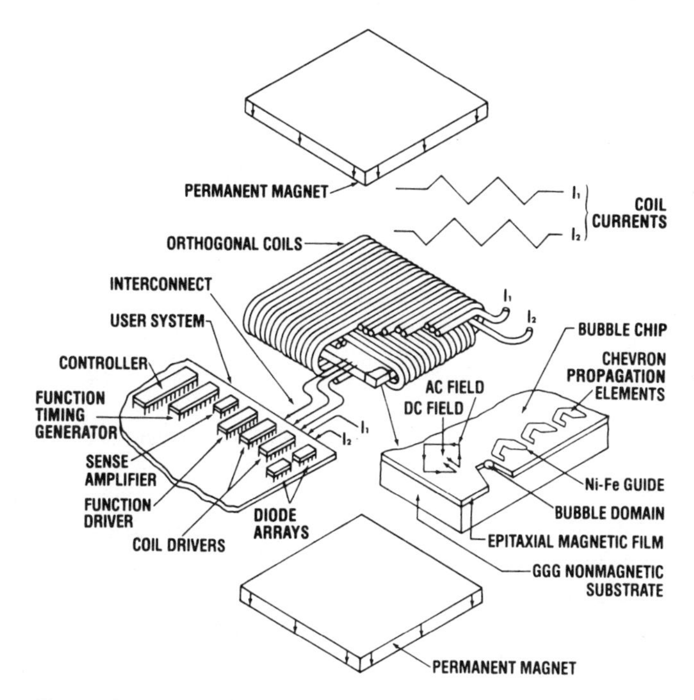

Figure 7.2 Exploded view of a magnetic bubble memory (MBM).

7.3 PROPAGATION AND CONTROL OF MAGNETIC BUBBLES

Having discussed how data may be stored in the form of magnetic bubbles, it now becomes obvious that control of the motion of these bubbles is required to allow access to stored information. It was decided very early in bubble memory technology that the only practical method of motion control was to line up all of the stored data in serial loop fashion and then cause all bubbles to shift one position under a magnetic stimulus applied in a precise manner. To realize this shift register-like motion, there had to be a way to hold the bubbles in known positions. In addition, provision had to be made to force them to jump from one known position to an adjacent known position.

It was discovered that the bubbles could be captured and held in a known location by spots of soft magnetic material in close proximity to the epitaxial film. (The permalloy patterns are deposited on top of the epitaxial film and are physically held in place by this bond.) The difficult part was finding out how to cause bubbles to jump from one known position to another, that is, to *propagate*. The answer was to provide a rotating magnetic field at the bubble chip and to design the shape of the permalloy pattern that would support propagation. The rotating magnetic field is supplied by the orthogonal coils illustrated in Fig. 7.2.

Figure 7.3 shows the shape of the first successful propagation element.

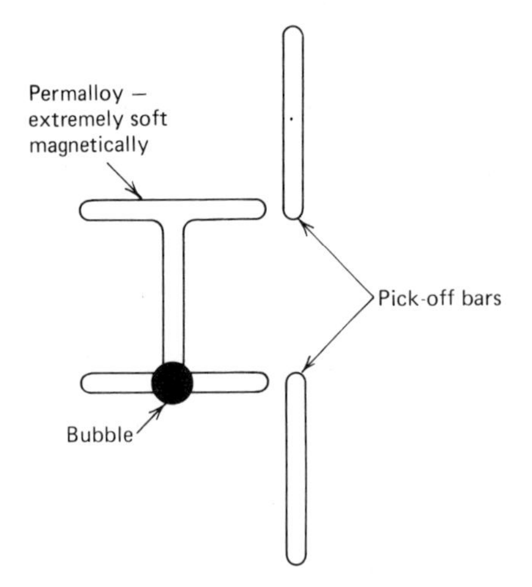

Figure 7.3 Basic T-BAR propagation element.

This shape is called the T-BAR and was used in the first commercial bubble memories. Propagation occurs because the rotating magnetic field causes the permalloy patterns to become tiny alternating magnetic dipoles that attract and repel the bubbles, forcing them to move one position with each field coil cycle. The pickup bars help the bubbles propagate from one T-BAR to the next.

It was soon discovered that a new shape, called the *asymmetric chevron* (Fig. 7.4), was a far superior propagation element than the T-BAR. Figure 7.5 illustrates the steps involved during propagation of an asymmetric chevron. The improvement of the chevron over the T-BAR is that the large end of the chevron develops a stronger magnetic dipole than the small end. This greatly helps the bubble to jump across the gap between chevrons. The asymmetric chevron is used in almost all MBM products on the market today.

With propagation understood, the next thing that must be learned is how bubbles are stored and retrieved from the shift register-like chains of propagation elements. This is accomplished through the use of special permalloy patterns, under which are aluminum conductor runs placed in defined positions. This combination is called a control element, or *gate*. Table 7.1 lists common types of these gates and their functions, as well as the basic shape of the permalloy and conductor patterns for each gate. The theory of their operation is beyond the scope of this chapter. It is essential, however, that the MBM system designer understand and remember the function of each type of gate.

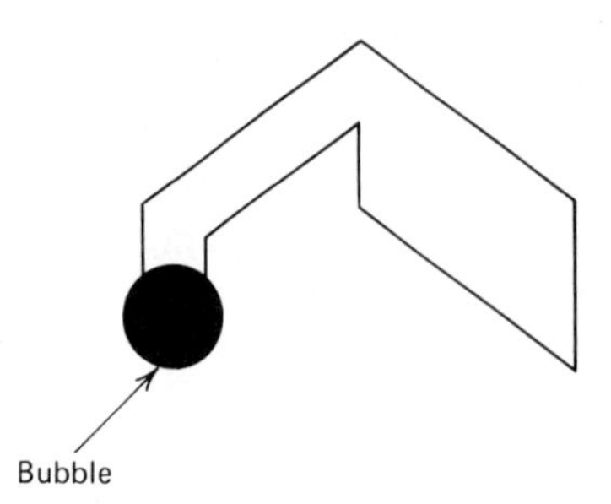

Figure 7.4 Asymmetric chevron propagation element.

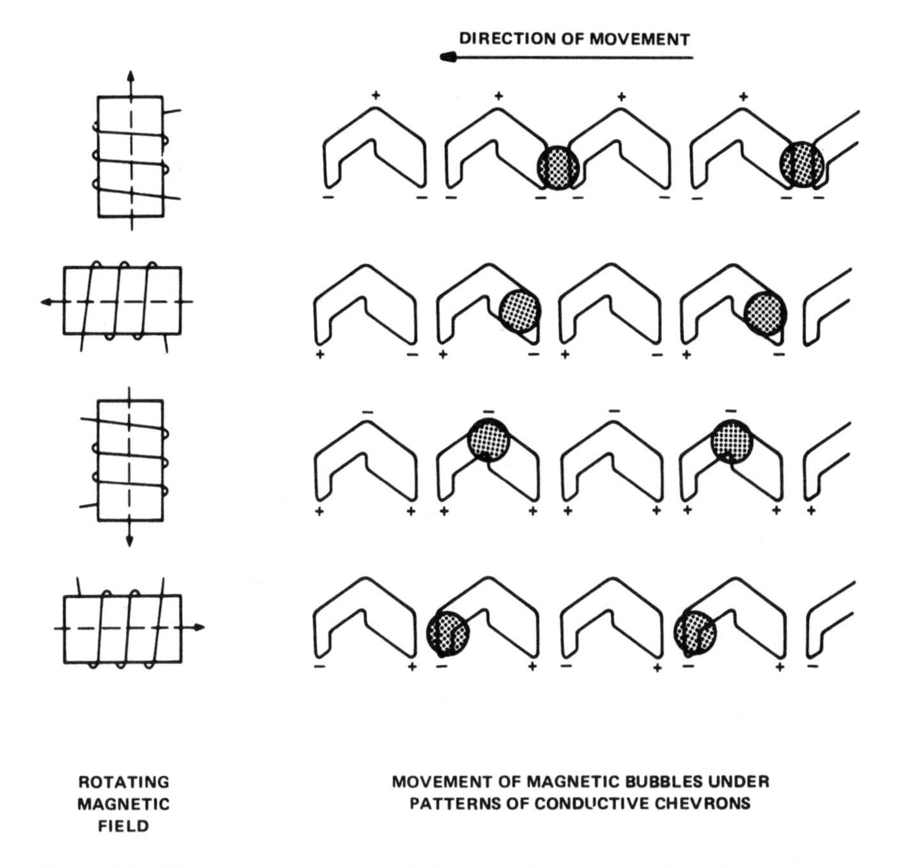

Figure 7.5 Movement of magnetic bubbles under patterns of conductive chevrons.

7.4 DEVICE ARCHITECTURE

The simplest MBM architecture is the *single serial loop* illustrated in Fig. 7.6. This architecture uses a minimum of control elements. The *generator* is pulsed to place bubbles in the loop and the *annihilator* is pulsed to place voids in the loop. A bubble is used to represent a "1" and a void to represent a "0." The *detector* detects the presence of a bubble, and the *replicator* is pulsed to place a copy of a bubble on the detector track. This type of architecture is acceptable only for very small MBM devices because extremely long loop lengths suffer tremendous yield losses and the cost of manufacturing such a device becomes unrealistic. Even if a long loop length could be manufactured, the access time of the device would be unacceptable.

The first practical architecture from both a cost and performance standpoint is the *major-minor loop,* illustrated in Fig. 7.7. Instead of a single storage loop, this approach used multiple parallel loops, called *minor loops,* which are connected to one common loop, called the *major loop*. Data was stored in this

Table 7.1 **Common control elements and their functions**

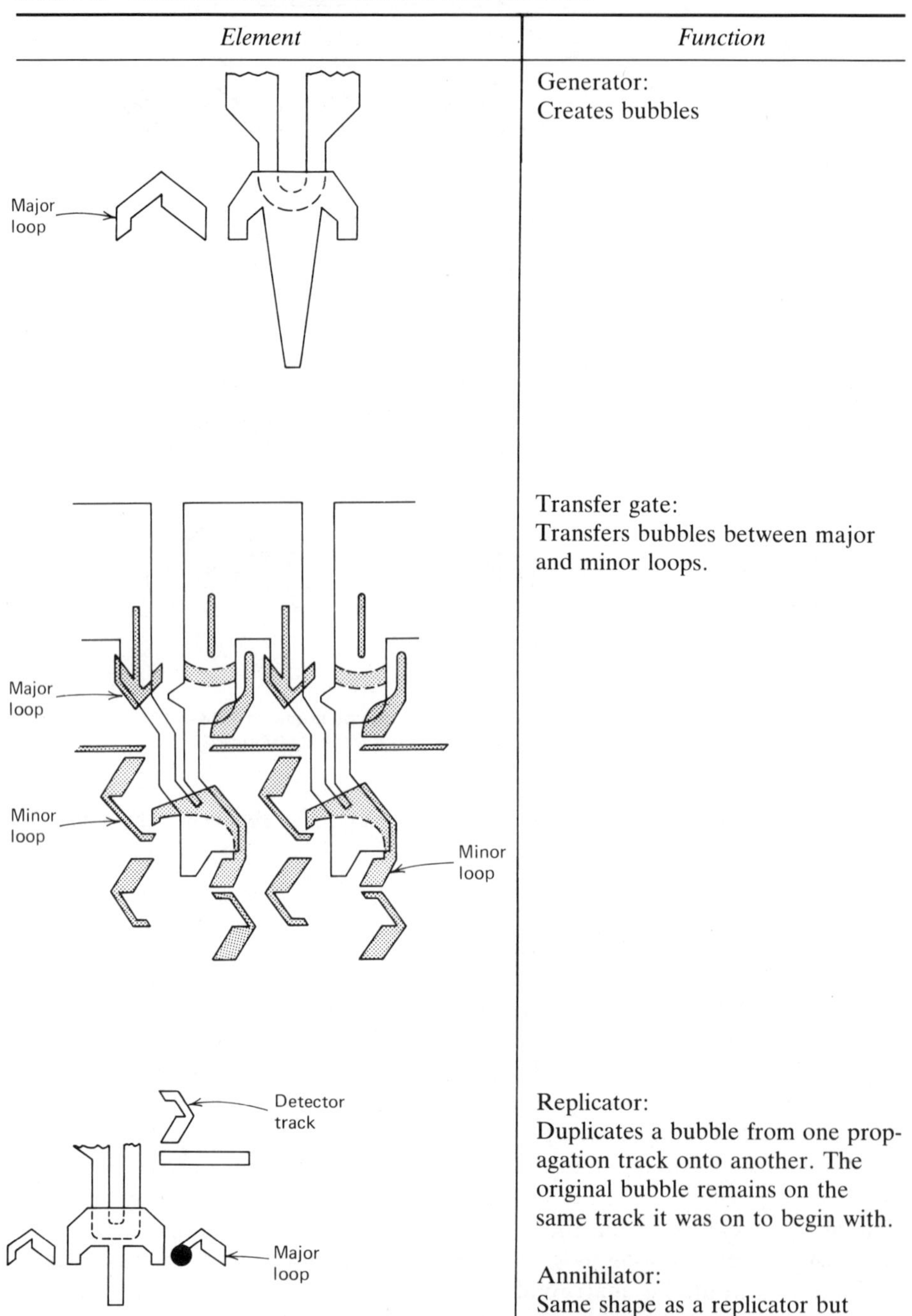

Element	*Function*
	Generator: Creates bubbles
	Transfer gate: Transfers bubbles between major and minor loops.
	Replicator: Duplicates a bubble from one propagation track onto another. The original bubble remains on the same track it was on to begin with. Annihilator: Same shape as a replicator but pulsed with a different current pulse. Used for getting rid of unwanted bubbles.

Table 7.1 **Common control elements and their functions (continued)**

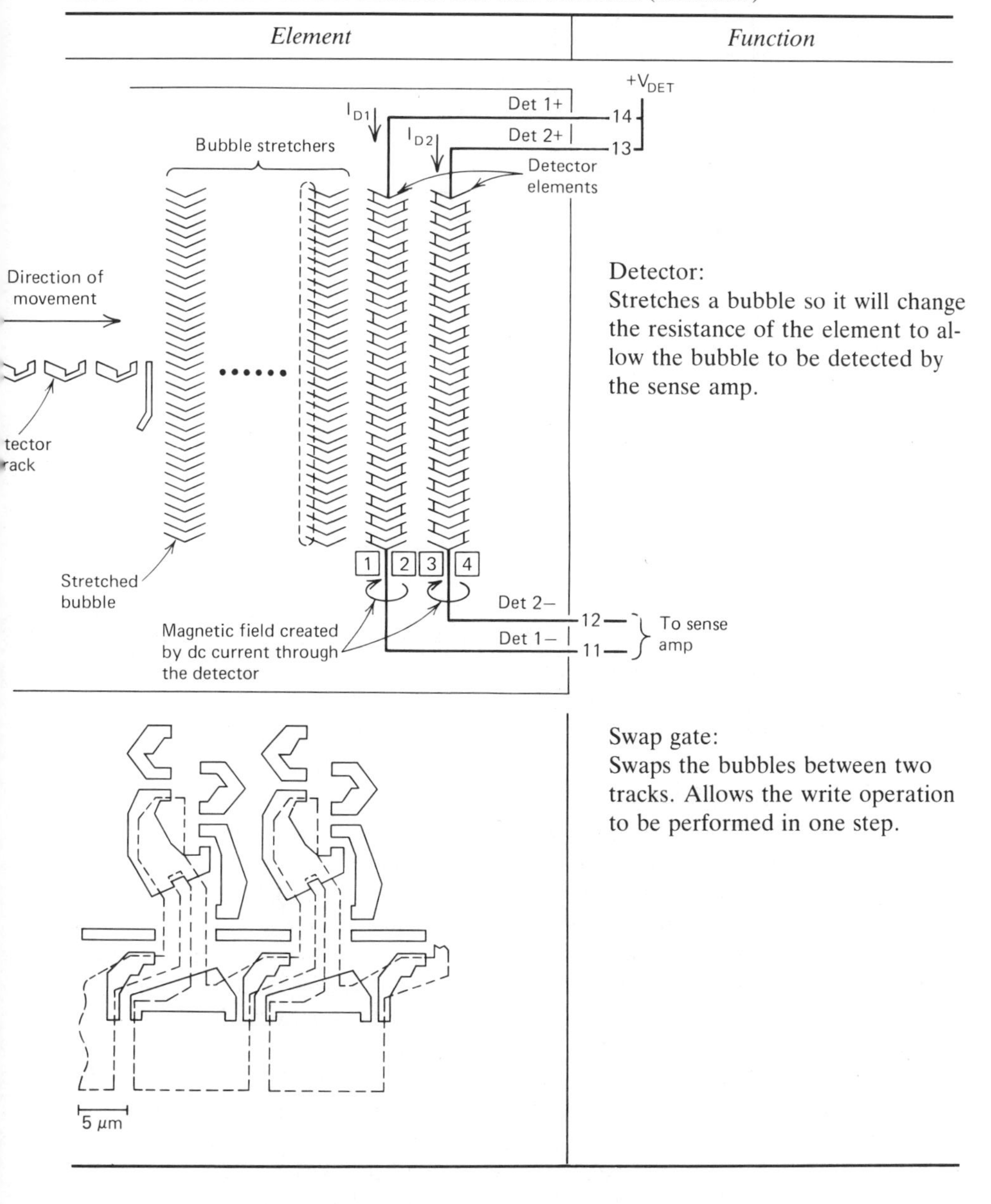

Element	*Function*
(detector drawing)	Detector: Stretches a bubble so it will change the resistance of the element to allow the bubble to be detected by the sense amp.
(swap gate drawing)	Swap gate: Swaps the bubbles between two tracks. Allows the write operation to be performed in one step.

device by pulsing the generator (or not pulsing it), and then it was propagated by the rotating magnetic field around the major loop until it lined up with the minor storage loops. Transfer gates were then pulsed and a string of length equal to the number of minor loops was transferred into the minor loops.

The string of data was called a *page*. To read a page of data, all bubbles in the device were propagated until the correct page was at the major loop.

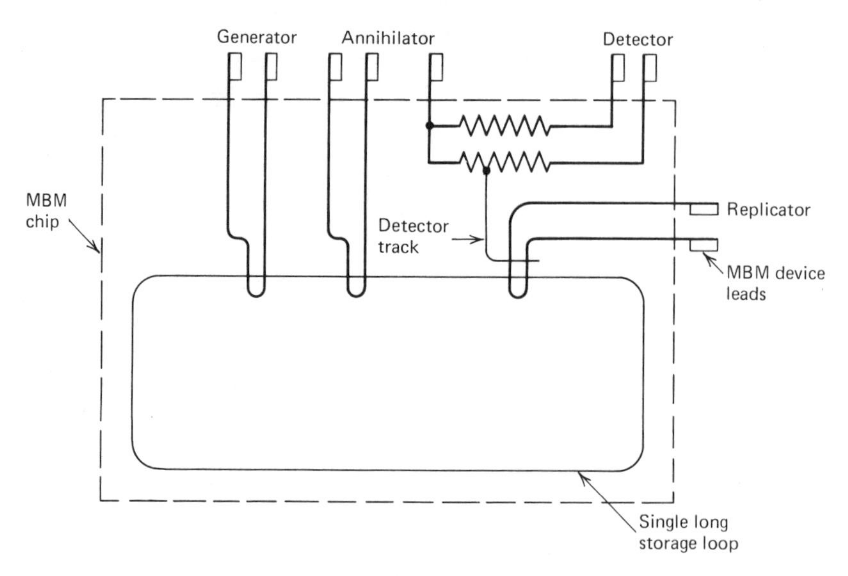

Figure 7.6 The simplest MBM architecture—the single serial loop.

Then the transfer gates were again pulsed bringing the data onto the major loop. Next, the data was rotated around the major loop and passed by the replicator where bubbles were split, allowing one to go to the detector. The others remain on the major loop to be rotated back around and transferred onto the minor loops at the same page it was removed from. As can be seen, the transfer operation is a destructive read of the minor loops, and thus the need to rewrite the page back after reading it.

The major-minor loop concept not only improved access time, but it also opened the door to a new approach in improving yields. Build more minor loops onto each chip than are actually needed and then let that number of extra loops be defective. These extra loops are called *redundant loops*.

All MBM devices now in production use yet another type of architecture. Illustrated in Fig. 7.8, this architecture is called *block swap/replicate*. It eliminates the destructive read problem mentioned in the previous paragraph. To write in this device, the generator is pulsed placing a page of data on the input, or write track. The bubbles in the device are then propagated until the page lines up with the minor loops. At that time, the swap gates are pulsed, swapping the new page of data with the data that had been stored on that page. The old page of data is then rotated to the guard rail where it is destroyed.

Reading a page back is accomplished by rotating the device until the desired page is at the output, or read track. Then the replicate gates are pulsed, placing a copy of the stored contents of that page on the output track. (This is a non-destructive read operation.) Next, the page of data is taken to the detector where it is read from the device. From Fig. 7.8, it can be seen that there is another type of loop present; the redundancy loop. This loop is for storage of the redundancy map that is used to mask out defective loops by the MBM system.

The block swap/replicate architecture retains all of the advantages of the major/minor loop approach. These include low access time, redundant loops, and ease of manufacture while solving the problem of the destructive read.

7.5 BUBBLE MEMORY SYSTEMS

From the preceding discussion, it is apparent that controlling all of the required steps involved with reading and writing a bubble memory device is not a simple matter. Indeed, it is extremely complex, requiring considerable sophistication on the part of the support circuitry. To realize this sophistication requires either massive quantities of standard "off the shelf" ICs (plus some transistors and other discrete components), or a family of custom support integrated circuits. The custom IC approach is the one taken by most MBM manufacturers. This

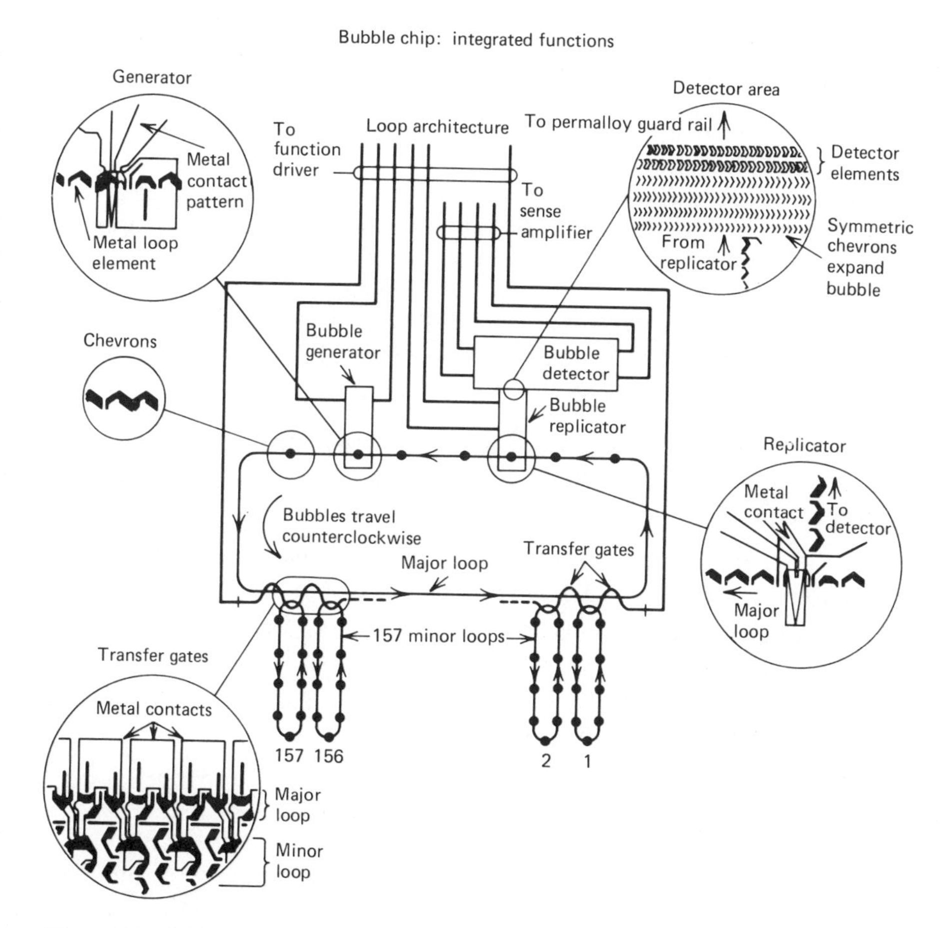

Figure 7.7 Major-minor loop architecture.

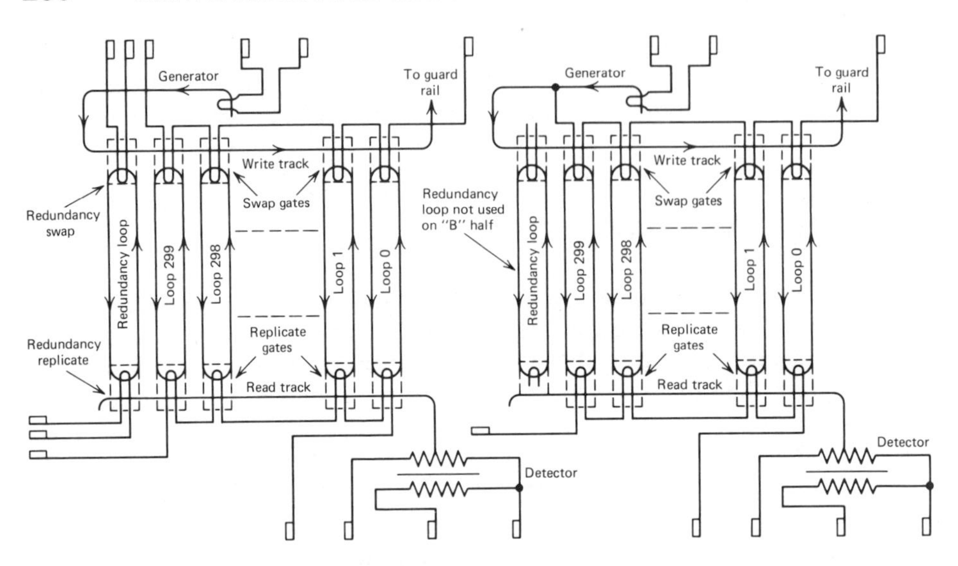

Figure 7.8 Block swap/replicate architecture.

approach is the only one that takes advantage of the high-density storage properties of the MBM device.

There are many ways to partition the required support circuitry in a bubble memory system. The block diagram of Fig. 7.9 illustrates one such partitioning scheme. Each block represents one custom integrated circuit.

The first thing that must be done in a bubble memory system is to provide conversion of the analog current-mode MBM signals to digital voltage levels that are easily controlled by a computer. The output of an MBM device can be thought of as digital in the sense that an analog signal is present to indicate a "1" and absent to indicate a "0." In Fig. 7.9, this is handled by the custom ICs that make up the MMU, or *modular memory unit*. Each MBM device must have its own set of MMU circuits. These circuits consist of two coil drivers, a diode array, a function driver, and a sense amplifier.

1 The coil drivers and diode array provide the triangular current waveshapes needed to produce the rotating magnetic field.

2 The function driver provides the precise current pulses that control the path that the magnetic bubbles take within the MBM device.

3 The sense amplifier detects the presence or absence of a bubble that has been sent to the detector site by the function driver.

The *bubble memory controller,* or BMC, is by far the most complex custom IC in a bubble memory system. The main function of this chip is to provide a relatively simple interface to the host CPU. To achieve this, the controller must be, essentially, a special purpose microsequencer with a variety of different operational modes. Modes must include reading and writing data to the MBM device, in addition to modes that initialize, reset, provide mask interrupts, and so on. Each controller on the market today has a different set of operational modes which should be carefully studied. The controller always dictates the MBM system structure.

The *function timing generator,* or FTG, is the source of all of the required timing signals in the bubble memory system. This IC operates in close conjunction with the controller, and indeed could have been made part of the chip if it had not been for two very important reasons. First, the signals to the rest of the MBM system must be extremely precise, and the NMOS process used for the controller will not run fast enough for this. The controller had to be designed using NMOS because of its complexity. The FTG, therefore, was designed as a separate device using standard low-power Schottky processing.

The second reason that FTG was not made part of the controller was to improve yields and keep costs at a minimum while still providing mask programmability to permit timing changes to be handled in the simplest way possible. Mask programmability permits an IC to be completely manufactured, except for the last step and packaging. The last step is where a series of constants are programmed into the device through the use of a mask.

Power Supply Considerations

Selection of an appropriate power supply is critical to the success of any bubble memory system design. Most system designers are accustomed to the rather forgiving nature of modern circuit components. They are used to selecting a power supply by simply checking the voltage and current requirements and then finding a supply that will fit in the allotted space. In bubble memory system design, however, several additional considerations are necessary.

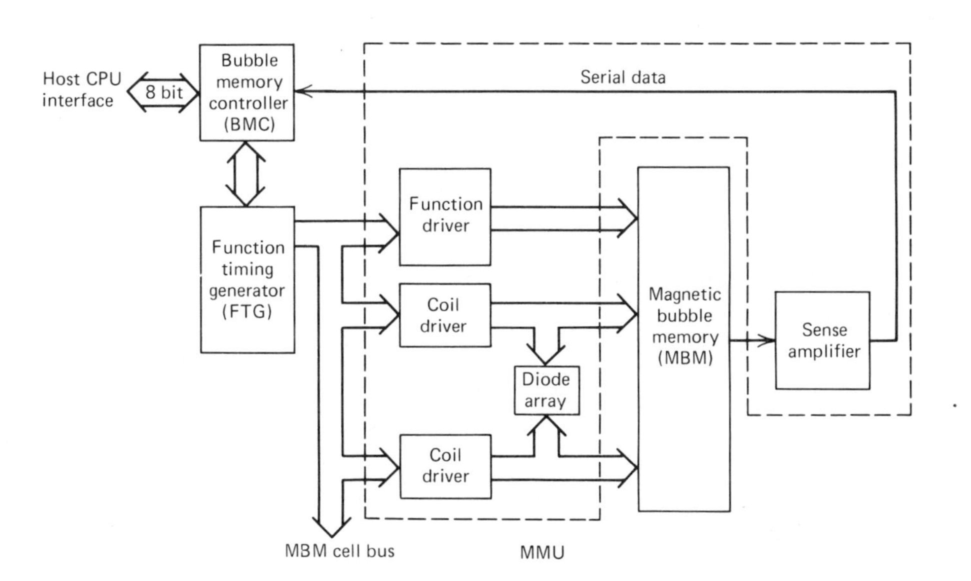

Figure 7.9 Block diagram of support circuitry for a bubble memory system.

There is one fundamental reason that makes the choice of the power supply so critical in an MBM system. This is *nonvolatility*. The nonvolatile nature of the MBM device must be carefully protected at the system level, and the starting point for this protection is the power supply.

In addition to being capable of supplying the correct amount of power in the operating mode, a power supply under consideration for use in an MBM system must have fully specified turn-on and turn-off with no overshoot or undershoot. It also must be a carefully regulated supply with *remote sensing* on all voltages.

In remote sensing, wires are connected from the load that a power supply is driving back to special inputs on the power supply. These lines draw very little current and the voltage drop in the connecting lines is, therefore, negligible. This provides an accurate "voltage at the load" indication to the power supply so that it can adjust its outputs to compensate for voltage drops in the high-current supply lines.

When power is applied to an MBM system, there must be some way for the controller to set up its internal page count register to the same address that the MBM device was left at when power was removed. This is called *reestablishing the address reference point*. One way to accomplish this is to ensure that the MBM device is always rotated to a predetermined page before shutdown; the system of Fig.7.9 uses this approach. After completion of any operation, the controller automatically rotates the MBM device to page zero. Then on power-up, the page count register is simply initialized to zero, and from that point on the controller increments the page count register for each field coil cycle.

This method seems very straightforward and, indeed, it is if the power supply never fails. The problem is how do you protect the data in the MBM device against sudden power loss? If the controller is in the middle of an operation and power goes off the MBM device will not get rotated back to page zero. Then, when power is restored later, the controller will not be able to

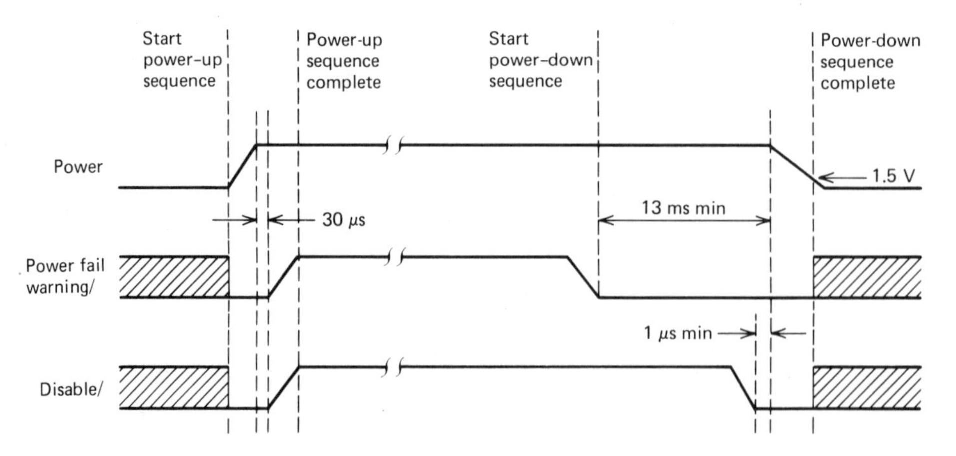

Figure 7.10 Power-up/power-down sequence for a typical major-minor loop system.

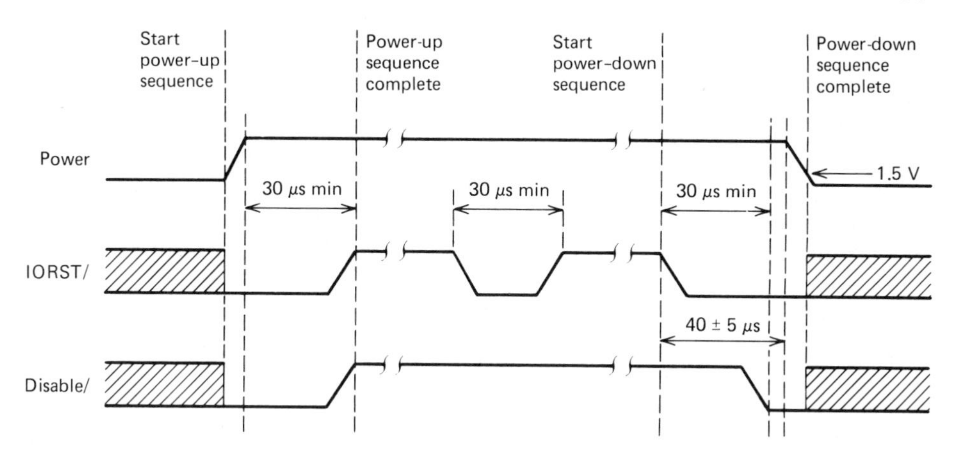

Figure 7.11 Power-up/power-down sequence for a typical block swap/replicate system.

restore its address reference point. If this occurs, there is generally no way to recover and, for all practical purposes, the data has been lost.

The answer to this problem again lies in the choice of the power supply. The power supply should have a fully specified power fail warning signal that is issued far enough in advance of power coming down for the controller to rotate the bubble device back to page zero. For the system of Fig. 7.9, the controller will automatically rotate the bubble device to page zero when its RESET input is activated. The time required for this is typically 13 ms worst case. In addition, after the bubble device has been rotated to page zero, a solid DISABLE signal must be applied to the coil and function drivers while the power is actually transitioning. The DISABLE inputs to these devices protect the bubble device from any spurious pulses that might be generated otherwise. Figure 7.10 shows the complete power-up/power-down sequence that must be observed in a typical major-minor loop system.

Another method to reestablish the page address reference point on power-up is to store a locator data pattern in the bubble device that the controller can search for. Then, when the controller finds this pattern, it can stop rotating the bubble device and set its page count register to the correct value. The obvious advantage of this approach is that the controller no longer has to rotate the bubble device back to page zero after completing an operation. The amount of power supply holdup time, therefore, is greatly reduced.

Figure 7.11 shows the entire power-up/power-down requirements of a typical block swap/replicate system. Note that the holdup time of the power supply is 30 μs as opposed to 13 ms for the previous system. The 30 μs is the length of time it takes the system to complete a field coil cycle it has started (10 μs), plus extra time to allow any coil ringing to subside. The locator data pattern on this device could be strings of ones followed by a zero stored in the redundancy loop.

Megabit Systems

Figure 7.12 shows a block diagram of a low-cost, one-megabit system. The bubble is a single one-megabit device. Each block represents a custom IC with the exception of the MBM device itself. Notice the similarities between this and the major-minor loop system. The MMU circuits consist of a dual-channel sense amplifier, two coil drivers, a diode array, a read function driver, and a write function driver. These circuits perform exactly the same functions as in the major-minor loop system described earlier. There is also a controller and a function timing generator that perform the same functions as their counterparts in the major-minor loop system.

The only fundamental difference between this megabit system and the major-minor loop system is in the way the redundancy mapping is handled. In the major-minor loop system, the map is stored in a PROM (see Chap. 6), and in the megabit system the map is stored in the redundancy loop inside the bubble device itself. Even though the controllers perform the same system function,

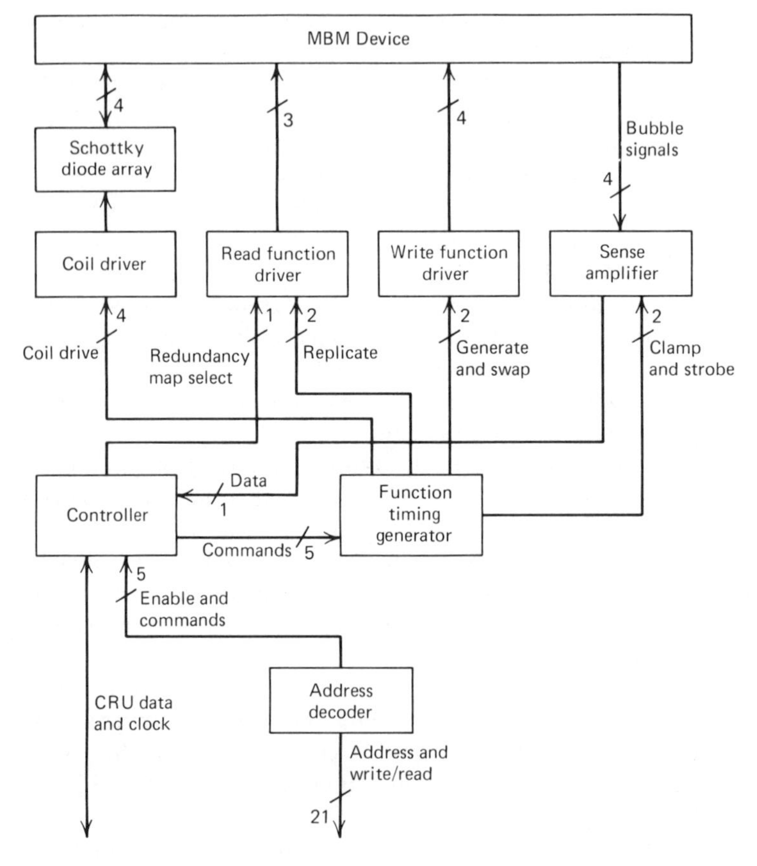

Figure 7.12 Block diagram of a 1 megabit system.

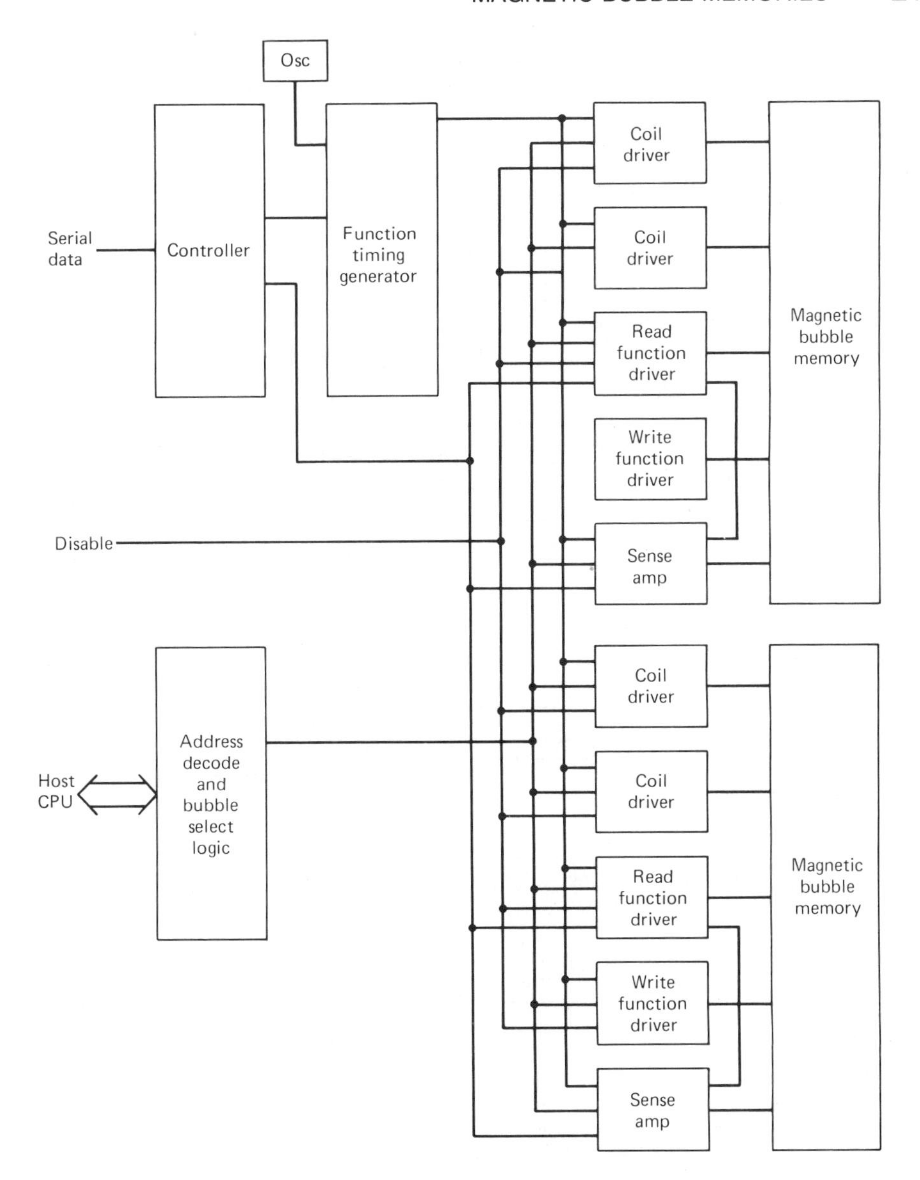

Figure 7.13 Larger megabit system using chip select expansion.

they operate totally differently because of the difference in the architecture of the major-minor loop compared to the block swap/replicate.

Larger megabit-based systems are easily designed by adding more MMUs (Fig. 7.13) and selecting between them with a standard chip select scheme. Providing the proper buffering rules are observed, there is virtually no limit to the amount of memory that can be added using this type of expansion.

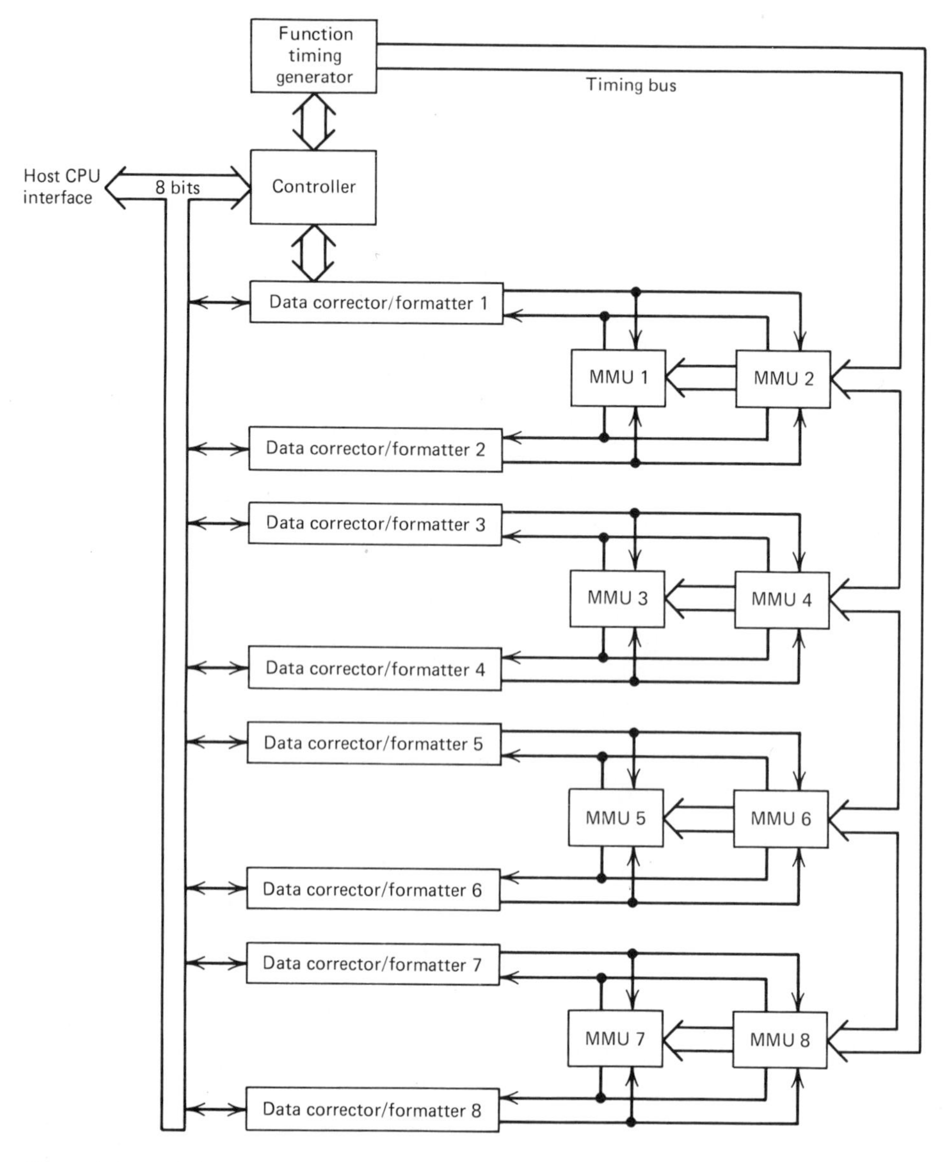

Figure 7.14 High-performance, multichannel, 8-megabit bubble memory system with error correction.

High-performance bubble memory systems basically require higher data rates than those systems mentioned thus far. To achieve higher data rates, either the field coil frequency must be increased, or multiple MBM channels must be run in parallel. Increasing the field coil frequency introduces several additional problems, such as changes in the field coil drive voltage and timing changes in the function timing generator. These types of changes require complete recharacterization of the MBM device at the new frequency, and this has not yet been done. Instead, the multiple-channel parallel approach has been taken. Running both channels of a block/swap replicate device in parallel, for example, will increase the data rate from 85 to 170 kbits/s (assuming a 100 kHz field rate). Running eight of these devices in parallel, or 16 channels, will increase the data rate to 1360 kbits/s.

To achieve parallel channel operation requires one major change to the basic megabit system block diagram. Figure 7.14 illustrates a high-performance, 8-channel system using four 1-megabit devices. Notice that each channel now has an additional custom IC called a *data corrector/formatter,* or DCF. This device is required for three reasons. The first, and most fundamental, is that there must be some on-line memory associated with each channel in which to store its redundancy map. Each DCF has a 300-bit RAM (see Chap. 6) for this purpose that gets loaded by the controller when it executes a RESTORE command. The second function of the DCF is to provide the formatting required to interface directly to the host CPU bus. Actually, the DCF and controller have been designed so that one, two, or four DCFs may be either connected to the CPU bus or to the controller. When connected to the controller, it formats the data into 8-bit bytes and presents it to the host CPU through the controller interface. The third function of the DCF is to provide error correction.

In single-channel systems, the basic error rate of the bubble device itself of approximately 10^{-10} errors per bit is usually acceptable. A typical single-channel, floppy disk replacement would make an error every 160 days, assuming a 50% read/write ratio and a 5% usage rate during an 8 hour day. A 16-channel, hard disk replacement is a different story. At the basic error rate of 10^{-10} errors per bit, this system would make an error every 19 hours, assuming a 50% read/write ratio and a 10% usage rate during a 24 hour day. The DCF will decrease the basic error rate to about 10^{-14}. The hard disk replacement now will make an error every 11 years.

Three basic types of bubble memory systems and several interconnect options have been considered. This brings the designer of an MBM system to the point where the merits of each system type and configuration must be examined closely and a choice made. Selecting which MBM system configuration is right for a particular application is largely a matter of *preventing over-design* to minimize overall system cost. For example, error correction or high data rates, if not required, simply add unneeded costs to an MBM system.

7.6 PRACTICAL MBM DESIGN

The following examples illustrate the steps taken by an MBM systems design team from the conception of a product through making it ready for production. The basic problem is to design a bubble memory system for a TM990 microcomputer that will be used for nonvolatile storage of files and data.

EXAMPLE 7.1 Define the steps required in designing an MBM system.

PROCEDURE

1 Choose the amount of bubble memory storage that will be required.
2 Determine the performance requirements of the MBM system.
3 Determine flexibility and expandability requirements.
4 Determine physical space limitations.
5 Determine the amount of power available from the host system power supply and whether or not a power fail warning signal is provided.
6 Study the system bus specification to understand how the host system will communicate with the MBM system. The type of host CPU must be known, and its I/O instructions fully understood.
7 Draw the MBM system block diagram. It is important to involve the software designers at this point, to be completely sure that the proposed hardware approach will not place too heavy a burden on the host CPU.
8 Draw the schematic. This schematic needs to show the interconnections as well as required decoupling or bypass capacitors. It is very important at this point to put a major amount of thought into the eventual layout.
9 Lay out the PC board and have prototype quantities built.
10 Debug the prototype and determine the best method of testing the production boards to follow.

EXAMPLE 7.2 Determine the amount of data storage required.

PROCEDURE

1 The TM990 microcomputer is a multipurpose device that was designed with widely varying industrial applications in mind. To serve a useful purpose in such a system, the bubble memory board had to be an extension of that philosophy. After examining traditional mass storage media such as a floppy disk, it was decided that the most sensible MBM system size would be one megabyte of storage of 8-bit bytes.
2 This amount of storage could not be sensibly handled with a major-minor loop; a megabit device, therefore, was chosen. Eight one-megabit devices will provide the one megabyte of storage desired.

Performance

Next, the required performance is considered. The performance of a floppy disk is used for comparison. Floppies, in general, have higher average data rates than the basic single-channel data rate of the one-megabit device; average random access times, however, are much lower on a floppy. Because both of these parameters are important to the overall performance of any memory system, some simple way to quickly compare systems is necessary. One way to do this is to define a *performance factor, P,* as:

$$P = (R \times RI)/(T \times TI) \text{ bits/s}^2 \qquad (7.1)$$

where R = average data rate (bits/s), RI = average data rate importance (numeric), T = average random access time (s), and TI = average random access time importance (numeric).

Terms RI and TI are weighting factors that are derived empirically. A system that transfers large blocks of data would have a larger RI than TI, because random access must only be performed once for each large block. A system that has a lot of random accessing, such as compiling, would usually have a higher TI factor.

Equation 7.1 gives a performance indication of any type of memory system. Variables RI and TI must be determined by careful system requirement study. For this MBM design, the importance of access time and data rate were judged to be equal, which causes RI and TI to cancel in Eq. 7.1 leaving:

$$P = R/T \text{ bits/s}^2 \tag{7.2}$$

EXAMPLE 7.3 Calculate the performance factor for a megabit device if R = 85,000 bits/s and $T = 11.2 \times 10^{-3}$ s. Assume $RI = TI$.

PROCEDURE

By Eq. 7.2,

$$P = 85{,}000/(11.2 \times 10^{-3}) = 7.59 \times 10^{6} \text{ bits/s}^2$$

Table 7.2 provides a comparison of performance factors for various types of memories, assuming equal importance between access time and data rate. Because the performance factor for a one-megabit MBM is nearly two orders of magnitude higher than the performance factor for the floppy disk, single-channel operation for this MBM board would be more than adequate.

Flexibility and Expandability Requirements

For a product such as the MBM system under consideration, it is always advisable to make it as flexible and expandable as possible. It was decided to design the system in such a way that one board could be set up as the controller board and

Table 7.2 **Performance factor comparisons of various memories**

Memory type	*Performance factor, P*
High-speed 1K × 8 PROM	5×10^{15}
250 ns 16K RAM	1.28×10^{14}
1 megabit MBM	7.59×10^{12}
Floppy disk	1.2×10^{11}
Cassette tape	55

the others would be set up as slave boards. This would provide a software transparent extension of an existing board. It was also decided to place address decode switches on the board so that system expansion could also be handled by adding additional boards as controllers. This would provide a method of expansion that was not transparent to the software, but rather, looked like an additional peripheral in the system.

Physical Dimensions Consideration

The design goal for this system was to put all eight one-megabit devices on a single board. Because only one channel would be operating at a time, there was no compelling reason to use an advanced, parallel, controller unless hardware error correction would be required. It was decided that, owing to the data rates involved, error correction would not be needed. This decision left the single-channel controller system as the only logical choice for this design.

Power Supply Estimates

The next step is to make some rough power supply estimates.

EXAMPLE 7.4 Estimate the power supply requirements for the MBM system under consideration.

PROCEDURE

1 The following values may be used for the estimate:

	100% duty cycle	Idle
Controller and FTG	1.1 W	1.1 W
	(220 mA, +5 V)	(220 mA, +5 V)
MMU	5.1 W	1.2 W
	(47 mA, +12 V)	(21 mA, +12 V)
	(256 mA, −12 V)	(41 mA, −12 V)
	(283 mA, + 5 V)	(99 mA, + 5 V)

2 For this board, the power supply must handle seven idle MMUs and one active MMU, in addition to the controller and FTG. The calculation procedure is as follows:

$$+12\text{ V: } 7 \times 21 + 47 = 194\text{ mA; } 0.194 \times 12 = 2.3\text{ W}$$

$$-12\text{ V: } 7 \times 41 + 256 = 543\text{ mA; } 0.543 \times 12 = 6.5\text{ W}$$

$$+5\text{ V: } 7 \times 99 + 283 + 220 = 1196\text{ mA; } 1.196 \times 5 = 6\text{ W}$$

3 Totaling the power figures shows that this board will dissipate about 15 W when being accessed, which is the value that must be used when selecting a power supply. Because this board plugs into a variety of configurations, however, a single power supply choice is not possible. A different size supply will be required for each configuration.

4 The matter of power-fail warning was next considered. It was decided to put a monitor on the board to detect when power was either coming on or going off. This meant that the power supply did not require a power fail warning signal and a considerable cost savings was thereby realized.

Interfacing

The bus for the TM990 is well defined, as are all popular buses. The bus specification indicates that the primary means of input/output is by the CRU (*communications register unit*). This is a high-speed serial link that greatly simplifies both hardware and software in the TM990 system. Hardware is simplified because only two interconnecting lines are required for each CRU peripheral, compared to 16 when parallel I/O is used. Software is simplified by the fact that the host CPU has a special set of I/O instructions specifically designed to support a CRU. With CRU interfacing understood, the controller data sheet must be examined to see how compatible it is with this method of I/O.

EXAMPLE 7.5 Define the system structure of this MBM system with a block diagram.

PROCEDURE

1 Figure 7.15 shows the block diagram. Notice the addition of several blocks that were not present in any of the previous block diagrams in this chapter. These blocks represent circuits that give this system additional features that make the board more appealing as an OEM product.

2 The interface of this board to the TM990 consists of the address bus lines that are labeled A1–A14, the CRU data lines labeled CRUOUT and CRUIN, and the CRU clock line labeled CRUCLK. The address decode block compares the address bus to the settings in the DIP SW1 (dual inline switch 1), and if they do compare, WMAPEN/, STAT/BUBEN/, BMCEN/, or MBMEN/ is set to logic 0.

These signal name abbreviations stand for Write MAP ENable, STATus BUBble ENable, Bubble Memory Controller ENable, and Magnetic Bubble Memory ENable. The slash (/) following the name is equivalent to the bar over the name on the block diagram, and means that the signal is active when low. These signals are used to enable the various circuits on the board and, one at a time, will go active according to which address is decoded (indicated in the ADDR versus FUNCTION chart on the diagram).

3 The BUB SEL (BUBble SELect) block generates an 8-bit bubble address that is fed through 3-state buffers to an 8-bit adder. Here this address is added with the setting of DIP SW2 to determine if the bubble being addressed is on this board or not. The LOGIC block uses the carry out of the ADDER to generate BOARDEN (BOARD ENable) if the bubble is on this board or OUTPUT STRREQ (SToRage REQuest) if the bubble is on another board. The DIP SW2 is set according to the number of bubbles on the board.

4 The MAP WRITER block consists of circuitry that allows rewriting the redundancy map in case of accidental loss. The WMAP switch is for disabling the map writer circuitry.

5 The RESET AND DISABLE CIRCUIT monitors the indicated TM990 bus signals to generate the RESET or DISABLE/ signals. RESET is sent directly to the BMC block to reset the controller. DISABLE is generated when power is transitioning, and is sent

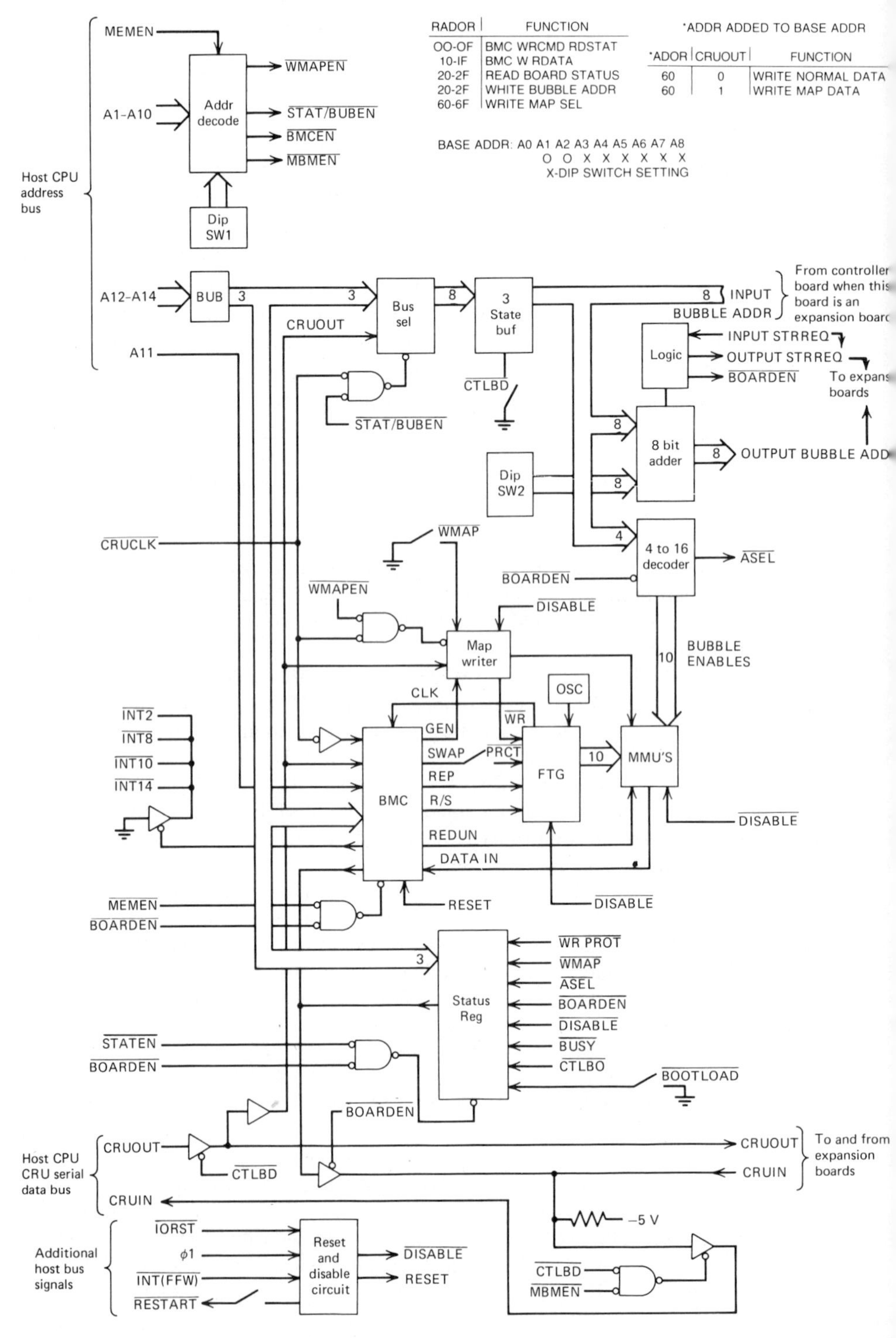

Figure 7.15 Block diagram of a practical 8-megabit bubble memory board.

to the FTG and MMUs to prevent any spurious pulses from being applied to the MBM devices that might alter data or cause an unwanted bubble rotation. The STATUS REGister is an 8-bit addressable latch that can be read by the host to determine the status of the board. The status lines are write protect, write map, A half select, board enable, disable, busy, controller board, and boot load.

The next step after the detailed block diagram is completed is to generate a schematic diagram of the system. It should follow very closely the block diagram from which it was derived. Careful attention to the grounding points around the MMU interface circuits is necessary.

PCB Layout

With the paper design completed, one is ready to build the board. First, the board is a printed circuit. The design team at Texas Instruments had absolutely no success with wire-wrapping or solder-wrapping techniques. The only successful approach to the layout problem has been to use either double-sided or multilayer PC boards with carefully chosen grounding points, decoupling capacitors, and as much ground plane as possible.

The basic reason that layout is so critical in a bubble memory system is that the bubble signal is very small (10 mA or so), and the coil drive signals are very large (800 mA). Figure 7.16 illustrates how a bubble signal looks on a scope for both a bubble present ("1") and a bubble not present ("0"). In addition, the function drive and sense amplifier reference inputs must be extremely quiet to achieve proper device operation. This means that the coil drivers, in general, must be as isolated as possible from the rest of the MMU interface circuits. This isolation must be both physical, by choice of component placement, and electrical, by choice of proper decoupling methods.

EXAMPLE 7.6 Develop a layout of a typical MBM cell.

PROCEDURE

Figure 7.17 shows the component placement on the MBM board. The layout observes the following guidelines:

1 Coil drivers.

a. *Etch runs.* The etch runs should be short, direct, and as wide as possible. Also, there should be a solid ground plane surrounding the entire coil drive area.

b. *Positioning.* The coil drivers should be located as close to the MBM device coil drive leads as possible and as far from the sense amplifier as possible.

c. *Capacitors.* Each coil driver should be decoupled from the power supplies by placing bypass capacitors on each voltage located as close as possible to the coil drivers. A 22 μF and a 1 μF capacitor should be placed in parallel on the −12 V supply, and a 1 μF capacitor should be placed on the +5 V supply. The 22 μF capacitor is for decoupling low-frequency noise, while the 1 μF capacitors are for decoupling high-frequency noise. These must be high-quality tantalum capacitors.

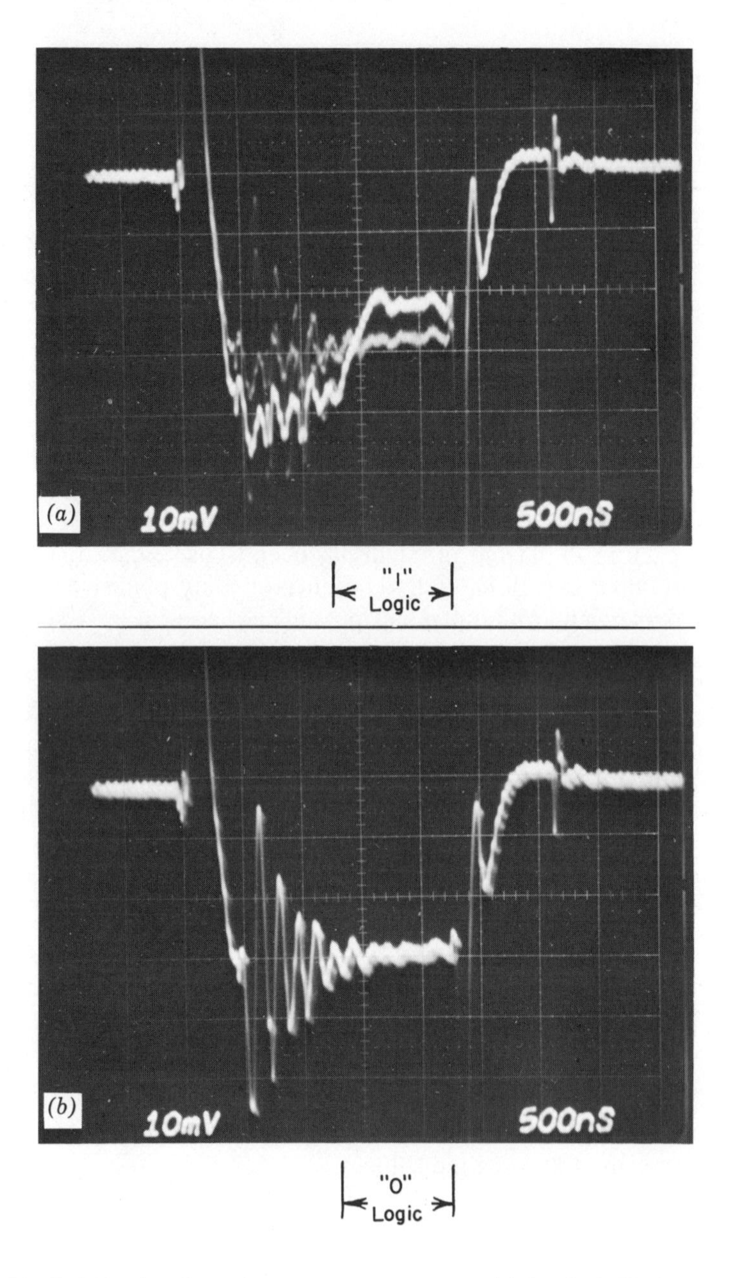

Figure 7.16 Bubble signals as they appear on a scope. (*a*) Logic 1. Notice some logic 0s also; these stem from the redundant masked-off loops. (*b*) Logic 0.

d. *Grounding*. The grounding points for the coil drivers and their associated decoupling capacitors should be as close together as possible.

2 Sense amplifiers.

a. *Etch runs*. The etch runs should be short and direct, especially on the detector

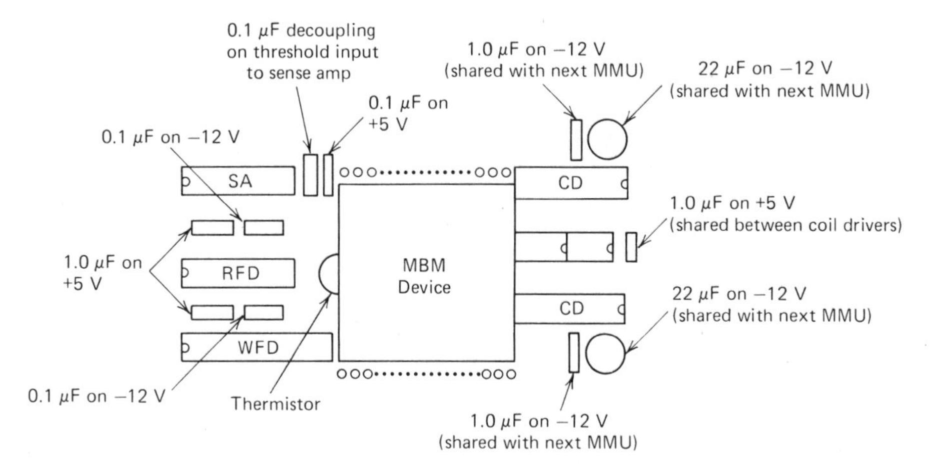

Figure 7.17 Proper cell layout including decoupling capacitors. This is the approximate layout used for the 8-megabit board described in the text.

leads. The detector leads should run parallel and be the same length to take advantage of the good common mode rejection ratio (CMRR) of the sense amplifier. Clocks, or other such signals, should not run near the detector leads or threshold input.

b. *Positioning.* The sense amplifier should be located as far as possible from the coil drivers or other noise-producing sources on the board. In multiple-board systems, location of the sense amplifier should also take into account the location of noise sources on adjacent boards.

c. *Capacitors.* The +5 V supply should be decoupled with a 0.1 μF capacitor located as close as possible to the sense amplifier. A 0.1 μF capacitor should be connected from the threshold input to ground. In extremely noisy environments, 150 pF capacitors may be required across the detector inputs to the sense amplifier.

d. *Grounding.* It is extremely important to ground the threshold capacitor at the ground pin of the sense amplifier. It is advisable to connect the +5 V decoupling capacitor ground at this point as well.

3 Function drivers.

a. *Etch runs.* Etch runs should be kept as short, direct, and isolated as possible. The runs to reference resistors are especially sensitive to noise, and must be kept as quiet as possible.

b. *Positioning.* The function drivers should be located as close as possible to the MBM device and as far as possible from the coil drivers.

c. *Capacitors.* The +5 V supply should be decoupled with a 1 μF capacitor, and the −12 V supply should be decoupled with a 0.1 μF capacitor. Generally, these two capacitors may be shared between the read and write function drivers.

d. *Grounding.* The ground point for the reference resistors must be the ground pin of the write function driver package. It also is advisable to ground the decoupling capacitors at the same point.

4 Function timing generator and controller.

These are not particularly sensitive to the layout and should be treated as any other digital device. Power supplies should be decoupled close to the package with 0.1 μF capacitors.

7.7 TESTING AND DEBUGGING

After fabricating the PCB for this MBM system and assembling it, the next step is debugging it. The simplest way to debug a bubble memory board is to write special test software that will exercise only a small portion of circuitry at a time.

EXAMPLE 7.7 Provide a testing and debugging procedure for an MBM system.

PROCEDURE

1 *Initialize controller.* This test will establish all of the initial values of the internal registers in the controller.
2 *Write redundancy map.* This test allows a user-selected pattern to be written in the redundancy RAM within the controller. The controller interface to the host system is the first step in debugging a bubble memory board.
3 *Read redundancy map.* This test will read back the contents of the redundancy map and display them for comparison with what was written in the redundancy RAM during step 2 above.
4 *Read status register.* This test will read the status register and display its contents. This is normally one of the first tests run on the board, because a good deal can be learned from this information about whether or not the board is functioning properly.
5 *Restore.* This test brings the redundancy map out of the bubble device into the redundancy RAM of the controller so that reading and writing to the MBM device may take place.
6 *Write.* This test allows a selected pattern to be written to a selected number of pages beginning at a selected page. Typically, patterns that are the hardest for the MBM device to handle are hexadecimal C3C3, F7F7, and other patterns involving strings of ones separated by a zero. Other test patterns always used in testing MBM devices are 0000, FFFF, 5555, and AAAA.
7 *Read.* This test permits a selected page, or pages, to read from a bubble. Two or more pages are always read for each write so that replication errors will not be missed. Sometimes a replicator failure will cause the replicator to act like a transfer gate, in which case the data will be correct the first time it is read, but not on subsequent reads.
8 *Write redundancy loop.* This test is for rewriting the redundancy map back into the bubble in case of accidental loss. This feature would seldom be used unless a large permanent magnet was brought too close to the board.

Magnetic bubble memory systems provide a nonvolatile storage medium that will withstand harsh industrial environments. Designing a magnetic bubble memory system is a totally new and different experience. It requires an unusually high level of understanding of the MBM device itself, as well as learning a fairly structured set of design rules. It requires that the designer understand TTL design rules (see Chap. 1), as well as PC board layout.

It is recommended that PC layouts be done by hand and not by automated layout programs. Automated layouts will not provide adequate ground and power planes and often run signal lines for sense amplifiers close to noise sources. In addition, the length of a line often cannot be controlled in an automatic layout.

The design of a typical MBM board has been covered from conception to tested product in an attempt to provide enough information to design one's own system. All bubble memory devices and systems on the market today are similar in nature. The design problems, layout considerations, and failure modes are essentially identical.

7.8 ERRORS

With the software package described in Ex. 7.7, testing and debugging the hardware is relatively straightforward. One area that does require close attention is the analysis of errors. There are two types of data errors that can occur in a bubble memory system: hard and soft errors.

A *hard error* is an error that occurs when a data bit within a minor storage loop gets changed. This type of error can be caused by many things, but is normally found to be the result of a defective MMU interface IC, most often a coil driver. At temperature extremes, however, hard errors can occur because of improper functioning of the MBM device itself. Self generation, or the appearance of a new bubble where one was not supposed to be, can happen if the magnets do not track properly. This would occur if the device were near the point of stripe out, which was discussed earlier. If the bubble devices were near the collapse point, they would disappear.

A *soft error* is an error that occurs once, but not again. This type of error is also known as a *read error* and can be caused by either a faulty sense amplifier or a faulty bubble device itself. If the sense amplifier's internal threshold changes slightly, it may start either picking up or dropping ones, depending on which way it drifted. Again, as with hard errors, soft errors can occur at temperature extremes owing to self-generation or collapse if the problem occurs on the read track.

7.9 REFERENCES

Chang, H., *Magnetic Bubble Memory Technology,* Marcel Dekker, New York, 1978.

Cox, G., "Storing Data in Magnetic Bubbles," *Machine Design,* p. 60, March 8, 1979.

Cox, G., "Bubble Memories—Mass Storage in the Palm of Your Hand," *Instrument and Control Systems,* p. 59, June 1979.

Cox, G., "Designing with Magnetic Bubble Memories," *Control Engineering,* p. 54, July 1979.

MacDonald, R., "Magnetic Bubble Memories for Large and Small System Nonvolatility," WESCON/80 paper 15/5, September 1980.

MacDonald, R., "Application Considerations of Bubble Memories," MIDCON/80 paper 13/3, November 1980.

CHAPTER 8

Microprocessors and Microcomputers

Dick Eden, President
Ellery W. Potash, Vice President
Intra Computer, Incorporated

8.1 INTRODUCTION

The term microprocessor (MPU, μP) refers to a central processing unit (CPU) consisting of an arithmetic and logic unit (ALU), a control block, and a register array, all designed on one or more chips. A microcomputer (MCU, μC), is a microprocessor combined with a memory storage system and peripheral input/output (I/O) chips (Fig. 8.1). The basic operations that a microprocessor can perform are listed in an instruction set, which can then be used to program the microcomputer for particular applications.

Though the terms "microprocessor" and "microcomputer" are sometimes used interchangeably, there is a distinction between the two that should be clear to the reader. First, the prefix "micro" hints at the smallness of the device. State of the art, LSI technology allows a complete system to be constructed upon a silicon wafer of only 30 to 40 thousand square mils in area. Second, these devices are characterized by a "micro" price; available to OEMs (equipment manufacturers) in thousand-piece quantities for under $10.00 each. This cost is contrasted with that of minicomputer systems that generally range from $2000 to $35,000, and mainframe computer systems that can cost millions of dollars. The microprocessor cost is quoted here in thousand-piece lots because, more often than not, microprocessors are being engineered into many mass-produced industrial and consumer products, including gasoline pumps (calculating price/gallon, causing automatic shut-off, and recording the purchase for automatic credit card billing), automatic sewing machines (allowing many different stitch patterns at

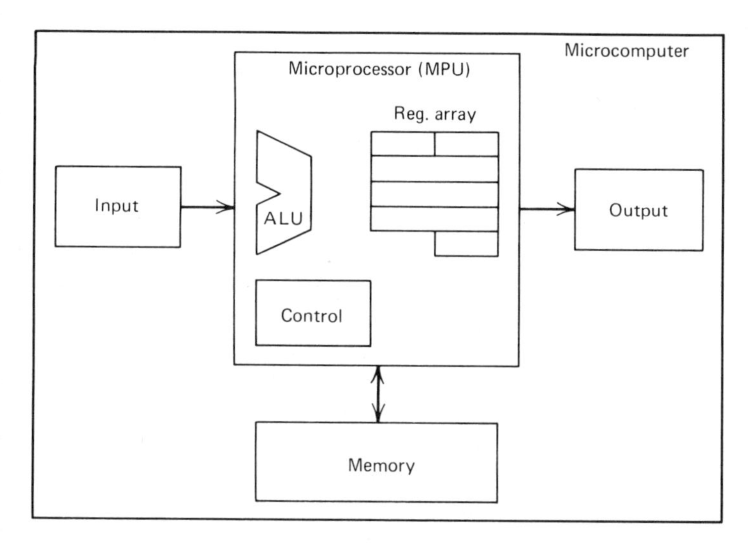

Figure 8.1 Simplified block diagram of a microcomputer.

the touch of a button), microwave ovens (selection of proper timing and temperature), digital communications, electronic games, and so on.

When we think of computers, images are brought to mind of a tool used to deliver numerical solutions to properly coded mathematical problems, for example, calculating the area of a circle using the classical formula $A = \pi r^2$. The term microprocessor more clearly describes the limited capabilities of the device as a *process controller*. Simple functions, such as multiplication and exponentiation, are accomplished only indirectly through complex and lengthy software algorithms.

A typical microprocessor application may be found in an automatic drill-press controller. After locating the piece to be drilled under the drill bit and locking it in place, the microprocessor is programmed to move the mounting platform (via stepping motors) so many units in the X and Y directions to locate the bit over the desired place for the first hole. The processor would then lower the bit onto the workpiece at the required speed and to the required depth for that hole. Next, the processor would raise the bit out of the hole and move the workpiece, then locate the coordinates of the second required hole, and repeat the process.

The computational limitations of a computer are most closely related to the operational word size of the unit. A *word* is a group of bits representing instructions or data acquired by the computer in a single fetch operation. With larger words, numbers of greater magnitude and instructions of higher complexity can be handled during a single machine cycle. Smaller words mean that more words and hence, more machine cycles are required to perform the same function as those machines with larger words. Large, mainframe computers generally operate on words of 16 to 32 bits. The word size of a minicomputer is usually 12 or 16 bits. Most microcomputers are *byte*-oriented machines, a byte being the term for an 8-bit word. The obvious conclusion is that microcomputers require more instructions and longer execution times to perform the same operations as their larger counterparts.

8.2 SYSTEM COMPONENTS

There are three basic building blocks common to all computers. One is the MPU (called the CPU in larger machines) which controls input/output operations, and performs basic arithmetic and logic operations on various forms of data. A quantity of memory words for program storage and storage of data from intermediate calculations is required for program execution. An interface between the MPU and peripheral devices (such as the stepping motors in the drill-press example, or a Teletype printer and keyboard for human interaction), allows program acquisition of data, or control of external devices by the computer.

All components of the microprocessor system (MPU, memory, peripherals) communicate with each other via three signal busses. (A *bus* is a group of signal-carrying wires or printed circuit traces having a common purpose.) The *address bus,* in most systems, is composed of 16 lines that carry the address of the memory location or an I/O device that the MPU will write data to, or read data from, as dictated by the program. Using 16 lines, the processor can directly access up to 65,536 (2^{16}) external memory locations. The address bus is *unidirectional,* that is, data flows only in one direction.

The *data bus* is eight lines wide (byte size), and is *bidirectional*; that is, data can flow toward or away from the processor over these lines, depending on whether a read or write operation is in progress.

The number of lines in the *control bus* varies from one processor to the next, but generally includes signals to interrupt, halt, or restart the processor, and status signals to indicate whether an instruction or data byte is being processed, or a read or write operation is occurring. Figure 8.2 illustrates typical bus connections. The control bus is bidirectional.

Except for the CMOS (complementary metal oxide semiconductor) families of microprocessors, all bus signals conform to TTL (Transistor Transistor Logic) interface specifications (logic 0 < 0.8 V, logic 1 > 2.4 V) with fan out capability of 1 TTL load and 130 pF (see Chap. 1). Considering this limitation,

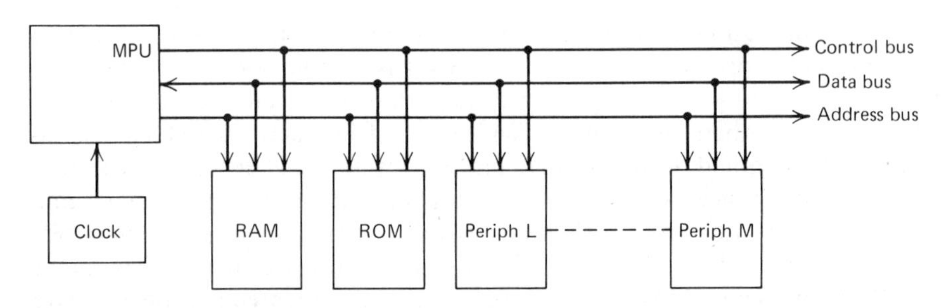

Figure 8.2 Typical microcomputer bus interconnections.

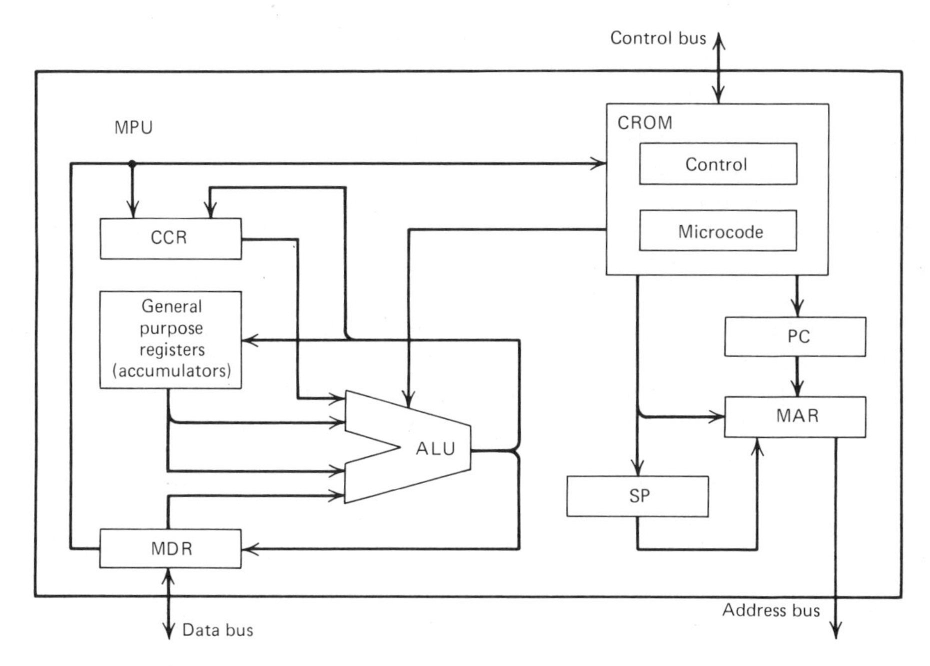

Figure 8.3 Block diagram of a generalized microprocessor.

we can attach approximately seven to 10 family devices to the busses without the need for special bus buffering circuitry. Exceeding this restriction without added buffering degrades signal quality and increases the chance of a malfunction.

8.3 INSIDE THE MPU

Within the microprocessor are the components that combine to function as the "brain" of a computer system (Fig. 8.3). A brief description of the basic units follows:

CROM The control and read-only memory section examines the words introduced to the MPU to determine whether they are program data or executable instructions, and directs the other MPU components to act on them accordingly. An internal microprogram code is used by the CROM to signal whether logic (AND, OR), arithmetic (ADD, SUB), shift, or I/O functions are to be performed. An externally or internally applied clock pulse synchronizes these operations.

ALU The arithmetic and logic unit is a three-port device used to perform both unary (one operand) and binary (two operand), arithmetic and logic operations on register or memory words. (An operand is a symbolic or numeric quantity upon which an operation is to be performed.) The ALU is supported by two special-purpose registers that are generally not under the control of the programmer. The MAR (memory address register) contains the current address in the main memory of the program *data* to be acted upon by the MPU. When the CROM detects an instruction that requires an action be taken upon the contents of a main memory location, that location's address is held in the MAR.

At the appropriate step in the operational sequence of the MPU, the data at the address pointed to by the MAR are loaded into the MDR (memory data register) to be processed by the ALU. Microcoded commands issued by the CROM direct the ALU to perform either unary functions, such as complement, shift, and rotate, or binary functions such as ADD, OR, XOR, AND, and INCREMENT (for which the second operand is the literal constant 1). The result of the operation performed is directed to the destination register or storage location specified in the program via a command issued by the CROM.

Register Array The registers are the temporary storage units inside the MPU, under direct control of the CROM. Because of their speed characteristics (relative to external RAM and ROM, discussed in the following pages and in Chap. 6) and their electrical proximity to the ALU, all MPU operations are usually carried out between registers, or between a register and a main memory word. Registers can either be general purpose types, or they may have specific uses such as an accumulator, program counter, condition codes register (status register), and stack pointer.

ACC The accumulator is the register that temporarily stores arithmetic or logic results of the ALU. Every MPU has to have at least one accumulator; some have several.

PC The program counter is a register that points to the next code location in main memory to be executed by the MPU. The contents of the PC will normally step sequentially through a program, unless altered by a branch or a jump instruction, or by a program interrupt sequence.

CCR The condition codes register contains a word composed of several FLAG bits. These flags are turned on or off depending upon the numerical result of the last instruction executed by the MPU. They represent whether the result of an operation was zero or negative, and whether a carry or borrow was required. The next sequential instruction may examine these flags to determine if any special action, such as a branch or jump, should be taken as a consequence.

SP A stack pointer serves to locate the next available location for data storage in a LIFO (Last In—First Out) STACK. This stack provides a convenient means of storage for register contents when these registers are to be used in subroutines or interrupt handling routines. In some microprocessors, both the stack and the stack pointer are internal to the MPU. Others maintain the stack within the main memory and enable modification of the stack pointer under program control.

IR Some microprocessors have special purpose registers called index registers. These are used to aid in specifying a main memory address for program access outside the program's immediate range of influence, or to access data stored in tabular form.

8.4 MEMORY*

Every computer system, including microprocessor-based systems, requires one or more forms of memory media for storage of program instructions, data, and/or a scratchpad work area for intermediate task results. These media are classified into three broad categories: read-only memory (ROM), sequential access memory (SAM), and random access memory (RAM). Virtually all memory types fall into one or more of these categories.

ROM Read only memory, as the name implies, contains data that can only be read by the computer. The computer cannot alter or delete data programmed in a ROM. ROMs will retain their data even when there is no power applied to the system. Examples of ROM media include punched cards, paper tape, and various forms of semiconductor arrays. There are some forms of semiconductor memories designed specifically to overcome some of the limiting characteristics of the ROM. These are the:

PROM Programmable read only memory can be programmed by the system designer using a portable programming machine that programs 1s or 0s into the memory by blowing fusable links, or depositing a charge on internal capacitive elements.

EPROM Erasable, programmable read only memory is used in the development stage of a microprocessor system. Data stored in the EPROM can be erased by exposure to ultraviolet (UV) light through a quartz window molded into the IC package for this purpose. The EPROM can then be reprogrammed as needed. Once the designer has verified that the software (program) functions correctly in the prototype system, there may be mass-produced copies of the memory data made by a masking process. This is done by the semiconductor manufacturer when the ROM is produced, rather than by the system designer.

SAM Sequential access memories usually require that their data be acquired in groups of words called *records* or *blocks* from units called *files*. As the name implies, the first word of the record must be accessed before the second, and the second before the third. SAM forms include punched cards and paper tape, magnetic tape, disks, and bubble memories. Unlike punched cards and paper tape, the magnetic media are also classified as read/write memory. The microprocessor can delete, create, or modify data by writing into these files, which can then be read back at a future time. Like ROMs, SAMs are nonvolatile, that is, they maintain the integrity of their data when shelved without power applied.

* Chapter 6 covers semiconductor memories in detail.

RAM Random access memories fall into one of three subcategories: magnetic core, semiconductor static, and semiconductor dynamic.

Core memory is constructed of millions of miniature "doughnuts" made of soft iron with select, write, and sense wires passing through each hole along three orthogonal axes. This tedious wiring is usually done by hand. Core memory is used almost exclusively in large installations, such as mainframe machines, because their cost, size, and nonportability preclude their use in most microprocessor applications. Core memory is nonvolatile.

Static memories are manufactured by any of several semiconductor processes, and feature small size, moderate cost, and solid-state portability. Data may be read from, or written into, any selected slot within the device. RAMs retain the written data only as long as power is supplied to the device. State of the art technology allows as many as 32K bits (4K bytes) to be manufactured on a single silicon chip.

Dynamic memories are generally less expensive and faster than static RAMs, as well as being capable of greater densities; up to 256K bits per IC. The distinguishing characteristic of dynamic RAMs is that they require a refreshing cycle. Every bit within the memory must be read on a regular basis whether or not the program requires the data therein. Additional support circuitry is generally required to refresh the dynamic RAMs without intruding on the normal memory access cycles of the microprocessor. Choice of either static or dynamic RAMs for a particular microprocessor application depends on a variety of factors, including allowable component count, cost, and available circuit board area, as well as speed (memory access time), and address decoding practices (see Chap. 6 for further details).

8.5 INTERFACING TO THE OUTSIDE WORLD*

To make the microprocessor a useful tool, there must be some way of interfacing its monitoring and control capabilities to the outside world. There are two basic methods of accomplishing this: either in parallel mode or in serial mode.

Parallel Mode

In parallel mode, data are transferred between the microprocessor and any external devices one byte at a time, using a control line strobe signal to indicate when data are valid. As 8 bits are transferred with a single read or write operation, data rates approaching 65 Kb/s (65,000 bytes/s) can easily be achieved in this manner. Parallel mode interfacing is generally restricted to short distance communications (for example, between the microprocessor and a local disk drive or high-speed printer), as the expense of running 10 pairs of shielded copper wire over distances greater than 15 feet can easily outpace the cost of the processor itself.

As an output device, the parallel port most resembles an 8-bit latch, made up of D-type flip flops with a common clock (Fig. 8.4). The D inputs are attached

* Chapter 16 covers interfacing in detail.

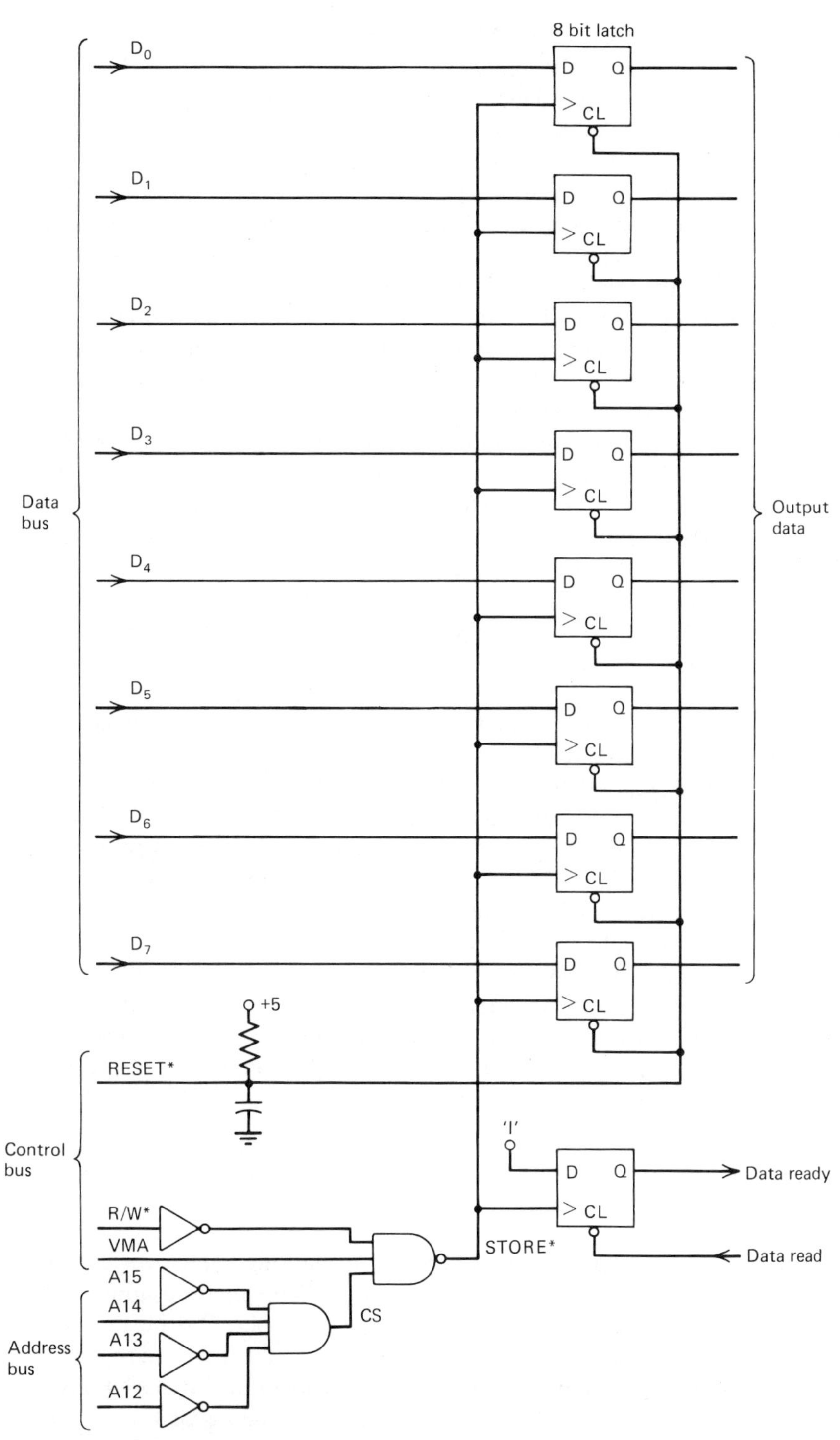

Figure 8.4 Parallel output latch.

to the processor's data bus while the clock strobe is derived from the address decode/select circuitry and the read/write control bus line. The VMA (valid memory address) control bus signal goes to logic 1 to indicate that the bits of the address bus have settled to their steady-state value. The R/W* signal turns to logic 0 to indicate that a write operation is taking place.

As an input device, the parallel port provides for the transfer of external data onto the bidirectional data bus of the processor system. Data from remote devices are latched into a parallel register by an externally supplied clock strobe. The availability of this data may be called to the attention of the processor by the exertion of a logic 0 signal on the IRQ* (Interrupt Request*) control bus line (see Fig. 8.5). Data held in the input latches are transferred to the data bus when the 3-state buffers (Fig. 8.5) are enabled. The enabling signal is derived from the address decode/select circuitry incorporating control bus lines VMA and R/W*. The signal, enabling the data to the bus, will also clear the IRQ* flip flop, acknowledging data receipt and enabling new data to enter. When disabled, the output of the 3-state buffers appears to the bus as a high impedance only. Enabling these buffers will place logic 0 or logic 1 data onto the bus, depending upon the state of the input.

The address decoding shown in the examples will select the parallel port whenever the hexadecimal address 4xxx is placed on the address bus by the processor. (xxx indicates the 12 don't care lines of the address bus not used in the example.)

In order to reduce the cost of implementing a parallel interface to the microprocessor bus, all components of the port (including bidirectional 8-bit latches, 3-state buffers, and status flip flops) have been integrated into a single circuit termed a peripheral interface adapter (PIA, Fig. 8.6). The PIA provides a universal means of interfacing peripheral equipment to the microprocesor. This device is capable of interfacing the MPU to peripherals through two 8-bit, bidirectional data ports and four control lines. No external logic is required for interfacing to most peripheral devices.

The functional configuration of the PIA is programmed by the MPU during system initialization. Each of the peripheral data lines can be programmed to act on an input or output, and each of the four control/interrupt lines may be programmed for one of several control modes. This allows a high degree of flexibility in the operation of the interface.

Serial Mode

In order to reduce the expense involved with transmitting bytes over extended lengths, it becomes necessary to serialize the data. This method allows information to be carried over a single pair of wires and permits the use of the telephone system for long-distance communication. Serialization and reconstruction are accomplished through the use of parallel load/unload shift registers. Serial communications can be done either in the synchronous or the asynchronous mode.

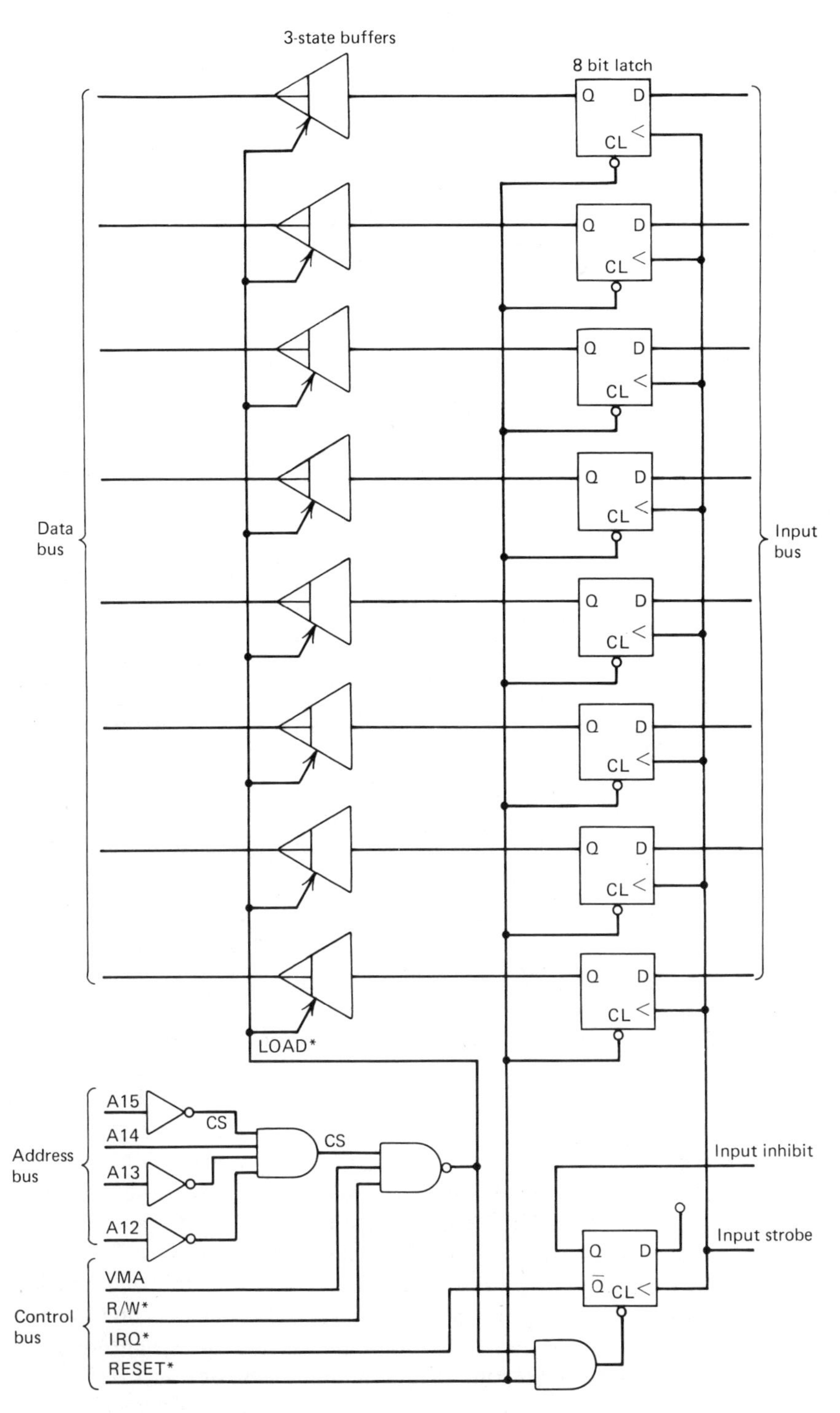

Figure 8.5 Parallel input latch.

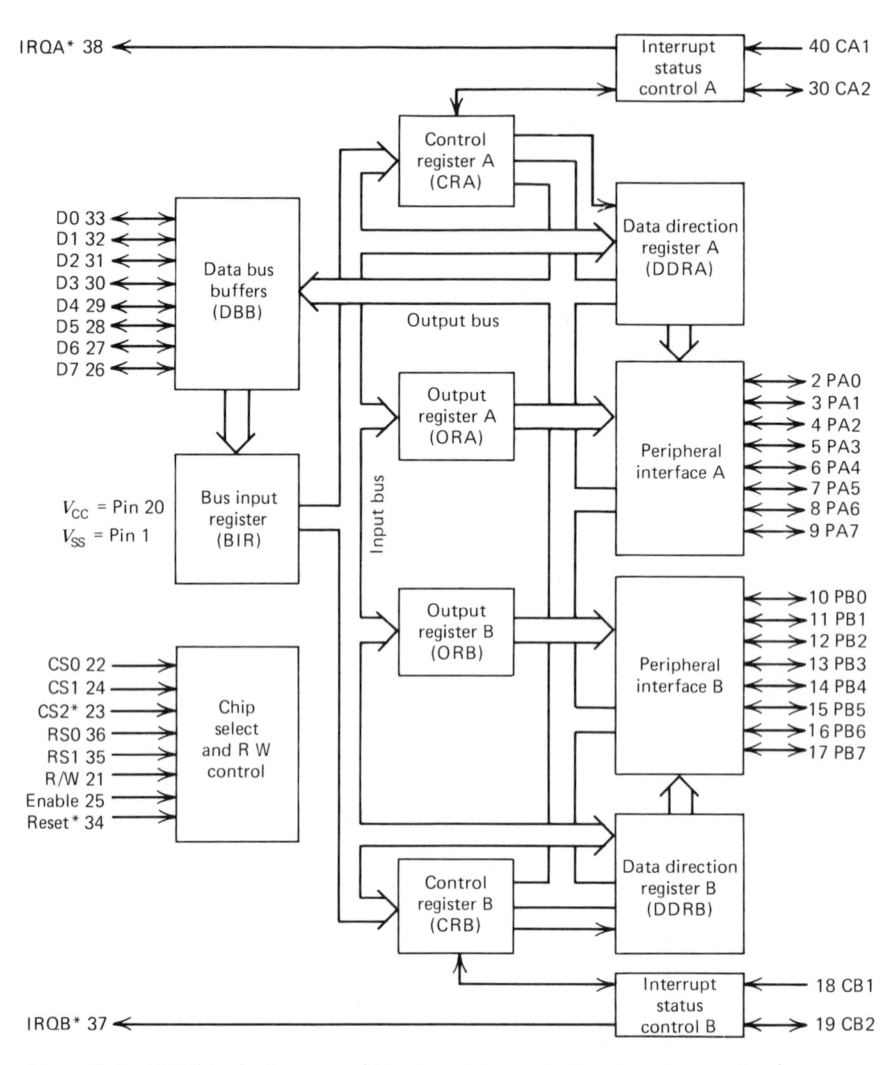

Figure 8.6 PIA block diagram. (Courtesy Motorola Semiconductor, Inc.)

Synchronous mode In the synchronous mode of serial transmission, a data byte is loaded into the shift register from the parallel data bus. This data is then shifted out of the register, one bit at a time, onto the transmission line (Fig. 8.7). The free-running clock signal, which strobes each bit out of the register in turn, is also transmitted to the receiving device over a separate pair of wires. The number of bits transmitted per second is called the *Baud rate*. This rate is dependent upon the clock frequency and is limited by the transmission line characteristics. Data rates exceeding 9600 Baud can be achieved using the synchronous mode. At the receiver, the clock input is used to synchronize the incoming data so that it can be reconstructed into parallel bytes by a system that functions in the opposite manner as the transmitter (Fig. 8.8). A detector circuit

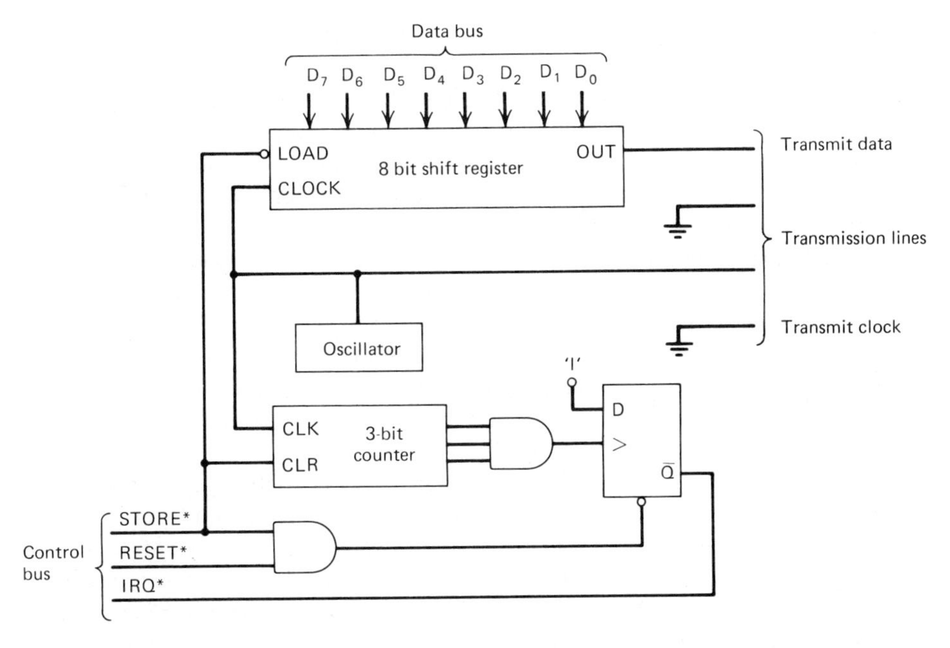

Figure 8.7 Synchronous transmitter. NOTE: STORE* signal is derived as for PIA selection.

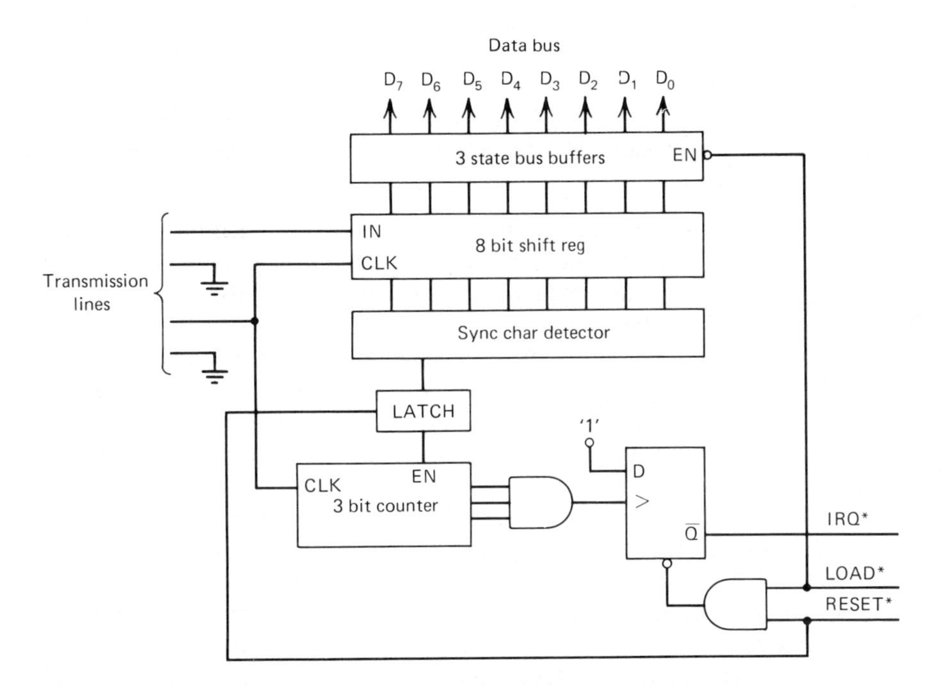

Figure 8.8 Synchronous receiver. NOTE: LOAD* signal is derived as for PIA selection.

compares the incoming bit stream with a predetermined synchronization character. When a match is recognized, a 3-bit counter is enabled to indicate the beginning of a character boundary. The counting circuit signals to the receiver that eight bits have been shifted in (an entire byte), and can be unloaded from the register in parallel form. At the transmitting end, a similar counter signals that eight bits have been shifted out and that the register is empty and ready to accept new input.

To boost economy, reliability, and functional efficiencies, the features necessary for synchronous communications have been integrated onto a single chip. The synchronous serial data adapter (SSDA) provides a bidirectional serial interface for synchronous data information interchange. It contains interface logic for simultaneously transmitting and receiving standard synchronous communications characters in bus organized microprocessor systems. The block diagram of the chip is provided in Fig. 8.9.

The bus interface of the SSDA includes select, enable, read/write, interrupt, and bus interface logic to allow data transfer over an 8-bit, bidirectional data bus. The parallel data of the bus system is serially transmitted and received by the synchronous data interface with synchronization, fill character, insertion/deletion, and error checking. The functional configuration of the SSDA is programmed via the data bus during system initialization. Built-in programmable registers provide control for variable word lengths, transmit control, receive control, synchronization control, and interrupt control. Status, timing, and control lines provide peripheral or modem (modulator/demodulator) control. A modem performs the function of an FM transmitter/receiver and allows serial digital data to be transferred over common telephone lines.

Typical applications of the SSDA include floppy disk controllers, cassette or cartridge tape controllers, data communications terminals, and numerical control systems.

Asynchronous mode Asynchronous techniques allow the transfer of serial data between two remote stations without the need of a synchronizing clock. In this mode, additional bits are appended to each character during transmission. These bits enable character synchronization to be established at the receiver.

At the beginning (before the least significant bit) of each character is placed a START bit that is a logic 0 (space) level. At the end of a character are placed one or two STOP bits that are at logic 1 (marking) level. These bits *frame* the character to indicate character limits and the occurrence of errors during transmission.

Data are usually transmitted in ASCII character format (Fig. 8.10). This format includes seven data bits, which allow 128 combinations for all the common printable characters, and an eighth bit, P, used for error detection, called *parity*.

The parity bit permits error detection by first having both the transmitter and receiver agree on whether to use odd or even parity. If even parity is selected, then for each character transmitted, the parity bit is set or cleared so that the total number of logic 1 bits between the START and STOP bits is an even quantity. The reverse is true for odd parity. If, for even parity transmission, an

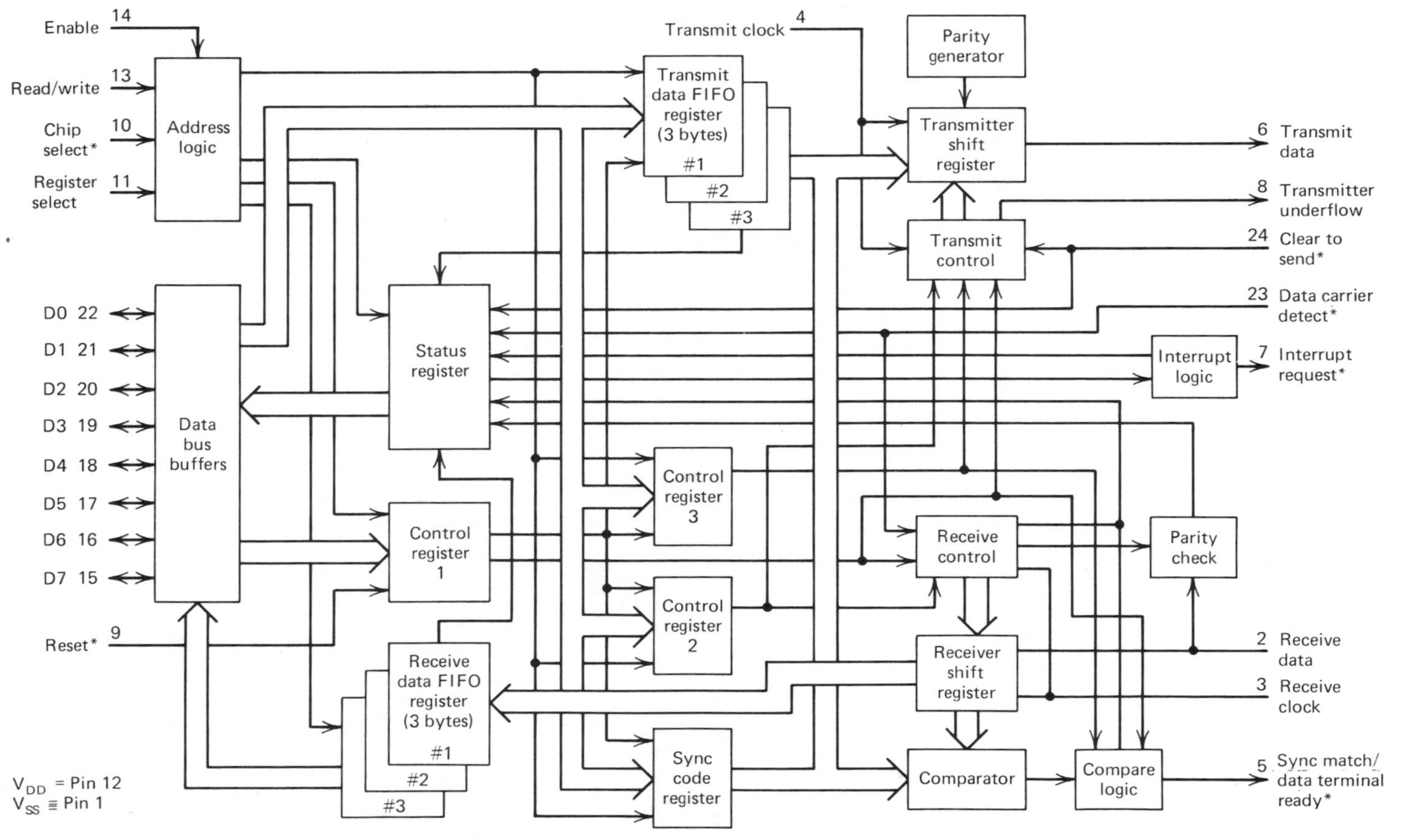

Figure 8.9 SSDA block diagram. (Courtesy Motorola Semiconductor, Inc.)

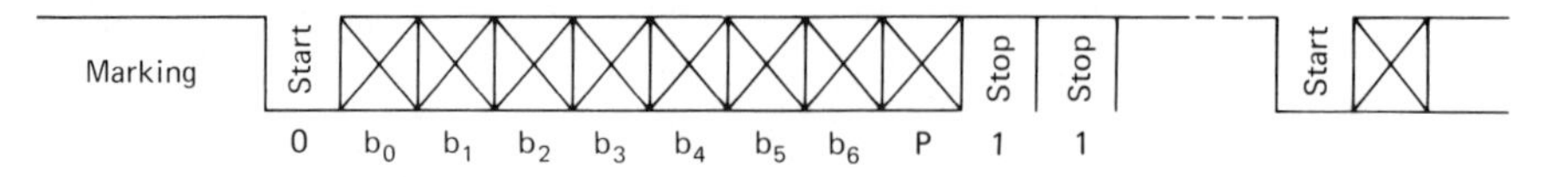

Figure 8.10 ASCII character format for asynchronous serial data transmission.

odd number of 1 bits is counted at the receiver, a PARITY ERROR is signaled and the receiver can ask for a retransmission of the message in error.

In addition to having both transmitter and receiver agree on the type of parity, they must also agree on the same Baud rate (number of bits being transferred per second), the number of information bits per character, and the number of STOP bits.

To provide character synchronization, the receiver monitors the incoming data line and waits for the first MARK to SPACE transition indicating the possible beginning of a start bit. An internal counter then causes a delay of ½ bit time, after which, the line level is checked again. If the line reads high (logic 1), then the signal was a false start probably caused by induced noise on the line. If the line continues to read low, then a true START bit is detected and the system is ready to accept the data bits. Once a valid start bit has been detected, the remaining bits are synchronized automatically and are shifted into the receiving shift register at their approximate ½ bit time. All internal operations, including time-delays for sampling at the bit midpoint and bit shifting, are controlled by a clock oscillator operating typically at 16 times the Baud rate.

The asychronous communications interface adapter (ACIA) provides the data formatting and control to interface serial asynchronous data communications information to microprocessor systems (Fig. 8.11). The bus interface of the ACIA includes select, enable read/write, interrupt, and bus interface logic to allow data transfer over an 8-bit, bidirectional data bus. The parallel data of the bus system are serially transmitted and received by the asynchronous data interface with proper formatting and error checking. The functional configuration of the ACIA is programmed via the data bus during system initialization. A programmable control register provides variable word lengths, clock division ratios, transmit, receive, and interrupt control. Three control lines are provided for peripheral, or modem control.

Asynchronous information throughput is generally slower than the throughput for synchronous systems. For each eight bits of real information, two or three bits of protocol padding (standard stops) are added to provide synchronization.

8.6 MEMORY MAPPING

A *memory map* is a graphical representation of how a computer's address space is allocated to the basic memory classes and peripheral devices. Most memories are mapped so that one or more forms of RAM respond to lower addresses. System software in ROM will respond to higher addresses on the bus. Peripheral

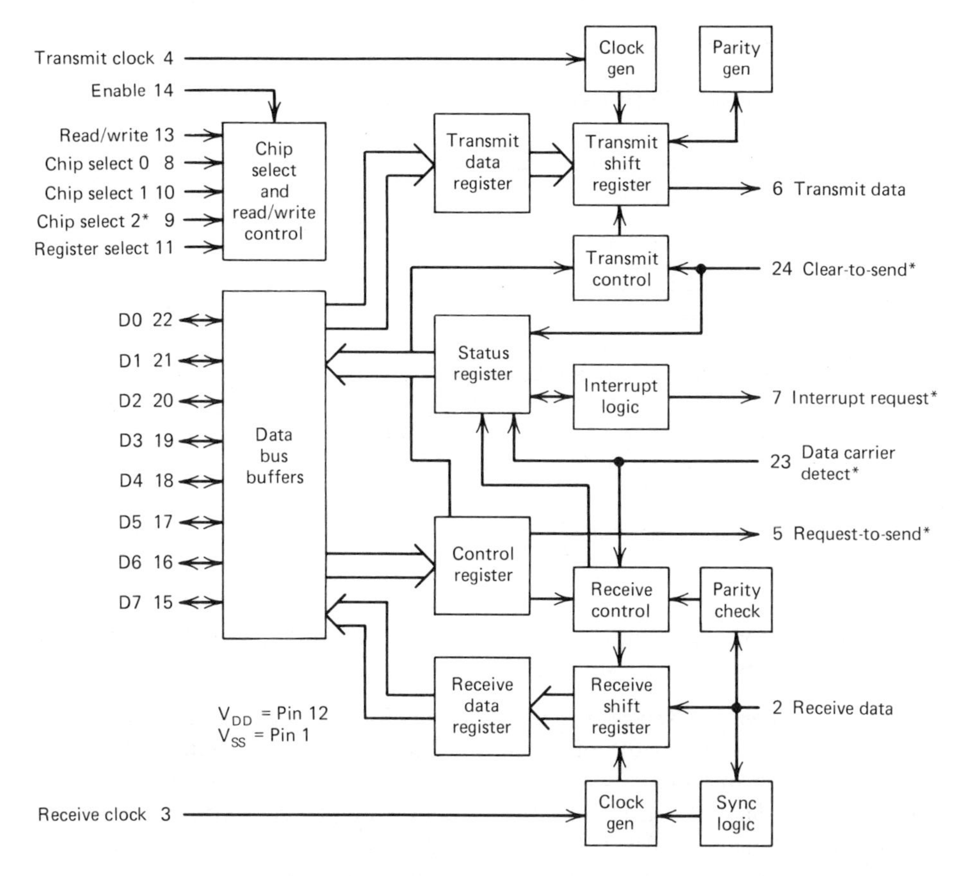

Figure 8.11 ACIA block diagram. (Courtesy Motorola Semiconductor, Inc.)

devices, including external memories, are located midrange. The address of a memory section or peripheral will depend upon how the system building blocks are wired together (Fig. 8.12).

8.7 PROGRAMMING

The computer program provides the means by which an ordered sequence of instructions is presented to the microprocessor for execution. The written program is referred to as *software,* as opposed to *hardware,* which is the term for the physical components (MPU, RAM, ROM, I/O) used to construct the system.

Those familiar with combinational logic (Chap. 1) recognize that any digital function can be realized using only AND, OR, and INVERT GATES as basic building blocks. The software equivalents of these functions form the basic INSTRUCTION SET of any digital computer. These simple instructions are augmented by ADD, SHIFT, ROTATE, LOAD, and STORE instructions and

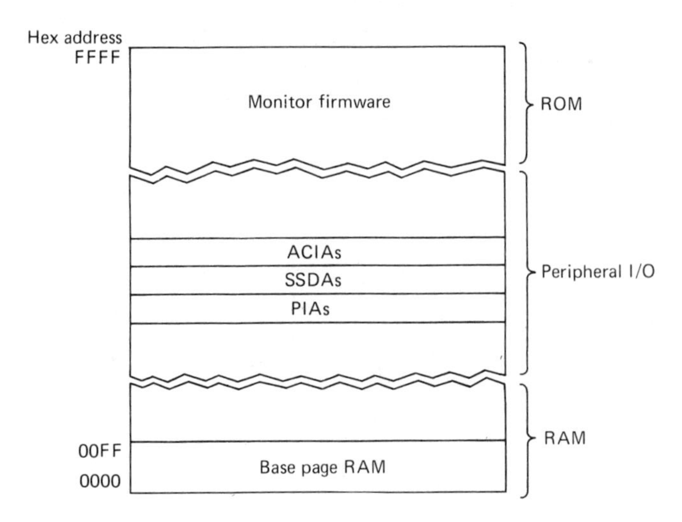

Figure 8.12 Microcomputer memory map.

operate on 8-bit words fetched from the microprocessor data bus. Programming a computer can be done on several levels:

Machine language A series of numeric codes for the various operations that the computer is capable of executing. The actual codes used will depend upon where in memory the program is to be loaded for execution.

Assembly language A series of alphabetic mnemonic codes corresponding to machine language instructions. A program written in assembly language requires an assembler that examines the assembly language mnemonics and translates them into machine language for execution. The assembler has the ability to RELOCATE assembly language program code to work in systems with different memory maps.

The following are four of the commonly used higher level languages:

BASIC A higher level language than assembly language, BASIC statements may include +, −, *, /, and exponentiation operators. A machine language program, called the BASIC interpreter, examines the program and compiles a series of machine codes for each BASIC statement. These codes can then be executed by the microprocessor.

FORTRAN Formula Translation is a higher level language than BASIC and is specifically designed for programming mathematical operations. A FORTRAN compiler generates machine code from FORTRAN program statements.

PL/M Program language for microprocessors is designed for scientific, control, and character processing applications. A machine language program, called the PL/M translator, converts PL/M source statements into machine code for microprocessor execution.

PASCAL Named for the famed French mathematician, this increasingly popular language promotes *structural programming,* which is a technique for making programs more readable and easier to debug. Larger programs are constructed from smaller, easy to understand modules.

8.8 ASSEMBLY LANGUAGE OVERVIEW

In cases where microprocessors are to be included in a high-volume, mass-produced product, they are most often programmed in assembly language. Assembly language is considered to be efficient. It allows the fewest number of instruction bytes to be used to accomplish a given task. This efficiency results in less required memory and faster execution of time critical operations. The disadvantage of using assembly language instead of one of the higher level languages is that the programmer using assembly language is required to specify a great many more simple instructions to the processor rather than a few comprehensive instructions to achieve the same result. Debugging and documenting an assembly language program also are generally more difficult for this low-level language.

Assembly language provides the programmer with an intuitive method of instructing the processor to perform a routine without having to remember each of the numeric codes that the processor decodes to perform a selected operation. It is considerably easier to remember that ABA means ADD B TO A than to remember the machine language code 1B, which has a meaning only to the processor. Similarly, ANDA, ORAB, and COMA are assembly language mnemonics for AND WITH ACCUMULATOR A, OR WITH ACCUMULATOR B, and COMPLEMENT A, respectively. Their hard-to-remember machine language codes are 84, CA, and 43, respectively. Assembly language mnemonics have been chosen by the chip manufacturer to simplify the programming task. The three or four letters that make up the code are generally the initials or the prefix of the operation they perform. It should be no surprise then that LD and ST, ADD and SUB, and SE and CL begin the assembly language codes for LOAD and STORE, ADD and SUBTRACT, and SET and CLEAR operations.

Tables 8.1 through 8.6 show a typical programming reference card supplied for a popular microprocessor, the 6800. Using this card, the assembly language programmer is made aware of all possible instruction types supported by the particular device. In Table 8.3, the first column of the card lists the operations and their mnemonic codes grouped by operation and in alphabetical order. The column headed "boolean/arithmetic operation" shows in standard notation exactly what function is performed when each instruction is executed. The columns on the right edge of the card describe the changes in the status flags of the condition code register (CCR) resulting from instruction execution.

The programming language provides a means for the programmer to instruct the computer to perform a series of tasks. Before writing a program, the programmer must have a clear understanding of which tasks are to be performed. It is often best to express the program tasks in a system flowchart.

***Table 8.1* Microprocessor instruction set—alphabetic sequence (courtesy Motorola Semiconductor, Inc.)**

ABA	Add Accumulators	INX	Increment Index Register
ADC	Add with Carry	JMP	Jump
ADD	Add	JSR	Jump to Subroutine
AND	Logical And	LDA	Load Accumulator
ASL	Arithmetic Shift Left	LDS	Load Stack Pointer
ASR	Arithmetic Shift Right	LDX	Load Index Register
BCC	Branch if Carry Clear	LSR	Logical Shift Right
BCS	Branch if Carry Set	NEG	Negate
BEQ	Branch if Equal to Zero	NOP	No Operation
BGE	Branch if Greater or Equal Zero	ORA	Inclusive OR Accumulator
BGT	Branch if Greater than Zero	PSH	Push Data
BHI	Branch if Higher	PUL	Pull Data
BIT	Bit Test	ROL	Rotate Left
BLE	Branch if Less or Equal	ROR	Rotate Right
BLS	Branch if Lower or Same	RTI	Return from Interrupt
BLT	Branch if Less than Zero	RTS	Return from Subroutine
BMI	Branch if Minus	SBA	Subtract Accumulators
BNE	Branch if Not Equal to Zero	SBC	Subtract with Carry
BPL	Branch if Plus	SEC	Set Carry
BRA	Branch Always	SEI	Set Interrupt Mask
BSR	Branch to Subroutine	SEV	Set Overflow
BVC	Branch if Overflow Clear	STA	Store Accumulator
BVS	Branch if Overflow Set	STS	Store Stack Register
CBA	Compare Accumulators	STX	Store Index Register
CLC	Clear Carry	SUB	Subtract
CLI	Clear Interrupt Mask	SWI	Software Interrupt
CLR	Clear	TAB	Transfer Accumulators
CLV	Clear Overflow	TAP	Transfer Accumulators to Condition Code Reg.
CMP	Compare	TBA	Transfer Accumulators
COM	Complement	TPA	Transfer Condition Code Reg. to Accumulator
CPX	Compare Index Register	TST	Test
DAA	Decimal Adjust	TSX	Transfer Stack Pointer to Index Register
DEC	Decrement	TXS	Transfer Index Register to Stack Pointer
DES	Decrement Stack Pointer	WAI	Wait for Interrupt
DEX	Decrement Index Register		
EOR	Exclusive OR		
INC	Increment		
INS	Increment Stack Pointer		

Flowcharts

The *flowchart* is a graphic representation of the sequence of operations to be performed. It allows a visual conception of the assignment and serves to document the process for status reporting as well as enabling several programmers on a project team to make their individual ideas known to all members.

There are several standard flowchart symbols recognized throughout the industry. The programmer may make up new ones as the need arises. The basic symbols are shown in Fig. 8.13. A directed arrow between flowchart symbols indicates direction of program flow. Information within the symbols describe the operand of the program step. The flowchart should be universal in nature, such that a programmer working in any of the available languages can code a program directly from the chart.

Writing Programs

By examining the completed flowchart, the assembly language programmer should be able to write source statements representing each chart block. There are *four*

Table 8.2 **Instruction addressing modes and associated execution times (times in machine cycles) (courtesy Motorola Semiconductor, Inc.)**

	(Dual Operand)	ACCX	Immediate	Direct	Extended	Indexed	Inherent	Relative
ABA		•	•	•	•	•	2	•
ADC	x	•	2	3	4	5	•	•
ADD	x	•	2	3	4	5	•	•
AND	x	•	2	3	4	5	•	•
ASL		2	•	•	6	7	•	•
ASR		2	•	•	6	7	•	•
BCC		•	•	•	•	•	•	4
BCS		•	•	•	•	•	•	4
BEA		•	•	•	•	•	•	4
BGE		•	•	•	•	•	•	4
BGT		•	•	•	•	•	•	4
BHI		•	•	•	•	•	•	4
BIT	x	•	2	3	4	5	•	•
BLE		•	•	•	•	•	•	4
BLS		•	•	•	•	•	•	4
BLT		•	•	•	•	•	•	4
BMI		•	•	•	•	•	•	4
BNE		•	•	•	•	•	•	4
BPL		•	•	•	•	•	•	4
BRA		•	•	•	•	•	•	4
BSR		•	•	•	•	•	•	8
BVC		•	•	•	•	•	•	4
BVS		•	•	•	•	•	•	4
CBA		•	•	•	•	•	2	•
CLC		•	•	•	•	•	2	•
CLI		•	•	•	•	•	2	•
CLR		2	•	•	6	7	•	•
CLV		•	•	•	•	•	2	•
CMP	x	•	2	3	4	5	•	•
COM		2	•	•	6	7	•	•
CPX		•	3	4	5	6	•	•
DAA		•	•	•	•	•	2	•
DEC		2	•	•	6	7	•	•
DES		•	•	•	•	•	4	•
DEX		•	•	•	•	•	4	•
EOR	x	•	2	3	4	5	•	•

	(Dual Operand)	ACCX	Immediate	Direct	Extended	Indexed	Inherent
INC		2	•	•	6	7	•
INS		•	•	•	•	•	4
INX		•	•	•	•	•	4
JMP		•	•	•	3	4	•
JSR		•	•	•	9	8	•
LDA	x	•	2	3	4	5	•
LDS		•	3	4	5	6	•
LDX		•	3	4	5	6	•
LSR		2	•	•	6	7	•
NEG		2	•	•	6	7	•
NOP		•	•	•	•	•	2
ORA	x	•	2	3	4	5	•
PSH		4	•	•	•	•	•
PUL		4	•	•	•	•	•
ROL		2	•	•	6	7	•
ROR		2	•	•	6	7	•
RTI		•	•	•	•	•	10
RTS		•	•	•	•	•	5
SBA		•	•	•	•	•	2
SBC	x	•	2	3	4	5	•
SEC		•	•	•	•	•	2
SEI		•	•	•	•	•	2
SEV		•	•	•	•	•	2
STA	x	•	•	4	5	6	•
STS		•	•	5	6	7	•
STX		•	•	5	6	7	•
SUB	x	•	2	3	4	5	•
SWI		•	•	•	•	•	12
TAB		•	•	•	•	•	2
TAP		•	•	•	•	•	2
TBA		•	•	•	•	•	2
TPA		•	•	•	•	•	2
TST		2	•	•	6	7	•
TSX		•	•	•	•	•	4
TXS		•	•	•	•	•	4
WAI		•	•	•	•	•	9

NOTE: Interrupt time is 12 cycles from the end of the instruction being executed, except following a WAI instruction. Then it is 4 cycles.

Table 8.3 **Accumulator and memory instructions (courtesy Motorola Semiconductor, Inc.)**

ACCUMULATOR AND MEMORY OPERATIONS	MNEMONIC	IMMED OP	IMMED ~	IMMED #	DIRECT OP	DIRECT ~	DIRECT #	INDEX OP	INDEX ~	INDEX #	EXTND OP	EXTND ~	EXTND #	INHER OP	INHER ~	INHER #	BOOLEAN/ARITHMETIC OPERATION (All register labels refer to contents)	5 H	4 I	3 N	2 Z	1 V	0 C
Add	ADDA	8B	2	2	9B	3	2	AB	5	2	BB	4	3				A + M → A	↕	•	↕	↕	↕	↕
	ADDB	CB	2	2	DB	3	2	EB	5	2	FB	4	3				B + M → B	↕	•	↕	↕	↕	↕
Add Acmltrs	ABA													1B	2	1	A + B → A	↕	•	↕	↕	↕	↕
Add with Carry	ADCA	89	2	2	99	3	2	A9	5	2	B9	4	3				A + M + C → A	↕	•	↕	↕	↕	↕
	ADCB	C9	2	2	D9	3	2	E9	5	2	F9	4	3				B + M + C → B	↕	•	↕	↕	↕	↕
And	ANDA	84	2	2	94	3	2	A4	5	2	B4	4	3				A · M → A	•	•	↕	↕	R	•
	ANDB	C4	2	2	D4	3	2	E4	5	2	F4	4	3				B · M → B	•	•	↕	↕	R	•
Bit Test	BITA	85	2	2	95	3	2	A5	5	2	B5	4	3				A · M	•	•	↕	↕	R	•
	BITB	C5	2	2	D5	3	2	E5	5	2	F5	4	3				B · M	•	•	↕	↕	R	•
Clear	CLR							6F	7	2	7F	6	3				00 → M	•	•	R	S	R	R
	CLRA													4F	2	1	00 → A	•	•	R	S	R	R
	CLRB													5F	2	1	00 → B	•	•	R	S	R	R
Compare	CMPA	81	2	2	91	3	2	A1	5	2	B1	4	3				A − M	•	•	↕	↕	↕	↕
	CMPB	C1	2	2	D1	3	2	E1	5	2	F1	4	3				B − M	•	•	↕	↕	↕	↕
Compare Acmltrs	CBA													11	2	1	A − B	•	•	↕	↕	↕	↕
Complement, 1's	COM							63	7	2	73	6	3				$\overline{M}$ → M	•	•	↕	↕	R	S
	COMA													43	2	1	$\overline{A}$ → A	•	•	↕	↕	R	S
	COMB													53	2	1	$\overline{B}$ → B	•	•	↕	↕	R	S
Complement, 2's (Negate)	NEG							60	7	2	70	6	3				00 − M → M	•	•	↕	↕	①	②
	NEGA													40	2	1	00 − A → A	•	•	↕	↕	①	②
	NEGB													50	2	1	00 − B → B	•	•	↕	↕	①	②
Decimal Adjust, A	DAA													19	2	1	Converts Binary Add. of BCD Characters into BCD Format	•	•	↕	↕	↕	③
Decrement	DEC							6A	7	2	7A	6	3				M − 1 → M	•	•	↕	↕	④	•
	DECA													4A	2	1	A − 1 → A	•	•	↕	↕	④	•
	DECB													5A	2	1	B − 1 → B	•	•	↕	↕	④	•
Exclusive OR	EORA	88	2	2	98	3	2	A8	5	2	B8	4	3				A ⊕ M → A	•	•	↕	↕	R	•
	EORB	C8	2	2	D8	3	2	E8	5	2	F8	4	3				B ⊕ M → B	•	•	↕	↕	R	•
Increment	INC							6C	7	2	7C	6	3				M + 1 → M	•	•	↕	↕	⑤	•
	INCA													4C	2	1	A + 1 → A	•	•	↕	↕	⑤	•
	INCB													5C	2	1	B + 1 → B	•	•	↕	↕	⑤	•
Load Acmltr	LDAA	86	2	2	96	3	2	A6	5	2	B6	4	3				M → A	•	•	↕	↕	R	•
	LDAB	C6	2	2	D6	3	2	E6	5	2	F6	4	3				M → B	•	•	↕	↕	R	•
Or, Inclusive	ORAA	8A	2	2	9A	3	2	AA	5	2	BA	4	3				A + M → A	•	•	↕	↕	R	•
	ORAB	CA	2	2	DA	3	2	EA	5	2	FA	4	3				B + M → B	•	•	↕	↕	R	•
Push Data	PSHA													36	4	1	A → M_{SP}, SP − 1 → SP	•	•	•	•	•	•
	PSHB													37	4	1	B → M_{SP}, SP − 1 → SP	•	•	•	•	•	•
Pull Data	PULA													32	4	1	SP + 1 → SP, M_{SP} → A	•	•	•	•	•	•
	PULB													33	4	1	SP + 1 → SP, M_{SP} → B	•	•	•	•	•	•
Rotate Left	ROL							69	7	2	79	6	3				M: C ← b7 ← b0 (rotate through C)	•	•	↕	↕	⑥	↕
	ROLA													49	2	1	A: C ← b7 ← b0	•	•	↕	↕	⑥	↕
	ROLB													59	2	1	B: C ← b7 ← b0	•	•	↕	↕	⑥	↕
Rotate Right	ROR							66	7	2	76	6	3				M: C → b7 → b0 (rotate through C)	•	•	↕	↕	⑥	↕
	RORA													46	2	1	A: C → b7 → b0	•	•	↕	↕	⑥	↕
	RORB													56	2	1	B: C → b7 → b0	•	•	↕	↕	⑥	↕
Shift Left, Arithmetic	ASL							68	7	2	78	6	3				M: C ← b7 ← b0 ← 0	•	•	↕	↕	⑥	↕
	ASLA													48	2	1	A: C ← b7 ← b0 ← 0	•	•	↕	↕	⑥	↕
	ASLB													58	2	1	B: C ← b7 ← b0 ← 0	•	•	↕	↕	⑥	↕
Shift Right, Arithmetic	ASR							67	7	2	77	6	3				M: b7 → b7 → b0 → C	•	•	↕	↕	⑥	↕
	ASRA													47	2	1	A: b7 → b7 → b0 → C	•	•	↕	↕	⑥	↕
	ASRB													57	2	1	B: b7 → b7 → b0 → C	•	•	↕	↕	⑥	↕
Shift Right, Logic	LSR							64	7	2	74	6	3				M: 0 → b7 → b0 → C	•	•	R	↕	⑥	↕
	LSRA													44	2	1	A: 0 → b7 → b0 → C	•	•	R	↕	⑥	↕
	LSRB													54	2	1	B: 0 → b7 → b0 → C	•	•	R	↕	⑥	↕
Store Acmltr	STAA				97	4	2	A7	6	2	B7	5	3				A → M	•	•	↕	↕	R	•
	STAB				D7	4	2	E7	6	2	F7	5	3				B → M	•	•	↕	↕	R	•
Subtract	SUBA	80	2	2	90	3	2	A0	5	2	B0	4	3				A − M → A	•	•	↕	↕	↕	↕
	SUBB	C0	2	2	D0	3	2	E0	5	2	F0	4	3				B − M → B	•	•	↕	↕	↕	↕
Subract Acmltrs.	SBA													10	2	1	A − B → A	•	•	↕	↕	↕	↕
Subtr. with Carry	SBCA	82	2	2	92	3	2	A2	5	2	B2	4	3				A − M − C → A	•	•	↕	↕	↕	↕
	SBCB	C2	2	2	D2	3	2	E2	5	2	F2	4	3				B − M − C → B	•	•	↕	↕	↕	↕
Transfer Acmltrs	TAB													16	2	1	A → B	•	•	↕	↕	R	•
	TBA													17	2	1	B → A	•	•	↕	↕	R	•
Test, Zero or Minus	TST							6D	7	2	7D	6	3				M − 00	•	•	↕	↕	R	R
	TSTA													4D	2	1	A − 00	•	•	↕	↕	R	R
	TSTB													5D	2	1	B − 00	•	•	↕	↕	R	R

programming fields to each assembly language statement. The first is an *optional label field* where the programmer can assign a symbolic name to the memory address of the statement. Labels are limited to five or six alphanumeric characters, the first of which must be alphabetic. Certain characters (such as blanks and other delimiters) are not allowed within the label field.

Table 8.4 **Index register and stack manipulation instruction (courtesy Motorola Semiconductor, Inc.)**

INDEX REGISTER AND STACK		IMMED			DIRECT			INDEX			EXTND			INHER				5	4	3	2	1	0
POINTER OPERATIONS	MNEMONIC	OP	~	#	OP	~	#	OP	~	#	OP	~	#	OP	~	#	BOOLEAN/ARITHMETIC OPERATION	H	I	N	Z	V	C
Compare Index Reg	CPX	8C	3	3	9C	4	2	AC	6	2	BC	5	3				$(X_H/X_L) - (M/M+1)$	•	•	(7)	↕	(8)	•
Decrement Index Reg	DEX													09	4	1	$X - 1 \rightarrow X$	•	•	•	↕	•	•
Decrement Stack Pntr	DES													34	4	1	$SP - 1 \rightarrow SP$	•	•	•	•	•	•
Increment Index Reg	INX													08	4	1	$X + 1 \rightarrow X$	•	•	•	↕	•	•
Increment Stack Pntr	INS													31	4	1	$SP + 1 \rightarrow SP$	•	•	•	•	•	•
Load Index Reg	LDX	CE	3	3	DE	4	2	EE	6	2	FE	5	3				$M \rightarrow X_H, (M+1) \rightarrow X_L$	•	•	(9)	↕	R	•
Load Stack Pntr	LDS	8E	3	3	9E	4	2	AE	6	2	BE	5	3				$M \rightarrow SP_H, (M+1) \rightarrow SP_L$	•	•	(9)	↕	R	•
Store Index Reg	STX				DF	5	2	EF	7	2	FF	6	3				$X_H \rightarrow M, X_L \rightarrow (M+1)$	•	•	(9)	↕	R	•
Store Stack Pntr	STS				9F	5	2	AF	7	2	BF	6	3				$SP_H \rightarrow M, SP_L \rightarrow (M+1)$	•	•	(9)	↕	R	•
Indx Reg → Stack Pntr	TXS													35	4	1	$X - 1 \rightarrow SP$	•	•	•	•	•	•
Stack Pntr → Indx Reg	TSX													30	4	1	$SP + 1 \rightarrow X$	•	•	•	•	•	•

Table 8.5 **Jump and branch instructions (courtesy Motorola Semiconductor, Inc.)**

JUMP AND BRANCH		RELATIVE			INDEX			EXTND			INHER				5	4	3	2	1	0
OPERATIONS	MNEMONIC	OP	~	#	OP	~	#	OP	~	#	OP	~	#	BRANCH TEST	H	I	N	Z	V	C
Branch Always	BRA	20	4	2										None	•	•	•	•	•	•
Branch If Carry Clear	BCC	24	4	2										C = 0	•	•	•	•	•	•
Branch If Carry Set	BCS	25	4	2										C = 1	•	•	•	•	•	•
Branch If = Zero	BEQ	27	4	2										Z = 1	•	•	•	•	•	•
Branch If ≥ Zero	BGE	2C	4	2										$N \oplus V = 0$	•	•	•	•	•	•
Branch If > Zero	BGT	2E	4	2										$Z + (N \oplus V) = 0$	•	•	•	•	•	•
Branch If Higher	BHI	22	4	2										$C + Z = 0$	•	•	•	•	•	•
Branch If ≤ Zero	BLE	2F	4	2										$Z + (N \oplus V) = 1$	•	•	•	•	•	•
Branch If Lower Or Same	BLS	23	4	2										$C + Z = 1$	•	•	•	•	•	•
Branch If < Zero	BLT	2D	4	2										$N \oplus V = 1$	•	•	•	•	•	•
Branch If Minus	BMI	2B	4	2										N = 1	•	•	•	•	•	•
Branch If Not Equal Zero	BNE	26	4	2										Z = 0	•	•	•	•	•	•
Branch If Overflow Clear	BVC	28	4	2										V = 0	•	•	•	•	•	•
Branch If Overflow Set	BVS	29	4	2										V = 1	•	•	•	•	•	•
Branch If Plus	BPL	2A	4	2										N = 0	•	•	•	•	•	•
Branch To Subroutine	BSR	8D	8	2											•	•	•	•	•	•
Jump	JMP				6E	4	2	7E	3	3				See Special Operations	•	•	•	•	•	•
Jump To Subroutine	JSR				AD	8	2	BD	9	3					•	•	•	•	•	•
No Operation	NOP										01	2	1	Advances Prog. Cntr. Only	•	•	•	•	•	•
Return From Interrupt	RTI										3B	10	1		(10)					
Return From Subroutine	RTS										39	5	1		•	•	•	•	•	•
Software Interrupt	SWI										3F	12	1	See special Operations	•	S	•	•	•	•
Wait for Interrupt	WAI										3E	9	1		•	(11)	•	•	•	•

A blank character is used to separate the label field from the OP-CODE (Assembly Mnemonic Operation Code) field. OP-CODES are three or four characters long, encompassing the machine's full instruction set. A blank character separates the OP-CODE from the operands field of the assembly language statement. A statement may have 2, 1, or no operands, depending upon the type of statement and the addressing mode. Implied, immediate direct, relative, indexed, and extended addressing modes tell the processor which memory accessing method will be used for a particular statement. (For details, refer to Motorola references in Sec. 8.12.) A blank character separates the operands field from the comments field. The comments field is ignored by the assembler and does not produce any executable machine (object) code. Comments are messages from the programmer to him or herself or to anyone else reading the program, explaining what the line of assembly code is trying to accomplish. Comments should correspond to blocks of the flowchart.

Having completed the writing of the assembly source text, the programmer directs the assembler to examine the text and produce a corresponding execut-

Table 8.6 **Condition code register manipulation instruction (courtesy Motorola Semiconductor, Inc.)**

CONDITIONS CODE REGISTER OPERATIONS	MNEMONIC	INHER OP	INHER ~	INHER =	BOOLEAN OPERATION	5 H	4 I	3 N	2 Z	1 V	0 C
Clear Carry	CLC	0C	2	1	0 → C	•	•	•	•	•	R
Clear Interrupt Mask	CLI	0E	2	1	0 → I	•	R	•	•	•	•
Clear Overflow	CLV	0A	2	1	0 → V	•	•	•	•	R	•
Set Carry	SEC	0D	2	1	1 → C	•	•	•	•	•	S
Set Interrupt Mask	SEI	0F	2	1	1 → I	•	S	•	•	•	•
Set Overflow	SEV	0B	2	1	1 → V	•	•	•	•	S	•
Acmltr A → CCR	TAP	06	2	1	A → CCR	⑫					
CCR → Acmltr A	TPA	07	2	1	CCR → A	•	•	•	•	•	•

LEGEND:

OP	Operation Code (Hexadecimal);	00	Byte = Zero;
~	Number of MPU Cycles;	H	Half-carry from bit 3;
=	Number of Program Bytes;	I	Interrupt mask
+	Arithmetic Plus;	N	Negative (sign bit)
−	Arithmetic Minus;	Z	Zero (byte)
•	Boolean AND;	V	Overflow, 2's complement
M_{SP}	Contents of memory location pointed to be Stack Pointer;	C	Carry from bit 7
		R	Reset Always
		S	Set Always
+	Boolean Inclusive OR;	↕	Test and set if true, cleared otherwise
⊕	Boolean Exclusive OR;	●	Not Affected
$\bar{M}$	Complement of M;	CCR	Condition Code Register
→	Transfer Into;	LS	Least Significant
0	Bit = Zero;	MS	Most Significant

CONDITION CODE REGISTER NOTES:

(Bit set if test is true and cleared otherwise)

1. (Bit V) Test: Result = 10000000?
2. (Bit C) Test: Result = 00000000?
3. (Bit C) Test: Decimal value of most significant BCD Character greater than nine? (Not cleared if previously set.)
4. (Bit V) Test: Operand = 10000000 prior to execution?
5. (Bit V) Test: Operand = 01111111 prior to execution?
6. (Bit V) Test: Set equal to result of N ⊕ C after shift has occurred.
7. (Bit N) Test: Sign bit of most significant (MS) byte of result = 1?
8. (Bit V) Test: 2's complement overflow from subtraction of LS bytes?
9. (Bit N) Test: Result less than zero? (Bit 15 = 1)
10. (All) Load Condition Code Register from Stack. (See Special Operations)
11. (Bit I) Set when interrupt occurs. If previously set, a Non-Maskable Interrupt is required to exit the wait state.
12. (ALL) Set according to the contents of Accumulator A.

able machine code. The assembler is a two-pass program. On the first pass, the assembler associates statement labels with memory locations and assigns values to literal constants. Every label within an assembly program must be unique. On the second pass, the assembler assembles machine code from the assembly source, completing memory reference instructions with first-pass label assignments.

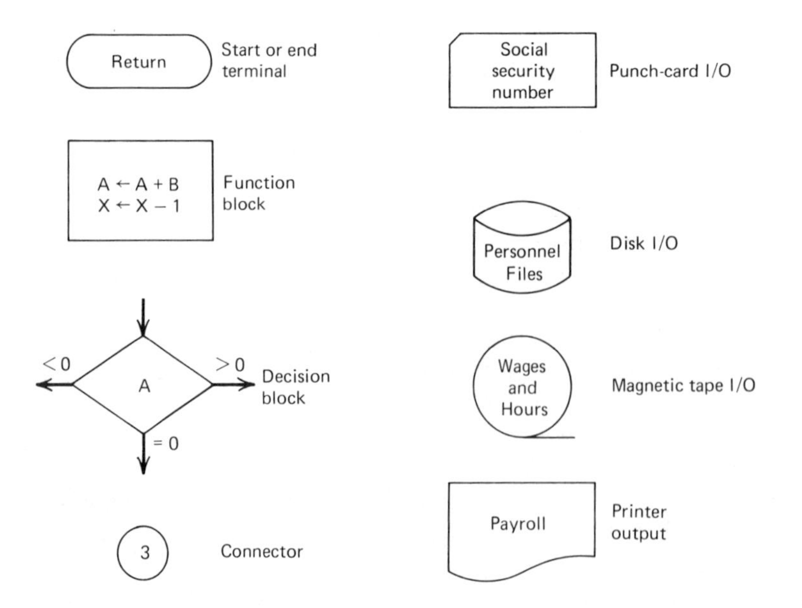

Figure 8.13 Frequently used flowchart symbols.

For short programs (under 20 or 30 statements), it is not difficult for programmers to assemble a machine code using the instruction set decoding chart provided by the microprocessor manufacturer. The middle columns of the programming card (Table 8.3) show the machine language OP codes for each assembly mnemonic in each of the possible addressing modes. Other pertinent information is provided, including the number of microprocessor clock cycles and the number of bytes required for the instruction. The method used is to convert the simplest (one byte) instructions first, the two byte immediate addressing instructions second, three byte immediate addressing instructions next, and direct, extended, indexed, and PC relative instructions last.

The machine language program can then be loaded into memory for execution. This can be done through the front panel switches of the microcomputer, a keyboard, or in systems with greater capabilities, using a LOADER program.

8.9 TYPES OF SYSTEMS

The variety of tools available to the microprocessor designer covers a broad spectrum in both sophistication and cost. The simplest of these tools is generally classified as educational trainers. They include a microcomputer circuit board with an MPU, a small quantity of RAM, an interface to binary switches or hexadecimal keyboard for program entry, and LEDs for program display. An on-board ROM based MONITOR program allows the user to enter a short program in machine code, verify the correctness of manual entry, and execute the loaded routine for program logic debugging. Some trainers allow machine language programs to be stored and retrieved from a common audio cassette recorder, eliminating the need to key-in the same program each time the device is repowered. Educational trainers usually cost between $100 and $300.

At the midrange of available equipment are the hobby computers, featuring an S-100, S-50, or independent support architecture. These specifications refer to the edge connector pin-outs that enable several plug-in PC boards in a system to communicate over a common back plane. They allow low-cost modular expansion of memory and I/O ports, and may include a 5-¼ inch, mini floppy disk drive for program storage and retrieval. These computers are usually programmed in BASIC and cost between $500 and $5000, with options for CRT terminals and medium-speed line printers.

The industrial requirements of ruggedness, high reliability, and interactive hardware/software debugging aids are features of the microprocessor development system. These systems include dual 8-inch floppy disc drives with an extensive software library of editors, assemblers, loaders, and high-level languages supported by a full featured disk operating system. Many of these systems use IN CIRCUIT EMULATION to permit debugging of microprocessor hardware designs concurrently with their operating software; a bus-buffering cable connects the circuit under test with the development system. With an intelligent console terminal (containing special editing keys and CRT formats) and a high-speed printer, these systems cost $7000 to $28,000.

8.10 APPLICATIONS

Extended Arithmetic Functions

Although microprocessors are most at home in a process control environment, occasionally they are called upon to perform nontrivial arithmetic functions. Included in this class of tasks are double precision (16 bits for an 8-bit MPU) addition, subtraction, multiplication, and division. Because the basic accumulator size of most microprocessors is one byte wide, the largest number that can be directly processed is 255 (decimal). When numbers greater than this must be arithmetically manipulated, extended precision routines are employed.

EXAMPLE 8.1 Write an assembly language routine that will add the 16-bit number found in the index register to another 16-bit number that has been loaded into the accumulators. The most significant byte of this number is in ACC A; the least significant byte is in ACC B. When complete, the result remains in the accumulator pair.

PROCEDURE

1 The assembly language routine of Fig.8.14 will perform double precision addition as required.
2 A RAM memory location is used to temporarily hold the index register contents during the calculation.
3 Lines 0010 and 0020 are the segment header that declare to the reader the purpose of the routine.
4 Program line 0030 stores the index in two sequential memory locations, the first of which is named TEMP.
5 Line 0040 adds together the two least significant bytes of the addend and augend, and stores the least significant byte of the sum in accumulator B.
6 The statement of line 0050 adds the two most significant bytes, plus any carry C that may have resulted from the addition of least significant bytes.

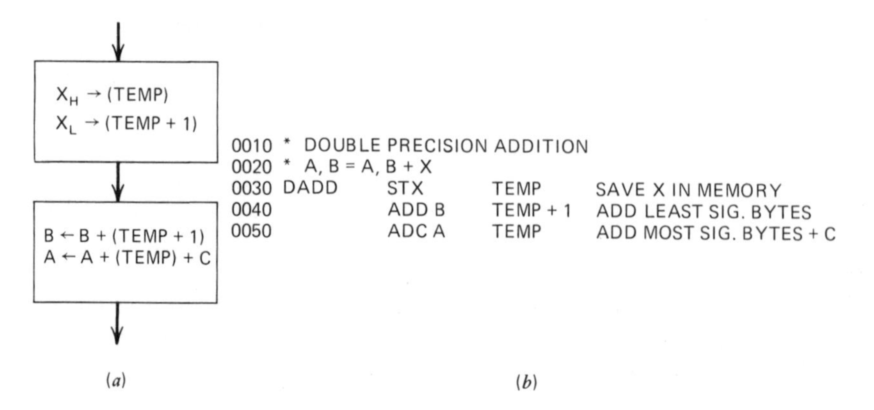

Figure 8.14 Double precision addition routine. (*a*) Flowchart. (*b*) Program.

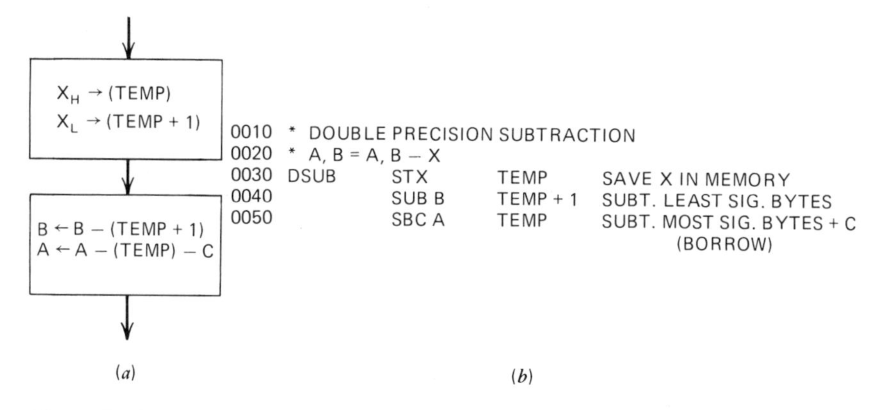

Figure 8.15 Double precision subtraction routine. (*a*) Flowchart. (*b*) Program.

7 The result of this instruction places the most significant byte of the sum in accumulator A. Thus, the carry bit (one of the flags in the condition code register) serves as a link between least and most significant bytes during the calculation.

To perform double precision substraction, the programmer need only replace the ADD and ADC instructions with SUB and SBC codes, as shown in Fig. 8.15.

EXAMPLE 8.2 Write an assembly language routine to multiply the 8-bit number in accumulator A by the 8-bit number in accumulator B. Note that a result as large as 16-bits wide may occur. Store this result in the AB accumulator pair.

PROCEDURE

1 One method of accomplishing arithmetic multiplication is by *repetitive addition*. This tactic is illustrated in Fig. 8.16.
2 Lines 0030, 0040, 0050, and 0080 load the multiplier (A) into the index register and place the multiplicand (B) into memory location TEMP+2.
3 Lines 0060 and 0070 initialize the product (A,B) to zero.
4 If the multiplier is nonzero (line 0090), the multiplicand is added to the partial product (line 0100).
5 Line 0110 links the least and most significant bytes to the result.
6 The multiplier is then decremented (line 0120) and the process repeated (line 0130) until the multiplier (in index register) reaches zero, at which time the program branches (line 0090) to the statement labeled DONE (line 0140).

EXAMPLE 8.3 Write a routine in assembly language to divide the 8-bit number in accumulator B by the contents of accumulator A. The quotient of the result is to be placed into A; the remainder in B.

PROCEDURE

1 Division may be easily accomplished by the method of *repetitive subtraction*.

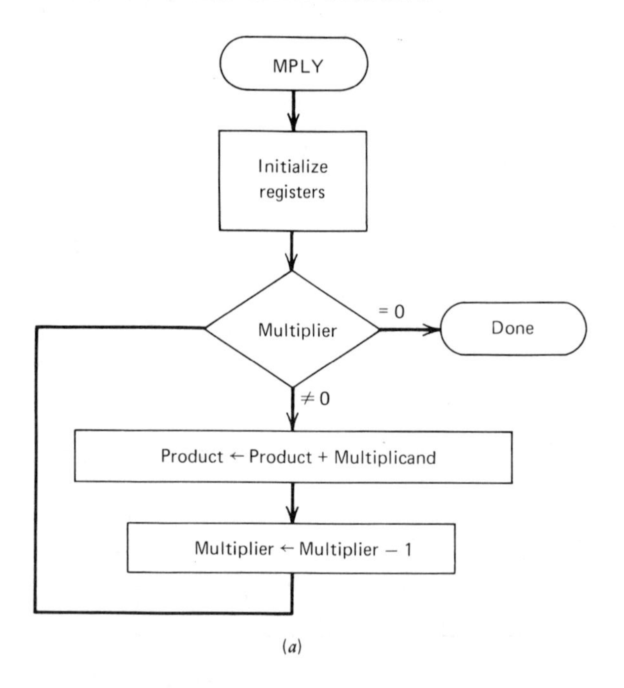

(*a*)

```
0010    * MULTIPLICATION BY REPEATED ADDITION
0020    * A, B = A * B
0030    MPLY    CLR     TEMP        CLEAR A MEMORY LOCATION
0040            STA A   TEMP + 1    STORE A IN THE NEXT
0050            STA B   TEMP + 2    STORE B IN THE ONE AFTER
0060            CLR A               CLEAR A
0070            CLR B               CLEAR B
0080            LDX     TEMP        MOVE TEMP, TEMP + 1 TO X
0090    TEST    BEQ     DONE        DONE IF X = 0
0100            ADD B   TEMP + 2    ADD MULTIPLICAND
0110            ADD A   # 0         ADJUST MOST SIG. BYTE
0120            DEX                 DECREMENT MULTIPLIER
0130            BRA     TEST        DO IT AGAIN
0140    DONE    EQU     *           CONTINUE
```

(*b*)

Figure 8.16 Multiplication by repetitive addition. (*a*) Flowchart. (*b*) Program.

2 As shown in Fig. 8.17, each time the divisor is subtracted from the dividend (line 0050), the result is tested for minus (line 0060).

3 If not negative, the quotient is incremented and the process repeated (lines 0070–0080); otherwise, there was one excessive subtraction performed.

4 This error is repaired in line 0090, and the routine then terminated in line 0100 with the final quotient and remainder in accumulators A and B.

Note that for correct operation, both dividend and divisor must each be positive values less than 127 (decimal), as bit 7 is used to represent a negative number (when set) in the format that is called two's complement notation. There

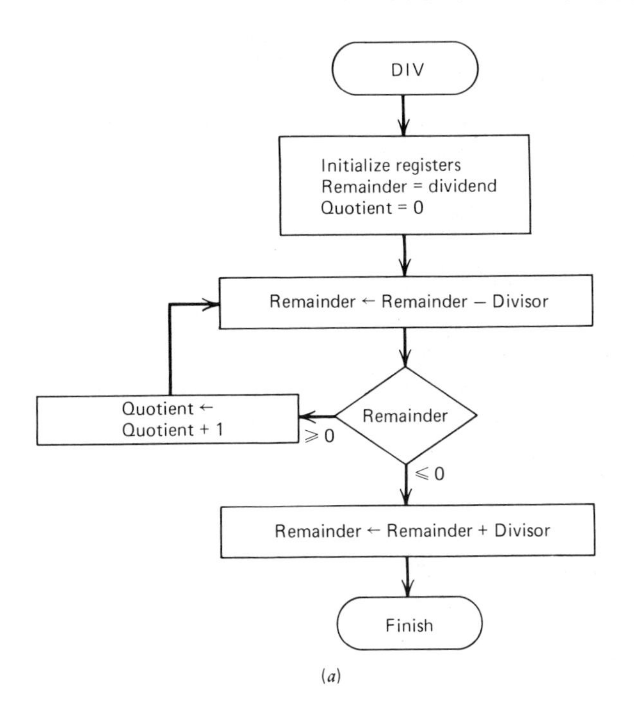

(a)

```
0010    * DIVISION BY REPEATED SUBTRACTION
0020    * A(QUOTIENT) B(REMAINDER) = B(DIVIDEND)/A(DIVISOR)
0030    DIV      STA A    TEMP     SAVE THE DIVISOR
0040             CLR A             ZERO THE QUOTIENT
0050    NEXT     SUB B    TEMP     SUBT DIVISOR FROM DIVIDEND
0060             BMI      FIXUP    IF RESULT MINUS, THEN TOO FAR
0070             INC A             RESULT OK. INCREMENT QUOTIENT
0080             BRA      NEXT     DO IT AGAIN
0090    FIXUP    ADD B    TEMP     RESTORE LAST VALUE REMAINDER
0100    FINISH   EQU      *        CONTINUE
```

(b)

Figure 8.17 Division by repetitive subtraction. (*a*) Flowchart. (*b*) Program.

are other, more complicated algorithms, that are more efficient and unrestricted. These, however, are beyond the scope of this chapter.

The ability of a microprocessor to satisfy the needs of a process control environment can best be illustrated by a few examples demonstrating common control functions.

Time Delay

Many applications require either logging the time at which a particular event has occurred (as with a fire reporting system), or performing an operation after a finite time has elapsed following a trigger event (as when heat settings are programmed for a microwave oven). The short routines below show how common time delays are implemented in software.

EXAMPLE 8.4 Using assembly language, write a program implementing a one-half second delay.

PROCEDURE

1 The flowchart and assembly language source code are provided in Fig. 8.18.
2 Assuming that our microprocessor is operating at an MPU clock frequency of 1 MHz, then each cycle of an instruction execution will take 1 μs, the MPU clock period.
3 From the programming card (Table 8.4), we see (under the column headed '~') that the number of cycles required for the "DEcrement indeX" statement on line 0020 is 4. Similarly, the number of cycles required for the "Branch if Not Equal to Zero" instruction of line 0030 can be found to be 4 as well. These two statements form a DELAY LOOP.
4 Statement 0020 subtracts one from a counter.
5 Statement 0030 asks if the counter has reached zero and, if not, returns control to statement 0020. The programmer's concern, then, is "How many times must the two statements of the loop be executed to delay the required amount of time?" If each statement requires four cycles, then together they require eight cycles or 8 μs considering the MPU clock period.
6 The number of passes through the loop can be calculated by dividing the desired delay time by the loop time:

$$\frac{0.5 \text{ s}}{8 \ \mu\text{s}} = \frac{0.5}{8 \times 10^{-6}} = \frac{0.5 \times 10^{6}}{8} = 62{,}500 \text{ passes.}$$

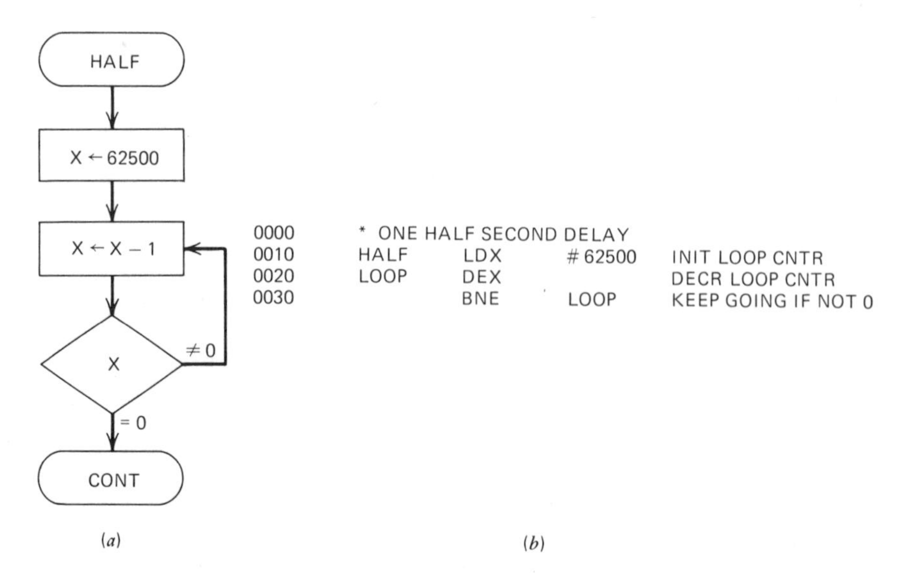

Figure 8.18 Realizing a one-half second delay. (*a*) Flowchart. (*b*) Program.

7 This calculated value is used to initialize the 16-bit loop counter by the "LoaD indeX" statement of program line 0010.
8 The real time error associated with this approach is related to the number of cycles required for execution of the initialization statement. From Table 8.4, under the column headed IMMED ("#" sign of statement 0010 indicated immediate addressing mode), the number of cycles for that statement not included in the calculation (under "~") is 3. The resultant error, then, is +3 μs or 0.0006%.

EXAMPLE 8.5 Using assembly language, program a one second delay. Note that the largest number representable with a 16-bit register is 65,535.

PROCEDURE

1 Simple time delay routines can be cascaded, as needed, to implement longer delays with high accuracy. As we see from Fig. 8.19, beginning execution at line 0040 will provide the half second delay previously discussed.
2 Entering the program at line 0010 will cause two half-second delays to execute in succession, yielding a full one-second delay.

EXAMPLE 8.6 Using assembly language, write a program to provide one minute of delay. Keep in mind the need for program efficiency to minimize memory costs.

PROCEDURE

1 Extended time delays can be achieved by multiplying the number of times that a simple routine is executed. This is done by embedding the simple delay loops in an external repeating loop (Fig. 8.20).
2 The new statements involved with the procedure appear in lines 0010, 0080, and 0090. The "LoaD Accumulator B" instruction of line 0010 initializes an 8-bit register to the value 60, the number of times the one second delay is to be executed. In line 0080, the "DECrement B" instruction subtracts one from this register for each second elapsed, while the BNE statement of line 0090 directs repetition until the minute has expired.
3 The error associated with the one minute delay routine stems from the cycle counts of statements 0010, 0020, 0050, 0080, and 0090, which are not directly involved with time delay loops and are considered overhead. From Table 8.4, the respective number of cycles for each are 2, 3, 3, 2, and 4, yielding a total error of +14 μs, or only 0.0000233%.

Polling Sensor Inputs

In a control environment, it is often necessary to be able to determine when any of several external inputs are signaling an active state or status change to the microprocessor. These inputs may be temperature threshold probes, tachometer velocity indicators, or simple switch closures.

Although from a programming standpoint, it would be easiest to assign each discrete input device a different address in the memory map, and to load and test each location for activity (as in Fig. 8.21), this method proves most inefficient from a hardware designer's point of view.

Because the microprocessor data bus is 8-bits wide, attaching only one sensor input per address would not only waste seven data lines, but would require

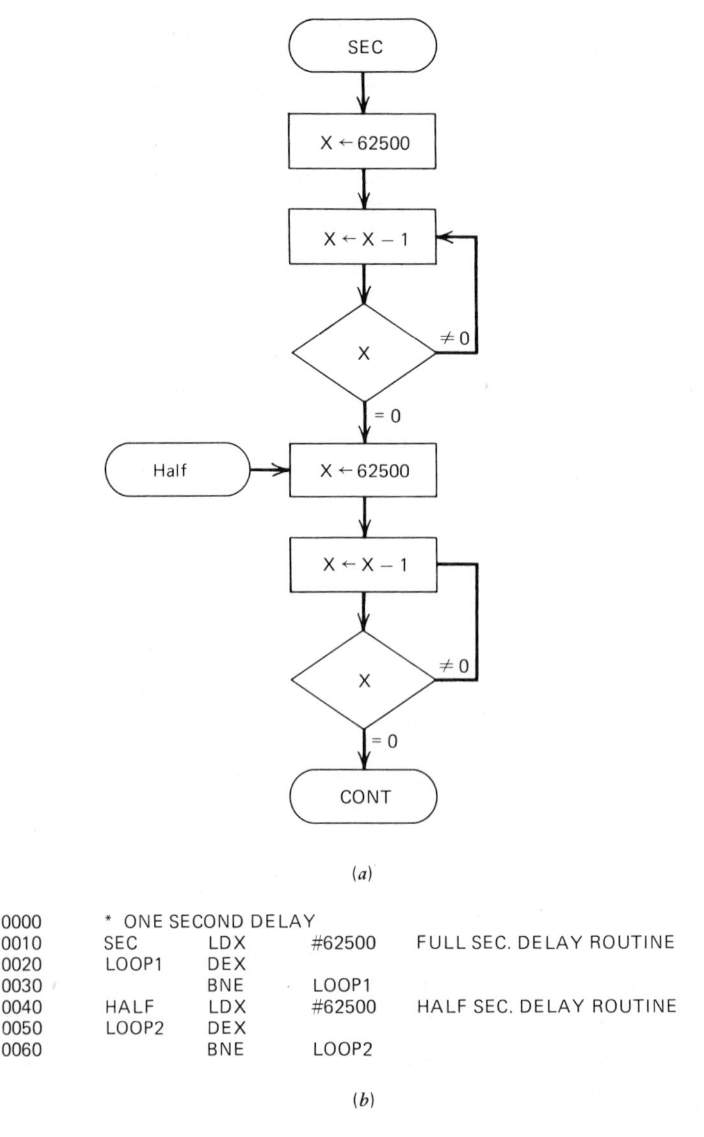


```
0000    * ONE SECOND DELAY
0010    SEC      LDX    #62500    FULL SEC. DELAY ROUTINE
0020    LOOP1    DEX
0030             BNE    LOOP1
0040    HALF     LDX    #62500    HALF SEC. DELAY ROUTINE
0050    LOOP2    DEX
0060             BNE    LOOP2
```

(b)

Figure 8.19 Realizing a one second delay. (*a*) Flowchart. (*b*) Program.

extensive decoding circuitry to assign each input a unique address. A more efficient method is to use a PIA to acquire eight status bits at a time and to write a *polling routine* (as in an election to see who votes yes and who votes no) to test each bit accordingly and to take appropriate action.

A polling routine of this type appears in Fig. 8.22. Because such a routine will most likely be executed many times during a program run, it may not be desirable for it to service an active status bit that had been reported active, and

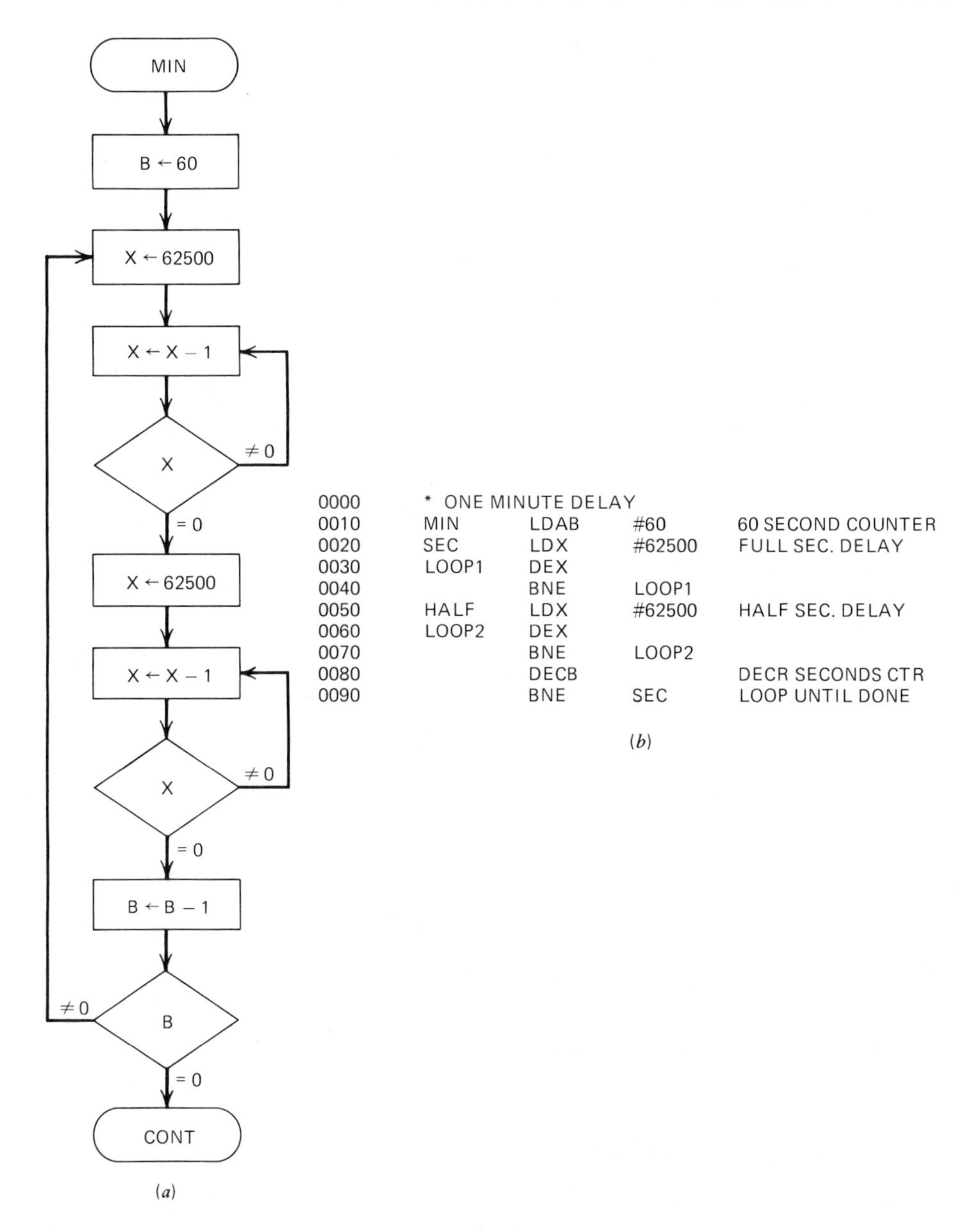

Figure 8.20 Realizing a one minute delay. (*a*) Flowchart. (*b*) Program.

that had been serviced during a previous scan of the inputs. A method is required whereby only zero-to-one transitions of the status bits are recognized and serviced by the routine, while bits that are zero, or that have been previously serviced, are ignored. A polling routine of this type can be programmed by substituting the code of Fig. 8.23 for the first statements of the previous example.

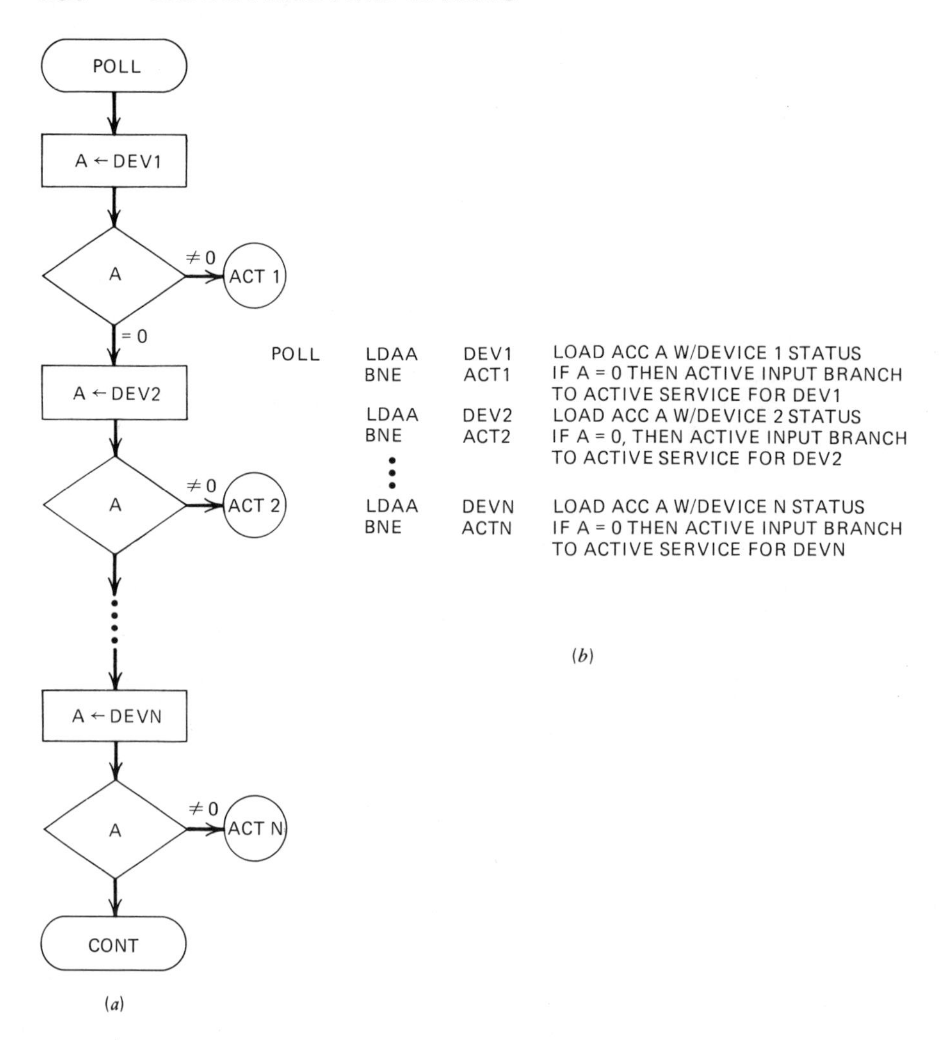

Figure 8.21 Device status polling routine. (*a*) Flowchart. (*b*) Program.

Setting/Clearing Control Outputs

Just as efficiency was achieved by accessing input signals as 8-bit parallel bytes rather than from discrete addresses, similar gains can be realized when applying that concept to output control. Some families of microprocessor products support instructions that can test, set, or clear bits on an individual basis. Those processors that do not, must achieve equivalent results through more complicated means if the parallel I/O implementation is to be successful.

To turn on a control output in systems without bit addressing capability, the output port register is logically ORed with an 8-bit pattern that has a 1 in the bit positions to be turned ON and zeros everywhere else. Turning OFF a

```
PIARA    EQU     $4000     ADDRESS OF PIA REG A IS HEX
                           4000
WAIT     LDAA    PIARA     LOAD A WITH 8 STATUS BITS
TEST     BMI     ACT7      IF BIT 7 ON, SERVICE DEV7
         ROLA              ROTATE LEFT A ONE POSITION
         BMI     ACT6      IF BIT 6 ON, SERVICE DEV6
         ROLA
         BMI     ACT5      IF BIT 5 ON, SERVICE DEV5
         ROLA
         BMI     ACT4      IF BIT 4 ON, SERVICE DEV4
         ROLA
         BMI     ACT3      IF BIT 3 ON, SERVICE DEV3
         ROLA
         BMI     ACT2      IF BIT 2 ON, SERVICE DEV2
         ROLA
         BMI     ACT1      IF BIT 1 ON, SERVICE DEV1
         ROLA
         BMI     ACT0      IF BIT 0 ON, SERVICE DEV0
         BRA     WAIT      NO INPUTS YET. KEEP WAITING
```

Figure 8.22 Efficient device status polling routine.

```
PIARA    EQU     $4000     ADDRESS OF PIA REG A
WAIT     LDAA    PIARA     LOAD A WITH PIA INPUT BITS
         STAA    NEW       SAVE IN TEMP LOCATION FOR
                           NEW INPUT
         EORA    LAST      EXCLUSIVE OR NEW AND OLD IN-
                           PUTS
*                          A NOW HAS ONES WHERE BITS
                           HAVE CHANGED
         LDAB    NEW       NEW INPUT FOR THIS PASS . . .
         STAB    LAST      BECOMES OLD INPUT FOR NEXT
                           PASS
         ANDA    NEW       LOGICAL AND BIT CHANGES WITH
                           NEW BITS
*                          A NOW HAS ONES ONLY FOR 0 TO
                           1 TRANSITION
TEST     BMI     ACT7      IF BIT7 ON SERVICE DEV7
* CONTINUE AS BEFORE
```

Figure 8.23 Polling modification to detect status changes only.

control output requires logical ANDing the output port with a bit pattern that has zeros in the bit positions to be switched off, and ones in all others.

Figure 8.24 shows two subroutines; one for turning a control bit on, the other to turn control bits off. These routines are designed so as not to affect any bits of the output port other than those designated for activation or deactivation. Each subroutine requires that, upon entry, accumulator A have a bit pattern with ones in positions to be affected.

Completing the time delay routines described in Figs. 8.18, 8.19, and 8.20 (half, sec, min) with ReTurn from Subroutine (RTS) statement, permits each to be used as a subroutine when required for sequencing the setting and the clearing of control outputs.

EXAMPLE 8.7 A control output sequence is required that will turn on outputs 0, 1, and 3; delay 1 minute, turn on output 2 and turn off output 1; delay 2-½ s; turn on outputs 1 and 4, and then stop.

PROCEDURE

1 The program of Fig. 8.25 illustrates a calling sequence for previously defined subroutines.

2 The Branch to SubRoutine instructions cause the program counter to be saved in the stack while control is transferred to the indicated subroutine.

3 When a ReTurn from Subroutine is encountered at subroutine end, the PC is automatically restored from the stack and the main routine continues.

The routines presented here have demonstrated that the microprocessor can be used as a cost-effective replacement for those monitor and control tasks previously associated with relay logic. Not only does the microprocessor offer solid-state reliability and high accuracy, it is more efficient in both power consumption and space utilization. Furthermore, the flexibility inherent to the stored program concept permits modifications in the sequence or duration of operations without any hardware alteration.

```
PIARB   EQU    $4002    ADDRESS OF PIA REG B
ON      ORAA   PIARB    OR OUTPUT WITH BIT PATTERN
                        IN A
        STAA   PIARB    SET NEW OUTPUT BITS
        RTS             RETURN TO MAIN
OFF     COMA            COMPLEMENT BIT PATTERN . . .
                        TO MAKE EFFECTED BITS ZERO
        ANDA   PIARB    AND OUTPUT WITH BIT PATTERN
                        IN A
        STAA   PIARB    CLEAR NEW OUTPUT BITS
        RTS             RETURN TO MAIN
```

Figure 8.24 Device control subroutines.

```
CONTROL  LDAA  #00001011%  BINARY BITS 0, 1, AND 3
         BSR   ON          BRANCH TO TURN ON SUBROU-
                           TINE
         BSR   MIN         BRANCH TO DELAY 1 MINUTE
                           SUB
         LDAA  #00000100%  BINARY BIT 2
         BSR   ON          BRANCH TO TURN ON SUBROU-
                           TINE
         LDAA  #00000010%  BINARY BIT 1
         BSR   OFF         BRANCH TO TURN OFF SUB
         BSR   SEC         DELAY 2.5 SECONDS
         BSR   SEC
         BSR   HALF
         LDAA  #00010010%  BINARY BITS 1 AND 4
         BSR   ON          BRANCH TO TURN ON SUBROU-
                           TINE
         WAI               WAIT FOR INTERRUPT
```

Figure 8.25 Industrial controller program.

Having inventoried the hardware building blocks of a microprocessor design and acquired a working knowledge of the processor's software instruction set, the system designer is ready to incorporate microprocessor technology into a product. A typical product design sequence is presented in Ex. 8.8.

EXAMPLE 8.8 Design and build a microprocessor-based electronic piano toy. It should be safe (operating at low voltage from flashlight batteries), reliable, and inexpensive to produce. Although it will cover three octaves, it should provide a one-octave piano style keyboard with eight notes of the basic musical scale and five keys for sharps and flats, 13 keys in total. As this is intended to be a children's toy, the piano will not be required to play chords (multiple notes) at the same time.

PROCEDURE

1 To design an electronic piano using microprocessor technology, the project is divided into separate hardware and software sections. This allows work to proceed in parallel. Hardware design determines which system components will be used. Software design and program code can then be written to make the I/O devices function as per the product specification.
2 The first step in hardware design is to define and diagram those system sections that are product independent. In this case, they are the MPU and the clock circuit for program sequencing and hardware timing. As the product is a toy and does not require highly accurate timing components, an RC controlled oscillator is used. The schematic diagram of Fig. 8.26 shows the MPU and clock sections of the microprocessor system.

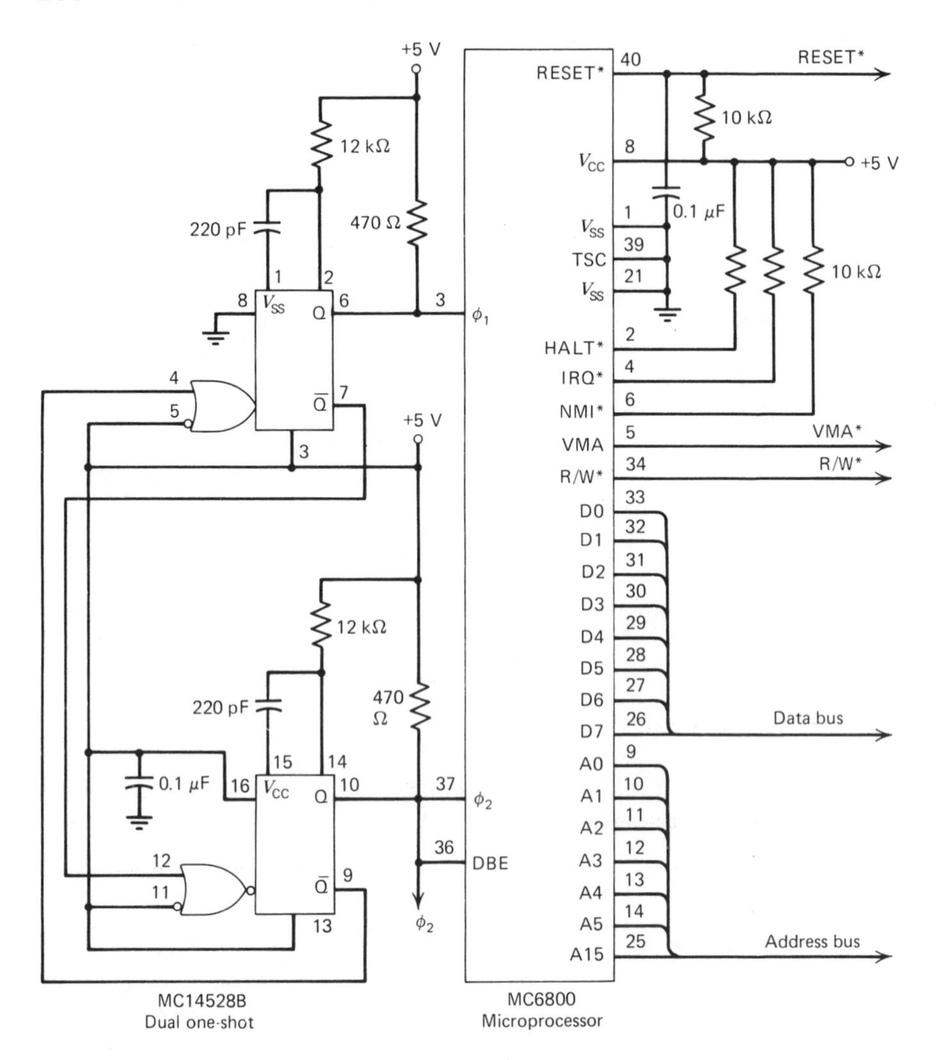

Figure 8.26 MPU and clock sections of a microcomputer.

3 The two-phase oscillator is made using a dual, one-shot (monostable multivibrator) with each half producing one phase pulse. The falling edge of ϕ_1 triggers ϕ_2 which, in turn, retriggers ϕ_1 on its falling edge. There are 470 Ω pull-up resistors on the ϕ_1 and ϕ_2 clocks to meet the MOS level requirements (logic 1 > 4.7 V) of the MPU inputs. Pulse widths are determined by the 220 pF capacitors and 12 kΩ resistors. The pulse width, *PW*, is:

$$PW = 0.2RC \ln (V_{CC}) \tag{8.1}$$

$$= 0.2\,(12 \times 10^3)(220 \times 10^{-12}) \ln 5$$

$$= 850 \text{ ns}$$

The clock period, *t*, is:

$$
\begin{aligned}
t &= 2(PW) \qquad (8.2)\\
&= 2(850 \times 10^{-9})\\
&= 1.7 \times 10^{-6} = 1.7\ \mu s
\end{aligned}
$$

The clock frequency, *f*, is:

$$
\begin{aligned}
f &= 1/t \qquad (8.3)\\
&= 1/(1.7 \times 10^{-6})\\
&= 5.88 \times 10^{5} = 588\ \text{kHz}
\end{aligned}
$$

4 For this product, MPU control bus signals TSC (3-state control), HALT*, IRQ* (interrupt request), and NMI (the priority input nonmaskable interrupt request line) are not required and are disabled by connecting them to their inactive logic levels. The usable signals of the control bus are ϕ_2 (for data transfer timing), system RESET*, VMA (valid memory address), and R/W* (read/write select). The entire data bus is used, but only a limited number of address bus lines are required.

5 To fulfill the requirements of the product specification, an I/O device is required that will interface 13 momentary action switches to the processor and provide an output to drive an acoustic speaker at controlled audio frequencies. A suitable component for this purpose is the PIA, which has two 8-bit parallel ports whose bits can be programmed as either inputs or outputs. Figure 8.27 illustrates how the PIA can be used to connect the required switch and audio devices to the microprocessor busses.

6 PIA inputs RS0 and RS1 are used for selecting the internal control and data registers and are connected to the two least significant bits of the address bus. CS0, CS1, and CS2* are the PIA chip select inputs. The PIA will be properly selected when VMA is active and address 15 is at a logic 0. These signals are attached to CS0 and CS2*. CS1 is not required and is permanently enabled by connection to V_{CC}.

7 The R/W* control bus signal controls the direction of data transfer between the MPU and PIA while ϕ_2 indicates when data on the bus has become stable and is safe for access.

8 The RESET* control bus line initializes all PIA internal registers to known initial conditions during power-on reset.

9 Keyboard switches are connected to the PIA I/O ports on PA0-PA7 and PB3-PB7, which each have 10 kΩ pull-up resistors to V_{CC}. Normally, closed switches keep these inputs at a logic 0 until a key is pressed. PB0 is programmed as an output bit and controls frequencies to the speaker through a transistor buffer that functions as a high-current driver. A diode is reverse biased across the speaker coil and is provided to suppress spikes caused by inductive switching.

10 Finally, selections must be made for program memory. As the program does not require the storage of intermediate data (other than those values that can remain in registers), no RAM is required.

Preliminary programming estimates indicate that a 64 byte PROM is sufficient to contain the system software. At the time of this writing, there are no 64 byte 3-state PROMs available that operate from a single power source. To satisfy system requirements, the necessary PROM size is achieved by using two 32-byte PROM arrays with proper selection decoded by ¾ of a quad 2-input NAND circuit and a

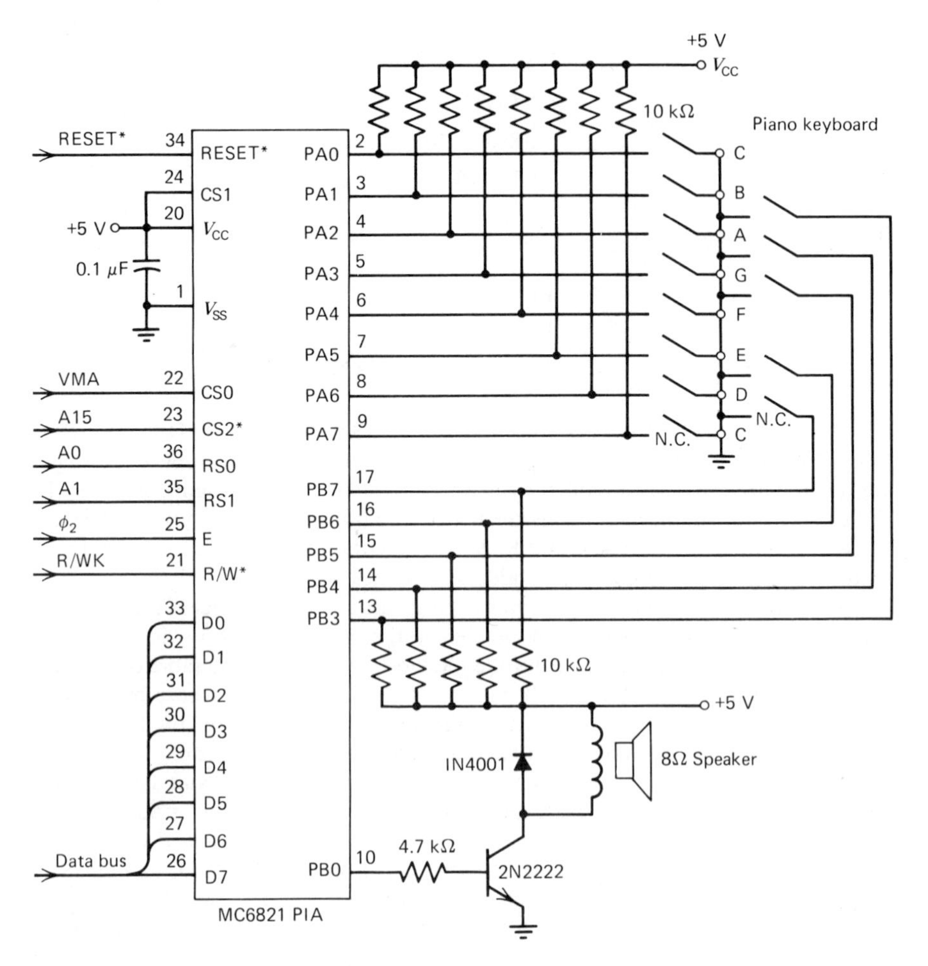

Figure 8.27 PIA configuration.

diode logic AND gate. Figure 8.28 shows how the selected PROM is interfaced to the microprocessor bus. Note that because this is a read-only, nonvolatile device, control bus signals RESET*, ϕ_2, and R/W* are not required.

11 To document the programming strategy and to visualize program flow, a flowchart is prepared. This makes the coding step, which follows, simpler. The required program may be divided into three sections: PIA initialization, keyboard polling, and frequency generation. Each section may be separately charted.

Figure 8.29 describes the operations required to configure the PIA to properly interface the speaker and keyboard switches as diagramed in Fig. 8.27.

12 The flowchart of Fig. 8.30 shows a variation of the polling routine illustrated in Fig. 8.22. As before, the inputs to be tested are loaded into an accumulator and are shifted so that each input passes through bit-7, a position that may be tested for 0 or 1 with the Branch if MInus (BMI) instruction. Greater efficiency is achieved here by performing the repetitive operation in a program loop. At the beginning of the loop, the

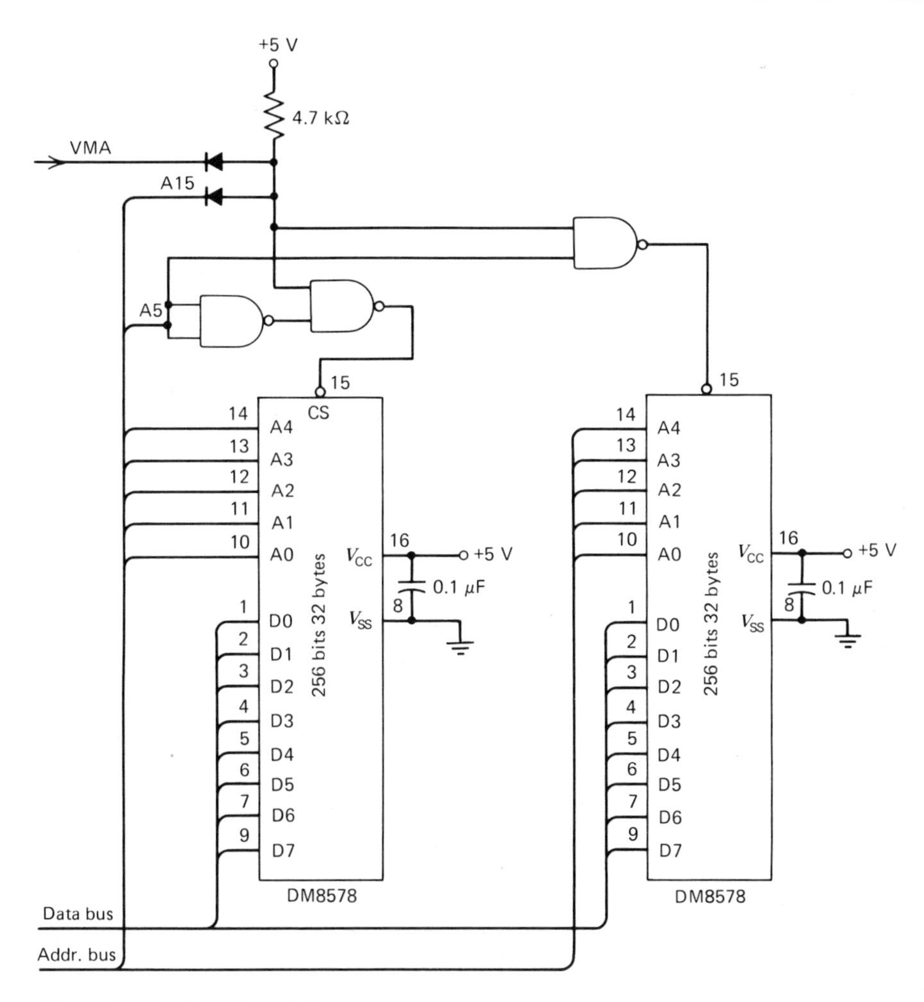

Figure 8.28 A 64-byte ROM memory system.

index register is loaded with the first address of a table or list of operands. As each bit to be tested passes through bit-7, the index register is incremented to point to the next entry in the table. When an active bit is polled, the loop will be exited with the index containing the address of the operand associated with the selected input.

13 A programmable audio frequency oscillator can easily be designed using the time delay techniques previously discussed. Because logic 0 and logic 1 are, in fact, represented in our system by TTL voltage levels 0.8 V and 2.4 V, an analog oscillator waveform can be achieved by turning a bit on; delaying one-half the desired period; turning the bit off; delaying again; and repeating the process. The general format of this oscillator is shown in Fig. 8.31. It should be obvious that altering the delay count will alter the period and frequency of the waveform. The delay count is directly proportional to the period, and inversely proportional to the frequency.

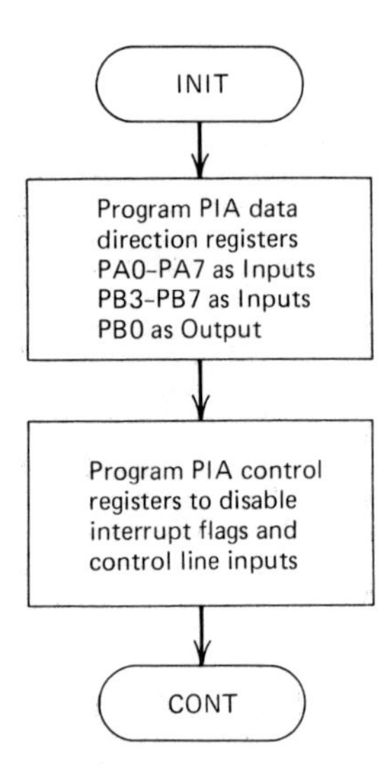

Figure 8.29 PIA initialization.

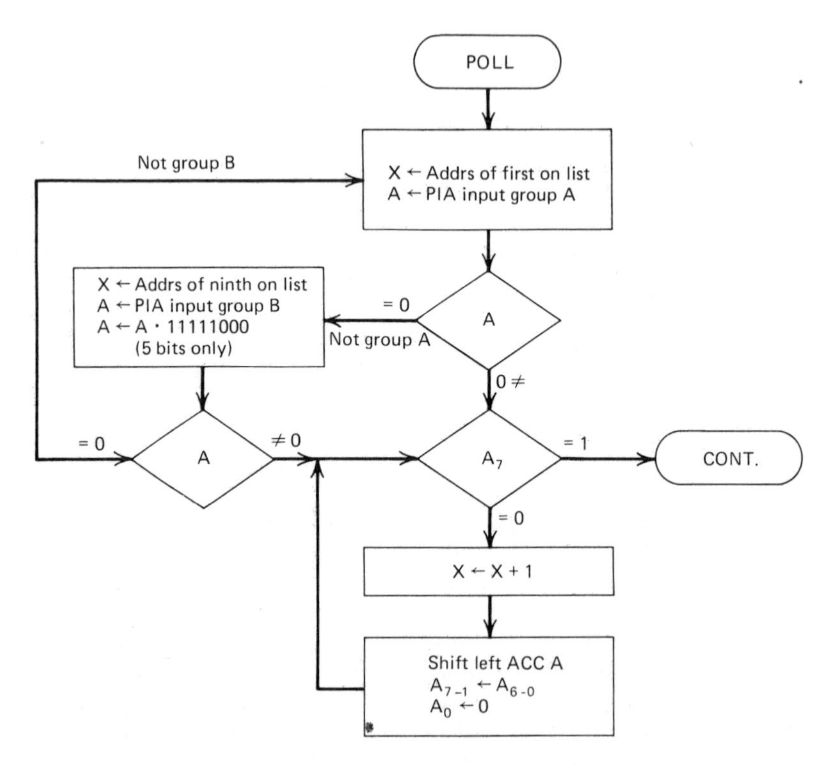

Figure 8.30 Keyboard polling.

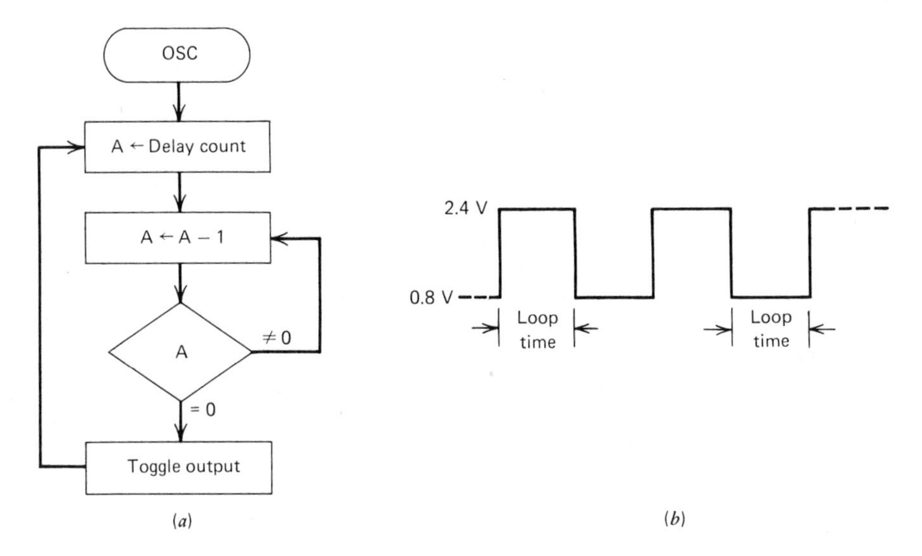

Figure 8.31 Audio frequency oscillator. (*a*) Flowchart. (*b*) Output waveform.

For example, the musical note C has a fundamental audio frequency of 256 Hz. Its period, t, then, is:

$$t = 1/f = 1/256 = 3.906 \text{ ms}$$

and the required software delay, d, is:

$$d = t/2 = (3.906 \times 10^{-3})/2 = 1.953 \times 10^{-3} \text{ s}$$

14 Recall from previous discussions that the timing loop requires eight clock cycles. The clock period of the MPU oscillator being used in our system is 1.7 μs. The delay count for note C can then be determined after calculating the time for one pass through the loop:

$$t_{\text{loop}} = 8(1.7 \times 10^{-6}) = 13.6 \times 10^{-6} \text{ s}$$

The delay count, DC, is:

$$\begin{aligned} DC &= d/t_{\text{loop}} \qquad (8.4) \\ &= (1.953 \times 10^{-3})/(13.6 \times 10^{-6}) \\ &= 143.6 \\ &\approx 143 \end{aligned}$$

(The remaining time will be closely matched by the overhead code.)

15 If we calculate the delay count for each of the thirteen notes we wish to reproduce, and store these values in sequential table locations corresponding to their keyboard switch positions, then the polling routine we have defined can exit with the index register containing the address of the delay count for the active key. The indexed addressing mode allows us to load an accumulator from the memory location pointed to by the index. The flowchart of Fig. 8.32 uses this technique to demonstrate a programmable oscillator. This routine must be repeated each half cycle and may alter its period by external manipulation of the index register.

16 Figure 8.33 shows the entire electronic piano program flowchart resulting from the linking of the three segments in Figs. 8.29, 8.30, and 8.32. Using this flowchart as a guide, assembly language source code can be written as a first step toward executable software. This source code is shown in Fig. 8.34 as it is prepared for processing by the assembler. Lines with asterisks (*) in the first column are comment lines and are inserted to clarify the purpose of program sections. Numbers on the left are line or record keys used for program editing purposes.

17 After visually verifying that there is correct program syntax (labels start in column 1, space between statement fields, legal op-codes, and proper number of operands for the desired addressing mode), the program is assembled. The complete assembly listing appears in Fig. 8.35.

The assembly listing shows, on the right-hand side, the original source code formatted by tab columns for easier reading. To the left of the source code are the machine codes to be loaded for program execution. The decimal numbers of the first column are line numbers corresponding to those produced by the editor. Line numbers followed by 'A' indicate that an Absolute (nonrelocatable) address has been calculated by the statement on that line.

The four hexadecimal digits of the second column indicate the address at which the first byte of that statement will be loaded for execution. Statements may occupy one, two, or three bytes. Column three of the listing shows the statement op-code

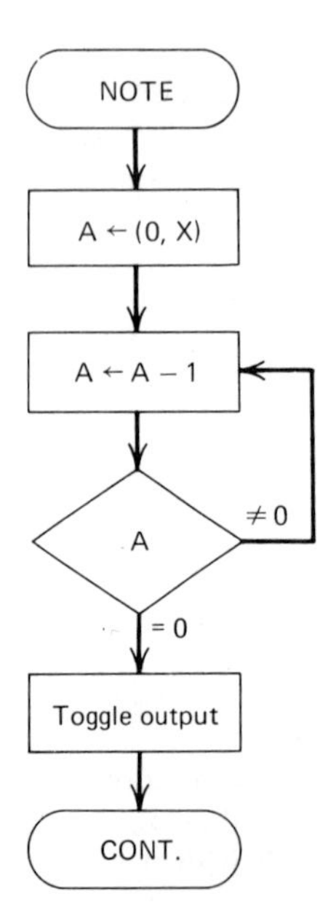

Figure 8.32 Flowchart for a programmable oscillator.

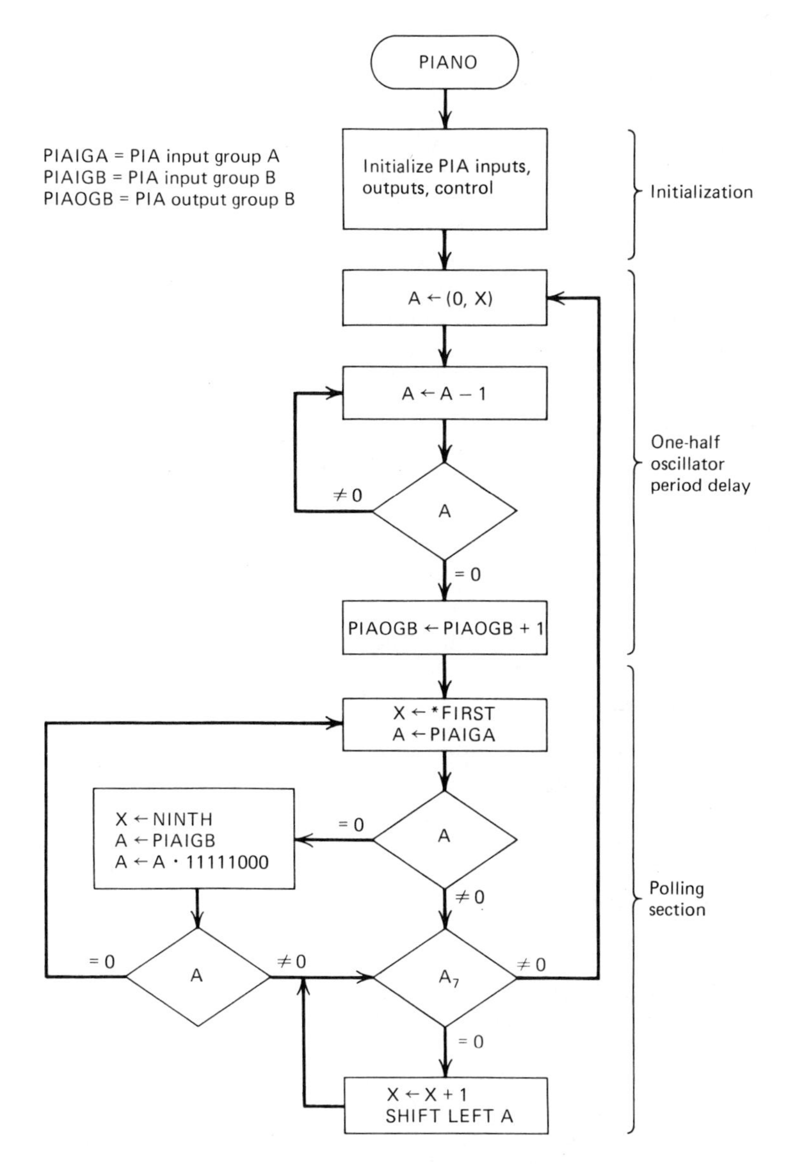

Figure 8.33 Electronic toy piano program flowchart.

(which can be verified against the program card for correctness). The next column shows the statement operands, if any, as calculated for the selected addressing mode. The fifth column contains a letter ("A" indicates that the operand is at an absolute address) or, for branch type instructions which use P.C. relative addressing, the hexadecimal target address of the statement. The columns of the source code are label, op-code, operands, and comments.

18 The first two lines of the program assign a program name and request the assembler to produce the symbol table at the end of the listing.

```
0010 NAM PIANO
0020 OPT S ASK FOR SYMBOL TABLE LIST
0030 * SYSTEM EQUATES
0040 PIADRA EQU $0000 PIA DATA REG A
0050 PIACRA EQU PIADRA+1 PIA CONTROL REG A
0060 PIADRB EQU PIADRA+2 PIA DATA REG B
0070 PIACRB EQU PIADRA+3 PIA CONTROL REG B
0080 PIAIGA EQU PIADRA PIA INPUT GROUP A
0090 PIAIGB EQU PIADRB PIA INPUT GROUP B
0100 PIAOGB EQU PIADRB PIA OUTPUT GROUP B
0110 ROMSIZ EQU 64 SIZE OF AVAILABLE ROM
0120 * STORE MACHINE CODE AT TOP OF MEMORY MAP
0130   ORG $FFFF-ROMSIZ+1
0140 * INITIALIZE THE PIA
0150 INIT CLR PIACRA DISABLE CONTROL INPUTS
0160   CLR PIACRB AND INTERRUPTS
0170   LDX #$0004 BITS 0-7 AS INPUTS
0180   STX PIADRA TO DATA & CONTROL REGS
0190   LDX #$0704 0-2 OUTPUTS. 3-7 INPUTS
0200   STX PIADRB TO DATA & CONTROL REGS
0202 * EXECUTE ONE-HALF OSCILLATOR PERIOD
0203 DELAY LDAA 0,X GET DELAY COUNT FROM LIST
0204 LOOP DECA TIME OUT. A=A-1
0205   BNE LOOP DELAY LOOP
0206   INC PIAOGB TOGGLE THE OUTPUT
0210 * BEGIN POLLING SECTION
0220 POLL LDX #FIRST POINT TO 1ST ON LIST
0230   LDAA PIAIGA READ PA0-PA7
0240   BEQ TESTB NO ACTIVE. TEST B GROUP
0250 NXTBIT BMI DELAY BRANCH IF BIT 7 ON
0260   INX MOVE POINTER. X=X+1
0270   ASLA SHIFT LEFT A
0280   BRA NXTBIT BRANCH TO TEST NEXT BIT
0290 TESTB LDX #NINTH POINT TO 9TH ON LIST
0300   LDAA PIAIGB READ PB0-PB7
0310   ANDA #%11111000 MASK OFF BOTTOM 3
0320   BNE NXTBIT CHK BIT 7 IF ANY ACTIVE
0330   BRA POLL NONE ACTIVE. CHK A SIDE
0400 * LIST OF DELAY COUNTS FOR NOTES
0410 * C - D - E - F - G - A - B - C
0420 FIRST FCB 143,127,113,107,95,85,75,71
0430 * C# - D# - F# - G# - A#
0440 NINTH FCB 137,122,103,91,82
0450 * POWER-ON RESET BRANCH VECTOR
0460   ORG $FFFE
0470 RESET FDB INIT
0480   END
```

Figure 8.34 Assembly language source code for toy piano.

19 Lines 3 through 11 permit symbolic names to be assigned to real values. These symbolic names can then be used throughout the program. If actual values were used, any hardware reconfiguration would cause the entire program to be reedited. If we use EQUates, only lines 4 through 11 would need revision. EQUs are called assembly directives. They do not appear on the program card and produce no executable code.

20 The ORG statement of line 13 is an assembly directive instructing the assembler where to begin addressing of the generated machine code (object code). We want the program to be PROMable at the top of the address map. The first address is calculated as the top address of the memory map minus the size of the PROM.

21 Statements 15 through 20 perform the function of PIA initialization. PIA internal registers referenced in this group have been previously defined in the EQU block. Operand codes used in lines 17 and 19 are selected in accordance with the PIA manufacturer's data sheet.

22 The oscillator block (lines 21 through 25) has been sequenced before the polling routine in the interest of program code efficiency. This location eliminates the need for a BRanch Always (BRA) instruction that would be required if the block were positioned elsewhere. Although this routine will be executed once, even before any polling is done to check for active switches, a one-half oscillator period will produce no audible sound and is therefore acceptable. An INCrement instruction is used to toggle the output instead of a COMplement (the more obvious choice) to gain an added design feature at no cost. By driving the speaker from either PB0, PB1, or PB2, a three-octave musical range is provided as each of these will toggle at half the frequency of the previous output.

23 Lines 26 through 38 implement the polling section which, in two phases, examines the input lines of the A and B PIA parallel channels and determines which, if any, inputs are active.

24 The form constant byte assembly directive (lines 41 and 43) permit the calculated delay count numbers to be entered into memory in list form. Note that the assembler automatically converts numbers of other bases to the hexadecimal format on the machine code side of the listing.

25 The power-on reset branch vector provides a means of telling the processor where to begin program execution when the device is first turned on. The processor will, at initialization, when RESET* is released, automatically load the PC from the top two bytes of the memory map. The form double byte assembly directive is used in line 46 to place the restart address (INIT = FFCO) into memory locations FFFE and FFFF.

26 An END statement is a required assembly directive used to define the limits of the source code.

At the bottom of the listing appears the error count and the symbol table that lists all program symbols and their associated values. This table (requested in line 2) may be helpful if program debugging is required.

Note that the total number of bytes required by the program is 63 (location FFFD is unused), one less than the PROM space allotted by estimation. PROMs can now be "blown" with the code produced by the assembler and inserted into the prototype hardware system for testing and verification. The complete system schematic appears in Fig. 8.36, and includes the octave select feature.

```
PAGE    001         PIANO

00001                                          NAM    PIANO
00002                                          OPT    S              ASK FOR SYMBOL TABLE LIST
00003                               * SYSTEM EQUATES
00004                   0000    A   PIADRA     EQU    $0000          PIA DATA REG. A
00005                   0001    A   PIACRA     EQU    PIADRA+1       PIA CONTROL REG A
00006                   0002    A   PIADRB     EQU    PIADRA+2       PIA DATA REG B
00007                   0003    A   PIACRB     EQU    PIADRA+3       PIA CONTROL REG B
00008                   0000    A   PIAIGA     EQU    PIADRA         PIA INPUT GROUP A
00009                   0002    A   PIAIGB     EQU    PIADRB         PIA INPUT GROUP B
00010                   0002    A   PIAOGB     EQU    PIADRB         PIA OUTPUT GROUP B
00011                   0040    A   ROMSIZ     EQU    64             SIZE OF AVAILABLE ROM
00012                               * STORE MACHINE CODE AT TOP OF MEMORY MAP
00013A  FFC0                                   ORG    $FFFF-ROMSIZ+1
00014                               * INITIALIZE THE PIA
00015A  FFC0    7F      0001    A   INIT       CLR    PIACRA         DISABLE CONTROL INPUTS
00016A  FFC3    7F      0003    A              CLR    PIACRB         AND INTERRUPTS
00017A  FFC6    CE      0004    A              LDX    #$0004         BITS 0-7 AS INPUTS
00018A  FFC9    DF      00      A              STX    PIADRA         TO DATA & CONTROL REGS
00019A  FFCB    CE      0704    A              LDX    #$0704         0-2 OUTPUTS. 3-7 INPUTS
00020A  FFCE    DF      02      A              STX    PIADRB         TO DATA & CONTROL REGS
00021                               * EXECUTE ONE-HALF OSCILLATOR PERIOD
00022A  FFD0    A6      00      A   DELAY      LDAA   0,X            GET DELAY COUNT FROM LIST
00023A  FFD2    4A                  LOOP       DECA                  TIME OUT. A=A-1
00024A  FFD3    26      FD  FFD2               BNE    LOOP           DELAY LOOP
00025A  FFD5    7C      0002    A              INC    PIAOGB         TOGGLE THE OUTPUT
00026                               * BEGIN POLLING SECTION
00027A  FFD8    CE      FFF0    A   POLL       LDX    #FIRST         POINT TO 1ST ON LIST
00028A  FFDB    96      00      A              LDAA   PIAIGA         READ PA0-PA7
00029A  FFDD    27      06  FFE5               BEQ    TESTB          NO ACTIVE. TEST B GROUP
00030A  FFDF    2B      EF  FFD0    NXTBIT     BMI    DELAY          BRANCH IF BIT 7 ON
00031A  FFE1    08                             INX                   MOVE POINTER. X=X+1
00032A  FFE2    48                             ASLA                  SHIFT LEFT A
00033A  FFE3    20      FA  FFDF               BRA    NXTBIT         BRANCH TO TEST NEXT BIT
```

```
00034A  FFE5  CE  FFF8      A  TESTB   LDX    #NINTH        POINT TO 9TH ON LIST
00035A  FFE8  96  02        A          LDAA   PIAIGB        READ PB0 – PB7
00036A  FFEA  84  F8        A          ANDA   #%11111000    MASK OFF BOTTOM 3
00037A  FFEC  26  F1  FFDF             BNE    NXTBIT        CHK BIT 7 IF ANY ACTIVE
00038A  FFEE  20  E8  FFD8             BRA    POLL          NONE ACTIVE. CHK A SIDE
00039                          * LIST OF DELAY COUNTS FOR NOTES
00040                          * C – D – E – F – G – A – B – C
00041A  FFF0  8F            A  FIRST   FCB      143,127,113,107,95,85,75,71
     A  FFF1  7F            A
     A  FFF2  71            A
     A  FFF3  6B            A
     A  FFF4  5F            A
     A  FFF5  55            A
     A  FFF6  4B            A
     A  FFF7  47            A
00042                          * C# – D# – F# – G# – A#
00043A  FFF8  89            A  NINTH   FCB      137,122,103,91,82
     A  FFF9  7A            A
     A  FFFA  67            A
     A  FFFB  5B            A
     A  FFFC  52            A
00044                          * POWER-ON RESET BRANCH VECTOR
00045A  FFFE                           ORG    $FFFE
00046A  FFFE      FFC0      A  RESET   FDB    INIT
00047                                  END
```

PAGE 002 PIANO

TOTAL ERRORS 00000

```
DELAY   FFD0  FIRST   FFF0  INIT    FFC0  LOOP    FFD2  NINTH   FFF8
NXTBIT  FFDF  PIACRA  0001  PIACRB  0003  PIADRA  0000  PIADRB  0002
PIAIGA  0000  PIAIGB  0002  PIAOGB  0002  POLL    FFD8  RESET   FFFE
ROMSIZ  0040  TESTB   FFE5
```

Figure 8.35 Assembler listing with machine code and symbol table for toy piano.

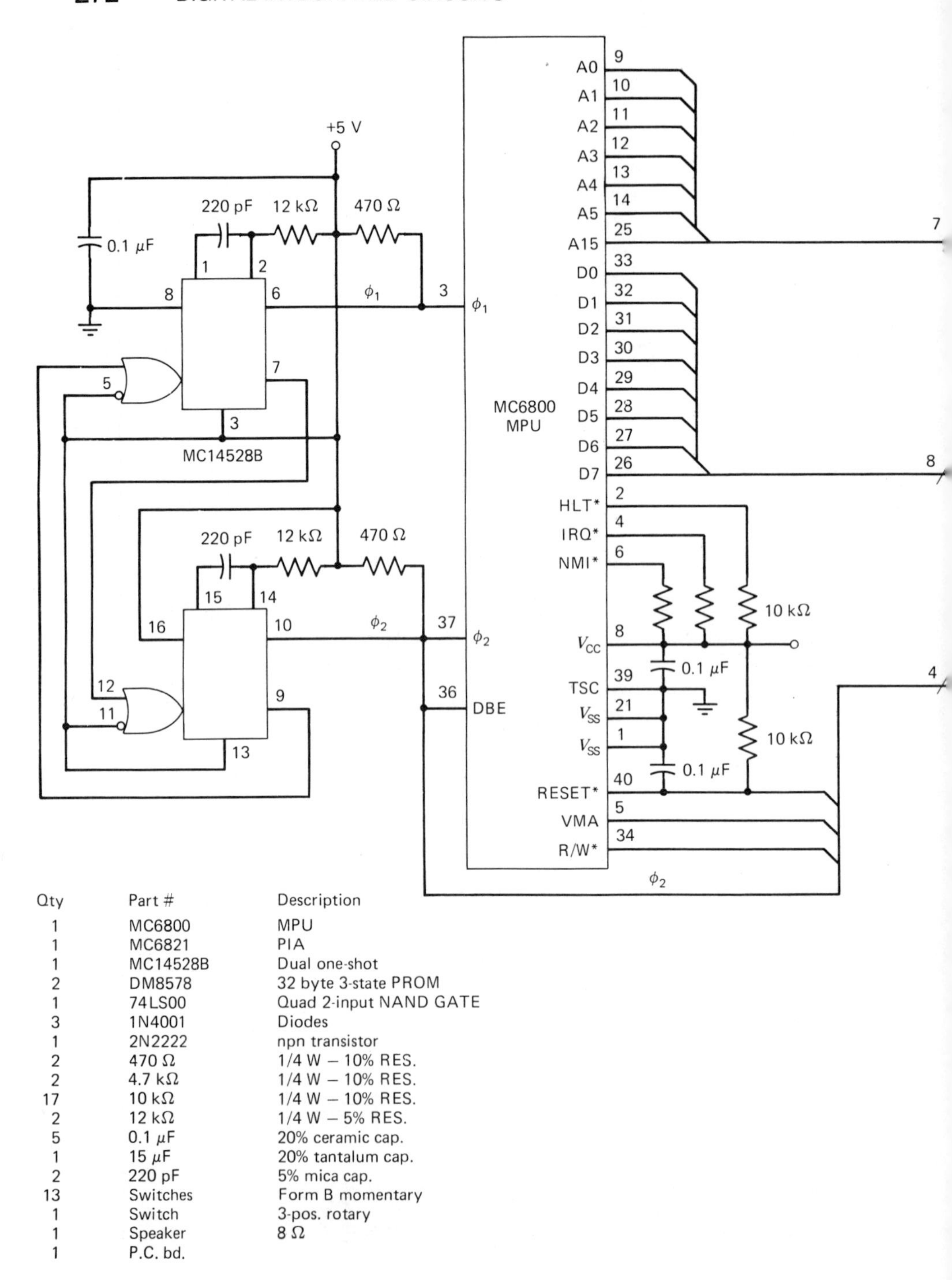

Qty	Part #	Description
1	MC6800	MPU
1	MC6821	PIA
1	MC14528B	Dual one-shot
2	DM8578	32 byte 3-state PROM
1	74LS00	Quad 2-input NAND GATE
3	1N4001	Diodes
1	2N2222	npn transistor
2	470 Ω	1/4 W – 10% RES.
2	4.7 kΩ	1/4 W – 10% RES.
17	10 kΩ	1/4 W – 10% RES.
2	12 kΩ	1/4 W – 5% RES.
5	0.1 μF	20% ceramic cap.
1	15 μF	20% tantalum cap.
2	220 pF	5% mica cap.
13	Switches	Form B momentary
1	Switch	3-pos. rotary
1	Speaker	8 Ω
1	P.C. bd.	

Figure 8.36 Schematic diagram of a microprocessor based three-octave toy piano.

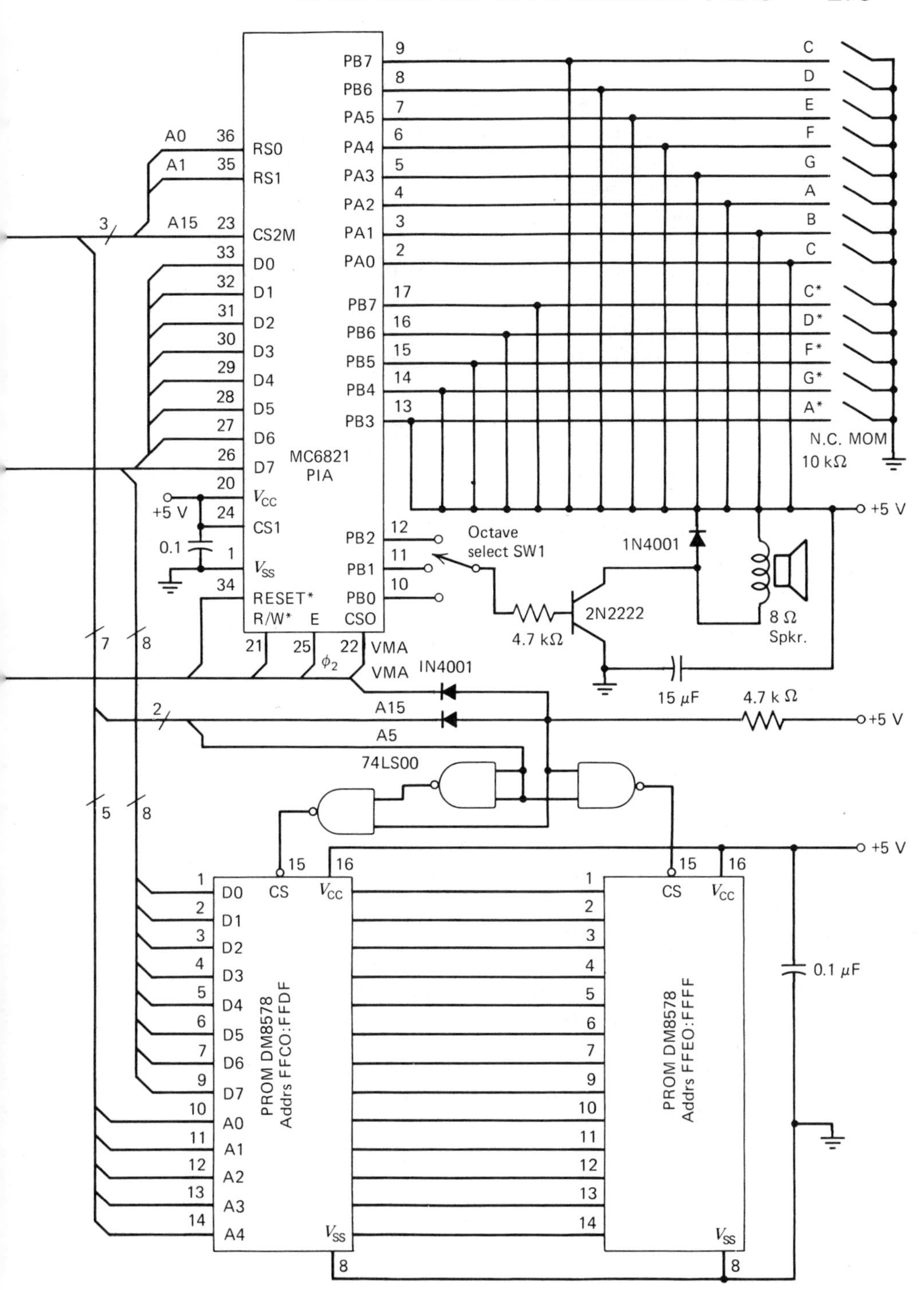
MC6821
PIA
RS0
RS1
CS2M
D0
D1
D2
D3
D4
D5
D6
D7
V_{CC}
CS1
V_{SS}
RESET*
R/W*
E
CSO
PB7
PB6
PA5
PA4
PA3
PA2
PA1
PA0
PB7
PB6
PB5
PB4
PB3
PB2
PB1
PB0
A0
A1
A15
+5 V
0.1
VMA
ϕ_2
C
D
E
F
G
A
B
C
C*
D*
F*
G*
A*
N.C. MOM
10 kΩ
+5 V
Octave
select SW1
1N4001
2N2222
4.7 kΩ
8 Ω
Spkr.
15 μF
IN4001
A15
A5
4.7 k Ω
74LS00
CS
V_{CC}
PROM DM8578
Addrs FFCO:FFDF
PROM DM8578
Addrs FFEO:FFFF
0.1 μF
V_{SS}

8.11 TRENDS IN MICROPROCESSING

Although the design examples in this chapter have featured the Motorola 6800 microprocessor and its family of peripheral components, the techniques presented here are applicable to systems using chips by Fairchild, Intel, National, Zilog, or by any of the other popular semiconductor manufacturers.

Recent trends in microprocessor products indicate a push toward greater power (16-bit data bus, multiply, divide, and block transfer-type instructions), functional integration (MPU, RAM, ROM, PIA, ACIA, TIMER, all on the same I.C.), and real-world interfacing (A/D and D/A converters built in) to allow a digital approach to the processing of analog signals.

In addition to the traditional use of microprocessors in industrial process control environments, microprocessors are having an ever increasing impact on our daily lives. The revolution in home computing and the availability of off-the-shelf hardware/software systems for small-business data processing has put the microprocessor within everyone's reach. Out of this growth comes new opportunities for engineers, programmers, technicians, repair service people, and others in related fields, from sales and marketing to space exploration and energy conservation.

8.12 REFERENCES

Bishop, R., *Basic Microprocessors and the 6800,* Hayden, Rochelle Park, NJ, 1979.

Frenzel, Jr., L. E., *The Howard W. Sams Crash Course in Microcomputers,* Sams, Indianapolis, 1980.

Greenfield, J. D. and W. C. Wray, *Using Microprocessors and Microcomputers: The 6800 Family,* Wiley, New York, 1981.

Kaufman, M. and A. H. Seidman, *Handbook of Electronics Calculations for Engineers and Technicians,* McGraw-Hill, New York, 1979. Chapter 17.

Klingman, E. E., *Microprocessor Systems Design,* Prentice-Hall, Englewood Cliffs, NJ, 1977.

Krutz, R. L., *Microprocessors and Logic Design,* Wiley, New York, 1980.

Motorola, *M6800 Microprocessor Applications Manual,* latest edition.

Motorola, *M6800 Microcomputer System Design Data,* latest edition.

Motorola, *M6800 Programming Reference Manual,* latest edition.

Osborne, A., *An Introduction to Microcomputers,* latest editions, vols. 0, 1, 2, and 3. Osborne/McGraw-Hill, Berkeley.

Tocci, R. and L. Laskowski, *Microcomputers and Microprocessors: Hardware and Software*, second edition, Prentice-Hall, Englewood Cliffs, NJ, 1982.

PART TWO

LINEAR INTEGRATED CIRCUITS

CHAPTER 9

The Operational Amplifier

Roger C. Thielking
Professor of Electrical Technology
Onondaga Community College

9.1 INTRODUCTION

The operational amplifier (op amp) is probably the most versatile and widely used type of linear integrated circuit. It is also often used with, or as a part of, other linear ICs (a number of examples appear in other chapters).

The name op amp refers to circuits that perform mathematical operations. These circuits need amplifiers with a differential input and very high gain, so that the overall circuit gain could be accurately controlled with precision external components. They were widely used in analog computers in the 1950s, when they were built entirely of discrete components, and were relatively bulky and expensive. IC fabrication, however, has reduced size and cost greatly and has improved performance as well. As a result, op amps have outgrown their name, and are now used in many types of circuits where amplification is needed.

Basic Operation

The schematic symbol for the op amp is shown in Fig. 9.1. There are two input terminals: the noninverting (+) and the inverting (−). The voltage difference

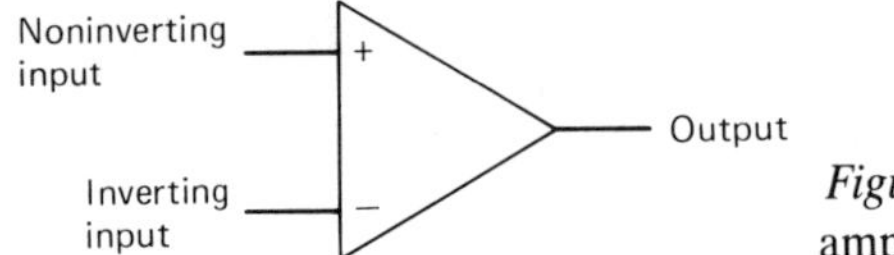

Figure 9.1 Electrical symbol for an operational amplifier (op amp).

between these terminals is amplified by the op amp *open loop gain*, A_{vol}, to produce the output signal.

A simplified equivalent circuit of an op amp is given in Fig. 9.2. If the input signal is applied to the noninverting input (with respect to the inverting input), then $V_{(-)} = 0$ V (ground), $V_d = V_{(+)}$, and the output, V_{out}, has the same polarity as $V_{(+)}$. However, the input may be applied instead to the inverting input, $V_{(-)}$ (with $V_{(+)}$ grounded). In this case, $V_d = -V_{(-)}$, and the output polarity (or phase, for an ac signal) is opposite to that of the input. The gain, A_{vol}, applies to dc as well as ac, and all voltages involved may include both ac and dc values.

This equivalent circuit shows the input impedance to be infinite, or an open circuit, with zero impedance in series with the output terminal. These assumptions are usually valid for analyzing practical circuits. At times, however, these and other op amp parameters may be important. They are described, with examples of their effects, in Secs. 9.3 and 9.4.

9.2 VOLTAGE LEVELS AND GAIN IN FEEDBACK AMPLIFIERS

Op amps have very large values of open loop gain A_{vol}, often over 100,000. For linear amplification, negative feedback is used to reduce circuit gain to the desired value and to make it largely independent of A_{vol}, which may vary considerably between op amps.

In the following examples, op amp characteristics, other than gain, are assumed to be ideal.

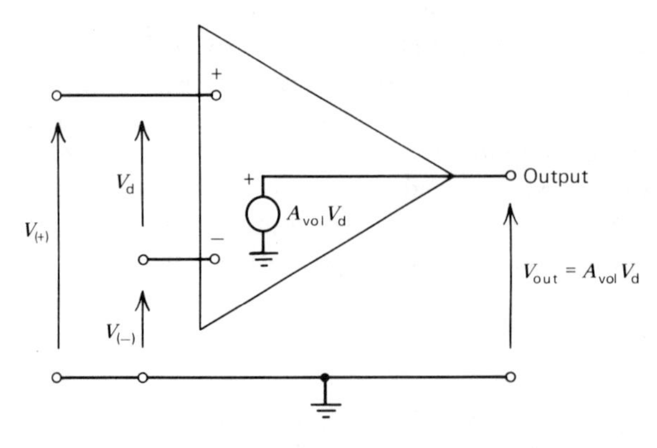

Figure 9.2 Simplified equivalent circuit of an op amp.

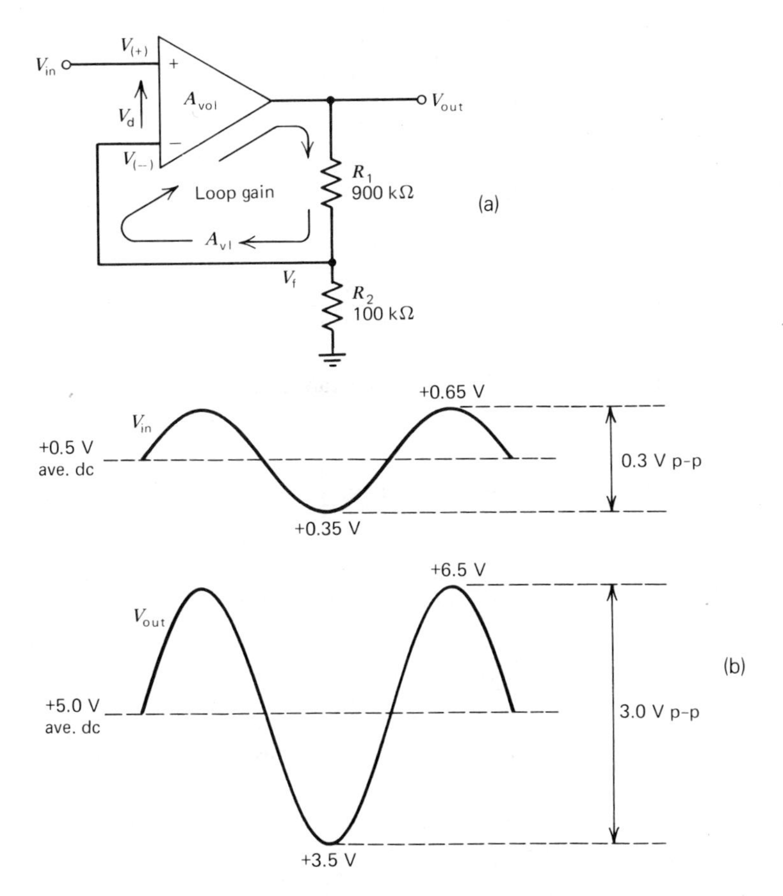

Figure 9.3 Noninverting amplifier. (*a*) Circuit diagram. (*b*) Input and output voltage waveforms (not drawn to scale).

Noninverting Amplifier

EXAMPLE 9.1 A noninverting amplifier is shown in Fig. 9.3*a*. Op amp gain A_{vol} = 50,000. Input voltage V_{in} is +0.5 V average dc, with an ac sine wave component of 0.3 V peak-to-peak (see Fig. 9.3*b*). Calculate the exact closed-loop circuit gain $A_{vcl(exact)}$, approximate $A_{vcl(approx.)}$ assuming infinite op amp gain, and the dc and ac values of V_{out}.

PROCEDURE

1 From the equivalent circuit of Fig. 9.2, $V_{out} = A_{vol} V_d$. From Fig. 9.3*a*, $V_d = V_{in} - V_f$. Combining these terms yields:

$$V_{out} = A_{vol}(V_{in} - V_f)$$

2 From the voltage divider relationship of R_1 and R_2, $V_f = V_{out}R_2/(R_1 + R_2)$. Substituting, we obtain

$$V_{out} = A_{vol}\left(V_{in} - \frac{R_2 V_{out}}{R_1 + R_2}\right)$$

3 Solving for the exact circuit gain $A_{vcl(exact)}$ yields:

$$A_{vcl(exact)} = \frac{V_{out}}{V_{in}} = \frac{1}{\dfrac{1}{A_{vol}} + \dfrac{R_2}{R_1 + R_2}} \tag{9.1}$$

4 The approximate gain, $A_{vcl(approx.)}$, is found by assuming an ideal op amp with $A_{vol} = \infty$. Equation 9.1 then becomes:

$$A_{vcl(approx.)} = \frac{R_1 + R_2}{R_2} = 1 + \frac{R_1}{R_2} \tag{9.2}$$

5 A concept that is often useful is the loop gain, A_{vl}, defined as the gain around the loop consisting of the op amp and its feedback components, given by:

$$A_{vl} = \frac{R_2 A_{vol}}{R_1 + R_2} = \frac{A_{vol}}{A_{vcl(approx.)}} \tag{9.3}$$

This may be solved for $A_{vol} = A_{vl} A_{vcl(approx.)}$. If we substitute this and Eq. 9.2 into Eq. 9.1,

$$A_{vcl(exact)} = \frac{1}{\dfrac{1}{A_{vl} A_{vcl(approx.)}} + \dfrac{1}{A_{vcl(approx.)}}}$$

$$= \frac{A_{vcl(approx.)}}{\dfrac{1}{A_{vl}} + 1} \tag{9.4}$$

6 Substituting values, we find that the approximate circuit gain from Eq. 9.2 is:

$$A_{vcl(approx.)} = \frac{900\ \text{k}\Omega + 100\ \text{k}\Omega}{100\ \text{k}\Omega} = 10.0$$

The loop gain from Eq. 9.3 is:

$$A_{vl} = \frac{50{,}000}{10} = 5000$$

and the exact circuit gain, from Eq. 9.4, is:

$$A_{vcl(exact)} = \frac{10}{\dfrac{1}{5000} + 1} = 9.998$$

7 For practical purposes, the exact and approximate values of A_{vcl} are equal, and the output voltages are calculated by multiplying input voltages by the gain:

The dc output component,

$$V_{OUT} = 10(+0.5\text{ V}) = +5.0\text{ V dc}$$

The magnitude of the ac output component,

$$v_{out} = |10(0.3\text{ V})| = 3.0\text{ V ac peak-to-peak}$$

Because the gain is positive (noninverting), the output waveform is in phase with the input, as shown in Fig. 9.3*b*.

Inverting Amplifier

EXAMPLE 9.2 An inverting amplifier is shown in Fig. 9.4*a*. Op amp gain $A_{vol} = 50{,}000$. Input voltage V_{in} is +0.5 V average dc, with an ac sine wave component of 0.3 V peak-to-peak (see Fig. 9.4*b*). Calculate the exact closed-loop circuit gain $A_{vcl(exact)}$, approximate $A_{vcl(approx.)}$ assuming infinite op amp gain, and the dc and ac values of V_{out}.

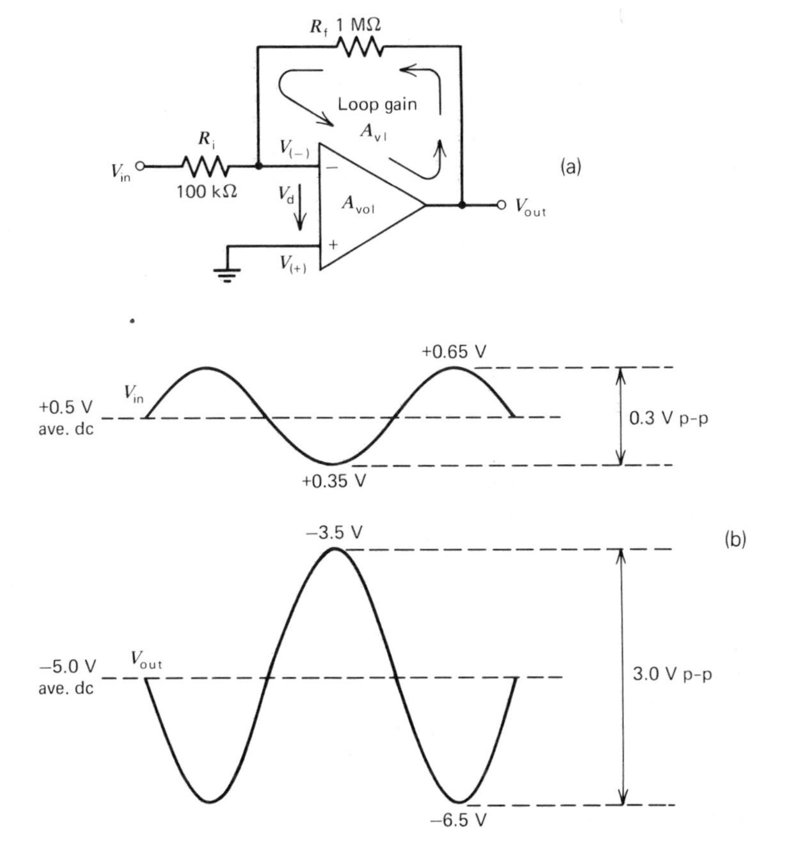

Figure 9.4 Inverting amplifier. (*a*) Circuit diagram. (*b*) Input and output voltage waveforms (not drawn to scale).

PROCEDURE

1 The approximate gain $A_{vcl(approx.)}$ is determined by assuming the ideal op amp gain, $A_{vol} = \infty$. Then, from Fig. 9.4*a*, $V_d = -V_{out}/A_{vol} = 0$ V, and $V_{(-)} = 0$ V. The same current must flow through R_i and R_f, so that $(V_{in} - 0 \text{ V})/R_i = (0 \text{ V} - V_{out})/R_f$. Solving for gain yields:

$$A_{vcl(approx.)} = \frac{V_{out}}{V_{in}} = -\frac{R_f}{R_i} \tag{9.5}$$

2 The loop gain A_{vl} around the feedback loop is:

$$A_{vl} = \frac{R_i A_{vol}}{R_f + R_i} \tag{9.6}$$

Closed-loop considerations are similar to those in the previous example for the noninverting amplifier. Equation 9.4 may be derived by a similar analysis, to show that it applies for the inverting amplifier as well.

3 From Eq. 9.5 the approximate gain is:

$$A_{vcl(approx.)} = -\frac{1 \text{ M}\Omega}{100 \text{ k}\Omega} = -10.0$$

From Eq. 9.6 the loop gain is:

$$A_{vl} = \frac{100 \text{ k}\Omega \cdot 50{,}000}{1 \text{ M}\Omega + 100 \text{ k}\Omega} = 4550$$

4 If we use these values in Eq. 9.4, the exact gain becomes:

$$A_{vcl(exact)} = \frac{-10}{\dfrac{1}{4550} + 1} = -9.9978$$

5 As for the noninverting circuit, exact and approximate values of A_{vcl} are equal for practical purposes. Output voltages are:

$$V_{OUT} = -10(+0.5 \text{ V}) = -5.0 \text{ V dc}$$
$$v_{out} = |-10(0.3 \text{ V})| = 3.0 \text{ V ac peak-to-peak}$$

The negative value of gain causes a phase reversal of the output ac signal with respect to the input, as shown in Fig. 9.4*b*.

The "Virtual Ground"

The preceding examples show that very large values of op amp open-loop gain A_{vol} cause the input difference voltage V_d to be very small; that is, $V_d \approx 0$ V (provided the circuit operates linearly, with no limiting of the output voltage). In an inverting amplifier circuit, as in Fig. 9.4*a*, where V_d is referenced to ground, the voltage at the inverting terminal is equal to V_d and also is very small. Although not connected to ground, the inverting terminal is at ground potential for most practical purposes and is therefore referred to as a *virtual ground*.

9.3 CHARACTERISTICS OF PRACTICAL OP AMPS

Characteristics are best described and understood by referring to the internal circuitry, and by knowing generally how the circuit operates. The type 741, with the circuit block diagram and simplified schematic shown in Fig. 9.5, is typical of the most widely used general-purpose devices. In other types, circuit details may be different, but the general form of the circuit and the way it operates are usually similar.

Circuit Operation

The various sections of the 741 circuit may be divided and categorized as shown in the block diagram, Fig. 9.5*a*.

Differential Input Amplifier Referring to Fig. 9.5*b*, we find that the feedback bias voltage, applied in common to Q_3 and Q_4, results in essentially the same emitter voltage at Q_1 and Q_2. If both input voltages are equal, the equally biased transistors will have equal collector currents. A very slight input difference voltage, however, unbalances the currents, resulting in an output at the collector of Q_4. The very high impedance of current source I_3 maximizes the voltage gain.

The basic circuit for IC current sources is shown in Fig. 9.6. Current I_{in}, through the diode-connected, forward-biased transistor Q_1, is fixed by the voltage and resistance values. The transistors are well matched and equally biased, so that $I_{out} = I_{in}$ in Q_2 regardless of the collector voltage.

In Fig. 9.5*c*, current source I_1 includes Q_8, Q_9, Q_{10}, and Q_{11}. Currents I_2 and I_3, controlled by Q_5, Q_6, and Q_7, are equal to each other and to the collector current of Q_3. This configuration reinforces the output signal and increases voltage gain still further.

Voltage Amplifier and Output Driver Transistors Q_{16} and Q_{20} are Darlington-connected to form a conventional high gain voltage amplifier. The collector load is the I_4 current generator, Q_{12} and Q_{13}. Resistors R_7, R_8, and their associated transistor produce a fixed voltage drop of about 1.1 V, for proper biasing of the output stage. High-frequency negative feedback, owing to C_1, limits high-frequency response to provide frequency compensation (see Sec. 9.5).

Output Amplifier Transistors Q_{14} and Q_{15} constitute a conventional complementary class B output stage. Because they are emitter followers, they have high current and power gain, but no voltage gain.

Transistors Q_{15} and Q_{22} are for current limiting. Normally cutoff, excessive

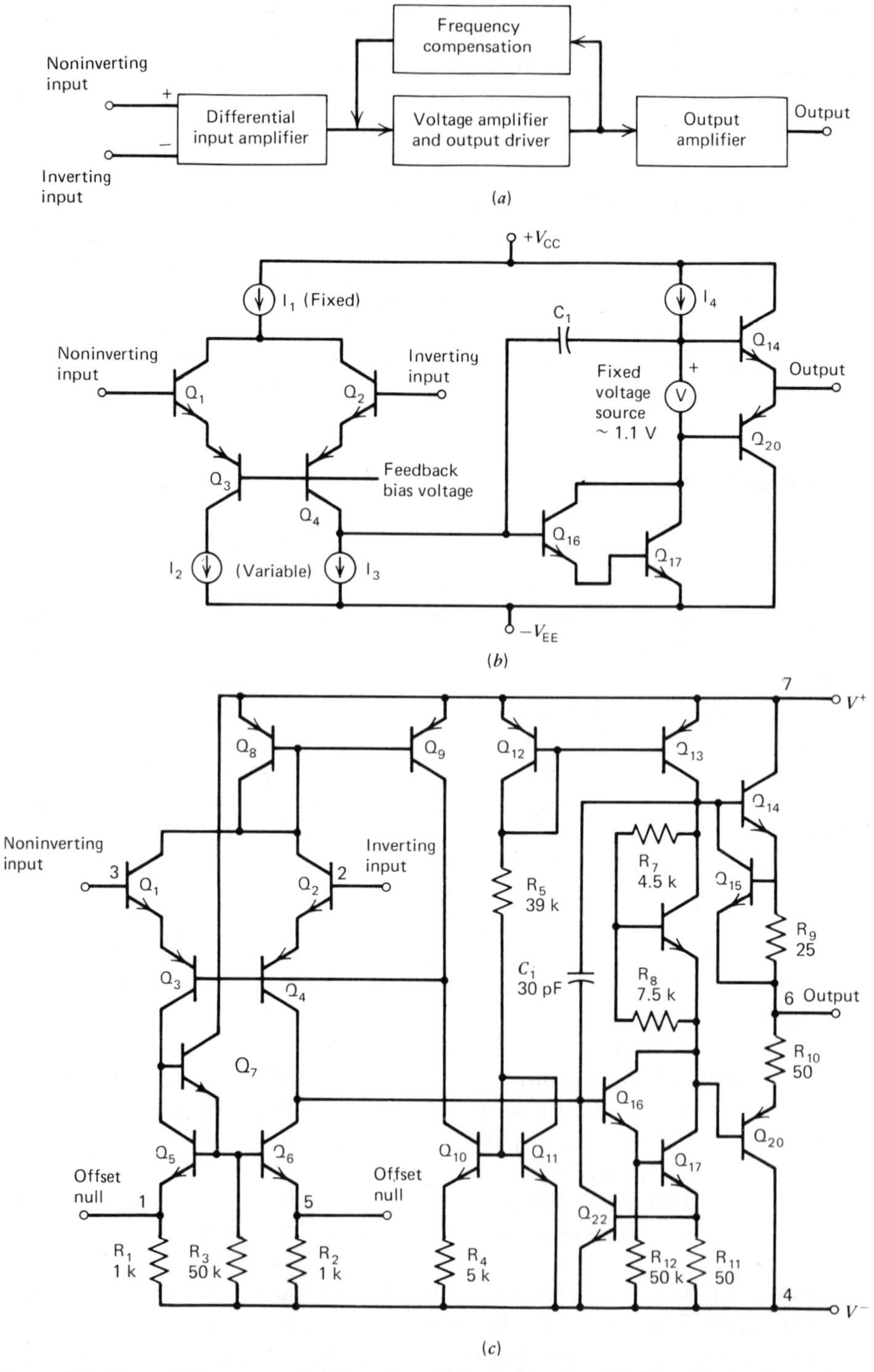

Figure 9.5 Type 741 op amp. (*a*) Functional block diagram. (*b*) Simplified schematic diagram. (*c*) Complete schematic diagram. (Courtesy National Semiconductor Corporation.)

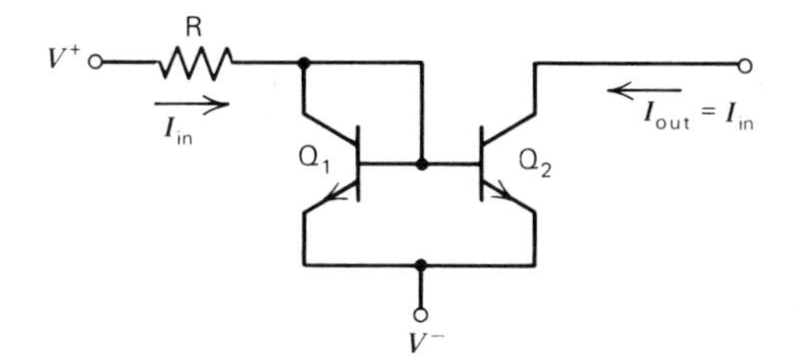

Figure 9.6 Integrated circuit current source.

current brings them into conduction so as to divert some of the base drive current to Q_{14} or Q_{17}.

Characteristics

Characteristics and parameters for the 741 op amp are listed in Table 9.1, together with ideal values. Both typical and worst limit values, if available, are shown. The worst limit is either a maximum or minimum, whichever has the greatest effect on circuit performance. (These values are for certain conditions including room temperature, 25°C. Manufacturers' specifications usually give additional data for other conditions.)

Table 9.1 **Characteristics of a type 741 op amp**

		Type 741 values		
Characteristic	*Symbol*	*Worst* limit*[a]	*Typical*[a]	*Ideal value*
Input offset voltage	V_{IO}, V_{OS}	±5 mV	±1 mV	0
Input bias current	I_{IB}, I_B	500 nA	80 nA	0
Input offset current	I_{IO}, I_{OS}	±200 nA	±20 nA	0
Input resistance	r_i, r_{in}	0.3 M	2 M	∞
Input capacitance	C_i, C_{in}		1.4 pF	0
Offset voltage adjustment range	V_{IOR}, V_{OSR}		±15 mV	
Large signal voltage gain	A_{vol}, A_v	50 V/mV	200 V/mV	∞
Output resistance	r_o		75 Ω	0
Common mode rejection ratio	CMRR	70 dB	90 dB	∞
Supply voltage rejection ratio	PSRR	150 μV/V	30 μV/V	0
Output voltage swing	V_O	±12 V	±14 V	
Output short-circuit current	I_{os}		20 mA	
Supply current	I_D, I_S	2.8 mA	1.7 mA	
Power consumption	P_C	85 mW	50 mW	
Transient response:				
Rise time	t_{TLH}		0.3 μs	0
Overshoot	*OS*		15%	0
Slew rate	*SR*		0.5 V/μs	∞

[a] These values are subject to certain conditions, and may vary among manufacturers. The manufacturer's specification should be consulted for guaranteed limits.

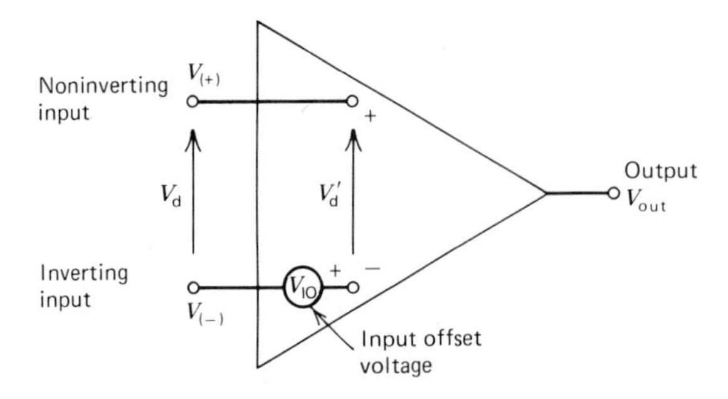

Figure 9.7 Input offset voltage equivalent circuit.

Input Offset Voltage (V_{IO}, V_{OS}) The dc value of V_d, the difference voltage at the inputs, that results in zero output voltage. It may be shown in an equivalent circuit (see Fig. 9.7) as a voltage source in series with one of the inputs. When $V_d = V_{IO}$, the internal input signal $V'_d = 0$, corresponding to $V_{out} = 0$. The offset voltage results from slight imbalances between the op amp input transistors Q_1, Q_2, Q_3, and Q_4 (Fig. 9.5*c*) and can be of either polarity, that is, $\pm V_{IO}$. Because an offset voltage causes an undesired change in output, very small values are best.

Input Bias Current (I_{IB}, I_B) The dc current at either input, as shown in Fig. 9.8. (In Fig. 9.5*c*, it is the base drive current for transistor Q_1 or Q_2.) Because this may load the input signal and feedback network significantly, the value should be very small.

Input Offset Current (I_{IO}, I_{OS}) The difference between the input bias currents at the two inputs. Referring to Fig. 9.8, $I_{IO} = I_{IB(+)} - I_{IB(-)}$. Good balance of the IC transistors results in $I_{IB(+)} \approx I_{IB(-)}$, so that I_{IO} is much smaller than I_{IB}.

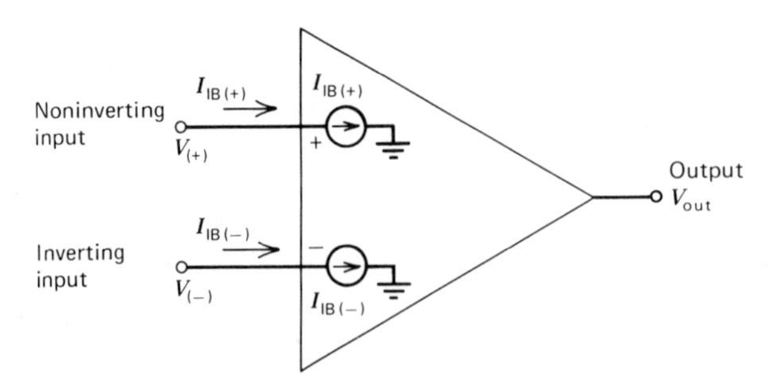

Figure 9.8 Input bias currents equivalent circuit.

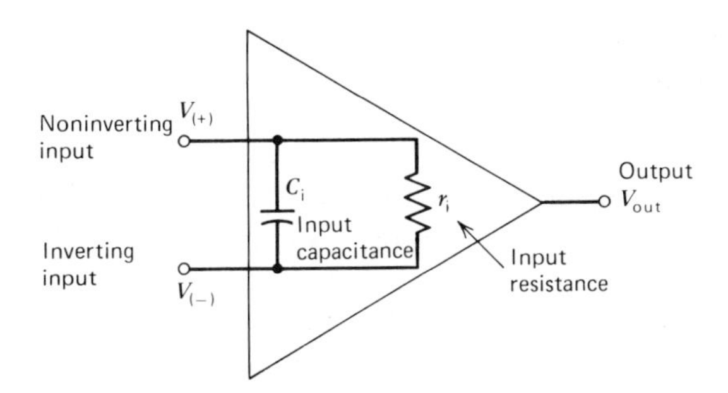

Figure 9.9 Input capacitance and resistance equivalent circuit.

Input Resistance (r_i, r_{in}) The voltage change divided by the current change, or the ac resistance, between the two input terminals as shown in the equivalent circuit, Fig. 9.9. This tends to load the inputs. However, in negative current feedback circuits the effective input resistance (with respect to ground) is r_i multiplied by loop gain A_{vl}. Typical A_{vl} values are very large (see Ex. 9.2), so that the effect of r_i is almost always insignificant.

Input Capacitance (C_i, C_{in}) As shown in Fig. 9.9, this is the internal capacitance between the inputs. It is rarely large enough to be significant, and is often not specified.

Offset Voltage Adjustment Range (V_{IOR}, V_{OSR}) If an op amp has terminals for offset null adjustment, a potentiometer may be added to change the offset voltage through the specified range (Sec. 9.4).

Large Signal Voltage Gain (A_{vol}, A_v) The open-loop voltage gain of the op amp, V_{out}/V_d (see Fig. 9.2), at dc and low frequencies. Small signal gain may be somewhat greater. For higher frequencies, the gain will decrease.

Output Resistance (r_o) The ratio of a change in output voltage to a change in output current (with no feedback connections). It has the effect of an equivalent resistance, as shown in Fig. 9.10. In practical circuits with negative feed-

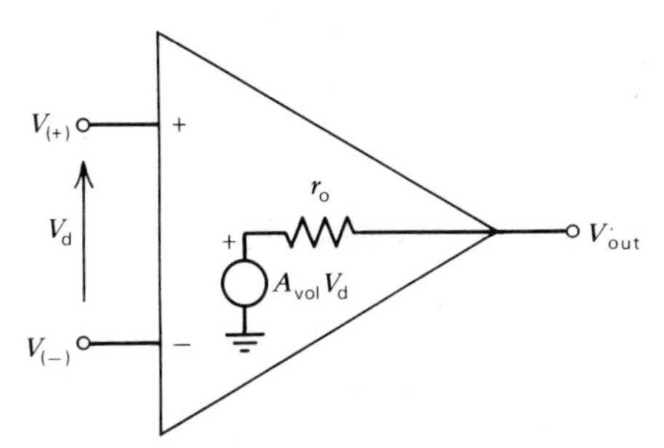

Figure 9.10 Output resistance equivalent circuit.

back, the output impedance of the circuit is r_o divided by A_{vl}, the loop gain. The very small result is rarely significant.

Common Mode Rejection Ratio (CMRR) Ideally, op amps respond only to a difference in input voltages, V_d, and not to a common mode voltage (an identical voltage or voltage change applied simultaneously to both inputs). The common mode voltage will, however, have a slight effect. Common mode rejection ratio, CMRR, is the ratio of the desired output voltage change (due to V_d) to the undesired output voltage change due to the common mode input voltage.

Supply Voltage Rejection Ratio (PSRR) A small change in power supply voltage should not affect the output. There is, however, a small effect, given by the power supply rejection ratio, PSRR, defined as the ratio of the change in output voltage divided by the power supply voltage change that causes it. The effect is negligible if power supplies are well regulated, and is usually insignificant even with unregulated supplies.

Output Voltage Swing (V_O) This is the maximum output voltage for a particular, specified output load resistance. The actual voltage limits may be greater, but op amp performance may degrade when the output voltage swing is exceeded.

Output Short-Circuit Current (I_{os}) Excessive output current might cause overdissipation or other damage to the op amp. To prevent this, internal protective circuitry limits current to the indicated value.

Supply Current and Power Consumption (I_D, P_C) These values are for no-signal, no-load conditions.

Transient Response: Rise Time and Overshoot (t_{TLH}, OS) These values apply only to a very small, specified output signal level. The transient response of most larger signals is limited by the slew rate (below). Rise time t_{TLH} may, however, be used for frequency response calculations.

Slew Rate (SR) The maximum rate of change of the output voltage. It is determined by C_1 (Fig. 9.5*b*) and its limiting charging and discharging currents. In uncompensated op amps (see Sec. 9.5) the slew rate may be determined by externally connected frequency compensation components.

9.4 OUTPUT EFFECTS OF OP AMP CHARACTERISTICS

Inverting and noninverting amplifier circuits with ideal op amp characteristics, and the effect of op amp voltage gain A_{vol}, were examined in Sec. 9.2. The practical op amp characteristics defined in Sec. 9.3 may cause deviations or errors in the output signal from the ideal. These effects are illustrated in the following examples.

Effect of Input Offset Voltage

EXAMPLE 9.3 A 741 op amp is used in the inverting amplifier circuit (Fig. 9.4*a*). Find the effect on output voltage V_{out} of the worst-limit value of offset voltage V_{IO}.

PROCEDURE

1 The amplifier circuit, including the equivalent circuit for the offset voltage (from Fig. 9.7), is shown in Fig. 9.11. Input voltage V_{in} is grounded; therefore, from Eq. 9.5, V_{out} should be 0.0 V. Thus, any voltage at the output is caused only by V_{IO}. Because the op amp is assumed ideal in other respects, the values are $V'_d = 0.0$ V, $V_d = V_{IO}$, and $V_{(-)} = -V_{IO}$. Then, from the voltage divider relationship, the result is:

$$V_{(-)} = \frac{R_i V_{out}}{R_i + R_f} = -V_{IO} \tag{9.7}$$

2 Solving for V_{out} yields:

$$V_{out} = -\frac{(R_i + R_f)V_{IO}}{R_i} \tag{9.8}$$

NOTE: The input offset voltage is always multiplied by the noninverting gain.

3 From Table 9.1, $V_{IO} = \pm 5$ mV (the offset voltage can be either polarity). Substituting values into Eq. 9.8 we obtain:

$$V_{out} = -\frac{(100\text{ k}\Omega + 1.0\text{ M}\Omega)(\pm 5\text{ mV})}{100\text{ k}\Omega} = \mp 55\text{ mV}$$

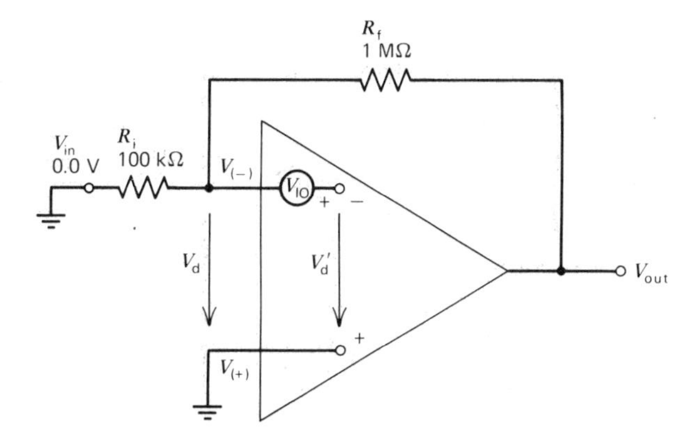

Figure 9.11 Inverting amplifier with the equivalent circuit of the input offset voltage included.

With higher gain, R_f is relatively larger and the effect of V_{IO} on output is greater.

EXAMPLE 9.4 A 741 op amp is used in the noninverting amplifier circuit of Fig. 9.3. Find the effect on output voltage V_{out} due to the worst limit value of the offset voltage V_{IO}.

PROCEDURE

1 The amplifier circuit, including the equivalent circuit for the offset voltage (from Fig. 9.7), is shown in Fig. 9.12. This circuit, with V_{in} grounded, is identical (except for the resistor designators and values) to Fig. 9.11.

2 Therefore, the same solution applies and Eq. 9.8 becomes:

$$V_{out} = -\frac{(R_1 + R_2)V_{IO}}{R_2} \tag{9.9}$$

3 Substituting values yields:

$$V_{out} = -\frac{(900\text{ k}\Omega + 100\text{ k}\Omega)(\pm 5\text{ mV})}{100\text{ k}\Omega} = \mp 50\text{ mV}$$

Combining Characteristics Effects with Signal Levels

In these examples, V_{in} is grounded to simplify the calculations. With signal voltages at V_{in}, the desired, or ideal, outputs may be calculated as in Ex. 9.1 or 9.2, and the offset effect as in Ex. 9.3 or 9.4. Using superposition, these are then added to determine the exact output signal. The effects of input currents and common mode voltage (Exs. 9.5, 9.7, 9.8, and 9.10) may similarly be added to signal levels.

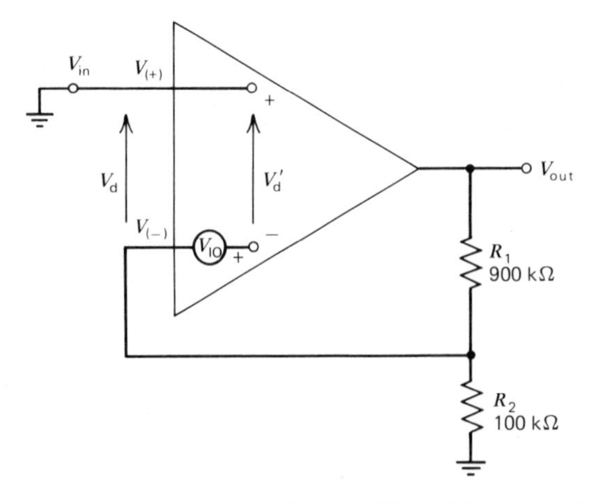

Figure 9.12 Noninverting amplifier with the equivalent circuit of the input offset voltage included.

Effect of Bias and Offset Currents

EXAMPLE 9.5 A 741 op amp is used in the inverting amplifier circuit of Fig. 9.4*a*. Find the effect on output voltage V_{out} owing to the worst limit value of input bias current I_{IB}.

PROCEDURE

1 The amplifier, including bias currents (see Fig. 9.8), is shown in Fig. 9.13*a*. Because V_{in} is grounded, any voltage at the output is caused by bias current. Otherwise, ideal op amp characteristics are assumed, so that $V_d = 0.0$ V and $V_{(-)} = V_{(+)} = 0.0$ V.

2 Resistance R_i has no voltage drop and no current; therefore, all of $I_{IB(-)}$ flows through R_f as shown. Then:

$$V_{out} - V_{(-)} = V_{out} = I_{IB(-)}R_f \tag{9.10}$$

3 From Table 9.1, worst limit I_{IB} is 500 nA. Using this for $I_{IB(-)}$ and substituting values yields:

$$V_{out} = 500 \text{ nA} \cdot 1 \text{ M}\Omega = 0.5 \text{ V}$$

EXAMPLE 9.6 In Fig. 9.13*b*, compensating resistor R_c has been added to minimize the effect of input bias current. Determine the optimum value for R_c.

PROCEDURE

1 Thevinize the inverting terminal input resistors, as shown in Fig. 9.13*c*. The optimum value of R_c will cause $V_{out} = 0.0$ V when $I_{IB(-)} = I_{IB(+)} = I_{IB}$. Then:

$$V_{TH} = \frac{V_{out}R_i}{R_i + R_f} = 0.0 \text{ V}$$

$$V_{(-)} = -I_{IB(-)}R_{TH} = -I_{IB} R_{TH}$$

$$V_{(+)} = -I_{IB(+)} R_c = -I_{IB} R_c$$

2 Because of the following:

$$V_d = 0 \text{ V}, \quad V_{(-)} = V_{(+)} = -I_{IB} R_{TH} = -I_{IB} R_c.$$

Solving for R_c yields:

$$R_c = R_{TH} = R_i \parallel R_f = \frac{R_iR_f}{R_i + R_f} \tag{9.11}$$

3 If we substitute values:

$$R_c = \frac{100 \text{ k}\Omega \cdot 1 \text{ M}\Omega}{100 \text{ k}\Omega + 1 \text{ M}\Omega} = 90.9 \text{ k}\Omega$$

(The value of R_c is not critical, and any value within about $\pm 10\%$ would work almost as well.)

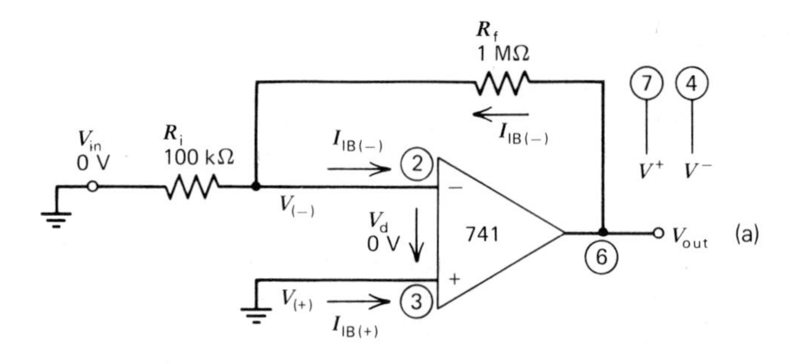

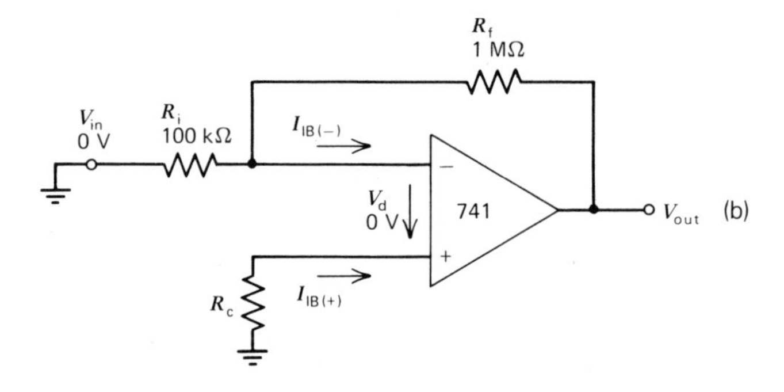

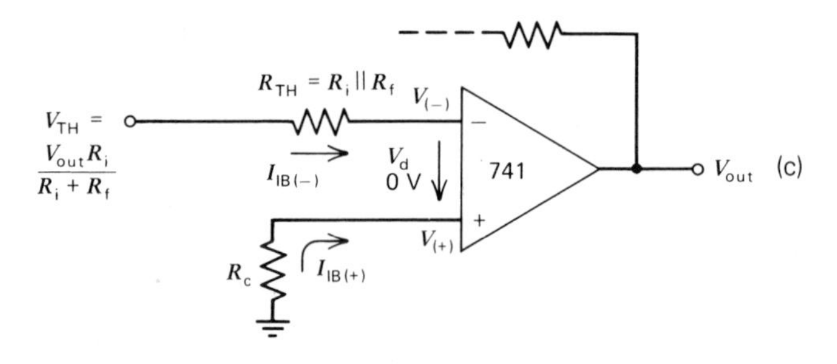

Figure 9.13 Inverting amplifier showing input bias currents. (*a*) Basic circuit. (*b*) Circuit with compensating resistor R_c. (*c*) Circuit with feedback resistors and voltages represented by their Thevenin equivalents.

EXAMPLE 9.7 A 741 op amp is used in the circuit of Fig. 9.13*b*. Find the effect on output voltage V_{out} owing to the worst-limit value of the input offset current I_{IO}. Assume $R_c = R_i \| R_f = 90.9$ k (as determined in Ex. 9.6).

PROCEDURE

1 Use the Thevenized circuit, Fig. 9.13*c*. Assume $V_d = 0$ V so that $V_{(-)} = V_{(+)}$. Then the results are:

$$V_{TH} - V_{(+)} = I_{IB(-)} R_{TH} = \frac{I_{IB(-)} R_i R_f}{R_i + R_f} \tag{9.12}$$

$$V_{TH} = \frac{V_{out} R_i}{R_i + R_f} \tag{9.13}$$

$$V_{(+)} = V_{(-)} = -I_{IB(+)} R_c = -\frac{I_{IB(+)} R_i R_f}{R_i + R_f} \tag{9.14}$$

2 Substituting Eqs. 9.13 and 9.14 into Eq. 9.12 yields:

$$\frac{V_{out} R_i}{R_i + R_f} - \frac{-I_{IB(+)} R_i R_f}{R_i + R_f} = \frac{I_{IB(-)} R_i R_f}{R_i + R_f}$$

3 Solving for V_{out}, we obtain:

$$V_{out} = (I_{IB(-)} - I_{IB(+)}) R_f$$

4 The difference current, $I_{IB(-)} - I_{IB(+)}$ is defined as the offset current, I_{IO}. Thus, the result is

$$V_{out} = I_{IO} R_f \tag{9.15}$$

5 Worst limit I_{IO}, from Table 9.1, is ± 200 nA. Substituting values yields:

$$V_{out} = \pm 200 \text{ nA} \cdot 1 \text{ M}\Omega = \pm 0.2 \text{ V}$$

(I_{IO} can assume either polarity, depending upon which input bias current is greater.)

EXAMPLE 9.8 For the noninverting amplifier of Fig. 9.3*a*, determine the expressions for the output effects of input currents, corresponding to Eqs. 9.10, 9.11, and 9.15.

PROCEDURE

1 When the input voltage V_{in} is grounded for these analyses (see Fig. 9.14), the circuits become identical to those of Fig. 9.13 except for the resistor designations. Resistances

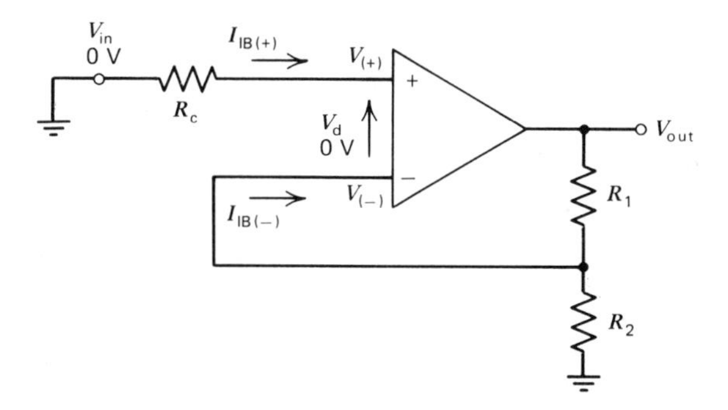

Figure 9.14 Noninverting amplifier with the input grounded.

R_f and R_i become R_1 and R_2 respectively, and R_c may be inserted at $V_{(+)}$, as shown in Fig. 9.14. Equation 9.10 for input bias current becomes:

$$V_{out} = I_{IB(-)}R_1 \tag{9.16}$$

2 Equation 9.11 for the compensating resistor becomes:

$$R_c = R_1 \parallel R_2 = \frac{R_1R_2}{R_1 + R_2} \tag{9.17}$$

3 Equation 9.15 for input offset current becomes:

$$V_{out} = I_{IO}R_1 \tag{9.18}$$

4 Equation 9.17 assumes that the source resistance of V_{in} is small. If it is not, the series sum of the source resistance and R_c should equal $R_1 \parallel R_2$.

Offset Null Adjustments

Some op amps, such as the 741, have terminals for offset null adjustments and a manufacturer-recommended circuit, as shown in Fig. 9.15*a*. If the op amp does not have offset null terminals, a circuit such as that in Fig. 9.15*b* may be used.

EXAMPLE 9.9 Adjust the null potentiometer R_n of either circuit in Fig. 9.15 to compensate for the effects of the input offset voltage and current.

PROCEDURE

1 Ground V_{in}, as shown.
2 Measure V_{out} with a sensitive dc voltmeter that will read to 1 mV or less.
3 Carefully adjust R_n for a null (0.0 V) at V_{out}.

Because input offset voltage and current drift with temperature, R_n must be readjusted after a temperature change in the op amp has occurred, if an exact null is needed.

Effect of Large Signal Voltage Gain

See Exs. 9.1 and 9.2.

Effect of Common Mode Voltage

EXAMPLE 9.10 In the circuit of Fig. 9.3*a*, find the effect on output voltage V_{out} because of the common mode voltage V_{CM} when the input voltage V_{in} changes by 1.0 V. Assume a 741 op amp and use the worst-limit value of CMRR.

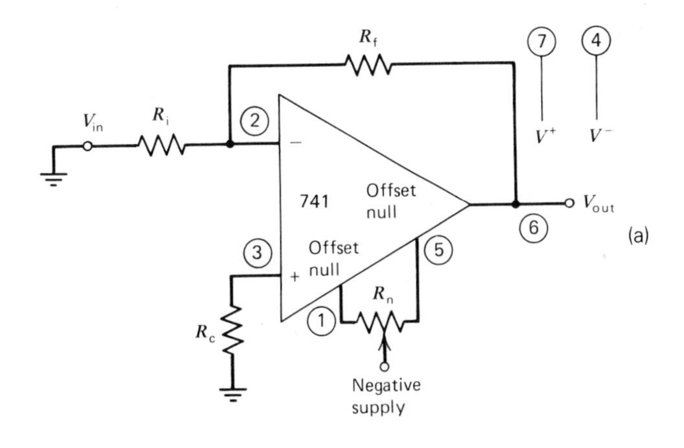

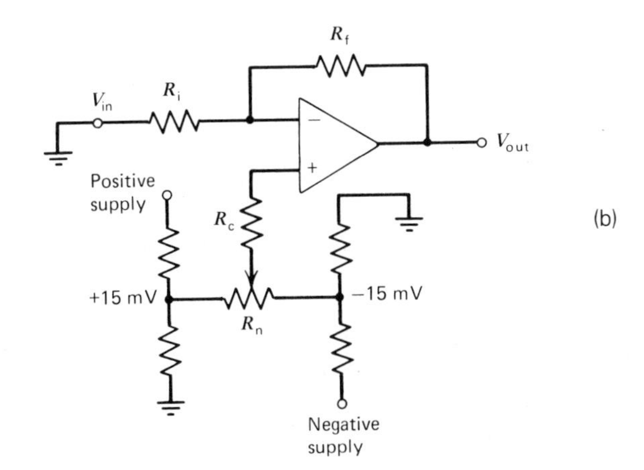

Figure 9.15 Op amps with offset null adjustment. (*a*) Type 741 with recommended null circuit. (*b*) Nulling circuit for an op amp lacking offset null terminals.

PROCEDURE

1 Common mode rejection ratio (CMRR) may be expressed by:

$$\mathrm{CMRR} = 20\log\left(\frac{V_{\mathrm{CM}}}{V_{\mathrm{d(CM)}}}\right)\ \mathrm{dB} \tag{9.19}$$

where $V_{\mathrm{d(CM)}}$ is the effective input difference voltage (V_{d}) resulting from V_{CM}.

2 Solving for $V_{\mathrm{d(CM)}}$ yields:

$$V_{\mathrm{d(CM)}} = \frac{V_{\mathrm{CM}}}{\mathrm{antilog}\left(\dfrac{\mathrm{CMRR}}{20}\right)} \tag{9.20}$$

3 Voltage $V_{d(CM)}$ has the same effect as input offset voltage V_{IO}, and may be substituted for it in Eq. 9.9 as follows:

$$V_{out} = \frac{(R_1 + R_2)V_{d(CM)}}{R_2}$$

$$= \frac{(R_1 + R_2)V_{CM}}{R_2 \text{ antilog}\left(\frac{\text{CMRR}}{20}\right)} \qquad (9.21)$$

4 CMRR, $V_{d(CM)}$ and V_{out} may be either polarity. Voltage V_{CM} is equal to the change in V_{in}, because this change is applied to the noninverting input and, through the feedback, to the inverting input of the op amp as well. From Table 9.1, worst-limit CMRR is 70 dB. Then, substituting values yields:

$$V_{out} = \pm\frac{(900\text{ k}\Omega + 100\text{ k}\Omega)\ 1.0\text{ V}}{100\text{ k}\Omega \cdot \text{antilog}\left(\frac{70\text{ dB}}{20}\right)} = \pm 3.16\text{ mV}$$

This effect, in noninverting amplifiers, is very small, usually insignificant. However, in differential amplifiers (see Chap. 10), common mode effects may be much greater.

Frequency Response Effects

EXAMPLE 9.11 Find the approximate open-loop frequency response for a 741 op amp.

PROCEDURE

1 The typical value of the open-loop gain, A_{vol}, from Table 9.1, is 200 V/mV or 200,000. This may also be expressed in decibels as $20\log(200{,}000) = 106$ dB. This value applies to very low frequencies only. At higher frequencies gain decreases, until it reaches unity at frequency f_T, calculated by:

$$f_T = \frac{0.35}{t_{TLH}} \qquad (9.22)$$

where t_{TLH} is the transient response rise time for a unity-gain amplifier.

2 From Table 9.1, t_{TLH} is typically 0.3 μs. Substituting yields:

$$f_T = \frac{0.35}{0.3\ \mu\text{s}} = 1.17\text{ MHz}$$

3 The attenuation stems from a single time constant, so that it is −20 dB/decade (or −6 dB/octave) with a phase shift of −90°, and the high-frequency gain is:

$$A_{vol(high\ freq.)} = -j\frac{f_T}{f} \qquad (9.23)$$

4 The gain at any frequency is:

$$A_{vol} = \frac{200{,}000}{1 + j\dfrac{f}{f_c}} \tag{9.24}$$

where f_c is the upper bandwidth limit of the op amp, which occurs when $A_{vol(high\ freq.)}$ from Eq. 9.23 equals 200,000. Thus, $f_T/f_c = 200{,}000 = 1.17\ \text{MHz}/f_c$ and solving, $f_c = 1.17\ \text{MHz}/200{,}000 = 5.85\ \text{Hz}$. Then, Eq. (9.24) becomes:

$$A_{vol} = \frac{200{,}000}{1 + j\dfrac{f}{5.85\ \text{Hz}}} \tag{9.25}$$

5 From Eq. 9.25, the Bode approximation of frequency response may be plotted, as in Fig. 9.16*a*. Frequency f_c also may be found graphically, by drawing the 20 dB/decade slope through $f_T = 1.17$ MHz at 0 dB, and determining where this line indicates a gain of 200,000 or 106 dB.

The Bode approximation graph is close to the actual values except at frequencies near f_c, where there is a 3 dB error and the actual gain is 103 dB. Exact values may be calculated by Eq. 9.25.

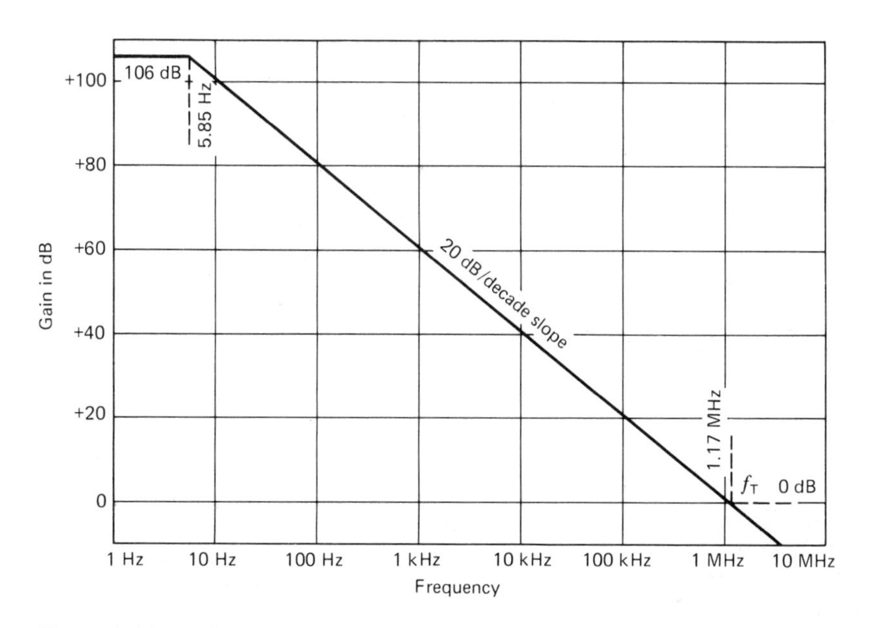

Figure 9.16 Bode approximations of the frequency response of a 741 op amp. (*a*) Open loop response, A_{vol}.

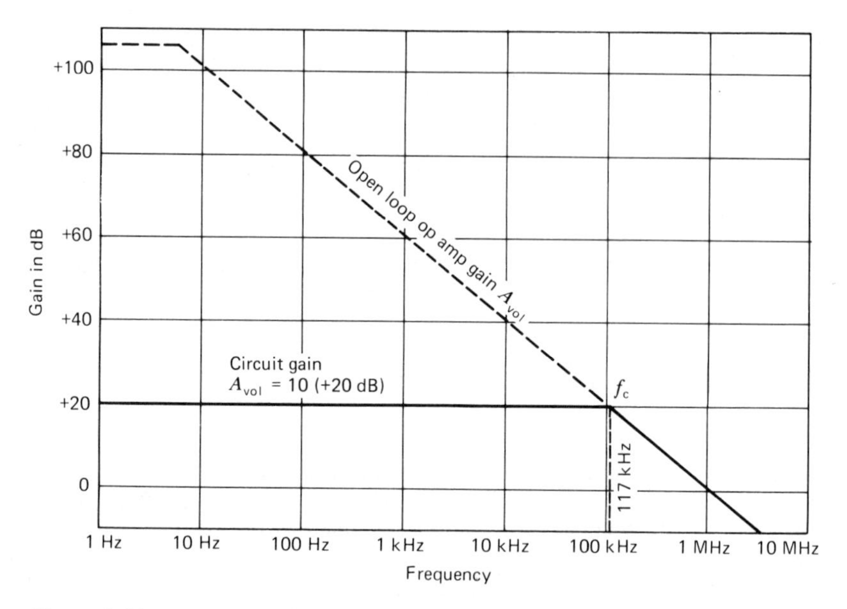

Figure 9.16 continued (*b*). Closed loop response, A_{vcl} (low freq.) = 10.

EXAMPLE 9.12 Find the bandwidth and frequency response for an inverting amplifier with closed-loop gain $A_{vcl} = -10$ using a 741 op amp.

PROCEDURE

1 Equations 9.3 and 9.4 apply, using A_{vcl} and $A_{vcl(low\ freq.)}$ in place of $A_{vcl(exact)}$ and $A_{vcl(approx.)}$, respectively. Then, substituting Eq. 9.3 into Eq. 9.4 yields:

$$A_{vcl} = \frac{A_{vcl(low\ freq.)}}{1 + \dfrac{A_{vcl(low\ freq.)}}{A_{vol}}}$$

2 Substituting Eq. 9.23 for high-frequency values, we obtain:

$$A_{vcl(high\ freq.)} = \frac{A_{vcl(low\ freq.)}}{1 + j\dfrac{f}{f_T} A_{vcl(low\ freq.)}} \tag{9.26}$$

3 The bandwidth, or upper frequency bandwidth limit, f_c, is defined as the frequency at which the gain decreases by 3 dB to 70.7%, with a phase shift of $-45°$. This occurs when the magnitudes of the real and imaginary parts of the denominator of Eq. 9.26 are equal:

$$\frac{f_c}{f_T} A_{vcl(low\ freq.)} = 1$$

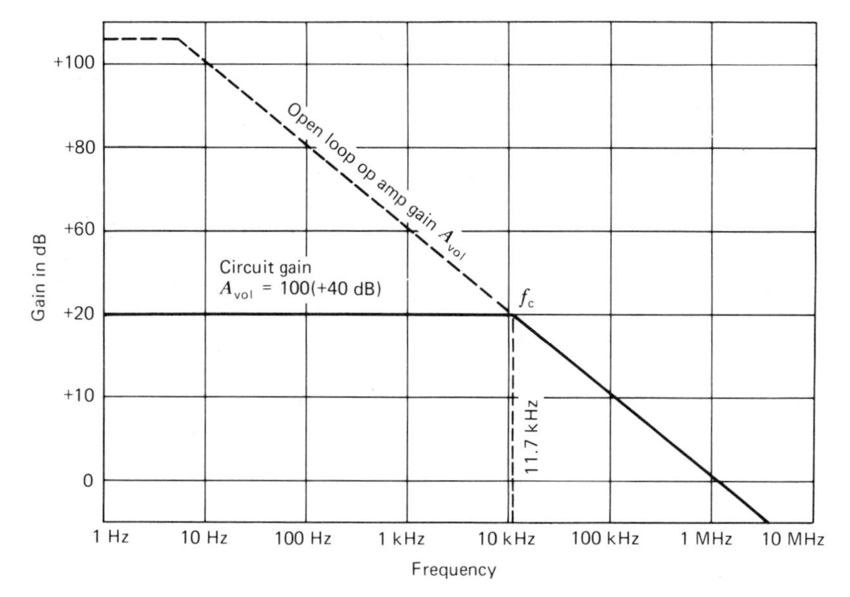

Figure 9.16 continued (*c*). Closed loop response, A_{vcl} (low freq.) = 100.

4 Solving for f_c yields:

$$f_c = \frac{f_T}{|A_{vcl(low\ freq.)}|} \tag{9.27}$$

5 From Ex. 9.11, f_T = 1.17 MHz. Substituting values gives us

$$f_c = \frac{1.17\ \text{MHz}}{10} = 117\ \text{kHz}$$

The Bode approximation of frequency response is shown in Fig. 9.16*b*. Exact values of gain at any frequency may be calculated by substituting values into Eq. 9.26.

EXAMPLE 9.13 A noninverting amplifier using a 741 op amp has a voltage gain A_{vcl} = 100 (or +40 dB). Calculate its bandwidth and frequency response.

PROCEDURE

1 Typical f_T for the 741, from Ex. 9.11, is 1.17 MHz. Substituting values into Eq. 9.27, we see that the upper bandwidth limit f_c becomes:

$$f_c = \frac{1.17\ \text{MHz}}{100} = 11.7\ \text{kHz}$$

2 The Bode approximation of the frequency response is shown in Fig. 9.16*c*. With increased circuit gain, bandwidth decreases and high-frequency response is more limited.

Transient Response Effects

EXAMPLE 9.14 An amplifier is used with a square wave input signal, as shown in Fig. 9.17*a*. The output amplitude is 10 V peak-to-peak. Find typical values for the rise time, t_r, and the fall time, t_f, of the output waveform.

PROCEDURE

1 Conventional methods of transient response analysis using bandwidth (as calculated in Exs. 9.11 and 9.12) may be used. However, if these show a slope in the output waveform that is greater than the slew rate, as in this example, then the slew rate is the limiting factor. Rise time t_r and fall time t_f are defined as the time duration for the signal voltage change, ΔV, from 10 to 90% of final value (as shown in Fig. 9.17*b*). During these times the signal slope follows the slew rate, SR. Then, $SR = \Delta V/t_r = \Delta V/t_f$. After solving for t_r and t_f, we obtain:

$$t_r = t_f = \frac{\Delta V}{SR} \tag{9.28}$$

2 The change in output voltage, ΔV, is 80% of the peak-to-peak output, or 8.0 V. From Table 9.1, SR is typically 0.5 V/μs for both positive and negative slopes. Substituting values yields:

$$t_r = t_f = \frac{8.0 \text{ V}}{0.5 \text{ V/μs}} = 16 \text{ μs}$$

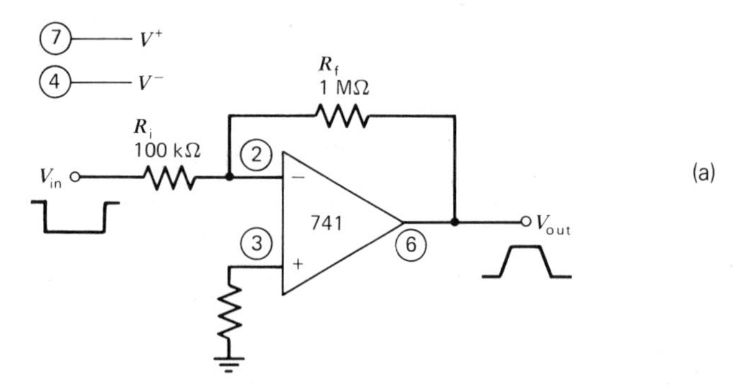

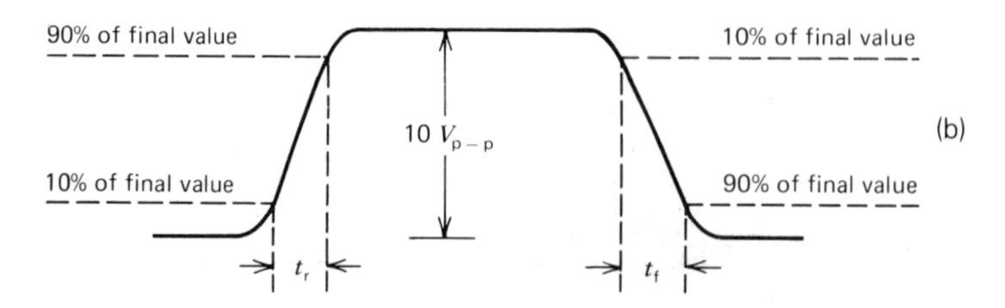

Figure 9.17 Amplifier square wave response. (*a*) Circuit with waveforms (not drawn to scale). (*b*) Detail of output voltage waveform showing rise (t_r) and fall (t_f) times.

3 For larger signals, these times will be longer. They will be shorter for smaller signals, until they become limited by the amplifier bandwidth.

EXAMPLE 9.15 An amplifier has a 20 kHz sinewave input signal. Find the largest amplitude that the output of the 741 op amp can be, without distortion owing to slew-rate limiting.

PROCEDURE

1 If the rate of change, or slope, of the sine waveoutput exceeds the slew rate of the op amp, the slope becomes limited to the slew rate, and distortion results as shown in Fig. 9.18. The maximum slope of a sinewave is $2\pi f V_p$, where V_p is the peak voltage. Setting this equal to the slew rate, we get $SR = 2\pi f V_p$. Solving for V_p then yields:

$$V_p = \frac{SR}{2\pi f} \tag{9.29}$$

2 From Table 9.1, SR is typically 0.5 V/μs. If we substitute values:

$$V_p = \frac{0.5 \text{ V}/\mu\text{s}}{2\pi \cdot 20 \text{ kHz}} = 3.98 \text{ V (peak)}$$

EXAMPLE 9.16 An amplitude of 2 V peak-to-peak, without distortion owing to slew rate, is needed at the output terminal of a 741 in an amplifier circuit. Find the highest frequency sinewave that can be used for a signal.

PROCEDURE

1 Solving Eq. 9.29 for f gives us:

$$f = \frac{SR}{2\pi V_p} \tag{9.30}$$

2 From Table 9.1, SR is typically 0.5 V/μs. The 2 V peak-to-peak output corresponds to $V_p = 1.0$ V. Substituting values, we obtain:

$$f = \frac{0.5 \text{ V}/\mu\text{s}}{2\pi \cdot 1.0 \text{ V}} = 79.6 \text{ kHz}$$

As frequency increases, the maximum undistorted amplitude decreases.

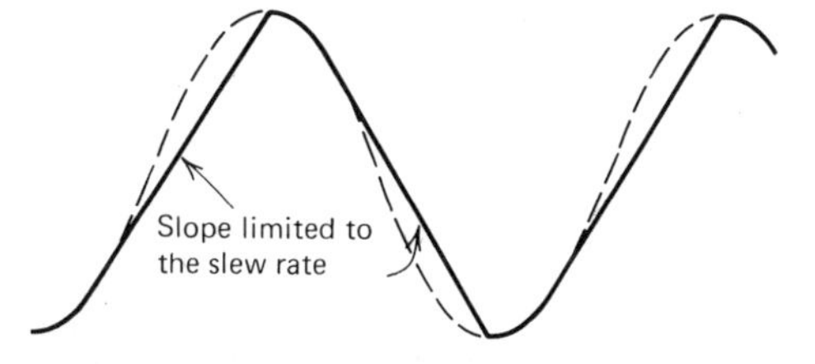

Figure 9.18 Sinewave output exhibiting slew rate distortion.

9.5 UNCOMPENSATED OP AMPS

Compensated op amps, such as the 741, have a frequency response characteristic that makes them suitable for any amplifier circuit gain down to unity. However, as illustrated by Exs. 9.12 and 9.13, when used in high-gain circuits the bandwidth is reduced in proportion to the gain, degrading the high-frequency response. Uncompensated op amps that, in general, exhibit better frequency response, require one or more additional components, that are externally connected, to provide frequency compensation. (Without compensation, there might be too much phase shift through the amplifier and around the feedback loop. If there is additional phase shift of 180°, making the total phase shift 360° at any frequency at which the loop gain is unity or more, oscillation will result.)

Optimum compensation components may be selected for the amplifier circuit gain, to give the best possible frequency response. Component values and connections are different for various op amps. Manufacturer's recommendations should be followed, to ensure stable operation under various conditions.

EXAMPLE 9.17 The inverting amplifier of Fig. 9.19 uses a 709 op amp. Circuit gain $A_{vcl} = 50$. Find values for the compensation components R_1, C_1, and C_2. What are the resulting bandwidth and slew rate?

PROCEDURE

1 Pertinent data from the 709 specifications are provided in Fig. 9.20. Compensation component values are given on the frequency response curves. The nearest curve is for 40 dB gain ($A_{vcl} = 100$). Note that the capacitor values are approximately inversely proportional to gain.
2 Therefore, for $A_{vcl} = 50$, the capacitor values should be multiplied by two. Then the values become:

$$R_1 = 1.5\ \text{k}\Omega$$

$$C_1 = 2 \times 100\ \text{pF} = 200\ \text{pF}$$

$$C_2 = 2 \times 3\ \text{pF} = 6\ \text{pF}$$

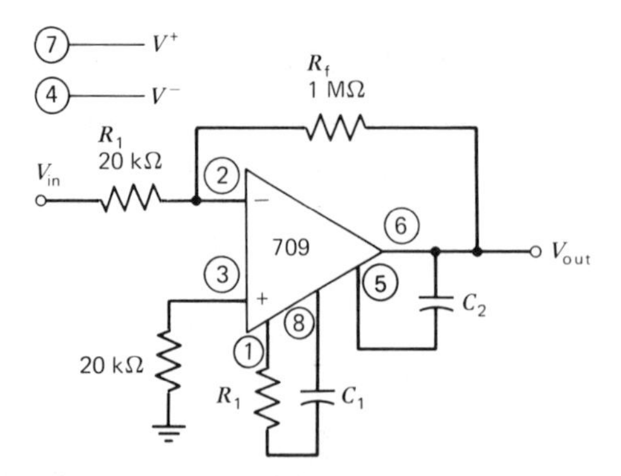

Figure 9.19 Amplifier using an uncompensated (709) op amp, with frequency compensation components.

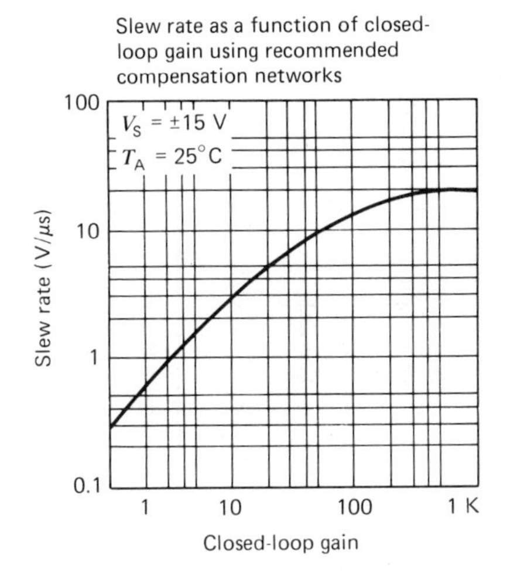

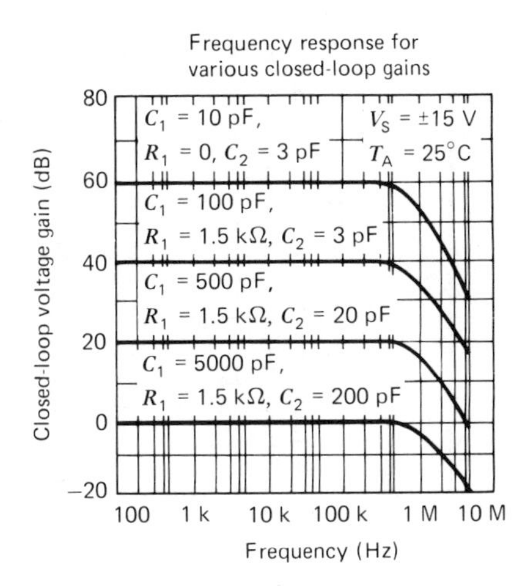

Figure 9.20 Excerpts from specifications for the type 709 op amp. (Courtesy National Semiconductor Corp.)

3 The curves show the bandwidth to be 1 MHz and the slew rate as 10 V/μs, approximately. These are much better than the 741, when operated at the same circuit gain.

9.6 OP AMPS WITH SPECIAL CHARACTERISTICS

In some applications, op amp parameters may produce undesirable effects on the output signal, as was considered in Sec. 9.4. Special op amp types are available, with improved characteristics to minimize some of these effects. Other special op amps for particular circuit constraints or for certain functions, are covered in Secs. 9.7 and 9.8. A number of devices are in more than one of the groups listed in these sections; for example, a single-supply op amp (see Sec. 9.7) may also be programmable and precision (see below). Special-characteristics op amps are summarized in Table 9.2.

Programmable Op Amps

The internal circuitry of IC op amps usually includes a number of current sources, as shown in Sec. 9.3, which are controlled by one or more internal resistors. In a programmable op amp, the current-determining resistor is external, allowing control of the internal currents over a wide range of values. Power consumption changes accordingly, and other parameters are also affected. At low current levels, input bias and offset currents are minimum, but frequency response and slew rate are degraded. Higher currents increase the bandwidth and slew rate, and also the input currents. Internally generated noise is also somewhat less with higher currents.

Table 9.2 **Op amps with special characteristics**

Type	*Example*[1] *(type number)*	*Important characteristics*	*When used*
Programmable	4250	Input bias current I_{IB} Input offset current I_{IO} Frequency response Slew rate SR Power consumption P_C	If it is desirable to control the parameter values, or if a single type is used for different parameter requirements. Also may be used for low power consumption or dissipation requirements.
Precision	357	Very low input bias and offset currents, I_{IB} and I_{IO}, down to pA values Low input offset voltage, V_{IO}, to <0.5 mV Low temperature coefficient, $\Delta V_{IO}/°C$, to <5 μV/°C	For precise dc output voltage levels. When input voltage source impedance is very large, with noninverting amplifiers.
High slew rate	357	Slew rate, SR, to >50 V/μs Large bandwidth, to >20 MHz	If signal has frequencies to the megahertz range. When good transient response of large signals is needed.
Low noise	157	Equivalent input noise voltage, e_n, to 10 $nV/\sqrt{Hz}$	Very low signal level, high-gain amplifiers.

[1] Type numbers may also have prefix and/or suffix letters, which vary depending upon the manufacturer.

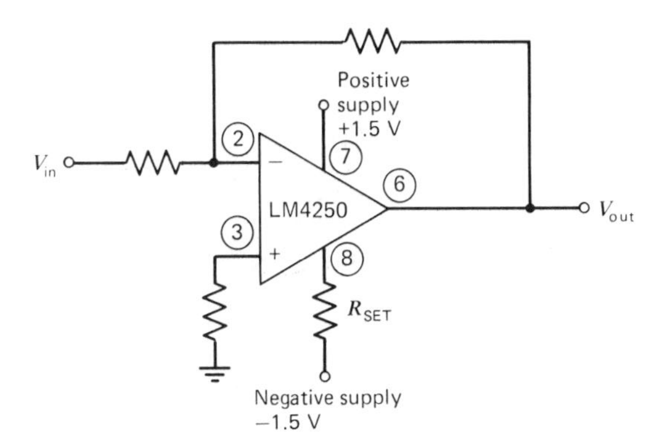

Figure 9.21 Inverting amplifier using an LM4250 programmable op amp.

Programmability makes the device more versatile, possibly allowing use of a single op amp type in various applications where two or more would otherwise be needed. In experimenting and in development work, the programming resistor may be made variable, to check the effect on circuit performance as the op amp characteristics are changed.

EXAMPLE 9.18 The programmable op amp circuit, using an LM4250, shown in Fig. 9.21, has ± 1.5 V supplies. It is in a small battery-operated instrument. Input bias currents are unimportant, but a typical slew rate of at least 0.1 V/μs must be maintained to avoid signal degradation. Find the value of R_{SET} that will minimize power dissipation, for longest battery life.

PROCEDURE

1. Excerpts of the LM4250 specifications are given in Fig. 9.22. Current I_{SET} should be as small as possible for low power dissipation.
2. The "slew rate vs. I_{SET}" shows that I_{SET} must be 6 μA or more, for a typical slew rate of at least 0.1 V/μs. Thus, 6 μA is the best value.
3. From the "R_{SET} vs. I_{SET}" graph, 6 μA corresponds to R_{SET} = 390 kΩ.

Precision Op Amps

Sometimes these are called *high performance, low drift, high input impedance, low offset,* or *low input current* amplifiers. All of them feature reduced values of input bias current and input offset current. Some also have reduced input offset voltage, and reduced drift (with temperature) of input offset voltage and current. These are the characteristics that can cause dc output voltage errors, illustrated in Exs. 9.3 through 9.8 in Sec. 9.4. Some precision op amps have low power dissipation, relatively poor high-frequency response, and low slew rate.

The best precision op amps usually have FET (field effect transistor) input stages, as shown in Fig. 9.23. This reduces input currents to extremely low values.

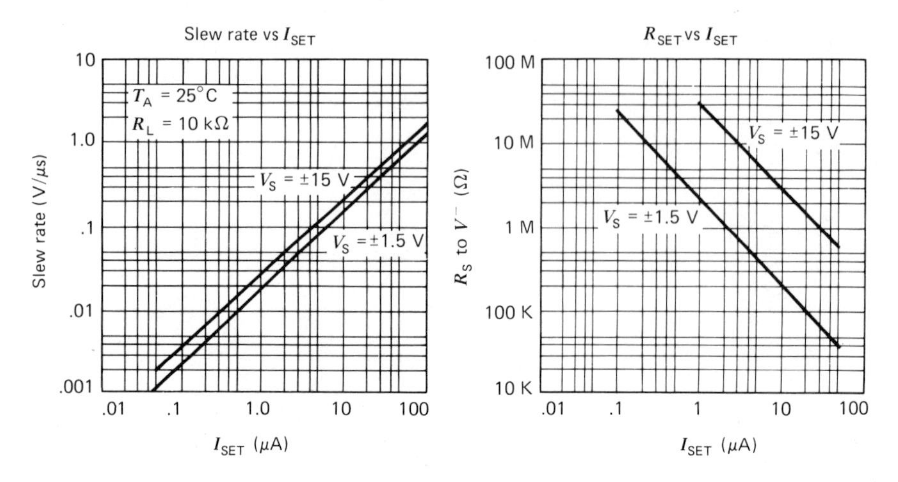

Figure 9.22 Excerpts from specifications for the LM4250 op amp. (Courtesy National Semiconductor Corp.)

High Slew Rate Op Amps

These are sometimes referred to as *wideband* or *high frequency.* Op amps with high slew rates usually also have improved high-frequency response. When signals contain high frequencies, wideband amplifiers may help overcome frequency response limitations, as illustrated in Exs. 9.12 and 9.13. Under some conditions, however, the slew-rate effect may degrade the output signal at frequencies within the bandwidth, as in Exs. 9.14 through 9.16. The slew-rate value may then be more important than the frequency response.

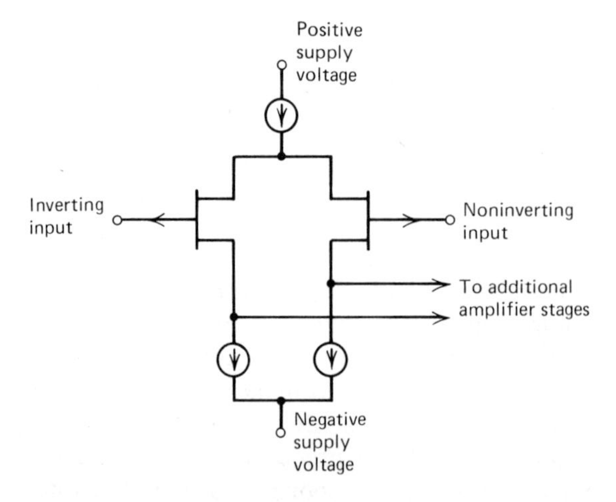

Figure 9.23 Simplified schematic of a field effect transistor (FET) input stage of a precision op amp.

In high-gain amplifier circuits, high slew rate and improved frequency response may be obtained by using an uncompensated op amp (see Sec. 9.5). There also are special types of op amps with exceptionally good high-frequency characteristics, suitable for video frequency amplification (see Secs. 9.8 and 9.9).

Low-Noise Op Amps

Internally generated op amp noise is not usually a problem. It will typically produce an output noise level well under a millivolt, which is usually insignificant. In circuits having extremely low signal levels, or very high gain (100 or more) where output noise may increase to several millivolts, low-noise op amps may be needed. They are available with equivalent input noise voltage, e_n, down to about 10 nV/$\sqrt{\text{Hz}}$. (Noise calculations involve other factors as well and are beyond the scope of this chapter.)

9.7 OP AMPS FOR SPECIAL CIRCUIT CONSTRAINTS

Special op amps are sometimes needed because of circuit-imposed constraints, such as available supply voltages or power restrictions. In other respects, these devices are usually similar to general purpose op amps. They are summarized in Table 9.3.

Table 9.3 **Op amps for special circuit constraints**

Type	*Example*[a] *(type number)*	*Circuit constraints*
High voltage	343	Power supplies greater than ±15 V, up to ±40 V.
Low voltage	4250	Power supplies less than ±5 V, down to ±1 V.
Low power	4250	Low power dissipation or consumption, down to 30 μW.
High current	13080	Output currents above 20 mA, up to 250 mA.
Single supply	358	Positive power supply only is available, and all signal levels are positive.

[a] Type numbers may also have prefix and/or suffix letters, which vary depending upon the manufacturer.

High-Voltage and Low-Voltage Op Amps

Most op amps work well over a range of voltages from ±5 to ±15 V. For higher or lower power supply voltages, special devices may be needed and are available for voltages as high as ±40 to as low as ±1 V. Some of these also will work well within the usual voltage range. At the lowest supply voltage levels, power consumption is usually small, and the frequency response and slew rate may be relatively low.

Low-Power Op Amps

If supply current or power dissipation must be limited, low-power op amps will operate at a supply current as low as about 10 μA. They may be programmable (Sec. 9.6). Frequency response and slew rate are usually poor at very low power levels.

High-Current Op Amps

Most op amps limit the output current to about 25 mA or less, to protect the device. If more current is needed, high-output current op amps provide up to about 250 mA (peak) to the load.

Single Supply Op Amps

These need no negative supply voltage, but they must be operated with positive voltages at the output and at both input terminals. When used in a noninverting amplifier with positive voltage signals, conventional feedback shown in Fig. 9.3 applies. For inverting amplification, a fixed positive voltage must be connected at $V_{(+)}$, as shown in Fig. 9.24.

Single supply op amps work well with input or output voltages very close to ground (within a few millivolts) level. They may also be used with two supplies, in the usual manner.

EXAMPLE 9.19 In the single-supply inverting amplifier of Fig. 9.24, V_{in} is a 0.3 V peak-to-peak 100 Hz sine wave at +10.5 V average dc. Find the ac and dc levels of V_{out}, assuming ideal op amp characteristics.

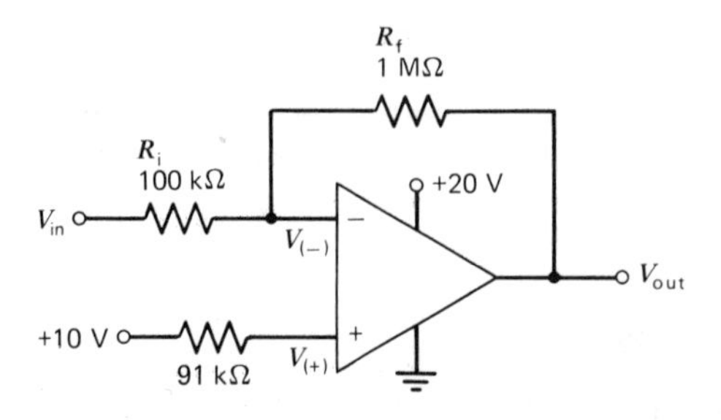

Figure 9.24 Inverting amplifier using a single-supply op amp.

PROCEDURE

1 If the input and output voltages V_{in} and V_{out} are referenced to $V_{(+)}$ instead of ground, then this example is similar to Ex. 9.2 where Eq. 9.5 applies. Substituting values yields:

$$A_{vcl} = -\frac{1\ \text{M}\Omega}{100\ \text{k}\Omega} = -10$$

2 The dc reference change does not affect ac values. Thus the value is:

$$v_{out}\ (\text{ac}) = |-10\ v_{in}| = |-10\ (0.3\ \text{V})| = 3.0\ \text{V peak-to-peak}$$

(The negative gain indicates that the output is out-of-phase with the input, as in Ex. 9.2).

3 For dc values, the change in reference is handled by substituting (V_{IN} − 10 V) for V_{in}, and (V_{OUT} − 10 V) for V_{out}. Thus the substitution becomes:

$$\begin{aligned} V_{OUT} - 10\ \text{V} &= -10\ (V_{IN} - 10\ \text{V}) \\ &= -10\ (10.5\ \text{V} - 10\ \text{V}) \\ &= -5.0\ \text{V dc} \end{aligned}$$

4 Solving for V_{OUT} yields:

$$V_{OUT} = +5.0\ \text{V dc}$$

5 Because the output voltage is also positive, the circuit will work properly. Waveforms would be as shown in Fig. 9.4*b*, but with dc values shifted by +10 V.

9.8 SPECIAL FUNCTION OP AMPS

Some op amps have special internal connections or terminals, and/or special characteristics, for a particular functional use.

Comparators

The comparator is an open-loop application of the op amp, as shown in Fig. 9.25. The high open-loop gain causes the output voltage V_{out} to go into limiting (saturation), to within about a volt of the positive or negative supply voltage. Voltage V_{ref} is a reference voltage or signal. If $V_{in} < V_{ref}$, V_{out} becomes positive; if $V_{in} > V_{ref}$, V_{out} becomes negative. The output polarities may be reversed by connecting V_{in} to $V_{(+)}$ and V_{ref} to $V_{(-)}$. Any general-purpose op amp may be used as a comparator.

Comparators are often used to interface with digital circuits. Special comparator op amps are usually designed to be compatible with various logic families. They usually have fast response time and high slew rate. A strobe input, if

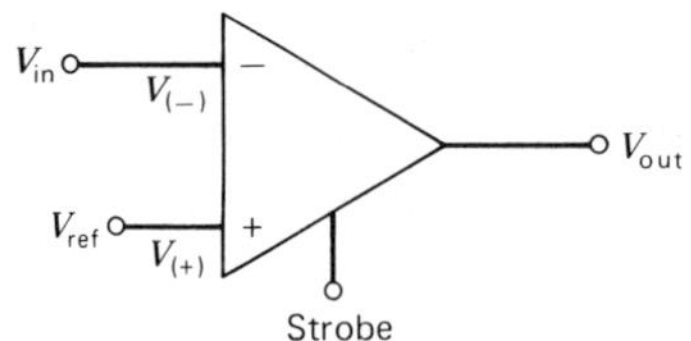

Figure 9.25 Op amp comparator with a strobe input.

Figure 9.26 Voltage follower.

provided, allows the comparator to become disabled, so that the output voltage is at a particular limit regardless of the value of V_{in}.

Voltage Followers

Also called a *buffer*, this type of op amp has the output connected directly back to the inverting input, as shown in Fig. 9.26. This gives noninverting, unity gain. Any general purpose op amp may be used this way. Voltage follower devices have the feedback internally connected, and usually have good high-frequency characteristics (see Ex. 10.3).

Instrumentation Amplifiers

For many instrumentation applications, op amps are used in differential amplifier circuits. Figure 9.27 shows the basic differential amplifier circuit. The difference between the input voltages, $V_{in(1)} - V_{in(2)}$, is amplified by the circuit closed-loop gain A_{vcl}.

In instrumentation amplifier op amps, the feedback network is internally provided. Gain is usually controlled with an external resistor, and input impedances are very high.

Video Amplifiers

Most op amps have a frequency response that is too limited for video signals. However, frequency response may be greatly extended with special design using

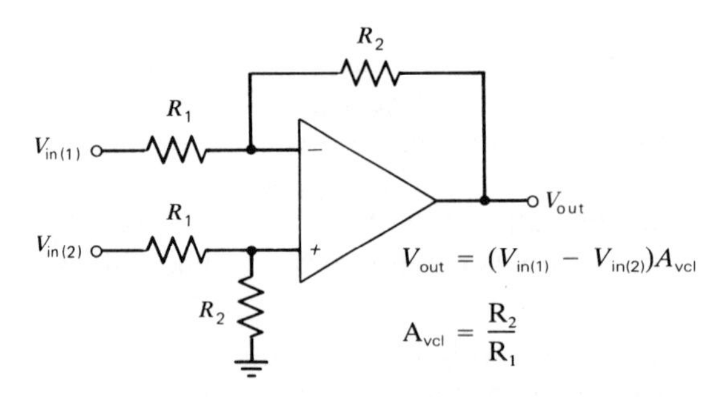

Figure 9.27 Basic differential (difference) amplifier.

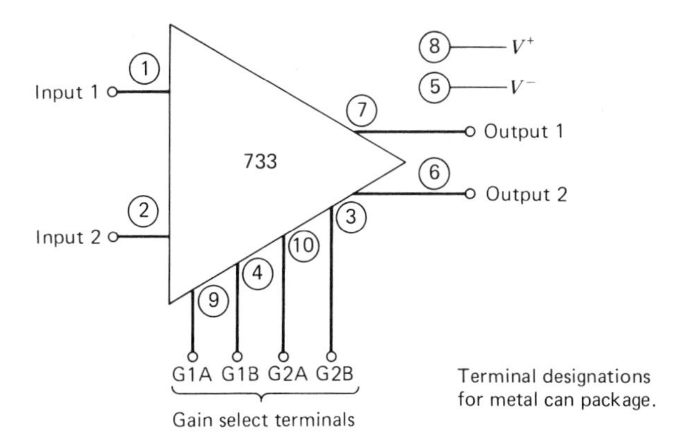

Figure 9.28 Type 733 video amplifier.

internal feedback networks. An example is the type 733, shown in Fig. 9.28. It has a bandwidth up to 120 MHz, and voltage gain of 10 to 400. Gain is set by making various connections between the gain select terminals. Inputs and outputs are both differential. The output responds to the difference signal between Input 1 and Input 2, and the outputs are of opposite polarity. (Refer to Chap. 14 for a detailed discussion of video and rf amplifiers.)

The current differencing, or Norton type amplifier (see Sec. 9.9), also may be used for video amplification.

9.9 CURRENT DIFFERENCING (NORTON) AMPLIFIERS

The symbol for the current differencing, or Norton type amplifier, is shown in Fig. 9.29*a*. The current generator symbol indicates that the positive-terminal current, $I_{in(+)}$, subtracts from the negative-terminal current, $I_{in(-)}$, at the negative input terminal. Thus, the net input current is the difference current, $I_{in(-)} - I_{in(+)}$. The input circuitry that accomplishes this is shown in Fig. 9.29*b*. Transistors Q_1 and Q_2 form a current source, sometimes called a *current mirror*, which reproduces $I_{in(+)}$ at the collector of Q_2 (see Sec. 9.3 and Fig. 9.6). The current difference is then the base drive to Q_3, which is the first amplifier stage. Usually a single positive supply is used, and input and output voltages must be kept positive at all times.

Current differencing amplifiers may be used at very high frequencies, having bandwidths up to 400 MHz and slew rates as high as 60 V/μs. They also have low noise, and may have additional terminals (not shown) for programming resistors and frequency compensation components.

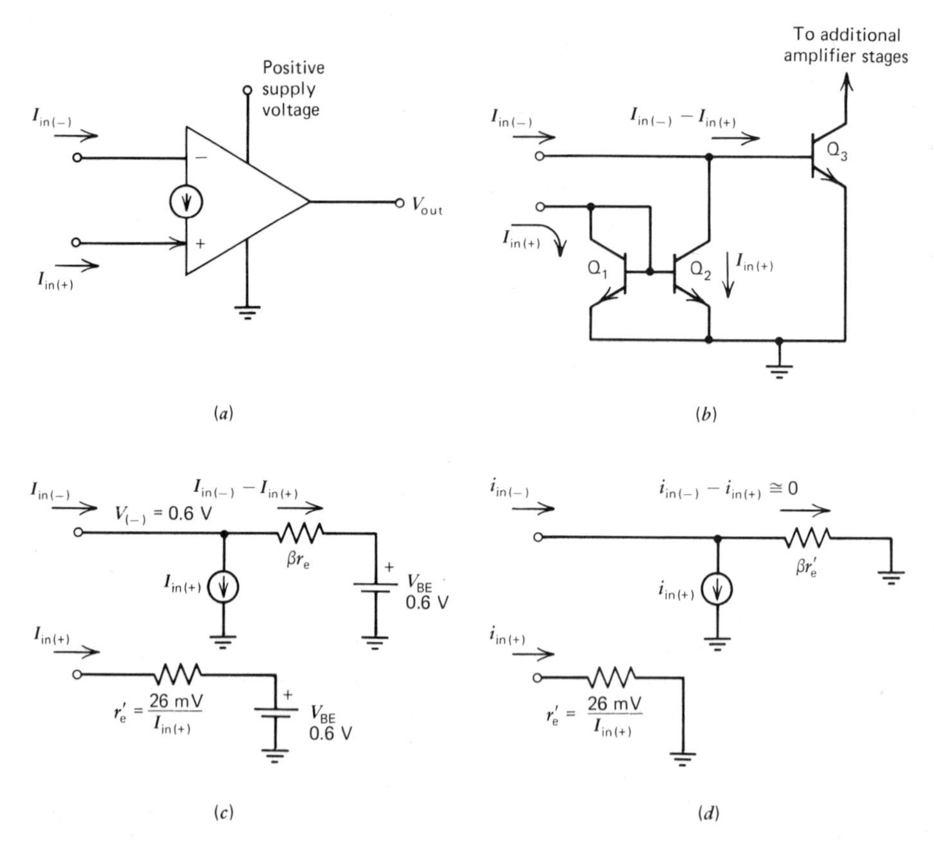

Figure 9.29 Current differencing (Norton) amplifier. (*a*) Electrical symbol (terminals for programming and frequency compensation not shown). (*b*) Schematic of the input stage. (*c*) The dc equivalent circuit of the input. (*d*) The ac equivalent circuit of the input.

EXAMPLE 9.20 A type LM159 current input amplifier is used as a noninverting amplifier, shown in Fig. 9.30. The input signal V_{in} is $+1.14$ V dc (average) and 50 mV peak-to-peak ac. Find the closed-loop gain A_{vcl}, and the dc and ac values of the output voltage V_{out}. Assume ideal op amp parameters and that the amplifier is properly programmed.

PROCEDURE

1 Equivalent circuits for the inputs, needed to analyze the circuits, are shown in Figs. 9.29*c* and *d*. The value of the internal diode resistance, $r'_e = 26\text{ mV}/I_{in(+)}$, follows from semiconductor theory. Ideal op amp assumptions include infinite gain, so that $I_{in(+)} - I_{in(-)} = 0$, or $I_{in(+)} = I_{in(-)}$.

2 There is no voltage drop across $\beta r'_e$; therefore, $V_{(-)} = +0.6$ V. Then, for dc values (Fig. 9.29*c*), the results are:

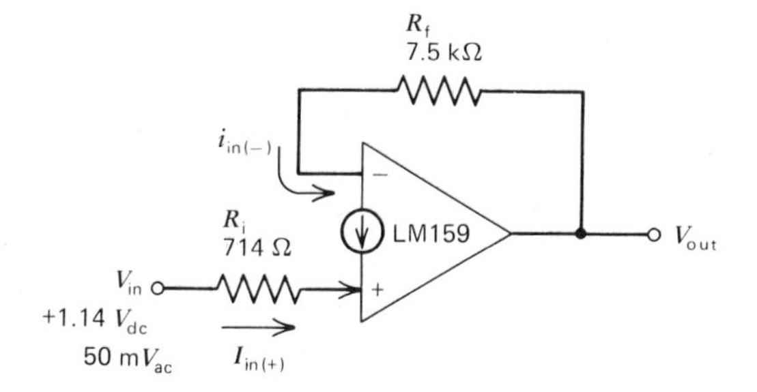

Figure 9.30 Noninverting amplifier using the current differencing op amp (programming and frequency compensation circuits not shown).

$$V_{OUT} = 0.6\text{ V} + I_{in(-)}R_f = 0.6\text{ V} + I_{in(+)}R_f$$

$$\begin{aligned} V_{IN} &= 0.6\text{ V} + r'_e I_{in(+)} + R_i I_{in(+)} \\ &= 0.6\text{ V} + \frac{26\text{ mV}}{I_{in(+)}} I_{in(+)} + R_i I_{in(+)} \\ &= 0.626\text{ mV} + R_i I_{in(+)} \end{aligned} \tag{9.31}$$

3 Solving for $I_{in(+)}$ yields:

$$I_{in(+)} = \frac{V_{IN} - 0.626\text{ V}}{R_i} \tag{9.32}$$

4 For ac values, using Fig. 9.29*d* yields:

$$v_{out} = i_{in(-)}R_f = i_{in(+)}R_f$$

$$v_{in} = i_{in(+)}(r'_e + R_i)$$

5 Dividing these equations, we see that the closed-loop circuit gain A_{vcl}:

$$A_{vcl} = \frac{v_{out}}{v_{in}} = \frac{R_f}{r'_e + R_i} \tag{9.33}$$

6 Substituting values in Eq. 9.31 yields:

$$I_{in(+)} = \frac{1.14\text{ V} - 0.626\text{ V}}{714\ \Omega} = 720\ \mu\text{A}$$

7 Then, $r'_e = 26\text{ mV}/720\ \mu\text{A} = 36\ \Omega$. If we substitute values in Eqs. 9.31 and 9.33, the solution is:

$$A_{vcl} = \frac{7.5\text{ k}\Omega}{36\ \Omega + 714\ \Omega} = 10$$

$$V_{OUT} = 0.6\text{ V} + 720\ \mu\text{A} \cdot 7.5\text{ k}\Omega = 6.0\text{ V dc (average)}$$

$$v_{out} = A_{vcl}v_{in} = 10 \cdot 50\text{ mV} = 500\text{ mV ac (peak-to-peak)}$$

9.10 REFERENCES

Bannon, E., *Operational Amplifiers: Theory and Servicing*, Reston, Reston, VA, 1977.

Coughlin, R. F. and F. F. Driscoll, *Operational Amplifiers and Linear Integrated Circuits,* second edition, Prentice Hall, Englewood Cliffs, NJ, 1982.

Gray, P. R. and R. G. Meyer, *Analysis and Design of Analog Integrated Circuits*, Wiley, New York, 1977.

Lenk, J. D., *Manual for Operational Amplifier Users*, Reston, Reston, VA, 1976.

Linear Applications Handbook, National Semiconductor Corporation, Santa Clara, CA, 1978.

Prensky, S. D. and A. H. Seidman, *Linear Integrated Circuits: Practice and Applications*, Reston, Reston, VA, 1981.

Smith, J. I., *Modern Operational Circuit Design*, Wiley-Interscience, New York, 1971.

Stout, D. F. and M. Kaufman (ed.), *Handbook of Operational Amplifier Circuit Design*, McGraw-Hill, New York, 1976.

Wong, Y. J. and W. E. Ott, *Function Circuits Design and Applications*, McGraw-Hill, New York, 1976.

CHAPTER 10

Circuits Using the OP AMP

Roger C. Thielking
Professor of Electrical Technology
Onondaga Community College

10.1 INTRODUCTION

The operational amplifier (op amp) is almost a "universal" device for linear integrated circuits. It can be used for various purposes in most linear circuit systems at frequencies from dc up to the low-megahertz range. In fact, the op amp is used in many of the circuits described in Chaps. 11 through 17. These include active filters, wave generators, D/A and A/D converters, communications circuits, voltage and current regulators, certain multifunction circuits, and phase-locked loops. In some, the op amps may be integrated with other circuitry in the chip, while other circuits use separate op amps. Because these applications are fully described in those chapters, most of them are not included here.

Circuits covered in this chapter include various basic amplifiers, and other important and representative circuit types. Circuit analysis considerations that relate to the interaction of the circuit with the op amp are emphasized.

Ideal Op Amp Characteristics

Op amp characteristics and their effects are described in Chap. 9. Usually the effects are slight, so that the circuit operates essentially as it would with ideal op amp characteristics. In this chapter, ideal op amp characteristics are assumed,

Table 10.1 **Ideal values of op amp characteristics**

Characteristic	*Symbol*	*Ideal value*
Input offset voltage	V_{IO}, V_{OS}	0 mV
Input bias current	I_{IB}, I_B	0 nA
Input offset current	I_{IO}, I_{OS}	0 nA
Input resistance	r_i, r_{in}	∞ MΩ
Large signal voltage gain	A_{vol}, A_v	∞
Output resistance	r_o	0 Ω
Common mode rejection ratio	CMRR	∞ dB
Transient response:		
Rise time	t_{TLH}	0 μs
Slew rate	*SR*	∞ V/μs

except as otherwise indicated. Some of the more important op amp characteristics, with ideal values, are summarized in Table 10.1.

A more complete list, with values for the type 741 op amp, appears in Chap. 9, Table 9.1.

Actual effects of op amp characteristics for various circuits, if needed, may usually be calculated by using the analysis techniques illustrated in the examples in Sec. 9.4, Chap. 9.

Circuit Equations

Certain equations derived in Chap. 9, for the basic inverting and noninverting amplifiers, are used widely for analysis of op amp circuits. They are repeated here, using complex impedances in place of resistors. The noninverting amplifier is shown in Fig. 10.1. From Eq. 9.2 the closed-loop circuit gain $A_{vcl} = V_{out}/V_{in}$ becomes:

$$A_{vcl} = \frac{Z_1 + Z_2}{Z_2} \tag{10.1}$$

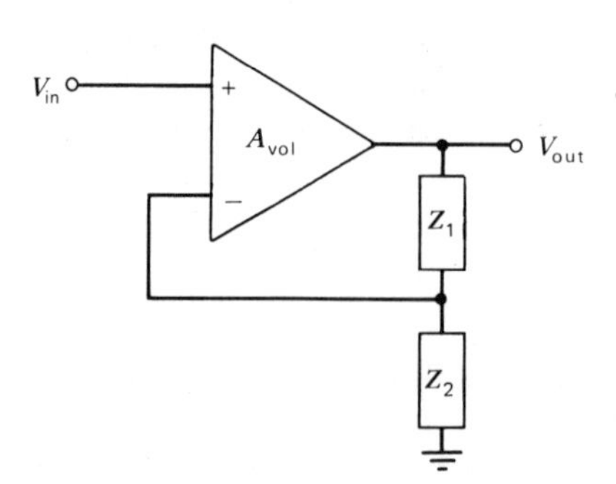

Figure 10.1 Noninverting amplifier.

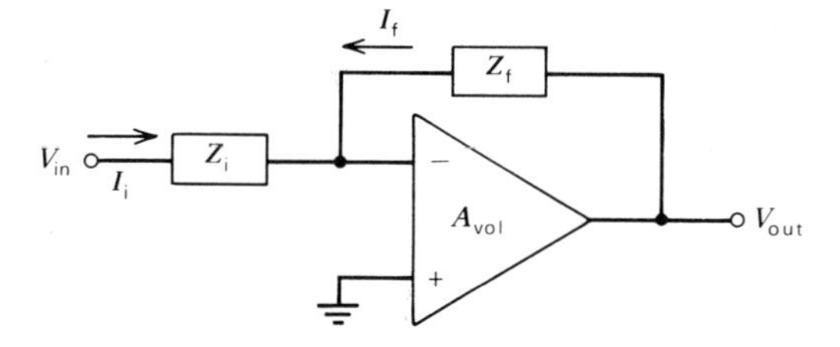

Figure 10.2 Inverting amplifier.

For the inverting amplifier of Fig. 10.2, from Eq. 9.5, the closed-loop gain is:

$$A_{vcl} = -\frac{Z_f}{Z_i} \tag{10.2}$$

Another very useful relationship is that of the currents in the feedback network. From Kirchhoff's current law, $I_f + I_{in} = 0$, or:

$$I_f = -I_{in} \tag{10.3}$$

10.2 BASIC AMPLIFIERS

Comparator

In comparator applications, the op amp is operated open loop, with no feedback or with a small amount of positive feedback. The very high circuit gain causes the output voltage to limit (saturate) at its high or low value (usually within a volt or so of the power supply voltages).

EXAMPLE 10.1 The MLM311 comparator in Fig. 10.3 has an open-loop gain A_{vol} = 200,000. Operated from a single +5 V supply, the output voltage V_{out} limits at about +0.3 and +4 V. Calculate the exact input voltage levels that cause limiting. Assume that the offset adjustment pot, R_s, is properly set.

PROCEDURE

1. The output voltage $V_{out} = V_d A_{vol}$, which is the input difference voltage times the open-loop gain. Because $V_d = V_{ref} - V_{in}$, the output voltage becomes: $V_{out} = (V_{ref} - V_{in})A_{vol}$.
2. Substituting values gives us, at the high limit, +4 V = (1.5 V − V_{in})200,000 and at the low limit, +0.3 V = (1.5 V − V_{in})200,000.
3. Solving for V_{in}, high-voltage limiting occurs when $V_{in} \leq +1.49998$ V and low-voltage limiting occurs when $V_{in} \geq +1.4999985$ V. (For practical purposes, both of these values are equal to +1.5 V.)

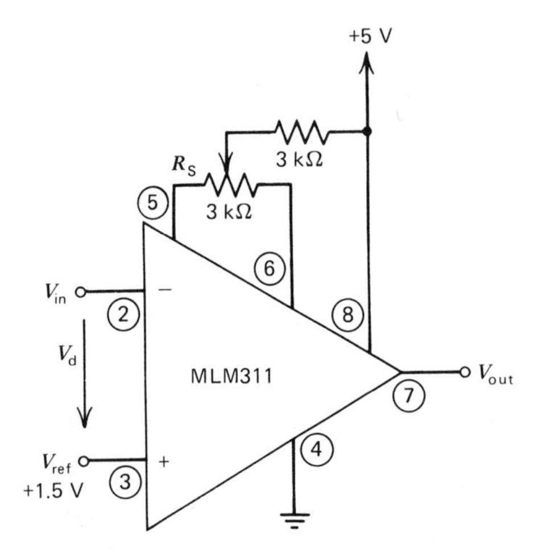

Figure 10.3 Comparator.

The voltage change needed for full switching is less than 20 μV. Signals greater than this value result from noise and other disturbances, which will cause the output to switch erratically between the limits if the input voltage is very close to +1.5 V. If this becomes a problem, positive feedback may be added as described in the next example.

EXAMPLE 10.2 The circuit of Ex. 10.1 is modified by adding the feedback resistors R_1 and R_2 to provide positive feedback, as shown in Fig. 10.4*a*. Calculate the input voltage levels that cause switching, and determine the output waveform for the input shown in Fig. 10.4*b*. (Assume inifinite op amp gain.)

PROCEDURE

1 Feedback resistors R_1 and R_2 are connected to the noninverting input, resulting in *positive,* or *regenerative,* feedback. When switching begins to occur, the feedback reinforces the change, causing a latching effect and eliminating the potential unstable condition described above. Also, the switching voltage for a positive-going input is slightly different than for a negative-going input.

2 Because an ideal op amp is assumed, switching occurs when $V_d = 0$ V, so that $V_{in} = V_{(+)}$. The voltage divider effect of R_1 and R_2 yields:

$$V_{(+)} = V_{ref} + (V_{out} - V_{ref}) \frac{R_1}{R_1 + R_2} \tag{10.4}$$

3 If V_{in} is initially low, $V_{out} = +4$ V, the positive limit. Substituting values yields:

$$V_{(+)} = 1.5 \text{ V} + (4 \text{ V} - 1.5 \text{ V}) \frac{100\ \Omega}{10\ \text{k}\Omega + 100\ \Omega} = 1.525 \text{ V}$$

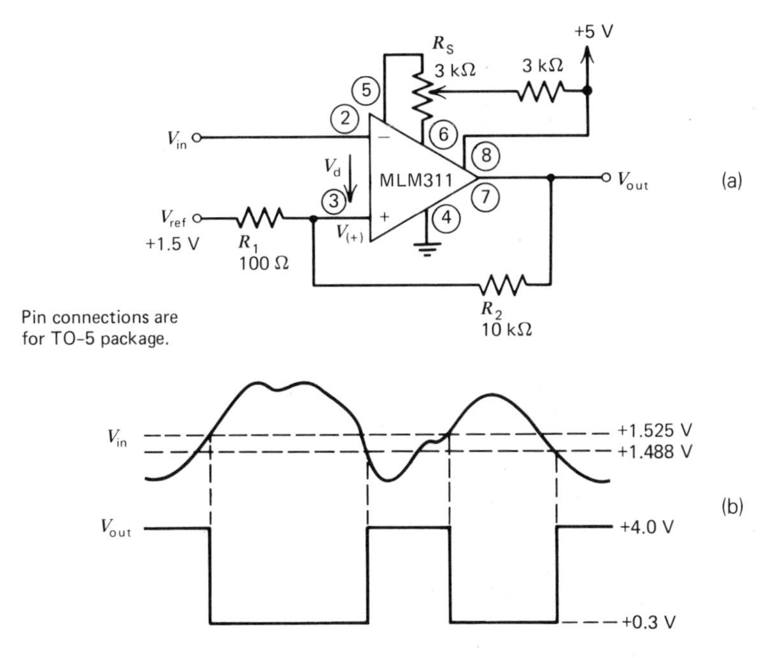

Figure 10.4 Comparator with positive feedback. (*a*) Circuit diagram. (*b*) Input and output voltage waveforms.

4 The other possibility is that V_{in} is initially high, so that $V_{out} = +0.3$ V, the low limit. Then the result becomes:

$$V_{(+)} = 1.5 \text{ V} + (0.3 \text{ V} - 1.5 \text{ V}) \frac{100 \ \Omega}{10 \text{ k}\Omega + 100 \ \Omega} = 1.488 \text{ V}$$

Thus, switching occurs when a positive-going voltage reaches 1.525 V and when a negative-going voltage drops to 1.488 V, as shown in the output voltage waveform in Fig. 10.4*b*. The *hysteresis,* or difference between switching levels, of 37 mV is probably greater than noise and other disturbances, so that erratic switching is prevented.

Unity Follower

EXAMPLE 10.3 A signal source V_s with a source resistance $R_s = 100$ kΩ must be applied to a load with resistance $R_L = 1$ kΩ. No appreciable attenuation or dc voltage change can be allowed.

PROCEDURE

1 Use an op amp as a unity (voltage) follower, as shown in Fig. 10.5. The output voltage V_{out} is fed back directly to the inverting terminal and compared with the source voltage, to ensure accurate reproduction.

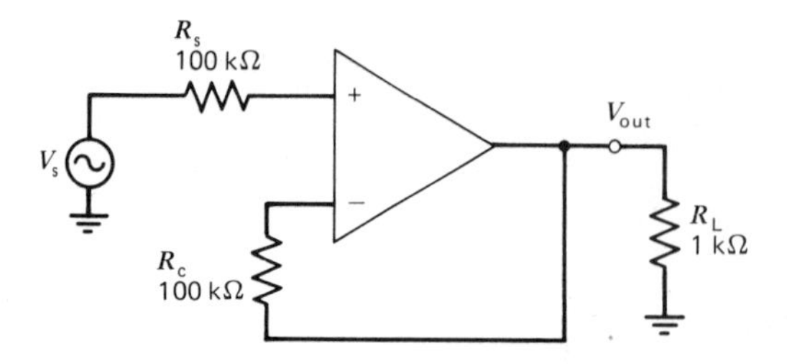

Figure 10.5 Unity (voltage) follower.

2 The current-compensating resistor R_c is made equal to the source resistance R_s, to minimize the error caused by input bias currents (see Chap. 9, Ex. 9.6).
3 The output voltage will reproduce the source voltage to within a few millivolts, using a general-purpose op amp.

For error calculations and ways to minimize errors, see Chap. 9, Secs. 9.4 and 9.6.

Noninverting Amplifier with Level Shift

EXAMPLE 10.4 In the noninverting amplifier of Fig. 10.6*a*, R_3 is added in the feedback network to change the dc level of the output. Calculate the output signal, including the amount of the level shift.

PROCEDURE

1 Analysis is simplified by Thevenizing R_2 and R_3, as in Fig. 10.6*b*. Then, the output values due to V_{in} and V_{TH} are determined separately and combined, using superposition.
2 To calculate V_{out} because of V_{in}, ground V_{TH}. Equation 10.1 applies, with $Z_1 = R_1$ and $Z_2 = R_{TH}$. Then, this component of V_{out} becomes:

$$V_{out} = A_{vcl}V_{in} = \frac{R_1 + R_{TH}}{R_{TH}} V_{in} \tag{10.5}$$

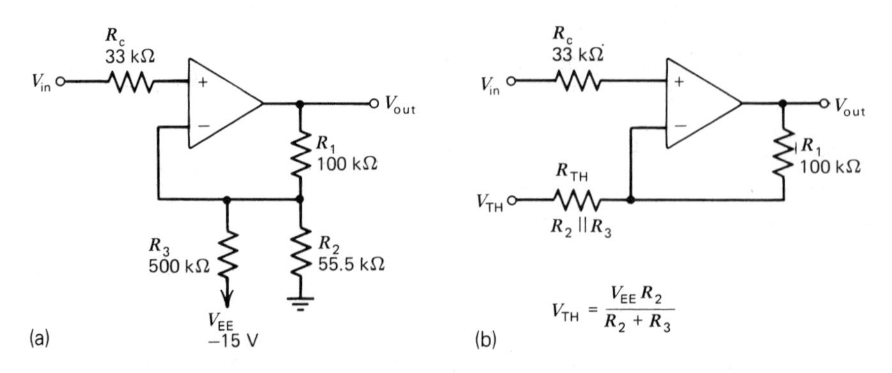

Figure 10.6 Noninverting amplifier with level shift. (*a*) Circuit diagram. (*b*) Circuit diagram with R_2 and R_3 Thevinized.

3 To calculate V_{out} owing to V_{TH}, ground V_{in}. This is the circuit for an inverting amplifier, and Eq. 10.2 applies, with $Z_f = R_1$ and $Z_i = R_{TH}$. Then, V_{out} (due to V_{TH}) becomes:

$$V_{OUT} = A_{vcl}V_{TH} = -\frac{R_1}{R_{TH}} V_{TH} \tag{10.6}$$

4 Substituting values, we obtain $R_{TH} = R_2 \| R_3 = 55.5\ \text{k}\Omega \| 500\ \text{k}\Omega = 50\ \text{k}\Omega$ and $V_{TH} = V_{EE}\ R_2/(R_2 + R_3) = (-15\ \text{V})(55.5\ \text{k}\Omega)/(500\ \text{k}\Omega + 55.5\ \text{k}\Omega) = -1.5$ V. Then, the output signal, by Eq. 10.5, is:

$$V_{out} = \frac{100\ \text{k}\Omega + 50\ \text{k}\Omega}{50\ \text{k}\Omega} V_{in} = 3\ V_{in}$$

5 The level shift, by Eq. 10.6, is:

$$V_{OUT} = -\frac{100\ \text{k}\Omega}{50\ \text{k}\Omega}(-1.5\ \text{V}) = +3\ \text{V}$$

Thus, the input is amplified by three and also shifted in level by +3 V dc at the output.

Inverting Amplifier with Level Shift

EXAMPLE 10.5 The inverting amplifier of Fig. 10.7 has a dc voltage at the noninverting input to shift the dc level of the output. Calculate the output voltage V_{out}, including the level shift.

PROCEDURE

1 The combined effect of the two inputs may be determined by superposition. With V_{dc} grounded, Eq. 10.2 applies with $Z_i = R_1$ and $Z_f = R_2$. Then, the result becomes:

$$V_{out} = -\frac{R_2}{R_1} V_{in} = -\frac{30\ \text{k}\Omega}{10\ \text{k}\Omega} V_{in} = -3\ V_{in}$$

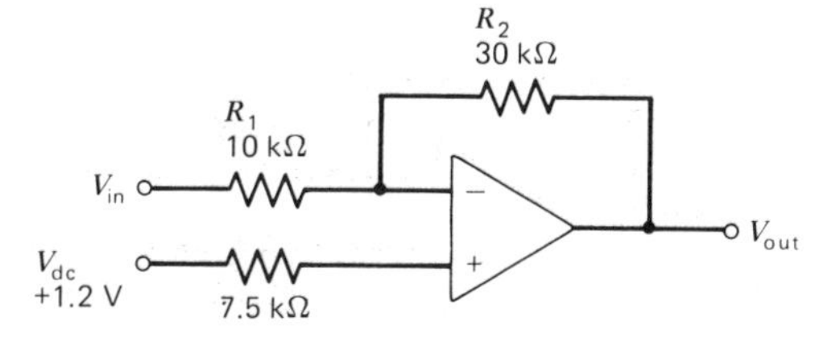

Figure 10.7 Inverting amplifier with level shift.

2 With V_{in} grounded, the circuit acts as a noninverting amplifier for the V_{dc} input. Equation 10.1 applies with $Z_1 = R_2$ and $Z_2 = R_1$, and thus yields:

$$V_{OUT} = \frac{R_2 + R_1}{R_1} V_{dc} = \frac{30\ k\Omega + 10\ k\Omega}{10\ k\Omega}(1.2\ V) = +4.8\ V\ dc$$

3 The total output voltage is $-3\ V_{in}$, added to the $+\ 4.8$ V dc level change.

Another way to do this is to use a summing amplifier (see Sec. 10.3, Ex. 10.13). One input is used for the input signal V_{in}, and a dc voltage is applied to the other input to cause the level shift.

The ac Coupled Inverting Amplifier

EXAMPLE 10.6 Find the frequency response of the ac coupled inverting amplifier shown in Fig. 10.8*a*. The gain-bandwidth product, f_T, of the compensated op amp is 1.0 MHz.

PROCEDURE

1 At mid and low frequencies (well below f_T), the gain is given by Eq. 10.2. Impedance $Z_i = R_1 - jX_C$ and $Z_f = R_2$, so that the results are:

$$A_{vcl} = \frac{-R_2}{R_1 - jX_C} = \frac{-R_2}{R_1 + \dfrac{1}{j\omega C_1}}$$

2 Multiplying numerator and denominator by $j\omega C_1$ yields:

$$A_{vcl} = \frac{-j\omega R_2 C_1}{1 + j\omega R_1 C_1} \tag{10.7}$$

3 At the low cutoff frequency, $f_{c(l)}$, the real and imaginary components in the denominator are equal, so that $\omega R_1 C_1 = 2\pi f_{c(l)} R_1 C_1 = 1$. Solving for $f_{c(l)}$ yields:

$$f_{c(l)} = \frac{1}{2\pi R_1 C_1} \tag{10.8}$$

4 For low frequencies $f < f_{c(l)}$, $\omega R_1 C_1 < 1$ and Eq. 10.7 becomes $A_{vcl} \approx -j\omega R_2 C_1$, so that the gain decreases by 20 dB/decade as frequency decreases. At midfrequencies where $f > f_{c(l)}$, $\omega R_1 C_1 > 1$ and Eq. 10.7 becomes:

$$A_{vcl(mid\ freq.)} \approx \frac{-j\omega R_2 C_1}{j\omega R_1 C_1} = \frac{-R_2}{R_1} \tag{10.9}$$

5 The high cutoff frequency $f_{c(h)}$ owing to the op amp bandwidth f_T given by Eq. 9.27,

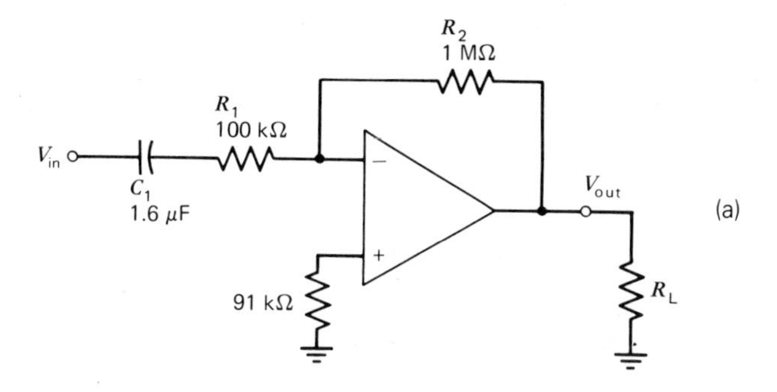

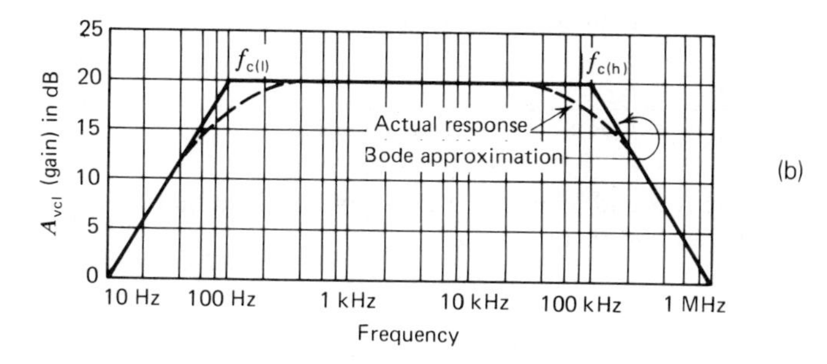

Figure 10.8 An ac coupled inverting amplifier. (*a*) Circuit diagram. (*b*) Frequency response.

Chap. 9 is:

$$f_{c(h)} = \frac{f_T}{|A_{vcl(mid\ freq.)}|} \tag{10.10}$$

At high frequencies $f > f_{c(h)}$, the response decreases by 20 dB/decade above $f_{c(h)}$ (see Ex. 9.12, Chap. 9).

6 Substituting values, by Eq. 10.8 yields:

$$f_{c(l)} = \frac{1}{2\pi \cdot 100\ \text{k}\Omega \cdot 1.6\ \mu\text{F}} = 99.5\ \text{Hz}$$

7 By Eq. 10.9 the result is:

$$A_{vcl(mid\ freq.)} = \frac{-1\ \text{M}\Omega}{100\ \text{k}\Omega} = -10\ (20\ \text{dB magnitude})$$

and by Eq. 10.10 the high cutoff frequency is:

$$f_{c(h)} = \frac{1\ \text{MHz}}{10} = 100\ \text{kHz}$$

8 The frequency response curve is given in Fig. 10.8*b*.

Low-Frequency Inverting Amplifier

EXAMPLE 10.7 In the inverting amplifier of Fig. 10.9*a*, capacitor C_1 is added in the feedback network to attenuate high frequencies. Find the frequency response.

PROCEDURE

1 Circuit gain is given by Eq. 10.2, with $Z_i = R_1$ and $Z_f = R_2 \| X_C$, so that,

$$A_{vcl} = -\frac{R_2 \| X_C}{R_1} = -\frac{1}{R_1}\frac{R_2/(j\omega C_1)}{R_2 + 1/(j\omega C_1)}$$

2 Multiplying numerator and denominator by $j\omega C_1$, we obtain:

$$A_{vcl} = -\frac{1}{R_1}\frac{R_2}{1 + j\omega R_2 C_1} = -\frac{R_2}{R_1}\frac{1}{1 + j\omega R_2 C_1} \tag{10.11}$$

At the cutoff frequency f_c, the real and imaginary values are equal, so that $\omega R_2 C_1 = 2\pi f_c R_2 C_1 = 1$. Solving for f_c, we find that

$$f_c = \frac{1}{2\pi R_2 C_1} \tag{10.12}$$

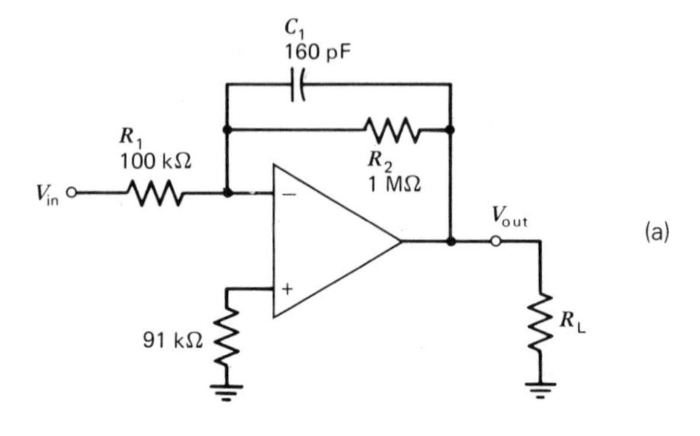

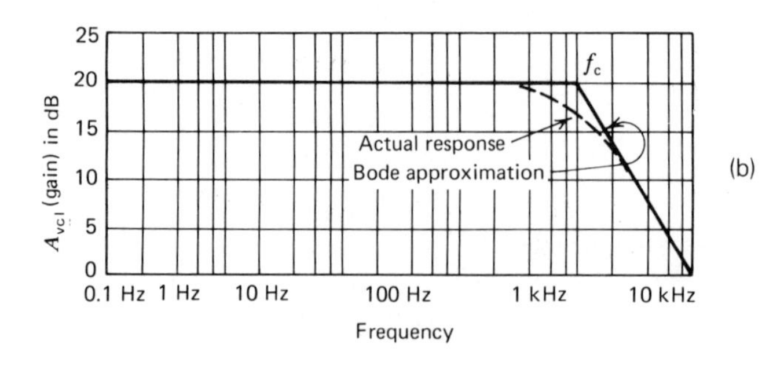

Figure 10.9 Low-frequency inverting amplifier. (*a*) Circuit diagram. (*b*) Frequency response.

3 At low frequencies where $f < f_c$, $\omega R_2 C_1 < 1$; Eq. 10.11 becomes:

$$A_{\text{vcl(low freq.)}} \approx -\frac{R_2}{R_1} \tag{10.13}$$

Above the cutoff frequency where $f > f_c$, $\omega R_2 C_1 > 1$; Eq. 10.11 becomes:

$$A_{\text{vcl(high freq.)}} \approx -\frac{R_2}{R_1}\frac{1}{j\omega R_2 C_1}$$

Thus, as frequency increases above f_c, gain decreases by 20 dB/decade.

4 The cutoff frequency calculated by Eq. 10.12 is:

$$f_c = \frac{1}{2\pi \cdot 1\ \text{M}\Omega \cdot 160\ \text{pF}} = 995\ \text{Hz}$$

5 Low-frequency gain, by Eq. 10.13, is:

$$A_{\text{vcl(low freq.)}} = -\frac{1\ \text{M}\Omega}{100\ \text{k}\Omega} = -10\ (20\ \text{dB magnitude})$$

6 The resulting frequency response is shown in Fig. 10.9*b*.

Inverting Amplifier with Power Buffer

EXAMPLE 10.8 An inverting amplifier with a gain $A_{vcl} = -4$ is needed to drive a load, requiring a current of up to 0.5 A, which is beyond the capability of available op amps. What should be added to the circuit to accomplish this?

PROCEDURE

1 Add a complementary power output stage to the op amp, as shown in Fig. 10.10. Feedback should be from the output, as shown.

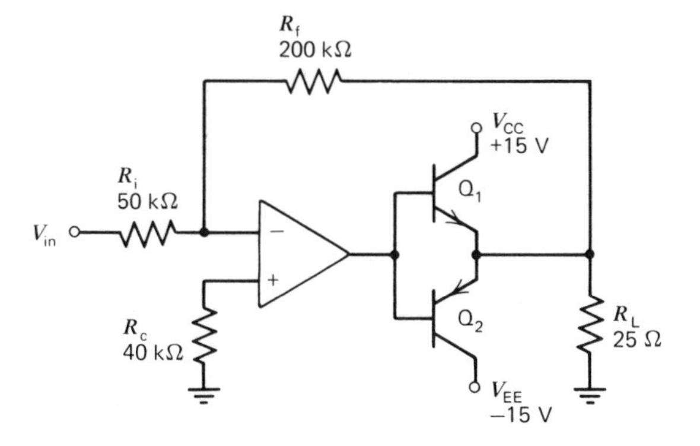

Figure 10.10 Inverting amplifier with power buffer.

2 The power output stage becomes in effect, part of the op amp. Transistors Q_1 and Q_2 should have a gain–bandwidth product, f_T, greater than that of the op amp to ensure stability (freedom from oscillations). Because most general-purpose op amps have gain–bandwidth products of about 1 MHz or less, 10 MHz would be sufficient for the transistors.

3 Gain Eq. 10.2 would apply, with $Z_f = R_f$ and $Z_i = R_i$.

Difference Amplifier

A difference amplifier has two inputs: one to the inverting terminal of the op amp and the other to the noninverting terminal, as illustrated in Fig. 10.11. The difference between these two inputs is amplified and becomes the output signal.

EXAMPLE 10.9 Find the circuit gain, $A_{vcl} = V_{out}/(V_{in(+)} - V_{in(-)})$, of the difference amplifier of Fig. 10.11.

PROCEDURE

1 Use superposition to find the contribution of each input to the output, then add to obtain the total output signal. With $V_{in(+)}$ grounded, the circuit is an inverting amplifier and Eq. 10.2 applies, with $Z_i = R_1$ and $Z_f = R_2$, so that the result is:

$$V_{out} \text{ (due to } V_{in(-)}) = A_{vcl}V_{in(-)} = -\frac{R_2}{R_1} V_{in(-)} \tag{10.14}$$

2 With $V_{in(-)}$ grounded, the circuit is a noninverting amplifier. Equation 10.1 applies, with $Z_1 = R_2$ and $Z_2 = R_1$. Then, V_{out} (due to $V_{in(+)}$) $= A_{vcl}V_{in(+)} = (R_1 + R_2)V_{(+)}/R_1$. By the voltage divider relationship, $V_{(+)} = V_{in(+)}R_2/(R_1 + R_2)$. Combining these equations yields:

$$V_{out} \text{ (due to } V_{in(+)}) = \frac{V_{in(+)}R_2}{R_1 + R_2}\,\frac{R_1 + R_2}{R_1} = \frac{R_2}{R_1} V_{in(+)} \tag{10.15}$$

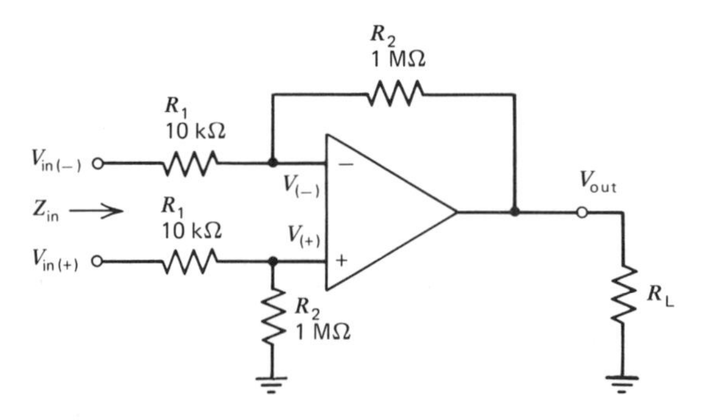

Figure 10.11 Difference amplifier.

3 Adding Eqs. 10.14 and 10.15 yields $V_{out} = (V_{in(+)} - V_{in(-)})R_2/R_1$.
4 When we solve for the gain,

$$A_{vcl} = \frac{V_{out}}{V_{in(+)} - V_{in(-)}} = \frac{R_2}{R_1} \tag{10.16}$$

5 Substituting values gives us $A_{vcl} = (1\ \text{M}\Omega)/(10\ \text{k}\Omega) = 100$. The equal-value resistors must be very closely matched to minimize common mode, output voltage error.

EXAMPLE 10.10 Find the input impedance, Z_{in}, of the difference amplifier in Fig. 10.11.

PROCEDURE

1 The feedback causes $V_{(+)} \approx V_{(-)}$, so that they are effectively shorted together as viewed from the input terminals.
2 Then, $Z_{in} = 2R_1 = 20\ \text{k}\Omega$.

EXAMPLE 10.11 In the difference amplifier of Fig. 10.11, the op amp has a common mode rejection ratio CMRR = 80 dB. Find the output error caused by a +10 V common mode signal at the inputs.

PROCEDURE

1 The common mode signal has the effect of producing an input offset voltage, which is analyzed by grounding the inputs. The circuit is then the same as in Chap. 9, Ex. 9.10 (except that R_1 and R_2 are interchanged), so that Eq. 9.21, with the resistor designators changed, may be used:

$$V_{out} = \frac{(R_1 + R_2)V_{CM}}{R_1 \text{ antilog}\left(\dfrac{\text{CMRR}}{20}\right)} \tag{10.17}$$

Where V_{CM} is the common mode voltage.
2 Substituting values yields:

$$V_{out} = \frac{(10\ \text{k}\Omega + 1\ \text{M}\Omega)\ 10\ \text{V}}{(10\ \text{k}\Omega) \text{ antilog}\left(\dfrac{80\ \text{dB}}{20}\right)} = 101\ \text{mV}$$

Comparison with Ex. 9.10 in Chap. 9 indicates that common mode effects may be much larger in difference amplifiers than in noninverting amplifiers.

Instrumentation Amplifier

Instruments often use a high-impedance transducer or sensor, in a bridge circuit, as an input signal source. This requires a difference amplifier with very high

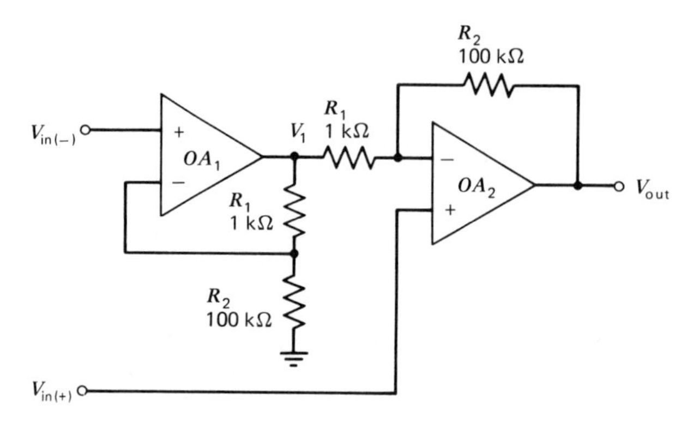

Figure 10.12 Instrumentation amplifier.

input impedance, such as the circuit illustrated in Fig. 10.12. If FET-input op amps are used, the input impedance is extremely high.

EXAMPLE 10.12 Find the circuit gain, $A_{vcl} = V_{out}/(V_{in(+)} - V_{in(-)})$, of the high input impedance difference amplifier shown in Fig. 10.12.

PROCEDURE

1 Op amp OA_1 is a noninverting amplifier. Equation 10.1 applies, with $Z_1 = R_1$ and $Z_2 = R_2$, so that:

$$V_1 = A_{vcl} V_{in(-)} = \frac{R_1 + R_2}{R_2} V_{in(-)} \tag{10.18}$$

2 Because OA_2 has two inputs, V_1 and $V_{in(+)}$, V_{out} may be determined by superposition. If we set V_1 to ground, Eq. 10.1 again applies, with $Z_1 = R_2$ and $Z_2 = R_1$:

$$V_{out} \text{ (due to } V_{in(+)}) = A_{vcl} V_{in(+)} = \frac{R_1 + R_2}{R_1} V_{in(+)} \tag{10.19}$$

3 With $V_{in(+)}$ grounded, OA_2 is an inverting amplifier to V_1. Equation 10.2 applies, with $Z_i = R_1$ and $Z_f = R_2$:

$$V_{out} \text{ (due to } V_1) = A_{vcl} V_1 = -\frac{R_2}{R_1} V_1 \tag{10.20}$$

4 Combining Eqs. 10.19 and 10.20 to sum the output components, we obtain:

$$V_{out} = \frac{R_1 + R_2}{R_1} V_{in(+)} - \frac{R_2}{R_1} V_1 \tag{10.21}$$

5 Substituting Eq. 10.18 into Eq. 10.21 yields:

$$V_{out} = \frac{R_1 + R_2}{R_1} V_{in(+)} - \frac{R_2}{R_1}\frac{R_1 + R_2}{R_2} V_{in(-)}$$

$$= \frac{R_1 + R_2}{R_1} (V_{in(+)} - V_{in(-)})$$

6 Solving for the gain gives us:

$$A_{vcl} = \frac{V_{out}}{V_{in(+)} - V_{in(-)}} = \frac{R_1 + R_2}{R_1} \tag{10.22}$$

7 Now we substitute values, $A_{vcl} = (1\ k\Omega + 100\ k\Omega)/(1\ k\Omega) = 101$. The equal-value resistors must be very closely matched, to minimize common mode error.

A number of commercial monolithic and hybrid instrumentation amplifiers are available that, in addition to providing high values of input impedance and CMRR, offer the feature of variable voltage gain. Representative examples include the *Analog Devices* AD520 and the *Burr–Brown* 3630 instrumentation amplifiers.

10.3 AMPLIFIERS FOR MATHEMATICAL OPERATIONS

The earliest widespread use of op amps was for performing mathematical operations in analog computers. These have now been replaced, generally, with digital technology. However, the various analog circuits are widely used for similar functions in many types of electronic equipment.

Addition and Subtraction

These functions are done with *summing amplifiers*. Inverters, or unity-gain inverting amplifiers, may also be used.

EXAMPLE 10.13 A basic two-input summing amplifier is shown in Fig. 10.13. Find the output function, V_{out}, in terms of the input voltages.

PROCEDURE

1 The inverting input, $V_{(-)}$, is the *summing junction*. By extension of Eq. 10.3 to include both inputs, $I_1 + I_2 + I_3 = 0$. Since $V_{(-)} = 0$ V, $I_1 = V_A/R_1$, $I_2 = V_B/R_2$, and $I_3 = V_{out}/R_3$.

2 Substituting, $V_A/R_1 + V_B/R_2 + V_{out}/R_3 = 0$. Solving for V_{out} yields:

$$V_{out} = -\frac{R_3}{R_1} V_A - \frac{R_3}{R_2} V_B \tag{10.23}$$

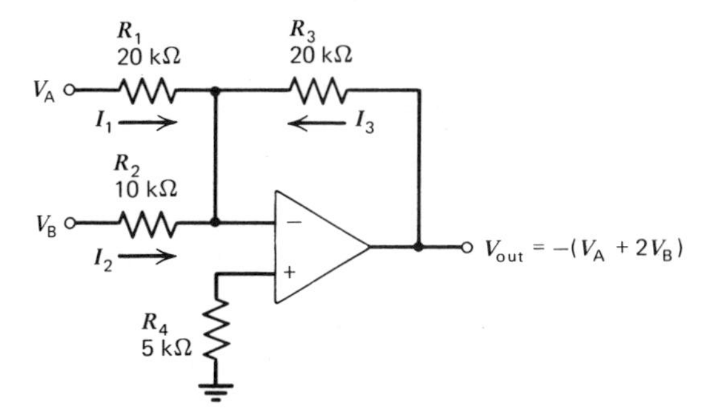

Figure 10.13 Summing amplifier.

3 Substituting values yields:

$$V_{\text{out}} = -\frac{20\text{ k}\Omega}{20\text{ k}\Omega}V_{\text{A}} - \frac{20\text{ k}\Omega}{10\text{ k}\Omega}V_{\text{B}} = -V_{\text{A}} - 2V_{\text{B}}$$

4 Any number of inputs may be similarly added. The weighting of each is equal to the feedback resistor (R_3) divided by the associated input resistor (R_1 or R_2).

EXAMPLE 10.14 Find the output voltage V_{out} of the adding-subtracting circuit shown in Fig. 10.14.

PROCEDURE

1 OA_1 and OA_3 have equal input and feedback resistors. By Eq. 10.2, $A_{\text{vcl}} = -1$; thus they are unity gain inverters with input and output voltages as indicated in Fig. 10.14.

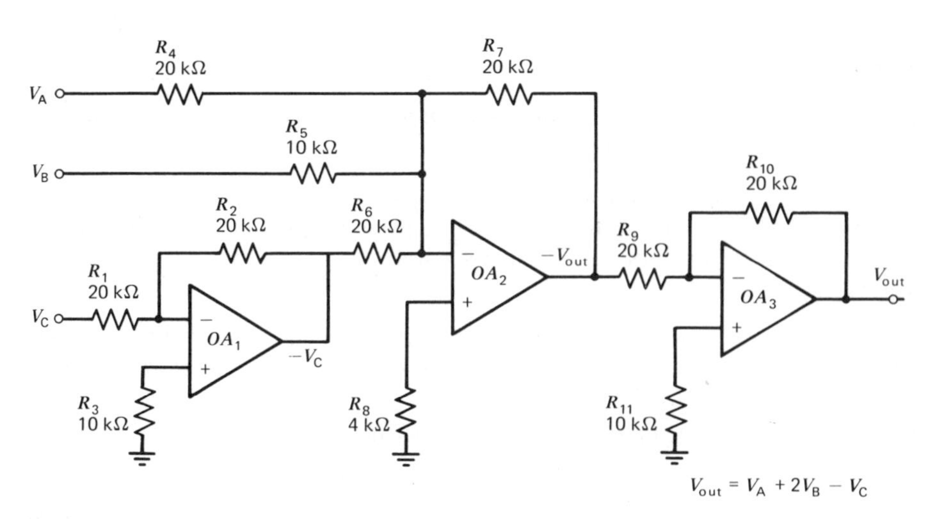

Figure 10.14 Adding-substracting circuit.

2 The summing amplifier is OA_2. Summing the inputs with weighting factors as in Ex. 10.13, and including the output inversion, we find that:

$$V_{\text{out}} = \frac{R_7}{R_4} V_A + \frac{R_7}{R_5} V_B - \frac{R_7}{R_6} V_C$$

3 Substituting values, $R_7/R_4 = R_7/R_6 = 20\ \text{k}\Omega/20\ \text{k}\Omega = 1$ and $R_7/R_5 = 2$, we obtain:

$$V_{\text{out}} = V_A + 2V_B - V_C$$

Logarithmic and Antilogarithmic Amplifiers

Some mathematical functions may involve the manipulation of logarithms. Conversion of values to and from logarithms is done with log and antilog amplifiers.

EXAMPLE 10.15 For the log amplifier of Fig. 10.15, calculate the output voltage V_{out}.

PROCEDURE

1 The output is the base-to-emitter junction voltage of a transistor, which, based on semiconductor theory, is:

$$-V_{\text{out}} = V_{\text{be}} = \frac{kT}{q} \ln I_c$$

where ln is the natural logarithm (base $e = 2.718$); k, Boltzmann's constant, $= 1.38 \times 10^{-23}$ J/°K (joules/degree kelvin); T is temperature in degrees kelvin (at room temperature, + 298°); and q, the charge on the electron $= 1.60 \times 10^{-19}$ C (coulomb).

2 Then, at room temperature the result is:

$$V_{\text{out}} = \frac{-(1.38 \times 10^{-23}\text{J/K})(298°\text{K})}{1.60 \times 10^{-19}\ \text{C}} \ln I_c = -0.0257 \ln I_c \qquad (10.24)$$

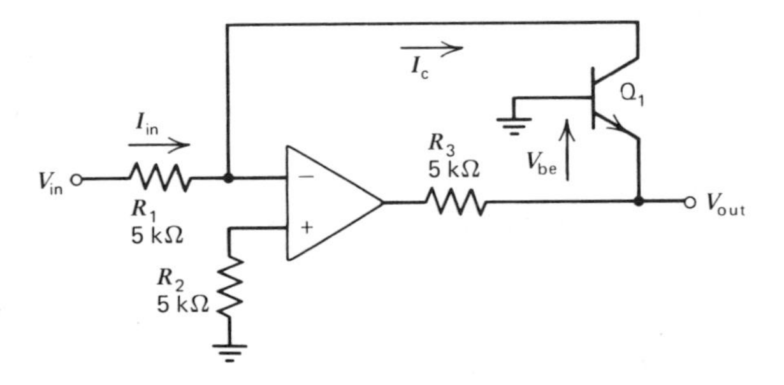

Figure 10.15 Logarithmic amplifier.

3 At the summing junction, by Eq. 10.3, $I_c = I_{in} = V_{in}/R_1$. Substituting yields:

$$V_{out} = -0.0257 \ln\left(\frac{V_{in}}{R_1}\right) = -0.0257 \ln\left(\frac{V_{in}}{5 \text{ k}\Omega}\right) \tag{10.25}$$

4 If the value is needed in base 10 logarithm, the conversion factor $\log x/\ln x = 0.4343$ is used (log is the base 10 logarithm). The result is:

$$V_{out} = -0.4343\left[0.0257 \log\left(\frac{V_{in}}{5 \text{ k}\Omega}\right)\right]$$

$$= -0.0112 \log\left(\frac{V_{in}}{5 \text{ k}\Omega}\right)$$

This circuit requires positive input voltages. Resistor R_3 is included to increase the gain to approximately unity at the op amp output. Op amps are usually not designed to operate at fractional gain values, and may oscillate if the closed-loop gain $A_{vcl} < 1$.

The output voltage will vary somewhat with temperature changes. Log amps may include temperature compensation circuitry to minimize this effect.

EXAMPLE 10.16 For the antilog amplifier of Fig. 10.16, calculate the output voltage V_{out}.

PROCEDURE

1 Equation 10.24, with V_{out} changed to V_{in}, applies. Collector current $I_c = I_1 = V_{out}/R_1$. Substituting yields:

$$V_{in} = -0.0257 \ln\left(\frac{V_{out}}{R_1}\right)$$

2 Solving for V_{out} gives us:

$$\ln\left(\frac{V_{out}}{R_1}\right) = -\frac{V_{in}}{0.0257} = -38.9\ V_{in}$$

$$\frac{V_{out}}{R_1} = -\ln^{-1}(38.9\ V_{in})$$

$$V_{out} = -R_1 \ln^{-1}(38.9\ V_{in}) = -(5 \text{ k}\Omega) \ln^{-1}(38.9\ V_{in}) \tag{10.26}$$

3 The input voltage must be negative. As with the log amplifier, the output varies somewhat with temperature unless compensating circuits are added.

Division and Multiplication

Because logarithms are exponents, they are added to multiply and subtracted to divide. The antilog then provides the desired result.

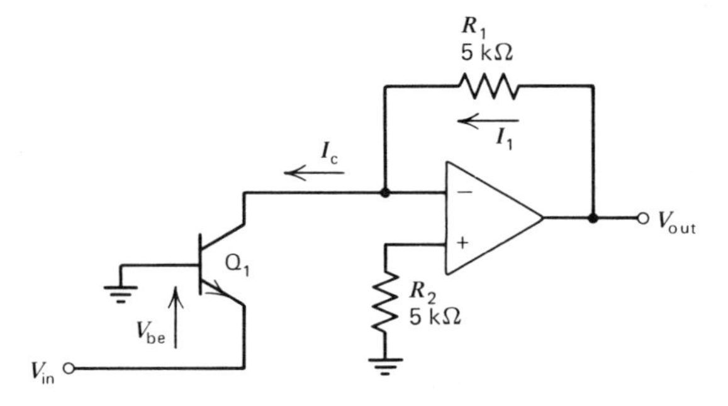

Figure 10.16 Antilogarithmic amplifier.

EXAMPLE 10.17 A basic multiplier-divider circuit is shown in Fig. 10.17. Find the output function, V_{out}.

PROCEDURE

1 Assume that R_5 is set so that $V_B = 0$ V. Then,

$$V_{\text{be}_1} + V_{\text{be}_2} = V_{\text{be}_3} + V_{\text{be}_4}$$

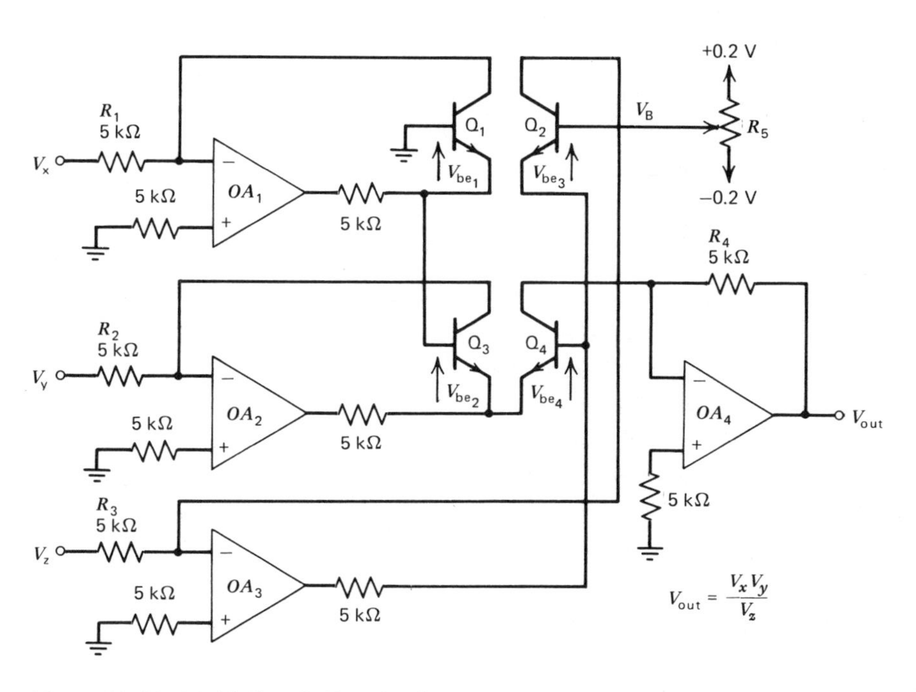

Figure 10.17 Multiplier-divider circuit.

2 Solving for V_{be_4} yields:

$$V_{be_4} = V_{be_1} + V_{be_2} - V_{be_3} \tag{10.27}$$

3 OA_1, OA_2, and OA_3 are log amps with outputs producing $-V_{be_1}$, $-V_{be_2}$, and $-V_{be_3}$. If we apply Eq. 10.25 to this circuit, we get $V_{be_1} = 0.0257 \ln (V_x/R_1)$, $V_{be_2} = 0.0257 \ln (V_y/R_2)$, and $V_{be_3} = 0.0257 \ln (V_z/R_3)$.

4 Substituting into Eq. 10.27 yields:

$$V_{be_4} = 0.0257 \left(\ln \frac{V_x}{R_1} + \ln \frac{V_y}{R_2} - \ln \frac{V_z}{R_3} \right) \tag{10.28}$$

V_{be_4} is the input voltage for the antilog amplifier OA_4.

5 Applying Eq. 10.26 gives us $V_{out} = -R_4 \ln^{-1} (38.9\ V_{be4})$. We substitute Eq. 10.28 to obtain:

$$V_{out} = -R_4 \ln^{-1} \left[38.9(-0.0257) \left(\ln \frac{V_x}{R_1} + \ln \frac{V_y}{R_2} - \ln \frac{V_z}{R_3} \right) \right]$$

$$= R_4 \ln^{-1} \left(\ln \frac{V_x}{R_1} + \ln \frac{V_y}{R_2} - \ln \frac{V_z}{R_3} \right)$$

6 When we take the antilog, logarithmic additions become multiplication and subtractions become division; therefore, the result is:

$$V_{out} = R_4 \frac{V_x}{R_1} \frac{V_y}{R_2} \frac{R_3}{V_z} = \frac{R_3 R_4}{R_1 R_2} \frac{V_x V_y}{V_z} \tag{10.29}$$

7 Substituting values yields:

$$V_{out} = \frac{5\ \text{k}\Omega \times 5\ \text{k}\Omega}{5\ \text{k}\Omega \times 5\ \text{k}\Omega} \frac{V_x V_y}{V_z} = \frac{V_x V_y}{V_z}$$

All inputs, and the outputs, must have positive voltage values to properly bias the transistors. Because transistor differences can cause significant errors, transistor pairs Q_1, Q_4 and Q_2, Q_3 should be well matched. Slight mismatches can be compensated for by adjusting R_5.

EXAMPLE 10.18 In the multiplier-divider circuit of Fig. 10.17, adjust R_5 to the optimum setting.

PROCEDURE

1 Apply the same midrange voltage, for example +5 V, to all three inputs V_x, V_y, and V_z.

2 As a result, $V_{out} = V_x V_y / V_z = (+5\ \text{V})(+5\ \text{V})/(+5\ \text{V}) = +5\ \text{V}$.

3 Adjust R_5 to obtain this exact value.

EXAMPLE 10.19 How can the circuit of Ex. 10.17 be used as a multiplier only, so that $V_{out} = V_x V_y$?

PROCEDURE

1 Apply the voltage to V_z that corresponds to the value of 1.
2 As a result, $V_{out} = V_x V_y/1 = V_x V_y$.

EXAMPLE 10.20 How can the circuit of Ex. 10.17 be used as a divider only, so that $V_{out} = V_y/V_z$?

PROCEDURE

1 Apply the voltage to V_x that corresponds to the value of 1.
2 As a result, $V_{out} = 1V_y/V_z = V_y/V_z$.

Analog Computation

Mathematical function circuits may be combined in various ways to perform almost any type of mathematical computation.

EXAMPLE 10.21 Devise a circuit to do the following computation:

$$V_{out} = \frac{(V_A + 2V_B - V_C)V_D}{2V_E}$$

PROCEDURE

1 Combine the circuits of the summing amplifier, Fig. 10.14, with the multiplier-divider, Fig. 10.17; the result is the circuit shown in Fig. 10.18.
2 The summing junction of op amp OA_2 is used for the $(V_A + 2V_B - V_C)$ summing operation. The output is given by Eq. 10.29, with these substitutions: $(V_A/R_1) + (V_B/R_2) - (V_C/R_3)$ for V_x/R_1, V_D/R_6 for V_y/R_2, R_7/V_E for R_3/V_z, and R_8 for R_4. The result is:

$$V_{out} = R_8\left(\frac{V_A}{R_1} + \frac{V_B}{R_2} - \frac{V_C}{R_3}\right)\frac{V_D}{R_6}\frac{R_7}{V_E}$$

3 Substituting values yields:

$$V_{out} = 5\text{ k}\Omega\left(\frac{V_A}{5\text{ k}\Omega} + \frac{V_B}{2.5\text{ k}\Omega} - \frac{V_C}{5\text{ k}\Omega}\right)\frac{V_D}{5\text{ k}\Omega}\frac{2.5\text{ k}\Omega}{V_E}$$
$$= \frac{(V_A + 2V_B - V_C)V_D}{2V_E}$$

4 The inverting function of OA_3 in Fig. 10.14 is performed by the antilog amplifier, OA_5.

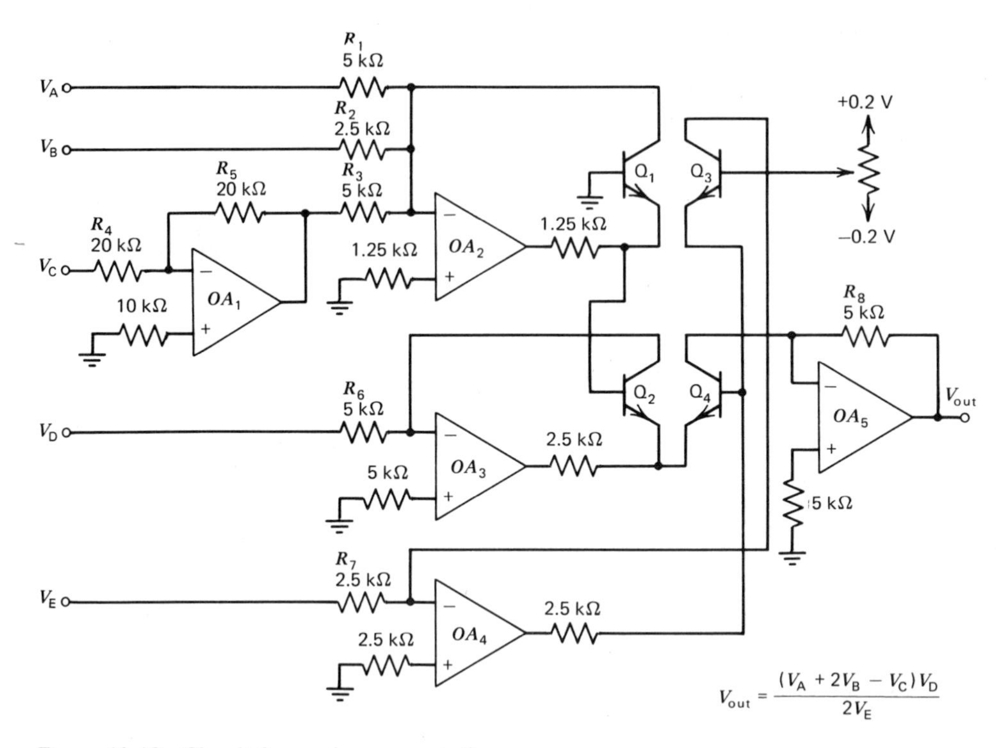

Figure 10.18 Circuit for analog computation.

Differentiation and Integration

The mathematical operations of differentiation and integration are calculus functions. However, the amplifiers for these functions are also used for related purposes, such as waveshaping.

EXAMPLE 10.22 The differentiator of Fig. 10.19*a* has a triangular, or ramp-type, input voltage as shown. Find the waveshape and amplitude of the output waveform.

PROCEDURE

1 Output voltage $V_{out} = -i_2R_2$. From Eq. 10.3, $i_2 = i_1$, so that $V_{out} = -i_1R_2$. Because i_1 also charges the capacitor, $i_1 = (\Delta V_C/\Delta t)C_1$. If the small voltage across R_1 is neglected, $V_C = V_{in}$.

2 Substituting yields:

$$V_{out} = -\frac{\Delta V_{in}}{\Delta t} C_1R_2 \tag{10.30}$$

3 Examining the input waveform (Fig. 10.19*b*), $\Delta V_{in}/\Delta t = 0$ V/μs before t_1, 8 V/40 μs = 0.2 V/μs during t_1, −8 V/80 μs = −0.1 V/μs during t_2, and 0 V/μs after t_2.

4 Substituting values into Eq. 10.30 yields:

$$\begin{aligned} V_{out} &= -0 \text{ V/}\mu\text{s} \times 5 \text{ nF} \times 10 \text{ k}\Omega = 0 \text{ V before } t_1 \\ &= -0.2 \text{ V/}\mu\text{s} \times 5 \text{ nF} \times 10 \text{ k}\Omega = -10 \text{ V during } t_1 \\ &= +0.1 \text{ V/}\mu\text{s} \times 5 \text{ nF} \times 10 \text{ k}\Omega = +5 \text{ V during } t_2 \\ &= -0 \text{ V/}\mu\text{s} \times 5 \text{ nF} \times 10 \text{ k}\Omega = 0 \text{ V after } t_2 \end{aligned}$$

5 These calculations result in the output waveform in Fig. 10.19*b*. As indicated by Eq. 10.30 and shown in the waveform, the output amplitude is proportional, at every point, to the slope, or rate of change, of the input.

Resistor R_1 is needed to limit high-frequency gain. Without it, the amplifier would have extremely high gain at high frequencies, and amplified high-frequency noise might be a problem.

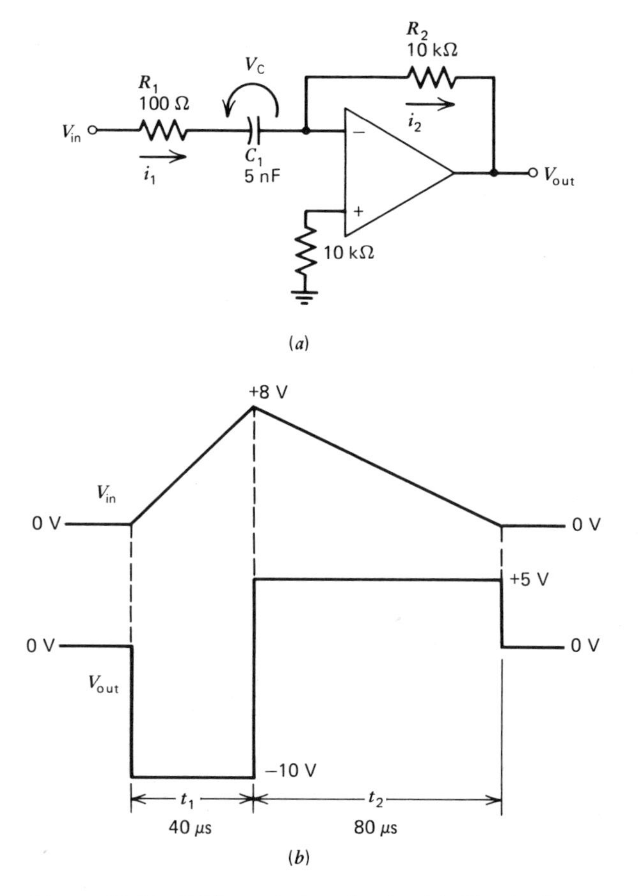

Figure 10.19 Differentiating circuit. (*a*) Circuit diagram. (*b*) Input and output voltage waveforms.

EXAMPLE 10.23 The integrator of Fig. 10.20*a* has a square wave input, as shown in Fig. 10.20*b*. Find the waveshape and amplitude of the output waveform.

PROCEDURE

1 The output voltage is the voltage across the capacitor, so that $\Delta V_{out}/\Delta t = \Delta V_C/\Delta t = i_2/C_1$. The resistor R_2 current is assumed negligible, so that $i_2 = i_1 = -V_{in}/R_1$. After substituting, we get $\Delta V_{out}/\Delta t = i_1/C_1 = -V_{in}/(R_1C_1)$. Now we solve for ΔV_{out}:

$$\Delta V_{out} = -\frac{V_{in}\Delta t}{R_1C_1} \tag{10.31}$$

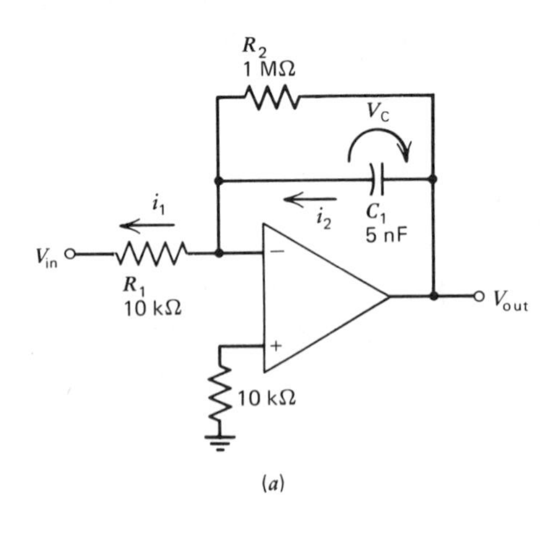

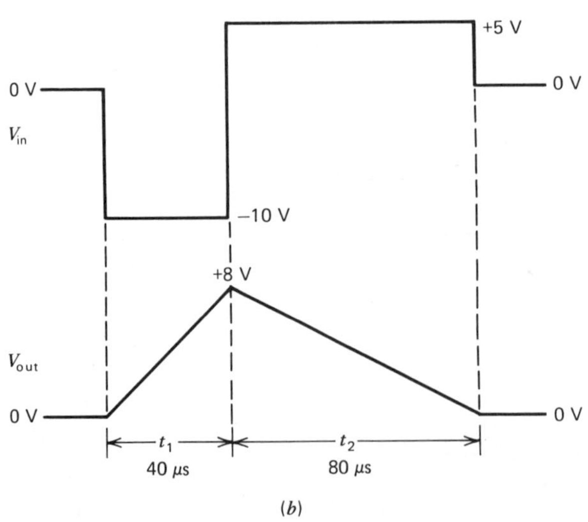

Figure 10.20 Integrating circuit. (*a*) Circuit diagram. (*b*) Input and output voltage waveforms.

2 Before time interval t_1, no V_{out} change occurs since V_{in} is 0. Then, by substituting values, we obtain:

$$\Delta V_{out} = -\frac{-10 \text{ V} \times 40 \text{ µs}}{10 \text{ k}\Omega \times 5 \text{ nF}} = +8 \text{ V during } t_1$$

$$= -\frac{5 \text{ V} \times 80 \text{ µs}}{10 \text{ k}\Omega \times 5 \text{ nF}} = -8 \text{ V during } t_2$$

3 After t_2, $V_{in} = 0$ so that the output voltage does not change further. The resulting output waveform is shown in Fig. 10.20*b*.

4 Resistance R_2 is needed for dc feedback, to bring V_{out} to 0 V when there is no input signal. Because $R_2 >> X_C$ for most frequencies, it has very little effect on the ac value of the output.

The circuit operation of the integrator is the opposite of the differentiator (Ex. 10.22). The ouptut voltage V_{out}, for the integrating amplifier, has a rate of change or slope that is proportional to the amplitude of the input voltage, V_{in}.

10.4 VOLTAGE AND CURRENT CONVERSION

These circuits are used when the output current, controlled by an input voltage, must be independent of the load, or when a load-independent output voltage is controlled by an input current.

EXAMPLE 10.24 In the voltage-controlled current source of Fig. 10.21, output current $I_{out} = 5$ mA, flowing into the load, is desired. Calculate the required input voltage V_{in}.

PROCEDURE

1 The input voltages to the op amp, by voltage division, are $V_{(+)} = V_{out}R_1/(R_1 + R_2)$ and $V_{(-)} = V_{in} + (V_1 - V_{in})R_1/(R_1 + R_2)$. Assume $V_d = 0$, so that $V_{(-)} = V_{(+)}$.

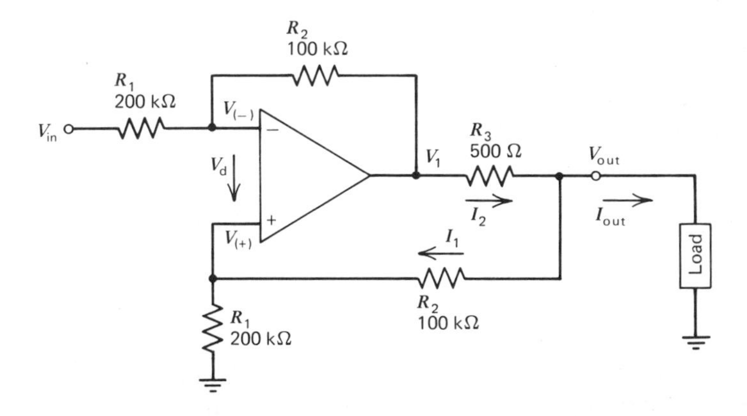

Figure 10.21 Voltage-controlled current source.

Then, the result is:

$$\frac{(V_1 - V_{in})R_1}{R_1 + R_2} + V_{in} = \frac{R_1 V_{out}}{R_1 + R_2}$$

2 But $V_1 = V_{out} + I_2 R_3$. Assume $I_1 << I_{out}$, then $I_2 = I_{out}$ and $V_1 = V_{out} + I_{out} R_3$. Substituting yields:

$$\frac{(V_{out} + I_{out} R_3 - V_{in})R_1}{R_1 + R_2} + V_{in} = \frac{R_1 V_{out}}{R_1 + R_2}$$

3 Multiplying through by $(R_1 + R_2)$ and solving for V_{in} gives us:

$$R_1 V_{out} + R_1 R_3 I_{out} - R_1 V_{in} + R_1 V_{in} + R_2 V_{in} = R_1 V_{out}$$

$$V_{in} = -\frac{R_1 R_3 I_{out}}{R_2} \tag{10.32}$$

Output voltage V_{out} drops out of the equation because it is fed back equally to both inputs. This causes the output current to be independent of the output voltage, controlled only by the input voltage.

4 Substituting values yields:

$$V_{in} = -\frac{5\text{ mA} \times 500\ \Omega \times 200\text{ k}\Omega}{100\text{ k}\Omega} = -5.0\text{ V}$$

EXAMPLE 10.25 In Ex. 10.24, it was assumed that $I_2 = I_{out}$. Actually, this is exactly true only when $V_{out} = 0$, so that no current flows in the feedback resistor. Find the error, and the exact output current I_{out}, when $V_{out} = +10$ V.

PROCEDURE

1 The feedback current $I_1 = V_{out}/(R_1 + R_2) = (10\text{ V})/(100\text{ k}\Omega + 200\text{ k}\Omega) = 33\ \mu\text{A}$. The feedback maintains the current through R_3 constant at exactly 5 mA, therefore, $I_{out} = 5\text{ mA} - 33\ \mu\text{A} = 4.967$ mA.

2 The error is $< 1\%$.

EXAMPLE 10.26 In the current-to-voltage converter of Fig. 10.22, calculate the value of R_1 needed for a conversion factor of $V_{out} = -10$ V/mA.

PROCEDURE

1 Because $V_{(-)} \approx 0$, $V_{out} = I_1 R_1$. From Eq. 10.3, $I_1 = -I_{in}$. If we substitute, the value becomes:

$$V_{out} = -I_{in} R_1 \tag{10.33}$$

2 Solving, we obtain $R_1 = -V_{out}/I_{in} = -(-10\text{ V/mA}) = 10\text{ k}\Omega$. The negative polarity of Eq. 10.33 indicates that current into the op amp produces a negative output voltage. The output may be made positive by reversing the direction of the input current.

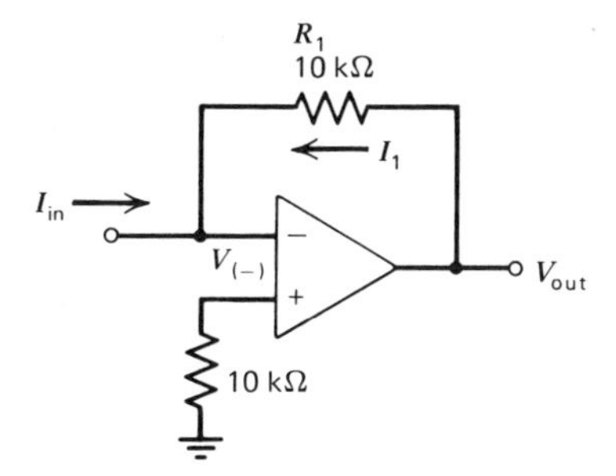

Figure 10.22 Current-to-voltage converter.

10.5 SIMULATED REACTANCES

Large-value capacitors and inductors are space consuming and expensive. Often their function can be performed with op amps and small, inexpensive components.

Capacitance Multiplier

EXAMPLE 10.27 For the capacitance multiplier of Fig. 10.23*a*, calculate the input impedance Z_{in} and the equivalent capacitance C'.

PROCEDURE

1 First, find V_{C1}. This is also the output of the unity follower op amp. From the voltage divider R_2 and C_1, the result is:

$$V_{C1} = \frac{X_{C1}V_{in}}{X_{C1} + R_2} = \frac{\dfrac{1}{j\omega C_1}V_{in}}{\dfrac{1}{j\omega C_1} + R_2} = \frac{V_{in}}{1 + j\omega R_1 C_1}$$

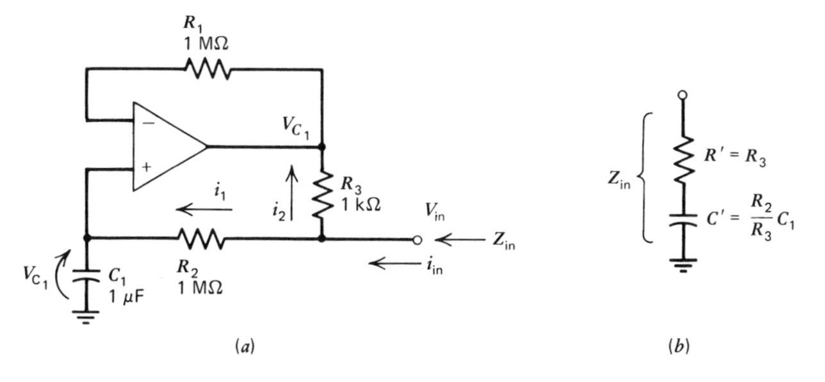

Figure 10.23 Capacitance multiplier. (*a*) Circuit diagram. (*b*) Equivalent circuit for Z_{in}.

2 In these circuits, $R_2 >> R_3$ and $i_2 >> i_1$, so that $i_{in} \approx i_2 = (V_{in} - V_{C1})/R_3$. Substituting yields:

$$i_{in} = \frac{1}{R_3}\left(V_{in} - \frac{V_{in}}{1 + j\omega R_2 C_1}\right) = \frac{1}{R_3}\frac{j\omega R_2 C_1}{1 + j\omega R_2 C_1}$$

3 Input impedance $Z_{in} = V_{in}/i_{in}$. Again substituting, we obtain:

$$Z_{in} = \frac{V_{in}}{\dfrac{1}{R_3}\dfrac{j\omega R_2 C_1 V_{in}}{1 + j\omega R_2 C_1}} = R_3 + \frac{R_3}{j\omega R_2 C_1}$$

This is a resistance in series with capacitive reactance. Let the equivalent components be R' and C', as shown in Fig. 10.23*b*. Then $Z_{in} = R' + 1/(j\omega C')$, $R' = R_3$, and the result becomes:

$$C' = \frac{R_2}{R_3} C_1 \tag{10.34}$$

4 Substituting values gives us $R' = 1\ \text{k}\Omega$ and $C' = (1\ \text{M}\Omega)/(1\ \text{k}\Omega)(1\ \mu\text{F}) = 1\ \text{mF}$.

For varying dc levels and very low frequencies, R' has little effect and the circuit acts essentially as a very large capacitor. An offset voltage or current at the op amp will cause the circuit to act as a current source, of perhaps several microamperes. If objectionable, this effect can be minimized with an offset null adjustment (see Chap. 9, Sec. 9.4).

Simulated Inductor

EXAMPLE 10.28 For the simulated inductor of Fig. 10.24*a*, calculate the input impedance Z_{in}, the equivalent inductance L', and the effective frequency range.

PROCEDURE

1 Because the op amp is a unity follower, its output voltage is $V_{(+)}$, as shown. $Z_{in} = V_{in}/I_{in} = V_{in}/(I_1 + i_2)$. $I_1 = (V_{in} - V_{(+)})/R_1$ and as a result:

$$i_2 = \frac{V_{in}}{R_3 + X_{C1}} = \frac{V_{in}}{R_3 + \dfrac{1}{j\omega C_1}} = \frac{V_{in}\, j\omega C_1}{1 + j\omega R_3 C_1}$$

2 Substituting I_1 and i_2 yields:

$$Z_{in} = \frac{V_{in}}{\dfrac{V_{in} - V_{(+)}}{R_1} + \dfrac{V_{in}\, j\omega C_1}{1 + j\omega R_3 C_1}}$$

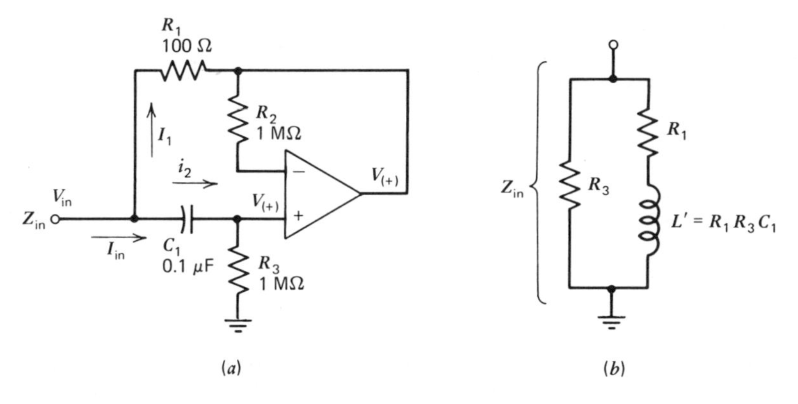

Figure 10.24 Simulated inductor. (*a*) Circuit diagram. (*b*) Equivalent circuit for Z_{in}.

Also,

$$V_{(+)} = i_2 R_3 = \frac{V_{in}\, j\omega R_3 C_1}{1 + j\omega R_3 C_1}$$

3 Substituting $V_{(+)}$ gives us:

$$Z_{in} = \frac{V_{in}}{\dfrac{V_{in}}{R_1} - \dfrac{1}{R_1}\dfrac{V_{in}\, j\omega R_3 C_1}{1 + j\omega R_3 C_1} + \dfrac{V_{in}\, j\omega C_1}{1 + j\omega R_3 C_1}}$$

Dividing by V_{in} drops the term out. Now we simplify:

$$Z_{in} = \frac{1}{\dfrac{1 + j\omega R_3 C_1 - j\omega R_3 C_1 + j\omega R_1 C_1}{R_1 + j\omega R_1 R_3 C_1}} = \frac{1}{\dfrac{1 + j\omega R_1 C_1}{R_1 + j\omega R_1 R_3 C_1}}$$

$$= \frac{R_1 + j\omega R_1 R_3 C_1}{1 + j\omega R_1 C_1} \qquad (10.35)$$

4 The frequency range of interest is $1/(R_1C_1) > \omega > 1/(R_3C_1)$. Then, $\omega R_1 R_3 C_1 > R_1$ and $\omega R_1 C_1 < 1$, so that both R_1 and $j\omega R_1 C_1$ may be neglected; Eq. 10.35 then becomes,

$$Z_{in} = j\omega R_1 R_3 C_1 \quad \text{(at midfrequencies)}$$

Let L' be the equivalent inductance, so that $Z_{in} = j\omega R_1 R_3 C_1 = j\omega L'$.

5 If we solve for inductance, we obtain:

$$L' = R_1 R_3 C_1 \qquad (10.36)$$

6 We substitute values: $L' = 100\ \Omega \times 1\ \text{M}\Omega \times 0.1\ \mu\text{F} = 10\ \text{H}$.

7 The limits of the frequency range for inductive Z_{in} are where the real and imaginary magnitudes of the denominator and numerator of Eq. 10.35 are equal. At these frequencies, f_1 and f_2, $R_1 = \omega_1 R_1 R_3 C_1 = 2\pi f_1 R_1 R_3 C_1$ and $1 = \omega_2 R_1 C_1 = 2\pi f_2 R_1 C_1$. Solving yields:

$$f_1 = \frac{R_1}{2\pi R_1 R_3 C_1} = \frac{1}{2\pi \times 1\ \text{M}\Omega \times 0.1\ \mu\text{F}} = 1.59\ \text{Hz}$$

$$f_2 = \frac{1}{2\pi R_1 C_1} = \frac{1}{2\pi \times 100\ \Omega \times 0.1\ \mu\text{F}} = 15.9\ \text{kHz}$$

8 Above and below these frequencies Z_{in} becomes resistive. The equivalent circuit for Z_{in} is shown in Fig. 10.24*b*. At low frequencies, the j terms become negligibly small and Eq. 10.35 becomes $Z_{in} = R_1 = 100\ \Omega$. At high frequencies the real terms, R_1 and 1, can be neglected so that $Z_{in} = j\omega R_1 R_3 C_1/(j\omega R_1 C_1) = R_3 = 1\ \text{M}\Omega$.

As with the capacitance multiplier, offsets in the op amp will cause the circuit to act as a small current source. Currents, which could reach 50 μA or so, may be greatly reduced by using a precision or low-offset type of op amp, or with an offset null adjustment, or both.

10.6 STORAGE CIRCUITS

Sample and Hold

This circuit samples the input voltage when a "sample" signal is applied. The sampled input voltage is reproduced at the output, and held there until the next sample signal.

EXAMPLE 10.29 In the sample-and-hold circuit of Fig. 10.25, the FET-input op amp, type LF355, has maximum bias current $I_B = 200$ pA. If the interval between sample pulses is 5 s, find the change in output voltage during the storage time.

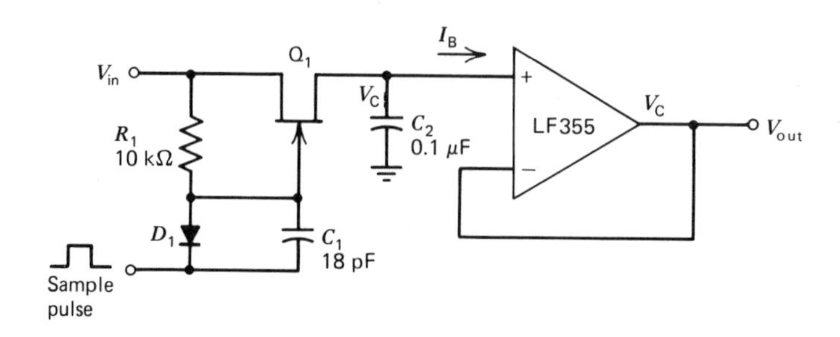

Figure 10.25 Sample-and-hold circuit.

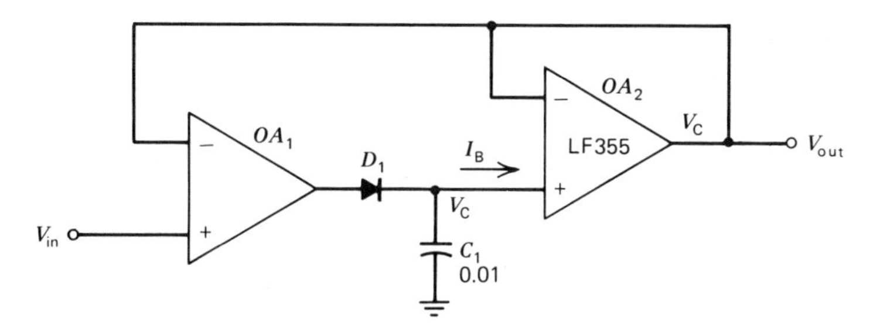

Figure 10.26 Peak detector.

PROCEDURE

1 Diode D_1 is normally forward biased by a voltage at the sample input, which is sufficiently negative to bias Q_1 into cutoff. The positive-going sample pulse reverse biases D_1 and Q_1 is now unbiased, gate to source, and conducts, so that $V_C = V_{in}$.
2 Upon termination of the sample pulse, D_1 again conducts, Q_1 becomes cutoff, and C_2 retains the sampled voltage value. The unity follower action of the op amp maintains $V_{out} = V_C$.
3 Capacitor C_2 will, however, lose some charge due to the leakage current of the cutoff transistor Q_1 and the bias current I_B of the op amp. The leakage curent of Q_1 is usually much smaller than I_B so that it can be neglected. The voltage change of the capacitor is then $\Delta V_C/\Delta t = I_B/C_2 = \Delta V_{out}/\Delta t$.
4 Solving, we get $\Delta V_{out} = I_B\Delta t/C_2$. When we substitute values, $\Delta V_{out} = 200\ \text{pA} \times (5\ \text{s})/(0.1\ \mu\text{F}) = 10\ \text{mV}$ change, between sample pulses. This is a maximum change, and will be less for smaller values of I_B.

Peak Detector

The peak detector maintains a dc output voltage equal to the most positive (or negative) point of the input signal or waveform.

EXAMPLE 10.30 For the peak detector of Fig. 10.26, calculate the period, or time between positive peak input voltage values, that results in an output voltage change or error of 50 mV. The input bias current I_B of the LF355 op amp is 200 pA. Assume the reverse leakage current of diode D_1 is negligibly small.

PROCEDURE

1 Op amp OA_2 is a unity follower, with $V_{out} = V_C$. A positive-going input voltage causes positive output at op amp OA_1, and conduction of D_1. Voltage V_C, which is fed back (through unity follower OA_2), increases until $V_{out} = V_C = V_{in}$. When V_{in} becomes negative-going, however, D_1 becomes reverse biased. The voltages across capacitor C, V_C, and V_{out} are maintained at their previous maximum positive values.
2 Between the positive peak input voltages, there is some change in V_{out} owing to I_B discharging C_1. $\Delta V_C/\Delta t = I_B/C_1 = \Delta V_{out}/\Delta t$.
3 Solving gives us $\Delta t = C_1\Delta V_{out}/I_B = 0.01\ \mu\text{F} \times (50\ \text{mV})/(200\ \text{pA}) = 2.5\ \text{s}$.

If it is necessary to change the input and output voltages quickly, to a less-positive peak value, switching circuitry may be added to discharge C_1 at the time that the conditions are changed.

10.7 REFERENCES

Bannon, E., *Operational Amplifiers: Theory and Servicing*, Reston, Reston, VA, 1977.

Coughlin, R. F. and F. F. Driscoll, *Operational Amplifiers and Linear Integrated Circuits*, second edition, Prentice Hall, Englewood Cliffs, NJ, 1982.

Gray, P. R. and R. G. Meyer, *Analysis and Design of Analog Integrated Circuits*, Wiley, New York, 1977.

Lenk, J. D., *Manual for Operational Amplifier Users*, Reston, Reston, VA, 1976.

Linear Applications Handbook, National Semiconductor Corporation, Santa Clara, CA, 1978.

Prensky, S. D. and A. H. Siedman, *Linear Integrated Circuits: Practice and Applications*, Reston, Reston, VA, 1981.

Smith, J. I., *Modern Operational Circuit Design*, Wiley-Interscience, New York, 1971.

Stout, D. F. and M. Kaufman (ed.), *Handbook of Operational Amplifier Circuit Design*, McGraw-Hill, New York, 1976.

Wong, Y. J. and W. E. Ott, *Function Circuits Design and Applications*, McGraw-Hill, New York, 1976.

CHAPTER 11

Active Filters

Lester J. Hadley Jr.
Strategic Applications Manager,
Standard Linear Circuits
Signetics Corporation

11.1 INTRODUCTION

The design of modern electronic systems often requires the shaping of signal response in the frequency domain. It may be desirable to limit the bandwidth of signals being received in order to improve the signal-to-noise ratio and thus increase system accuracy. Or, signals may be crowded into a continuous spectrum and it becomes necessary to select individual signals, thereby eliminating adjacent channel interference.

The signal to be filtered may be sinusoidal, square wave, sawtooth, or a repetitive pulse. Each of these signal types requires special consideration regarding filter characteristics in order that the filter itself does not cause distortion in the retrieved signal. In order to successfully design a filter, it becomes clear that the output signal quality is very important. It is for this reason that active filter design must begin with the signal characteristics and progress to the proper selection of a filter type that most nearly and accurately reproduces the signal.

An active filter, as opposed to a passive filter, consists of R and C elements connected in either the feedback or forward paths, or both, of an operational amplifier (op amp). The gain of the op amp is controlled by the external elements. This type of configuration provides predictably stable gain over the full frequency range of the filter. (See Chaps. 9 and 10 for a detailed discussion of op amps.)

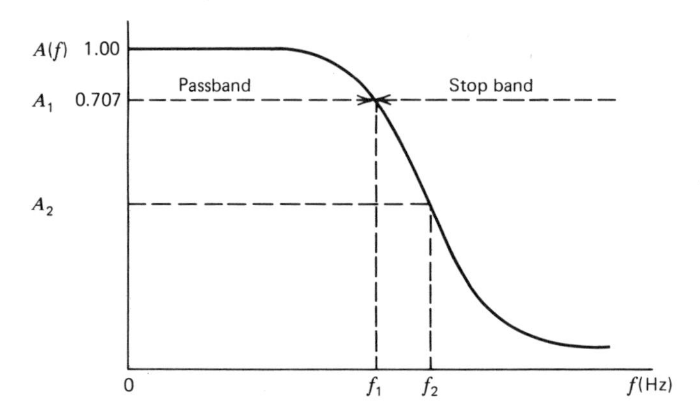

Figure 11.1 Typical low-pass filter response.

11.2 FILTER TYPES

The Low-Pass Filter

The purpose of the low-pass filter is to selectively differentiate frequency domain signals based on a designer-assigned break frequency. Signals below the break frequency are propagated through the filter with little or no attenuation. Signals above the break frequency are rejected or nulled (Fig. 11.1).

Note that the passband of a low-pass filter extends from zero to f_1 hertz. The *break frequency,* f_1, occurs at the frequency corresponding to an amplitude which has decreased to a value that is 0.707 times its value near zero hertz. This is commonly referred to as the filter *3 dB frequency* and shall be referred to in this way in the chapter. The passband is then equal to the 3 dB bandwidth. The amplitude response below the 3 dB frequency is of primary importance in a low-pass filter. The slope of the response between f_1 and f_2, where $f_2 = 2f_1$, determines the *order* of the desired filter response. The slope of the filter response is expressed in dB/octave:

$$\text{dB/octave} = [(A_1 - A_2)\text{dB}]/[(\log f_2 - \log f_1)/\log 2] \tag{11.1}$$

Another way of expressing the slope is in dB/decade, where a decade is 10 times f_1, that is, $f_2 = 10f_1$. Slope may, therefore, be expressed in either dB/octave or dB/decade. For example, 6 dB/octave is equal to 20 dB/decade.

Some filters, such as the Butterworth (to be described later), exhibit characteristics similar to those described as follows:

Response		
dB/octave	**dB/decade**	**Order *n***
6	20	1
12	40	2
18	60	3

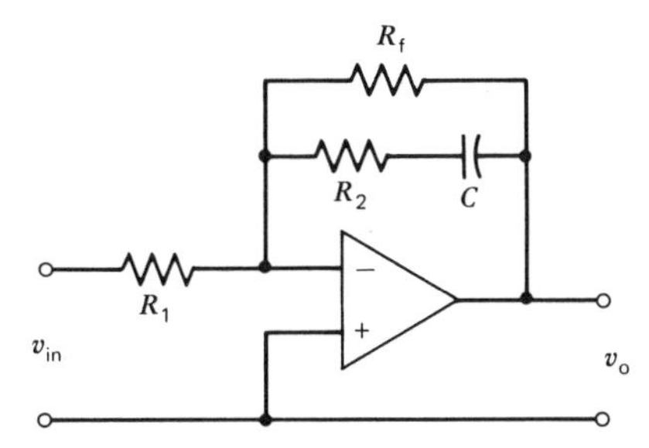

Figure 11.2 Example of a first-order, low-pass filter stage.

Combinations of active filter stages are used to obtain higher-order filters of $n = 3, 4, 5$, and so on.

First-order filter stage This filter is used to obtain an odd-order filter such as $n = 1, 3, 5$, and so on. In effect, a single-RC network is oriented around an op amp. Figure 11.2 shows an example of a first-order stage for low-pass operation. Gain in the passband is adjusted by changing the ratio R_f/R_1; that in the stop band by R_2/R_1 for $R_2 << R_f$.

High-Pass Filter

The high-pass filter is the mirror image of the low-pass filter. Signals *below* the 3 dB frequency are attenuated and those above are passed (Fig. 11.3). The passband of the ideal active high-pass filter extends to infinity. Operational amplifiers, however, have upper-frequency limits (see Chap. 9) and the op amp therefore restricts the high-frequency response.

High-Pass Filter with Gain At this point it is necessary to call attention to the fact that active filters have *gain* in the passband (Fig. 11.4). Thus, desired output levels must be considered in the design procedure.

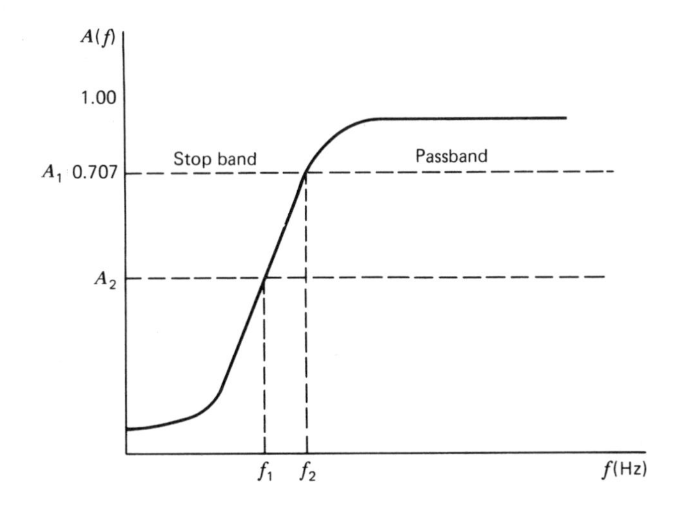

Figure 11.3 Typical high-pass filter response.

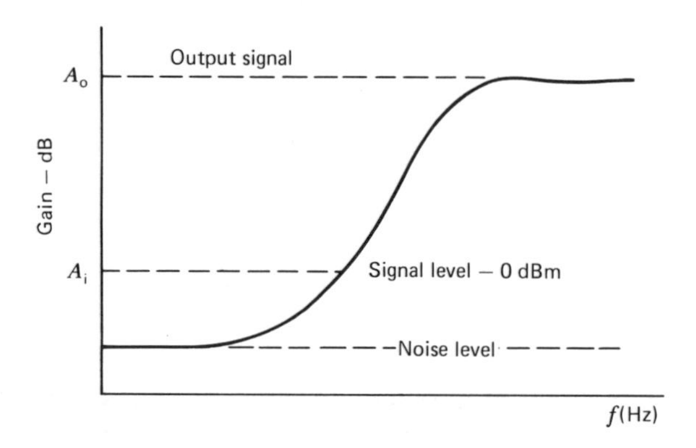

Figure 11.4 Gain of a filter in its passband.

Active filter gains may be measured in decibels milliwatt (dBm), decibels relative to a power level of 1 mW into a 600 Ω load or decibels volts (dBV), 1 V rms into a 600 Ω load. It is necessary to consider signal levels when using active filters in order to stay within the dynamic range of the op amp. For example, if an active filter has a gain of 10 in the passband, a 1 V p-p signal at the input will produce a 10 V p-p signal at the output. The amplifier must be capable of 10 V p-p operation at the cutoff frequency or distortion will occur at the output.

Bandpass Filter

Assume that a pair of filters, one low pass and the other high pass, is placed in series. The break frequency of the low-pass filter is at f_{cL} and for the high-pass filter is at f_{cH}. A bandpass filter results whose bandwith is equal to $f_{cL} - f_{cH}$ (Fig. 11.5). In order for a passband to exist, $f_{cH} < f_{cL}$. Bandpass active filters

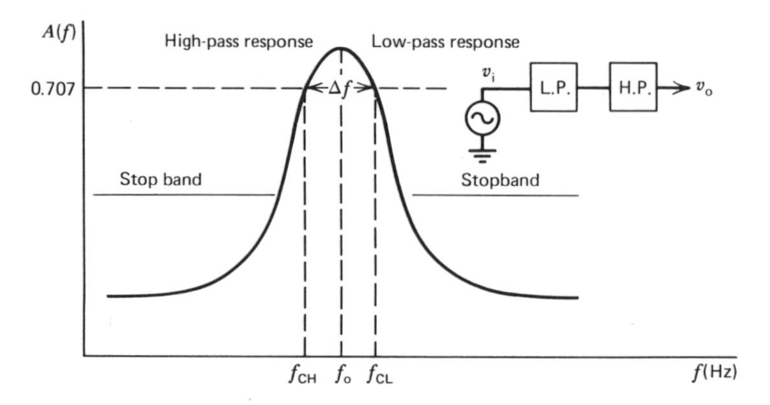

Figure 11.5 Bandpass filter realized by cascading low-pass (L.P.) and high-pass (H.P.) filter stages.

may be realized using the function building approach illustrated in Fig. 11.5 where dual or quad op amp packages are used, or a single op amp, as will be shown later in the chapter.

The higher the order of the filter, the sharper the rolloff of the resulting bandpass response. That is, amplitude reduction versus frequency occurs more rapidly as the order, *n*, increases. A measure of the effectiveness of a bandpass filter in terms of rolloff is the circuit Q. It is expressed by:

$$Q = f_o/\Delta f \tag{11.2}$$

where f_o is the center frequency of the filter and Δf is the 3 dB bandwidth.

Band-Reject Filter

As illustrated in Fig. 11.6, a band-reject (band-stop) filter is the inverse of a bandpass filter. It can be derived from low-pass and high-pass filter sections in parallel. The passbands of the filter sections do not overlap except in the reject region. In this case, the breakpoint of the low-pass filter is below that of the high-pass filter such that $f_{cL} < f_{cH}$.

A band-reject filter's effectiveness is judged by the depth of the notch and is measured by the ratio of the passband signal to the minimum signal in the stop-band, or notch, in decibels. Typically, attenuation in the stop band is greater than 50 dB.

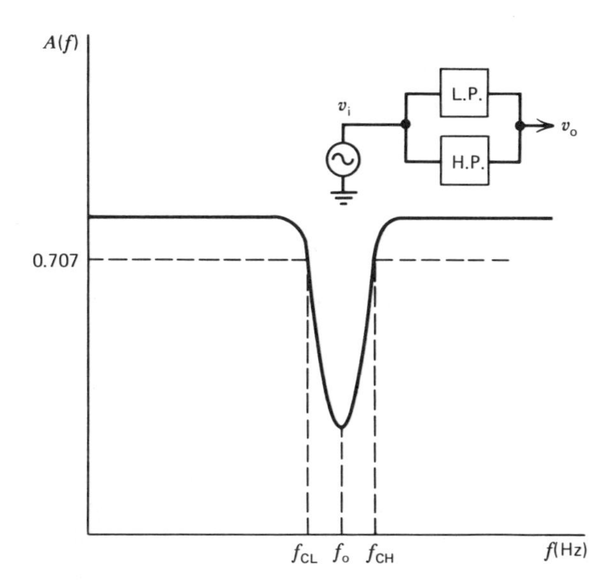

Figure 11.6 Band-reject (band-stop) filter realized by paralleling L.P. and H.P. sections.

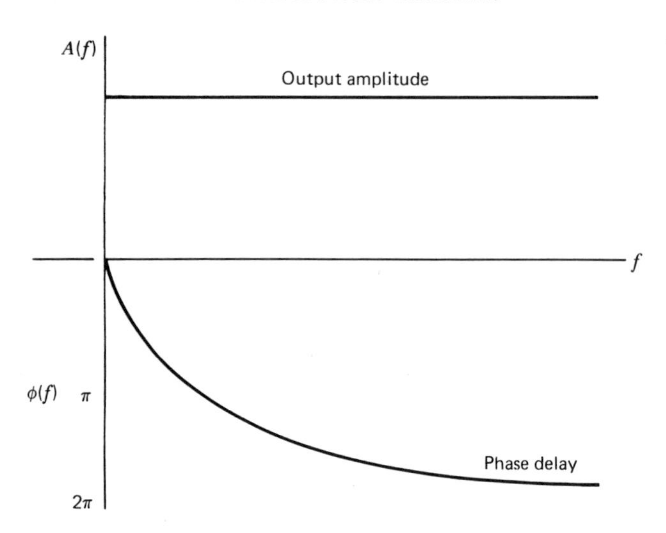

Figure 11.7 All-pass (phase-shift) filter characteristics.

All-Pass Filter

The all-pass, or phase-shift, filter produces a frequency-dependent phase shift and, theoretically, the signal amplitude is constant (Fig. 11.7). *Phase delay* is equivalent to *time delay* and for this reason the all-pass filter may be thought of as an incremental time delay filter. That is, for an incremental shift in signal frequency, a predictable change in time delay results as the signal traverses the filter from input to output. A phase shift of 90° will delay the signal by one-quarter of a cycle. Certain kinds of signal correlation and recognition analysis require frequency-time delays which are suited to the use of the all-pass filter.

11.3 FILTER RESPONSE

Butterworth

The Butterworth is one of the most popular filter types having good amplitude characteristics near zero frequency. The sharpness of the transition frequency increases with the order of the filter, as illustrated in Fig. 11.8. Also, the phase response is more linear with frequency than other similar type filters.

Chebyshev

This type filter is capable of a much sharper rolloff rate at the break frequency than is the Butterworth. Higher-order Chebyshev filters are, therefore, able to differentiate more efficiently between frequencies at the edge of the passband and those in the stop band. One major difference between the Butterworth and a same order Chebyshev is that the amplitude response of the latter has *ripple* in its passband (Fig. 11.9). The ripple width may be controlled by the selection of filter parameters (covered later in the chapter). The phase response of the Chebyshev filter is less linear than that of the Butterworth filter.

Bessel

The Bessel filter is also known as a *constant time delay* filter. The significance of constant time delay is particularly useful in circuits for processing variable frequency signals, such as tone bursts, pulse trains, and other unique systems of information transfer.

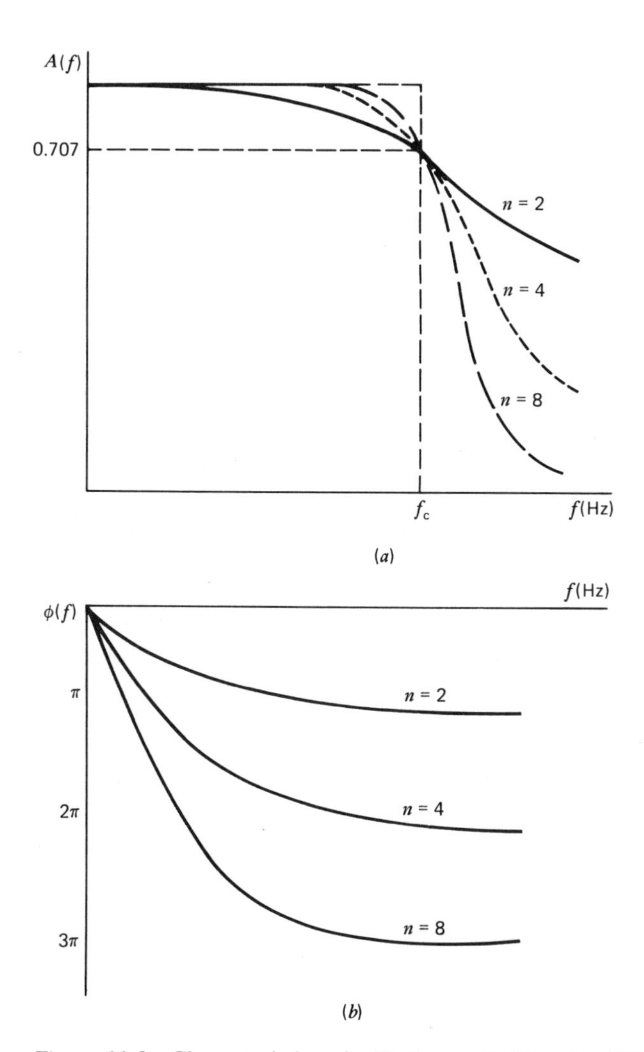

Figure 11.8 Characteristics of a Butterworth filter for different orders, *n*. (*a*) Amplitude response. (*b*) Phase response.

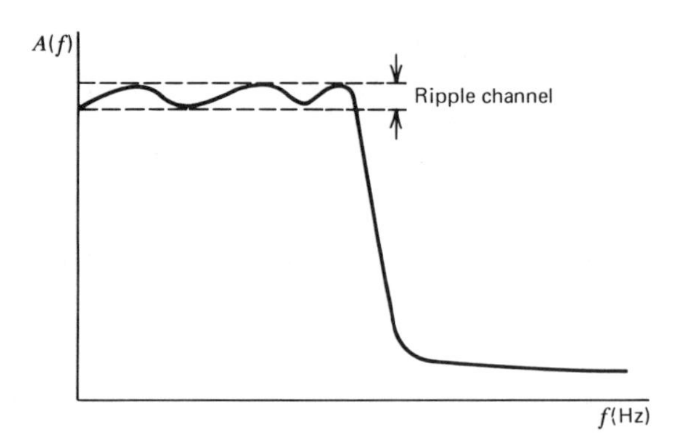

Figure 11.9 Typical amplitude response of a Chebyshev filter.

Time delay distortion is a nonlinear characteristic that results in various frequencies being phase shifted selectively or differentially. A Bessel filter, which also has constant delay, passes all frequencies below the cutoff frequency with equal time delay.

A constant time delay device is said to be *nondispersive*. An example of a nondispersive circuit is a wideband coaxial cable which has been carefully terminated and passes all signal frequencies *without delay distortion*. If delay distortion were present, the cable would be dispersive.

Negative Impedance Converter

A circuit that uses an op amp to reflect a mirror image from the output to the input is the *negative impedance converter* (NIC). Basically, the circuit is capable of converting a low-pass to a high-pass function, and vice versa, relative to the input and output ports of an NIC. If a high-pass network is connected to the input of the op amp (Fig. 11.10), there will be a low-pass impedance at the output. The circuit, therefore, yields a first-order, low-pass response with a rolloff of 6 dB/octave (20 dB/decade) above the break frequency. By paralleling a low-pass network from output to input (Fig. 11.11), a second-order filter is obtained. If the break frequencies for both networks are the same, the rolloff will be 12 dB/octave.

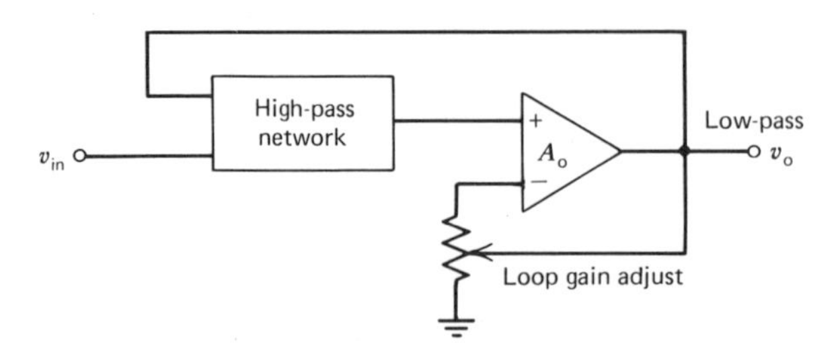

Figure 11.10 Using an NIC to realize a first-order, low-pass filter.

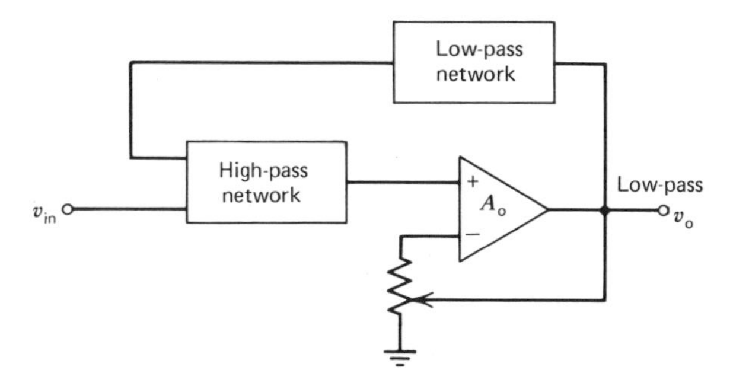

Figure 11.11 The addition of a low-pass network in the feedback path results in an NIC second-order, low-pass filter.

An adaptation of the NIC is the *gyrator* which rotates, or gyrates a capacitive impedance at the output of an NIC to an inductive impedance at the input. The NIC may then be used to simulate an inductor.

Because the NIC involves the combined use of positive and negative feedback, balancing the two effects is, therefore, critical to stability.

Multiple-Stage Filters

As mentioned previously, the sharpness of a filter's response is determined by its order, n. The higher the order, the sharper the transition region. By cascading low-order filters, a higher-order filter may be realized. Thus, if three second-order, low-pass filters are cascaded, all having the same cutoff frequency, a sixth-order filter is obtained (Fig. 11.12).

11.4 DESIGNING ACTIVE FILTERS

Complex mathematical forms may be used to derive filter characteristics and component values. A more straightforward procedure, however, is to make use of *design tables*. This will be the method used in this section.

Design Procedure

The following is an outline of the procedure for designing an active filter, such as a low pass, high pass, or bandpass, having a Chebyshev or Butterworth response characteristic.

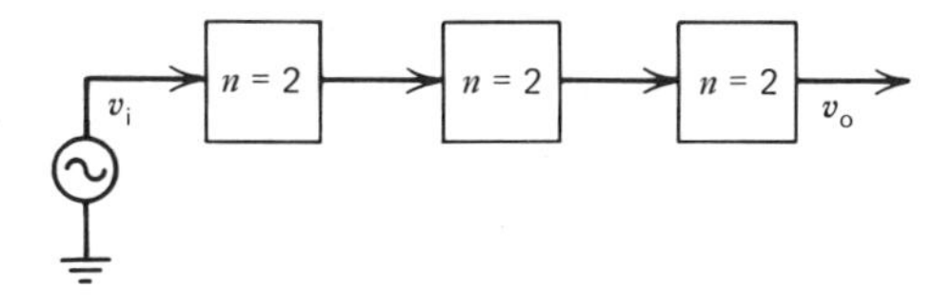

Figure 11.12 Cascading three second-order filters results in a sixth-order filter.

1 Define type filter: for example, low-pass, high-pass, and so on.
2 Select response: Butterworth, Chebyshev, and so on.
3 Specify cutoff frequency or bandwidth.
4 Specify order of filter: $n = 2, 4$, and so on.
5 Select the best circuit configuration for application.
6 Specify the total gain of filter. For cascaded stages, the total gain is equal to the product of the individual stage gains.
7 Select a value of C in microfarads. For the following frequency ranges, a general rule is:

100 Hz to 1 kHz	$C = 0.1\ \mu$F
1 kHz to 10 kHz	0.01 μF
10 kHz to 100 kHz	0.001 μF

8 Use an appropriate table to determine the values of the remaining components.

Filter Circuit Characteristics

Three commonly used circuit configurations for the realization of filters are:

1 Voltage Controlled Voltage Source (VCVS), also called the Sallen and Key filter: Provides noninverting gain which is adjustable.
2 Infinite Gain Multiple Feedback (MFB): Provides inverting gain with high stability and low output impedance.
3 Biquad: Highly stable with ease of tuning and good cascading. Exhibits a high Q.

EXAMPLE 11.1 Design a low-pass filter having a break frequency of 10 kHz and a gain of one. The circuit configuration to be used is the VCVS. To ensure fast rolloff, a fourth-order Chebyshev response is selected.

PROCEDURE

1 A dual op amp, such as the NE5532 which is internally compensated at unity gain, is chosen for the active element.
2 A multiplier constant, K, must now be computed. The expression for K is:

$$K = 100/(f_c C') \tag{11.3}$$

where f_c is the cutoff frequency and C' is in microfarad (μF).
3 Select $C = 0.01\ \mu$F. Substitution of values in Eq. 11.3 yields:

$$K = 100/(10^4 \times 0.01) = 1$$

4 A 0.5 dB ripple in the passband is acceptable. Referring to Table 11.1, we obtain:

Table 11.1 **Fourth-order low-pass Chebyshev cascaded VCVS filter design (0.5 dB)***

	Circuit Element Values[a]						
Gain	1	2	6	10	36	100	Stage
R_1	7.220	4.538	1.303	0.808	0.808	0.547	
R_2	12.219	0.525	1.828	2.948	1.474	2.177	
R_3	Open	10.126	4.696	4.695	2.738	3.027	1
R_4	0	10.126	9.393	18.781	13.692	27.240	
C_1	0.027C	C	C	C	2C	2C	
R_1	2.994	2.994	1.880	1.880	1.033	0.773	
R_2	5.050	0.050	3.781	3.781	3.440	4.596	
R_3	Open	Open	11.321	11.321	5.368	5.966	2
R_4	0	0	11.321	11.321	26.838	53.695	
C_1	0.47C	0.47C	C	C	2C	2C	

[a] Resistances in kilohms for a K parameter of 1.

Stage 1	*Use*
$R_1 = 7.220 \times 1 = 7.22\ \text{k}\Omega$	7.5 kΩ
$R_2 = 12.219 \times 1 = 12.219\ \text{k}\Omega$	12 kΩ
$R_3 = \text{Open}$	
$R_4 = 0$	
$C_1 = 0.027 \times 0.01\ \mu\text{F} = 270\ \text{pF}$	270 pF
Stage 2	
$R_1' = 2.994 \times 1 = 2.994\ \text{k}\Omega$	3 kΩ
$R_2' = 5.050 \times 1 = 5.05\ \text{k}\Omega$	5.1 kΩ
$R_3' = \text{open}$	
$R_4' = 0$	
$C_1' = 0.47 \times 0.01\ \mu\text{F} = 0.0047\ \mu\text{F}$	0.0047 μF

5 The resulting design is given in Fig. 11.13 and the measured frequency response of the filter is shown in Fig. 11.14.

EXAMPLE 11.2 Design a low-pass VCVS filter having a break frequency of 2 kHz and a fourth-order Butterworth response. An overall gain of 10 is required.

PROCEDURE

1 Let the gain of the first stage equal 10 and the gain of the second stage equal 1. Not only does the second stage act as a voltage follower with a low output impedance, but the best signal-to-noise ratio is achieved in the combined filter.

* Tables 11.1 to 11.9 are from *Rapid Practical Design of Active Filters,* D. Johnson and J. Hilburn.

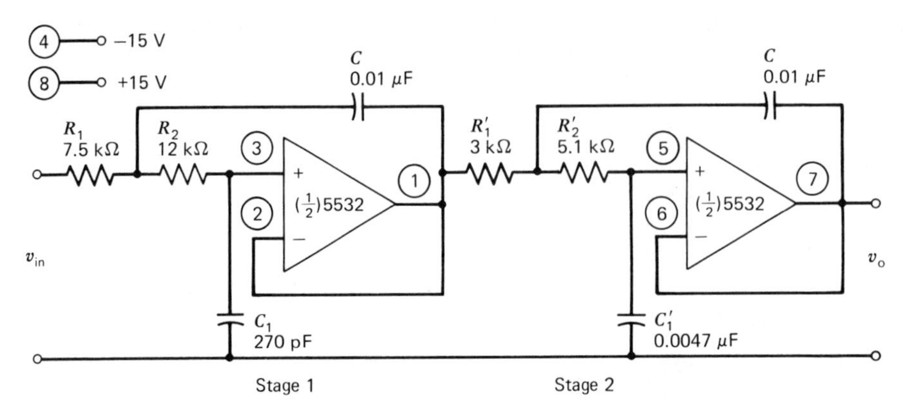

Figure 11.13 A fourth-order VCVS low-pass Chebyshev filter (Ex. 11.1).

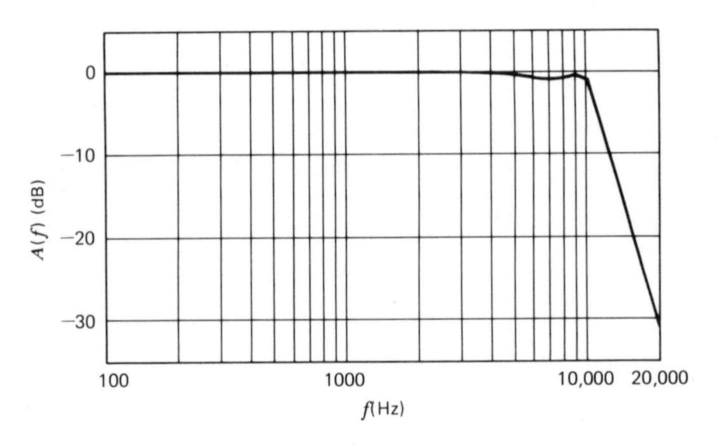

Figure 11.14 Measured frequency response of the low-pass Chebyshev filter of Fig. 11.13.

2 Selecting $C = 0.01\ \mu\text{F}$, by Eq. 11.3, gives us

$$K = 100/(2 \times 10^3 \times 0.01) = 5$$

3 Referring to Table 11.2, we obtain:

Stage 1 (gain = 10)	*Use*
$R_1 = 0.738 \times 5 = 3.69\ \text{k}\Omega$	3.6 kΩ
$R_2 = 3.432 \times 5 = 17.17\ \text{k}\Omega$	18 kΩ
$R_3 = 5.213 \times 5 = 26.06\ \text{k}\Omega$	27 kΩ
$R_4 = 20.851 \times 5 = 104.26\ \text{k}\Omega$	100 kΩ
$C_1 = 0.01\ \mu\text{F}$	0.01 μF
Stage 2 (gain = 1)	
$R_1' = 1.048 \times 5 = 5.24\ \text{k}\Omega$	5.1 kΩ

Table 11.2 **Fourth-order low-pass Butterworth cascaded VCVS filter designs**

	Circuit Element Values[a]						
Gain	1	2	6	10	36	100	Stage
R_1	2.661	2.079	1.095	0.738	0.738	0.521	
R_2	9.521	1.218	2.313	3.432	1.716	2.432	
R_3	Open	6.595	5.112	5.213	2.945	3.281	1
R_4	0	6.595	10.225	20.851	14.725	29.527	
C_1	0.1C	C	C	C	2C	2C	
R_1	1.048	1.048	0.861	0.861	0.551	0.427	
R_2	4.833	4.833	2.941	2.941	2.297	2.965	
R_3	Open	Open	7.604	7.604	3.418	3.769	2
R_4	0	0	7.604	7.604	17.092	33.924	
C_1	0.5C	0.5C	C	C	2C	2C	

[a] Resistances in kilohms for a K parameter of 1.

$R_2' = 4.833 \times 5 = 24.16\ \text{k}\Omega$ 24 kΩ

$R_3' = \text{Open}$

$R_4' = 0$

$C_1' = 0.5 \times 0.01\ \mu\text{F} = 0.005\ \mu\text{F}$ 0.005 μF

4 Using a dual op amp (NE5512), we show the final cascaded circuit in Fig. 11.15. The response of the filter is given in Fig. 11.16.

EXAMPLE 11.3 Design a fourth-order, high-pass filter having a gain of six and a break frequency of 3 kHz. Assume a VCVS configuration and a 1 dB ripple Chebyshev response.

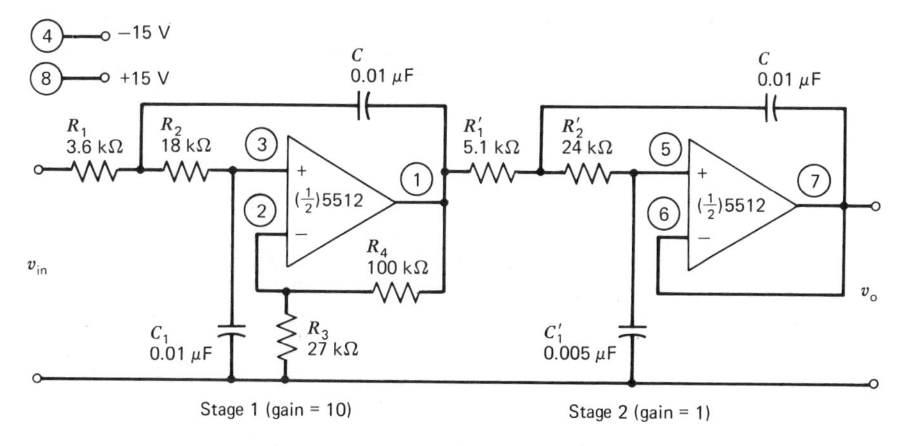

Figure 11.15 A fourth-order VCVS low-pass Butterworth filter (Ex. 11.2).

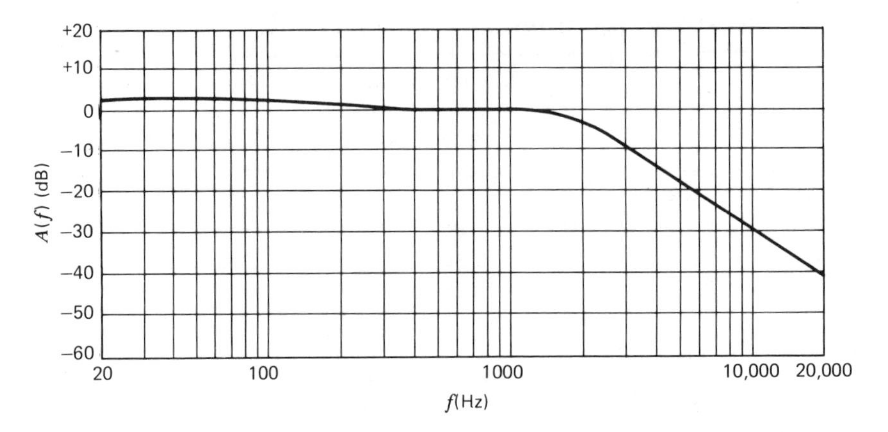

Figure 11.16 Measured frequency response of low-pass Butterworth filter of Fig. 11.15.

PROCEDURE

1 When we select $C = 0.01\ \mu F$, by Eq. 11.3,

$$K = 100/(3 \times 10^3 \times 0.01) = 3.33$$

2 Referring to Table 11.3, we obtain:

Stage 1	*Use*
$R_1 = 1.461 \times 3.33 = 4.86\ k\Omega$	4.7 kΩ
$R_2 = 1.710 \times 3.33 = 5.69\ k\Omega$	5.6 kΩ
$R_3 = 2.89 \times 3.33 = 9.62\ k\Omega$	10 kΩ
$R_4 = 4.189 \times 3.33 = 13.9\ k\Omega$	13 kΩ

Table 11.3 **Fourth-order high-pass Chebyshev cascaded VCVS filter designs (1 dB)**

	Circuit Element Values[a]						
Gain	1	2	6	10	36	100	Stage
R_1	0.222	0.839	1.461	1.758	2.613	3.466	1
R_2	11.252	2.979	1.710	1.421	0.956	0.721	
R_3	Open	10.170	2.890	2.078	1.148	0.801	
R_4	0	4.212	4.189	4.494	5.738	7.209	
R_1	0.536	0.735	1.033	1.183	1.625	2.073	2
R_2	1.320	0.962	0.685	0.598	0.436	0.341	
R_3	Open	3.286	1.158	0.875	0.523	0.379	
R_4	0	1.361	1.679	1.892	2.613	3.415	

[a] Resistances in kilohms for a K parameter of 1.

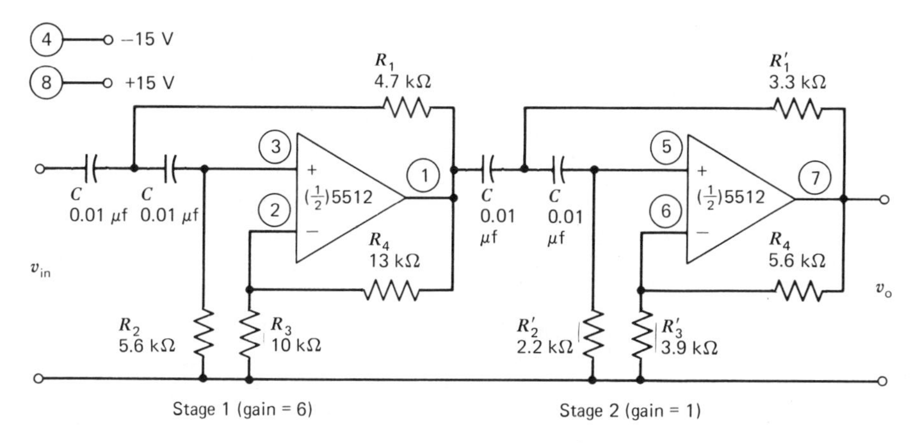

Figure 11.17 A fourth-order VCVS high-pass Chebyshev filter (Ex. 11.3).

Stage 2

$R_1' = 1.033 \times 3.33 = 3.44$ kΩ	3.3 kΩ
$R_2' = 0.685 \times 3.33 = 2.28$ kΩ	2.2 kΩ
$R_3' = 1.15 \times 3.33 = 3.83$ kΩ	3.9 kΩ
$R_4' = 1.679 \times 3.33 = 5.59$ kΩ	5.6 kΩ

3 The circuit is shown in Fig. 11.17 and the response is plotted in Fig. 11.18.

The response of a high-pass, fourth-order Butterworth filter is illustrated in Fig. 11.19. Note that the rolloff is not as sharp as that for the Chebyshev response of Fig. 11.18.

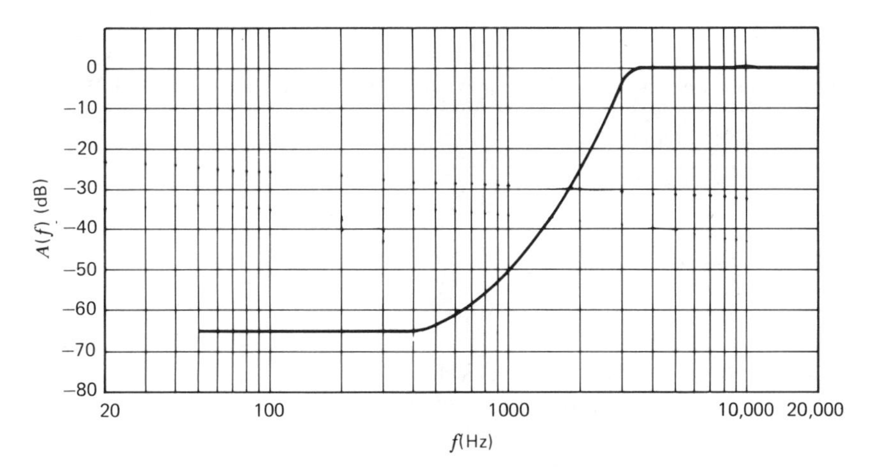

Figure 11.18 Measured frequency response of high-pass Chebyshev filter of Fig. 11.17.

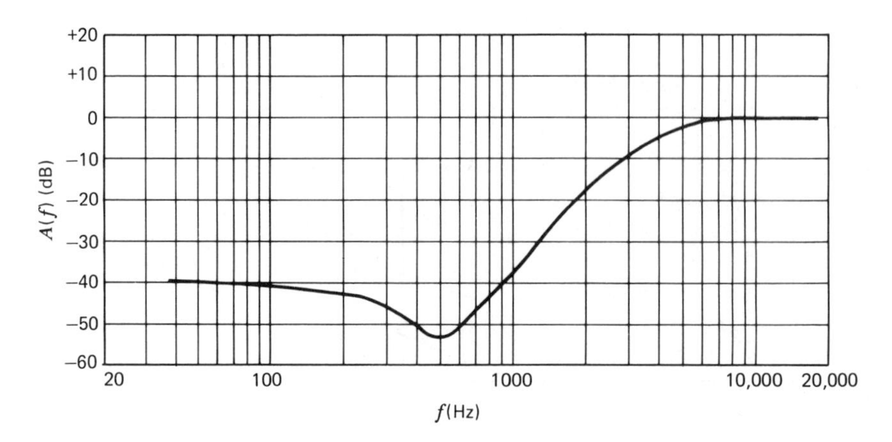

Figure 11.19 Measured frequency response of a fourth-order, high-pass Butterworth filter. Note that the rolloff for this filter type is not as steep as the rolloff for the Chebyshev of Fig. 11.18.

EXAMPLE 11.4 Design a high-pass filter having a break frequency of 1 kHz. A fourth-order response, a gain of 25, low output impedance, and good stability are desired. A Chebyshev response with 0.5 dB ripple is specified.

PROCEDURE

1 To meet the requirements of low output impedance and good stability, an MFB configuration is selected.
2 Select $C = 0.01\ \mu F$. By Eq. 11.3,

$$K = 100/(10^3 \times 0.01) = 10$$

Table 11.4 **Fourth-order high-pass Chebyshev cascaded MFB filter designs (0.5 dB)**

	Circuit Element Values[a]				
Gain	1	4	25	100	Stage
R_1	0.186	0.223	0.254	0.266	
R_2	14.479	24.132	53.090	101.354	1
C_1	C	0.5C	0.2C	0.1C	
R_1	0.449	0.539	0.613	0.642	
R_2	2.010	3.350	7.370	14.069	2
C_1	C	0.5C	0.2C	0.1C	

[a] Resistances in kilohms for a K parameter of 1.

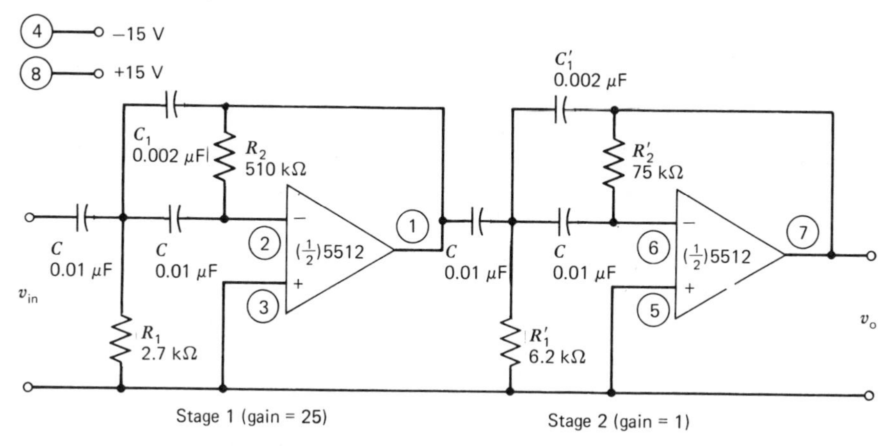

Figure 11.20 A fourth-order MBF high-pass Chebyshev filter (Ex. 11.4).

3 Referring to Table 11.4, we obtain:

Stage 1	*Use*
$R_1 = 0.254 \times 10 = 2.54\ \text{k}\Omega$	2.7 kΩ
$R_2 = 53.09 \times 10 = 531\ \text{k}\Omega$	510 kΩ
$C_1 = 0.2 \times 0.01\ \mu\text{F} = 0.002\ \mu\text{F}$	0.002 μF

Stage 2	*Use*
$R'_1 = 0.613 \times 10 = 6.13\ \text{k}\Omega$	6.2 kΩ
$R'_2 = 7.37 \times 10 = 73.7\ \text{k}\Omega$	75 kΩ
$C'_1 = 0.2 \times 0.01\ \mu\text{F} = 0.002\ \mu\text{F}$	0.002 μF

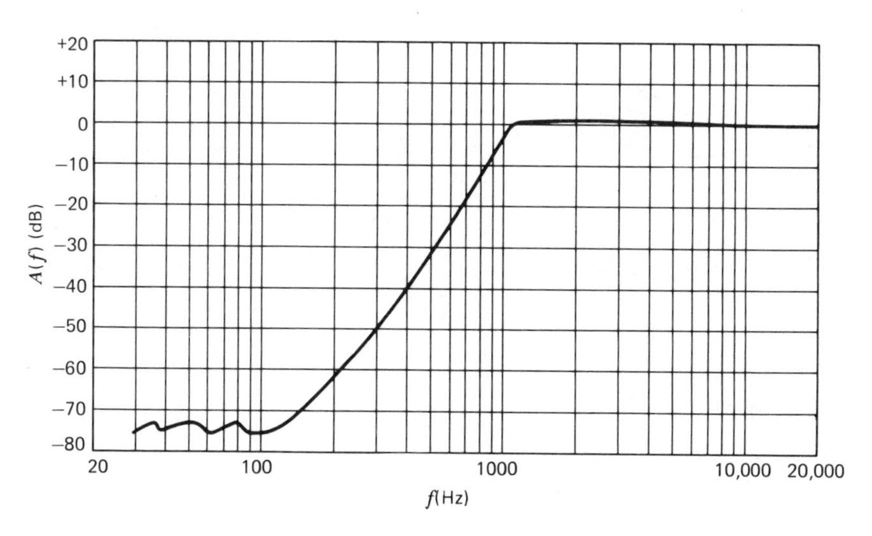

Figure 11.21 Measured frequency response of fourth-order, high-pass MBF filter of Fig. 11.20.

Table 11.5 **Fourth-order bandpass Chebyshev cascaded biquad filter designs (1 dB)**

	Circuit Element Values[a]									
Q	2	4	6	8	10	20	30	40	50	Stage
R_1	$3.032/G$	$6.063/G$	$9.095/G$	$12.126/G$	$15.158/G$	$30.315/G$	$45.473/G$	$60.630/G$	$75.788/G$	
R_2	4.753	10.436	16.192	21.970	27.756	56.725	85.712	114.705	143.699	1
R_3, R_4	1.017	1.272	1.371	1.423	1.455	1.522	1.545	1.556	1.563	
R_1	$3.032/G$	$6.063/G$	$9.095/G$	$12.126/G$	$15.158/G$	$30.315/G$	$45.473/G$	$60.630/G$	$75.788/G$	
R_2	7.438	13.054	18.798	24.571	30.355	59.321	88.308	117.300	146.295	2
R_3, R_4	2.491	1.991	1.848	1.780	1.741	1.664	1.640	1.628	1.620	

[a] Resistances in kilohms for a K parameter of 1. G = stage gain at the center frequency of the filter. The overall gain is the product of the stage gains. R_4 = 1.592 in all cases.

4 The two-stage filter is shown in Fig. 11.20 and the response curve is provided in Fig. 11.21. Note that the stop band is down 80 dB from the passband response.

EXAMPLE 11.5 Design a bandpass filter with the following characteristics: center frequency $f_0 = 440$ Hz, minimum $Q = 30$, Chebyshev response with 1 dB ripple, and an overall gain of 10.

PROCEDURE

1 By cascading two second-order Biquad bandpass filters a relatively high Q response may be obtained with inexpensive components.

2 Select $C = 0.1$ μF. The value of K is calculated by:

$$K = 100/(f_0C') \tag{11.4}$$

$$= 100/(440 \times 0.1) = 2.3$$

3 The gain of the first stage will be 10 and that of the second stage, 1. Referring to Table 11.5, we obtain,

Stage 1	*Use*
$R_1 = 45.473 \times 2.3/10 = 10.5$ kΩ	10 kΩ
$R_2 = 85.712 \times 2.3 = 198$ kΩ	200 kΩ
$R_3 = 1.545 \times 2.3 = 3.7$ kΩ	3.6 kΩ
$R_4 = 1.592 \times 2.3 = 3.7$ kΩ	3.6 kΩ

Stage 2	*Use*
$R_1' = 45.473 \times 2.3/1 = 105$ kΩ	100 kΩ
$R_2' = 88.308 \times 2.3 = 203$ kΩ	200 kΩ
$R_3' = 1.640 \times 2.3 = 3.77$ kΩ	3.6 kΩ
$R_4' = 1.592 \times 2.3 = 3.7$ kΩ	3.6 kΩ

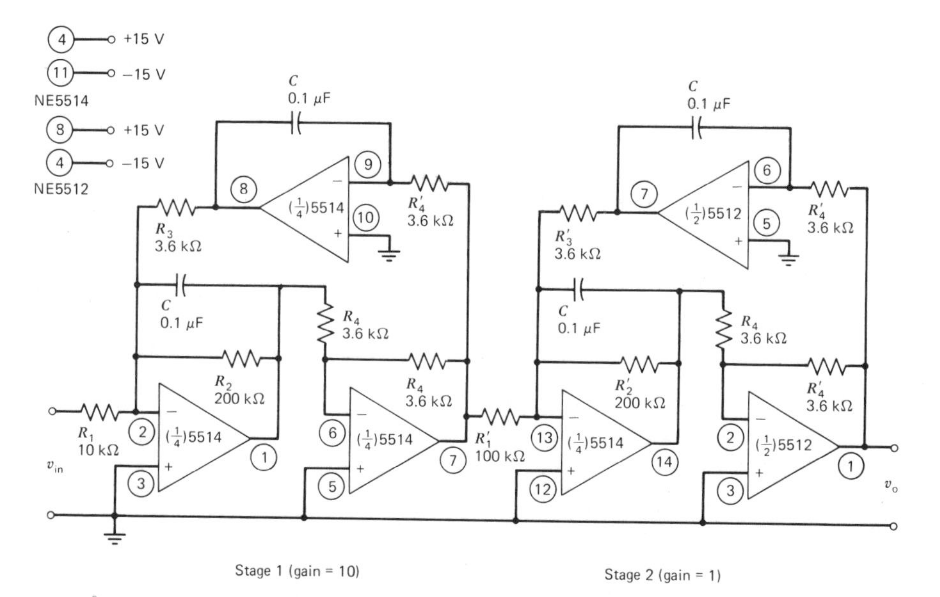

Figure 11.22 A fourth-order Biquad bandpass Chebyshev filter (Ex. 11.5).

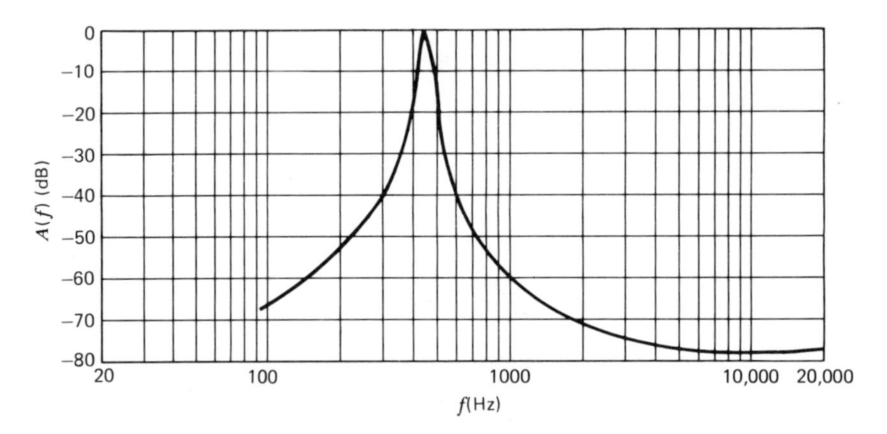

Figure 11.23 Measured frequency response of fourth-order bandpass Chevyshev filter of Fig. 11.22.

4 Referring to Fig. 11.22, we see that the first stage incorporates a quad op amp (NE5514) with the remaining stage combined with a dual NE5512 op amp for the second stage. The overall response curve is given in Fig. 11.23.

5 Voltage gain may be adjusted by varying R_1 and Q may be adjusted by varying R_2. Resistor R_3 affects the center frequency.

EXAMPLE 11.6 Design a fourth-order bandpass filter having a center frequency of 10 kHz, a gain of 100, and a Q of 10.

PROCEDURE

1 Significantly simpler than the Biquad, the VCVS bandpass filter is suited for a Q of ten or less.

2 For the first stage, $C = 0.01\ \mu\text{F}$ is selected. By Eq. 11.4,

$$K = 100/(10^4 \times 0.01) = 1$$

3 Each stage will have a gain of 10. Referring to Table 11.6, we obtain:

Table 11.6 **Second-order VCVS bandpass filter designs (Q = 10)**

	Circuit Element Values[a]					
Gain	1	2	4	6	8	10
R_1	31.831	15.915	7.958	5.305	3.979	3.183
R_2	2.251	2.332	2.502	2.684	2.876	3.078
R_3	1.167	1.166	1.160	1.148	1.131	1.110
R_4, R_5	4.502	4.664	5.004	5.368	5.752	6.156

[a] Resistances in kilohms for a K parameter of 1.

Stage 1	*Use*
$R_1 = 3.183 \times 1 = 3.183$ kΩ	3 kΩ
$R_2 = 3.078 \times 1 = 3.078$ kΩ	3 kΩ
$R_3 = 1.110 \times 1 = 1.11$ kΩ	1.1 kΩ
$R_4 = R_5 = 6.156 \times 1 = 6.156$ kΩ	6.2 kΩ

4 For the second stage, $C' = 0.0047$ μF is selected. By Eq. 11.4,

$$K = 100/(10^4 \times 0.0047) = 2.13$$

Therefore, for the second stage, we obtain:

Stage 2	*Use*
$R'_1 = 3.183 \times 2.13 = 6.8$ kΩ	6.8 kΩ
$R'_2 = 3.078 \times 2.13 = 6.6$ kΩ	6.8 kΩ
$R'_3 = 1.11 \times 2.13 = 2.36$ kΩ	2.4 kΩ
$R'_4 = R'_5 = 6.156 \times 2.13 = 13.1$ kΩ	13 kΩ

5 The resulting filter is shown in Fig. 11.24 and its response curve in Fig. 11.25. The combined Q for the two-stage filter may be calculated as follows:

$$Q = 1.55Q_1$$

where Q_1 is the single stage Q. Therefore, the overall Q is increased by cascading identical filters.

EXAMPLE 11.7 Design a second-order VCVS band-reject filter. The given specifications are: $f_o = 1$ kHz, $Q = 30$, and the gain is equal to unity.

PROCEDURE

1 Select $C = 0.001$ μF. The value of K, from Eq. 11.4, is:

$$K = 100/(10^3 \times 0.001) = 100$$

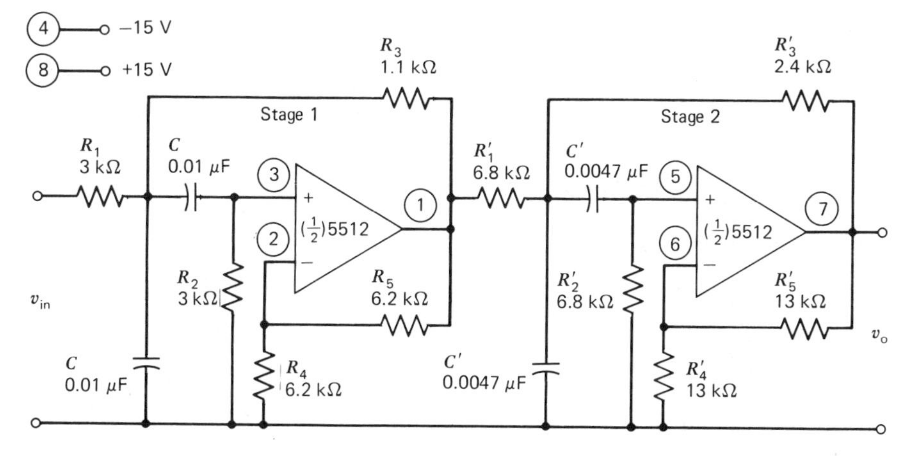

Figure 11.24 A fourth-order VCVS bandpass filter (Ex. 11.6).

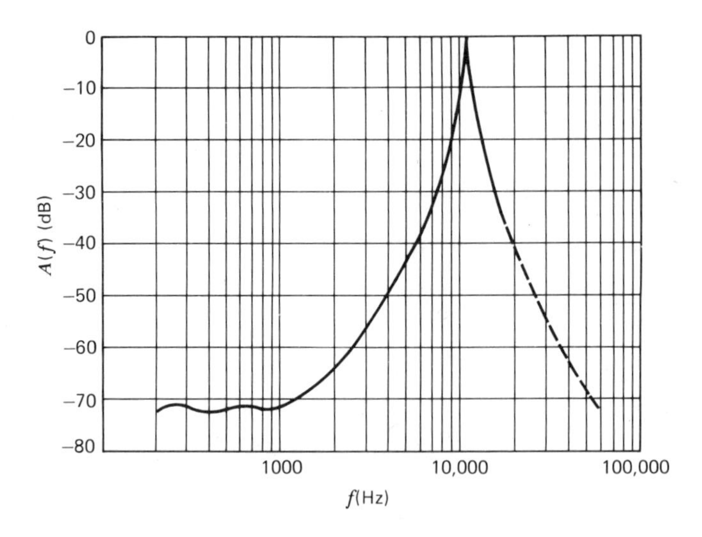

Figure 11.25 Measured frequency response curve of bandpass filter of Fig. 11.24.

2 Referring to Table 11.7, we obtain:

	Use
$R_1 = 0.796 \times 100/30 = 2.65\ \text{k}\Omega$	2.7 kΩ
$R_2 = 3.183 \times 100 \times 30 = 9.6\ \text{M}\Omega$	10 MΩ
$R_3 = (9.6\ \text{M})/(4 \times 30^2 + 1) = 2.66\ \text{k}\Omega$	2.7 kΩ

3 The final circuit using an NE535 is drawn in Fig. 11.26 and its response curve is given in Fig. 11.27. By trimming the value of $2C$ for a null, the output can be minimized at the notch frequency and a maximum attenuation of 54 dB attained. The final value of $2C$ was found to be close to 0.0024 μF.

Second-order Biquad Band-Reject Filter

The Biquad band-reject filter has several advantages over the VCVS design of Ex. 11.7. For one, gain is adjustable over a limited range. In addition, the center

***Table 11.7* Second-order VCVS band-reject filter designs**

Circuit Element Values[a]	
R_1	$0.796/Q$
R_2	$3.183Q$
R_3	$R_2/(4Q^2 + 1)$

[a] Resistances in kilohms for a K parameter of 1, gain is 1, quality factor is Q.

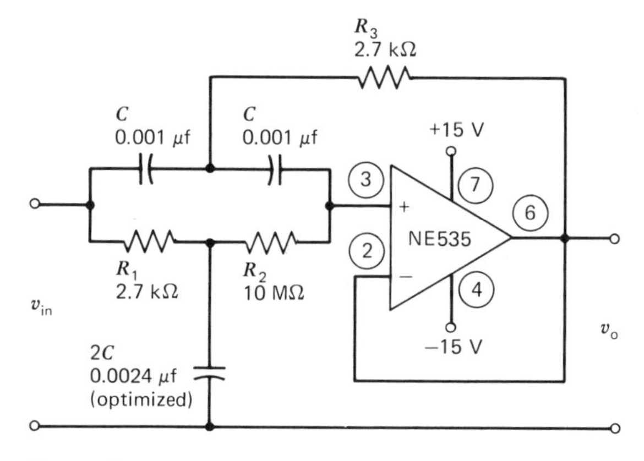

Figure 11.26 A second-order VCVS band-reject filter (Ex. 11.7).

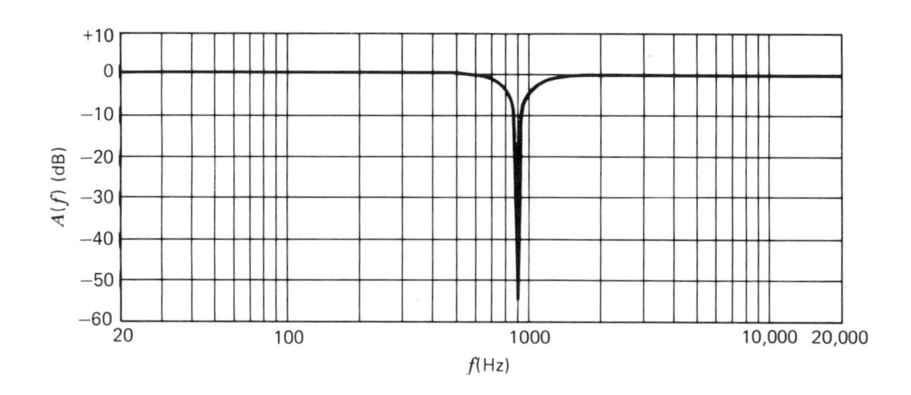

Figure 11.27 Measured frequency response curve of band-reject filter of Fig. 11.26.

Table 11.8 **Second-order biquad band-reject filter designs**

Circuit Element Values[a]	
R_1	$1.592Q/G$
R_2	$1.592Q$
R_3, R_4, R_5	1.592
R_6, R_7	$1.592/G$

[a] Resistances in kilohms for a *K* parameter of 1, *Q* and *G* are quality factor and gain.

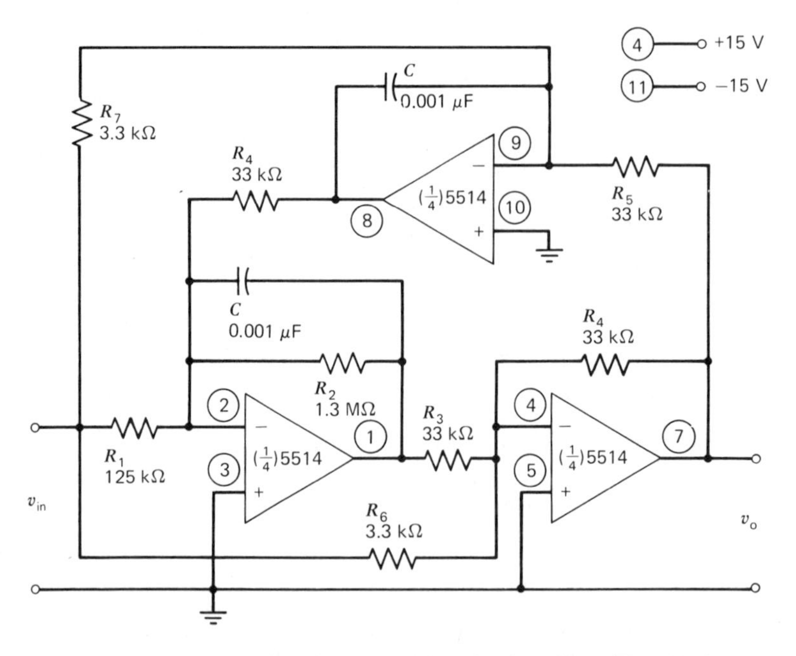

Figure 11.28 A second-order Biquad band-reject filter (Ex. 11.8).

frequency and Q may be adjusted, which is important in tracking two stages of a fourth-order filter. It also is capable of a much higher Q than a VCVS design.

EXAMPLE 11.8 Design a second-order, Biquad band-reject filter having a 5 kHz center frequency. The $Q = 40$ and the filter has a Chebyshev 1 dB ripple response; gain = 10.

PROCEDURE

1 We select $C = 0.001\ \mu\text{F}$. By Eq. 11.4:

$$K = 100/(5 \times 10^3 \times 0.001) = 20$$

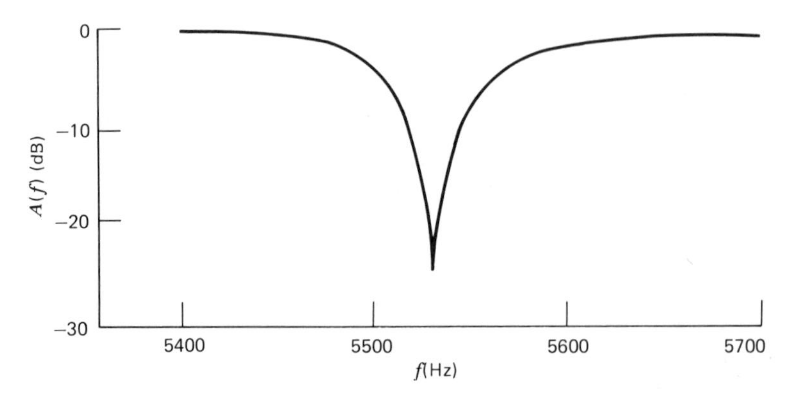

Figure 11.29 Measured frequency response of band-reject filter of Fig. 11.28.

2 Referring to Table 11.8, we obtain:

Single Stage	*Use*
R_1 = 1.592 × 20 × 40/10 = 127.4 kΩ	125 kΩ
R_2 = 1.592 × 20 × 40 = 1273.6 kΩ	1.3 MΩ
R_3 = R_4 = R_5 = 1.592 × 20 = 31.84 kΩ	33 kΩ
R_6 = R_7 = 1.592 × 20/10 = 3.184 kΩ	3.3 kΩ

3 The resulting filter is shown in Fig. 11.28 where a quad op amp, such as a NE5514, is used. The response curve is provided in Fig. 11.29. The measured Q = 78 and the notch depth = 24 dB.

State Variable Filter

A unique circuit, known as a *state variable filter,* provides low-pass, bandpass, and high-pass functions simultaneously. An example of such a filter is illustrated in Fig. 11.30. The center frequency, f_o, is determined by R_3:

$$f_o = \frac{0.159}{RC}\sqrt{\frac{R_{3a}}{R_{3b}}} \tag{11.5a}$$

If $R_{3a} = R_{3b}$, then

$$f_o = \frac{0.159}{RC} \tag{11.5b}$$

For the given values of C = 0.01 μF and R = 16 kΩ in Fig. 11.30, f_o = 1 kHz.

The responses of the state variable filter for low-pass, bandpass, and high-pass are provided in Fig. 11.31*a*, *b*, and *c*, respectively.

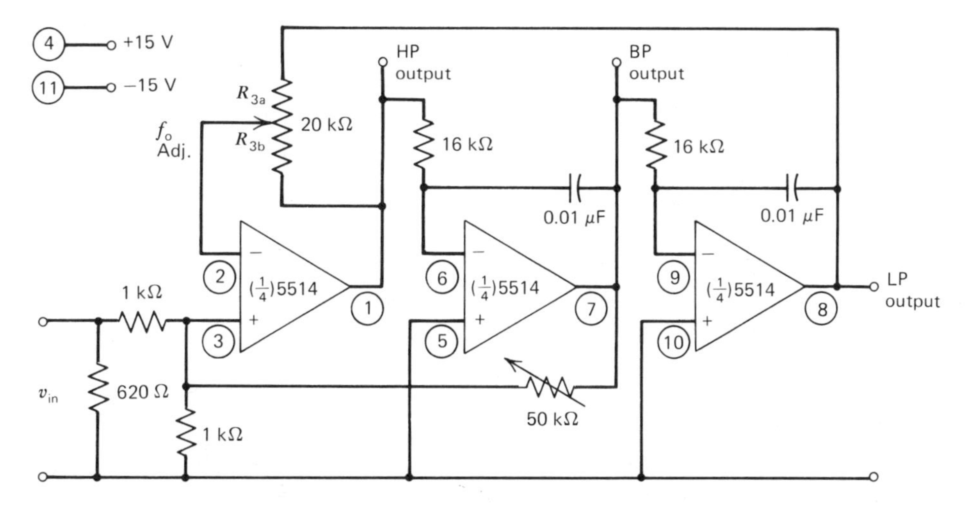

Figure 11.30 An example of a state variable filter.

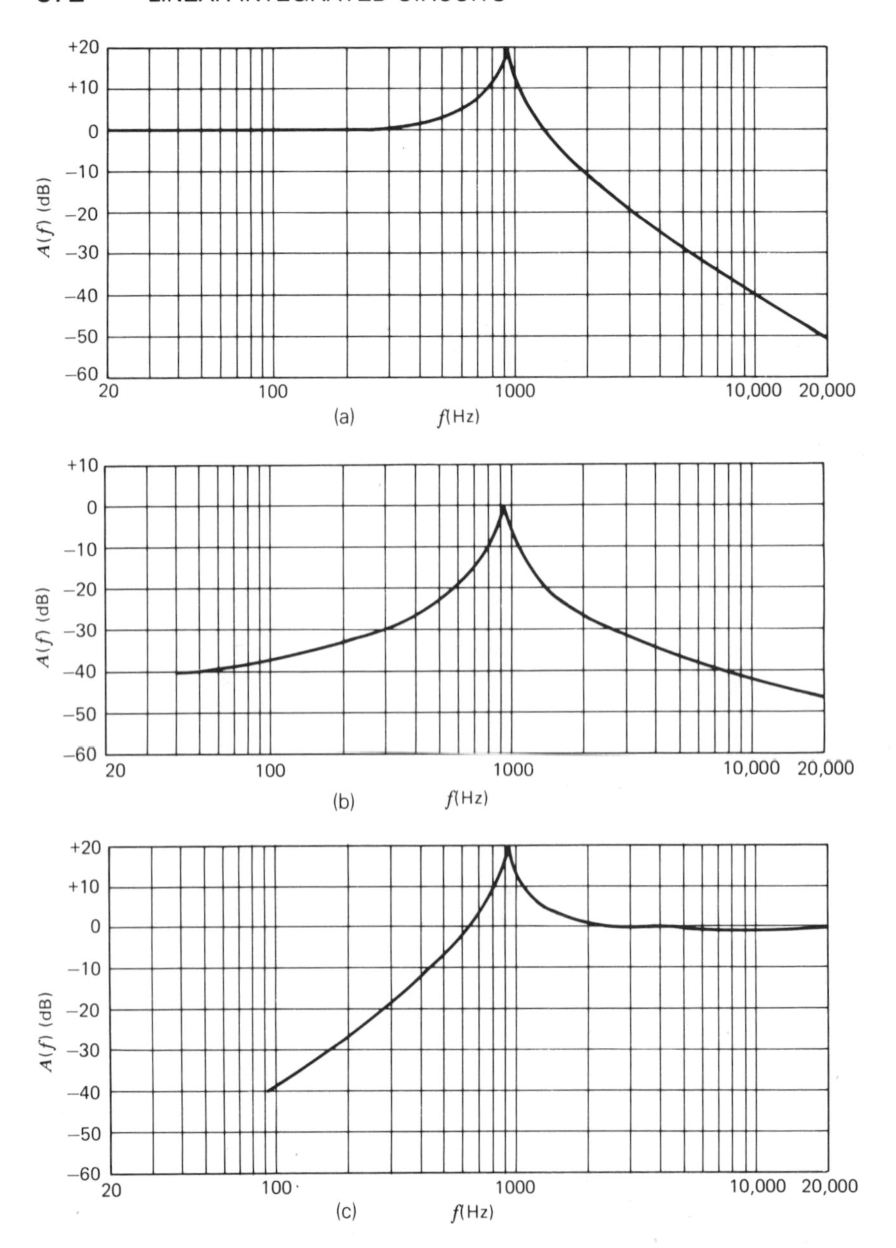

Figure 11.31 Frequency response curves of state variable filter of Fig. 11.30. (*a*) Low pass. (*b*) Bandpass. (*c*) High pass.

Time Delay Filter

The constant time delay (all-pass) filter considered in this section is configured from a Biquad circuit using all four op amps in an NE5514 package. As mentioned earlier, constant delay filters exhibit a uniform time delay, T_d, over a range of frequencies from zero to the cutoff frequency, f_c:

$$f_c = 0.159/T_d \tag{11.6}$$

Time delay T_d is the rate of change in phase with respect to frequency and is a constant.

EXAMPLE 11.9 Design an all-pass filter for a cutoff frequency of 10 kHz. The phase shift at 10 kHz = $-90°$ and T_d = 15 μs for $f << 10$ kHz.

PROCEDURE

1 Select C = 0.001 μF. By Eq. 11.4,

$$K = 100/(10^4 \times 0.001) = 10.$$

2 Referring to Table 11.9, we obtain:

Component Values	*Use*
$R_1 = R_2 = R_3 = 0.984 \times 10 = 9.84$ kΩ	10 kΩ
$R_4 = 0.492 \times 10 = 4.92$ kΩ	4.7 kΩ
$R_5 = R_6 = 1.592 \times 10 = 15.92$ kΩ	16 kΩ

3 The circuit is given in Fig. 11.32 and the resulting time delay waveforms are shown in Fig. 11.33.

11.5 CONSTRUCTION TECHNIQUES

Methods used to construct active filter circuits are similar to the general good practices employed in other linear amplifier circuits. The following recommendations, however, should prove to be useful.

Power Supply

When single supply operation is required, the main consideration will be the biasing of the noninverting input to the op amp. If the filter is also noninverting, the divider resistor values in the bias circuit must be sufficiently high so as not

Table 11.9 **All-pass (phase-shift) biquad filter designs ($-65°$ to $-90°$)**

	Circuit Element Values[a]					
Shift	$-65°$	$-70°$	$-75°$	$-80°$	$-85°$	$-90°$
R_1, R_2, R_3	0.774	0.819	0.863	0.904	0.945	0.984
R_4	0.387	0.410	0.431	0.452	0.472	0.492
R_5, R_6	1.592	1.592	1.592	1.592	1.592	1.592

[a] Resistances in kilohms for a K parameter of 1, gain is 1 (inverting).

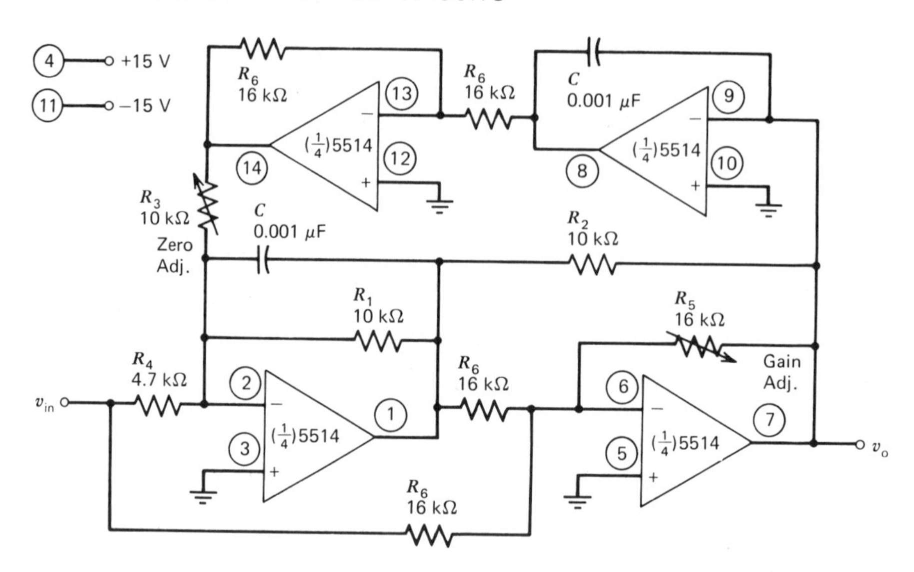

Figure 11.32 Biquad all-pass filter (Ex. 11.9).

to load the filter elements. Also, capacitive coupling will be required to prevent dc flow to or from the source. Special testing may then be necessary to ensure that filter degradation does not occur.

For example, consider the second-order VCVS filter of Fig. 11.34 operating from a single 15 V supply. With proper biasing as shown in the figure, the common mode voltage at the output is one-half the supply, or 7.5 V. The output signal voltage will now swing a maximum of +2 to +13 V before limiting occurs.

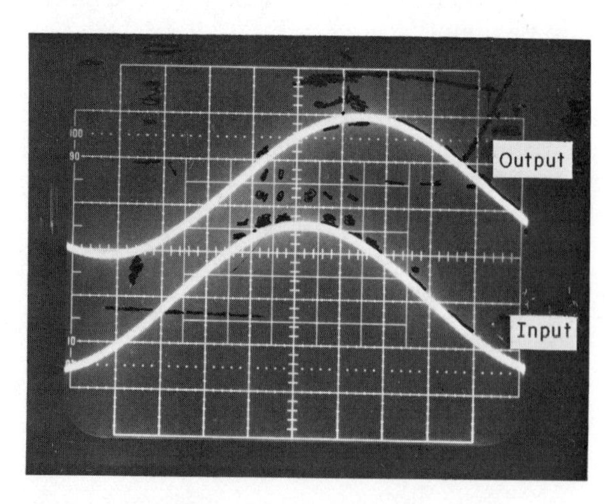

Figure 11.33 Response of all-pass filter of Fig. 11.32. The delay between input and output waveforms is 15 μs. The calibration for the time axis is 10 μs/cm and 0.5 V/cm for the voltage axis.

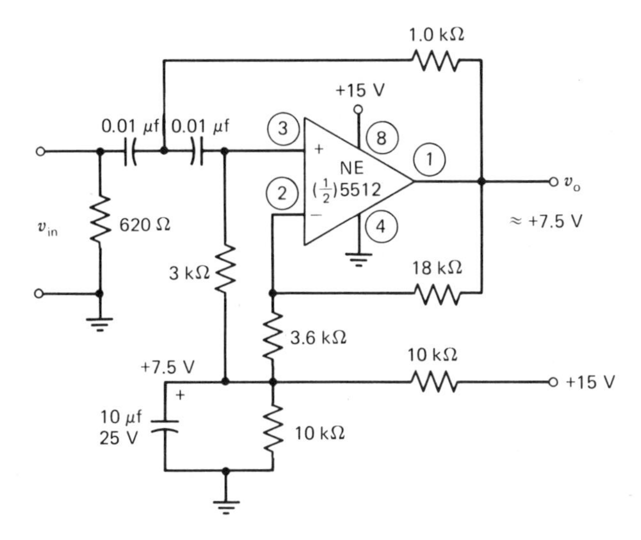

Figure 11.34 Example of a second-order VCVS filter operating from a single 15 V power supply.

When dual supplies are available the situation is quite simplified. The major considerations are power supply regulation and current requirements. Common mode ripple and noise within the supply rejection specification of the op amp will be greatly attenuated.

For active filters with gain and large signal inputs, the maximum undistorted output signal swing may be an important consideration. Here again, the op amp specification sheet will be the guide to follow in evaluating the maximum swing capability.

Power Supply Precautions

Power supply lines should be carefully decoupled, especially in wideband filter circuits. Bypass capacitors must be ceramic to reject high frequencies and wideband noise, such as interference from nearby pulse sources. In the event that the overall gain of several stages is high, then circuit layout and ground return lead distribution must be considered. In all cases, the first choice for good noise rejection and stability is the ground plane printed circuit board.

Noise

Noise at the output of an active filter is a function of several variables. First, there is the wideband and 1/f noise generated in the input stage of an op amp. The total root mean square noise from this source is a function of the noise gain and noise bandwidth of the amplifier. Noise bandwidth, based on a square response function and fitted to a standard first-order filter which rolls off at 6 dB/octave, will be 1.57 times the 3 dB bandwidth. As the filter order increases, the noise bandwidth will decrease until it finally approaches the 3 dB bandwidth of the filter.

Other sources of noise may arise from adjacent pulse circuits, Zener diodes, the supply line, and from resistors in the circuit. The latter noise may be minimized by the use of carbon film or metal film resistors.

11.6 REFERENCES

Burr-Brown, *Handbook of Operational Amplifier Active RC Circuits*, Burr-Brown Research Corporation, Tucson, 1966.

Irvine, R. G., *Operational Amplifier Characteristics and Applications*, Prentice-Hall, Inc., Englewood Cliffs, NJ, 1981.

Johnson, D. E. and J. L. Hilburn, *Rapid Practical Designs of Active Filters*, Wiley, New York, 1975.

Roberge, J. K., *Operational Amplifiers Theory and Practice*, Wiley, New York, 1975.

Signetics, *Analog Applications Manual*, Signetics Corp., Sunnyvale, CA, 1979.

Signetics, *Analog Data Manual*, Signetics Corp., Sunnyvale, CA, 1981.

Stout, D. F. and M. Kaufman, *Handbook of Operational Amplifier Circuit Design*, McGraw-Hill, New York, 1976.

Williams, A. B., *Electronic Filter Design Handbook*, McGraw-Hill, New York, 1981.

CHAPTER 12

Waveform Generation

David L. Terrell
Evening Dean
ITT Technical Institute

John M. Tondra
Development Engineer
Cryogenic Associates

12.1 INTRODUCTION

The primary objective of this chapter is to provide the reader with circuit techniques used for the generation of a variety of waveforms. Most of the circuits presented are quite simple in operation and design and can be used for many waveform generator applications. If your particular application is unique, or extremely critical in some way, then you may still glean some practical conceptual ideas from reading the chapter that can be adapted to a specific design.

12.2 WIEN BRIDGE OSCILLATOR

Figure 12.1 shows a Wien bridge sine wave oscillator. The R_2C_2 input and R_1C_1 feedback networks attenuate the zero-phase shift oscillation frequency by a factor of three. The amplifier gain must, therefore, be equal to three to realize the unity gain positive feedback required for sustained oscillations. A tungsten lamp may be used as a variable resistor to ensure stable operation. As current through the filament changes, the resistance varies directly with current because tungsten has a positive temperature coefficient.

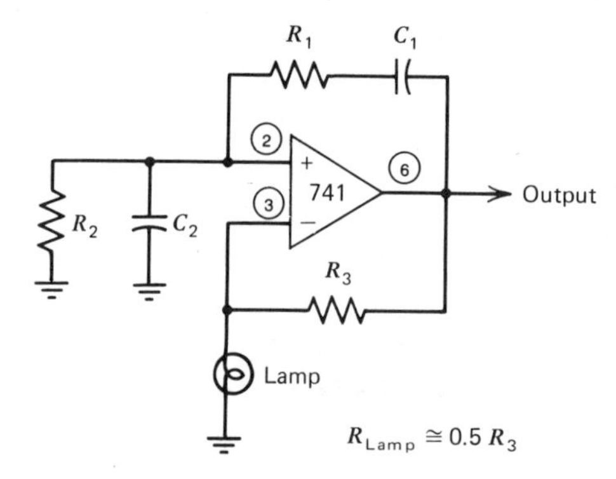

Figure 12.1 Basic circuit for a Wien bridge sine wave oscillator. A tungsten lamp is employed to ensure stable operation.

The circuit as shown is classic, but tends to be nonrepeatable in production quantities owing to the inconsistencies between various lamps. Figure 12.2 shows a similar circuit that employs the dynamic impedance of a p-channel FET (Q_1) to provide the trimming resistance needed to maintain the circuit forward gain at three.

Important design equations for the Wien bridge oscillator are:

$$R_1 = R_2 = R \tag{12.1a}$$

$$C_1 = C_2 = C \tag{12.1b}$$

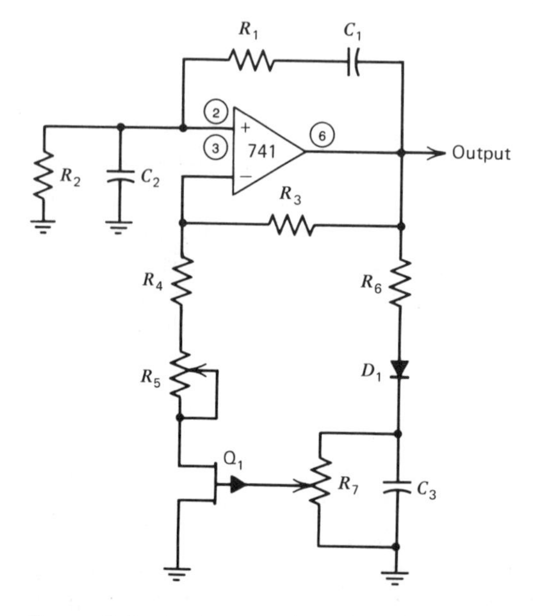

Figure 12.2 An improved version of a Wien bridge sine wave oscillator using a p-channel FET, in place of a tungsten lamp.

Frequency of oscillation, f, is:

$$f = 0.159/RC \tag{12.1c}$$

The value of capacitance C is selected.

EXAMPLE 12.1 Design an oscillator that will generate a low-distortion sine wave at a frequency of 9 kHz.

PROCEDURE

1 The circuit of Fig. 12.2 will be used.
2 From Eq. 12.1c,

$$RC = 0.159/f$$
$$= 0.159/9000 = 17.7\ \mu s$$

A reasonable value for C is 1000 pF. Hence, $R = (17.7\ \mu s)/(1000\ pF) = 17.7\ k\Omega$.
3 Resistors R_4 and R_5 should have a combined resistance of approximately $0.5R_3 - 1000\ \Omega$.
4 Optimum values for R_6, R_7, and D_1 depend upon the particular FET chosen for Q_1. These components sample the output and allow trimming of the feedback to Q_1. One set of typical components for a 2N5020 FET are: $R_6 = R_7 = 1\ M\Omega$. Diode D_1 is a 1N914A.

Practical Considerations

a. Resistor R_5 does not have to be a potentiometer; but a pot allows for variations between FETs.
b. Capacitor C_3 filters the rectified feedback signal and should be selected to provide Q_1 with a smooth dc.

12.3 CMOS CRYSTAL OSCILLATOR

A simple, but reliable and stable rectangular waveform source, can be designed around a CMOS inverter package and an inexpensive quartz crystal. Figure 12.3 shows one possible configuration for this circuit. The operating frequency, f, is primarily determined by the crystal and can range from 10 kHz to as high as 1 MHz. The power supply voltage can fall anywhere within the normal operating range of CMOS inverters. The actual component values are relatively noncritical, but can be trimmed for optimum performance with a particular crystal and a given circuit board layout.

The basic design equations for the circuit are:

$$R_1 = 5 \times 10^6 e^{-(10 \times 10^{-6} f)} \tag{12.2a}$$

$$R_2 = 0.12R_1 \tag{12.2b}$$

$$R_3 = R_2/(0.3V_{CC} - 0.5) \tag{12.2c}$$

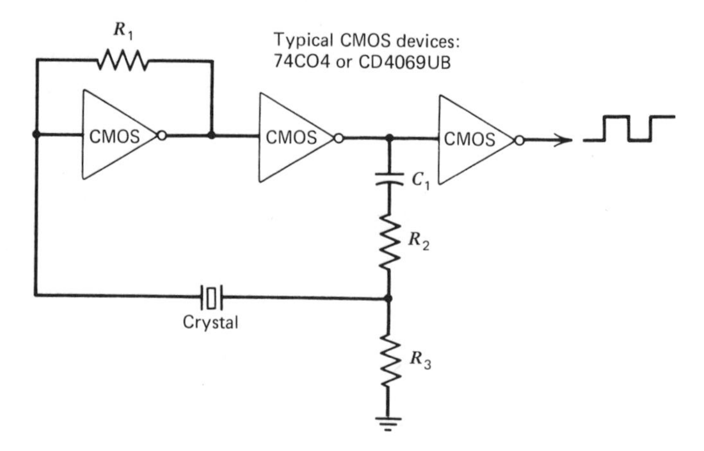

Figure 12.3 An example of a crystal oscillator employing CMOS inverters.

The value of capacitor C_1 is selected. The value of R_1 should be between 100 kΩ and 5 MΩ.

EXAMPLE 12.2 Design a crystal rectangular waveform generator using CMOS inverters (Fig. 12.3) that will oscillate at 32.768 kHz and operate from a 10 V dc power supply.

PROCEDURE

1 By Eq. 12.2*a*,

$$R_1 = 5 \times 10^6 e^{-(10 \times 10^{-6} \times 32{,}768)}$$
$$= 3.6\ \text{M}\Omega$$

2 By Eq. 12.2*b*,

$$R_2 = 0.12 R_1$$
$$\simeq 430\ \text{k}\Omega$$

3 By Eq. 12.2*c*,

$$R_3 = R_2/(0.3\ V_{CC} - 0.5)$$
$$= (430\ \text{k}\Omega)/(0.3 \times 10 - 0.5) \simeq 180\ \text{k}\Omega$$

4 For its initial value, select $C_1 = 10$ pF.

Practical Considerations

a. The circuit board layout can have a significant effect on oscillator performance. The board should be designed to minimize stray capacitance across the crystal.

b. One method for selecting an optimum value for R_1 is to remove the crystal and allow the circuit to operate at its natural frequency. With the crystal removed, R_1

can be adjusted to cause the natural free-running frequency to be approximately double the desired crystal frequency.

c. The value of R_3 may be adjusted to vary the drive level to the crystal. Insufficient drive may result in poor oscillator starting, as well as dropout and jitter. Excessive drive may reduce the life of the crystal.

d. The power supply lines to the oscillator should be decoupled as close as possible to the inverters with a 0.01 to 1 μF ceramic capacitor.

12.4 CRYSTAL OSCILLATOR USING A COMPARATOR

Another useful and inexpensive crystal oscillator can be designed around an analog comparator IC. One possible advantage of this circuit over the CMOS crystal oscillator is the possibility of extending the upper frequency limits and the availability of more current drive. In Fig. 12.4, resistors R_3 and R_4 are equal so that the comparator switches symmetrically about $(+V)/2$ V. The specific time constant R_2C is noncritical and is determined primarily by the availability of component values. If the time constant R_2C is greater than $3/f$, where f is the frequency of oscillation, then a 50% duty cycle is obtained by maintaining a dc voltage at the inverting input equal to the absolute average value of the output waveform.

Resistor R_1, a pull-up resistor, is optional. For a 50% duty cycle,

$$R_2C > 3/f \tag{12.3a}$$

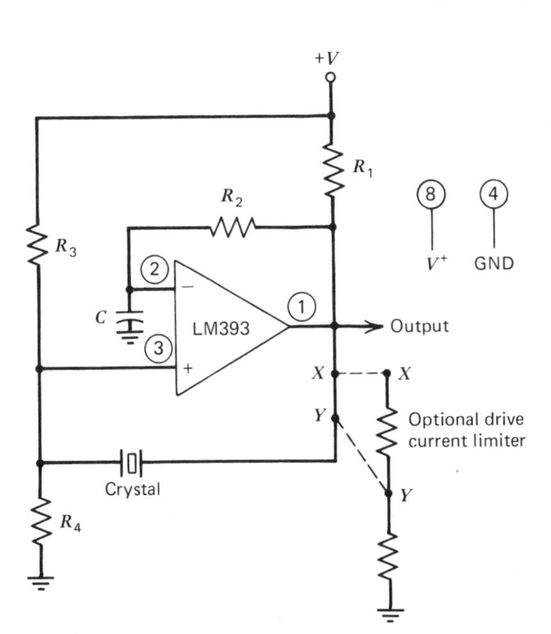

Figure 12.4 A crystal oscillator using a comparator.

Also,

$$10 \text{ k}\Omega < R_3 < 100 \text{ k}\Omega \tag{12.3b}$$

and

$$R_4 = R_3 \tag{12.3c}$$

The value of C is selected.

EXAMPLE 12.3 Design a crystal oscillator using a comparator (Fig. 12.4) for generating a 32.768 kHz sine wave.

PROCEDURE

1 Choose $R_1 = R_2 = 100$ kΩ for low current consumption.
2 By Eq. 12.3*a*, the result is:

$$R_2C > 3/f$$

$$3/32{,}768 = 92 \ \mu\text{s}$$

3 Select $C = 0.01$ μF; $R_2 = 100$ kΩ was chosen in step 1. (These values are readily available.) The resulting time constant is $10^{-8} \times 10^5 = 1$ ms which is greater than 92 μs.
4 Select $R_3 = 47$ kΩ. By Eq. 12.3*c*, $R_4 = 47$ kΩ, which is a midrange value based on Eq. 12.3*b*.

Practical Considerations

a. The circuit board should be designed to minimize stray capacitance.
b. The crystal for this circuit should be of the series resonant type.
c. If drive current to the crystal needs to be limited, a divider network can be used from the output to ground, as illustrated in Fig. 12.4. The specific drive current requirement will be specified by the crystal manufacturer.
d. More precise values for R_1, R_3, and R_4 can be obtained by taking the exact IC parameters such as bias current, offset voltage, and drive capabilities into account. For many applications, however, these considerations will have a minimal effect on the desired performance.

12.5 SQUARE WAVE GENERATOR USING A COMPARATOR

An inexpensive and fairly stable square wave generator can be designed around an op amp, or comparator. The circuit of Fig. 12.5 employs an RC feedback network to the inverting input and resistive feedback to the noninverting input of an op amp. The ratio of R_1 to the sum of R_1 and R_2 defines the threshold for switching. When the output is at +V, the timing capacitor C charges from the negative threshold to the positive threshold. When the positive threshold is

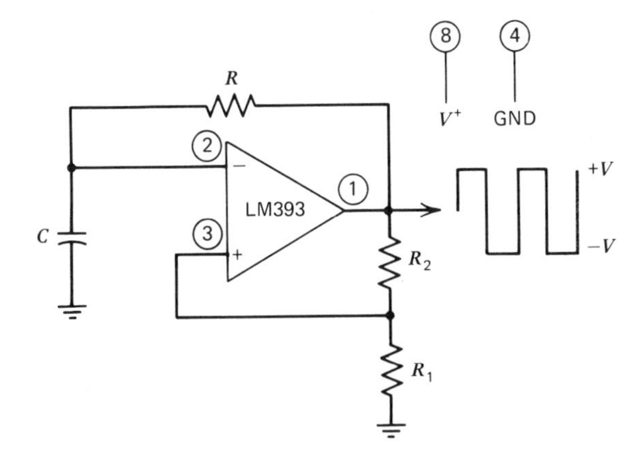

Figure 12.5 A square wave generator designed around a comparator.

reached, the output switches to −V and the capacitor begins charging from the positive threshold back to the negative threshold.

If $R_1 = 0.86R_2$, the frequency of oscillation, *f*, is:

$$f = \frac{1}{2RC} \qquad (12.4a)$$

If $R_1 \neq 0.86R_2$, then the frequency of oscillations becomes:

$$f = \frac{1}{2RC \ln (2R_1/R_2 + 1)} \qquad (12.4b)$$

The values of R_2 and C are selected.

EXAMPLE 12.4 Based on the circuit of Fig. 12.5, design a square wave generator operating at a frequency of 22.7 kHz. The application is relatively precise, but does not demand crystal accuracy.

PROCEDURE

1. Select $R_2 = 120\ \text{k}\Omega$, a reasonable value for minimizing the effects of input bias current and loading of the comparator.
2. Assume $R_1 = 0.86R_2 = 0.86 \times 120\ \text{k}\Omega = 103\ \text{k}\Omega$.
3. By Eq. 12.4*a*,

$$2RC = 1/22{,}700 = 44\ \mu\text{s}$$

Based on availability and cost, the timing capacitor value is usually selected first and the resistor value R calculated. If the timing capacitor is selected as 0.1 μF, then the timing resistor value, R, is:

$$R = (44\ \mu\text{s}/2)/(0.1\ \mu\text{F}) = 220\ \Omega$$

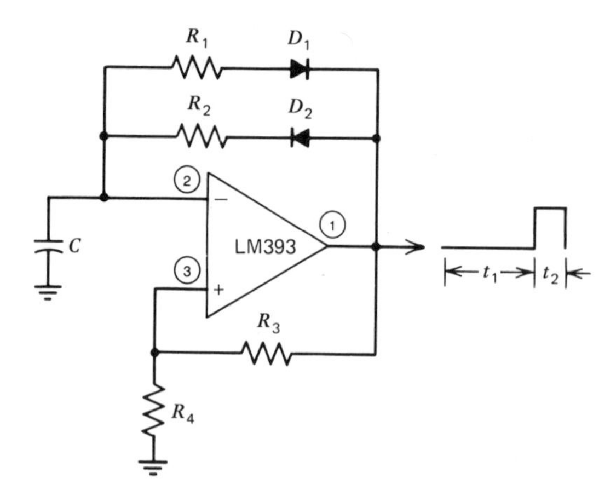

Figure 12.6 A rectangular wave generator.

Practical Considerations

a. A symmetrical square wave is not always the desired output. Figure 12.6 shows the circuit modified to generate rectangular waves by varying the duty cycle. This is accomplished by providing separate charge and discharge paths for the timing capacitor, C. The capacitor charging through R_2 and D_2 determines T_2. The discharge path through R_1 and D_1 determines T_1. Resistors R_3 and R_4 define the circuit hysteresis. The amount of hysteresis depends upon the particular application.

b. For the circuit of Fig. 12.6, the frequency of oscillation, f, is:

$$f = 1/(T_1 + T_2) \tag{12.5a}$$

If $R_4 = 0.86R_3$, then:

$$T_1 = R_1C \tag{12.5b}$$

$$T_2 = R_2C \tag{12.5c}$$

$$T = R_1C + R_2C \tag{12.5d}$$

The value of C should be selected so that:

$$T/10^6 < C < T/10^3 \tag{12.5e}$$

Select a value for R_3 that is small enough to reduce the effects of input bias current and yet large enough to minimize loading of the output.

EXAMPLE 12.5 Based on the circuit of Fig. 12.6, design an oscillator that generates a 100 μs pulse at a frequency of 1 kHz.

PROCEDURE

1 Select $R_3 = 150$ kΩ; then, $R_4 = 0.86 \times 150$ kΩ $= 129$ kΩ.
2 100 μs $= R_2C$. Select $C = 0.1$ μF; then $R_2 = (100\ \mu s)/(0.1\ \mu F) = 1000\ \Omega$.

3 1 kHz = $1/(T_1 + 100\ \mu s)$; therefore, $T_1 = 900\ \mu s$.
4 900 μs = R_1C; $R_1 = (900\ \mu s)/(0.1\ \mu F) = 9\ k\Omega$.

12.6 CMOS SQUARE WAVE GENERATOR

A square wave generator can be designed using CMOS inverters. Referring to Fig. 12.7, the frequency is primarily determined by the charge and discharge times of capacitor C.

The design equations for the circuit are:

$$R_2 = \frac{1}{2.2fC} \tag{12.6a}$$

$$R_1 = 10R_2 \tag{12.6b}$$

The value of C is selected.

EXAMPLE 12.6 Design a square wave generator using CMOS inverters (Fig. 12.7). The nominal frequency is 2 kHz.

PROCEDURE

1 Select $C = 0.01\ \mu F$.
2 By Eq. 12.6*a*,

$$R_2 = \frac{1}{2.2fC}$$
$$= 1/(2.2 \times 2000 \times 10^{-8}) = 22\ k\Omega$$

3 By Eq. 12.6*b*,

$$R_1 = 10 \times 22\ k\Omega = 220\ k\Omega$$

Practical Considerations

a. The value of C should be selected so R_2 falls into the range of 5 kΩ to 5 MΩ. These values are not ultimate limits; however, they will tend to minimize the effects of leakage currents, imperfect capacitors, and variations between inverters.

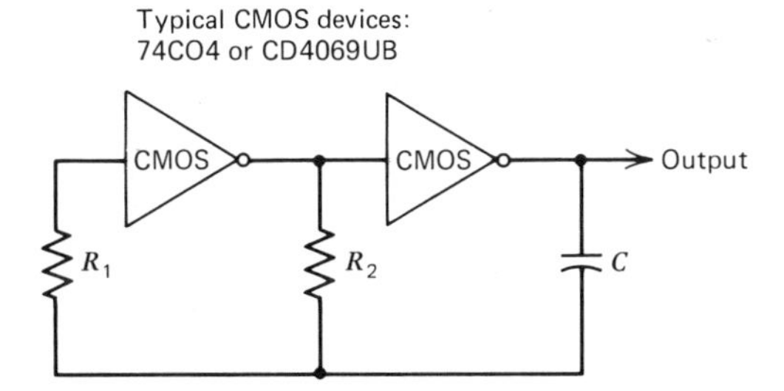

Figure 12.7 A square wave generator employing CMOS inverters.

b. Capacitor C should be nonpolarized with fairly low leakage. A ceramic disc capacitor will work well.

c. The frequency range is very broad and extends from well into the fractional hertz range to values approaching a megahertz.

d. The switching point of CMOS inputs are typically specified in the range of ±40% from the ideal 50% point. The circuit, therefore, needs some individual trimming if a precise frequency is necessary.

e. The circuit is relatively immune to temperature and supply voltage variations.

12.7 STAIRCASE GENERATOR

In the staircase generator of Fig. 12.8, the counter can be any practical number of bits and would, generally, be a standard binary counter. The D/A converter (see Chap. 13) can be selected based on the number of bits in the counter and the desired performance of the overall circuit.

As the counter is incremented, the input to the D/A converter increases and, therefore, the analog output also increases. Because the input changes in finite increments, the output obtained is the familiar staircase waveform. The relationship between the input frequency, and the number of output steps, is:

$$f_{in} = f_o S \tag{12.7a}$$

where f_{in} is the input frequency, f_o is the output frequency, and S is the number of output steps.

The minimum number of bits in the counter and converter, N, is:

$$N = 3.32 \log S \tag{12.7b}$$

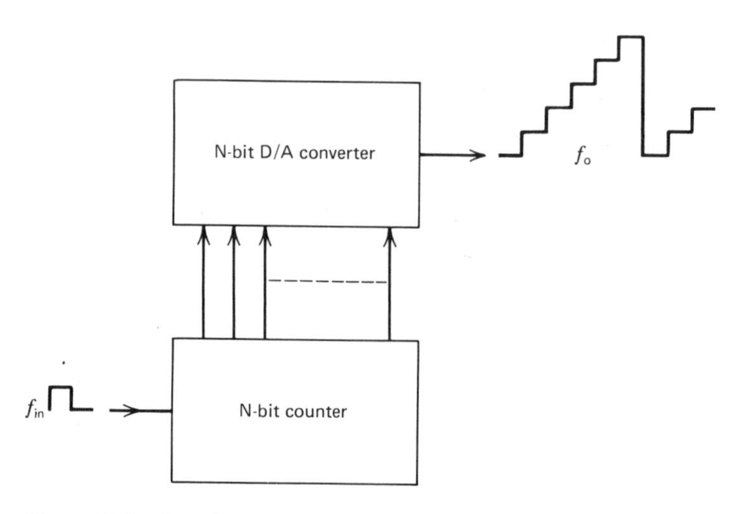

Figure 12.8 A staircase generator.

EXAMPLE 12.7 Based on Fig. 12.8, design a staircase generator to produce a waveform consisting of 256 steps at a repetition frequency of 100 Hz.

PROCEDURE

1 By Eq. 12.7*a*, the result is:

$$f_{in} = f_o S$$
$$= 100 \times 256 = 25.6 \text{ kHz}$$

2 By Eq. 12.7*b*,

$$N = 3.32 \log S$$
$$= 3.32 \times \log 256 = 8$$

3 The preceding calculations tell us that we need an 8-bit counter and D/A converter. The counter will be clocked by a 25.6 kHz pulse.

Practical Considerations

a. The practical limits of this circuit primarily stem from the rapid increase of f_{in} as S increases.

b. Selection of technology, such as CMOS or TTL, and choice of a particular D/A converter, are determined by the applications and are based on standard considerations for these devices (Chap. 13).

12.8 ARBITRARY WAVEFORM GENERATOR

There are certain applications, such as testing of analog systems, where it is necessary to generate nonstandard periodic waveform. Figure 12.9 is a block diagram of a circuit that can generate nearly any desired waveform. The practicality of the circuit depends primarily upon frequency and resolution. The D/

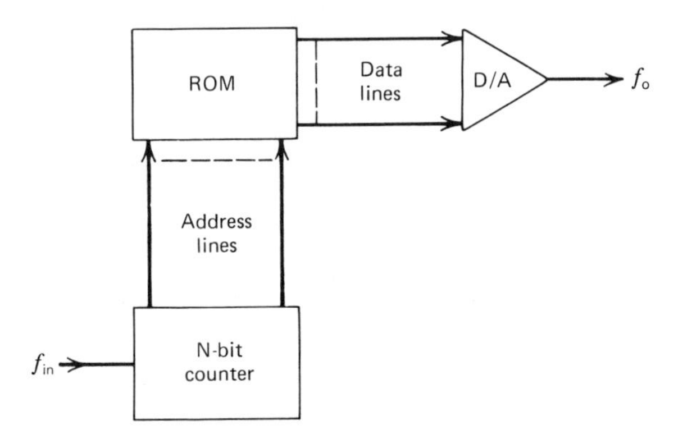

Figure 12.9 Block diagram of a circuit that generates arbitrary waveforms.

A converter can range from a simple R–$2R$ ladder to a high-accuracy monolithic converter (see Chap. 13). The counter can vary from 2 to N bits, depending on the desired resolution. Likewise, the ROM length and width are determined by the desired horizontal and vertical resolutions, respectively.

The key design equations for this circuit are:

$$f_{in} = 2^N f_o \tag{12.8a}$$

where N is the number of bits in the counter, f_{in} is the clock frequency, and f_o is the desired output frequency.

$$\text{Vertical (amplitude) resolution} = 1 \text{ part in } 2^W \tag{12.8b}$$

where W is the ROM width.

$$\text{Horizontal (time) resolution} = 1 \text{ part in } 2^L \tag{12.8c}$$

where L is the ROM length.

EXAMPLE 12.8 Design a circuit to generate the waveform of Fig. 12.10. The circuit of Fig. 12.9 will be used. Resolution required is 16 parts for both the vertical and horizontal axes. The output frequency is 1 kHz.

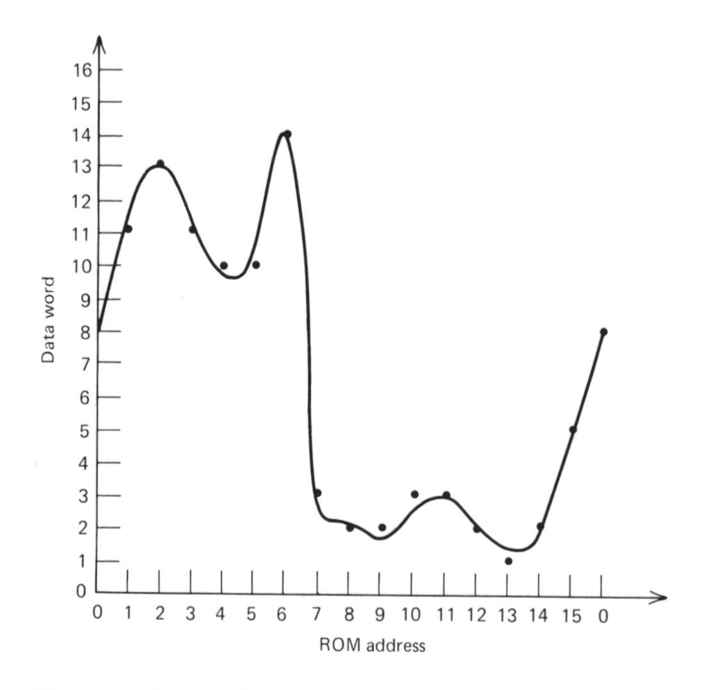

Figure 12.10 Waveform to be generated in Ex. 12.8.

Table 12.1 **ROM contents for generating the waveform of Fig. 12.10**

ROM address	*Data word*
0	1000 – 8
1	1011 – 11
2	1101 – 13
3	1011 – 11
4	1010 – 10
5	1010 – 10
6	1110 – 14
7	0011 – 3
8	0010 – 2
9	0010 – 2
10	0011 – 3
11	0011 – 3
12	0010 – 2
13	0001 – 1
14	0010 – 2
15	0101 – 5

PROCEDURE

1 The plotted points in Fig. 12.10 show a "best fit" plot with the stated resolutions. The required ROM data words for any given address can easily be determined directly from the graph. Table 12.1 is a summary of the ROM contents in truth table form.

2 For 16-part resolution, $N = 4$.

3 The required clock frequency, by Eq. 12.8*a,* is:

$$f_{in} = 2^N f_0$$
$$= 2^4 \times 1 \text{ kHz} = 16 \text{ kHz}$$

4 The vertical and horizontal resolution dictate a ROM have a 4×4 configuration (see Chap. 6). The actual memory device may be larger and can be a ROM, PROM, EPROM, or EAROM.

Practical Considerations

a. The vertical resolution can be increased to at least 1 part in 256 for a very nominal increase in ROM cost. Horizontal resolution likewise, can easily be increased to 1 part in 2048 or more, but the required input frequency can quickly become impractical.

b. The overall waveform distortion can be significantly reduced by following the D/A converter with a low-pass filter tuned to pass the highest desired variations, but rejecting the high frequencies associated with the step voltages between consecutive addresses.

12.9 555 MONOSTABLE MV

The 555 timer is a versatile building block for generating rectangular waveforms. It can operate in either the monostable or astable modes at pulse widths (lengths) of microseconds to minutes and frequencies of 100 kHz to 0.0008 Hz.

The standard connection for use as a monostable multivibrator (MV), or one-shot, is shown in Fig. 12.11. The output is normally low and goes high when the trigger voltage on pin 2 falls below $V_{CC}/3$. Once triggered, the output remains high, independent of the trigger. The width (length) of the pulse, T, is:

$$T = 1.1R_1C_1 \tag{12.9a}$$

Practical values for R_1 range from 2 kΩ to well into the megohm range. The minimum value of the timing capacitor, C_1, is approximately 500 pF; its maximum value is determined by the quality of the capacitor used, but is generally less than 1000 μF. The power supply voltage range for V_{CC} is from 4.5 to 15 V.

The maximum value of R_1, $R_{1(max)}$, is:

$$R_{1(max)} \leq V_{CC}/(0.75\ \mu A) \tag{12.9b}$$

The minimum value of R_1 is 1000 Ω.

Bypass capacitors C_2 and C_3, which are ceramic or tantalum types, range in value from 0.01 to 10 μF. The range of values for the pullup resistor, R_2, is between 10 Ω and 10 MΩ. The exact value is determined by the drive capability of the triggering circuit.

EXAMPLE 12.9 Design a 555 monostable MV to generate a 10 s wide pulse. The value of V_{CC} is 12 V.

PROCEDURE

1 Select $C_1 = 150\ \mu F$.
2 From Eq. 12.9*a*,

$$R_1 = T/1.1C_1$$
$$= 10/(1.1 \times 150 \times 10^{-6}) = 60.6\ k\Omega$$

3 Verify the value of R_1 is acceptable; from Eq. 12.9*b*,

$$(12\ V)/(0.75\ \mu A) = 16\ M\Omega$$

Hence, $R_1 = 60.6\ k\Omega < 16\ M\Omega$ and $R_1 > 1000\ \Omega$.
4 If the input pulse is capacitor coupled, R_2 can serve as part of a differentiator to shorten the trigger pulse. If the input is direct coupled and the high-level trigger voltage is greater than $V_{CC}/3$, then R_2 may be eliminated.
5 Select $C_2 = 0.1\ \mu F$ as a ceramic disc and $C_3 = 10\ \mu F$ as a tantalum capacitor.

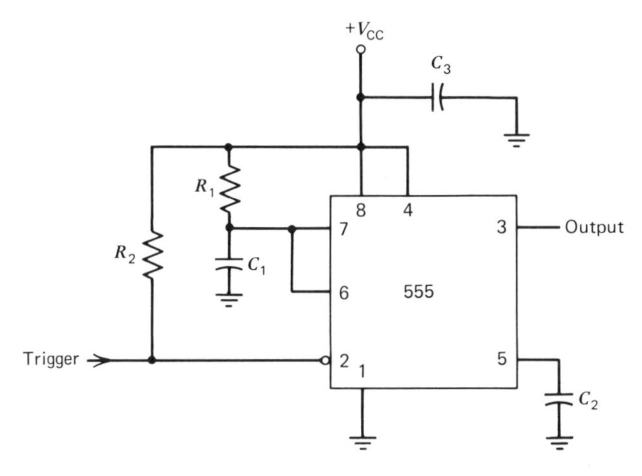

Figure 12.11 A monostable (one-shot) MV based on the 555 timer.

12.10 555 ASTABLE MV

An astable (free running) MV using the 555 timer is illustrated in Fig. 12.12. Timing capacitor C_1 charges through R_1 and R_2 toward V_{CC} until it reaches a value of $2V_{CC}/3$ V. At that time, pin 7 is internally returned to ground which allows C_1 to discharge through R_2 toward zero. The cycle is repeated when C_1 has discharged to $V_{CC}/3$ V.

Power supply voltage V_{CC} can be between 4.5 and 15 V. The time interval for the high-level output, T_{HI}, is:

$$T_{HI} = 0.693(R_1 + R_2)C_1 \tag{12.10a}$$

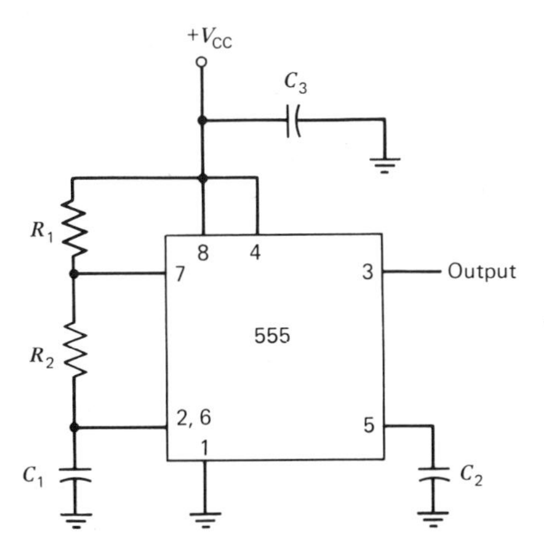

Figure 12.12 An astable (free running) MV based on the 555 timer.

The time interval for the low-level ouput, T_{LO}, is:

$$T_{LO} = 0.693R_2C_1 \tag{12.10b}$$

The period, T, of the output waveform is:

$$T = 0.693(R_1 + 2R_2)C_1 \tag{12.10c}$$

and the frequency, f, is equal to $1/T$:

$$f = \frac{1.44}{(R_1 + 2R_2)C_1} \tag{12.10d}$$

where $0.00002 \text{ Hz} < f < 100 \text{ kHz}$.

Duty cycle, D, is expressed by:

$$\begin{aligned} D &= T_{HI}/T \\ &= (R_1 + R_2)/(R_1 + 2R_2) \end{aligned} \tag{12.10e}$$

Generally, D is selected to be between 0.6 and 0.9.

The value of the timing capacitor, C_1, is between 500 pF and 1000 μF. Timing resistor, R_1, is:

$$R_1 = R_2(2D - 1)/(1 - D) \tag{12.10f}$$

Generally, $1 \text{ k}\Omega < R_1 < 3.3 \text{ M}\Omega$.
Timing resistor R_2 is:

$$R_2 = 1.44(1 - D)/fC_1 \tag{12.10g}$$

where $1 \text{ k}\Omega < R_2 < 3.3 \text{ M}\Omega$.

Bypass capacitors C_2 and C_3 range between 0.1 and 10 μF. Disc ceramic or tantalum capacitors are recommended.

EXAMPLE 12.10 Design a 555 astable MV to operate at 2.5 kHz. The duty cycle $D = 0.75$.

PROCEDURE

1 Select $C_1 = 0.01$ μF.
2 By Eq. 12.10*g*,

$$\begin{aligned} R_2 &= 1.44(1 - D)/fC_1 \\ &= 1.44(1 - 0.75)/(2.5 \times 10^3 \times 10^{-8}) = 14.4 \text{ k}\Omega \end{aligned}$$

3 Verify the value of R_2:

$$1\ \text{k}\Omega < 14.4\ \text{k}\Omega < 3.3\ \text{M}\Omega$$

4 By Eq. 12.10*f*,

$$R_1 = R_2(2D - 1)/(1 - D)$$
$$= (14{,}400)\ (2 \times 0.75 - 1)/(1 - 0.75) = 28.8\ \text{k}\Omega$$

5 Verify the value of R_1:

$$1\ \text{k}\Omega < 28.8\ \text{k}\Omega < 3.3\ \text{M}\Omega$$

6 Select $C_2 = 0.1\ \mu\text{F}$ disc ceramic and $C_3 = 10\ \mu\text{F}$ tantalum.

Practical Considerations

- **a.** The output may be forced low at any time by connecting pin 4 to ground.
- **b.** In the monostable mode, pin 2 should return to a voltage that is greater than $V_{CC}/3$ before the end of the timing period.
- **c.** Duty cycles from 0.05 to 0.95 can be attained by placing a diode in series with R_2 and bypassing the diode-R_2 combination with a second diode.
- **d.** Output current capability is 200 mA.
- **e.** The timing capacitor should be silver mica, polystyrene, tantalum, or mylar for best results.
- **f.** Timing capacitor leakage should be much less than available charging current.
- **g.** Pin 5 can be used to vary the timing (or frequency) as a function of voltage. As a guide, $0.45V_{CC} <$ pin 5 $< 0.9V_{CC}$.
- **h.** A dual version of the 555 is available as a 556.

12.11 IC FUNCTION GENERATOR

The 8038 precision waveform generator is a multipurpose IC that can be connected to generate sine, square, and triangular waveforms. These outputs may be varied over a frequency range from less than a hertz to hundreds of kilohertz. In addition, the IC can be operated as a voltage controlled oscillator (VCO) for frequency modulation or sweeping applications (see Chap. 17).

The simplest connection for the chip is shown in Fig. 12.13. Frequency f is controlled by variable resistor R_C and capacitor C according to:

$$f = 0.15/R_C C \qquad (12.11a)$$

In this simple connection, the duty cycle is approximately 50%. Resistor R_1 is part of a biasing network for the internal sine converter and should be approximately 82 kΩ in this simple configuration. Resistor R_2 is a pullup resistor for the open-collector output on pin 9. Because the output is open collector, the

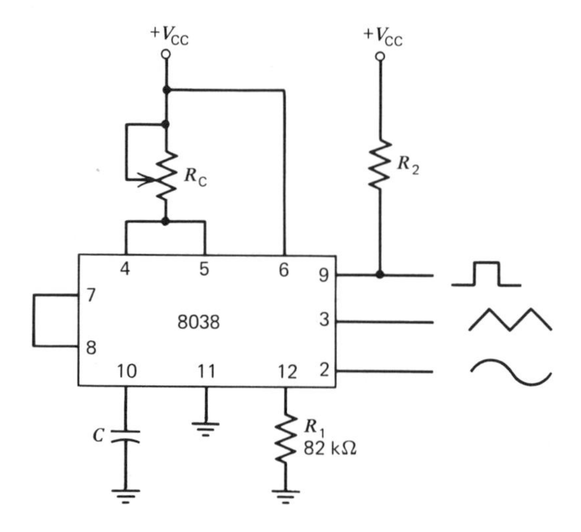

Figure 12.13 The 8038 multipurpose IC connected to generate sine, square, and triangular waveforms.

load resistor may be connected to some voltage that is higher or lower than V_{CC}. The upper voltage limit is 30 V while the lower limit is below 5 V.

To improve the performance of the 8038, the circuit of Fig. 12.14 may be implemented. The variable resistor R_C in Fig. 12.13 has now been replaced by two variable resistors R_{C1} and R_{C2}. This allows independent adjustment of both halves of an output waveform. Resistor R_{C1} controls the rising portion of a triangular, the 270 to 90° portion of a sine, and the high-level state of a square

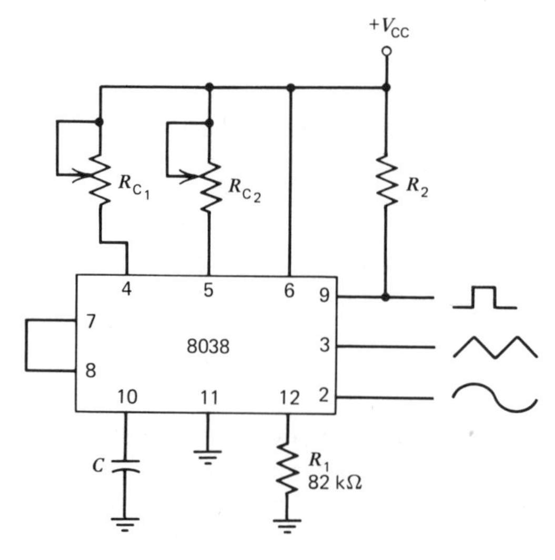

Figure 12.14 An improved version of the circuit shown in Figure 12.13.

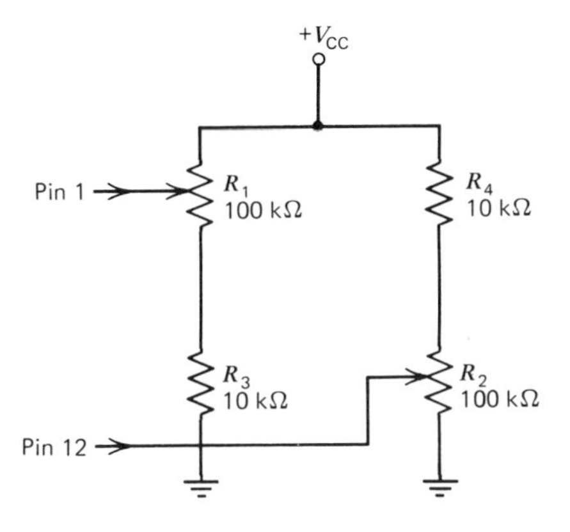

Figure 12.15 An adjustable bias network that may be connected between pins 1 and 12 of the 8038 chip.

wave. Resistor R_{C2} controls the timing on the opposite half of the wave, although R_{C1} will also interact with the adjustment. The frequency of the waveform is:

$$f = \frac{1}{1.66R_{C1}C\{1 + R_{C2}/(2R_{C1} - R_{C2})\}} \tag{12.11b}$$

For minimum sine wave distortion, the 82 kΩ bias resistor connected to pin 12 can be replaced by an adjustable bias network shown in Fig. 12.15. The positive-going (270–90°) portion of the sine wave is improved by adjusting R_1 and the falling (90–270°) is improved by adjusting R_2. Although there may be interaction in these two adjustments, sine wave distortion of less than 1% may be realized.

In addition to Eqs. 12.11*a* and *b,* other useful equations for designing with the 8038 chip are:

The positive slope of the triangular wave, the 270–90° on the sine wave, and the high-level pulse output on the square wave, $T+$, is:

$$T+ = 1.66R_{C1}C \tag{12.11c}$$

For the negative slope of the triangular waveform, (90–270°) of the sine wave, and the low level of the square wave, $T-$:

$$T- = (1.66R_{C1}R_{C2}C)/(2R_{C1} - R_{C2}) \tag{12.11d}$$

The amplitude of the triangular waveform, V_T, is:

$$V_T = 0.3V_{CC} \tag{12.11e}$$

The sine wave amplitude, V_s, is:

$$V_s = 0.2V_{CC} \tag{12.11f}$$

The value of the pulse pullup resistor, R_2, is:

$$R_{2(min)} = (V_{CC} - V_{OL})/(2 \text{ mA} \pm I_{LL}) \qquad (12.11g)$$

and

$$R_{2(\max)} = (V_{CC} - V_{OH})/(1 \text{ }\mu\text{A} \pm I_{LH}) \qquad (12.11h)$$

where I_{LL} is the pin-9 load current corresponding to a low (V_{OL}) output and I_{LH} is the pin-9 load current corresponding to a high (V_{OH}) output. Both of these currents are positive if current flows from V_{CC} through R_2 to the load. For opposite current flow, I_{LL} and I_{LH} are negative. The value of V_{OL} is between 0.1 and 1.25 V; V_{OH} is determined by the application.

Timing resistor and duty cycle control, R_{C1}, is:

$$R_{C1} = T+/1.66C \qquad (12.11i)$$

where $V_{CC}/(5 \text{ mA}) < R_{C1} < V_{CC}/(50 \text{ }\mu\text{A})$

For R_{C2},

$$R_{C2} = 2R_{C1}T-/(1.66\, R_{C1}C + T-) \qquad (12.11j)$$

where $V_{CC}/(5 \text{ mA}) < R_{C2} < V_{CC}/(50 \text{ }\mu\text{A})$.

Timing capacitor C is initially selected as large as possible.

EXAMPLE 12.11 Design a waveform generator using an 8038 chip. The frequency of oscillation is 10 kHz and the duty cycle is 50%. For $V_{CC} = 5$ V, typical values for the 8038 are: $V_{OH} = 3.6$ V, $V_{OL} = 0.2$ V, $I_{LL} = -1.6$ mA, and $I_{LH} = 40$ μA.

PROCEDURE

1 A 50% duty cycle yields:

$$T+ = T- = 1/(10{,}000 \times 2) = 50 \text{ }\mu\text{s}$$

2 Select $C = 0.01$ μF

3 By Eq. 12.11*i*,

$$\begin{aligned} R_{C1} &= T+/1.66C \\ &= (50 \times 10^{-6})/(1.66 \times 10^{-8}) = 3 \text{ k}\Omega \end{aligned}$$

4 By Eq. 12.11*j*,

$$\begin{aligned} R_{C2} &= 2R_{C1}T-/(1.66R_{C1}C + T-) \\ &= (2 \times 3000 \times 50 \times 10^{-6})/(1.66 \times 3000 \times 10^{-8} + 50 \times 10^{-6}) \\ &= 3 \text{ k}\Omega \end{aligned}$$

---

5 Verify the values of R_{C1} and R_{C2}:

$$V_{CC}/(5\text{ mA}) = (5\text{ V})/(5\text{ mA}) = 1\text{ k}\Omega$$
$$V_{CC}/(50\ \mu\text{A}) = (5\text{ V})/(50\ \mu\text{A}) = 100{,}000\ \Omega$$

Hence,

$$1000 < 3000 < 100{,}000\ \Omega$$

6 By Eq. 12.11*g*,

$$R_{2(\min)} = (V_{CC} - V_{OL})/(2\text{ mA} \pm I_{LL})$$
$$= (5 - 0.2)/(2\text{ mA} - 1.6\text{ mA}) = 12\text{ k}\Omega$$

7 By Eq. 12.11*h*,

$$R_{2(\max)} = (V_{CC} - V_{OH})/(1\ \mu\text{A} \pm I_{IL})$$
$$= (5 - 3.6)/(1\text{ A} + 40\ \mu\text{A}) = 34\text{ k}\Omega$$

Practical Considerations

a. All connections shown as being returned to ground may be connected to the $-V_{CC}$ terminal if a dual supply is used.

b. Pins 7 and 8 can be used for FM or sweep applications by applying a control voltage in the range of: $2V_{CC}/3 < V_C < V_{CC}$ volts.

12.12 REFERENCES

Coughlin, R. F. and F. E. Driscoll, *Operational Amplifiers and Linear Integrated Circuits,* Prentice-Hall, second edition, Englewood Cliffs, NJ, 1982.

Exar, *Applications Data Book,* latest edition.

Intersil, "Application Bulletin No. A012-A."

Jung, W. G., *IC Op-Amp Cookbook,* second edition, Sams, Indianapolis, 1980.

National Semiconductor, *Linear Applications Handbook,* latest edition.

O'Neil, B., "Application Bulletin No. A013," Intersil, Inc.

Prensky, S. and A. H. Seidman, *Linear Integrated Circuits: Practice and Applications,* Reston, VA, 1981.

Signetics, *Analog Applications Manual,* latest edition.

Statek, "Technical Note No. TN-14."

Stout, D. F. and M. Kaufman, *Handbook of Operational Amplifier Circuit Design,* McGraw-Hill, New York, 1976.

Young, T., *Linear Integrated Circuits,* Wiley, New York, 1981.

CHAPTER 13

Data Conversion

Dr. Ramesh Gaonkar
Professor, Electrical Technology
Onondaga Community College

13.1 INTRODUCTION

Physical quantities, such as temperature and pressure, are electronically measured by converting these quantities into corresponding electrical signals using devices called *transducers*. (For example, a thermocouple is a transducer which converts temperature into a corresponding electrical signal.) These electrical signals, called *analog* signals, are measurable and continuously variable, corresponding to the physical quantities they represent. On the other hand, *digital* signals are binary and discontinuous. They have two states, 0 and 1, which can be represented by two discrete voltages, such as 0 V and 5 V, respectively.

Even though an analog signal may represent physical parameters accurately, it is difficult to process or store for later use without introducing considerable error. An analog signal is not convenient to read or interpret, especially in a noisy environment. On the other hand, digital data are easily stored, transmitted, processed, and are not subject to ambiguity (for example, digital voltmeter readings).

To take advantage of digital processing and storage, analog data must be converted into digital data with an *analog-to-digital,* or *A/D converter* (ADC). Similarly, digital data are converted into an analog signal to represent a physical parameter using a *digital-to-analog,* or *D/A converter* (DAC). Both A/D and D/A converters are known as *data converters.*

Applications of data converters range from a simple digital voltmeter to complex high-speed video digitizing systems. In some systems, accuracy of conversion is important; in others, speed of conversion is critical. Data converters are available with a wide range of characteristics and specifications.

13.2 NUMBER SYSTEMS AND CONVERTER CODES

Digital signals (input/output) are generally in a binary format (code). The binary format, however, is not appropriate in all data converter applications. Therefore, converters are available in such various input/output codes as binary, BCD, 2's complement, and offset binary.

Number Systems

The commonly used number system is decimal (or base 10) which has 10 digits, 0–9. There exist other number systems, such as binary (base 2), octal (base 8), and hexadecimal (base 16) which in some instances are more appropriate than decimal for mathematical manipulation in electronic circuits.

Numbers, in general, are represented in the positional system, meaning that the digit carries a definite weight according to its position in the number. Any number, regardless of its base, can be expressed as:

$$N = A_n B^n + A_{n-1} B^{n-1} + \cdots + A_1 B^1 + A_0 B^0 \tag{13.1}$$

where N = number, B = base of number system, and A = coefficient which can be any digit in the base system. For example:

Decimal number: $(1475)_{10} = 1 \times 10^3 + 4 \times 10^2 + 7 \times 10^1 + 5 \times 10^0$
Binary number: $(1011)_2 = 1 \times 2^3 + 0 \times 2^2 + 1 \times 2^1 + 1 \times 2^0$
Octal number: $(7523)_8 = 7 \times 8^3 + 5 \times 8^2 + 2 \times 8^1 + 3 \times 8^0$
Hexadecimal number: $(1625)_{16} = 1 \times 16^3 + 6 \times 16^2 + 2 \times 16^1 + 5 \times 16^0$

By adding these numbers, the decimal equivalent of each number can be obtained.

Binary numbers This number system has two digits, 0 and 1. A binary number is converted into a decimal number as follows:

$$10010101_2 = 1 \times 2^7 + 0 \times 2^6 + 0 \times 2^5 + 1 \times 2^4 + 0 \times 2^3 + 1 \times 2^2 + 0 \times 2^1 + 1 \times 2^0 = 149_{10}$$

Octal numbers This number system has eight digits, 0 through 7. An octal number is converted into a decimal number as follows:

$$225_8 = 2 \times 8^2 + 2 \times 8^1 + 5 \times 8^0 = 149_{10}$$

Hexadecimal numbers This number system has sixteen digits with values of 0 through 15. The values of 10 through 15 are designated A through F, respectively, to avoid confusion with decimal numbers. Thus, A = 10, B = 11, C =

12, $D = 13$, $E = 14$, and $F = 15$. A hexadecimal number is converted to decimal as follows:

$$95_{16} = 9 \times 16^1 + 5 \times 16^0 = 149_{10}$$

From the above, it is clear that all three numbers—binary, octal, and hexadecimal—are equivalent and easily converted as shown in Table 13.1 and Ex. 13.1.

Table 13.1 **Binary, octal, and hexadecimal conversions for decimal numbers 0 to 15_{10}**

Decimal	*Binary*	*Octal*	*Hexadecimal*
0	0 0 0 0	00	0
1	0 0 0 1	01	1
2	0 0 1 0	02	2
3	0 0 1 1	03	3
4	0 1 0 0	04	4
5	0 1 0 1	05	5
6	0 1 1 0	06	6
7	0 1 1 1	07	7
8	1 0 0 0	10	8
9	1 0 0 1	11	9
10	1 0 1 0	12	A
11	1 0 1 1	13	B
12	1 1 0 0	14	C
13	1 1 0 1	15	D
14	1 1 1 0	16	E
15	1 1 1 1	17	F

EXAMPLE 13.1

a. Convert the binary number 1110_2 into octal and hexadecimal equivalent numbers.
b. Convert the hexadecimal number $6A_{16}$ into its binary and decimal equivalents.

PROCEDURE

1 To convert a binary number into its equivalent octal number, group the binary digits in groups of three, starting from the right. Then, find the octal equivalent for each group. The conversion from an octal to a binary number is the reverse process.
2 To convert a binary number into its equivalent hexadecimal number, the binary digits are grouped in fours, starting from the right. Then, find the hexadecimal equivalent for each group. The conversion from hexadecimal to binary is the reverse process.
3 a. $1110 = 1{,}110 = 16_8$
$1110 = 1110 = E_{16}$ (which is decimal 14)
b. $6A_{16} = 0110{,}1010 = 01101010_2$

$$6A_{16} = 6 \times 16^1 + A \times 16^0 = 106_{10}$$

Codes (Unipolar)

The analog signal input to an ADC can be either unipolar (positive or negative) or bipolar (positive and negative). The digital output, therefore, should be able to represent the polarity of the input signal. In addition, the output of a converter is used in such applications as decimal display, signal encoding, and mathematical manipulation. These various applications require different types of digital coding.

Straight Binary This is the most commonly used code for A/D converters interfacing with computers. Binary code is very efficient for digital processing. By Eq. 13.1, any binary number may be represented by:

$$N = A_n 2^n + A_{n-1} 2^{n-1} + \ldots + A_0 2^0$$

where coefficient A is either 0 or 1. In conversion, however, 1 is used as the normalized full-scale value and the binary code is represented as a fraction:

$$N = A_1 2^{-1} + A_2 2^{-2} + \ldots + A_n 2^{-n} \qquad (13.2)$$

where $A_1 2^{-1}$ is the most significant bit (MSB), $A_n 2^{-n}$ is the least significant bit (LSB), and coefficient A is either 0 or 1.

For example, a fraction can represent a 3-bit binary number as:

$$N = A_1 2^{-1} + A_2 2^{-2} + A_3 2^{-3}$$

If all coefficients are 1, then, $0.111 = \frac{1}{2} + \frac{1}{4} + \frac{1}{8} = \frac{7}{8}$. In this representation, the following points should be noted:

1 The MSB is $\frac{1}{2}$ the full-scale value.

2 The LSB is $\frac{1}{2^3} = \frac{1}{8}$ (which depends on the number of bits).

3 The maximum digital value $= \frac{7}{8}$, 1 LSB less than the full scale.

In converters, the normalized convention with fractional representation is assumed and the binary point is normally dropped.

EXAMPLE 13.2

a. Calculate the fractional value of the 4-bit number 1 1 1 1. **b.** Show that the difference between the maximum fraction represented by a 4-bit number is equal to the difference between the full-scale value and the LSB.

PROCEDURE

1 $1\,1\,1\,1 = \frac{1}{2} + \frac{1}{4} + \frac{1}{8} + \frac{1}{16}$

$= \frac{15}{16}$

2 LSB $= 2^{-4}$ (where 4 is the number of bits)

$= \frac{1}{16}$

Full-scale $-$ LSB $= 1 - \frac{1}{16}$

$= \frac{15}{16}$

This is the maximum fraction that can be represented by a 4-bit number.

The preceding result is based on a normalized full-scale value of 1. If an analog input voltage is 0 to 10 V full scale, the values need to be multiplied by 10 V to find actual voltages. Table 13.2 summarizes 4-bit fractional codes. However, the binary code needs to be modified if voltages are bipolar. The offset binary code is then used (see Codes, (Bipolar) below).

Table 13.2 **Fractional binary and Gray codes**

Decimal fraction	*Binary code*	*Gray code*
0	0 0 0 0	0 0 0 0
$\frac{1}{16}$	0 0 0 1	0 0 0 1
$\frac{2}{16}$	0 0 1 0	0 0 1 1
$\frac{3}{16}$	0 0 1 1	0 0 1 0
$\frac{4}{16}$	0 1 0 0	0 1 1 0
$\frac{5}{16}$	0 1 0 1	0 1 1 1
$\frac{6}{16}$	0 1 1 0	0 1 0 1
$\frac{7}{16}$	0 1 1 1	0 1 0 0
$\frac{8}{16}$	1 0 0 0	1 1 0 0
$\frac{9}{16}$	1 0 0 1	1 1 0 1
$\frac{10}{16}$	1 0 1 0	1 1 1 1
$\frac{11}{16}$	1 0 1 1	1 1 1 0
$\frac{12}{16}$	1 1 0 0	1 0 1 0
$\frac{13}{16}$	1 1 0 1	1 0 1 1
$\frac{14}{16}$	1 1 1 0	1 0 0 1
$\frac{15}{16}$	1 1 1 1	1 0 0 0

Binary Coded Decimal (BCD) This is another popular code primarily used for decimal numeric displays. It represents the decimal digits 0 to 9 in a 4-bit binary format. Each digit in a decimal number is coded in groups of 4 bits. For example, the decimal 15 is represented in BCD as

$$15_{10} = (0\,0\,0\,1 \qquad 0\,1\,0\,1)_{\text{BCD}}.$$

Although the BCD code is appropriate for decimal displays, it is inefficient for digital processing. For decimal numbers greater than 9, more bits are needed for BCD than for straight binary.

EXAMPLE 13.3 The output voltage of a 4-bit A/D converter is given in binary code as 1 0 1 0 with a full-scale input voltage 10 V. Calculate the voltage represented by the binary code and convert the voltage in BCD code.

PROCEDURE

1 The binary code 1 0 1 0 $= \frac{1}{2} + 0 + \frac{1}{8} + 0$

$= \frac{5}{8}$ full scale

2 Voltage represented by the code $= \left(\frac{5}{8}\right) 10\text{V}$

$= 6.25\text{V}$

3 To represent this three-digit number in BCD it requires 12 bits:

$$6.25\text{ V} = 0\,1\,1\,0 \qquad 0\,0\,1\,0 \qquad 0\,1\,0\,1$$

Overranging Many A/D converters include an additional bit to the left of the MSB. The additional bit has a weight equal to full scale and it is used to indicate overrange. The overrange bit is typically used in digital voltmeters and panel meters to indicate that the input has exceeded the full-scale value and the reading may be erroneous.

Complementary Binary and BCD Complementary codes are used in digital systems were the logic 1 level is a negative voltage. These codes are commonly used in interfacing with minicomputers. The codes are obtained by complementing binary and BCD codes, so that 1 becomes 0 and 0 becomes 1.

Gray Code This code is used in shaft encoders to eliminate false readings when the shaft moves to different positions. This code is suitable for the parallel converter (explained in Sec. 13.11). The conversion from Gray to binary is performed by using exclusive OR gates.

In the Gray code, only one bit is allowed to change from one number to the next, and the bit position does not have binary numerical weighting. Gray code values from 0 to 15 are provided in Table 13.2.

Codes (Bipolar)

When voltages include both negative and positive values, bipolar codes are used in the conversion process. Commonly used codes are *offset binary, two's complement,* and *sign-magnitude*.

Offset binary In a D/A Converter, when a voltage range such as 0 to 10 V is changed, say to +5 V to −5 V, the offset binary code is generally used. It is implemented by adding a fixed, constant offset value to the output analog signal. However, no change is required in binary coding except to shift the representations so that 10 V becomes −5 V. For example, a 4-bit code will have the following values:

$$1\,1\,1\,1 = +4.375\ (+5\text{ V full scale})$$
$$1\,0\,0\,0 = 0\text{ V}$$
$$0\,0\,0\,0 = -4.375\ (-5\text{ V full scale})$$

Two's complement This code is primarily used for digital arithmetic in dealing with negative numbers. It is difficult to implement subtraction using logic cicuits. Therefore, subtraction is usually performed through addition using the two's complement method. Two's complement of a binary number is obtained by complementing the number and adding 1 to the complement.

EXAMPLE 13.4 Subtract 0 1 0 1 (5_{10}) from 1 1 1 1 (15_{10}) using the method of two's complement.

PROCEDURE

1 Find the two's complement of the number to be subtracted:

Number to be subtracted	= 0 1 0 1
Complement	= 1 0 1 0
Add 1	+ 1
Two's complement	= 1 0 1 1

2 Add the two's complement to the second number.

	15	= 1 1 1 1
		+
two's complement of	5	= 1 0 1 1
		1 1 0 1 0

Drop the carry of the most significant bit; therefore, the answer is 1 0 1 0 = 10_{10}

Sign-Magnitude This code uses the MSB to represent the sign of the number. If the MSB is 0, the number is positive; if it is 1, the number is negative. In a 4-bit code, +7 = 0 1 1 1 and −7 = 1 1 1 1.

Table 13.3 summarizes the bipolar codes. The various codes explained above will be used in the A/D converters discussed later in the chapter.

Table 13.3 **Bipolar codes**

Decimal fraction				
Positive reference	*Negative reference*	*Sign + magnitude*	*Two's complement*	*Offset binary*
$+\frac{7}{8}$	$-\frac{7}{8}$	0 1 1 1	0 1 1 1	1 1 1 1
$+\frac{6}{8}$	$-\frac{6}{8}$	0 1 1 0	0 1 1 0	1 1 1 0
$+\frac{5}{8}$	$-\frac{5}{8}$	0 1 0 1	0 1 0 1	1 1 0 1
$+\frac{4}{8}$	$-\frac{4}{8}$	0 1 0 0	0 1 0 0	1 1 0 0
$+\frac{3}{8}$	$-\frac{3}{8}$	0 0 1 1	0 0 1 1	1 0 1 1
$+\frac{2}{8}$	$-\frac{2}{8}$	0 0 1 0	0 0 1 0	1 0 1 0
$+\frac{1}{8}$	$-\frac{1}{8}$	0 0 0 1	0 0 0 1	1 0 0 1
0+	0−	0 0 0 0	0 0 0 0	1 0 0 0
0−	0+	1 0 0 0	(0 0 0 0)	(1 0 0 0)
$-\frac{1}{8}$	$+\frac{1}{8}$	1 0 0 1	1 1 1 1	0 1 1 1
$-\frac{2}{8}$	$+\frac{2}{8}$	1 0 1 0	1 1 1 0	0 1 1 0
$-\frac{3}{8}$	$+\frac{3}{8}$	1 0 1 1	1 1 0 1	0 1 0 1
$-\frac{4}{8}$	$+\frac{4}{8}$	1 1 0 0	1 1 0 0	0 1 0 0
$-\frac{5}{8}$	$+\frac{5}{8}$	1 1 0 1	1 0 1 1	0 0 1 1
$-\frac{6}{8}$	$+\frac{6}{8}$	1 1 1 0	1 0 1 0	0 0 1 0
$-\frac{7}{8}$	$+\frac{7}{8}$	1 1 1 1	1 0 0 1	0 0 0 1
$-\frac{8}{8}$	$+\frac{8}{8}$		(1 0 0 0)	(0 0 0 0)

13.3 DIGITAL-TO-ANALOG (D/A) CONVERTERS

Digital-to-analog converters transform a given digital signal into its equivalent analog signal. D/A converters can be broadly classified in three categories: current output, voltage output, and multiplying types.

Basic Concepts and Transfer Function

Conceptually, the D/A conversion process can be viewed as finding the equivalent weight of an object (less than one unit) with weights in geometrically proportional units, such as $\frac{1}{2}$, $\frac{1}{4}$ and $\frac{1}{8}$. By using these weights in various combi-

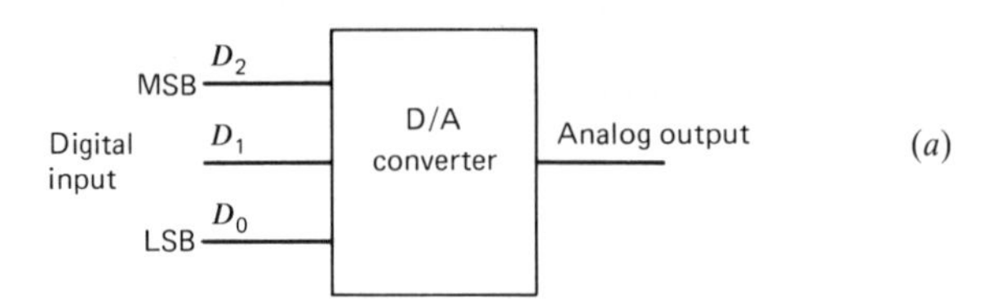

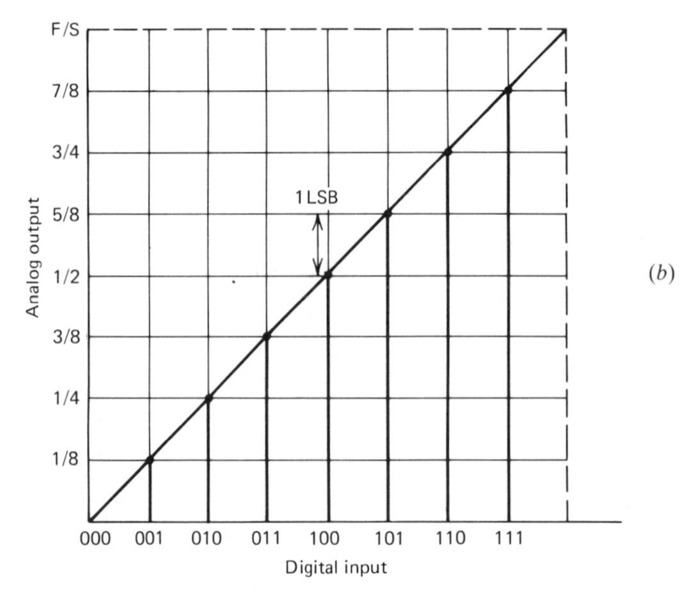

Figure 13.1 A 3-bit D/A converter. (*a*) Block diagram. (*b*) Ideal transfer function.

nations, eight different measurements ranging from 0 unit to $\frac{7}{8}$ unit can be obtained. A 3-bit D/A converter (Fig. 13.1*a*) has three digital inputs and one analog output. By various combinations of these three digital inputs, 2^3, or eight equivalent analog output levels can be obtained.

Assuming the straight binary code is used for the input signal, the equivalent output analog signal can be calculated for each digital combination at the input. The D_2 input is called the *most significant bit* (MSB) and the D_0 bit is called the *least significant bit* (LSB). These digital inputs can be either 1 or 0 (on or off). The fractional binary weighting of these inputs can be represented as 2^{-1} ($\frac{1}{2}$), 2^{-2} ($\frac{1}{4}$), and 2^{-3} ($\frac{1}{8}$) for D_2, D_1, and D_0, respectively. Thus, the input signal 001 is equal to $\frac{1}{8}$ and 100 is equal to $\frac{1}{2}$ of the full-scale analog signal.

EXAMPLE 13.5 Calculate the fraction of the full-scale signal if the input digital signal is 101.

PROCEDURE

$$
\begin{aligned}
101_2 &= 1 \times 2^{-1} + 0 \times 2^{-2} + 1 \times 2^{-3} \\
&= \frac{1}{2} + 0 + \frac{1}{8} \\
&= \frac{5}{8} \text{ of full scale}
\end{aligned}
$$

By convention, the binary point in the input signal (0.101) is assumed. Table 13.4 shows eight different inputs and equivalent analog outputs in fraction and in 10 V full-scale formats.

Table 13.4 3-bit digital input and analog output

Digital input	*Normalized analog output as a fraction*	*Analog Output (V) for 10 V full scale (V)*
0 0 0	0	0
0 0 1	$\frac{1}{8}$	1.25
0 1 0	$\frac{1}{4}$	2.50
0 1 1	$\frac{3}{8}$	3.75
1 0 0	$\frac{1}{2}$	5.00
1 0 1	$\frac{5}{8}$	6.25
1 1 0	$\frac{3}{4}$	7.50
1 1 1	$\frac{7}{8}$	8.75

Figure 13.1*b* shows the 3-bit DAC transfer function as a graphical representation of the binary input and resulting analog output for 1 V full scale. It is displayed as a bar graph because no other binary values exist for a 3-bit DAC. By examining the graph, the following points should be noted:

1 The 3-bit D/A converter has eight ($2^n = 8$) possible input combinations (n is number of input bits), ranging from 000_2 to 111_2.

2 If the full-scale analog voltage is 1 V, the smallest unit or LSB (001) is equivalent to $\frac{1}{8}$ V. (No voltage or a step smaller than $\frac{1}{8}$ V can be identified by a 3-bit converter.)

3 The MSB (100) has the equivalent value equal to $\frac{1}{2}$ V, 50% of the full scale.

4 For the maximum input signal (111), the analog output is $\frac{7}{8}$ V, which is $\frac{1}{8}$ V less than the full-scale value. This can be stated as:

$$V_{o(\max)} = (\text{full-scale value}) - (1 \text{ LSB}) \tag{13.3}$$

where $V_{o(\max)}$ is the maximum analog output voltage

D/A Converter Circuits

The D/A conversion process with straight binary code, described by the transfer function of Fig. 13.1*b* involves associating input signals with appropriate weights. The familiar circuit that performs this function is a summing operational amplifier (see Chaps. 9 and 10). These circuits may use binary weighted resistors, or an *R*/2*R* ladder network.

Binary Weighted Resistors This is the simplest type D/A converter, and uses the principle of the summing operational amplifier. Figure 13.2 shows a 3-input inverting op amp with input resistors R_1, R_2, and R_3 in binary weighted proportion. Each has double the value of the previous resistor. All three inputs D_2, D_1, and D_0, are 1 V. The total current, I_T, is:

$$\begin{aligned} I_T &= I_1 + I_2 + I_3 \\ &= \frac{V_{in}}{R_1} + \frac{V_{in}}{R_2} + \frac{V_{in}}{R_3} \\ &= \frac{1}{2} + \frac{1}{4} + \frac{1}{8} \\ &= 0.5 + 0.25 + 0.125 \\ &= 0.875 \text{ mA} \end{aligned}$$

The current contribution by each input is in binary proportion; therefore, voltage output V_o is also in the binary proportion:

$$\begin{aligned} V_o &= -R_F I_T \\ &= -(1 \text{ k}\Omega)(0.875 \text{ mA}) \\ &= -0.875 \text{ V} \\ &= -\frac{7}{8} \text{ V} \end{aligned}$$

The circuit of Fig. 13.2 simulates the binary digital input and the analog output. Assume the three inputs are connected to switches and can be turned on and off. If all switches are "off," the input represents the binary word 000

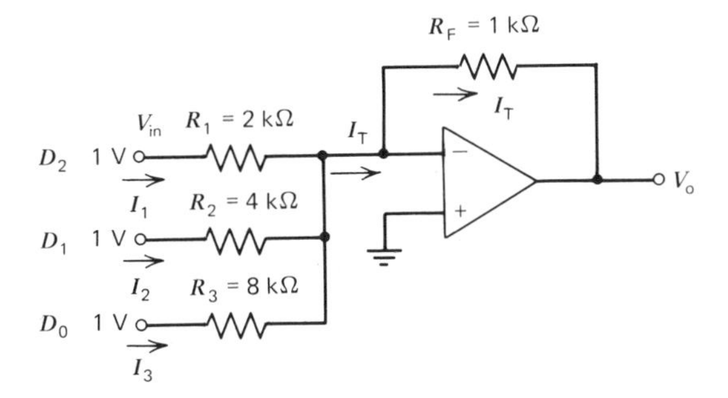

Figure 13.2 Summing amplifier connected to binary weighted input resistors.

and if all switches are "on" the input represents the binary word 111 (D_2 is the MSB and D_0 is the LSB). By using various combinations of switches, the input can assume any of eight binary (2^3) combinations, and the output will be the corresponding analog voltage.

EXAMPLE 13.6 In Fig. 13.2, calculate the output voltage if the inputs are $D_2 = 0$, $D_1 = 0$, and $D_0 = 1$ (001) and verify that the analog output corresponds to the LSB.

PROCEDURE

1 Current $I_T = 0 + 0 + \dfrac{1\text{ V}}{8\text{ k}\Omega}$

$= 0.125$ mA

2 Output voltage $V_o = -(1\text{ k}\Omega)(0.125\text{ mA})$

$= -\ 0.125\text{ V} = -\dfrac{1}{8}\text{ V}$

The output, $\frac{1}{8}$ V, corresponds to the LSB of a 3-bit DAC, as shown in the graph of its transfer function (Fig. 13.1*b*).

A major drawback in designing the binary weighted DAC is the requirement for various precision resistors of increasing values. The *R*/2*R* ladder approach described in the next section uses only two values of resistors for any number of bits. This eliminates the impractability of having many values of precision resistors.

R/2R Ladder D/A Converter The basic principle of an *R*/2*R* ladder DAC is similar to the binary weighted DAC except that individual input resistors are replaced with a "ladder" network of resistors of only two values. The resistors are connected in such a way that for any input, the effect becomes equivalent to that of the binary weighted resistor. Figure 13.3 shows a 3-bit input *R*/2*R* ladder DAC. The inputs D_0, D_1, and D_2 can either be grounded or connected to a 1 V supply through switches.

Assume $D_2 = 0$, $D_1 = 0$, and $D_0 = 1$ (001). The circuit is simplified by

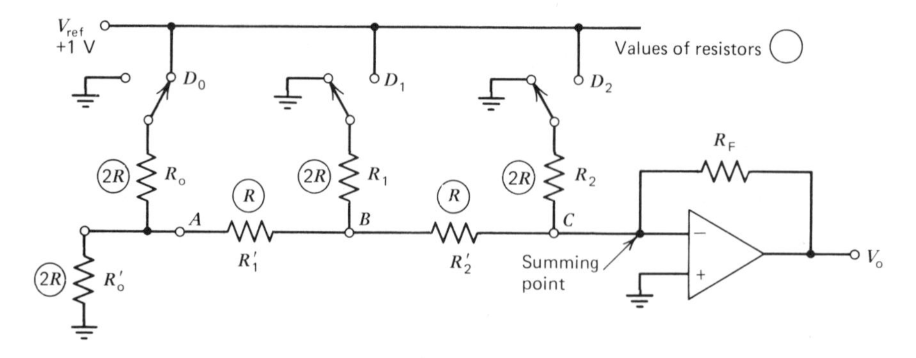

Figure 13.3 Summing amplifier connected to an R/2R ladder input network.

drawing its Thevinin equivalent. The Thevenin voltage becomes 0.5 V with R and R'_o replaced with their Thevinin equivalent resistance, $R_{To} = R_o // R'_o$ (Fig. 13.4*a*). The circuit is further simplified by similarly Thevinizing two more times as shown in Figs. 13.4*b* and 13.4*c*.

In the final Thevenized circuit, the voltage source is 0.125 V with R_{T2} as a driving resistance. If R_{T2} is assumed to be 1 kΩ, the current $I_o = 0.125/1 = 0.125$ mA and the output voltage = 0.125 V. This corresponds to the input 001.

In this circuit, the currents from various inputs are added in binary proportion at the summing point, and this current is the analog equivalent of the input signals. This current can be used as the output to drive a load, as in a *current output DAC,* or can be converted into an equivalent voltage, as in a *voltage output DAC.* In a voltage output DAC, an op amp with a feedback resistor is used, as shown in Fig. 13.3. Bipolar operation is obtained by adding bias current at the op amp summing point, or the output can be made equivalent to the product of the input signal and the reference voltage; this forms a *multiplying type DAC.*

The following points should be noted:

1 The basic circuit blocks included in D/A converters are a resistor network to provide appropriate weighting of the input signal, switches, and a reference voltage.

2 The output can be a current, or converted into a voltage by using an op amp.

3 The accuracy of an analog output signal is primarily dependent on the accuracy of the resistors.

4 Time required for conversion (see definition of Settling Time, Sec. 13.4) is dependent on the response time of the switches and the output op amp (for a voltage output DAC).

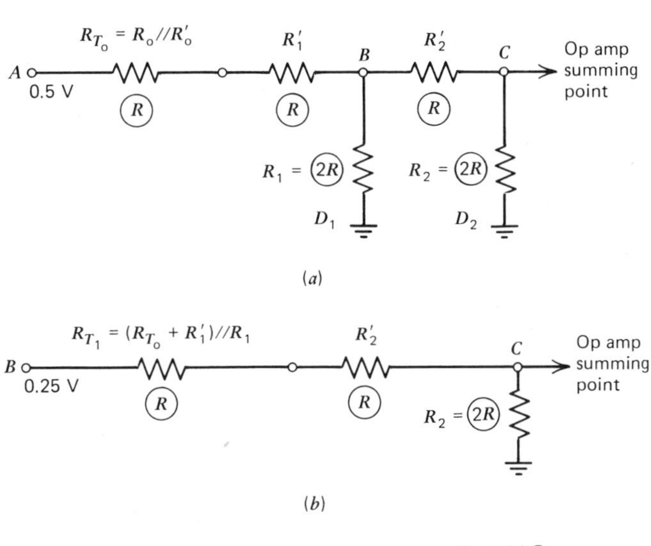

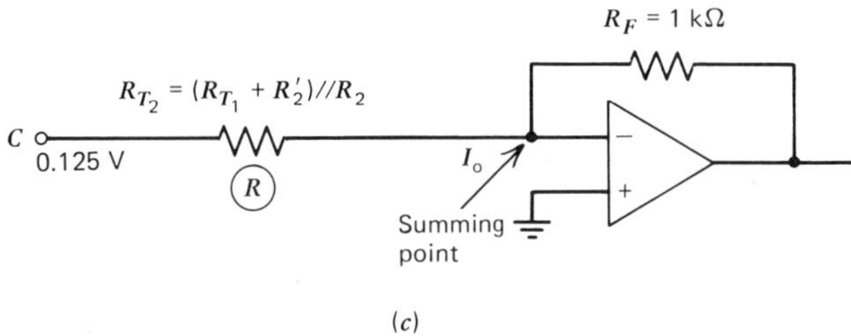

Figure 13.4 Developing Thevenin equivalent circuit of Fig. 13.13. Thevenized circuit at points (*a*) *A*, (*b*) *B*, and (*c*) *C*.

13.4 D/A CONVERTER SPECIFICATIONS

Commercial D/A converters are available in modules, hybrid, and monolithic versions. Modules and hybrid types, using several components in a single package, are able to combine the best features of various technologies for optimum performance. The monolithic version, with all circuitry on one chip, is small in size and low in cost. The ideal transfer function for a D/A converter is shown in Fig. 13.1*b*. However, the actual performance of a DAC deviates from the ideal transfer function, owing to changes in various characteristics and parameters. Important terms used to characterize a DAC are:

Resolution The smallest incremental change in output voltage of a D/A converter. An n-bit binary converter should be able to provide 2^n distinct analog output values corresponding to 2^n combinations of binary input. Resolution is defined as $1/2^n$ where n is the number of input bits. The resolution can also be stated as 1 part in 2^n, or as percent of full scale.

EXAMPLE 13.7 Calculate the resolution in percent and in millivolts for an 8-bit D/A converter with 10 V full-scale range.

PROCEDURE

1 Resolution $= 1/2^n$ (13.4)

$$= 1/2^8 = \frac{1}{256} = 0.39\%$$

2 For 10 V full scale, the resolution $= 10 \text{ V} \times \frac{1}{256} = 39 \text{ mV}$

Settling Time The time required, after a code transition, for a DAC output to reach final value within specified limits (usually $\pm \frac{1}{2}$ LSB). The settling time of a voltage output DAC is longer than that of a current output type owing to the slow response of the output op amp.

Absolute Accuracy How well a data converter follows the theoretical ideal transfer function taking into account all errors. Errors include *zero offset, gain,* and *linearity*.

Zero Offset or Offset Error The error when the digital input code calls for zero output (Fig. 13.5*a*). This error may be adjustable by setting the output to zero when the input bits are zero.

Gain Error The difference in slope between the actual transfer function and the ideal transfer function (Fig. 13.5*b*). The gain error may be adjustable by setting the output to full scale less 1 LSB when all input bits are 1s.

Linearity Error or Relative Accuracy The deviation from the ideal transfer function after adjusting zero offset and gain error (Fig. 13.5*c*).

Monotonicity The requirement that a converter analog output increase with increasing input (Fig. 13.5*d*).

Differential Linearity Error This measures the difference between any two adjacent steps (Fig. 13.6). If the differential linearity error is greater than 1 LSB, it leads to *nonmonotonicity*.

Zero Temperature Coefficient (TC) The zero shift over a specified temperature range. It is expressed as μV/°C or ppm/°C (parts per million) of full scale.

Gain TC The change of gain with temperature, causing the slope to change; it is expressed in ppm/°C.

Linearity TC The change in linearity due to temperature. It is expressed as ppm/°C of full scale.

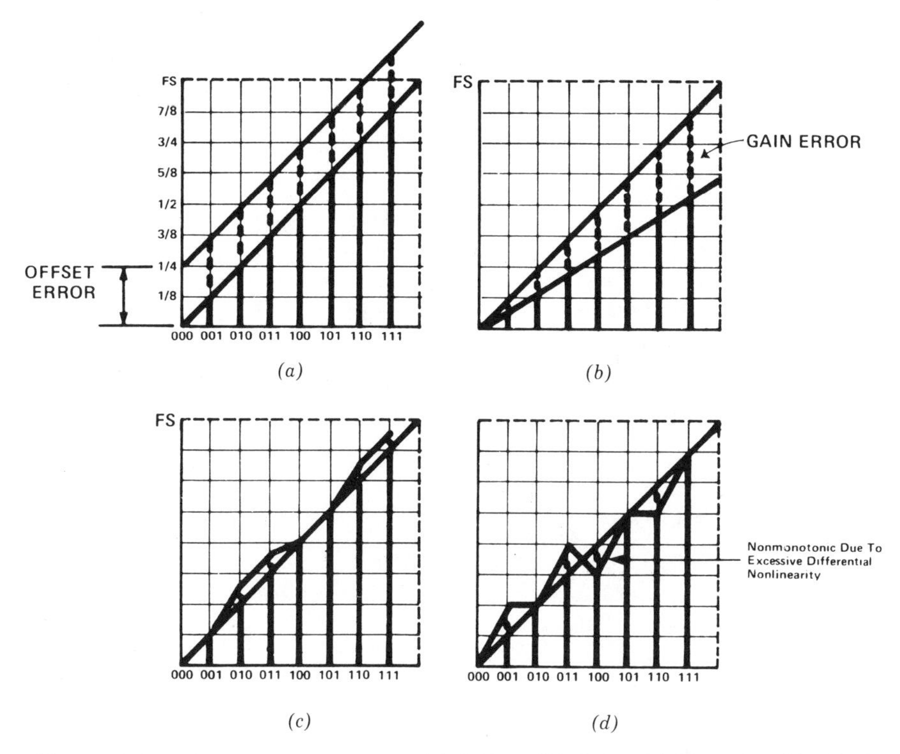

Figure 13.5 D/A conversion errors. (*a*) Offset. (*b*) Gain. (*c*) Linearity (bit 2 is $\frac{1}{2}$ LSB high). (*d*) Nonmonotonicity (bit 3 is 1 LSB high, bit 1 is 1 LSB low).

Differential Linearity TC The relative change in bit weights with temperature and is a measure of when a converter can be expected to go nonmonotonic. It is expressed as ppm/°C of full scale.

Power Supply Sensitivity The change in output stemming from variations in power supply voltages. It is normally expressed as output voltage change in percent (of full scale) or fraction of an LSB, for a 1% dc change in the power supply voltage.

Error Estimation

The major concern in selecting a D/A converter is the accuracy of analog output relative to the input over the required operating range of temperature. Monotonicity is essential for many control applications. Therefore, it is necessary to determine these errors and verify that the total error does not exceed 1 LSB.

EXAMPLE 13.8

a. Calculate the value of 1 LSB in a 12-bit D/A converter. Express the value in ppm and in millivolts if the full-scale range is 10 V.

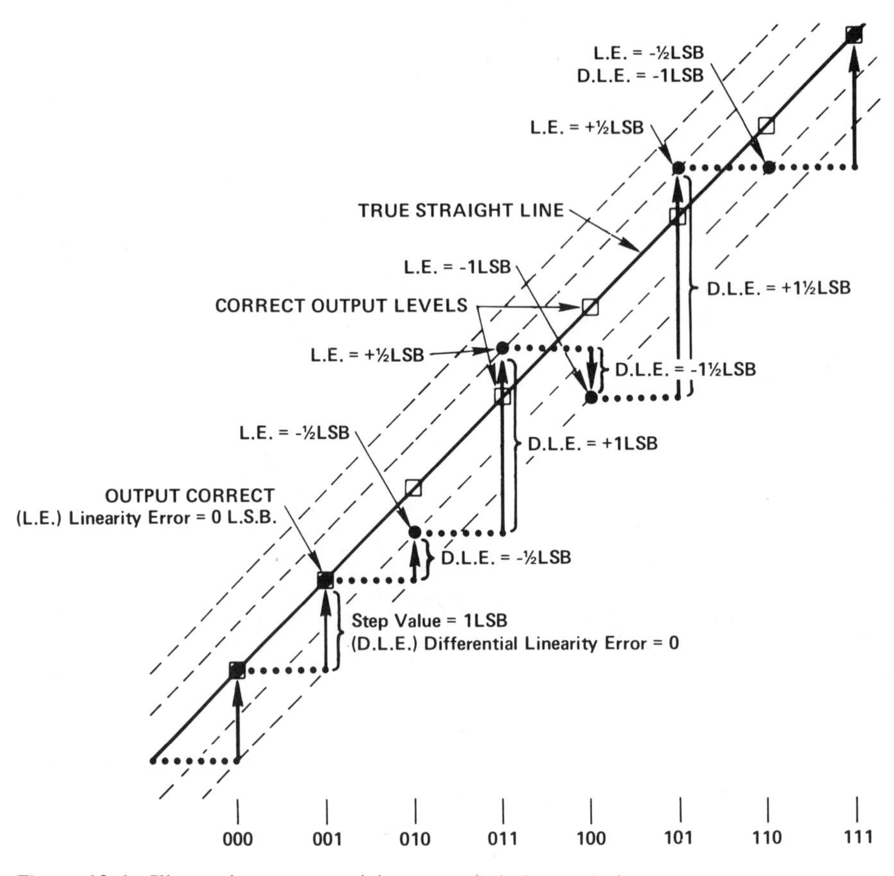

Figure 13.6 Illustrating monotonicity error (missing codes).

b. Calculate the total error over the temperature range of 0 to 100°C (with reference to 25°C) if the initial nonlinearity is $\frac{1}{2}$ LSB and the differential nonlinearity temperature coefficient is 2.5 ppm/°C.

PROCEDURE

1 1 LSB $= 1/2^{12}$
$= 2.4 \times 10^{-4}$
$= 244$ ppm
$= 2.44$ mV for 10 V full scale

2 Temperature Difference $= 100°\text{C} - 25°\text{C}$
$= 75°\text{C}$

Error due to differential nonlinearity $= 2.5 \text{ ppm} \times 75°\text{C}$
$= 187.5$ ppm

$$\text{Initial nonlinearity} = \frac{1}{2}\text{ LSB}$$
$$= 122 \text{ ppm}$$

Total Error = Initial nonlinearity + differential nonlinearity due to TC
= 187.5 + 122
= 309.5 ppm

This error exceeds 1 LSB (244 ppm); the output, therefore, is likely to be nonmonotonic over 100°C.

13.5 CURRENT OUTPUT DAC

The current output DAC is a commonly used converter and is faster than the voltage output DAC. For a general-purpose, 8-bit current DAC the settling time is about 300 ns. In high-speed current DACs, the settling time is around 25 ns. However, in many applications the output current is converted into a voltage signal with an op amp. Thus, the settling time is considerably affected by the response of the op amp.

The next section describes a general purpose D/A converter, the MC1408. D/A converters similar to the MC1408, or improved versions, are also available from several manufacturers. Typical applications include panel meters, digital voltmeters, peak detectors, programmable gain and attenuation, speech compression and expansion, and stepping motor drive.

The MC1408 DAC

This is a representative, low-cost, monolithic 8-bit D/A converter that is TTL and CMOS compatible and has a settling time of 300 ns (Fig. 13.7*a*). It requires two power supplies: $V_{CC} = +5$ V and $V_{EE} = -5$ V. Supply V_{EE} can range from -5 to -15 V. The reference current required is 2 mA for full-scale input. The pin diagram is provided in Fig. 13.7*b*.

Circuit Description The 1408 DAC consists primarily of eight high-speed current switches, an *R*/2*R* ladder network, and a reference current amplifier (Fig. 13.7*a*). The input pins are designated A1 through A8, MSB through LSB, respectively. (This is the reverse of the usual digital circuit convention.)

Figure 13.8 shows a typical circuit operation. Resistor R_{14} and voltage V_{ref} determine the total reference current. Resistor R_{15} is generally equal to R_{14} to match the input impedance of the reference amplifier. The reference current is switched through the *R*/2*R* ladder network in binary proportion according to the input bits. The output current, I_o, which is also proportional to the input value, flows into the circuit. It is equal to:

$$I_o = \left(\frac{V_{ref}}{R_{14}}\right)\left(\underset{\text{MSB}}{\frac{A1}{2}} + \frac{A2}{4} + \frac{A3}{8} + \frac{A4}{16} + \frac{A5}{32} + \frac{A6}{64} + \frac{A7}{128} + \underset{\text{LSB}}{\frac{A8}{256}}\right) \tag{13.5}$$

The output current can be used directly, as shown in Fig. 13.9*a,* in a panel meter read out or converted into a corresponding voltage using an op amp, as in Fig. 13.9*b*.

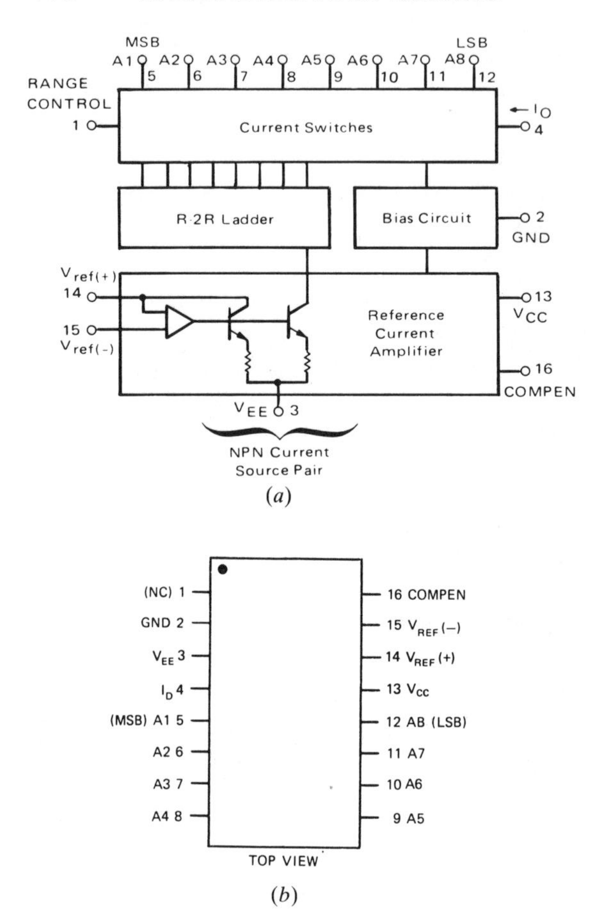

(*a*)

(*b*)

Figure 13.7 The MC1408 DAC. (*a*) Functional block diagram. (*b*) Pin connections. (Courtesy Motorola, Inc.)

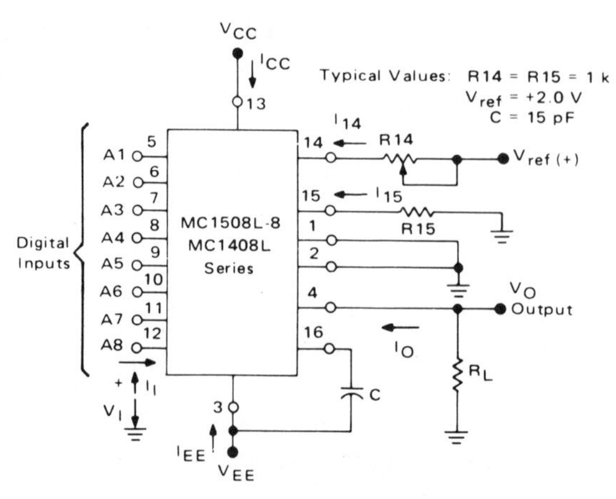

Figure 13.8 Typical circuit operation of the MC1408. (Courtesy Motorola, Inc.)

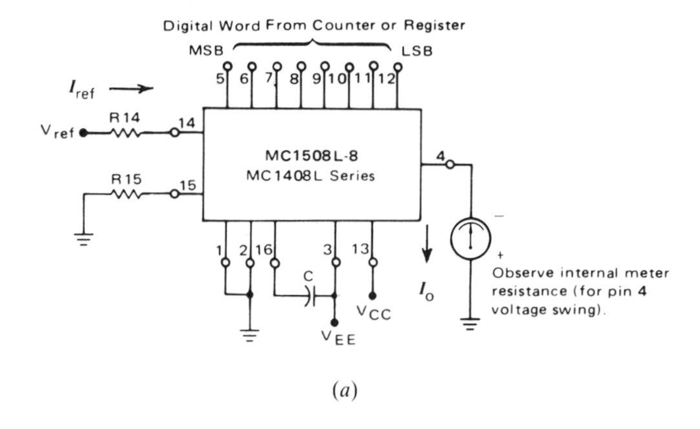

(a)

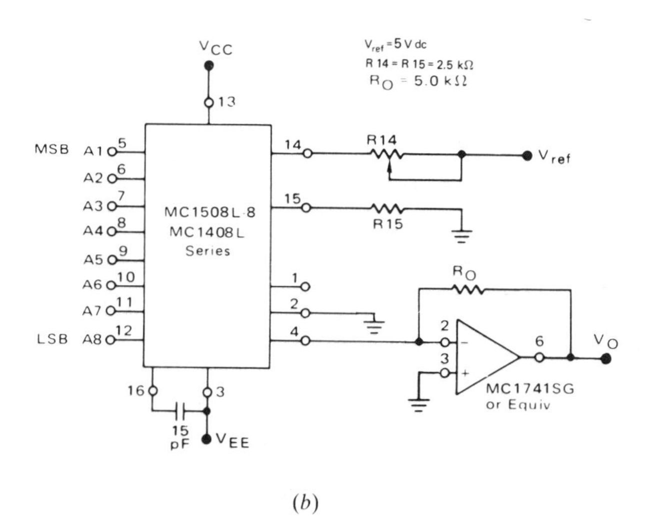

(b)

Figure 13.9 Examples of MC1408 applications. (*a*) Current meter. (*b*) Current to voltage converter. (Courtesy Motorola, Inc.)

EXAMPLE 13.9

a. In Fig. 13.9*a*, select R_{14} and calculate the reference current, I_{ref}; $V_{ref} = 5$ V.

b. Calculate the output current, I_o, when the inputs are 11111111 and 00000001 (LSB).

PROCEDURE

1 Resistor R_{14} is selected as 2.5 kΩ because the typical reference current specified is 2 mA for the maximum input (all bits are 1s). The maximum limit on I_{ref} for the 1408 DAC is 5 mA.

The reference current, therefore, is:

$$I_{ref} = V_{ref}/R_{14}$$
$$= (5 \text{ V})/(2.5 \text{ k}\Omega) = 2 \text{ mA}$$

2 The output current for the two different input words is calculated by Eq. 13.5:

$$I_o(A0\text{–}A8) = (V_{ref}/R_{14})\left(\frac{A1}{2} + \frac{A2}{4} + \cdots + \frac{A8}{256}\right)$$

$$= (2\,\text{mA})\left(\frac{1}{2} + \frac{1}{4} + \frac{1}{8} + \frac{1}{16} + \frac{1}{32} + \frac{1}{64} + \frac{1}{128} + \frac{1}{256}\right)$$

$$= 1.992\,\text{mA}$$

$$I_o(\text{LSB}) = (V_{ref}/R_{14})\left(\frac{A8}{256}\right)$$

$$= (2\,\text{mA})\left(\frac{1}{256}\right) = 7.8\,\mu\text{A}$$

EXAMPLE 13.10 Referring to Fig. 13.9*b*, calculate the output voltage V_o when the input words are 11111111 and 00000001.

PROCEDURE

1 From Ex. 13.9, I_o $(A0\text{–}A8 = 1) = 1.992$ mA; therefore, $V_o = 5 \times 1.992 = 9.961$ V.
2 I_o (LSB) $= 7.8\ \mu$A and $V_o = 39$ mV.

D/A Converter with Bipolar Voltage Range

In the previous example, the analog output voltage was restricted to positive voltages, 0 to 10 V. The output voltage can be made bipolar, however, by offsetting the current input to the op amp. The output current then is $I_o - I_{ref}$ (2 mA) for the maximum input signal. The output current can be reduced to 1 mA by connecting resistor R_B between the reference voltage and output pin 4 (Fig. 13.10).

EXAMPLE 13.11

a. Calculate resistance R_B in Fig. 13.10 to obtain the analog output voltage from +5 to −5 V.

b. Calculate the output voltage when the input word = 10000000.

PROCEDURE

1 To offset the output current by 1 mA, the current through resistor R_B should be 1 mA in the opposite direction:

$$R_B = \frac{5}{1} = 5\ \text{k}\Omega$$

2 The output voltage V_o for the input word 10000000 is:

$$I_o = \left(\frac{V_{ref}}{R_{14}}\right)\left(\frac{A1}{2}\right) - \frac{V_{ref}}{R_B}$$

$$= \frac{5}{2.5} \times \frac{1}{2} - \frac{5}{5} = 0\ \text{mA}$$

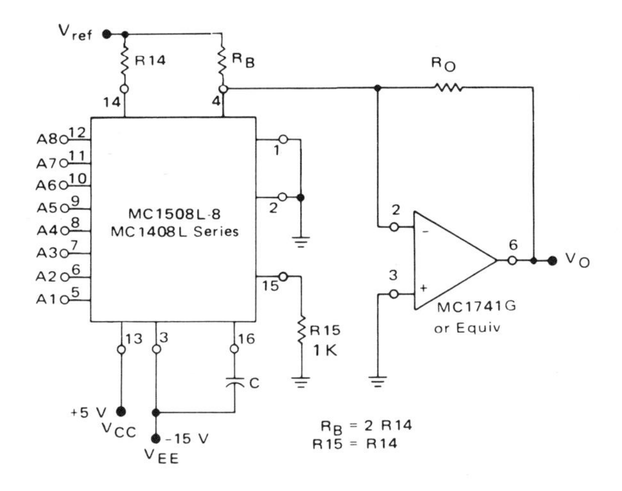

$$V_O = \frac{V_{ref}}{R14}(R_O)\left[\frac{A1}{2}+\frac{A2}{4}+\frac{A3}{8}+\frac{A4}{16}+\frac{A5}{32}+\frac{A6}{64}+\frac{A7}{128}+\frac{A8}{256}\right] - \frac{V_{ref}}{R_B}(R_O)$$

Figure 13.10 The MC1408 connected for bipolar operation. (Courtesy Motorola, Inc.)

Therefore,

$$V_o = R_o I_o = 5 \times 0 = 0 \text{ V}$$

13.6 VOLTAGE OUTPUT DAC

The voltage output type D/A converter is basically a current output DAC with an added op amp to convert output current into voltage. The basic principle was illustrated in Ex. 13.10. The voltage output converter has a longer settling time because of the op amp response time and stray capacitance. As an example, Fig. 13.11 shows a voltage output D/A converter (Datel–Intersil type DAC-UP8B). It is an 8-bit monolithic DAC with a high-speed output amplifier. It has an internal register that makes it suitable for interfacing with a microprocessor. Typically, the settling time is 2 μs.

When the line $\overline{\text{LOAD}}$ goes high, the input is latched and the data are retained until it goes low. When $\overline{\text{LOAD}}$ is low, the register is transparent, and the output changes according to the input word.

13.7 MULTIPLYING DAC

The D/A converter can be used as a multiplier by varying the reference voltage. The analog output of any DAC is proportional to the product of the reference

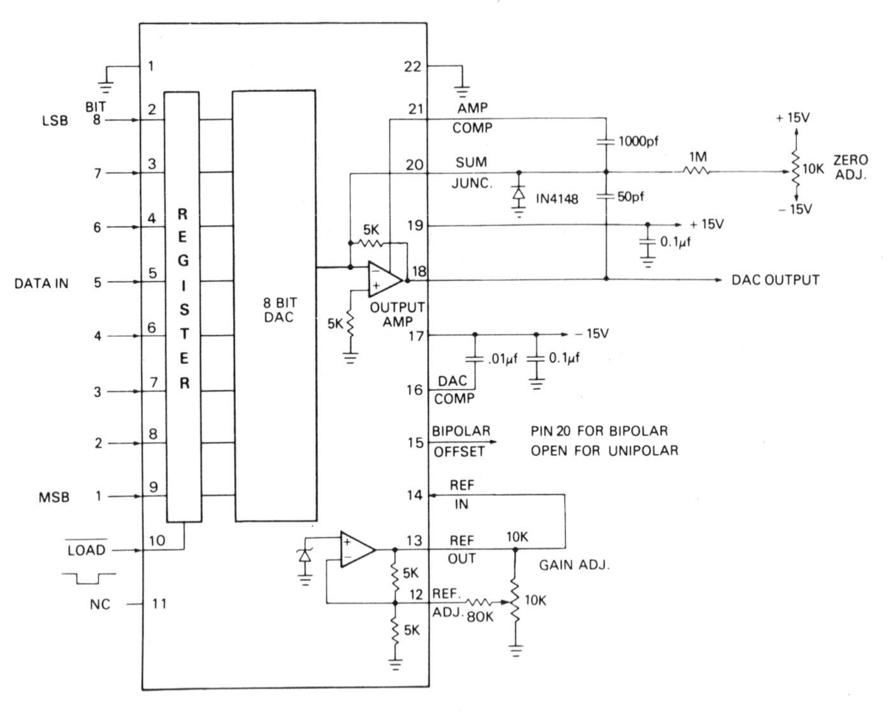

Figure 13.11 An example of a voltage output DAC. (Courtesy Datel–Intersil.)

voltage and the digital input value. In Fig. 13.8*b*, if the reference voltage is reduced from 5 to 2.5 V, the output V_o for the largest input word (11111111) is reduced to 4.98 V, half the output when the reference voltage is 5 V. Therefore, in some ways every DAC can be viewed as a multiplying device. However, with a general purpose DAC, the multiplying range is limited. DACs which are designed for multiplication provide a linear output over a greater range of reference voltage values.

Multiplying DACs are of two types: *two quadrant* or *four quadrant*. In a two-quadrant multiplying converter, one of the inputs, either the reference voltage or the digital input, can be bipolar and the output is in either of two quadrants. In a four-quadrant multiplying converter, both the reference and input voltages can be bipolar, yielding a four-quadrant output.

8-Bit Multiplying D/A Converter AD7523

As an example, the AD7523 is a low-cost, monolithic CMOS multiplying DAC packaged in a 16-pin DIP (Fig. 13.12*a*). It requires one power supply between +5 and +16 V; its settling time is less than 150 ns.

Figure 13.13*a* shows a typical application of unipolar binary operation. The reference voltage can assume positive and negative values, for two-quadrant operation. If the reference voltage is +10 V and if the MSB is one, 0.5 mA flows through switch S_1 (Fig. 13.12*b*). This current (OUT 1) flows into the summing junction of the op amp, and results in a contribution to the output voltage of $-5V$ (0.5 mA flowing through the 10 kΩ feedback resistor).

Figure 13.13*b* illustrates bipolar operation. In this configuration, current (OUT 2) which is the complement of (OUT 1), is inverted and added to current (OUT 1). This reduces the resolution to half and extends the output range in both polarities, as summarized in Table 13.5.

13.8 ANALOG-TO-DIGITAL (A/D) CONVERTERS

Analog-to-digital converters (ADCs) are used to interface analog signals with digital circuits or computers. Various approaches are used, depending upon the accuracy and the speed required, to convert an analog signal into its digital

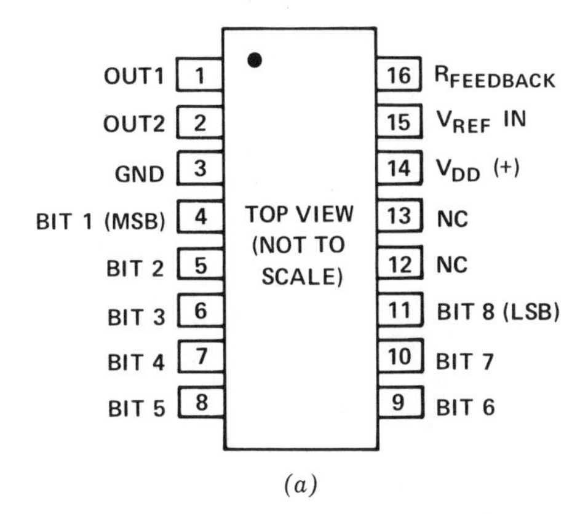

(*a*)

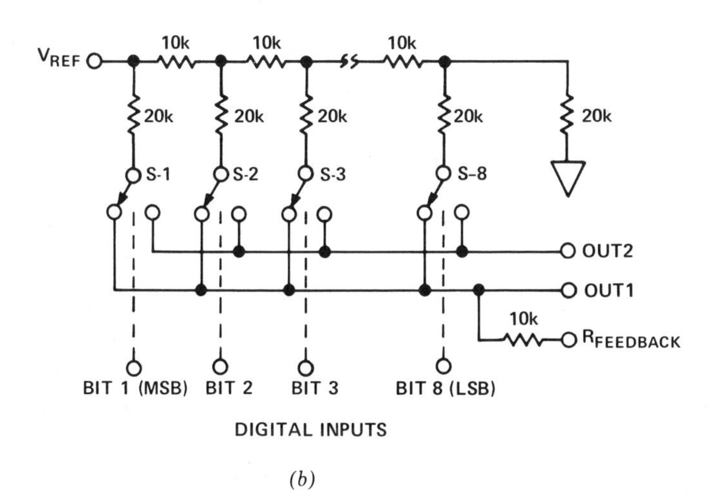

(*b*)

Figure 13.12 The A7523 multiplying DAC. (*a*) Pin connections. (*b*) Internal R/2R ladder network. (Courtesy Analog Devices.)

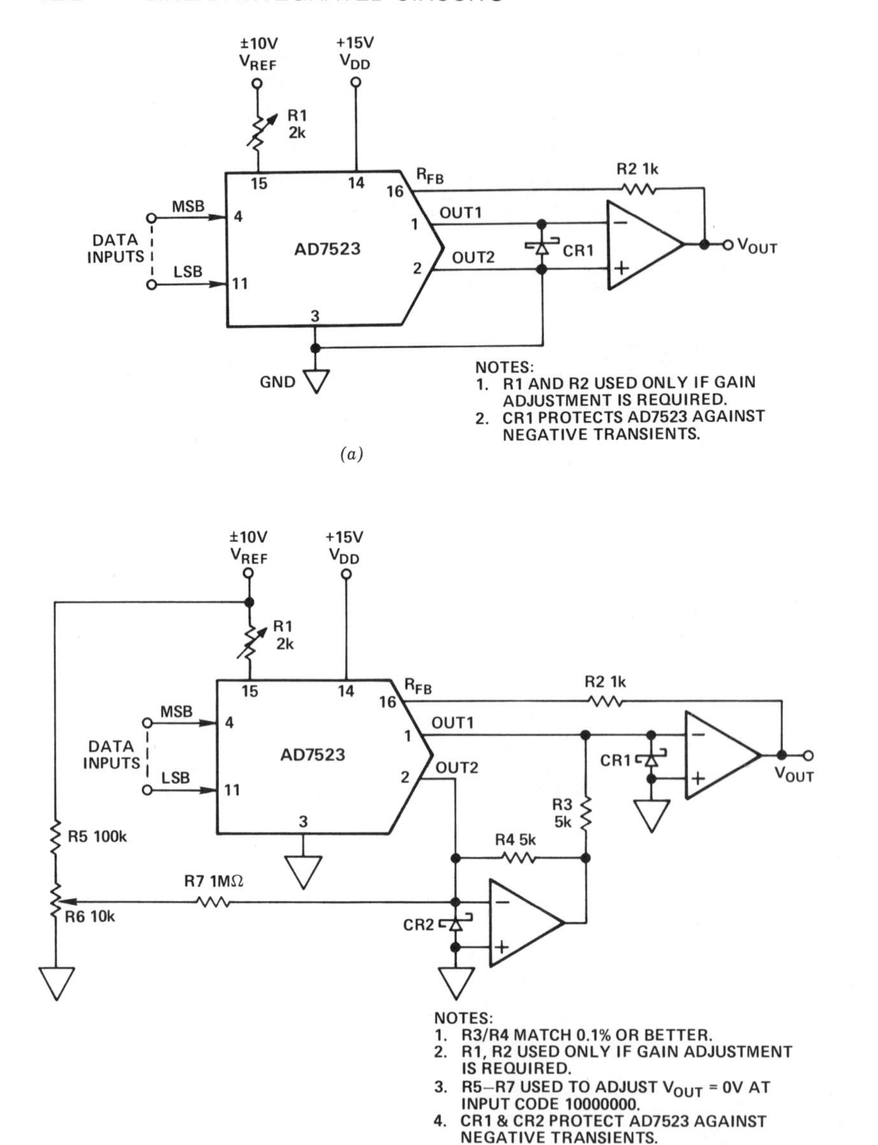

Figure 13.13 Multiplying DAC. (*a*) Unipolar operation. (*b*) Bipolar operation. (Courtesy Analog Devices.)

equivalent. Three such approaches are *successive approximation, integrating,* and the *parallel* (*flash*) types.

A/D conversion is essentially a quantizing process, whereby an analog signal is represented by discrete states. These states can be assigned appropriate codes such as straight binary, BCD, or two's complement.

Table 13.5 **Digital input and analog output for two- and four-quadrant multiplying DACs**

Digital input		*Analog output*	
		Two-quadrant multiplying	*Four-quadrant multiplying*
1 1 1 1	1 1 1 1	$-V_{ref}$ (255/256)	$-V_{ref}$ (127/128)
1 0 0 0	0 0 0 0	$-V_{ref}$ (128/256)	0
0 0 0 0	0 0 0 1	$-V_{ref}$ (1/256)	$+V_{ref}$ (127/128)
0 0 0 0	0 0 0 0	0	$+V_{ref}$ (128/128)

Transfer Function

Figure 13.14*a* shows a 3-bit A/D converter. Its transfer function, with 1 V full-scale input, is provided in Fig. 13.14*b*. A 3-bit converter has eight (2^3) discrete output states, 000_2 to 111_2, corresponding to eight segments ($\frac{1}{8}$ V each) of the analog input. In examining the ideal transfer function of Fig. 13.14*b,* note that:

1 The 3-bit A/D converter has $2^n = 2^3 = 8$ output states where *n* is the number of output bits.

2 The value of the first discrete level, or LSB, is obtained by dividing the full-scale input voltage into the number of discrete states.

$$1 \text{ LSB} = \frac{\text{Full-Scale Range (FSR)}}{2^n} = \frac{1}{8} \text{ V}$$

This is the *resolution* of the converter.

3 The analog input values shown (0.125 V, 0.25 V, and so on) represent the center points of the analog values for each output word, with transition points $\pm \frac{1}{2}$ LSB from the center points. The *quantizing error,* or *uncertainty,* is thus $\frac{1}{2}$ LSB.

4 The MSB $(100)_2$ output corresponds to 0.5 V, which is half of the input full-scale voltage. The largest output word 111_2 corresponds to $\frac{7}{8}$ V, and not the full-scale voltage which can be stated as:

$$V_{max} = \text{Full-Scale Range} - 1 \text{ LSB}$$
$$= 1 \text{ V} - \frac{1}{8} \text{ V} = \frac{7}{8} \text{ V}$$

These points are summarized in Table 13.6* and are further clarified in the following example.

* Note that Tables 13.4 and 13.6 are the same, except for the input and output reversal.

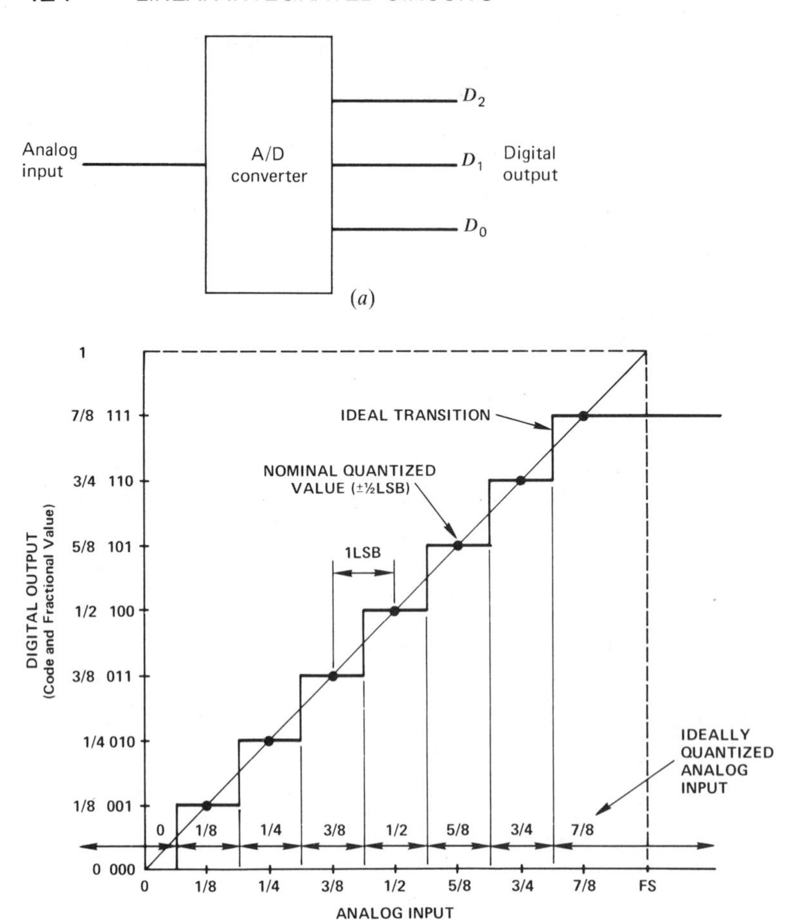

Figure 13.14 A 3-bit A/D converter. (*a*) Block diagram. (*b*) Ideal transfer function.

EXAMPLE 13.12 Calculate analog voltages corresponding to the LSB, and the maximum output for an 8-bit A/D converter (straight binary code) if the input range is 0 to 10 V.

PROCEDURE

1 For 8 bits, $2^8 = 256$ output states.

2 Because the full-scale input voltage is 10 V, the LSB voltage $V_{LSB} = \dfrac{10\text{ V}}{256} = 39\text{ mV}$.

3 The voltage corresponding to the maximum output word V_{max} = Full-Scale Voltage $- V_{LSB} = 10\text{ V} - 0.039\text{ V} = 9.961\text{ V}$

Table 13.6 **Analog input and binary output for a 3-bit A/D converter**

Analog input (normalized fraction)	*Analog input (V) (10 V full scale)*	*Binary digital output*
0	0	0 0 0
$\frac{1}{8}$	1.25	0 0 1
$\frac{1}{4}$	2.50	0 1 0
$\frac{3}{8}$	3.75	0 1 1
$\frac{1}{2}$	5.00	1 0 0
$\frac{5}{8}$	6.25	1 0 1
$\frac{3}{4}$	7.50	1 1 0
$\frac{7}{8}$	8.75	1 1 1

Sampling Concepts

An A/D converter requires a certain amount of time, called *conversion time,* to change an analog signal into the corresponding digital signal. If the analog signal changes during the conversion time, the converter output may be in error. To prevent this, a *sample-and-hold circuit* is used to sense the analog signal at the start of conversion, and store it on a capacitor during the remaining conversion time.

Figure 13.15 shows a simplified schematic of a sample-and-hold circuit. Amplifier A_1 is an input buffer with a high-input impedance and A_2 is an output amplifier with a low-output impedance. Switch S is generally a rapidly switching FET circuit. When S is closed, capacitor C charges rapidly to the input voltage. When S opens, the initial voltage is maintained. However, if a sampling rate is much slower than changes in the analog signal, information will be lost. Therefore, the minimum sampling rate, as defined by sampling theory, should be $2f_c$ samples/s, where f_c is the highest frequency component in the analog signal.

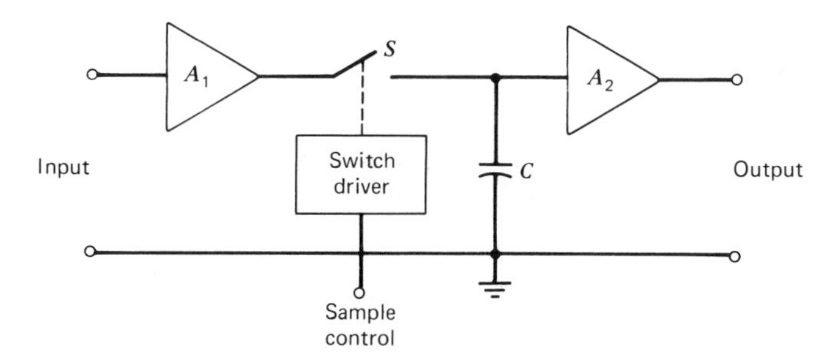

Figure 13.15 Basic sample-and-hold circuit.

Characteristics of A/D Converters

Figure 13.14*b* shows the ideal transfer function for an A/D converter. However, the actual performance of an ADC deviates from the ideal transfer function. These deviations are determined by the characteristics defined below, which are specified by the manufacturer.

Conversion Time The total time required to completely convert an analog input voltage to a digital output. It is affected by the propagation delay in various circuits. It may also be specified as a number of conversions per second.

This is one of the most important specifications considered in selecting an ADC. Converters having more output bits usually have a longer conversion time.

Resolution The amount of input voltage change required to increment the output by one LSB. The resolution is determined by the number of output bits. It is specified either in terms of the number of output bits, or as one part per number of output states. Thus, the resolution of an 8-bit converter is specified either as 8 bits, or 1 out of 256 $\left(\frac{1}{2^8} = \frac{1}{256}\right)$.

Quantizing Error or Uncertainity The inherent error associated with dividing a continuous signal into a finite number of discrete states. It is equal to $\pm \frac{1}{2}$ LSB and represents the uncertainity at the transition points (see Fig. 13.14*b*).

Absolute Accuracy The total error between the ideal and the actual analog input voltages required to produce a given output code. It consists of quantization error, gain error, zero error, and nonlinearities. However, it is rarely specified on data sheets.

Relative Accuracy The deviation of the analog voltage from its ideal value expressed as a fraction of full scale, with gain and offset errors adjusted to zero. Relative accuracy is a function of linearity and is usually specified as $\pm \frac{1}{2}$ LSB maximum. The factors that affect the relative accuracy are: differential linearity, offset and gain error, temperature coefficients, and power supply sensitivity. (Definitions of these characteristics, given for D/A converters, apply here as well.)

No Missing Code A digital output code should correspond to a quantum of analog input values 1 LSB apart (2^{-n} of full scale, for an n-bit converter). Any deviation of a step from the ideal width is called *differential nonlinearity*. A differential nonlinearity greater than 1 LSB can lead to a missing code (digital number). The *no missing code* characteristic indicates that this situation will not occur.

13.9 SUCCESSIVE APPROXIMATION ADCs

The successive approximation technique is one of the most widely used methods for general-purpose A/D converters. It involves comparing the unknown input analog voltage with the output of an internal D/A converter. The digital input to the D/A converter is also internally generated. When the DAC output matches the analog input voltage, the internally generated signal is equivalent to the input and becomes the digital output.

The procedure for comparing an unknown analog voltage with the output of the DAC is similar to weighing material less than 1 gm on a chemist's balance with a set of fractional weights, such as $\frac{1}{2}$ gm, $\frac{1}{4}$ gm, $\frac{1}{8}$ gm, and $\frac{1}{16}$ gm. The weighing procedure is to start with the heaviest weight, $\frac{1}{2}$ gm, and if not sufficient, subsequent weights are added one by one, until the balance is tipped. The weight that tips the balance is removed, and the successive weights are added in decreasing value until the last weight.

Figure 13.16*a* shows the block diagram of a successive approximation ADC. The diagram includes three major components: comparator, DAC, and SAR (successive approximation register). The process starts with a conversion command which clears the previous data. The MSB of the DAC (which has an output equal to $\frac{1}{2}$ of the full scale) is then turned on. If this output is insufficient, the successive bits of the DAC are turned on until the output of the DAC exceeds the analog input. The last bit, which causes the excess voltage, is turned off and the process continues until the LSB.

Figure 13.16*b* shows the comparison process graphically for a 4-bit DAC. When the MSB (D_3) is turned on, the DAC output is less than the analog input voltages; therefore, the MSB remains on. When bit D_2 is turned on, the output exceeds the input, and bit D_2 is turned off. Similarly, D_1 is turned off and D_0 remains 1. The digital output thus becomes 1001.

The successive approximation ADC has the following features:

1 Moderate to high speed.
2 High resolution (up to 16 bits).
3 Fixed conversion time which is independent of the input voltage.
4 Each conversion output is independent of the previous conversion output. This condition exists because the previous output is cleared before the start of the next conversion.

Its limitations are:

1 If the input changes during a conversion, the digital output no longer represents the input voltage.
2 For fast changing inputs, an additional sample- and hold-circuit is necessary.
3 Linearity and speed are primarily dependent on the properties of the DAC. If the DAC is nonmonotonic, the output may have missing codes.

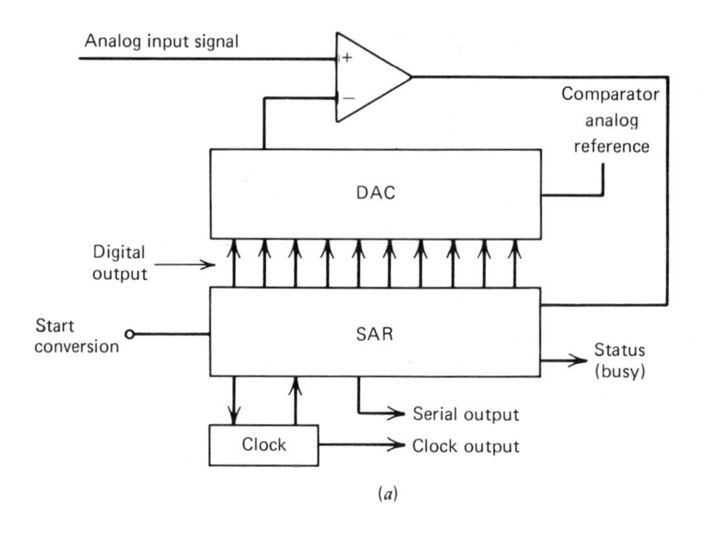

(a)

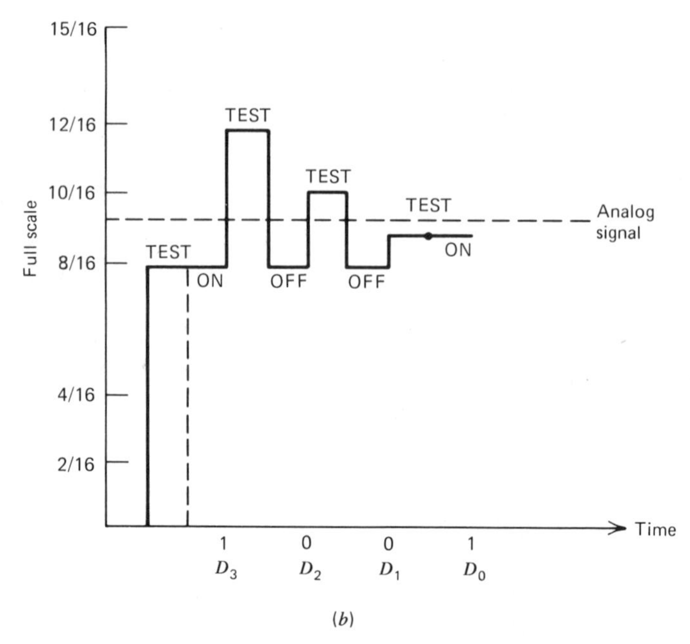

(b)

Figure 13.16 Successive approximation DAC. (*a*) Block diagram. (*b*) Internal process for determining output for a given input.

A representative IC converter using successive approximation is the 10-bit Analog Device type AD571. Conversion time is less than 25 μs; no external components are required; it has 10-bit accuracy with no missing codes; accepts either unipolar (0 to +10 V) or bipolar (−5 to +5 V) analog input voltages; and requires +5 and −15 V power supplies.

Circuit Description Figure 13.17*a* shows the functional block diagram of the AD571, which includes a 10-bit current output DAC, successive approximation register (SAR), tri-state buffer, and comparator. It is available in a dual-in-line package (DIP) with 18 pins, of which two pins, $\overline{\text{DATA}}$ $\overline{\text{READY}}$ and BLANK and $\overline{\text{CONVERT}}$, are control functions (Fig. 13.17*b*). There are separate grounds for the analog input and digital output signals. In most applications, all that is necessary is the connection of the power supplies (+5 V and −15 V), the analog input signal with a driving resistance R_{in} = 15 to 50 Ω, and the conversion start pulse.

When a low signal is applied to pin 11 (BLANK and $\overline{\text{CONVERT}}$), the conversion begins. The current output of the DAC is sequenced by the SAR from the MSB to the LSB. The input analog signal and the output of the DAC are compared, testing each bit. The SAR then contains a 10-bit binary word equivalent to the input signal and $\overline{\text{DATA}}$ $\overline{\text{READY}}$ becomes active low. The $\overline{\text{DATA}}$ $\overline{\text{READY}}$ signal also activates the tri-state output buffer, connecting the SAR content to the output. When pin 11 is made high (BLANK), the tri-state buffer disconnects the output and the SAR is prepared for another cycle. Before using the circuit, *zero-offset* and *full-scale adjustments* are necessary.

Zero Offset The zero point of the AD571 can be adjusted by grounding the analog input, and inserting an offset voltage between pin 14, ANALOG COMMON, and the external analog signal common (or ground). The circuit is then adjusted for zero output.

Full-Scale Calibration To calibrate the device for a nominal 10 V full scale, the analog input voltage is applied to pin 13 through a 15 Ω series resistance. To obtain full-scale output (all digits 1), we calculate the analog voltage required for 10 V full-scale operation, as shown in Ex. 13.13.

EXAMPLE 13.13 If the nominal full-scale input voltage is 10 V, calculate the actual analog voltage required for the AD571 10-bit A/D converter to obtain the maximum output (all 1's).

PROCEDURE

1 The full-scale analog input voltage required is:

$$\begin{aligned} V_{in} &= \text{full-scale range} - 1 \text{ LSB} \\ &= 10 \text{ V} - \frac{10 \text{ V}}{2^{10}} \\ &= 10 - 0.01 = 9.99 \text{ V} \end{aligned}$$

2 The input resistance, 15 Ω, is recommended because with the 9.99 V input, the resulting input current matches the current produced by the internal DAC for full-scale output. The 15 Ω resistor (along with the internal 5 kΩ input resistor) maintains the full-scale calibration error within ±2 LSB (or 0.2%). For higher accuracy, a 50 Ω trimmer is recommended.

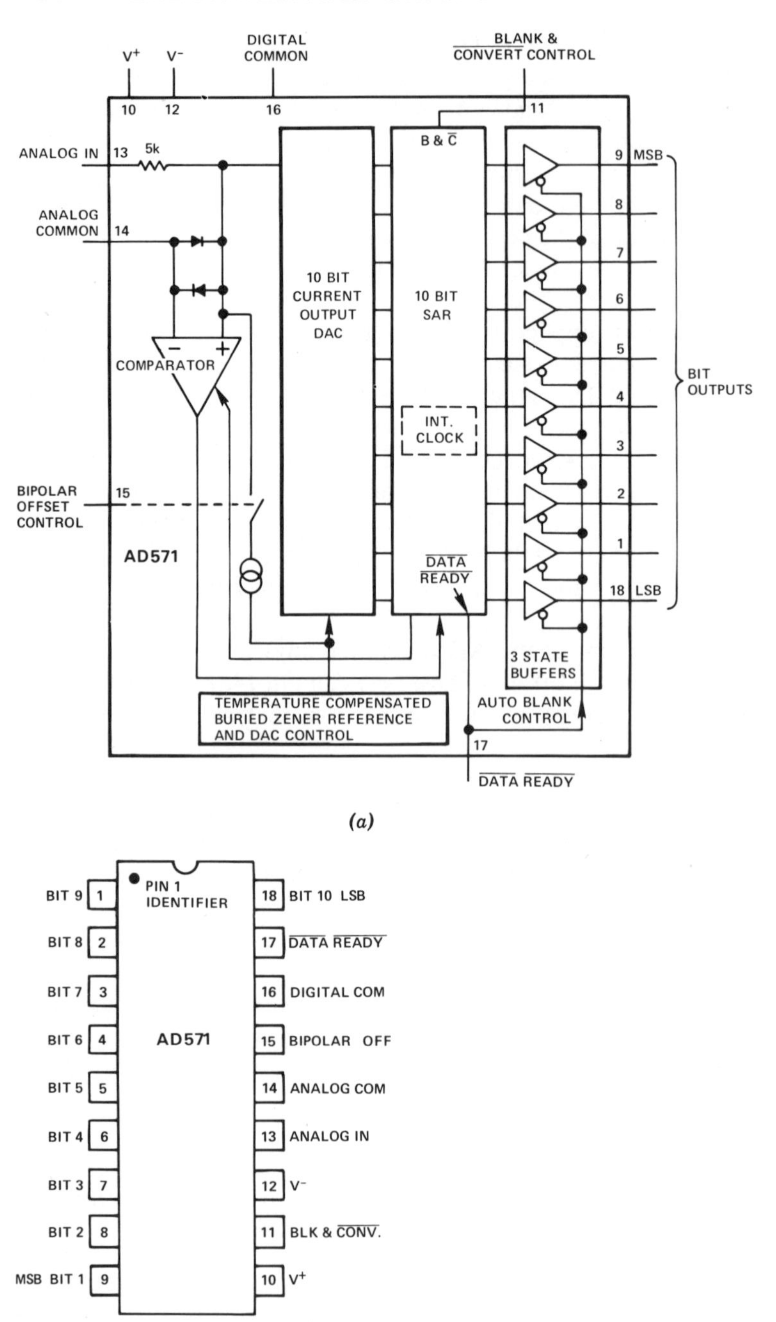

(a)

(b)

Figure 13.17 The AD571 successive approximation ADC. (*a*) Block diagram. (*b*) Pin connections. (Courtesy Analog Devices.)

Operating with Sample-and-Hold Amplifier

In a high-speed, data-acquisition system, or for a rapidly changing input signal, a sample-and-hold amplifier (SHA) is required in front of the AD571 (Fig. 13.18). The function of the SHA is to acquire the fast changing signal and hold it until the conversion is completed.

The analog input is connected to the SHA (AD582). The SHA has a control gate connected to the conversion start pulse and the $\overline{\text{DATA}}$ $\overline{\text{READY}}$ line of the AD571. When the conversion start pulse is low, the SHA goes into the hold mode, and the conversion begins. The high $\overline{\text{DATA}}$ $\overline{\text{READY}}$ line keeps the SHA in hold during the conversion. At the end of the conversion, the $\overline{\text{DATA}}$ $\overline{\text{READY}}$ goes low and the SHA is placed in the sample mode.

Successive approximation ADCs are preferred in systems that require moderate to high speed (or fast conversion time), and accuracy does not have utmost importance. Applications include high-speed data acquisition, pulse-code-modulation, automatic testing, and digital process control.

13.10 INTEGRATING ADCs

Integrating type A/D converters are based on an indirect conversion method. The analog input voltage is first converted into a time period which is then

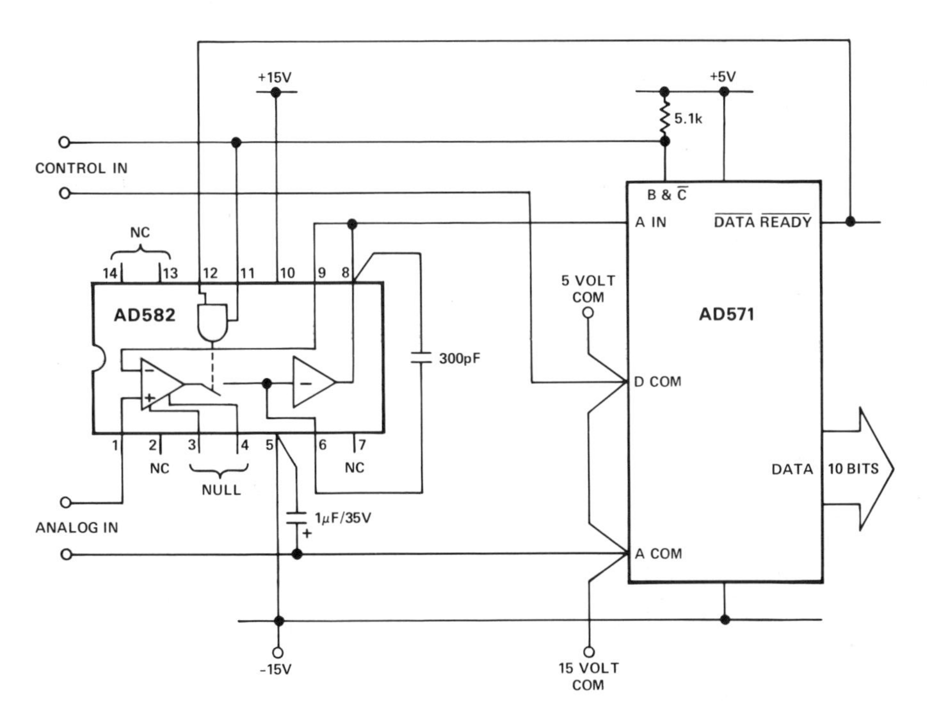

Figure 13.18 Sample-and-hold amplifier (AD582) connected to the AD571 ADC. (Courtesy Analog Devices.)

converted into a digital number by using a counter. There are a number of variations of this technique, the most widely used methods being *dual-slope* and *quad-slope*. The operation of the dual-slope type A/D converter is described below; the quad-slope type is an extension of the dual-slope ADC.

Dual-Slope A/D Converters

The circuit consists of an integrator, comparator, counter, and a reference voltage, as shown in Fig. 13.19*a*. When the switch closes, the analog input voltage V_{in} causes the charging current I_C to charge the capacitor, resulting in a linear ramp at the integrator output. This continues for a fixed period of time T_1, established by the counter (Fig. 13.19*b*). Then, the control circuit switches the integrator input to the negative reference votlage $-V_{ref}$ and the counter again begins to count from zero. Discharge current I_D flows, as the capacitor discharges with a ramp of opposite slope, for a period T_2, until the ramp crosses the zero level. The counter output becomes the digital output equivalent to the analog input voltage.

The total charge in the capacitor while charging must be equal to the charge while discharging because of the constant slope; therefore, $I_C T_1 = I_D T_2$. Because I_C and I_D are proportional to voltages V_{in} and V_{ref}, $V_{in} T_1 = V_{ref} T_2$. Solving for T_2 yields:

$$T_2 = \frac{V_{in}}{V_{ref}} T_1$$

Because T_1 and V_{ref} are constant, the counter output, which measures the period T_2, is proportional to the input voltage V_{in}.

The dual-slope ADC has the following features:

1 High-conversion accuracy, which is independent of changes in the clock rate and the capacitor value.

2 High-noise rejection.

3 Slow speed.

4 Responds to the average input level during the charging time.

BCD Converter

Dual-slope converters are well suited for numeric display applications. An example is the high-resolution A/D converter module, type ADC171. It has a 17-bit output, 16 bits for 4 BCD digits and one sign bit; hence, it is termed a $4\frac{1}{2}$ BCD converter. The ADC171 integrates the analog input for one ac line period, thus minimizing ac hum noise. Total conversion time is less than 40 ms.

Circuit Description Figure 13.20 shows the functional block diagram of the ADC171. The analog input voltage range is -10 to $+10$ V, with 20% overrange. Input impedance is 180 kΩ. Supplies of $+15$, -15 and $+5$ V are required.

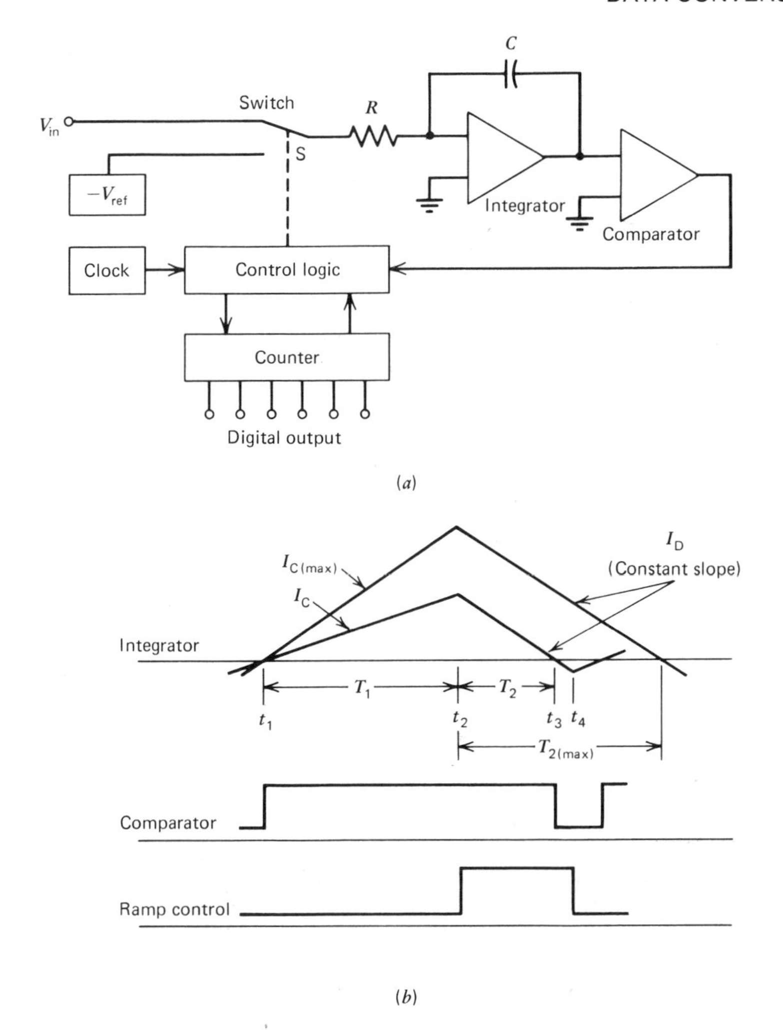

Figure 13.19 Dual-slope ADC. (*a*) Block diagram. (*b*) Ramp signal charging and discharging cycle.

A positive pulse, of 100 ns to 100 μs duration, starts the conversion cycle. The leading edge of the pulse resets previous data and the trailing edge initiates the conversion. First, the counter gates the 720 kHz clock on and the ramp signal rises for $16\frac{2}{3}$ ms with a slope proportional to the analog input signal. Then, the counter is reset and counts again during the negative ramp, until zero voltage is reached. Upon completion, the counter output is the digital value of the analog input voltage.

The timing diagram (Fig. 13.21) shows the signals available at the $\overline{\text{STATUS}}$, $\overline{\text{RAMP UP}}$, $\overline{\text{RAMP DOWN}}$, polarity, and overload terminals. The following are the manufacturer suggestions for effective use of the device:

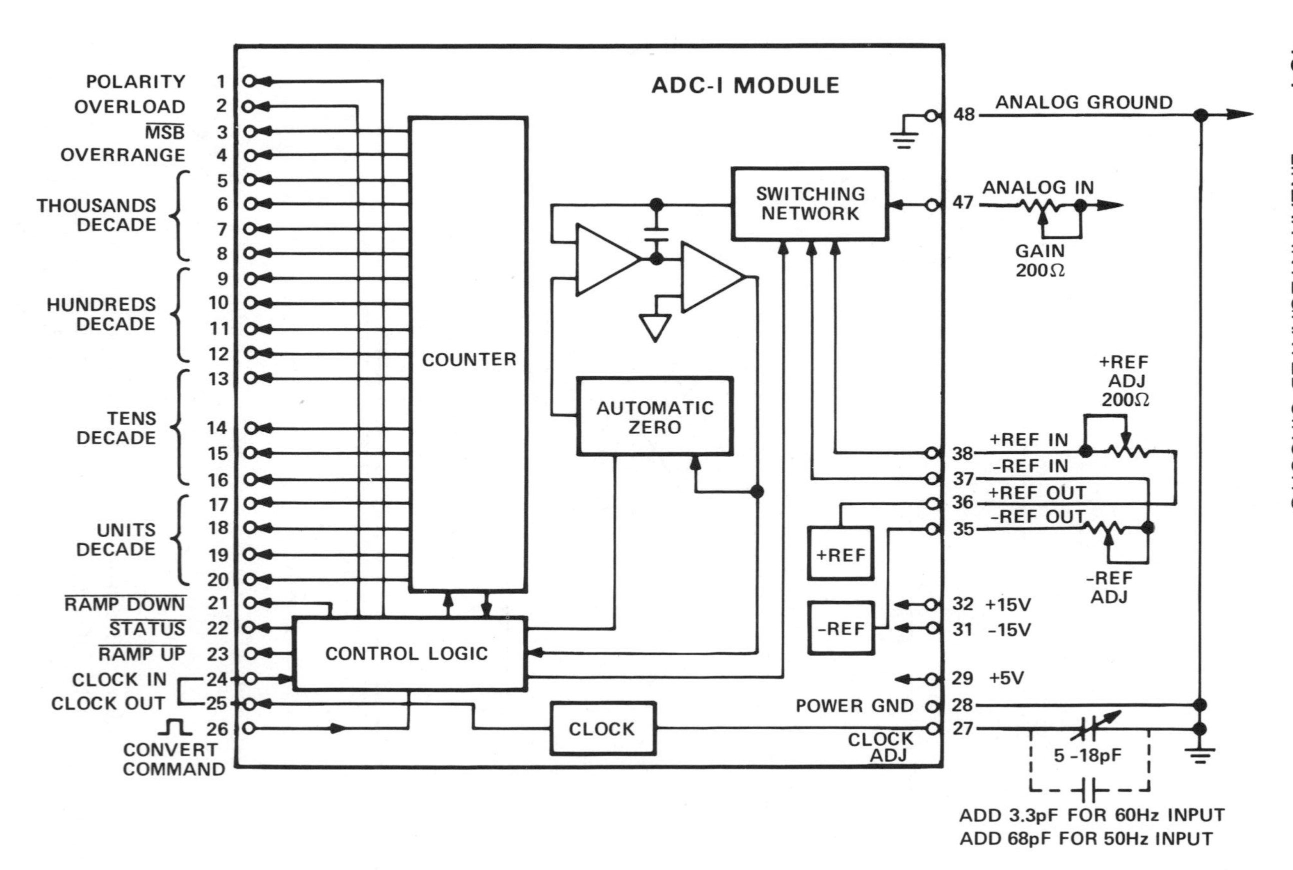

Figure 13.20 Functional block diagram of the ADC171 $4\frac{1}{2}$ BCD converter. (Courtesy Analog Devices.)

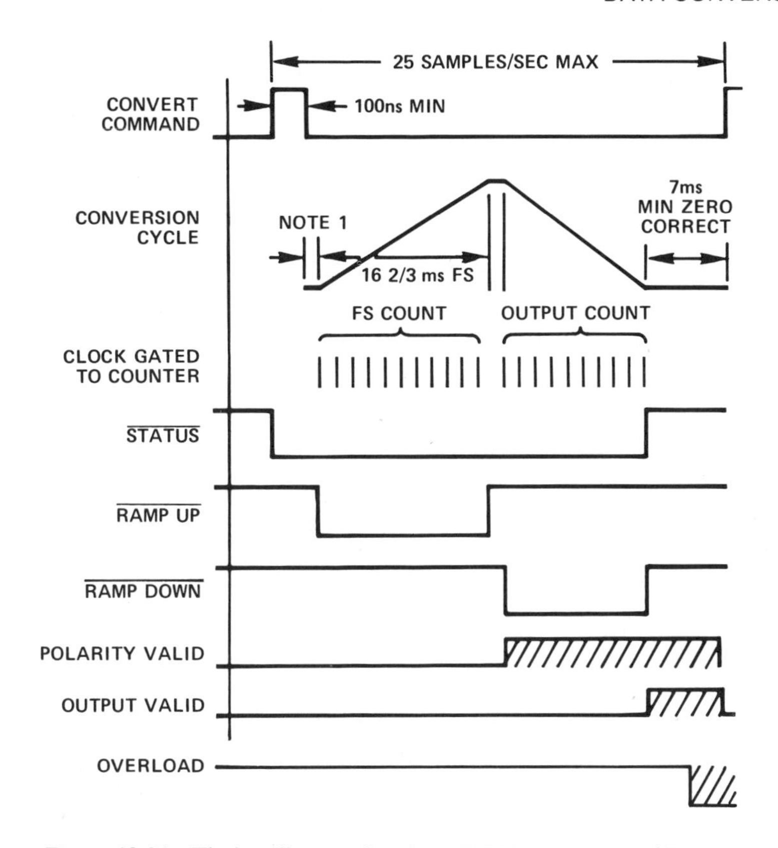

Figure 13.21 Timing diagram for the ADC171 converter. (Courtesy Analog Devices.)

Ground connections Analog ground and power ground are not internally connected and must be connected externally.

Clock operation An external clock may be used by connecting pins 24 and 25. For 60 Hz line frequency, the internal clock is adjusted to 720 kHz by using an external trimmer capacitor between pins 27 and 28. For 50 Hz line frequency, the clock should be adjusted to 600 kHz with a 68 pF capacitor in parallel with the trimmer capacitor (the ramp-up period becomes 20 ms).

Dual-slope A/D converters are used in systems where accuracy, rather than speed, is important. They are commonly employed in digital voltmeters, displays, laboratory measurements, slow-speed data-acquisition systems, monitoring systems, and measurements in high-noise environments.

13.11 THE PARALLEL (FLASH) ADC

Where ultra-high speed is required, the parallel A/D conversion technique is ideally suited. In Fig. 13.22, the circuit uses a series of comparators in parallel,

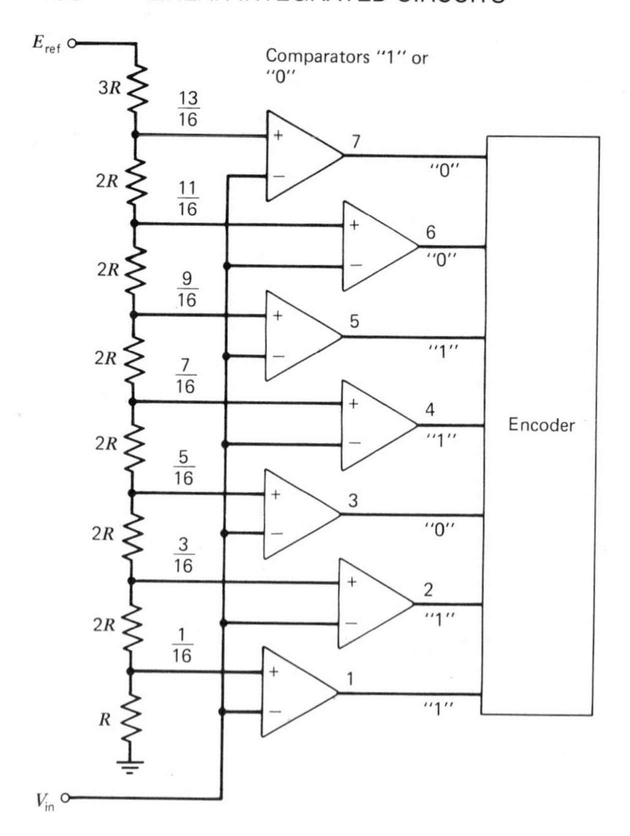

Figure 13.22 Basic parallel (flash) ADC.

with one terminal of the comparators connected to the analog input voltage. The second terminal of each comparator is biased in binary proportion, using the reference voltage and a voltage divider. Each of these terminals is 1 LSB from the next. For a given analog voltage, all comparators biased below the input voltage are turned on and the others remain off. The conversion takes place in one step; however, additional decoding is required to convert the output to binary code. A disadvantage of this method is the large number of comparators needed for high resolution. For example, an 8-bit A to D converter requires 255 ($2^n - 1$) comparators.

40 ns Converter

An example of a parallel converter is the Datel–Intersil type ADC-UH4B, a single-chip, 4-bit output device operating at ultra-high speeds up to 25 MHz (40 ns). The analog input voltage range is 0 to −2.56 V; ±15 V power supplies are needed. The available output is either straight binary, for unipolar operation, or offset binary, for bipolar operation.

Circuit Description The circuit (Fig. 13.23) consists of 15 parallel comparators, a 15 to 4 line encoder, a 4-bit storage register, and control logic circuitry. The comparators are biased 1 LSB apart by a precision resistor network.

There are two control signals: an input for START CONVERSION and an output for *EOC* (END OF CONVERSION). When an analog input voltage is applied to input pin 17, comparators biased below the input level are turned on and the others are turned off. The output of the 15 converters are decoded to a 4-bit binary and stored in the output register. The input to the converter is an inverted analog input; thus, the maximum output (digital 1111) is obtained for an input voltage of -2.56 V.

Parallel A/D converters are used in systems where ultra-high speed is critical. Applications include radar signal processing, digitizing video information, and high-speed instrumentation.

13.12 HOW TO SELECT CONVERTERS

Selecting a data converter that meets the requirements of the system under consideration is difficult. First, the design objectives must be defined as accurately as possible and then the specifications of various products examined. Here is a suggested check list of the factors to be considered in selecting converters:

1 *Input/Output.* The input-output requirements should be defined in terms of:
 a. Input signal range: full-scale value and polarity.
 b. Digital code: straight binary, offset binary, two's complement, Gray, BCD, and so on.

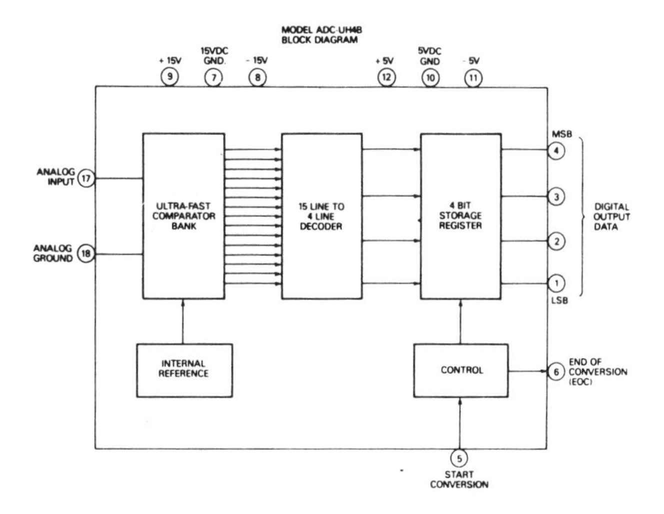

Figure 13.23 Functional block diagram of a parallel ADC (ADC-UH4B). (Courtesy Datel–Intersil.)

c. Input and output impedances: consider the input source impedance and the output load requirements.
d. Logic level compatibility: TTL, CMOS, and logic polarity.
e. Output signal (DAC): Current or voltage.

2 *Accuracy.* This is an important characteristic, and the requirements vary considerably from one system to another. It should be specified in terms of:
a. Resolution: number of bits.
b. Sources of errors: relative accuracy, nonlinearity, gain, and offset errors.

3. *Speed.* The *conversion time* in A/D converters range from nanoseconds to milliseconds. In many instances, speed determines the conversion method. In case of D/A converters, the speed is specified in terms of *settling time*.

4 *Environmental conditions.* These include temperature range, noise level, and power supply sensitivity. Each of these adds to the total error.

5 *Microprocessor interfacing.* The dedicated use of microcomputers as control and data loggers is now becoming common. Converters must often interface with them. For these applications, ADC *tri-state output* and DAC *input latch* may be important features.

6 *Cost.* The present cost of converters varies from less than $10.00 to a few hundred dollars. As the number of bits and performance demands increase, the cost also increases.

These factors should all be considered. However, the final selection is always a compromise between three parameters: speed, accuracy, and cost. Tables 13.7 and 13.8 summarize the selection factors for A/D and D/A converters.

13.13 INTERFACING WITH MICROPROCESSORS

Applications of microprocessor based systems are becoming widespread, replacing hard-wired digital systems. They often need data converters for input/output (I/O) functions. In addition, a microprocessor may be an integral part of an ADC, functioning as a counter or successive approximation register.

When interfacing converters, it is necessary to consider both their compatibility with microprocessors and the characteristics of microprocessors themselves. (See Chap.8 for a discussion of microprocessors and microcomputers.)

Interfacing DACs

Figure 13.24 shows a typical example of interfacing the widely used 1408 D/A converter (previously described) with the 8085 μP. The circuit includes the converter, a parallel 8-bit latch (74LS273), and an output op amp. From the

Table 13.7 **Selection guides for ADCs**

Type of converter	*Description*	*Conversion time*			*Features*	*Limitations*	*Applications*
		8 bits	*12 bits*	*16 bits*			
Integrating dual slope	Slow	20 ms	40 ms	250 ms	Inherent accuracy	Relatively slow	Digital meters
	Medium	1.8 ms	24 ms	—	Excellent noise rejection		Digital panel meters
	Fast	0.3 ms	5 ms	—	Noncritical components No missing codes Relatively low cost		Laboratory measurements Monitoring systems
Successive approximation	General purpose	30 μs	50μs	—	High speed	Several critical components	
	High performance	10 μs	20 μs	400 ns		Requires sample-and-hold circuit	High-speed data-acquisition systems
	Fast	4 μs	13 μs	—		Errors can exceed $\pm\frac{1}{2}$ LSB	Pulse-code modulation systems
	Very fast	2 μs	6 μs	—		No noise rejection	Automatic test systems
	Ultra fast	0.8 μs	2 μs	—			Digital process control
Parallel (flash)	Ultra fast	100 ns			Ultra high speed	Limited resolution	Video digitizing Radar signal processing
	Video speed	50 ns					

Table 13.8 Selection guides for DACs

Type of converters	*Description*	*Settling time*			*Applications*
		8 bit	10 bit	12 bit	
Current output	High speed—Fast	150 ns	250 ns	400 ns	Digital voltmeter, programmable gain and attenuator, CRT character generation
	Ultra fast	25 ns	25 ns	50 ns	CRT vector display, Video information, medical instruments
Voltage output	Slow	5 μs		20 μs	Process control Programmable power supplies
	Fast	2 μs		3 μs	Ramp generator Fast CRT display
Multiplying CMOS type	4 Quadrant	150 ns		500 ns	Battery operated equipment Digitally controlled gain and attenuator circuits
	Precision	1 μs		5 μs	Systems where high precision is required

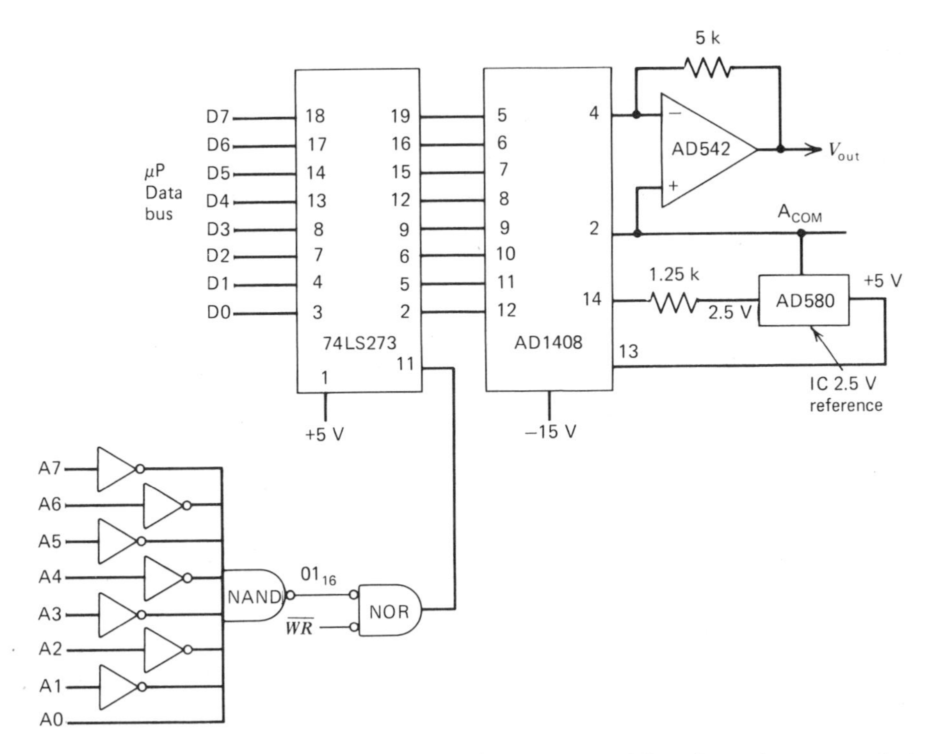

Figure 13.24 Interfacing a DAC with a microprocessor. The schematic assumes that the address bus and the data bus of the 8085 μP are demultiplexed.

μP, 8 address lines and one control signal $\overline{\text{WR}}$ are used. An 8-input NAND gate with inverters and a 2-input NOR gate are used for identifying the output port. The circuit displays the analog voltage corresponding to the digital input from the μP. The μP starts with a 00_{16} input, increasing the binary input by 1 count for each program cycle until it reaches the final count FF_{16} (all 1s). It then continuously repeats the process for a stable oscilloscope display.

Circuit Description When address 01_{16} is placed on the address bus, the output of the NAND gate which is the input to the NOR gate go low. The output of the NOR gate then goes high when the $\overline{\text{WR}}$ signal arrives from the μP, enabling latch 74LS273. The data bus information is stored for input to the converter until the next input arrives. If the response of the converter is slower than the input rate of the μP, a delay loop must be added in the program.

The D/A converter is used as an output port, with the port number 01_{16}. In this particular I/O technique, the μP can communicate with 256 different output ports from 00_{16} to FF_{16} and also with 256 different input ports.

Program A program in 8085 mnemonics for the circuit is as follows:

Labels	*Mnemonics*	*Comments*
	MVI A, 00_H	Load the accumulator with the first input 00_{16}.
D/A	OUT, 01_H	Output the accumulator content to the D/A converter at port number 01.
	MVI B, Data	B is a delay counter. Load register B with some data for delay.
DELAY	DCR B	Decrement B by one count.
	JNZ Delay	If B is not zero, go back to decrement B and stay in the loop until B is zero.
	INRA	Increment the accumulator by one.
	JMP D/A	Jump to location D/A for the next input display.

This program stays in a continuous loop and can display a D/A staircase output.

Microprocessor Compatible Second Generation Converters

The 1408 DAC requires a latch to interface with a μP. This increases the chip count and is undesirable. However, such newer converters as the AD7524, include an internal latch. Figure 13.25 shows a circuit interfacing the AD7524. The converter has pins for the control signal $\overline{WR}$ and for *chip select* ($\overline{CS}$) for the decoded address. The circuit description and the program are similar to that described above.

Interfacing a 12 Bit DAC with an 8-bit μP

In many applications, 12-bit DACs are needed for higher resolution. However, the 8-bit μP has only 8 data lines. Therefore, to transfer 12 bits, the 8-bit μP

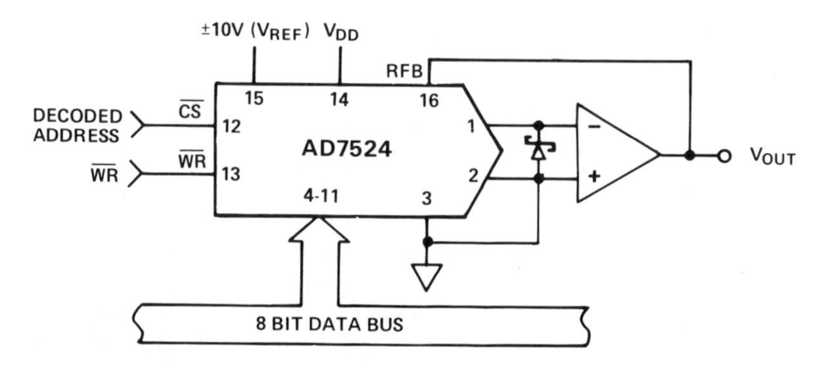

Figure 13.25 Interfacing a DAC (AD7524) with an 8085 microprocessor. (Courtesy Analog Devices.)

data bus is time shared by using two output ports: one for the first 8 bits, and the other for the remaining 4 bits.

Figure 13.26 shows the circuit with the output port address 06_{16} for low-order (least significant) data bits and the second output port (address 07_{16}) for the remaining high-order (most significant) data bits. The first instruction sends the least significant 8 bits to output port 06_{16}, and they are latched by the

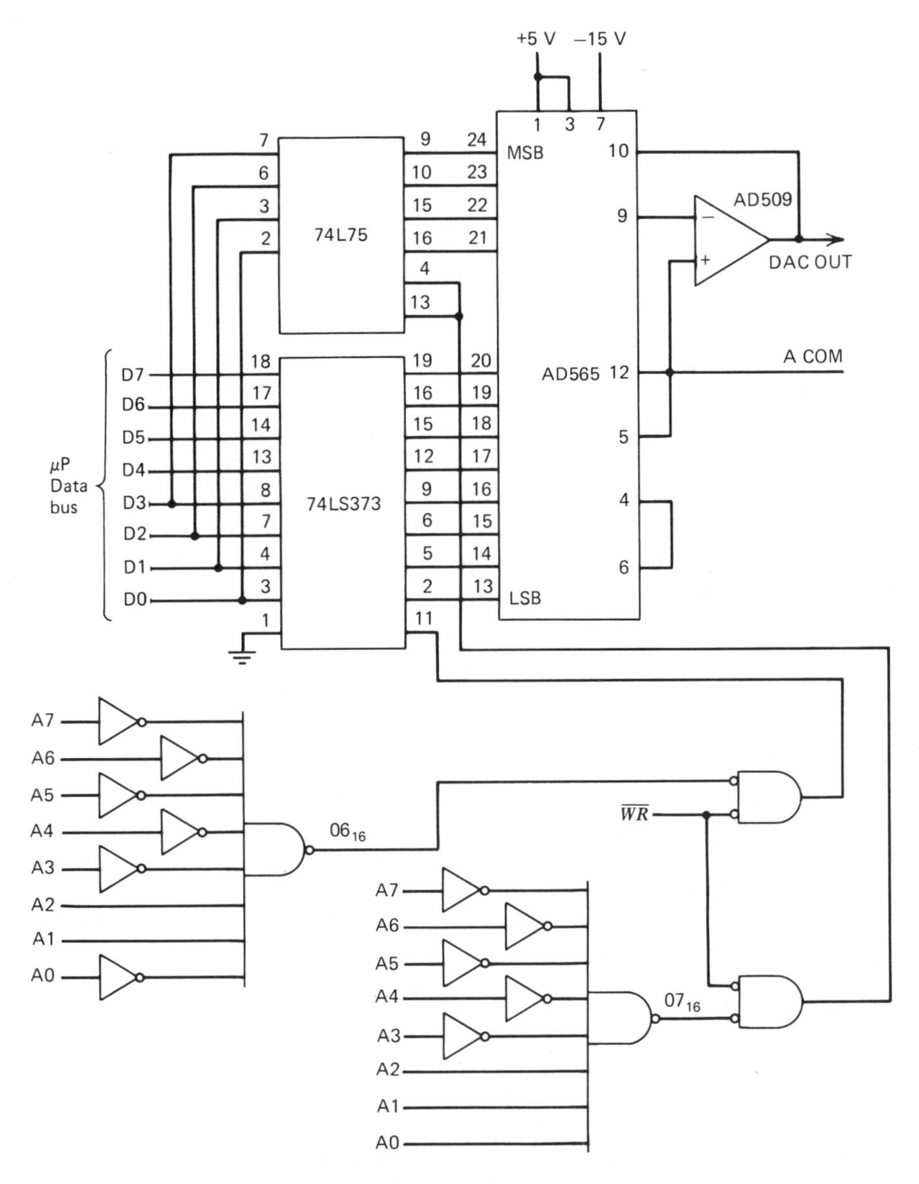

Figure 13.26 Interfacing a 12-bit ADC with an 8-bit microprocessor.

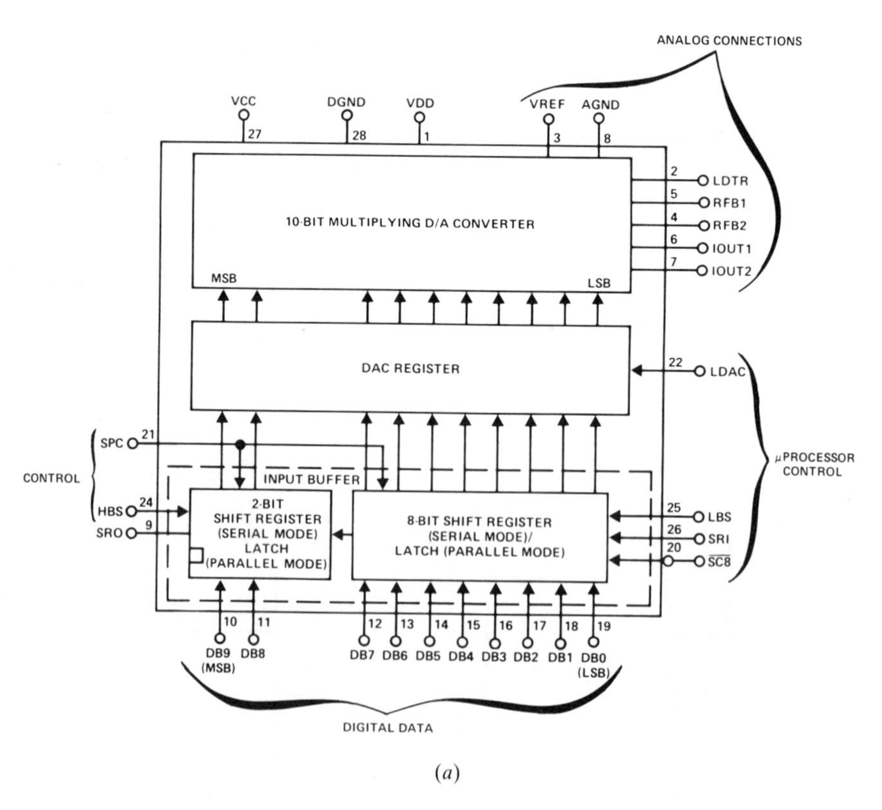

(*a*)

Figure 13.27 The AD7522 CMOS 10-bit DAC. (*a*) Block diagram. (*b*) Interfacing with a microprocessor. (Courtesy Analog Devices.)

74LS373. The four MSBs are sent by the next instruction to output port 07_{16}, using the data lines D0–D3. A disadvantage of this method is that the DAC output assumes an intermediate value between the two output operations. This is eliminated by using a double-buffered DAC (for example, AD7522).

The AD7522 (Fig. 13.27*a*) is a CMOS 10-bit D/A converter having an input buffer with shift and holding registers. The technique of loading 10 bits through an 8-bit data bus is similar, with data loaded into the input register in two steps (8 bit and 2 bit). After loading, however, they are switched into a holding register for the conversion by enabling the line LDAC through a third pot (Fig. 13.27*b*).

Interfacing an ADC

In interfacing an input A/D converter, the μP needs to perform three functions. The first is to provide an output command to the ADC to initiate conversion; the second is to check the data status ($\overline{\text{DATA READY}}$ line) input until the conversion is completed; and the third is to read input digital data.

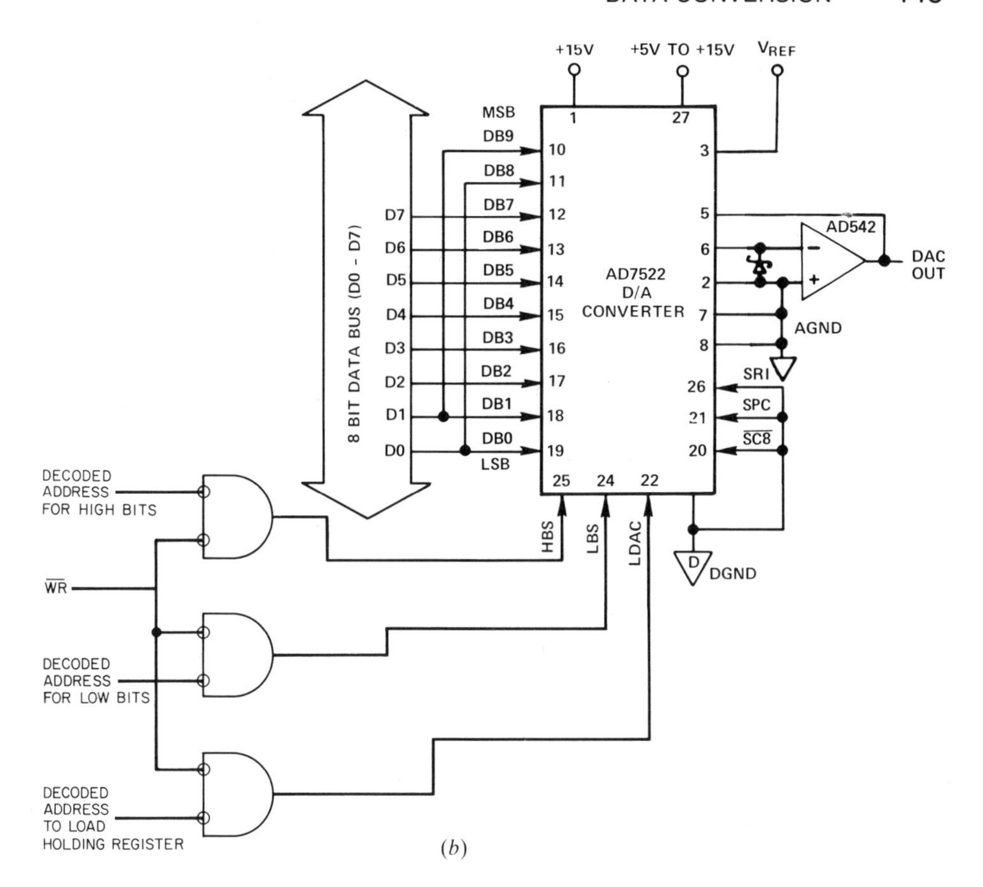

(*b*)

Figure 13.28 shows a typical circuit for interfacing an A/D converter to a microprocessor. The circuit includes an AD570 converter, two tri-state buffers 74LS244 and 74125, and address decoding circuits. This is a memory mapped I/O and uses memory-related control signals, *memory write* $\overline{\text{MEMW}}$ and *memory read* $\overline{\text{MEMR}}$. There are two port addresses: 4000_{16} for the data status bit and 8000_{16} for the A/D converter.

Circuit Description The A/D converter is assigned the address 8000_{16} by using address line A_{15} through an inverter; lines A_{14} to A_0 are "don't care." The status port is assigned the address 4000_{16} by using address line A_{14} through an inverter; again the other lines are assumed to be "don't care."

To initiate conversion, the μP sends a WRITE signal to the status port 4000_{16}. When the address 4000_{16} is placed on the address bus and the $\overline{\text{MEMW}}$ signal is sent, the output of gate G_1 goes high which is fed to NAND gate G_2. The other input of gate G_2 is already high because the data ready $\overline{\text{DR}}$ signal is

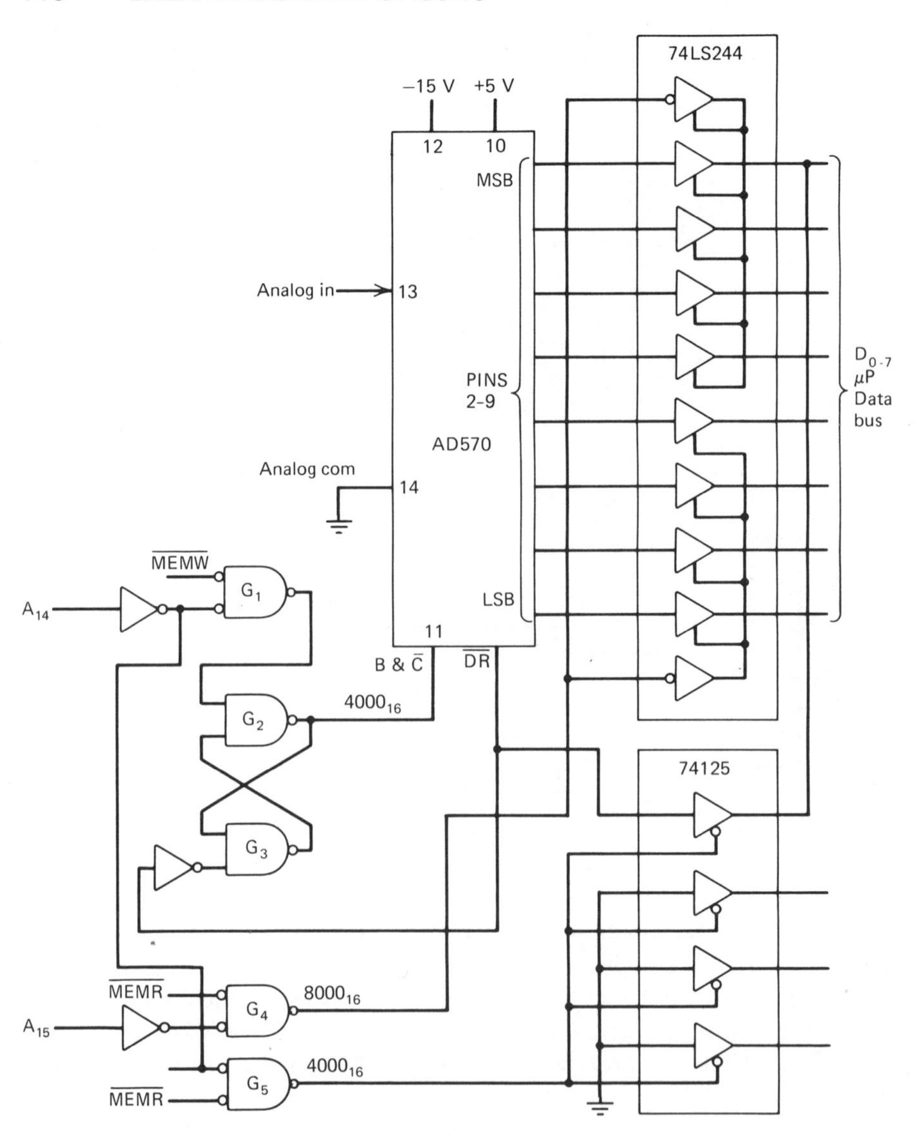

Figure 13.28 Interfacing an ADC with microprocessor.

high. Therefore, its output goes low and initiates the conversion process. Then the μP checks the status by reading the MSB, which is connected to the data ready $\overline{DR}$ line through a tri-state buffer. The μP repeatedly checks the status by executing the "TEST" loop (see the program, below) until the line $\overline{DR}$ goes low. Conversion is then complete, and the μP reads the port 8000 and accepts the digital output from the A/D converter.

Program The instructions in 8085 mnemonics are:

Label	*Mnemonics*	*Comments*
	LXI B, 4000H	Load B–C register with the address of status port.
	LXI H, 8000H	Load H–L register with A/D converter address.
	STAX B	Initiate conversion.
TEST	LDAX B	Read status port.
	RLC	Rotate left to place status bit in carry flag.
	JC TEST	If status bit is high, go back and check again.
	MOV B, M	Read data from A to D converter and store in register B.

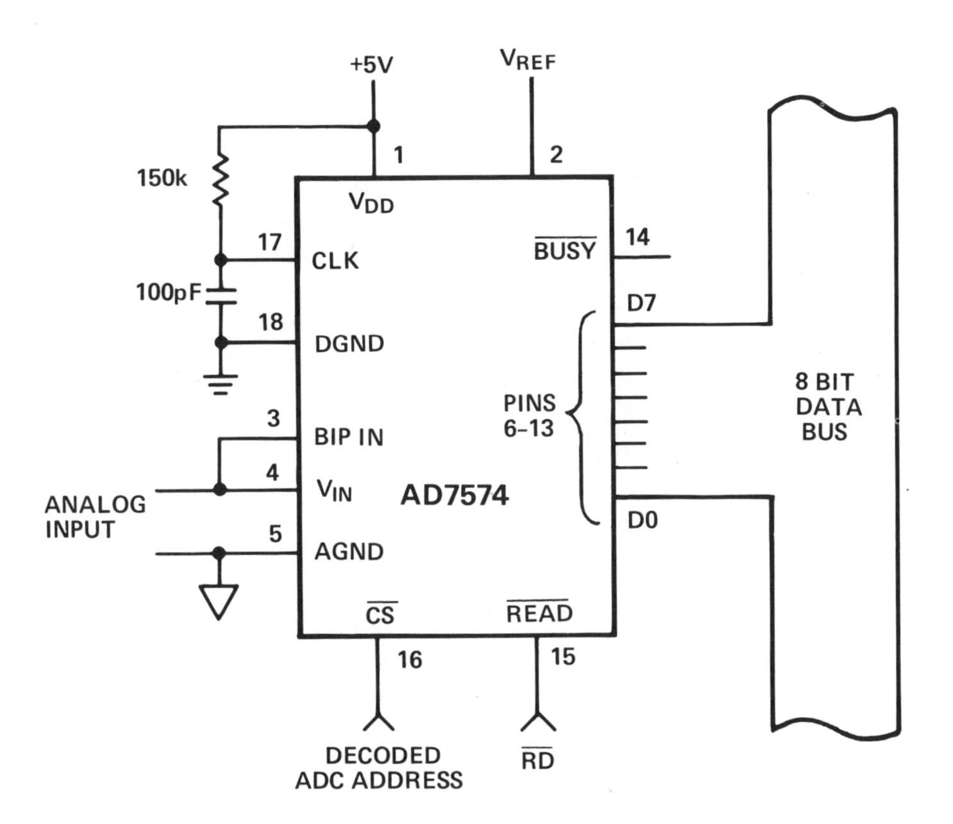

Figure 13.29 Microprocessor compatible ADC.

Microprocesor Compatible A/D Converter In the previous circuit, an 8-input, tri-state buffer is necessary to interface the ADC570. However, ADCs such as the ADC7574 include a tri-state buffer. Figure 13.29 shows the interfacing; the $\overline{\text{BUSY}}$ signal is used to check the conversion status.

Microprocessor Controlled Successive Approximation A/D Converter A successive approximation A/D converter requires a D/A converter, successive approximation register (SAR), and a comparator. In a microprocesor controlled ADC, the function of the SAR is accomplished with software.

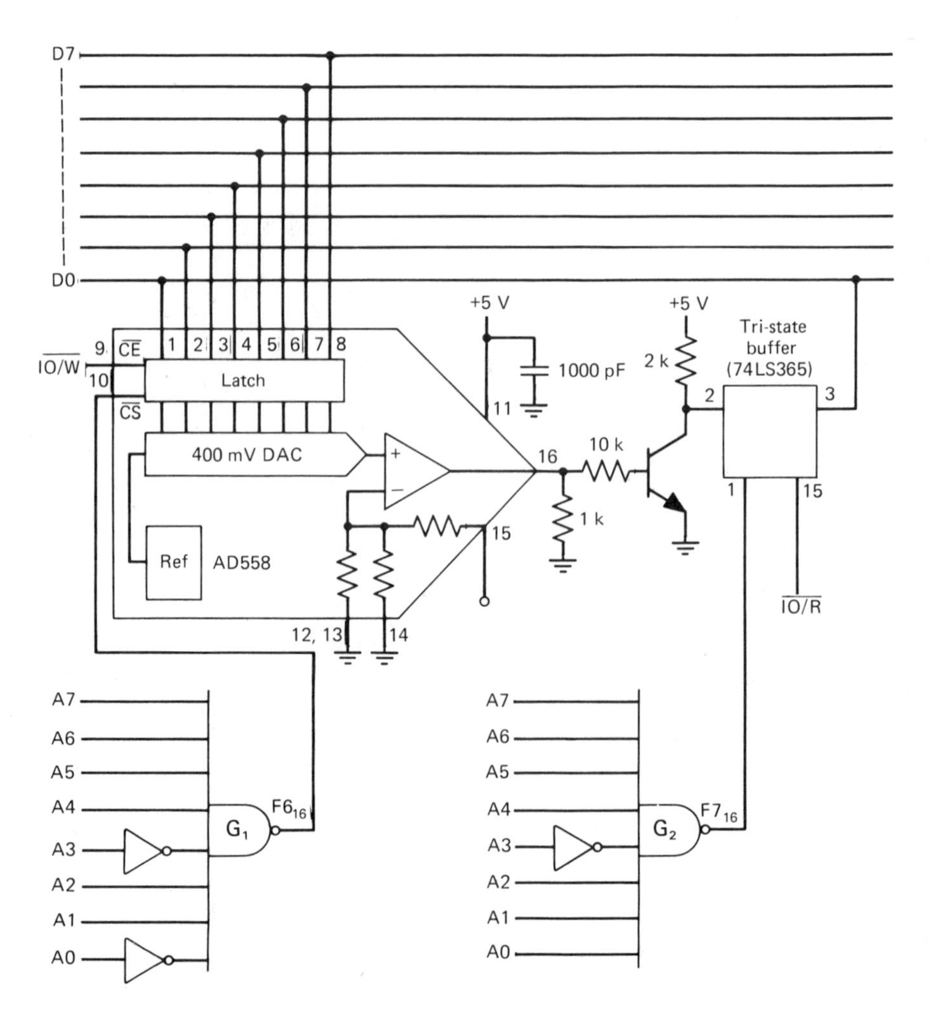

Figure 13.30 Microprocessor controlled successive approximation ADC. (Courtesy Analog Devices.)

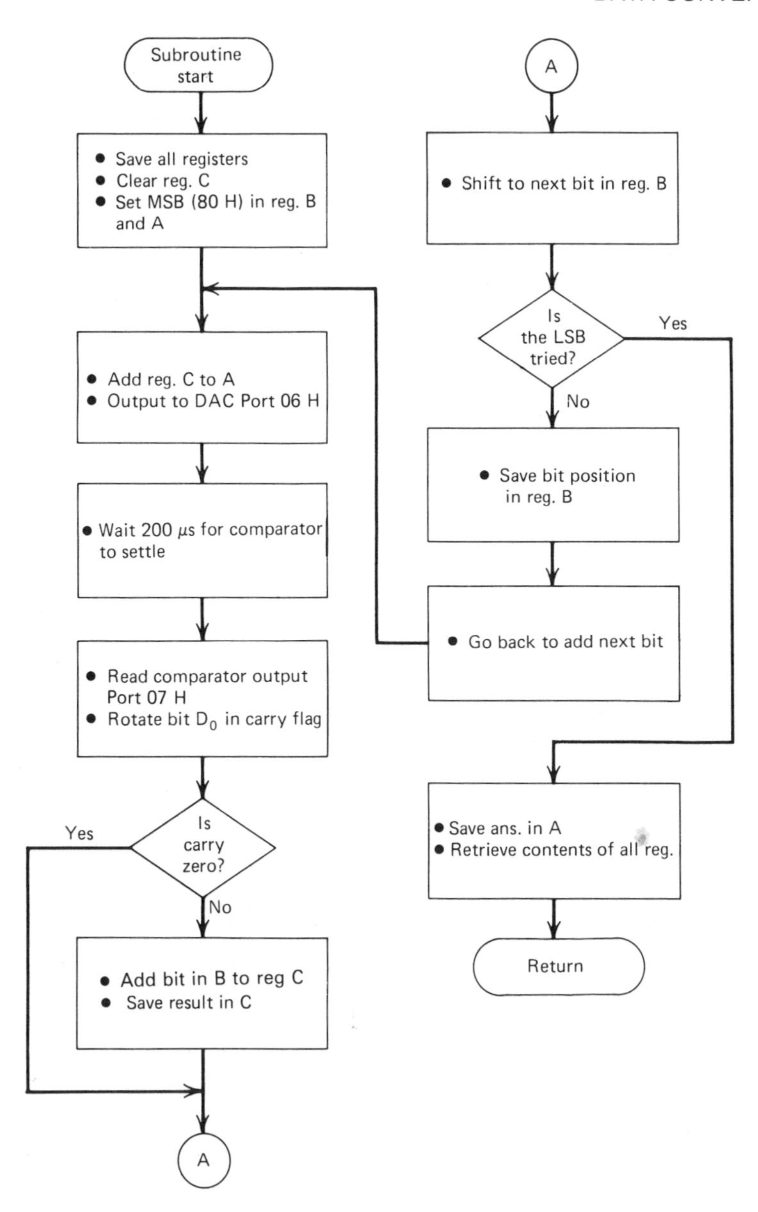

Figure 13.31 Flowchart for the successive approximation ADC.

The microprocessor controlled A/D converter is functionally the same as the conventional hardware-based, successive approximation ADC. The software algorithm starts with the MSB as an input to the DAC, and the output of the DAC is compared with the input analog signal. This process is repeated for successive bits until the LSB is reached. The bits that cause the DAC output to exceed the analog signal are turned off.

Circuit Description Figure 13.30 shows the circuit of an 8085 controlled successive approximation A/D converter. The circuit includes an AD558 D/A converter, a tri-state buffer, control signals $\overline{IO/R}$ and $\overline{IO/W}$, and two 8-input decoders. The circuit employs the direct I/O technique. The D/A converter is assigned port address $F6_{16}$ and the comparator is assigned address $F7_{16}$, by using two 8-input NAND gates. When the μP places $F6_{16}$ on the address bus, the output of NAND gate 1 is low, activating the chip select $\overline{CS}$. Control signal $\overline{IO/W}$ enables the ADC chip. When the μP places $F7_{16}$ on the address bus, the low output of NAND gate 2, and the control signal $\overline{IO/R}$ enable the tri-state buffer. The algorithm of the successive approximation technique is shown in Fig. 13.31.

13.14 REFERENCES

Analog Devices, *Application Guide to CMOS Multiplying D/A Converters,* 1978.

Analog Devices, *Data Acquisition Components and Subsystems,* 1980.

Analog Devices, *Data Acquisition Products Catalog,* 1978.

Analog Devices, *Integrated Circuit Converters, Data Acquisition Systems and Analog Conditioning Components,* 1979.

Datel–Intersil, *Data Acquisition Component Handbook,* 1980.

Grant, D., "Interfacing the AD558 Dacport to Microprocessors," Application Note, Analog Devices, 1980.

Henry, T., "Successive Approximation A/D Conversion," Application Note AN-716, Motorola, 1974.

Motorola, *Linear Interface Circuits,* 1979.

Renschler, E., "Analog-to-Digital Conversion Techniques," Application Note AN-471, Motorola, 1974.

Sheingold, D. H. (ed.), *Analog-Digital Conversion Notes,* Analog Devices, 1980.

CHAPTER 14

ICs for Communication Systems

Richard R. Kent

Assistant Professor, Electrical Technology
Onondaga Community College

14.1 INTRODUCTION

Linear integrated circuits (LICs) cover a broad range of functions, from control circuits to complex communication systems. The emphasis will be on the latter in this chapter. Useful practical information, whether it be for a simple IF amplifier or a complete AM receiver on a single chip, will be provided for the reader.

14.2 AM RECEIVER SUBSYSTEMS

The concept of an AM subsystem is generally categorized as a "receiver on a chip." Most AM subsystems include a converter, IF amplifier stage(s), a detector, and an audio preamplifier. The implications of such an LIC chip are most significant for the automotive and consumer industries, although their applications in commercial and military systems also are important.

Most LICs operate from a wide range of supply voltages, with some chips containing an internal zener diode for voltage regulation. In these cases, care must be exercised to avoid an excessive zener current to flow which could destroy the device. Particular attention should be given to the manufacturer's data sheets in this regard. Typical zener currents range from 20 to 30 mA, depending on the IC, with nominal zener voltages in the range of 6 to 12 V.

Most AM subsystems provide two IF stages to ensure adequate gain, leaving the designer only to add the necessary interstage transformers and associated capacitance. The IF input and output connections to the chip, therefore, allow wide flexibility in individual receiver design. Connections also are provided for an external automatic gain control (AGC) line for optional RF stages when desired. Internal AGC is provided for the first IF amplifier in most AM subsystems. Figure 14.1*a* is a block diagram of a representative (CA3088) AM chip.

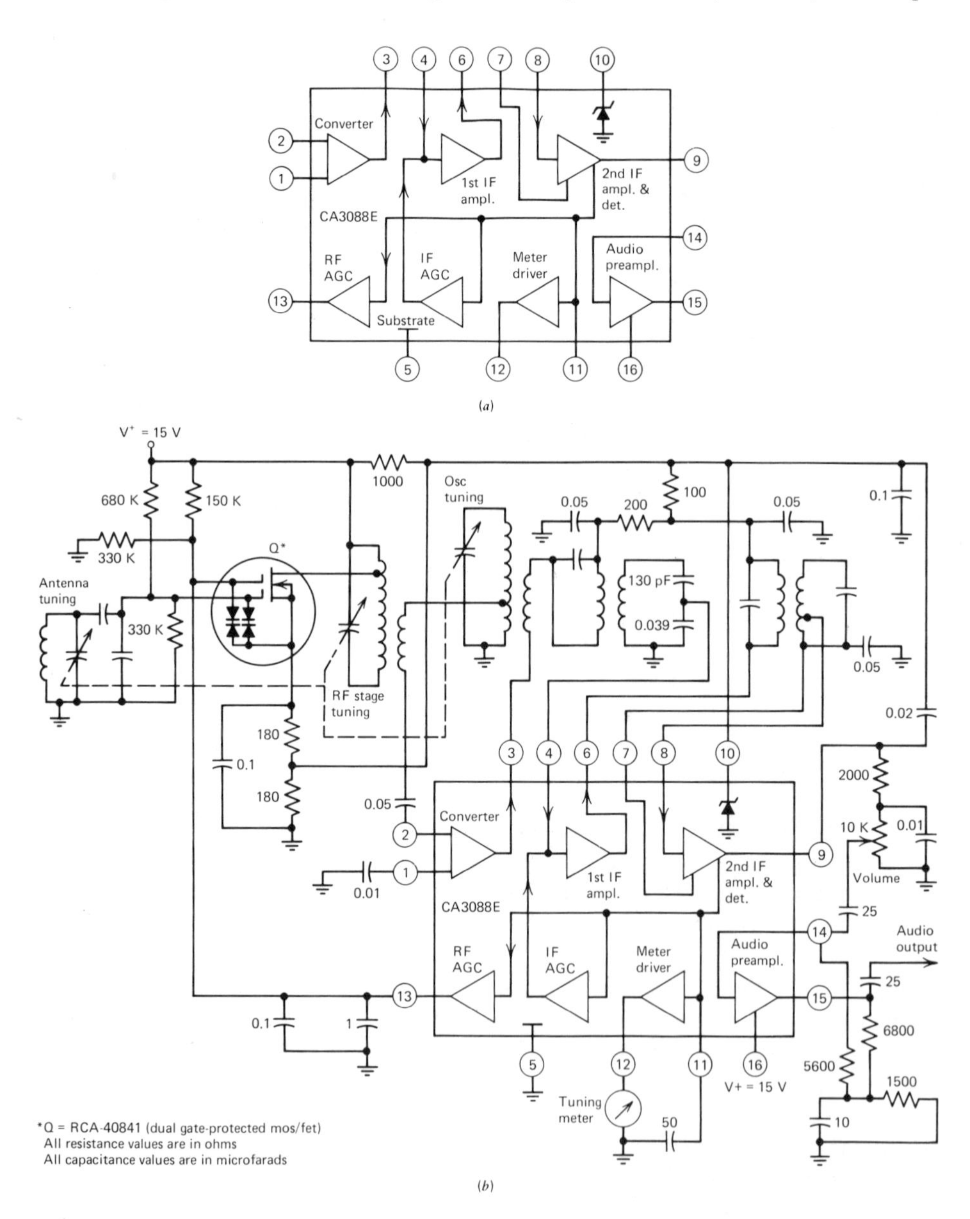

Figure 14.1 The CA3088 AM receiver subsystem. (*a*) Functional block diagram. (*b*) Typical AM receiver using the 3088 chip with an optional RF amplifier. (Courtesy RCA Corp.)

It should be noted that not all functions need be utilized in a particular application. For example, the CA3088 incorporates a meter driver for a tuning indicator. Not all applications, however, employing the chip require the addition of a metering circuit. Also, the use of tone controls may or may not be required. It is possible, if desired, to incorporate this feature should it be necessary into most AM subsystems. Typically, the tone circuit is connected between the audio preamplifier input and detector output. Any conventional circuit may be employed.

The zener diode in Fig. 14.1*a* regulates the second IF amplifier, AGC, and tuning meter circuits. Zener regulation may be used on any, or all, stages.

It is possible to provide the RF input directly from the antenna through an appropriate LC network to the converter stage. A more practical approach, to obtain higher gain, would be the addition of an RF stage, as in Fig. 14.1*b*. The RF amplifier in this case is a dual-gate MOSFET. When using a linear device such as a MOSFET, the ability to improve cross modulation, dynamic range, and intermodulation products are superior to the conventional bipolar junction transistor. Note that the output from pin 13 is used to control the gain of the RF stage, thereby eliminating overload problems in strong signal areas. The delayed AGC signal provided by the CA3088 is particularly useful in automotive applications where varying signal strengths are commonly encountered.

External filter and bypass capacitance should be used where indicated to maintain proper operation of the chip.

A complete AM broadcast receiver using a CA3123 IC is shown in Fig. 14.2*a*. A tuning meter has been included to indicate the necessary connections if this function is desired. The CA3123 is similar to the 3088 with the major addition being an RF amplifier. This eliminates the need for an external RF stage, thereby simplifying the receiver design. Figure 14.2*b* provides a functional block diagram with appropriate terminal designations for the CA3123.

The IF amplifier is tuned to 262 kHz (Fig. 14.2*a*). Ganged tuning is used to cover the broadcast band of 535 to 1650 kHz. The output from the detector is followed by an audio preamplifier; the signal from the detector is approximately 10 mV. Tone control circuitry may be added at this point if desired. Generally, an audio output stage operating class AB is used to provide sufficient drive for a speaker.

In summary, both types of AM subsystems offer distinct advantages depending upon the intended application. For many, a type such as the CA3088 (or the SK3146) will be the logical choice because they incorporate detectors and audio preamplifiers. However, because detector and audio preamplifier circuits used in AM are relatively straight forward and simple, the designer may elect to employ a chip such as the CA3123 or SK3171 containing an RF amplifier stage.

14.3 FM RECEIVER SUBSYSTEMS

Most FM subsystems are very similar to their AM counterparts in terms of function. Major differences include increased operating frequency, necessitating

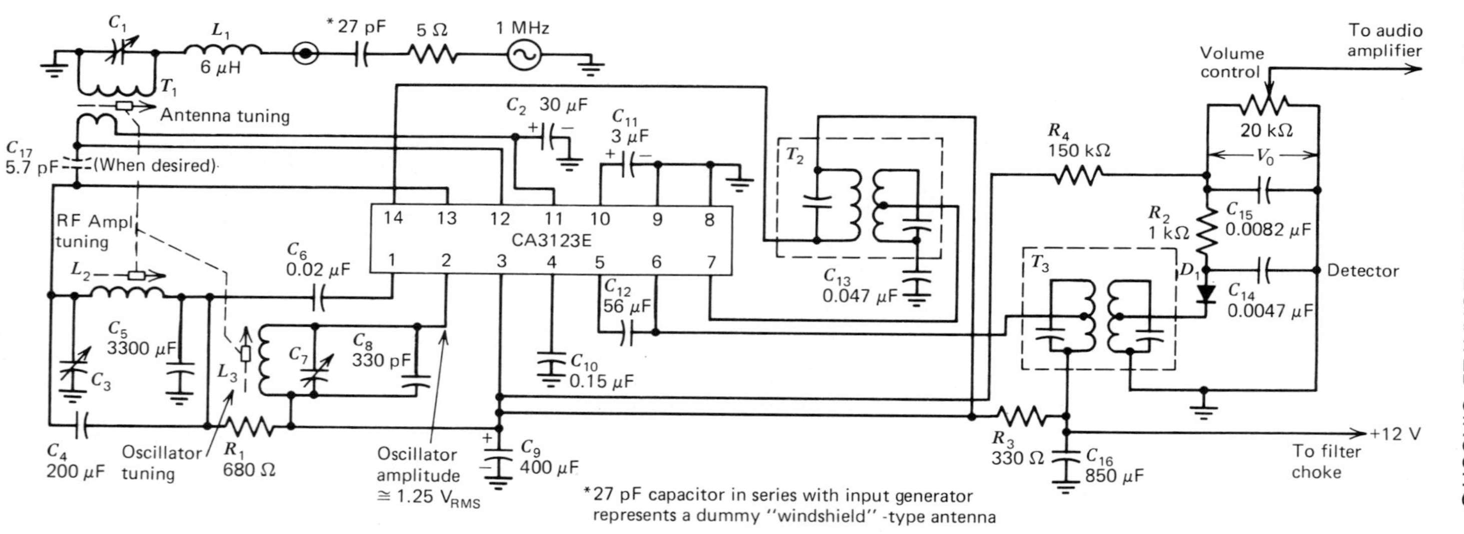

Transformer		Symbol	Frequency	Inductance μh (≈)	Capacitance pF (≈)	Q (≈)	Total turns to tap turns ratio	Coupling
First IF:	Primary Secondary	T_2	262 kHz	2840 2840	130 130	60 60	none 30:1 or 31:1	critical ≈ 0.017 ≈ 1/Q
Second IF:	Primary Secondary	T_3	262 kHz	2840 2840	130 130	60 60	8.5:1 8.5:1	critical ≈ 0.017 ≈ 1/Q
Antenna:	Primary	T_1	1 MHz	195	(C_1)–130	65		
	Secondary		Adjusted to an impedance of 75 Ω with primary resonant at 1 MHz. Coupling should be as tight as practical. Wire should be wound around end of coil away from tuning core.					
Coils		L_1 L_2 L_3	7.9 MHz 1 MHz 1.262 MHz	6 55 41		50 50 40		

(a)

Figure 14.2 The CA3123 AM receiver subsystem containing an RF stage. (*a*) AM receiver using the chip. (Courtesy RCA Corp.)

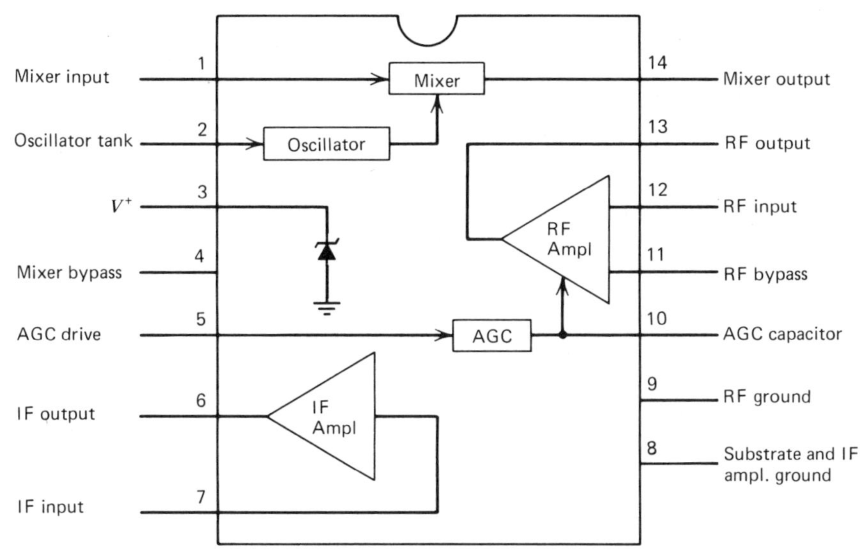

Figure 14.2 (*b*) Functional block diagram. (Courtesy RCA Corp.)

careful layout and construction practices, and the addition of stereo decoding, which will be discussed in Sec. 14.7.

As a guide, one can consider the typical FM subsystem to include the following:

1 IF amplifier(s) with limiter.

2 FM detector.

3 Audio preamplifier.

4 Internal zener regulator.

One need only include an RF stage, proper tuning of the IF for selectivity (a ceramic, crystal, or mechanical filter may be employed), a stable frequency source, and audio amplification to drive a speaker.

For example, the CA3041 (or SK3101), although designed for TV receiver sound, lends itself to FM receiver systems for either two-way or broadcast applications. This IC, supplied in a 14-pin DIP, features internal zener regulation at approximately 11.2 V, wide-frequency capability (from less than 100 kHz to more than 20 MHz), low-harmonic distortion, high sensitivity, and good AM rejection. The functions of IF amplification, limiting, detection, and audio preamplification are provided on the chip.

To better illustrate the specifications that must be considered in FM receiver design, a data sheet page for the CA3041 with typical specifications is provided in Fig. 14.3. Device dissipation is approximately 250 mW at room temperature. Care should be exercised not to exceed this limit, particularly when one considers the adverse conditions often encountered in mobile equipment.

The zener regulating voltage is generally the nominal knee voltage to which the chip will attempt to stabilize external voltage fluctuations. Manufacturers

CA3041

ELECTRICAL CHARACTERISTICS, *at an Ambient Temperature, T_A, of 25°C, and a DC Supply Voltage, V_{CC}, of +140 Volts applied to Terminal 14 through a resistance of 6.2 kΩ, unless otherwise indicated. Any other combination of DC Supply Voltage and Series Resistance which will not cause the Maximum Dissipation Limit or any of the Maximum Voltage or Current Limits for the CA3041 to be exceeded may be used.*

CHARACTERISTICS (See Page 7 for Definitions of Terms)	SYMBOLS	TEST CONDITIONS: SETUP AND PROCEDURE Fig.	TEST CONDITIONS	SPECIAL CONDITIONS	LIMITS TYPE CA3041 Min.	Typ.	Max.	Units	TYPICAL CHARACTERISTICS CURVES Fig.
Total Device Dissipation	P_T	11		T_A = 0°C	220	245	270	mW	2
				T_A = +25°C	225	250	275	mW	
				T_A = +85°C	230	255	280	mW	
Zener Regulating Voltage (DC Supply Voltage at Terminal 14)	V_{14}	-			10.5	11.2	12.3	V	-
Quiescent Operating Current (into Terminal 11)	I_{11}	11			0.25	0.63	1	mA	-
9-Volt Current Drain (Quiescent Operating Current into Terminal 14)	I_{14}	11		V_{CC} = +9 V applied directly to Terminal 14	7	11	16	mA	-
Input-Impedance Components: Parallel Input Resistance	R_i	3	f = 4.5 MHz		-	11	-	kΩ	-
Parallel Input Capacitance	C_i	3			-	5	-	pF	-
Output-Impedance Components: Parallel Output Resistance	R_o	-			-	100	-	kΩ	-
Parallel Output Capacitance	C_o	-			-	4	-	pF	-
Input Limiting Voltage (Knee)	$V_{i(lim)}$	7			-	150	200	μV (rms)	4
Amplitude-Modulation Rejection	AMR	10			45	58	-	dB	9
IF-Amplifier Voltage Gain	$A_{(IF)}$	5			-	67	-	dB	4
Recovered AF Voltage: 1. At FM-Detector Output	$V_o(af)$	-		R_L = 50 kΩ, Δf = ±25 kHz THD = 0.7% (typ.)	-	250	-	mV (rms)	-
2. At AF-Driver Output in Test Setup		-		THD < 5%	8	9	-	V (rms)	-
Total Harmonic Distortion	THD	7		$V_{o(af)}$ = 8 V(rms)	-	1.5	5	%	-
Discriminator Output Resistance	$R_{o(dis)}$	-	f = 1 kHz		-	10	-	kΩ	-
AF-Amplifier Input Resistance	$R_{i(af)}$	-			-	100	-	kΩ	-
AF-Amplifier Output Resistance	$R_o(af)$	-			-	30	-	kΩ	-
AF-Driver Voltage Gain	A_{af}	6			-	41	-	dB	8

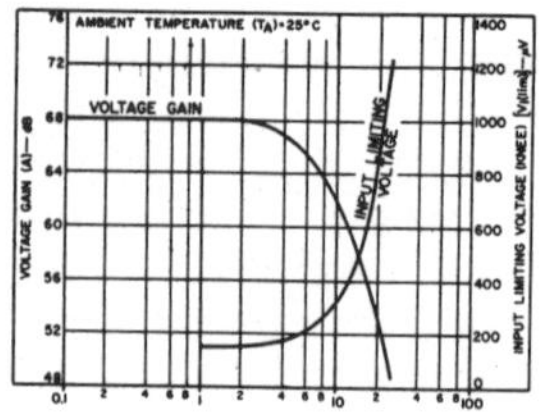

Fig.4 - Typical IF-amplifier voltage gain and input-limiting voltage (knee) characteristics.

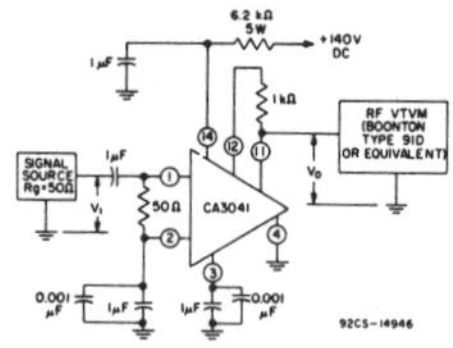

PROCEDURE:

A - Voltage Gain:

1) Set input frequency at desired value, v_i = 100 μV rms.
2) Record v_o.
3) Calculate Voltage Gain A from A = 20 $\log_{10} v_o/v_i$
4) Repeat Steps 1, 2, and 3 for each frequency and/or for temperature desired.

Fig.5 - Test setup for measurement of IF-amplifier voltage gain.

PROCEDURES:

Recovered AF Voltage:

1. Set Input Signal Generator as follows:
   Output frequency = 4.5 MHz
   Modulating frequency = 1 kHz
   Deviation = ± 25 kHz
   Output level for V_{in} = 100 mV rms
2. Set volume control for maximum af output.
3. Measure af output voltage and record as Recovered AF Voltage.

Total Harmonic Distortion:

1. Adjust volume control for an af output voltage of 300 mV rms.
2. Measure Total Harmonic Distortion of the output signal in accordance with the Operating Instructions for the Distortion Analyzer.

Input Limiting Voltage (Knee):

1. Decrease V_{in} until the af output voltage is 3 dB less than the value set in Step 1 of the procedure for measurement of Total Harmonic Distortion (300 mV - 3 dB = 210 mV)
2. Measure resulting value of V_{in} and record as Input Limiting Voltage (Knee).

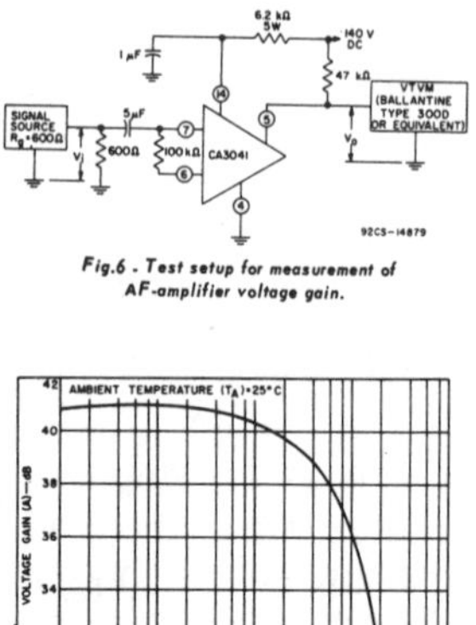

Fig.6 - Test setup for measurement of AF-amplifier voltage gain.

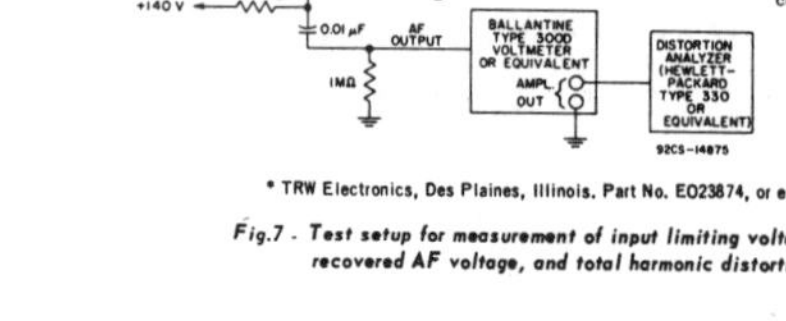

* TRW Electronics, Des Plaines, Illinois. Part No. EO23874, or equivalent.

Fig.7 - Test setup for measurement of input limiting voltage (Knee), recovered AF voltage, and total harmonic distortion.

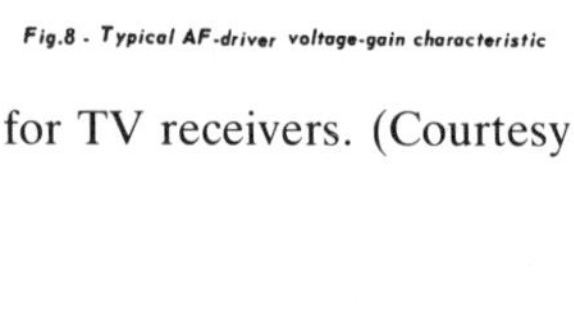

Fig.8 - Typical AF-driver voltage-gain characteristic

Figure 14.3 Data page for a CA3041 sound subsystem for TV receivers. (Courtesy RCA Corp.)

typically choose a voltage less than 12 V, thereby making the chip suitable for mobile systems where a 12 V electrical system is almost universal.

To maintain noise-free reception, AM rejection should be high; a safe figure is 40 to 50 dB. The input-limiting voltage, or knee (the point above which limiting occurs), is an indication of how strong the input signal must be in order

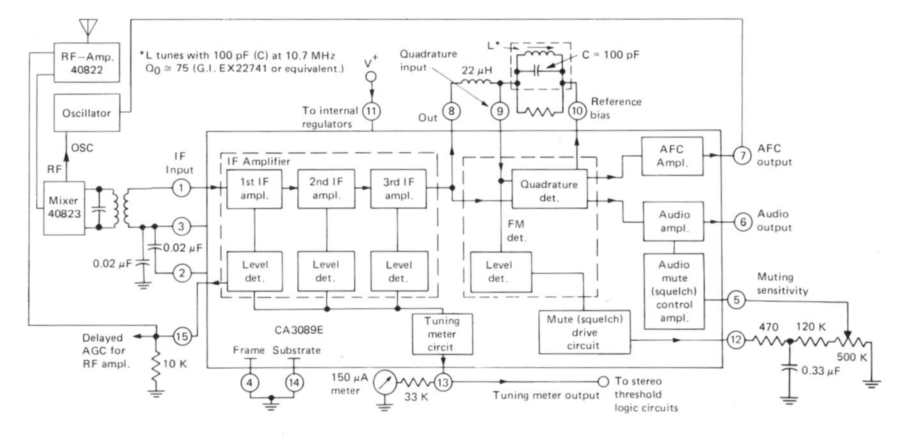

Figure 14.4 An example of an FM tuner designed around the CA3089E chip. (Courtesy RCA Corp.)

to push the amplifier into limiting. A low figure is desirable because this allows a weak signal to provide a fair degree of *quieting*. (Quieting is the amount of signal required to remove background hiss or noise. Full quieting provides dead silence under zero-signal conditions.)

Figures for the knee range from 50 to 500 μV. Where weak signals are anticipated, a low-limiting voltage is desirable; however, IF amplifier gain also must be considered. Depending on the design, these figures may exceed 100 dB, thereby ensuring adequate gain for even the most demanding requirements.

Another important factor often overlooked is that of audio preamplifier voltage gain. When space and weight are important, the greater internal voltage gain allows a reduction in succeeding amplifier stages, thereby conserving space and weight. Voltage gains in excess of 35 dB are common.

EXAMPLE 14.1 Design an FM tuner that provides suitable output for driving an audio amplifier for the 88 through 108 MHz FM broadcast band. Employ discrete RF amplification and an external mixer with it's associated local oscillator.

PROCEDURE

1 The CA3089E (or SK3147) chip shown in Fig. 14.4 is selected.

2 A 40822 (or SK3065) dual-gate MOSFET is acceptable for the RF amplifier and a 40823 for mixer action with the local oscillator signal.

3 A single-tuned circuit tunes a quadrature detector. If required, a tuning meter circuit is included on the chip. The CA3089E also incorporates a mute (squelch) feature to minimize noise when not tuned to a station or, when set to a higher level, to reject weak signals.

4 The audio output may be followed by an appropriate audio amplifier. Note that the signal is monophonic because no stereo decoder follows the CA3089E.

5 A ceramic filter can provide greater selectivity in crowded areas. The filter may be switched out of the circuit for improved frequency response of the audio when this feature is not needed.

Table 14.1 **Examples of ICs for AM/FM receivers**

Chip	*Function*
CA3002	IF amplifier
CA3004	RF amplifier
CA3012	Wideband amplifier
MC1350	IF amplifier
MC1355	Limiting FM IF amplifier
MC3310	Wideband amplifier

14.4 IF AMPLIFIERS

Often it is desirable to employ an IC that provides a *specific function* only, rather than a group of functions as provided in the AM and FM subsystems chips previously considered. Generally, in the case of FM IF amplifiers, the differential amplifier configuration (Chap. 9) is used because it provides symmetrical limiting over a wide range of input voltages.

For example, the CA3012 (or SK3129) wideband amplifier of Fig. 14.5 is well suited for IF amplification. The minimum supply voltage V^+ is 5.5 V with

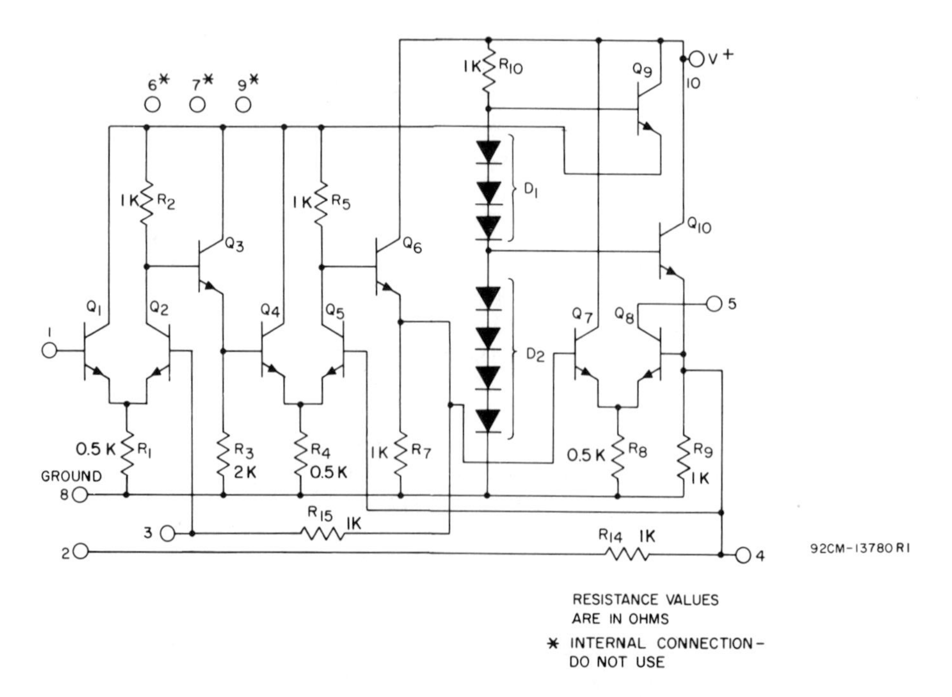

Figure 14.5 Schematic diagram of the CA3012 wideband amplifier. (Courtesy RCA Corp.)

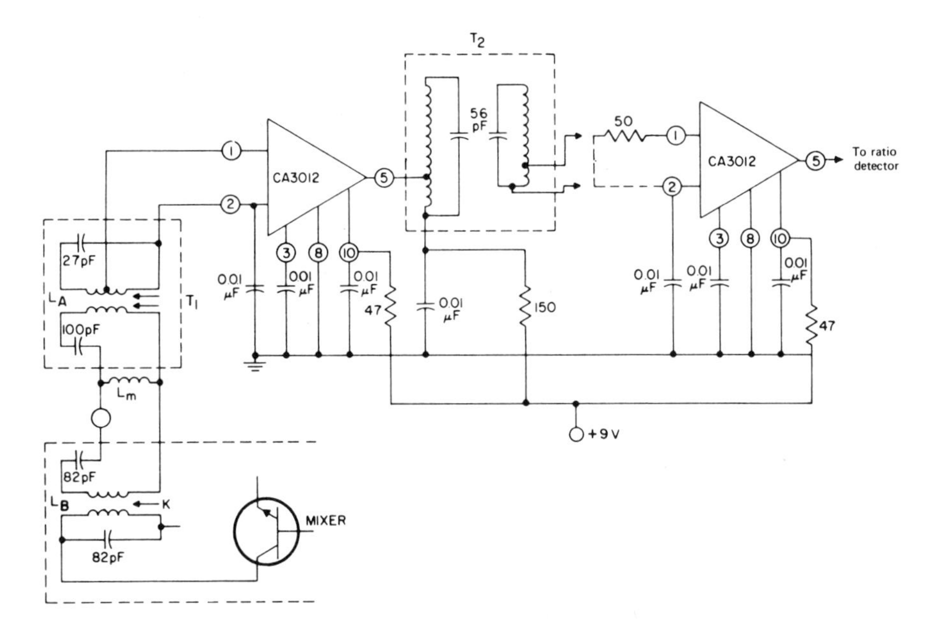

Figure 14.6 The CA3012 employed in a 10.7 MHz IF strip. (Courtesy RCA Corp.)

an upper limit of 13 V. No internal zenering of the supply exists in this IC. Basically, the CA3012 serves as an IF amplifier/limiter used in conjunction with an external FM detector. Figure 14.6 shows a typical 10.7 MHz IF strip employing the CA3012 feeding a ratio detector.

The CA3012 is not the only chip suitable for IF amplifiers. Table 14.1 lists several ICs that lend themselves for similar applications. Many of the chips fall under the category of general-purpose, high-gain amplifiers; their use as IF amplifiers in AM or FM systems is only one application.

If it is desired to include detection as well as IF amplification on a chip, the possibility of further reducing the number of discrete components is increased. Table 14.2 lists several choices available, all of which include some type of FM detector. Obviously these are not complete subsystems because RF amplifiers, mixers, and the like are not included. Some ICs even incorporate an audio preamplifier, leaving an output stage(s) to be added.

EXAMPLE 14.2 The CA3005 IC is to be used in an FM IF strip. The mixer output is 10.7 MHz. Provide a design of the circuit.

PROCEDURE

1 Figure 14.7 provides the proper configuration for 10.7 MHz operation. The mixer output is applied to pin 7. Stage gain is approximately 25 dB.

Table 14.2 **Typical FM chips containing detectors**

Chip	*Internal zener*	*Discriminator*	*Quadrature detector*	*Audio pre-amplifier*	*Limiter*
CA1190			X	X	X
CA2111			X		X
CA3013		X		X	X
CA3014		X		X	X
CA3041	X	X		X	X
CA3042	X	X		X	X
CA3075	X		X	X	X
MC1351	X	X		X	X
MC1357	X		X		X
TDA1190Z	X	X		X	X

2 Transformer T_2 is an interstage transformer used to prevent the collector output of the CA3005 from saturating. Therefore, under large input voltage swings, the bandpass is limited.

Another method of IF amplification, particularly at lower frequencies, involves the use of high-frequency differential amplifiers (Fig. 14.8*a*), which can be operated from dc to frequencies in excess of 100 kHz. This makes the chip suitable for AM and FM IF applications.

EXAMPLE 14.3 Design a 455 Hz IF strip having a bandwidth of 3 kHz.

PROCEDURE

1 The differential amplifier is the heart of the system. Detector schemes are of the standard variety used for AM (Fig. 14.8*b*).

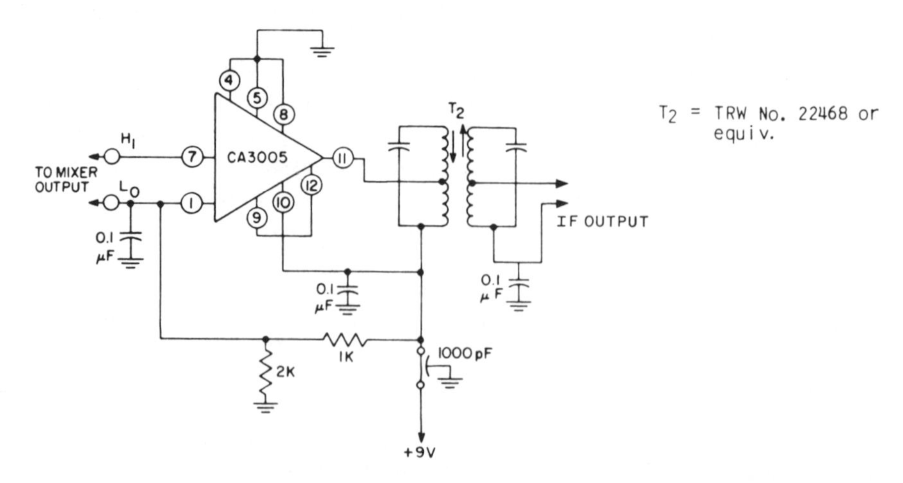

Figure 14.7 The CA3005 used as an IF amplifier. (Courtesy RCA Corp.)

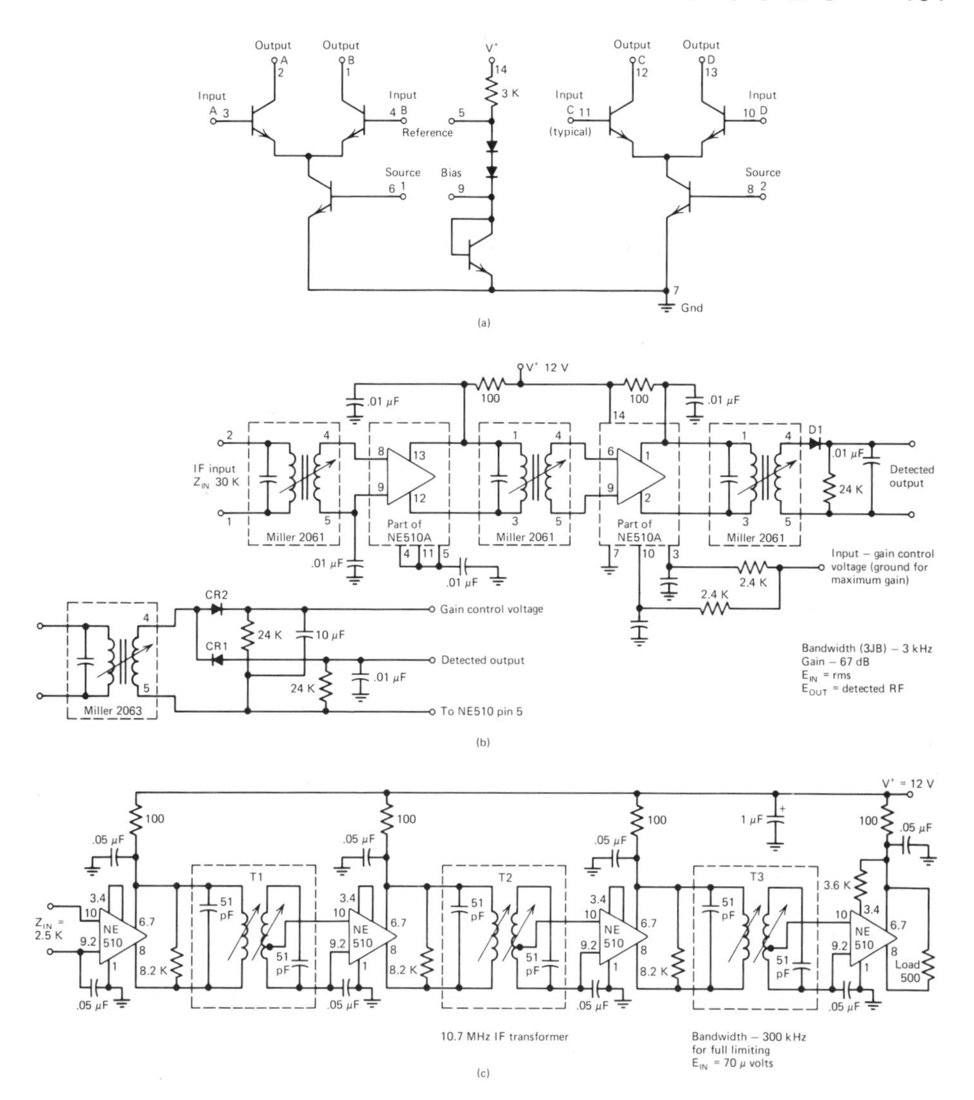

Figure 14.8 The NE510 dual high-frequency differential amplifier. (*a*) Equivalent schematic of circuit. (*b*) Employed in a 455 kHz IF strip. (*c*) Employed in a 10.7 MHz IF strip. (Courtesy Signetics Corp.)

2 The amplifier may also operate at 10.7 MHz and be used as a limiting amplifier with limiting occurring as low as 70 μV (Fig. 14.8*c*).

14.5 RF AMPLIFIERS

Many differential amplifiers function at higher frequencies, making them useful as RF amplifiers. The CA3028 is an example of such a chip. It provides high gain, good sensitivity, and a relatively low-noise figure.

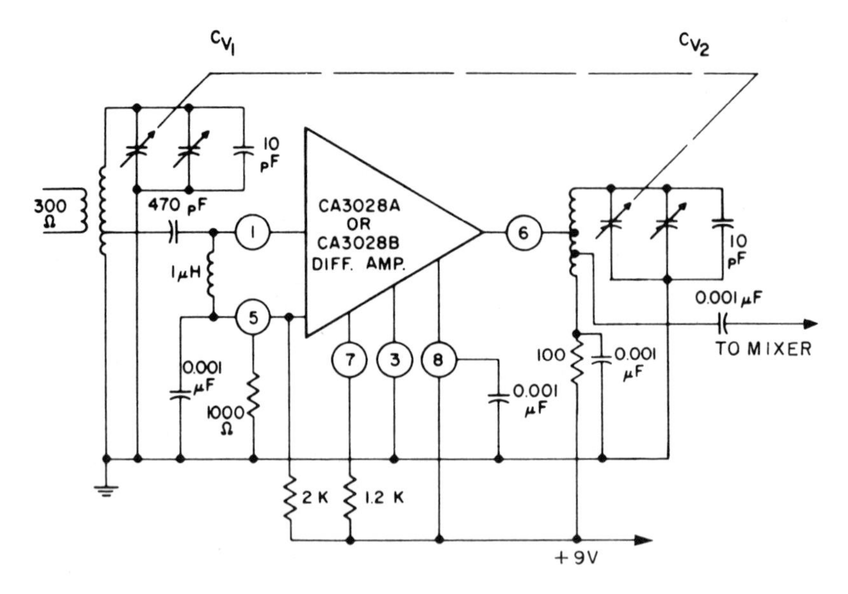

Figure 14.9 The CA3028 IC used as an RF amplifier. (Courtesy RCA Corp.)

EXAMPLE 14.4 A 50 MHz front end is to be designed using the CA3028 chip. Impedance matching must be provided to terminate the antenna into 50 Ω. The RF output is to be applied to a mixer stage. Design the circuit.

PROCEDURE

1 Figure 14.9 is a typical circuit using the CA3028. Input matching is realized at the RF antenna coil.

2 The output from pin 6 is applied to a tuned circuit and coupled to the mixer stage for further amplification and heterodyning action.

14.6 FM STEREO DECODERS

The trend in recent years has been toward more sophisticated, yet compact, stereo systems capable of ruler flat-frequency response, high-channel separation, low distortion, and high sensitivity. The chips discussed in this section offer premium performance for decoding stereo broadcasts at low cost.

The μA758, CA758E, MC1311P, ULX2244, and LM1800 are equivalent FM stero phase locked loop (Sec. 14.9) decoders that are pin compatible. These decoders also offer excellent subsidiary carrier authorization (SCA) suppression.

A minimum of external components is required for normal operation. Specifications for the CA758E are provided in Table 14.3 and a block diagram of the chip is given in Fig. 14.10. Right- and left-channel outputs, and an output for stereo indication and a single oscillator adjustment on pin 15, to simplify alignment, are available on the chip.

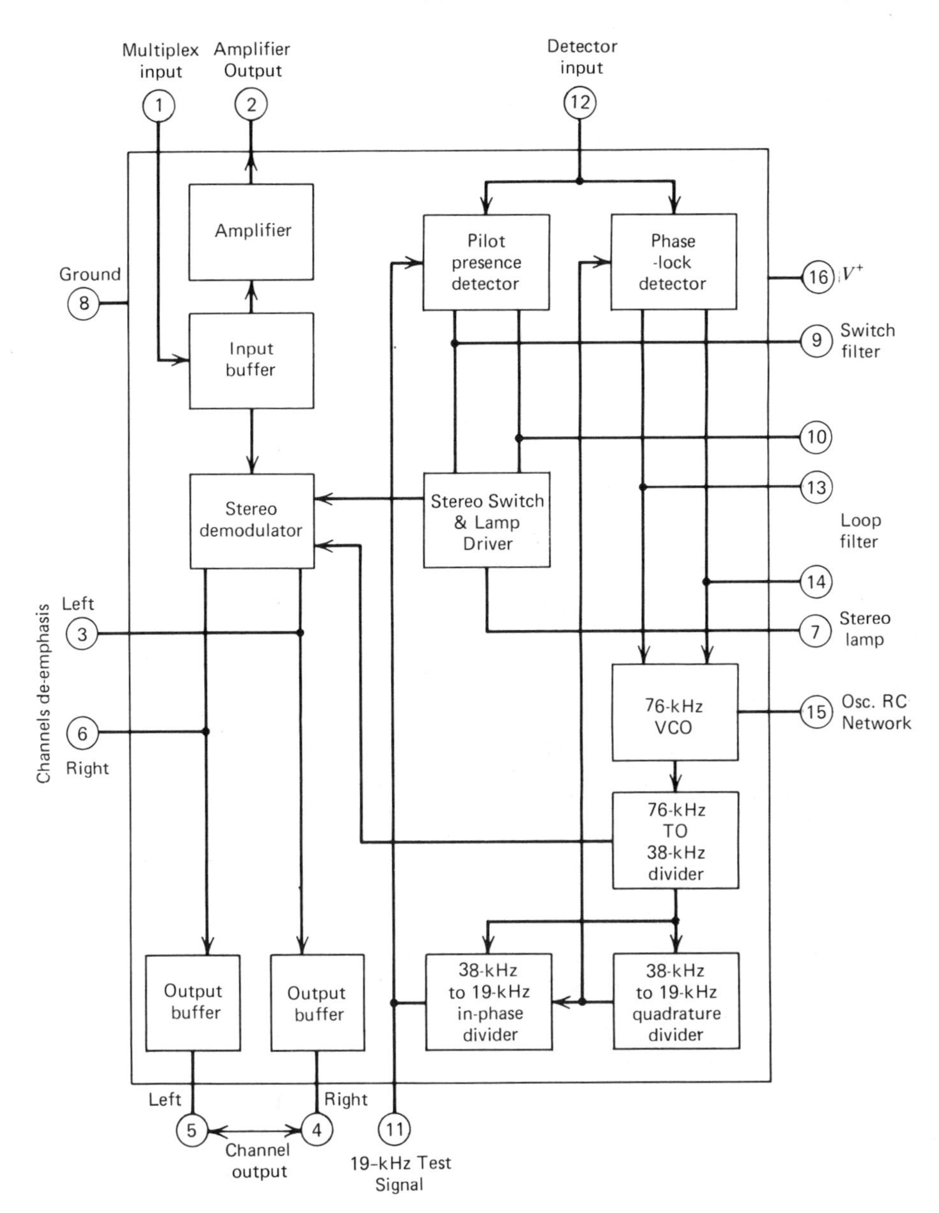

Figure 14.10 Functional block diagram of the CA758E phase locked loop stereo decoder. (Courtesy RCA Corp.)

EXAMPLE 14.5 An FM composite multiplexed signal is to be decoded into left- and right-channel outputs. Using the CA758E, design a circuit to accomplish this function. A stereo indicator is to be included.

Table 14.3 **Specifications for the CA758E PLL stereo decoder (Courtesy RCA Corporation)**

MAXIMUM RATINGS, *Absolute-Maximum Values at T_A = 25°C*

DC Supply Voltage	+18 V
DC Supply Voltage (for ⩽a 15-second period)	+22 V
DC Voltage at Term. 7 (Lamp Driver Circuit with Lamp "OFF")	+22 V
Device Dissipation:	
Up to T_A = 70°C	730 mW
Above T_A = 70°C derate linearly	9.1 mW/°C
Ambient Temperature Range:	
Operating	−40 to +85°C
Storage	−65 to +150°C
Lead Temperature (During soldering):	
At a distance not less than 1/32" (0.79 mm) from case for 10 s max.	+265°C

ELECTRICAL CHARACTERISTICS

CHARACTERISTIC	TEST CONDITIONS (Referenced to Fig. 3 unless otherwise specified) V^+ = 12 V, T_A = 25°C Multiplex Input Signal (L=R, pilot "OFF") = 300 mV RMS 19-kHz Pilot Level = 30 mV RMS f (modulation) = 400 Hz or 1 kHz	LIMITS Min.	Typ.	Max.	UNITS
Static Characteristics					
Total Current	Lamp "OFF"	–	26	35	mA
Maximum Available Lamp Current		75	150	–	mA
DC Voltage at Term. 7 (Lamp Driver)	I (Lamp) = 50 mA	–	1.3	1.8	V
DC Voltage Shift at either Term. 4 or 5 (Output)	Stereo-to-Mono Operation	–	30	150	mV
Dynamic Characteristics					
Power Supply Ripple Rejection	For a 200-Hz, 200-mV RMS Signal	35	45	–	dB
Input Resistance		20	35	–	kΩ
Output Resistance		0.9	1.3	2.0	kΩ
Channel Separation (Stereo)	At f = 100 Hz	–	40	–	dB
	f = 400 Hz	30	45	–	dB
	f = 10 kHz	–	45	–	dB
Channel Balance (Monaural)		–	0.3	1.5	dB
Voltage Gain	At f = 1 kHz	0.5	0.9	1.4	V/V
Pilot Input Level:					
19-kHz Input	Lamp "ON"	–	15	20	mV RMS
19-kHz Input	Lamp "OFF"	2.0	7.0	–	mV RMS
Hysteresis	Lamp "OFF"	3.0	7.0	–	dB
Capture Range (Deviation from 76-kHz Center Frequency)		±2.0	±4.0	±6.0	%
Total Harmonic Distortion	Multiplex Input Signal = 600 mV RMS (Pilot "OFF")	–	0.4	1.0	%
19-kHz Rejection		25	35	–	dB
38-kHz Rejection		25	45	–	dB
SCA (Storecast) Rejection	Measured Composite Signal: 80% Stereo, 10% Pilot, 10% SCA	–	70	–	dB
Voltage-Controlled Oscillator (VCO) Tuning Resistance	Total Resistance (Term. 15 to 8) required to set f_{REF} = 19 kHz ± 10 Hz (Term. 11)	21.0	23.3	25.5	kΩ
Voltage-Controlled Oscillator Frequency Drift	0° ⩽ T_A ⩽ 25°C	–	+0.1	±2	%
	25° ⩽ T_A ⩽ 70°C	–	−0.4	±2	%

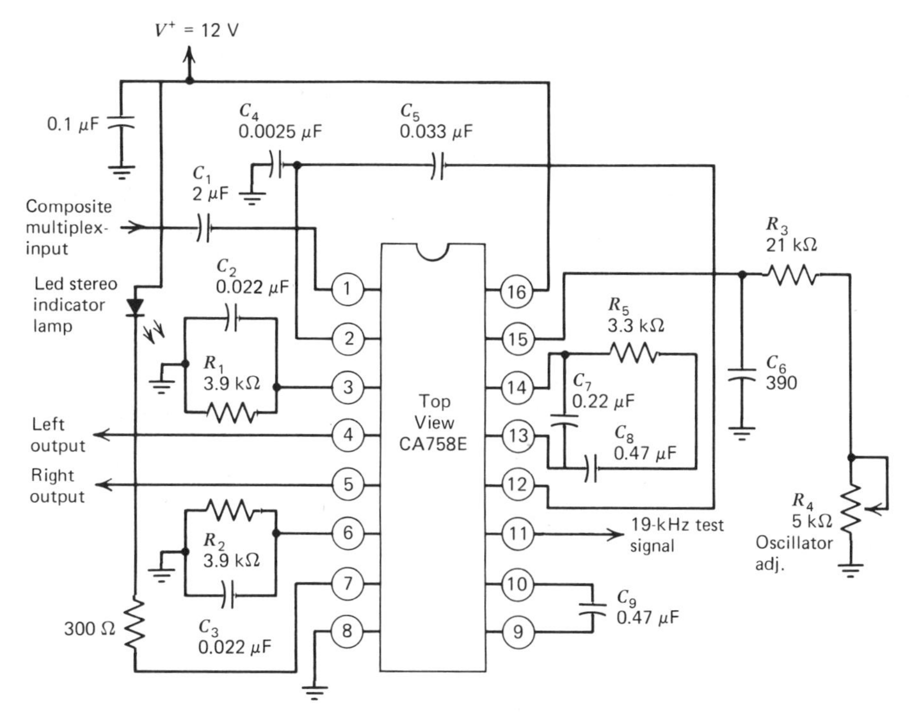

Notes:

Tolerance on resistors is ±5% and tolerance on capacitors is ±20%, unless otherwise specified.

C_1 = +100%, −20%

C_6 = ±1% in test circuit and ±5% in typical application

R_3 = ±1%

R_4 = ±10%

R_1 and R_2 = ±1% in test circuit and ±5% in typical application.

Figure 14.11 A stereo decoder employing a CA758E chip. (Courtesy RCA Corp.)

PROCEDURE

1 The complete circuit is provided in Fig. 14.11. An LED is employed for the stereo indicator.

2 Voltage V^+ is bypassed with a 0.1 μF capacitor connected to pin 16. The composite stereo signal is coupled to pin 1. Deemphasis for the left channel is provided by R_1, C_2 while right-channel deemphasis is accomplished by R_2, C_3. The external RC network connected to pin 15 determines the oscillation frequency and is trimmed by R_4. For test purposes, a 19 kHz test signal is connected to pin 11.

Pin 7 is a pulldown for the stereo LED when the 19 kHz pilot is present. A 300 Ω resistor provides current limiting. Output buffers (which follow the external RC circuits for deemphasis) provide the left- and right-channel audio outputs.

Another stereo decoder designed for use in high fidelity stereo tuners/receivers, as well as in auto stereo systems, is the TCA4500 chip. This IC operates over a supply voltage range of from 8 to 16 V. Manual stereo defeat is also possible. The TCA4500 can directly drive a stereo indicator lamp with capabilities up to 100 mA. Typical specifications for the chip are given in Table 14.4.

Variable separation control of the right and left channels is also possible with the TCA4500. Generally, separation is controlled by the AGC signal that is a function of the received signal strength. A common practice is to reduce the separation as the signal becomes weaker, thereby reducing noise without the need for total mono operation. Variable separation control can be manually overriden in the TCA4500. The circuit automatically switches to the mono mode when the 19 kHz pilot disappears.

Table 14.4 **Specifications for the TCA4500 stereo decoder (Courtesy Motorola, Inc.)**

MAXIMUM RATINGS (T_A = +25°C unless otherwise noted)

Rating	Value	Unit
Power Supply Voltage	16	Volts
Power Dissipation (Package limitation) Derate above T_A = +25°C	1800 15	mW mW/°C
Operating Temperature Range (Ambient)	-40 to +85	°C
Storage Temperature Range	-65 to +150	°C
Lamp Drive Voltage (Max. voltage at pin 7 with lamp "off")	30	Volts
Lamp Current	100	mA
Blend Control Input Voltage (pin 11)	10	Volts

ELECTRICAL CHARACTERISTICS Unless otherwise noted: V_{CC} = +12 Vdc, T_A = 25°C, 2.5 Vp-p standard multiplex composite signal with L or R channel only modulated at 1.0 kHz and with 10% pilot level, using circuit of Figure 1.

Characteristic	Min.	Typ.	Max.	Unit
Stereo Channel Separation: Unadjusted Optimised on other channel[1]	30 40	– –	– –	dB
Monaural Voltage Gain[1]	0.8	1.0	1.2	
THD at 2.5 Vp-p Composite Input Signal at 1.5 Vp-p Composite Input Signal	– –	– 0.2	0.3 –	%
Signal/Noise Ratio RMS 20 Hz - 15 kHz	–	90	–	dB
Ultrasonic Frequency Rejection 19 kHz 38 kHz	– –	31 50	– –	dB
Stereo Switch Level (19 kHz input level for lamp "on") Hysteresis	12 –	16 6.0	20 –	mVrms dB
Quiescent Output Voltage Change with Mono/Stereo Switching	–	5.0	20	mVdc
Stereo Blend Control Voltage (pin 11) 3 dB Separation (see Fig. 2) 30 dB Separation	– –	0.7 1.7	– –	V V
Minimum Separation (pin 11 at 0 V)	–	–	1.0	dB
Monaural Channel Imbalance (pilot tone off)	–	–	0.3	dB
ARI 57 kHz Pilot Tone Influence on THD[2]	–	–	0.5	%
Sub-carrier Harmonic Rejection 76 kHz 114 kHz 152 kHz	– – –	45 50 50	– – –	dB
Supply Ripple Rejection	–	50	–	dB
Input Impedance	–	50	–	KΩ
Output Impedance	–	100	–	Ω
Blend Control Current[1]	–	–	-300	μA
Capture Range	–	±5.0	–	%
Operating Supply Voltage	8.0	–	16	V
Current Drain (lamp off)	–	35	–	mA

Notes: [1] See Applications Information and Circuit Description
[2] ARI Test – Input signal: 1.5 Vp-p standard composite signal, 1 kHz modulation added to a CW 50 mVrms signal at 57.3 kHz.

In Fig. 14.12 the TCA4500 consists of three subsections: phase locked loop (PLL), stereo switch, and decoder. In order for this or any decoder to operate, the composite signal must be available. All other receiver functions must precede the decoder IC.

In the PLL section, the 228 kHz oscillator is inputted to a 3-stage Johnson (ring) counter after being divided by two. A train of 19 kHz square waves are thereby generated (Fig. 14.13*a*). These waveforms provide modulation functions as shown in Fig. 14.13*a*.

The stereo switch waveforms of Fig. 14.13*b* activate the external stereo indicator when a stereo broadcast is tuned in. Under mono signal conditions, no 19 kHz pilot is present and the TCA4500 operates in the mono mode. It also can operate in the mono mode when the pilot is present by connecting pin 9 to ground. This defeats the 228 kHz signal. Under weak signal conditions, the TCA4500 switches to the mono mode to reduce noise.

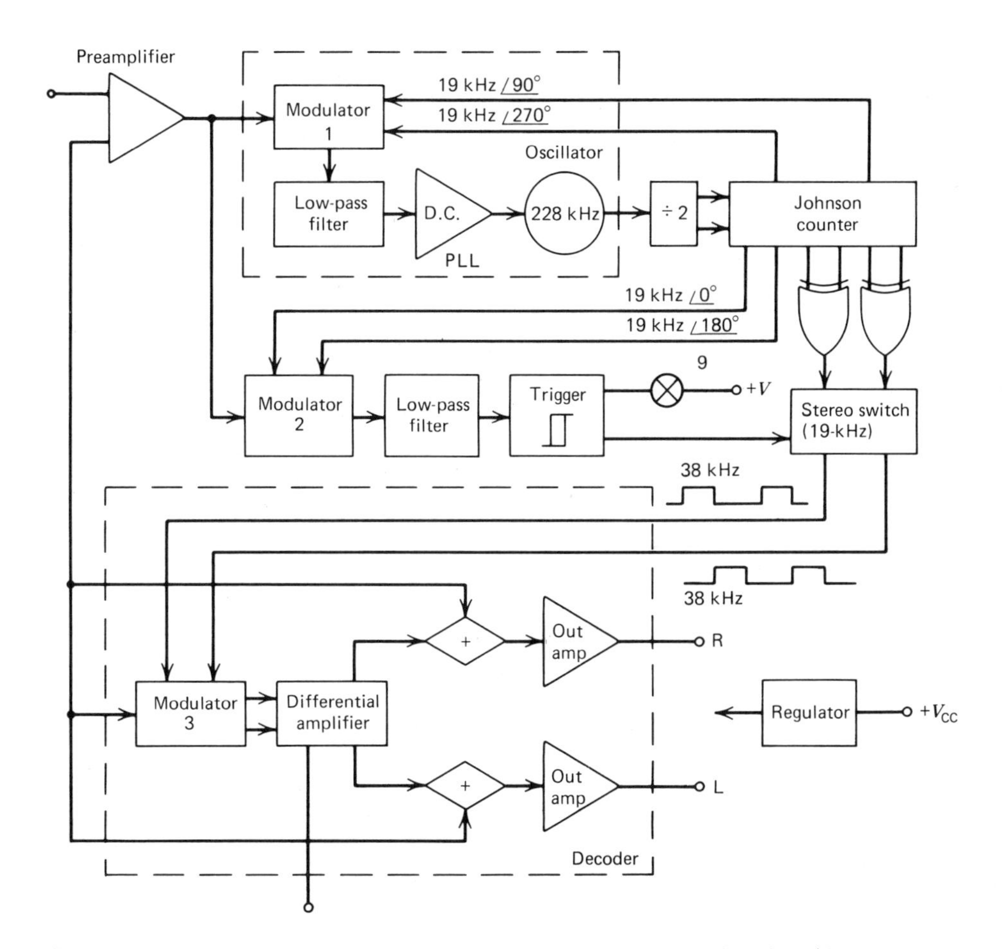

Figure 14.12 Functional block diagram of the TCA4500 stereo decoder. (Courtesy Motorola, Inc.)

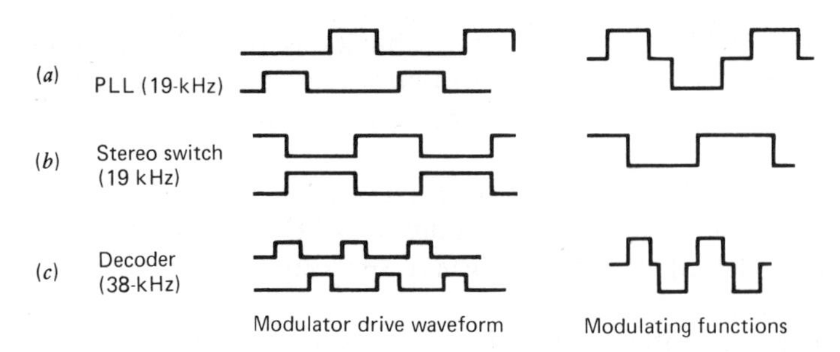

Figure 14.13 Waveforms generated in the TCA4500 chip from the (*a*) Phase locked loop. (*b*) Stereo switch. (*c*) Decoder. (Courtesy Motorola, Inc.)

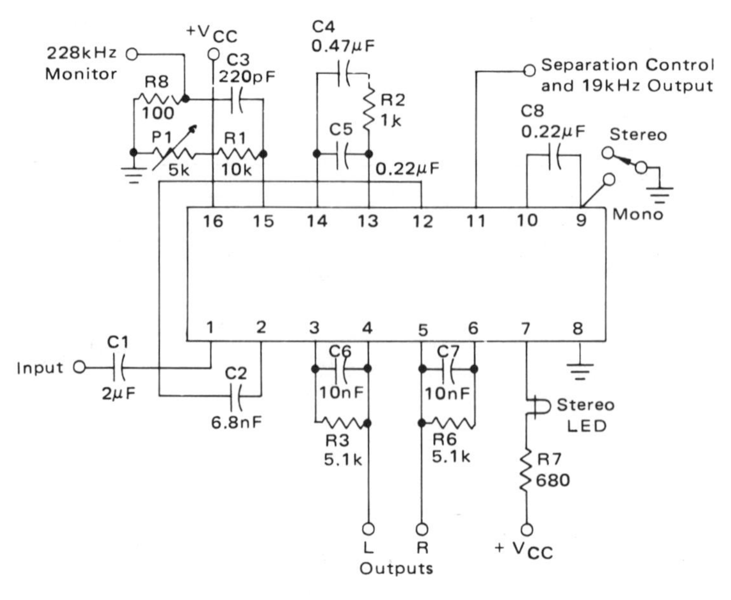

PIN FUNCTIONS

1 – Input
2 – Preamplifier output
3 – Left amplifier input
4 – Left channel output
5 – Right channel output
6 – Right amplifier input
7 – Stereo indicator Lamp
8 – Ground
9 – Stereo switch filter
10 – Stereo switch filter
11 – 19 kHz output/blend
12 – Modulator input
13 – Loop filter
14 – Loop filter
15 – Oscillator RC network
16 – V_{CC}

Figure 14.14 The TCA4500 employed in a stereo decoder circuit. (Courtesy Motorola, Inc.)

The waveforms of Fig. 14.13*c* drive the decoder. The outputs are the noninverted and inverted channel difference signals that are applied to the output amplifiers by a blend circuit. RC networks are employed to control deemphasis and gain, as shown in Fig. 14.14. For $R_3 = R_6 = 5.1$ kΩ, unity gain is achieved.

If pin 11 is unconnected, the 19 kHz has a dc component of approximately 4 V. The blend circuit is nonfunctional, thereby providing maximum separation of the right and left channels. If it is desired to reduce the separation, it is only necessary to reduce the potential at pin 11. When pin 11 is at 3.2 V, the 19 kHz pilot is eliminated. At 2.3 V, blending begins and separation decreases. When pin 11 is not used for blending, the 19 kHz signal can be used to indicate the oscillator frequency. When the blend system is employed, and the drive circuit inhibits access to the pilot signal, the oscillator frequency must be directly measured and tuned to 228 kHz. A test point, as indicated in Fig. 14.15, must be used.

EXAMPLE 14.6 Design an FM stereo receiver using the TCA4500 and CA3089E chips. The intermediate frequency is 10.7 MHz. A tuning meter, stereo indicator light, and a muting (squelch) circuit to eliminate interstation hiss should be included in the design.

PROCEDURE

1 Refer to Fig. 14.16 for the design. The CA3089E includes a 3-stage FM IF amplifier/limiter, a doubly balanced quadrature detector, and an audio preamplifier with mute capability.

2 Delayed AGC is provided by the CA3089E for RF amplifer control. Direct connection to the tuning meter is provided at pin 13 by a 33 kΩ resistor.

3 The use of a ceramic filter between the mixer and the CA3089E input enhances selectivity. The composite output from the CA3089E is coupled through C_1 to the input of the TCA4500 chip. When desired, this signal may be applied directly to an audio amplifier if stereo operation is not required.

14.7 ICs FOR TELEVISION APPLICATIONS

Over the years, IC technology has progressed to the point where high-density chips have replaced discrete components in various TV receiver sections. This has resulted in a reduction of assembly costs, size, and complexity, while achieving a level of performance hitherto not thought possible. An important circuit

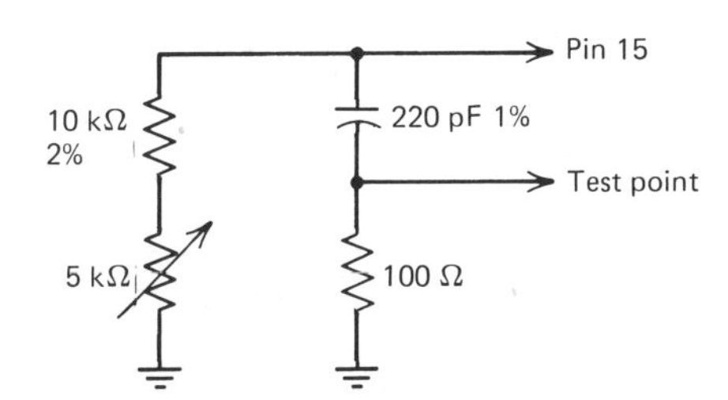

Figure 14.15 Test point for measuring the 228 kHz oscillator frequency.

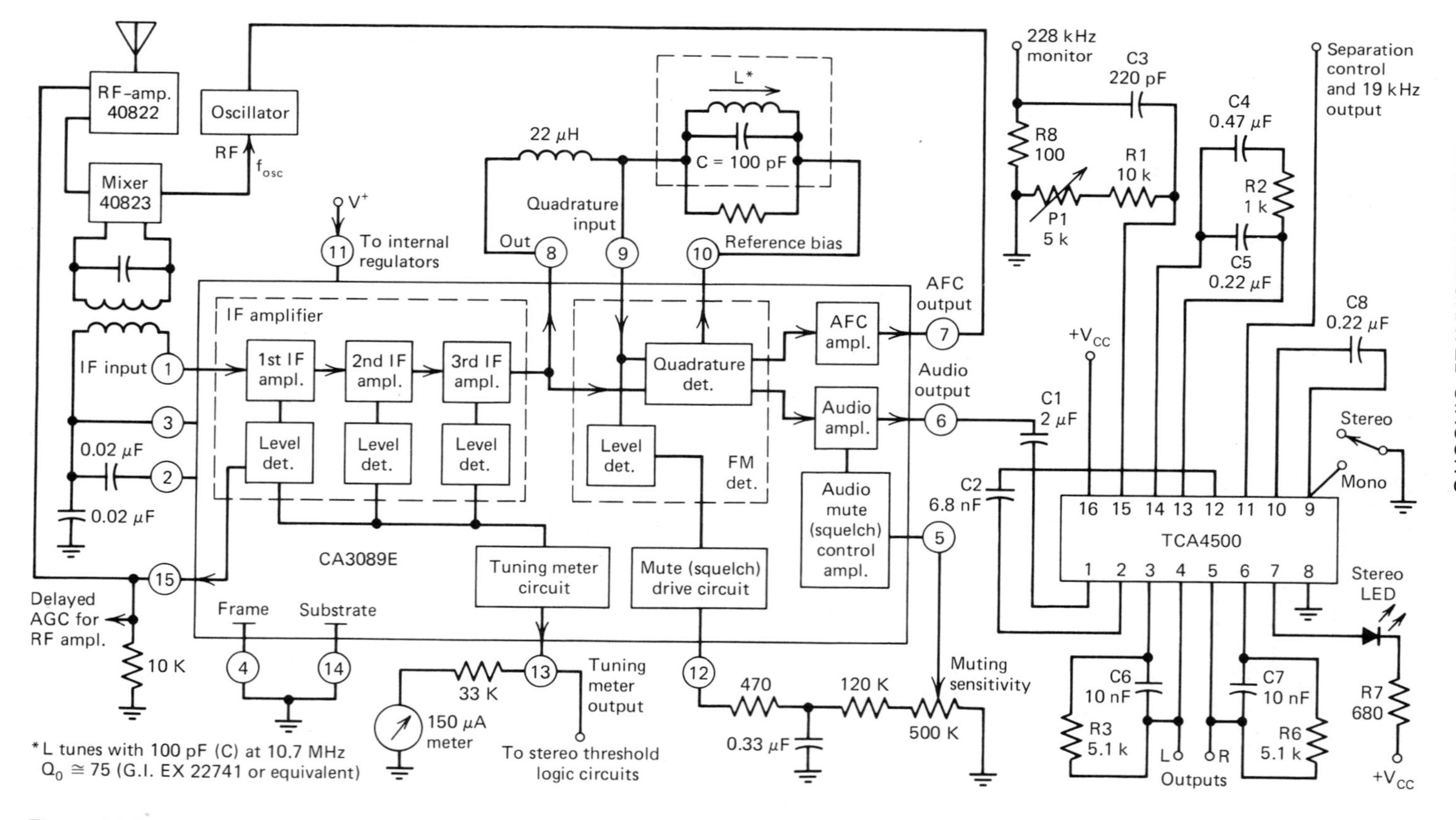

Figure 14.16 An FM stereo receiver using the CA3089E and TCA4500 chips.

performing various functions on one chip is the picture IF subsystem. All of the following functions are generally included on the chip:

1 Video IF amplification.

2 Linear detection.

3 Video output amplification.

4 Keyed AGC.

5 Delayed AGC for tuner control.

6 Sound carrier detection and amplification.

The CA3068 is an example of an IC performing the picture IF function, whether it be color or monochrome. Its block diagram is provided in Fig. 14.17 and its electrical characteristics are listed in Table 14.5.

A complete IF subsystem can be easily designed with the CA3068 requiring only the input from the tuner and a key pulse from the horizontal block. Outputs are provided to drive the video output stage, delay line, and sync separator. To provide sound IF amplification and detection, a CA3065 sound IF subsystem, for example, may be connected (through an appropriate filter) to pin 2 of the CA3068. The CA3065 contains the IF amplifier/limiter function, detector, and

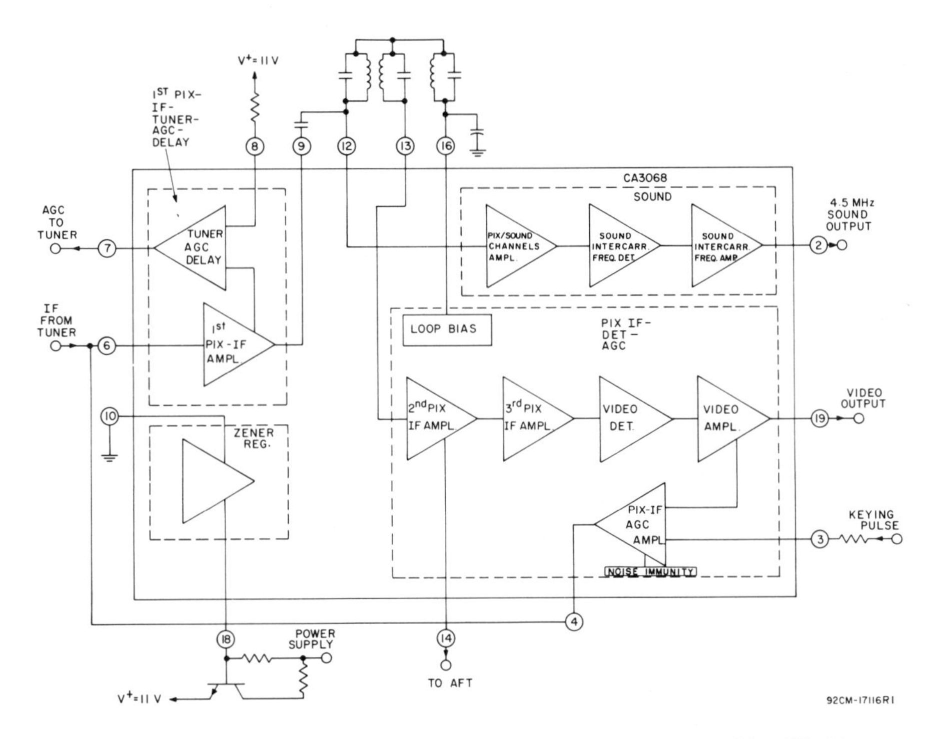

Figure 14.17 Functional block diagram of the CA3068 television video IF chip. (Courtesy RCA Corp.)

Table 14.5 **Electrical characteristics of the CA3068 video IF chip (Courtesy RCA Corporation)**

ELECTRICAL CHARACTERISTICS at $T_A = 25°C$

CHARACTERISTIC	SYMBOL	TEST CONDITIONS	CIRCUIT Fig. No.	LIMITS Min.	Typ.	Max.	UNITS
Static (DC) Characteristics							
Quiescent Circuit Current	I_{15}	–	3	15	–	45	mA
DC Voltages:							
Terminal 2 (Sound)	V_2	–	5	–	6	–	V
Terminal 3 (Keying Input)	V_3	–	3	6.4	–	10	V
Terminal 7 (1) (AGC)	V_7	–	3	16	–	21	V
Terminal 7 (2) (AGC)	V_7	–	4	–	1	–	V
Terminal 8 (AGC Delay)	V_8	–	4	–	4	–	V
Terminal 9 (Cascode Collector)	V_9	–	3	–	8.5	–	V
Terminal 16 (Bias)	V_{16}	–	3	1.1	–	2.3	V
Terminal 18 (Zener)	V_{18}	$V_5 = V_{17} = 0$ V, $I_{18} = 1$ mA	–	10.6	11.9	13.2	V
Terminal 19 (White Level)	V_{19}	–	5	6	–	10	V
Dynamic Characteristics							
Video Sensitivity	e_I	$f_o = 45.75$ MHz, Mod. (AM) = 85% at 400 Hz; Adjust e_I for 4 $V_{p\text{-}p}$ at Term. 19	6	40	100	200	μV
Sync. Tip Level Voltage	V_{19}	$f_o = 45.75$ MHz, e_I(CW) = 10 mV	6	0.4	0.8	1.6	V
Automatic Fine Tuning (AFT) Drive Level Voltage	V_{14}		6	–	15	–	mV
Delay Bias Voltage:	V_7	$f_o = 45.75$ MHz, e_I(CW) = 20 mV; Adjust R_1 for $V_7 = 14$ V	6				
At $e_I = 10$ mV				16	–	–	V
At $e_I = 30$ mV				0.5	–	2	V
3.58 MHz Chroma Output Voltage	V_{19}	$f_o = 45.75$ MHz, e_I(step mod.) = 10 mV; $f_1 = 42.17$ MHz, e_I(step mod.) = 3.33 mV	6	0.5	0.8	–	V
4.5-MHz Sound Output Voltage	V_2	$f_o = 45.75$ MHz, e_I(step mod.) = 10 mV; $f_2 = 41.25$ MHz, e_I(step mod.) = 2.5 mV	6	50	200	–	mV
Parallel Input Impedance:		$f_o = 45.75$ MHz Impedance and Admittance measured at bias conditions as developed by circuit shown in Fig. 7					
Resistance at Term. 6	R_{I-6}		7	4	–	–	kΩ
Capacitance at Term. 6	C_{I-6}			–	2	–	pF
Resistance at Term. 12	R_{I-12}		7	–	4.5	–	kΩ
Capacitance at Term. 12	C_{I-12}			–	4	–	pF
Resistance at Term. 13	R_{I-13}		7	–	5	–	kΩ
Capacitance at Term. 13	C_{I-13}			–	4	–	pF
Parallel Output Impedance:							
Resistance at Term. 9	R_{O-9}		7	30	–	–	kΩ
Capacitance at Term. 9	C_{O-9}			–	3	–	pF
Cascode Transfer Characteristics: Magnitude of Forward Transadmittance	$\lvert y_f \rvert$		7	–	50	–	mmho
Reverse Transfer Capacitance	C_r		7	–	0.001	–	pF

an audio driver stage which may directly drive an external npn device. A differential peak detector is employed, thereby requiring only a single-tuned circuit.

Accurate automatic frequency tuning (AFT) is required for good reception of TV, particularly color signals. The use of a well-designed AFT circuit minimizes picture deterioration under variable signal conditions, such as multipath. The CA3064, used as an AFT block, provides tuner control in conjunction with the AFT output drive from the CA3068 picture IF subsystem. (A block diagram of the CA3064 is provided in Fig. 14.18.) When varactor diodes are employed for tuner control, a dc error signal is produced from the CA3064 to shift the local oscillator frequency in the UHF or VHF tuner, thereby maintaining optimum picture quality. In order for this function to exist, a portion of the picture IF signal is applied to the CA3064 input from the CA3068 IF subsystem. The desired 45.7 MHz signal is supplied by the tuner when the correction signal is zero. A feedback error signal is, therefore, applied to the tuner ensuring "lock" at all times.

Color Processors

In color receivers the use of two ICs provide the basis for the color demodulation system. Several systems are possible; the discussion here, therefore, is limited to a particular pair of ICs that are capable of providing the necessary color (chroma) signal processing and color demodulation functions.

EXAMPLE 14.7 Analyze the operation of the CA3070 signal processor and CA3121 color demodulator.

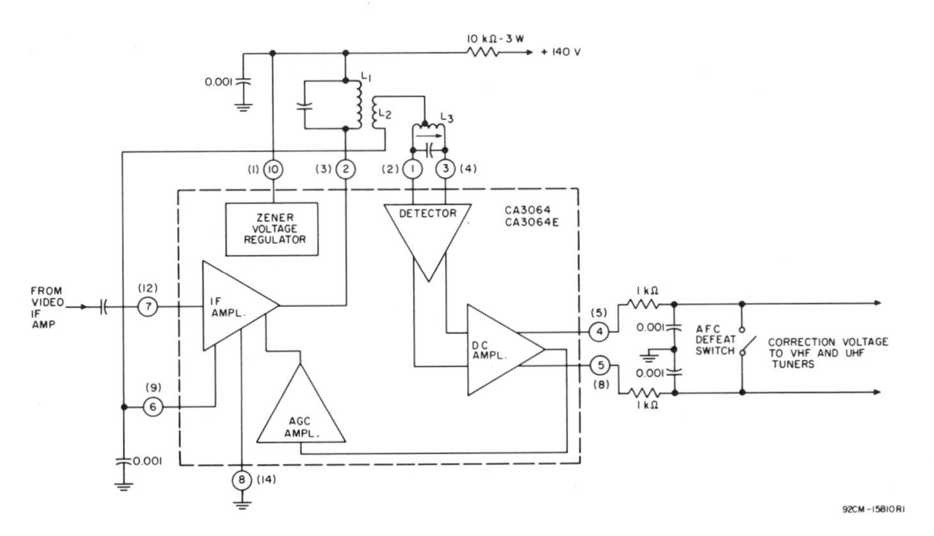

Figure 14.18 Block diagram of the CA3064 AFT circuit. (Courtesy RCA Corp.)

PROCEDURE

1 Figure 14.19*a* is a simplified block diagram of the CA3070 processor. The amplified color (chroma) signal from the CA3121 of Fig. 14.19*b* is applied to the APC and ACC terminals of the CA3070. These are the automatic phase and chroma controls, respectively. Both controls are keyed by the horizontal pulse. The ACC employs synchronous detection to provide an error voltage at pins 15 and 16. (In synchronous detection, a carrier signal is inserted in the detector to recover the chrominance signal.) The signal is then applied to inputs terminals 1 and 16 of the demodulator.

2 Synchronous detection is also performed in the APC block where a correction voltage is produced and internally supplied to the 3.58 MHz reference oscillator.

3 For phase requirements, an RC phase shifter is employed from terminals 13 and 14 and the color input. The feedback path of the oscillator is from terminals 7 and 8 to 6.

4 The frequency of the reference oscillator is varied by the phase of the feedback path.

5 Control of the oscillator stage is provided at pins 2 and 3 by the hue control input to terminal 1. External L, C, and R components control hue phase shift and couple the oscillator to the demodulator input.

6 Referring to Fig. 14.19*b*, the first amplifier output (terminal 3) is coupled to the CA3070 for ACC and APC action, as previously described. It also provides a signal to the second amplifier (terminal 4) which contains color gain control circuitry. The gain of amplifier 2 is reduced when required.

7 The demodulator input (terminal 13) receives its signal through a filter from terminal 14. The R–Y and B–Y subcarrier signals are applied to the demodulator from the CA3070.

8 A matrix recovers the R–Y, G–Y, and B–Y chroma difference signals, which are applied to the respective output stages, including the luminance signal.

14.8 MODULATOR AND DEMODULATOR CIRCUITS

The previous sections stressed receiving systems. For the most part, transmitting systems are usually composed of discrete components owing to the high powers encountered. Applications, however, exist for modulator and frequency doubler circuits operating at low power levels. The MC1596 balanced modulator/demodulator is an example of an IC designed for this function. It may also be used as a product detector, phase/frequency detector, or mixer in receivers.

Generally, the MC1596 is used in double-sideband suppressed carrier modulation systems. A typical AM modulator circuit is shown in Fig. 14.20*a*. Note that dual supplies are indicated; it is possible, however, to operate the chip from a single 12 V source.

As we see in Fig. 14.20*b*, the MC1596 circuit is basically a differential amplifier (Q_5, Q_6) driving a dual differential amplifier (Q_1, Q_2, Q_3, Q_4). Transistors Q_7 and Q_8 serve as constant current sources for the differential amplifier composed of Q_5 and Q_6. (See Chap. 9 for a detailed discussion of differential amplifiers and constant current sources.) Full wave balanced multiplication of the two input voltages, V_C and V_S, occurs because of the cross coupling of the output collectors of the dual differential amplifier. The output signal, V_o, is the product of the two input signals multiplied by a constant.

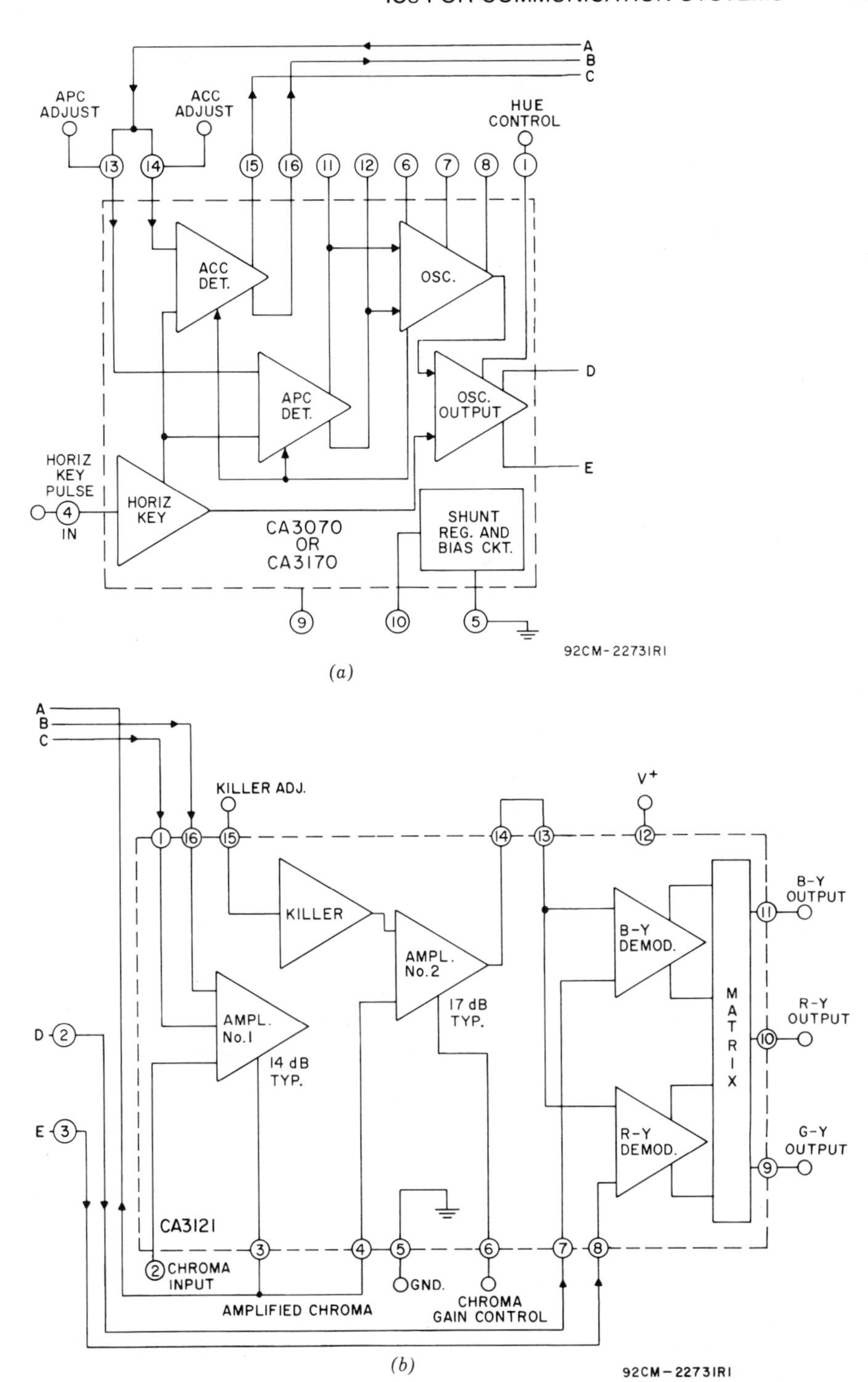

Figure 14.19 Block diagrams of the (*a*) CA3070 chroma signal processor and (*b*) CA3121E chroma amplifier/demodulator. (Courtesy RCA Corp.)

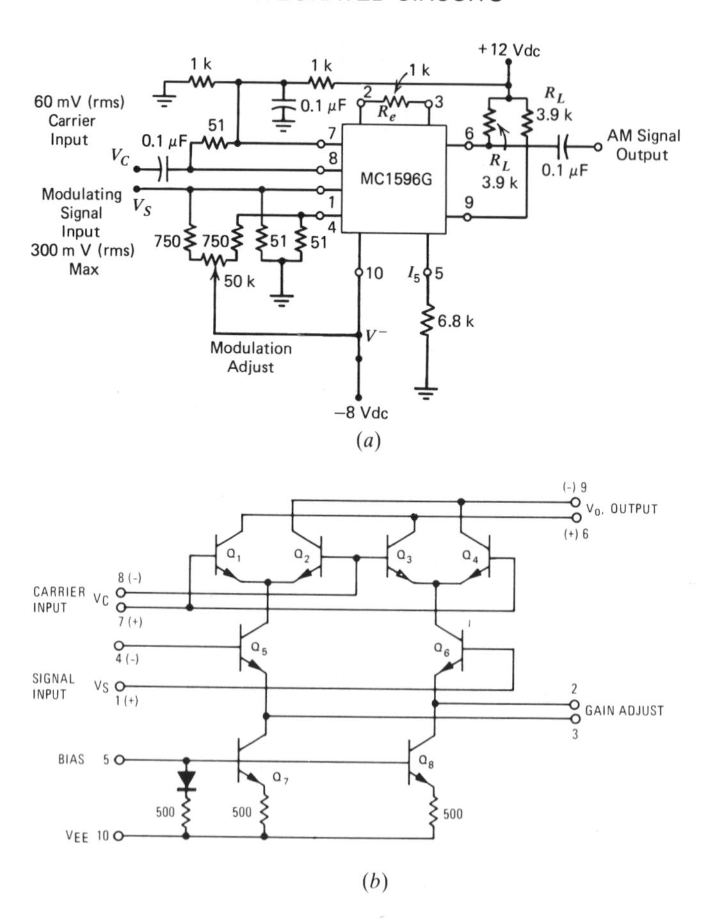

Figure 14.20 The MC1596 balanced modulator/demodulator. (*a*) Used as an AM modulator. (*b*) Circuit schematic. (Courtesy Motorola, Inc.)

For low-level signals at both inputs, the output will contain sum and difference components having an amplitude dependent upon the product of the amplitudes of the input signals. High-level operation at the carrier input and linear operation at the signal (modulating) input, results in an output containing sum and difference components of the modulating frequency, in addition to the fundamental carrier frequency and its odd harmonics.

Balanced Modulator

Figure 14.21 shows the MC1596 operating as a balanced modulator. Recommended input levels are 60 mV for the carrier and 300 mV for the modulating signal. In Fig. 14.20*b*, the carrier is applied to the upper differential amplifiers which are operated at the point of saturation. Linear operation is performed by the lower differential amplifier by keeping the modulating signal sufficiently low to prevent the generation of harmonics. For modulating voltages less than 300 mV, spurious sideband suppression is typically 55 dB at a carrier frequency of 500 kHz. When the carrier differential amplifiers are operated in their linear region, (15 mV or less), spurious outputs are reduced.

When employing the filter method of single-sideband (SSB) generation, one can rely on the filter to remove all spurious outputs with the exception of spurious sidebands of the carrier. In this case, operating with a high-level carrier is the logical choice to achieve maximum gain and ensure that the desired sideband does not contain spurious variations in amplitude.

Frequency Doubler

The output from the MC1596 contains sum and difference frequency components: $f_1 \pm f_2$. Therefore, the application of the same signal frequency, f_1, to both inputs can result in a single frequency at the output equal to $2f_1$. The output circuit may be broadbanded or tuned circuits may be used if desired.

EXAMPLE 14.8 A 300 MHz output signal is required from a 150 MHz oscillator. Apply the MC1596 in the doubler configuration, keeping spurious outputs 20 dB below the 300 MHz output level.

PROCEDURE

1. Figure 14.22 shows the MC1596 employed in a doubler circuit.
2. Because spurious outputs are to be kept 20 dB below the desired 300 MHz output, a tuned filter is employed.

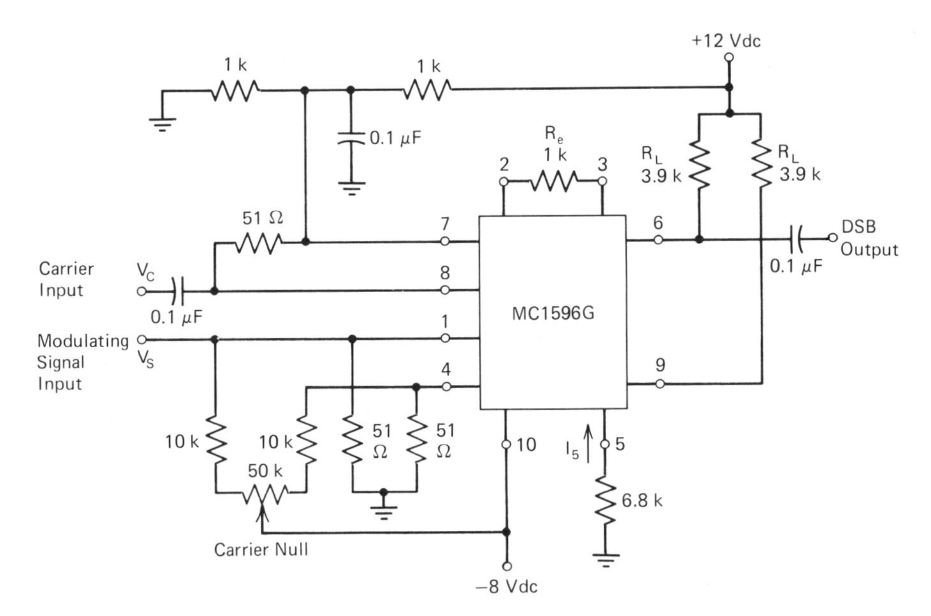

Figure 14.21 The MC1596 used as a balanced modulator providing a double sideband (DSB) output signal. (Courtesy Motorola, Inc.)

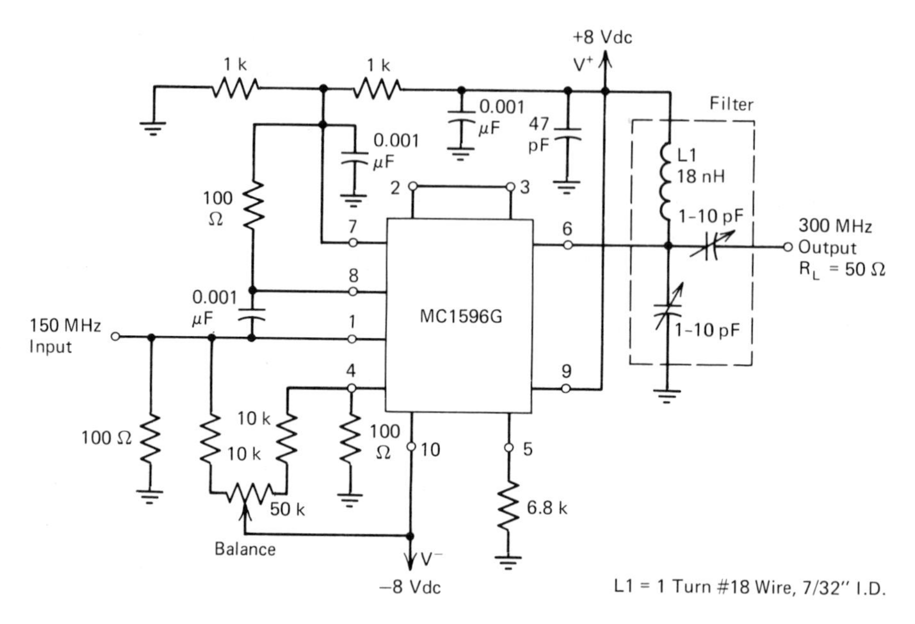

Figure 14.22 A 150 to 300 MHz doubler using the MC1596 chip. (Courtesy Motorola, Inc.)

Product Detector

A natural application of the MC1596 is the SSB product detector. In this case, all frequencies lie in the RF spectrum except for the recovered audio. No transformers or tuned circuits are required, thereby simplifying circuit design.

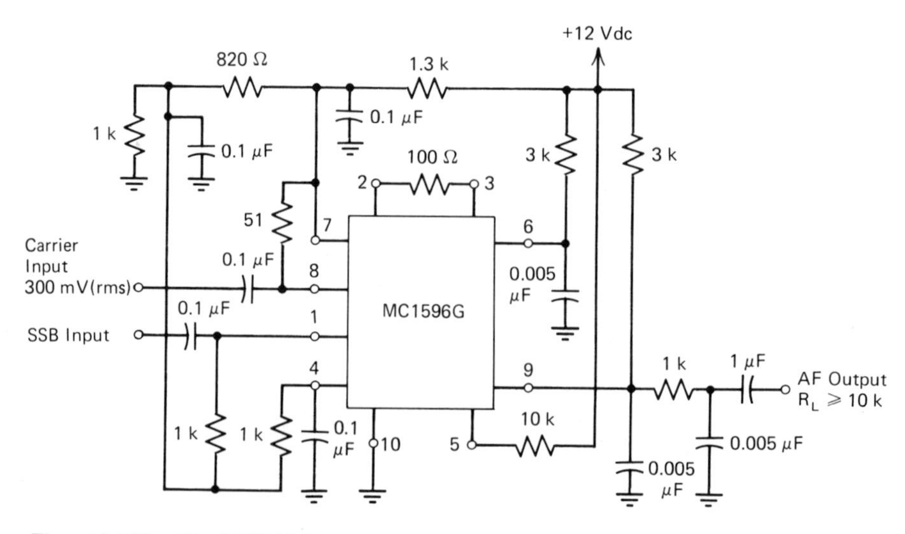

Figure 14.23 The MC1596 employed as a product detector. (Courtesy Motorola, Inc.)

EXAMPLE 14.9 Design an SSB product detector operating from a single 12 V supply. The carrier injection level is 300 mV while the SSB input signal level ranges from 3 μV to 100 mV.

PROCEDURE

1 Refer to Fig. 14.23. The circuit is, in effect, a mixer. The SSB signal is applied to the lower differential amplifier of the MC1596 (see Fig. 14.20*b*). Linear operation must be maintained.
2 The carrier injection signal is applied to the upper pair of differential amplifiers. Dual outputs (terminals 9 and 6) from the product detector can be used for audio amplification and AGC, if desired.

14.9 PHASE LOCKED LOOPS*

Phase locked loops (PLLs) for communication systems have gained wide popularity in recent years. This is particularly apparent in the CB area, because the use of PLLs removes the necessity for a separate receive and transmit crystal for each channel. In a 40-channel system, dramatic cost reductions are therefore possible. The PLL synthesizer also enjoys widespread use in amateur radio and consumer stereo applications, owing to its ability to generate and maintain stable signals.

Generally, a PLL is employed where a large number of frequencies with close spacings are to be generated. This however, is only a fraction of today's PLL applications. The PLL may be categorized as general purpose, dedicated, and for use in tone decoders. Table 14.6 lists typical general purpose PLLs. A block diagram of a PLL is provided in Fig. 14.24. Its basic components are a phase detector, low-pass filter, amplifier, and a voltage controlled oscillator (VCO), in a feedback loop. The VCO synchronizes with an incoming signal; this condition is referred to as *locking*. If the input frequency changes, the output voltage of the phase detector changes so the locked condition is maintained.

Special-purpose PLLs are listed in Table 14.7. The μA758 and it's equivalents were discussed in Sec. 14.6. Another popular PLL is the 8X08 AM/FM frequency synthesizer. This bipolar/MOS PLL, operating from a single 5 V

* For a detailed discussion of phase locked loop operation, refer to Chap. 17.

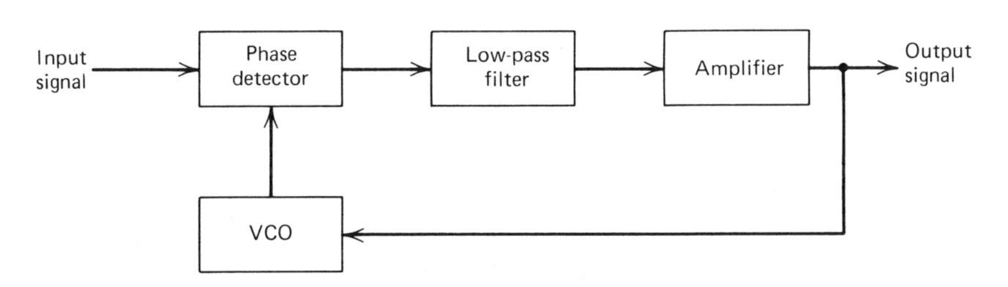

Figure 14.24 Block diagram of a basic phase locked loop.

Table 14.6 **Examples of general-purpose PLLs (Courtesy Signetics Corporation)**

NE564	High-speed modems, FSK transmitters and receivers, TTL and ECL compatible outputs.
NE565	Low frequency FSK.
NE567[1]	Touch-tone® decoder, AM demodulation, communications paging, TTL compatible output.
NE566[2]	FSK transmitters, signal generators.

NOTES
1. Although the NE567 is a special purpose PLL intended for tone decoder applications, it is included here since it serves many general purpose applications related to tone decoding.
2. While the NE566 is not a PLL but a precision voltage-controlled waveform generator, it is included here since it lends itself well to general purpose PLL applications.

supply, requires only one external crystal and a minimum number of discrete components to provide frequency synthesis of 200 AM channels (10 kHz spacing) and 2000 FM channels (100 kHz spacing). A scan feature may be easily implemented, in addition to memory storage of 10 stations for future recall. Countless applications in the consumer stereo industry are possible.

EXAMPLE 14.10 Using the 8X08 frequency synthesizer, design an AM/FM tuner. Assume 10 kHz spacing for AM and 100 kHz spacing for FM.

PROCEDURE

1 In Fig. 14.25, the 8X08 is used for generating the AM/FM local oscillator signals. The reference frequency is 3.6 MHz which is trimmed by a variable capacitor.
2 Two VCOs are employed: one for AM and the other for FM. Choice of AM/FM operation is provided by applying a voltage to the desired VCO.
3 The LM124 is a low power quad op amp.

Table 14.7 **Examples of special-purpose PLLs (Courtesy Signetics Corporation)**

DEVICE	PROCESS TECHNOLOGY	FEATURES AND APPLICATIONS
8X08 AM/FM Frequency Synthesizer	Combination bipolar and MOS	Uses digital PLL techniques to synthesize AM and FM local oscillator frequencies. Capable of digitally programming 200 AM channels with 10kHz spacing and 2000 FM channels with 100kHz spacing. Operates on a single 5 volt supply with input frequencies up to 80MHz. Requires only one crystal, one capacitor, and two resistors as external components. Easily adapted for electronically scanning both AM and FM bands, reversing scan directions, and storing up to 10 station locations in memory.
μA758 FM Stereo Multiplexer Decoder	Bipolar	Decodes FM stereo multiplex signal into left and right audio channels while suppressing Subsidiary Carrier Authorization (SCA) information that may be present in the input signal. Internal functions provide for automatic monaural-stereo mode switching and drive capability for an external lamp to indicate stereo mode operation. Requires no external tuning coils and only one potentiometer adjustment for complete alignment.

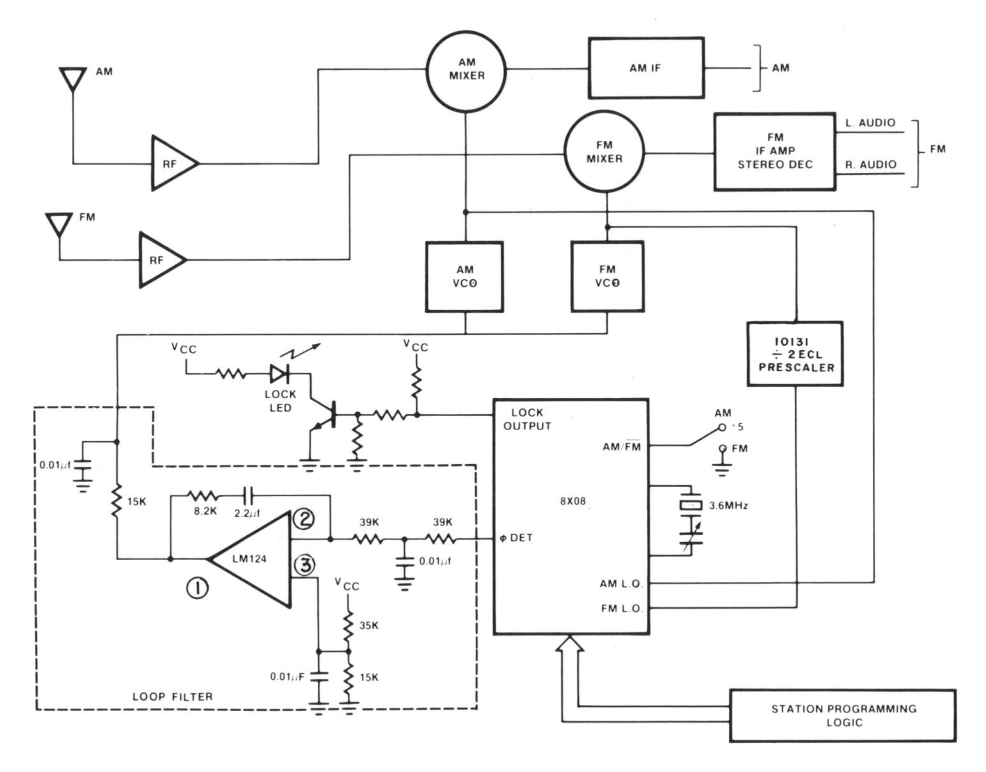

Figure 14.25 The 8X08 AM/FM frequency synthesizer used in an AM/FM tuner. (Courtesy Signetics Corp.)

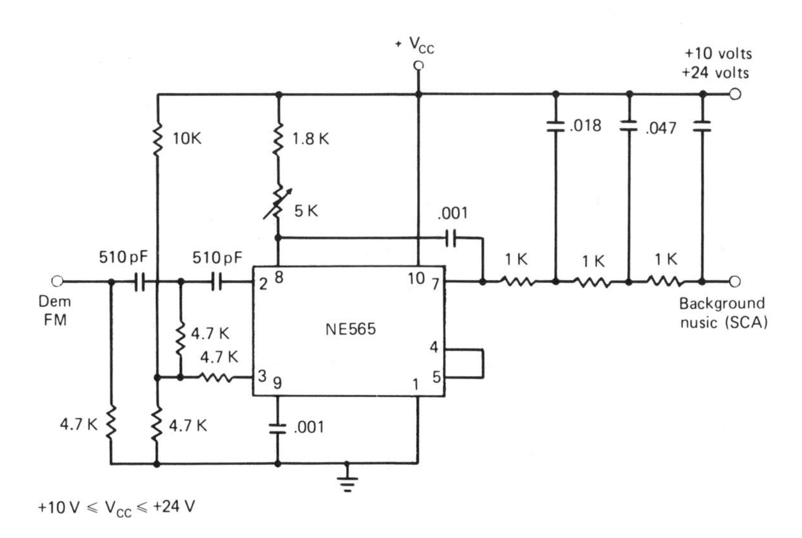

Figure 14.26 An example of a SCA decoder employing a NE565 phase locked loop chip. (Courtesy Signetics Corp.)

Most stereo decoders have a high suppression of the SCA; therefore, another method is required to detect this subcarrier of the main channel. The SCA is a 67 kHz frequency modulated subcarrier. This is well above the normal L–R composite signal of 23 through 25 kHz. The use of a NE565 PLL (Fig. 14.26) allows decoding of the SCA signal, which is then applied to an audio amplifier for further amplification.

CB Frequency Synthesizer

A low-power PLL, the MM55106, is employed by several CB transceiver manufacturers. A block diagram of the chip is shown in Fig. 14.27. The 10.24 MHz reference oscillator is divided by two to produce a 5.12 MHz signal. An offset frequency is provided when the 5.12 MHz signal is multiplied by three and mixed with the VCO frequency.

The 5.12 MHz signal is then divided by 512 or 1024, depending on the logic level as the frequency select (FS) terminal. A 10 kHz reference frequency is produced for the phase detector when the divider chain is set to 512 or, correspondingly, 5 kHz when programmed for 1024.

Program inputs are designated as P0 through P8. The most significant bit is 256 and the least significant bit is 1. Maximum division is by 511. Program inputs incorporate on board pulldown resistors for channel selection, thereby requiring only logic high inputs (Fig. 14.28).

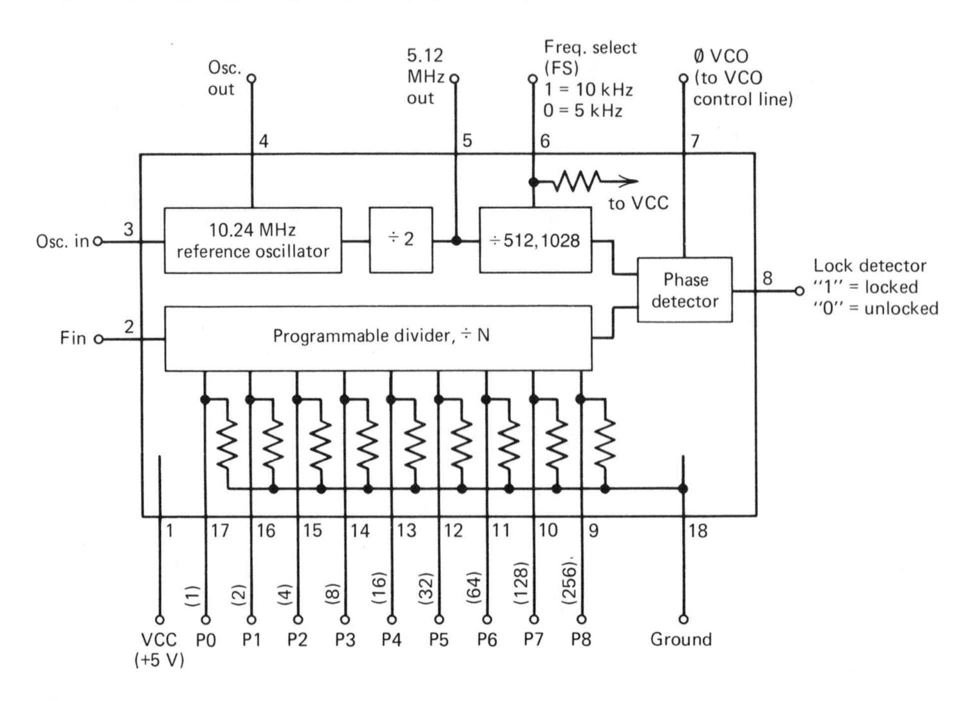

Figure 14.27 Block diagram of the MM55106 PLL chip. (Courtesy National Semiconductor.)

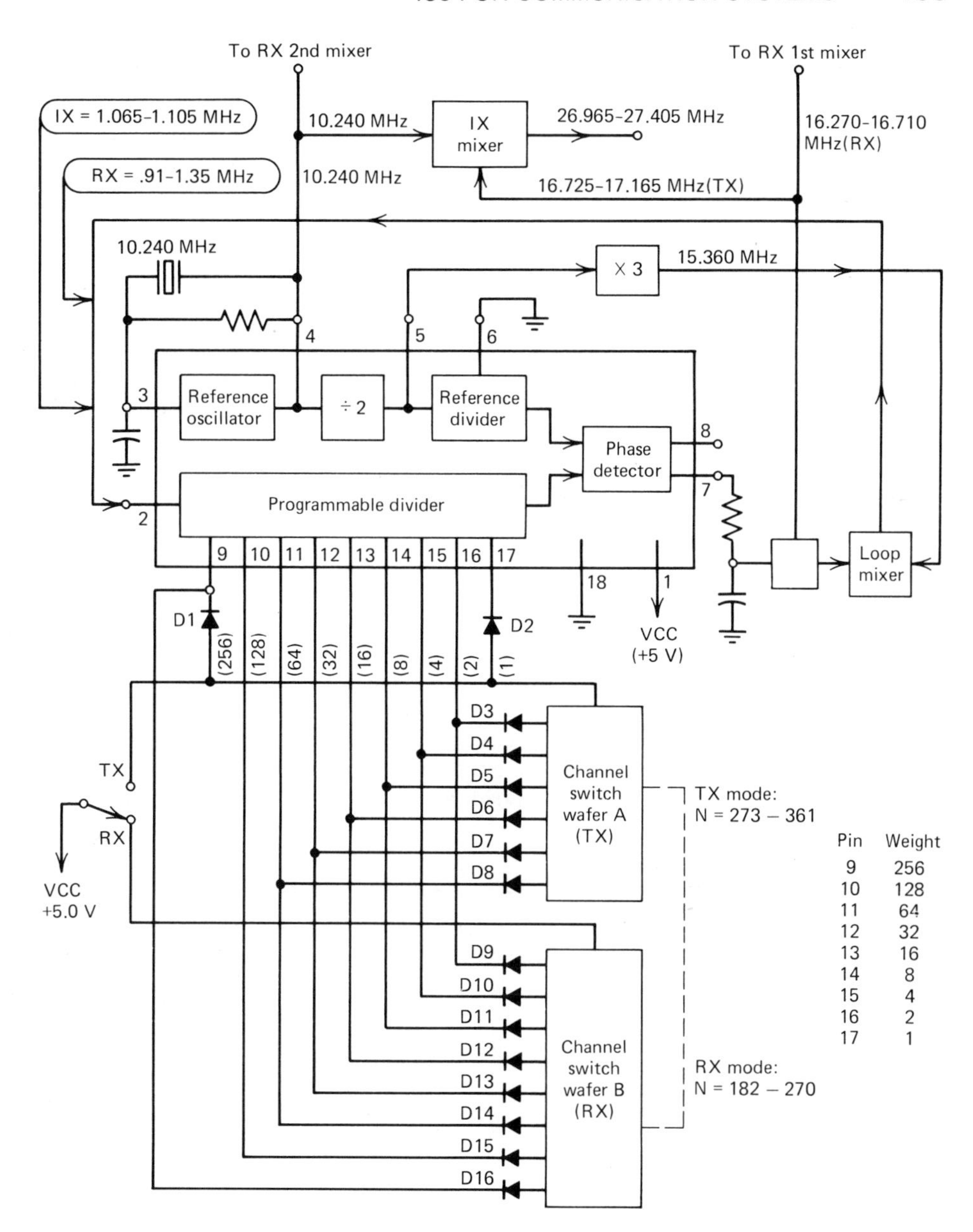

Figure 14.28 The MM55106 used as a frequency synthesizer for a CB transceiver. (Courtesy National Semiconductor.)

Only one crystal (10.24 MHz) is required with the MM55106 PLL. The VCO is, therefore, required to supply difference frequencies for receive and transmit. A common method for achieving this involves changing the division ratio in the transmit mode. A partial listing for channels 1 through 5 with the associated VCO frequency and program pin levels is provided in Table 14.8. Levels for both transmit and receive are included.

EXAMPLE 14.11 Analyze the MM55106 PLL connected in Fig. 14.28 as a frequency synthesizer used for a CB transceiver.

PROCEDURE

1 Assume channel 1 (26.965 MHz) is to be received. Division is set to 182 by the divider chain. In order to provide locking, the output frequency of the divider must be 5 kHz. This requires an input to the divider of 5 kHz $\times$ 182 = 0.91 MHz. The 5.12 MHz signal appearing at pin 5 is multiplied by three, yielding 15.36 MHz. This signal, in addition to the VCO signal, is fed to the loop mixer.

2 An offset frequency appears at pin 2 (the divider input) that is equal to the frequency of the VCO, f_{VCO}, minus 15.36 MHz. The phase detector output pushes the VCO up to $N \times 5$ kHz; therefore,

$$f_{VCO} = 15.36 + N(0.005)$$
$$= 15.36 + 182 \times 0.005 = 16.27 \text{ MHz}$$

3 This signal is applied to the receiver mixer stage. Mixer action recovers an IF of 10.695 MHz.

4 The IF is amplified, filtered, and applied to a second mixer where it is heterodyned with a 10.240 MHz signal provided by the reference oscillator. A difference frequency of 455 kHz results and is filtered, amplified, and detected in the normal manner.

Table 14.8 **Program pin levels and VCO frequencies covering channels 1–5 for the MM55106 frequency synthesizer chip.**

	CH	N	PRG CD-PN 9–17									VCO Frequency
			Weight 256	128	64	32	16	8	4	2	1	MHz
			PIN 9	10	11	12	13	14	15	16	17	
	1	182	0	1	0	1	1	0	1	1	0	16.270
	2	184	0	1	0	1	1	1	0	0	0	16.280
Receive	3	186	0	1	0	1	1	1	0	1	0	16.290
	4	190	0	1	0	1	1	1	1	1	0	16.310
	5	192	0	1	1	0	0	0	0	0	0	16.320
	CN	N	PRG CD-PN 9–17									VCO
			Weight 256	128	64	32	16	8	4	2	1	FREQ
			PIN 9	10	11	12	13	14	15	16	17	MHz
	1	273	1	0	0	0	1	0	0	0	1	16.725
	2	275	1	0	0	0	1	0	0	1	1	16.735
Transmit	3	277	1	0	0	0	1	0	1	0	1	16.765
	4	281	1	0	0	0	1	1	0	0	1	16.765
	5	283	1	0	0	0	1	1	0	1	1	16.775

Courtesy National Semiconductor.

5 The transmit mode of operation is similar except for a change in frequency division, which becomes 273. The programmable divider output continues to be 5 kHz to maintain locked conditions. This requires an input to the divider equal to:

$$273 \times 5 \text{ kHz} = 1.365 \text{ MHz}$$

and

$$f_{\text{VCO}} = 15.36 + 273 \times 0.005 = 16.725 \text{ MHz}$$

6 All that remains is to apply this signal to a mixer where it is heterodyned with a 10.24 MHz signal from the reference oscillator. The sum and difference frequencies result. Proper filtering allows recovery of the sum of the mixed signals resulting in the desired 26.965 MHz for the channel 1 transmit frequency. This signal is then amplified and applied to the antenna for transmission.

PLL Tone Decoders

Tone decoders, such as the NE567 special function PLL, are dedicated to control systems employing dual-tone multifrequency (DTMF) touch-tone dialing. In the block diagram of the 567 in Fig. 14.29, the chip contains a phase detector, amplifiers, a quadrature phase detector, and a current controlled oscillator. When the signal level at the lock frequency is sufficiently high, the driver am-

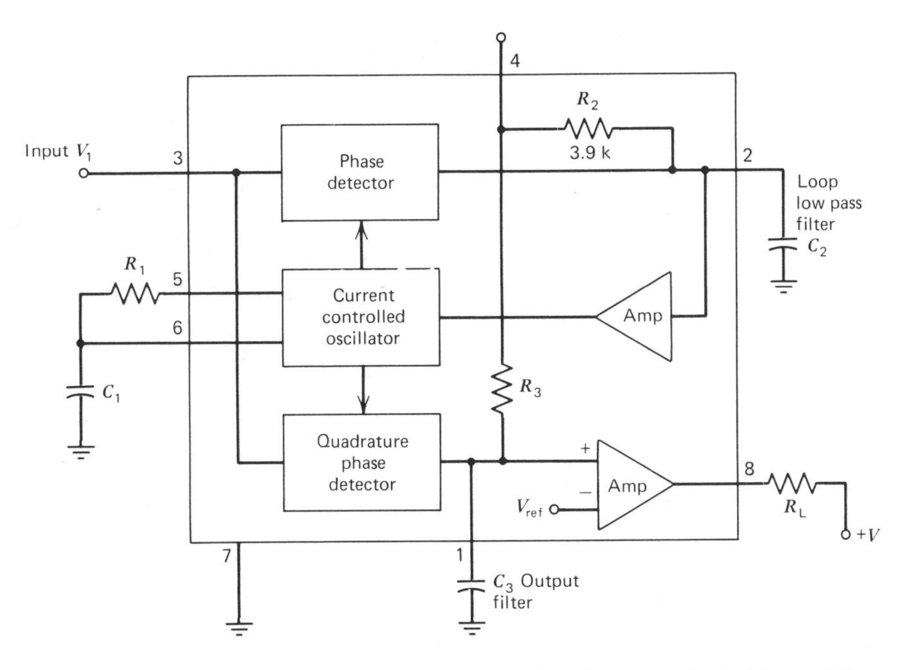

Figure 14.29 Block diagram of the NE567 tone decoder phase locked loop. (Courtesy Signetics Corp.)

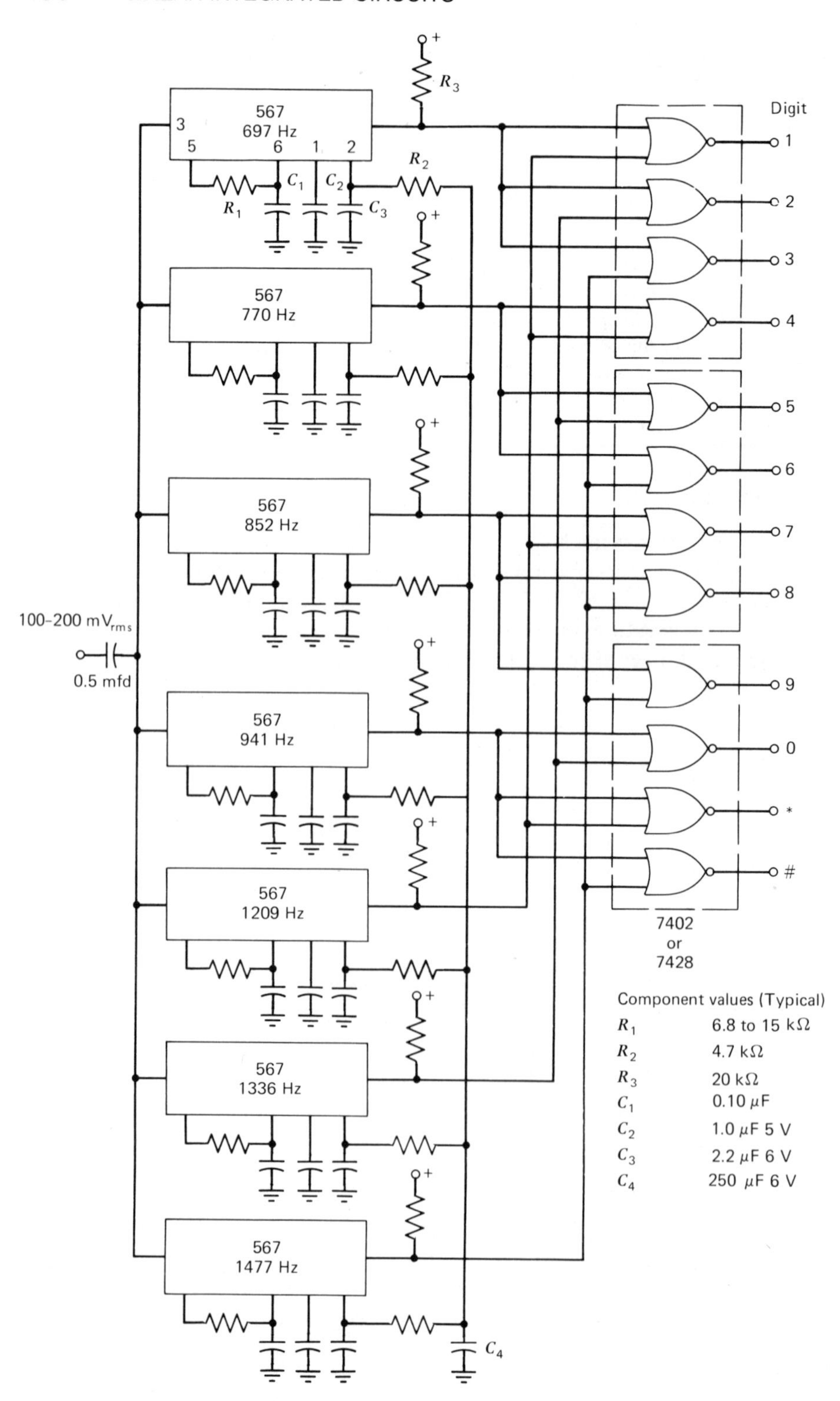

567
697 Hz
3
5
6
1
2
R1
C1
C2
C3
R2
R3
567
770 Hz
567
852 Hz
567
941 Hz
567
1209 Hz
567
1336 Hz
567
1477 Hz
C4
100-200 mV rms
0.5 mfd
Digit
1
2
3
4
5
6
7
8
9
0
*
#
7402
or
7428
Component values (Typical)
R1 6.8 to 15 kΩ
R2 4.7 kΩ
R3 20 kΩ
C1 0.10 μF
C2 1.0 μF 5 V
C3 2.2 μF 6 V
C4 250 μF 6 V

plifier turns on, providing load drive capability up to 200 mA. When an "inband" tone is present above the required minimum level, the 567 will provide an output whose value depends upon the supply voltage.

Several applications immediately come to mind. The transmission of DTMF tones from a mobile or portable amateur or commercial transmitter to a receiving site equipped with tone decoding circuitry allows easy access to a landline telephone system. This permits calls to be made, through an appropriate interface, while away from a standard telephone. The transmitting station need only contain a tone encoder capable of generating standard dial-tone signals, such as those present from a tone pad. These signals are used to modulate the carrier where they are then detected at the receiver and used to control various functions as required. Proper interfacing may then be used to inject the received tones into the telephone network for proper routing.

Because each digit generates two tones, one high and one low in frequency, two decoders, such as the 567, are required. Since a matrix is employed, seven decoders are needed to provide complete decoding of all 12 digit AND/OR functions. Proper interfacing with the logic results in each digit or function to provide control such as line on, line dump, redial, system down, and so on. Almost any function that can be controlled by a logic 1 or 0 can be accomodated.

EXAMPLE 14.12 Using the NE567 chips, design a simplified touch-tone decoder requiring 200 mV of detected signal to provide system control with digital 1 and 0 and special functions # and *.

PROCEDURE

1 The circuit is given in Fig. 14.30. Each decoder is tuned to the appropriate frequency for each tone.
2 The time constant R_1C_1 for each decoder provides the proper inband selected tone frequency.
3 Resistor R_2 reduces bandwidth to less than 10% at 100 mV and capacitor C_4 provides decoupling.
4 The outputs from the decoder are processed by the 12 NOR gates to provide the required logic commands.

14.10 REFERENCES

Egan, W. F., *Frequency Synthesis by Phase Lock,* Wiley-Interscience, New York, 1981.

Motorola, *Linear IC's,* latest edition.

Prensky, S. and A. H. Seidman, *Linear Integrated Circuits: Practice and Applications,* Reston, Reston, VA, 1981.

RCA, *Linear Integrated Circuits,* latest edition.

Signetics, *Analog Data Manual,* latest edition.

Young, T., *Linear Integrated Circuits,* Wiley, New York, 1981.

Figure 14.30 The NE567 chip employed in a touch tone decoder. (Courtesy Signetics Corp.)

CHAPTER 15

IC Voltage Regulators

Dr. David Cave

Design Manager, Voltage Regulators
Motorola Inc.
Semiconductor Products Sector

15.1 INTRODUCTION

The basic function of a voltage regulator is to provide a precise output voltage under varying load conditions and temperature from an unknown, and possibly changing, input voltage. Ideally, the output of a voltage regulator would behave as a zero output impedance battery. In reality, however, the regulator must be modeled as a voltage source, V, in series with positive output impedance, Z_o (Fig. 15.1). The value of V is not constant; rather, it varies with changes in input voltage, V_{in}, and with the IC junction temperature, T_j. (The junction temperature refers to the temperature of the p-n junctions that form the IC transistors which, for practical purposes, is the die temperature.) The regulator output voltage is also affected by the voltage drop across Z_o, resulting in variations in V_o with changes in the load current. Finally, the initial value of V is dependent on IC processing and varies from unit to unit by $\pm 5\%$ (for inexpensive regulators) to better than $\pm 1\%$.

It is clear then that an IC, or for that matter a discrete regulator, is not perfect. In fact, each regulator comes with a list of specifications on a data sheet that puts maximum limits on these imperfections. Because they are sometimes confusing, the next section is devoted to voltage regulator definitions and specifications.

15.2 SPECIFICATIONS IN A DATA SHEET

The regulator data sheet contains two sets of important specifications: *maximum ratings* and *electrical characteristics*. Maximum ratings are nonparametric; that

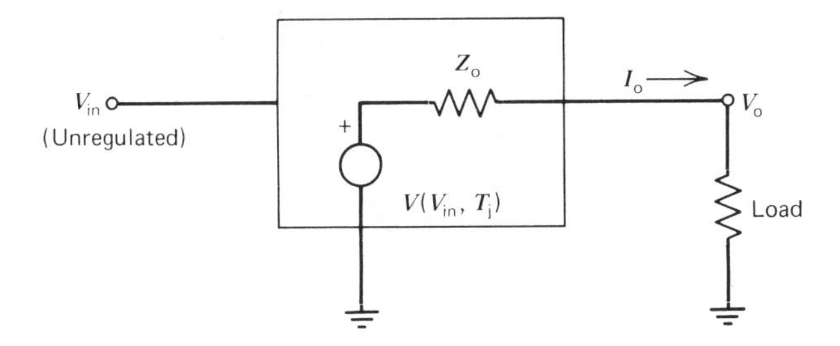

Figure 15.1 Model of a 3-terminal voltage regulator.

is, performance is not guaranteed. When the regulator is stressed to these limits, permanent damage will not result. Typically listed under maximum ratings are maximum input voltage, maximum power dissipation, maximum input-to-output differential voltage, and maximum output current. Operation beyond any of these specified limits *may result in permanent damage* (Table 15.1).

Under electrical characteristics, the ac and dc voltage and current specifications of the device are given. Unless otherwise stated, in each test the operating conditions are specified at the top of the table, as in Table 15.1. In the case of the MC7805C voltage regulator, those test conditions are $V_{in} = 10$ V, $I_o = 500$ mA, and the die, or junction, temperature is between 0 and 125°C.

1 *Output voltage.* ($T_j = +25$°C). Under the conditions at the head of the table, except that the junction temperature is maintained at 25°C, the output voltage is guaranteed to be between 4.8 and 5.2 V.

2 *Input regulation.* This specification relates changes in the output voltage (mV) to changes in the input voltage (V). Four conditions are provided, all at $T_j = 25$°C: two are at $I_o = 100$ mA, and two are at $I_o = 500$ mA. For example, with $T_j = 25$°C, $I_o = 100$ mA, changing the input voltage from 7 to 25 V causes the output to change 7 mV typically, but possibly up to 50 mV maximum.

3 *Load regulation.* As the load current, I_o, changes, the output voltage will also change owing to the regulator's finite output impedance, Z_o. For the MC7805C, two cases are cited: the first at $T_j = 25$°C and I_o from 5 mA to 1.5 A; the second, over the entire temperature range and I_o from 250 to 750 mA. It should be noted that the major contributor to Z_o is the wire bond connecting the IC to its package and, as such, is very small.

4 *Output voltage.* This is the all-encompassing output voltage specification that includes the full range of V_{in}, I_o, and temperature. Thus, under all normal operating conditions, the output of an MC7805C will fall between 4.75 and 5.25 V.

5 *Quiescent current.* The regulator requires a small current I_B from V_{in}, returned to ground, to operate its internal control circuitry and voltage reference. For the MC7805C, the quiescent current is typically 4.3 mA with a maximum of 8 mA at room temperature.

Table 15.1 **Partial data sheet for the MC7805, 5-V, three terminal voltage regulator. (Courtesy Motorola, Inc.)**

MC7800C Series MAXIMUM RATINGS (T_A = +25°C unless otherwise noted.)

Rating	Symbol	Value	Unit
Input Voltage (5.0 V - 18 V)	V_{in}	35	Vdc
(24 V)		40	
Power Dissipation and Thermal Characteristics			
Plastic Package			
T_A = +25°C	P_D	Internally Limited	Watts
Derate above T_A = +25°C	$1/\theta_{JA}$	15.4	mW/°C
Thermal Resistance, Junction to Air	θ_{JA}	65	°C/W
T_C = +25°C	P_D	Internally Limited	Watts
Derate above T_C = +95°C (See Figure 1)	$1/\theta_{JC}$	200	mW/°C
Thermal Resistance, Junction to Case	θ_{JC}	5.0	°C/W
Metal Package			
T_A = +25°C	P_D	Internally Limited	Watts
Derate above T_A = +25°C	$1/\theta_{JA}$	22.5	mW/°C
Thermal Resistance, Junction to Air	θ_{JA}	45	°C/W
T_C = +25°C	P_D	Internally Limited	Watts
Derate above T_C = +65°C (See Figure 2)	$1/\theta_{JC}$	182	mW/°C
Thermal Resistance, Junction to Case	θ_{JC}	5.5	°C/W
Storage Junction Temperature Range	T_{stg}	−65 to +150	°C
Operating Junction Temperature Range	T_J	0 to +150	°C

MC7805C ELECTRICAL CHARACTERISTICS (V_{in} = 10 V, I_O = 500 mA, 0°C < T_J < +125°C unless otherwise noted.)

	Characteristic	Symbol	Min	Typ	Max	Unit
1	Output Voltage (T_J = +25°C)	V_O	4.8	5.0	5.2	Vdc
2	Input Regulation	Reg_{in}				mV
	(T_J = +25°C, I_O = 100 mA)					
	7.0 Vdc ≤ V_{in} ≤ 25 Vdc		–	7.0	50	
	8.0 Vdc ≤ V_{in} ≤ 12 Vdc		–	2.0	25	
	(T_J = +25°C, I_O = 500 mA)					
	7.0 Vdc ≤ V_{in} ≤ 25 Vdc		–	35	100	
	8.0 Vdc ≤ V_{in} ≤ 12 Vdc		–	8.0	50	
3	Load Regulation	Reg_{load}				mV
	T_J = +25°C, 5.0 mA ≤ I_O ≤ 1.5 A		–	11	100	
	250 mA ≤ I_O ≤ 750 mA		–	4.0	50	
4	Output Voltage (7.0 Vdc ≤ V_{in} ≤ 20 Vdc, 5.0 mA ≤ I_O ≤ 1.0 A, P ≤ 15 W)	V_O	4.75	–	5.25	Vdc
5	Quiescent Current (T_J = +25°C)	I_B	–	4.3	8.0	mA
6	Quiescent Current Change	ΔI_B				mA
	7.0 Vdc ≤ V_{in} ≤ 25 Vdc		–	–	1.3	
	5.0 mA ≤ I_O ≤ 1.0 A		–	–	0.5	
7	Output Noise Voltage (T_A = +25°C, 10 Hz ≤ f ≤ 100 kHz)	V_N	–	40	–	μV
8	Long-Term Stability	$\Delta V_O/\Delta t$	–	–	20	mV/1.0 k HRS
9	Ripple Rejection (I_O = 20 mA, f = 120 Hz)	RR	–	70	–	dB
10	Input-Output Voltage Differential (I_O = 1.0 A, T_J = +25°C)	$V_{in}-V_O$	–	2.0	–	Vdc
11	Output Resistance (I_O = 500 mA)	R_O	–	30	–	mΩ
12	Short-Circuit Current Limit (T_J = +25°C)	I_{SC}	–	750	–	mA
13	Average Temperature Coefficient of Output Voltage I_O = 5.0 mA, 0°C ≤ T_A ≤ +125°C	TCV_O	–	-1.0	–	mV/°C

6 *Quiescent current change.* Symbol ΔI_B represents the change in quiescent current caused by changes in input voltage and load current. Note that the input voltage and load current exceed the range (stated in top of Table 15.1) in specification 5 and, therefore, under these conditions I_B may exceed 8 mA.

7 *Output noise voltage.* At a junction temperature of 25°C, the output noise of the regulator is typically 40 μV rms when band limited between 10 Hz and 100 kHz.

8 *Long-term stability.* The change in output voltage is measured in millivolts per 1000 hours of operation.

9 *Ripple rejection.* This test is similar to the test for input regulation (test 2) except that the input is sinusoidally changed at 120 Hz, the full rectified line frequency. Because ripple rejection is given in decibels, the level of input change is not specified; however, it is normally tested by the manufacturer using a 1 V rms ripple signal. The ripple rejection, *RR,* is expressed by:

$$RR = 20 \log_{10} [V_{o(rms)}/V_{in(rms)}]$$

10 *Input-output voltage differential.* Typically, the device requires the input voltage to be 2 V greater than the output voltage. Output voltage (test 4) limits the *maximum* differential in that at $V_{in} = 7$ V, the output must be at least 4.75 V.

11 *Output resistance.* A measure of Z_o (Fig. 15.1).

12 *Short-circuit current limit.* As the load impedance is decreased, the regulator delivers load current up to the short-circuit current limit, I_{SC}, under the specified operating conditions. Beyond this, that is, if the load impedance is continually decreased, the voltage regulator begins to act as a current source delivering I_{SC}. In order to protect the power device, I_{SC} is made to decrease with increasing input-to-output differential voltage.

13 *Average temperature coefficient of output voltage.* For each degree increase in temperature, the output voltage typically decreases 1 mV.

It should be noted that those parameters relying on feedback (all except quiescent current and input-output voltage differential) degrade with increasing output voltage. For example, when the input regulation of an MC7805 is compared to an MC7815, a threefold increase is found.

Finally, high-power tests, such as input regulation, load regulation, and short-circuit current limit, are pulse tested. By using short pulses, typically 1 ms in length, changes in V_o owing to heating are eliminated. Manufacturers normally indicate a pulse test by specifying a specific junction temperature, T_j. Some new voltage regulators have added the specification *thermal regulation* (%/W), which limits the changes in V_o caused by power dissipation (changes in die temperature).

The function of voltage regulation is accomplished using one of three basic configurations: *shunt, series pass,* or *switch mode.* IC regulators are, with few exceptions, series pass or control circuits for switch mode operation.

15.3 SHUNT REGULATOR

Figure 15.2 shows a conceptual 3-terminal shunt regulator. The unregulated input voltage, V_{in}, is applied to an external series dropping resistor, R_s. Regulation is accomplished in this conceptual model by varying resistance R_p such that the voltage drop across R_s provides the desired output. This, of course, implies some yet undefined control circuit for controlling R_p. The voltage drop

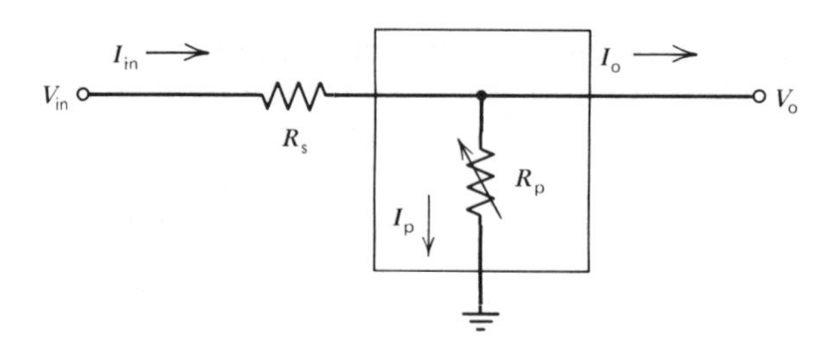

Figure 15.2 A conceptual 3-terminal shunt voltage regulator.

across R_s is the product of R_s and input current I_{in}. For a constant input voltage, I_{in} must remain constant and be at least equal to the peak load current. Current I_p is the difference between the peak load current and the operating load current, I_o. From Fig. 15.2, the output voltage, V_o, is:

$$V_o = V_{in} - R_s(I_o + I_p) \tag{15.1}$$

Because the source must at all times supply the peak load current, shunt regulators are generally inefficient. For example, if a shunt regulator is designed to deliver 1.0 A of current under full load, I_{in} must be 1.0 A even if the load current is 0.1 A. The remaining current, $I_p = 0.9$ A, is shunted through R_p.

Calculating the efficiency, η, we find:

$$\eta = \frac{P_{out}}{P_{in}} = \frac{V_o I_o}{V_{in} I_{o(max)}} \tag{15.2}$$

where $I_{o(max)}$ is the peak load current. Unfortunately, the lost power is dissipated in the IC and may possibly exceed the maximum ratings of power dissipation and junction temperature. The power dissipated in the IC, P_{IC}, is:

$$P_{IC} = V_o I_o \, [I_{o(max)}/I_o - 1] \tag{15.3}$$

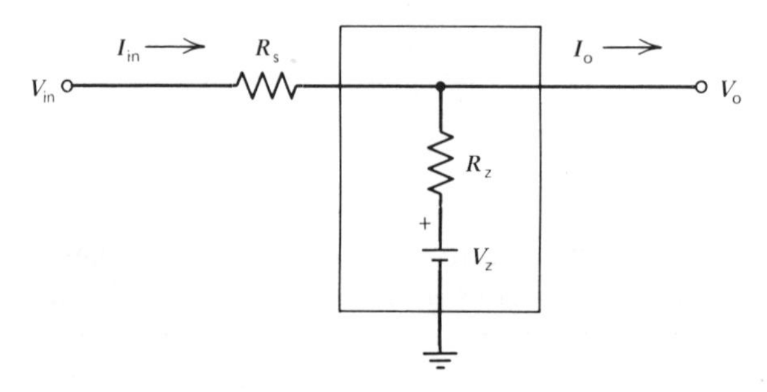

Figure 15.3 Model of a shunt 3-terminal voltage regulator.

The purpose of R_p is to produce a voltage drop across R_s, resulting in the desired output voltage. In actual practice, R_p is replaced with a voltage source, often a zener diode. A zener diode may be modeled as a voltage source, V_z, with series resistance, R_z. The resulting shunt regulator of Fig. 15.3 has an output voltage of:

$$V_o = \frac{V_z}{1 + R_z/R_s} + \frac{V_{in}}{1 + R_s/R_z} - \frac{I_o R_s}{1 + R_s/R_z} \qquad (15.4)$$

EXAMPLE 15.1 Find R_s, load and line regulation, the temperature coefficient, *TC,* of V_o, and the IC power dissipation at $I_o = 0.5$ A for a 5 V shunt regulator similar to that of Fig. 15.3. The following conditions are given: $V_{in(min)} = 30$ V, $V_{in(max)} = 34$ V, $I_{o(max)} = 1.0$ A, $V_z = 5$ V, and $R_z = 1\ \Omega$.

PROCEDURE

1 Analysis of Eq. 15.4 reveals that best performance is achieved when R_z is small and R_s is large. Thus, the maximum value of R_s, $R_{s(max)}$, is:

$$R_{s(max)} = \frac{V_{in(min)} - V_o}{I_{o(max)}} \qquad (15.5)$$

$$= \frac{30 - 5}{1} = 25\ \Omega$$

2 From Eq. 15.4, the change in output voltage, ΔV_o, for a change in input voltage, ΔV_{in}, is:

$$\Delta V_o = \frac{\Delta V_{in}}{1 + R_s/R_z} \qquad (15.6)$$

$$= \frac{4}{1 + 25/1} = 154 \text{ mV}$$

3 In a similar manner, the change in output voltage for a change in load current, ΔI_o, is:

$$\Delta V_o = \frac{-\Delta I_o R_s}{1 + R_s/R_z} \qquad (15.7)$$

$$= \frac{(-0.5)(25)}{1 + 25/1} = -481 \text{ mV}$$

4 By Eq. 15.3, the IC power dissipation is:

$$P_{IC} = 5 \times 0.5 \times (1/0.5 - 1) = 2.5 \text{ W}$$

5 The temperature coefficient, TC, of the regulator is determined essentially by V_z (neglecting any change in R_z with temperature), which is the first term of Eq. 15.4. The TC of zener diodes fabricated in IC form is given in Fig. 15.4. From the curve, the TC of a 5 V zener diode is estimated to be 0.7 mV/°C.

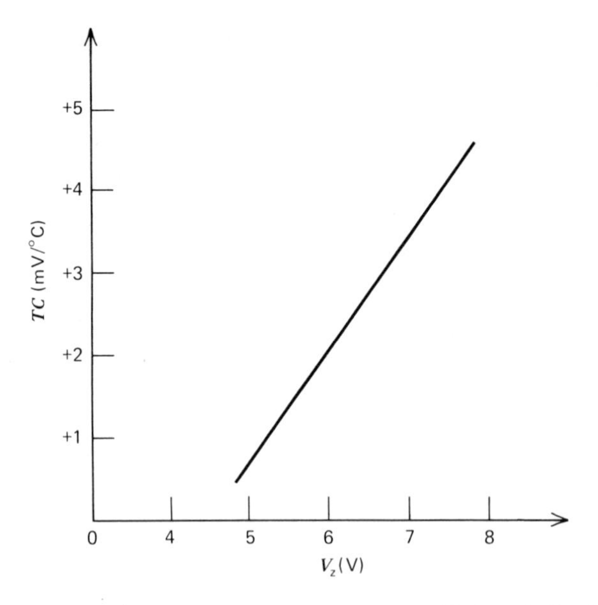

Figure 15.4 Average temperature coefficient (TC) of an IC zener diode versus zener voltage.

The rather poor load and line regulation of the shunt regulator in Ex. 15.1 may be corrected by replacing the passive zener with an active feedback network. As pointed out earlier, however, with few exceptions IC regulators are of the series pass or switch mode type. Therefore, we dispense with a discussion of improved shunt regulators. These exceptions (for example, Motorola MC1500, National LM136, and Texas Instruments TL431) are generally considered voltage references rather than voltage regulators. Their maximum output current is less than 100 mA and typically less than 10 mA.

Before leaving the subject of shunt regulators, their two inherently positive features should be mentioned. Because of the relatively constant input current, I_{in}, load current fluctuations are not passed on to the source and, owing to the external resistance R_s, the circuit is unharmed by shorted loads.

15.4 SERIES PASS REGULATOR

A conceptual series pass regulator is shown in Fig. 15.5. In operation, R_p is adjusted by a control circuit to pass the required load current. In effect, R_p is continuously adjusted to produce a voltage drop equal to the difference of the input voltage and the desired output voltage. The series pass regulator has an improved overall efficiency of:

$$\eta = \frac{V_o I_o}{V_{in} I_o} = \frac{V_o}{V_{in}} \tag{15.8}$$

It should be noted that the efficiency of a shunt regulator (Eq. 15.2) approaches that of a series pass type only when I_o approaches $I_{o(max)}$. The power dissipation in the IC is reduced in the series pass regulator to:

$$P_{IC} = (V_{in} - V_o)I_o \tag{15.9}$$

EXAMPLE 15.2 Calculate the maximum efficiency and associated power dissipation for the 5 V MC7805 series regulator. The input ripple is 10 V and the load current is 1 A.

PROCEDURE

1 From Eq. 15.8, maximum efficiency is attained for minimum input voltage. For this chip, the manufacturer recommends that a minimum of 7.5 V be added to the ripple voltage, (Table 15.1, test 4). Given a 10 V ripple, the input voltage is, therefore, set at 17.5 V.

2 By Eq. 15.8,

$$\eta = 5/17.5 = 0.29 = 29\%$$

3 By Eq. 15.9,

$$P_{IC} = (17.5 - 5) \times 1 = 12.5 \text{ W}$$

In actual practice, resistance R_p in Fig. 15.5 is replaced by a *series pass transistor,* Q_1 (Fig. 15.6). The control circuit contains an output sampling network, a voltage reference, and an error amplifier. The sampling network is a voltage divider composed of R_1 and R_2 in series that provides a voltage, V_s, equal to the reference voltage, V_{ref}, when V_o is the desired value. The error amplifier compares V_s to V_{ref}. If, for example, V_s is above V_{ref} (V_o is too high), the error amplifier reduces the drive to the base of Q_1 and the output voltage is thereby lowered. Each of the components in Fig. 15.6 is briefly described in the following sections.

Error Amplifier

The voltage regulator of Fig. 15.6 has been redrawn incorporating Q_1 inside an operational amplifier (Fig. 15.7). Now the op amp has, in addition to a good

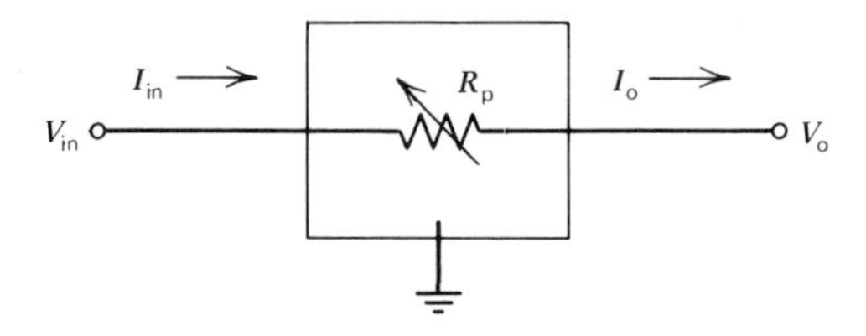

Figure 15.5 A conceptual 3-terminal series pass voltage regulator.

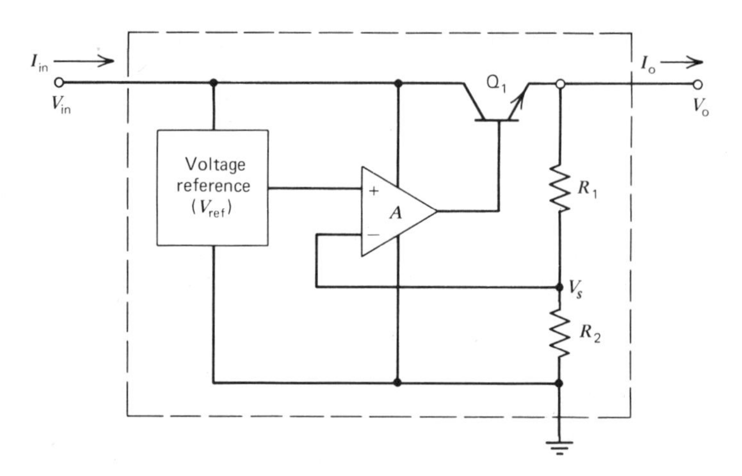

Figure 15.6 A 3-terminal series pass IC voltage regulator.

common mode rejection ratio (CMRR), power supply rejection ratio (PSRR), and open loop gain, the added capability of large output currents. (See Chap. 9 for an indepth discussion of op amps.) The output voltage may be expressed by:

$$V_o = [V_{ref} + V_{cm}/\text{CMRR} + V_{in}/\text{PSRR} + V_{io}]\,[1 + R_1/R_2] \quad (15.10)$$

for sufficiently large open loop gain. Voltge V_{cm} is the common mode voltage and V_{io} is the input offset voltage.

The output of this series pass regulator depends upon the reference voltage, power supply, common mode voltage, and the input offset voltage. In addition, there is a temperature dependence in the TC of V_{io} and V_{ref} of the error amplifier and a supply dependence on V_{ref}.

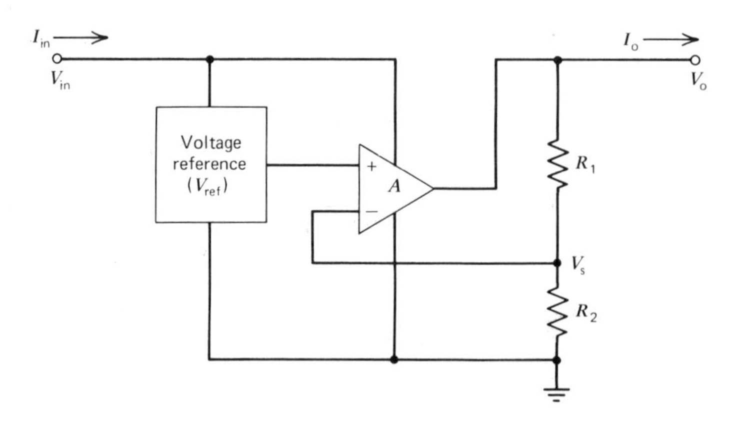

Figure 15.7 Transistor Q_1 in Fig. 15.6 is incorporated in the op amp.

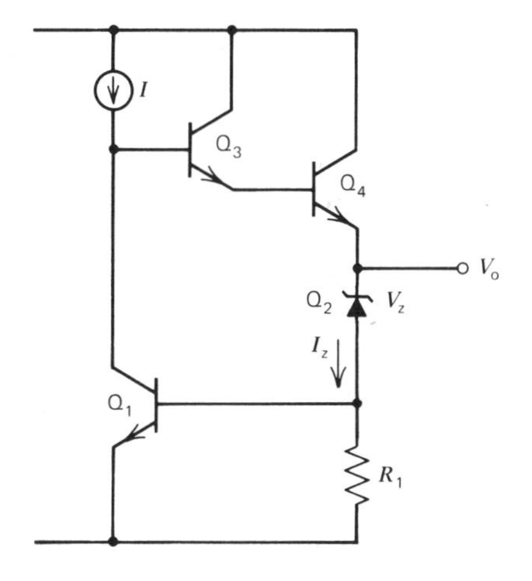

Figure 15.8 Simplified zener voltage reference, which is similar to that used in the μA723 regulator.

Voltage Reference

The voltage reference in a typical IC voltage regulator is one of two types: zener or bandgap. The zener is easier to process and control but requires larger input voltages and, until recently, was rather noisy and tended to drift with time. The bandgap reference is considered later in the chapter.

A simplified zener reference, similar to that found in the industry standard μA723 voltage regulator, is shown in Fig. 15.8. In operation, the current source provides base current for Q_3, the excess current being conducted to ground by Q_1. The emitter current of Q_4 is equal to the load plus the zener bias current, I_z. Zener bias current I_z is expressed by:

$$I_z = V_{BE(Q_1)}/R_1 \tag{15.11}$$

The output voltage is:

$$V_o = V_{BE(Q_1)} + V_z \tag{15.12}$$

If, for example, the output voltage attempts to increase, the base-emitter voltage of Q_1 increases, which in turn increases the collector current of Q_1. This forces the voltage at the base of Q_3 to decrease, causing the desired reduction in output voltage.

EXAMPLE 15.3 For the reference circuit of Fig. 15.8, determine the zener voltage and output voltage necessary to produce a zero-temperature coefficient, OTC, output. The TC of a pn junction is −2.2 mV/°C.

PROCEDURE

1 The TC of $V_{BE(Q_1)}$ is given as −2.2 mV/°C; therefore, OTC output is realized when the TC of the zener diode is +2.2 mV/°C.

2 From Fig. 15.4, the desired zener voltge is found to be approximately 6.3 V.
3 Finally, assuming V_{BE} of a transistor is 0.7 V, the total output voltage, by Eq. 15.12, is:

$$V_o = 0.7 + 6.3 = 7.0 \text{ V}$$

The recent introduction of ion-implanted zener diodes has moved the breakdown from a surface phenomenon, where it was subject to noise and drift, to a subsurface effect. This has resulted in much improved zener references. It should be noted, however, that to maintain regulation, the input voltage of even the most innovative circuit must be greater than the zener voltage, often an unacceptably high value.

The bandgap reference of Fig. 15.9 can inherently operate at lower input voltages and is, therefore, becoming one of the most common IC references. In operation, the mirror circuit of Q_3 and Q_4 forces equal currents in the collectors of Q_1 and Q_2. The emitter area of Q_1, however, is N times larger than that of Q_2, resulting in a ΔV_{BE} voltage across R_1.

Voltage V_1 is equal to the base-emitter voltage of Q_2, $V_{BE(Q_2)}$, plus $2IR_2$ (Fig. 15.9); hence,

$$\begin{aligned} V_1 &= 2IR_2 + V_{BE(Q_2)} \\ &= (2R_2/R_1)(kT/q) \ln (N) + V_{BE(Q_2)} \end{aligned} \tag{15.13}$$

where $kT/q = 26$ mV at room temperature. By carefully selecting the value of N and the ratio R_2/R_1, the positive TC inherent in the first term can be made to exactly cancel the negative TC inherent in the second term in Eq. 15.13. This will happen when V_1 is close to 1.25 V. The output voltage is adjusted upward from 1.25 V by the addition of resistors R_3 and R_4.

Power Device

The power device becomes the dominant factor in IC die size (well over half) for voltage regulators capable of delivering currents in excess of 1 A. Because these high-current regulators are able to dissipate large amounts of power, three safety features are commonly built in: *short-circuit current limit, high-voltage current reduction,* and *thermal limit.*

Figure 15.10 shows the output current, I_o, as a function of the input-output differential voltage, $V_{in} - V_o$, for a typical series pass regulator. Below a differential voltage of 3 V, the maximum output current is limited by the saturation resistance of the power device. From 3 to 7 V the output current is limited at $I_{o(max)}$ to protect the power device and wire bonds. Above 7 V, two phenomena must be considered: the absolute maximum power dissipation of the device and

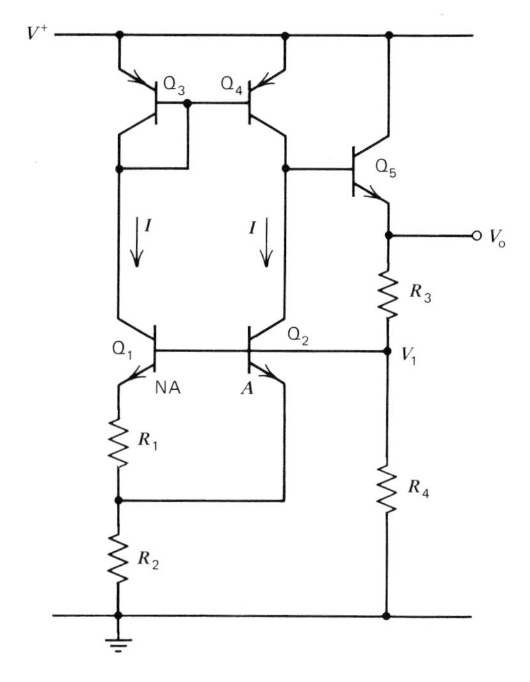

Figure 15.9 Simplified bandgap reference.

an abrupt reduction in breakdown voltage at high current, known as *second breakdown*. Both effects require that the maximum current be reduced.

In order to maintain long-term reliability and short-term parametric performance, die temperatures are typically not allowed to exceed 175 °C. When thermal limit is reached, internal circuitry shuts off the power device, thereby

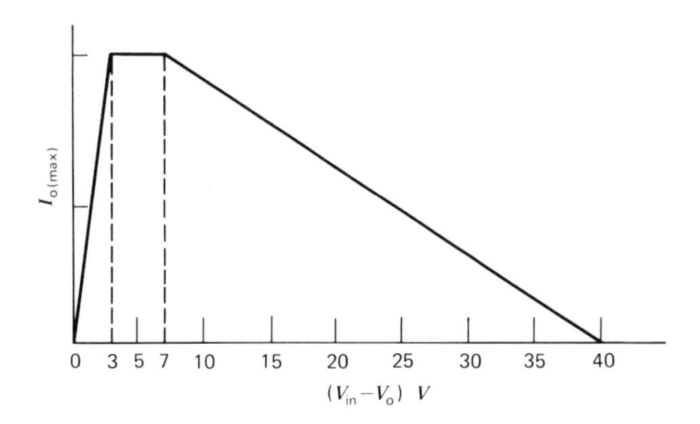

Figure 15.10 Typical plot of load current, I_o, as a function of the differential input-to-output voltage of an IC voltage regulator.

reducing the power dissipation to near zero. The die temperature, T_j, is related to power dissipation P, the thermal resistance from junction to ambient, θ_{jA}, and the ambient temperature T_A by an equation for heat transfer:

$$T_j = T_A + \theta_{jA}P \tag{15.14}$$

The thermal resistance from junction to ambient consists of the sum of the resistance from junction to case, θ_{jc}, case to heat sink, θ_{cs}, and heat sink to ambient, θ_{sA}:

$$\theta_{jA} = \theta_{jc} + \theta_{cs} + \theta_{sA} \tag{15.15}$$

EXAMPLE 15.4 Calculate the required heat sink thermal resistance, θ_{sA}, for an MC7805 regulator in a metal package. The desired maximum junction temperature, $T_{j(max)} =$ 125°C, $T_{A(max)} =$ 70°C, and $P = 2$ W.

PROCEDURE

1 From Table 15.1, θ_{jc} for an MC7805 is found to be 5.5°C/W. Using an insulator with a heat sink compound, Table 15.2 gives θ_{cs} for a TO-3 package as 0.4 °C/W.

2 From Eqs. 15.14 and 15.15,

$$T_j = T_A + P(\theta_{jc} + \theta_{cs} + \theta_{sA}) \tag{15.16}$$

Substituting the given values in Eq. 15.16,

$$125 = 70 + 2(5.5 + 0.4 + \theta_{sA})$$

3 Solving for θ_{sA},

$$\theta_{sA} = 21.6 \text{ °C/W}$$

The maximum die temperature of 125°C for this example was picked to be well below 175°C. This ensures that the regulator will not thermally shut down during load or line transients, when power dissipation is increased.

Table 15.2 **Thermal resistance from case to heat sink, θ_{cs}, for various packages and mountings. (Courtesy Motorola, Inc.)**

	θCS			
	METAL-TO-METAL*		USING AN INSULATOR*	
CASE	DRY	With Heatsink Compound	With Heatsink Compound	Type
TO-3	0.2 C W	0.1 C W	0.4 C W 0.35 C W	3 mil MICA Anodized Aluminum
TO-66	1.5 C W	0.5 C W	2.3 C W	2 mil MICA
199**	2 C W	1.6 C W	2.6 C W	3 mil MICA

*Typical values; heatsink surface should be free of oxidation, paint, and anodization
*Thermal data for Case 253 approximately the same.
**Thermal data for Case 313, 314 approximately the same.

EXAMPLE 15.5 Determine the maximum allowable T_A value for an MC7805 in a plastic package without a heat sink. The desired maximum junction temperature, $T_{j(max)}$ = 125°C and P = 0.25 W.

PROCEDURE

1 From Table 15.1, we find θ_{jA} = 65 °C/W
2 By Eq. 15.14 we obtain,

$$125 = T_A + 65 \times 0.25$$

3 Solving for T_A gives us,

$$T_A = 109°\text{C}$$

15.5 APPLICATIONS OF SERIES PASS REGULATORS

This final section on series pass regulators contains three examples which are intended to show how easily voltage regulation can be attained using modern ICs.

EXAMPLE 15.6 Design a 15 V regulator capable of delivering 1 A using an MC7815 in a plastic TO-220 package. Assume T_j = 25°C.

PROCEDURE

1 The simplicity of regulator design is revealed in this example. The front panel of the MC7800 data sheet, Fig. 15.11, shows design to be a matter of selecting an appropriate socket and application of a proper dc input voltage.
2 From Fig. 15.12, the appropriate dc input is found to be between approximately 18 V ($V_{in} - V_o = 18 - 15 = 3$) and 40 V ($V_{in} - V_o = 40 - 15 = 25$). These limits avoid low-voltage saturation and high-voltage current limit reduction.
3 If operated at high temperatures, heat sinking is recommended.

EXAMPLE 15.7 Design a 5 V regulator using the 3-terminal adjustable LM117 chip.

PROCEDURE

1 The LM117 is a 3-terminal regulator that provides 1.25 V between the output pin and the adjust pin (Fig. 15.13). If the adjust pin is grounded, the regulator behaves as a normal 1.25 V fixed 3-terminal voltage regulator. However, the IC has been designed so that the normal quiescent operating current exciting the circuit through the adjust pin is very low and relatively constant over all operating conditions ($I_{adj} \simeq 50\ \mu$A). This allows for the design of an adjustable regulator, as illustrated in Fig. 15.14.
2 The current I in R_1 is:

$$I = 1.25/R_1 \tag{15.17}$$

3-TERMINAL POSITIVE VOLTAGE REGULATORS

These voltage regulators are monolithic integrated circuits designed as fixed-voltage regulators for a wide variety of applications including local, on-card regulation. These regulators employ internal current limiting, thermal shutdown, and safe-area compensation. With adequate heatsinking they can deliver output currents in excess of 1.0 ampere. Although designed primarily as a fixed voltage regulator, these devices can be used with external components to obtain adjustable voltages and currents.

- Output Current in Excess of 1.0 Ampere
- No External Components Required
- Internal Thermal Overload Protection
- Internal Short-Circuit Current Limiting
- Output Transistor Safe-Area Compensation
- Output Voltage Offered in 2% and 4% Tolerance

THREE-TERMINAL POSITIVE FIXED VOLTAGE REGULATORS

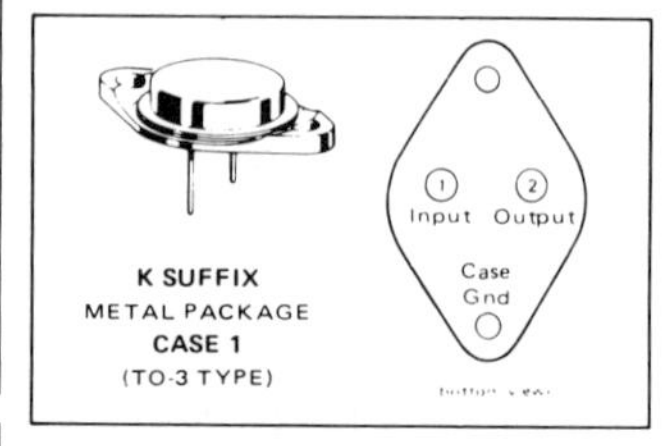

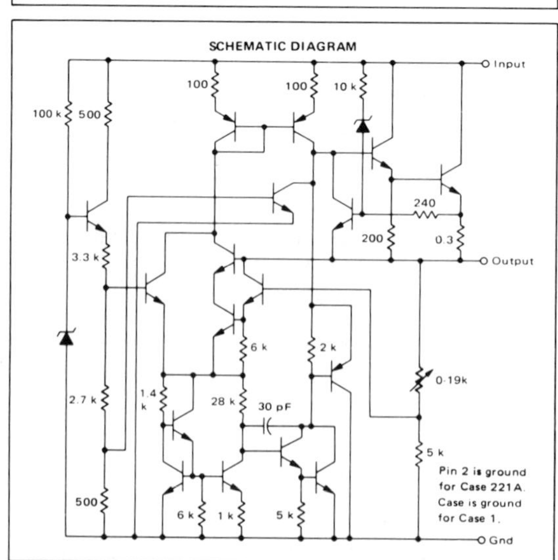

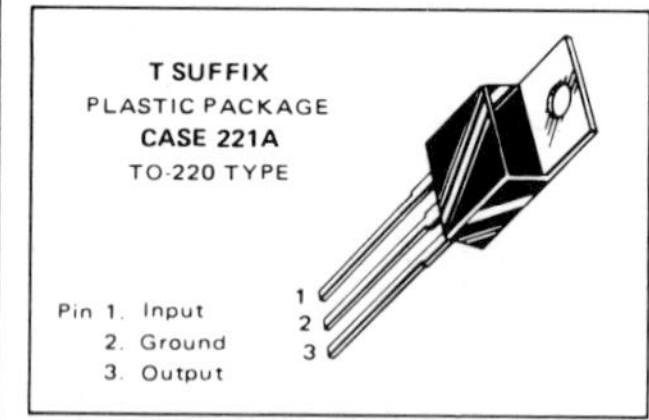

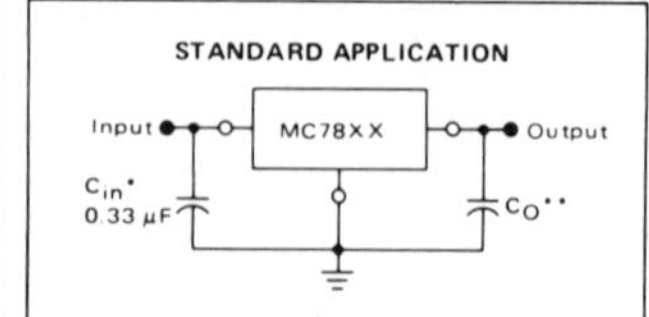

A common ground is required between the input and the output voltages. The input voltage must remain typically 2.0 V above the output voltage even during the low point on the input ripple voltage.

XX = these two digits of the type number indicate voltage.

* = C_{in} is required if regulator is located an appreciable distance from power supply filter.

** = C_O is not needed for stability; however, it does improve transient response.

XX indicates nominal voltage

ORDERING INFORMATION

Device	Output Voltage Tolerance	Temperature Range	Package
MC78XXK MC78XXAK	4% 2%	-55 to +150°C	Metal Power
MC78XXBK	4%	-40 to +125°C	
MC78XXCK MC78XXACK	4% 2%	0 to +125°C	
MC78XXCT MC78XXACT	4% 2%		Plastic Power
MC78XXBT	4%	-40 to +125°C	

TYPE NO./VOLTAGE			
MC7805	5.0 Volts	MC7815	15 Volts
MC7806	6.0 Volts	MC7818	18 Volts
MC7808	8.0 Volts	MC7824	24 Volts
MC7812	12 Volts		

Figure 15.11 Front panel of an MC7800 data sheet. (Courtesy Motorola, Inc.)

If R_1 is sufficiently small such that the current in the adjust pin can be ignored ($I >> I_{adj}$), then the voltage across R_2, V_2, is:

$$V_2 = IR_2$$
$$= 1.25R_2/R_1 \tag{15.18}$$

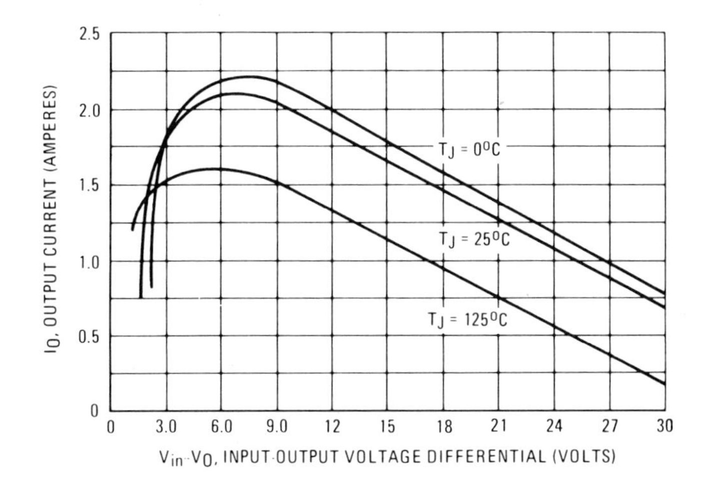

Figure 15.12 Peak output current as a function of input-output differential voltage for the MC7800 voltage regulator. (Courtesy Motorola, Inc.)

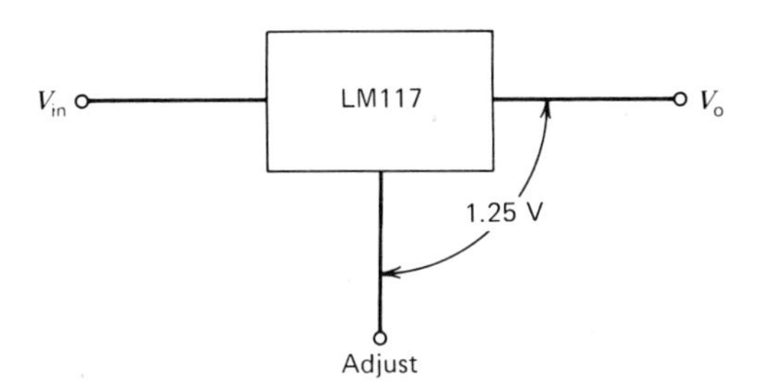

Figure 15.13 Block diagram of a 3-terminal (LM117) adjustable voltage regulator.

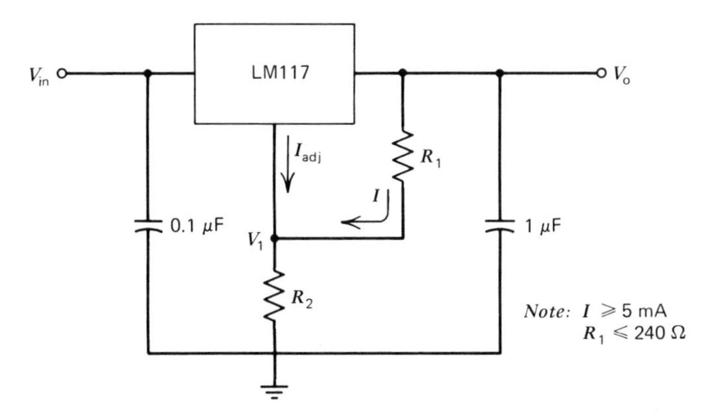

Figure 15.14 Circuit connection of a 3-terminal adjustable voltage regulator.

The total output voltage is:

$$V_o = 1.25\,(1 + R_2/R_1) \tag{15.19}$$

3 Selecting a value of 240 Ω for R_1 yields, by Eq. 15.17, I approximately $100I_{adj}$. Solving Eq. 15.19 for R_2 for an output voltage of 5 V yields $R_2 = 740\ \Omega$.
4 A capacitor is provided at the input to offset any inductance stemming from long leads to the raw dc power. Also, an output capacitor is provided to improve transient response of high-frequency load fluctuations.

EXAMPLE 15.8 Design a 15 V regulator using the industry standard 723 multipin voltage regulator.

PROCEDURE

1 Figure 15.15 shows a front panel from a 723 data sheet. Of significance, transistors Q_4 through Q_8 constitute the voltage reference, Q_{12} through Q_{15} the error amplifier, and Q_{16} and Q_{17} the output (Darlington pair) power device.
2 For operation above the reference voltage of 7 V, the output of the reference, pin 4, is connected to the noninverting input of the error amplifier, pin 3 (see Fig. 2 of the front panel). A capacitor to ground at pin 3 may be added to improve ripple rejection. The collector of the output power device, pin 7, is normally connected to V_{in}, pin 8. The output voltage, pin 6, is returned to the inverting input of the error amplifier by feedback resistors R_1 and R_2.
3 From Eq. 15.10, ignoring PSRR, CMRR, and V_{io}, we see that the output voltage is:

$$\begin{aligned} V_o &= V_{ref}(1 + R_1/R_2) \\ &= (7\ \text{V})\,(1 + R_1/R_2) \end{aligned} \tag{15.20}$$

If we are to provide a 15 V output, the ratio of R_1 to R_2 should be equal to 1.14.
4 Short-circuit current limit is provided by inserting a *sense resistor,* R_{sc}, in series with the output and connecting the base and emitter of Q_{18}, pins 10 and 1, respectively, across the resistor. The output is then limited to:

$$\begin{aligned} I_{sc} &= V_{BE(Q18)}/R_{sc} \\ &= 0.66/R_{sc} \end{aligned} \tag{15.21}$$

It should be noted that the voltage drop across R_{sc} is inside the feedback loop and, therefore, is not noticed at the output.
5 Frequency compensation is provided by inserting a 100 pF capacitor across pins 2 and 9.

15.6 SWITCHING REGULATOR

A switching regulator operates by chopping the unregulated dc supply voltage using a saturated transistor switch and then filtering the chopped voltage. By varying the duty cycle of the transistor, the output voltage can be regulated. A

MONOLITHIC VOLTAGE REGULATOR

The MC1723 is a positive or negative voltage regulator designed to deliver load current to 150 mAdc. Output current capability can be increased to several amperes through use of one or more external pass transistors. MC1723 is specified for operation over the military temperature range (-55°C to +125°C) and the MC1723C over the commercial temperature range (0 to +70°C)

- Output Voltage Adjustable from 2 Vdc to 37 Vdc
- Output Current to 150 mAdc Without External Pass Transistors
- 0.01% Line and 0.03% Load Regulation
- Adjustable Short-Circuit Protection

FIGURE 1 – CIRCUIT SCHEMATIC

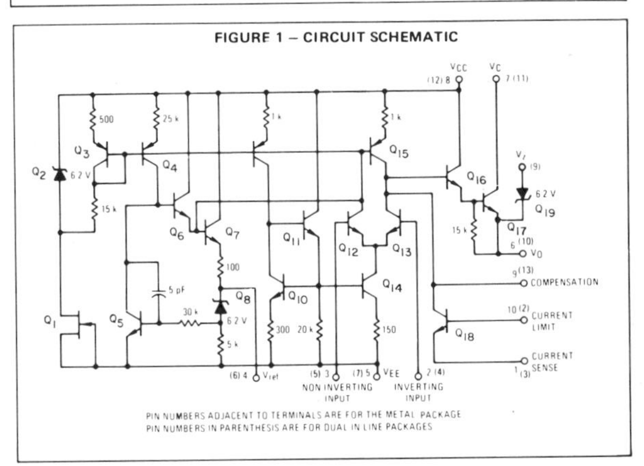

FIGURE 2 – TYPICAL CIRCUIT CONNECTION

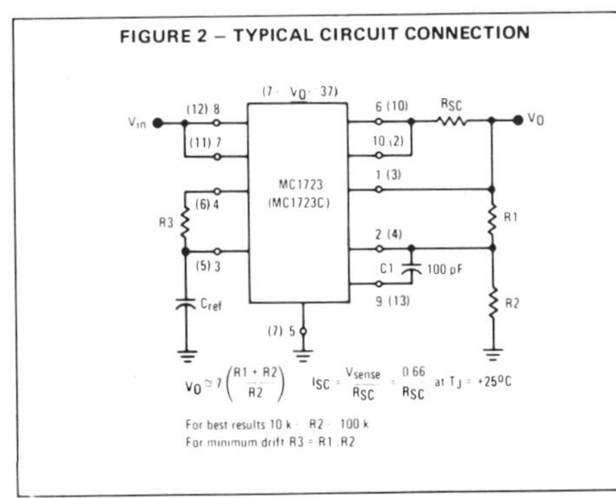

FIGURE 3 – TYPICAL NPN CURRENT BOOST CONNECTION

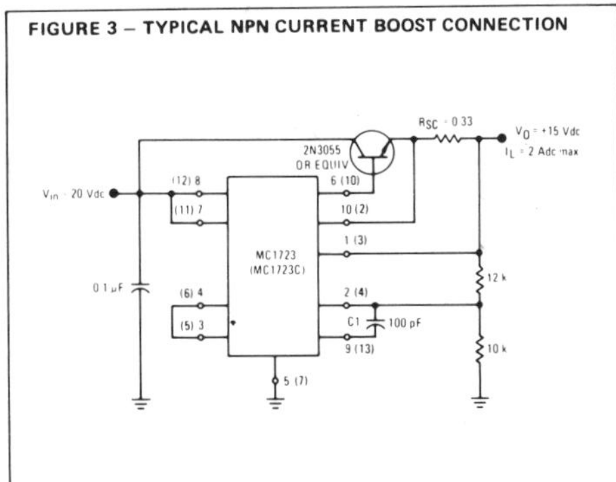

Figure 15.15 Front panel of a typical 723 voltage regulator. (Courtesy Motorola, Inc.)

block diagram of a switching regulator appears in Fig. 15.16. Because the regulating element (transistor) is operated either in saturation (collector-emitter voltage is approximately zero) or blocking (collector current is approximately zero), power losses are kept to a minimum. This results in higher efficiencies and minimized heat sinking requirements.

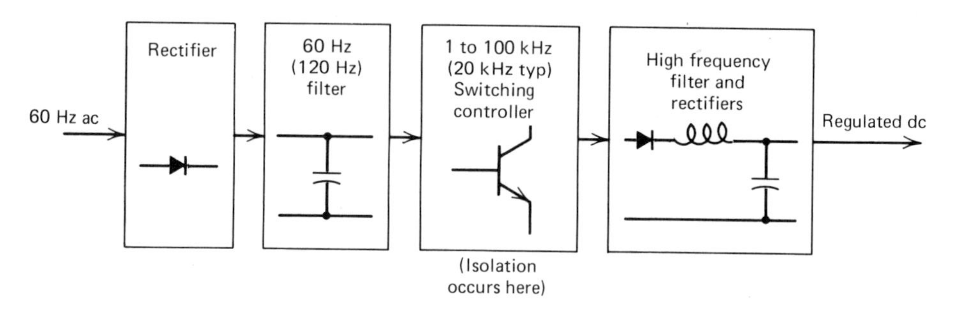

Figure 15.16 Block diagram of a switching regulator.

In addition, because the switching regulator can often operate directly off the ac line, the 60 Hz step-down transformer used in the series pass regulator is eliminated. In its place, a smaller and lighter high-frequency ferrite core transformer is used to transform the chopped dc voltage. Therefore, the switching regulator is lighter, smaller, and more efficient than a series pass type.

The switching regulator, however, has several disadvantages. Because it operates by switching at a high frequency, noise, both on the output and induced in the ac input lines, can be troublesome, thus requiring heavy filtering. The switching regulator also is limited in its transient response by the switching frequency and is usually considerably slower than a linear supply (milliseconds versus microseconds). In addition, the switching regulator is more complex than linear types. For these reasons, the use of switching regulators is usually limited to power levels above 100 W.

Although the complexity of switching regulators makes the design of such systems beyond the scope of this chapter, a short discussion follows on the filter section, commonly called the magnetics, and the control section. A comprehensive study of switch-mode systems may be found in some of the references listed in Sec. 15.7.

The switching and filtering functions in the circuit of Fig. 15.16 are accomplished using one or combinations of three basic configurations: buck, boost, and dc-dc transformer.

Buck

The buck regulator family all share the common characteristics of pulsating input current, continuous output current, and the input voltage being greater than the output voltage. Figure 15.17 illustrates the basic buck configuration. When the transistor is switched ON, full input voltage is applied to the LC filter; when the transistor is OFF, the input voltage is zero. With the transistor turned ON and OFF for equal amounts of time (*50% duty cycle*), the dc output voltage will be half the input voltage. The output voltage, V_o, will always equal the input voltage, V_{in}, times the duty cycle, D:

$$V_o = DV_{in} \tag{15.22}$$

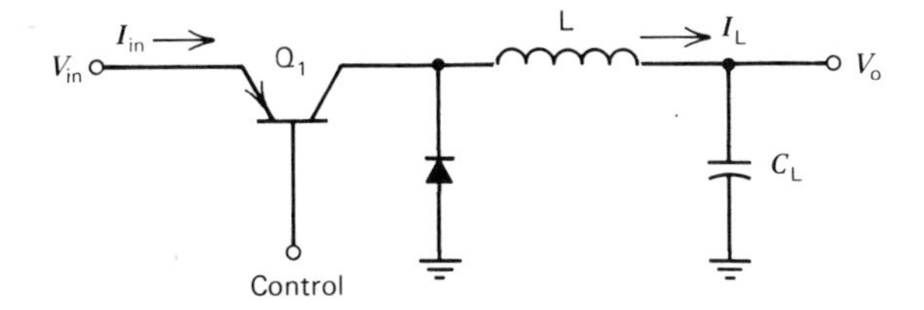

Figure 15.17 Basic buck regulator.

Varying the duty cycle will, therefore, compensate for changes in the input voltage. This technique is used to obtain a regulated output voltage.

Repetitive operation of the switching transistor at a fixed duty cycle produces the steady-state waveforms shown in Fig. 15.18. With the switch closed, inductor current I_L flows from the input voltage V_{in} to the load. The difference between the input and output voltages, $V_{in} - V_o$, is applied across the inductor. During this time, I_L increases.

With the switch open, stored energy in the inductor forces I_L to continue to flow in the load and return through the diode. The inductor voltage is now reversed and is approximately equal to V_o. During this time, I_L decreases.

The average current through the inductor is equal to the load current. Because the capacitor keeps V_o constant, the load current also will be constant. When the load current increases above its nominal value, the capacitor charges; when the load current decreases, the capacitor discharges. These waveform inflection points are indicated in Fig. 15.18.

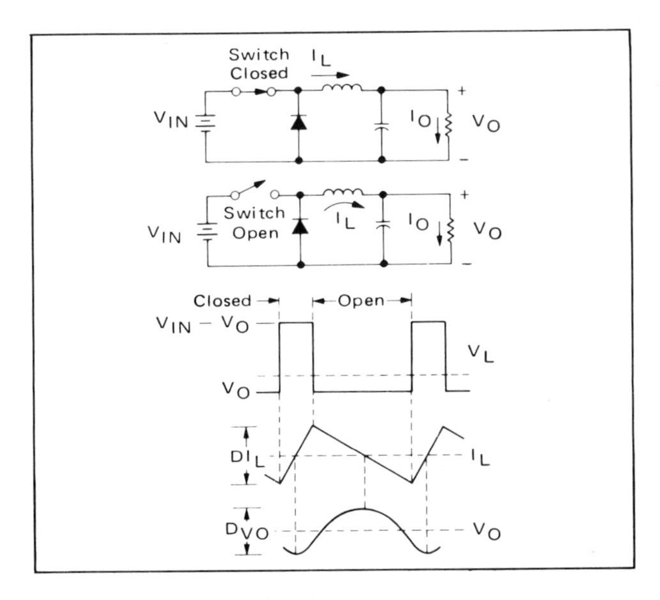

Figure 15.18 Theoretical waveforms of a buck regulator. (Courtesy Motorola, Inc.)

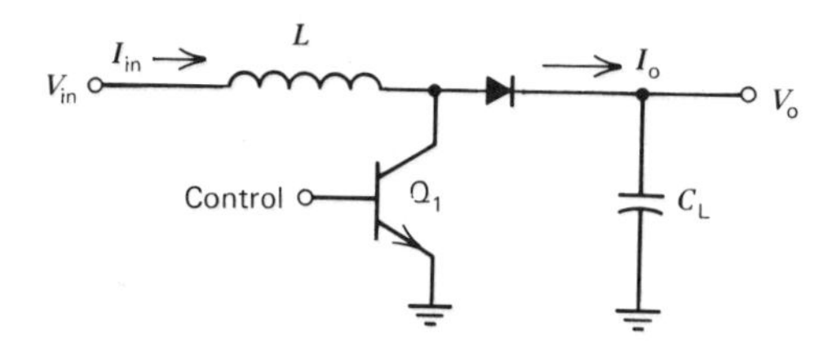

Figure 15.19 Basic boost regulator.

In summary, under steady-state operation:

1 The average inductor voltage is zero, but a wide variation from $V_{in} - V_o$ to V_o will be experienced.

2 The dc flowing in the inductor equals the load current. A small amount of sawtooth ripple also will be present.

3 The dc voltage across the capacitor is equal to the load voltage. A small amount of ripple (quasi sine wave) also will be present across the capacitor.

Transient operation must consider changes in V_{in} and I_o. Input voltage changes are automatically compensated for by appropriate duty cycle variations in a closed loop system. Input regulation and ripple rejection depend on loop gain, but are generally adequate.

Changes in I_o are more difficult to compensate for and load transient response is generally poor. Changes in I_o are compensated for with temporary duty cycle changes. For example, a change in load current from half to full will result in the following:

1 Duty cycle increases to its maximum; the transistor may just remain ON.

2 The inductor current takes many cycles to increase to its new dc value.

3 Duty cycle returns to its original value.

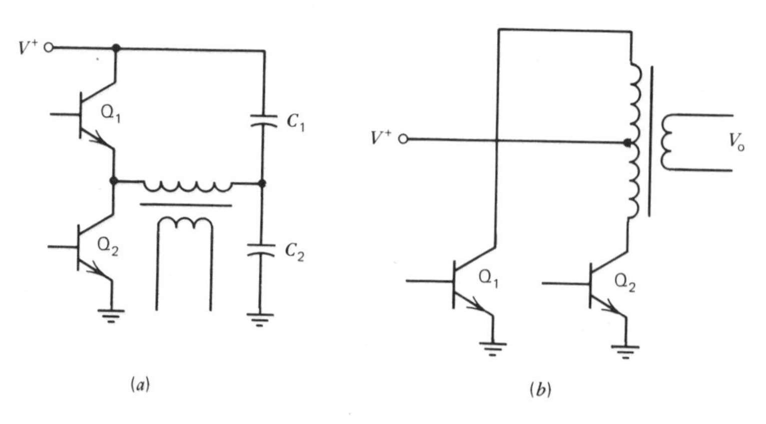

Figure 15.20 Two dc-dc transformers. (*a*) Bridge. (*b*) Quasi square wave.

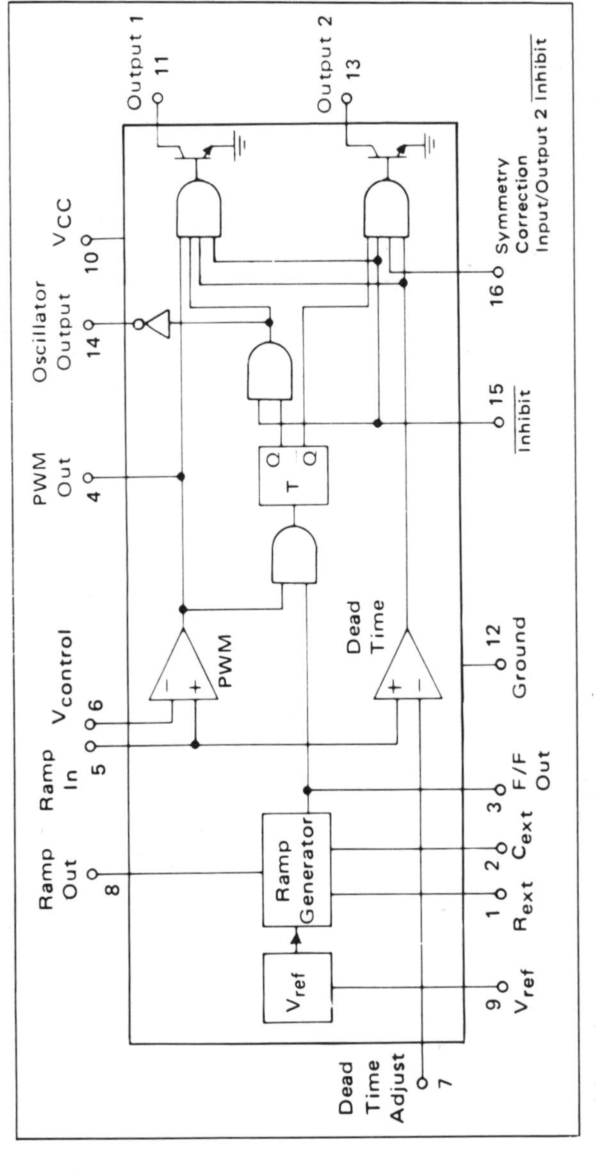

Figure 15.21 Block diagram of an MC3520 switching mode controller. (Courtesy Motorola, Inc.)

Boost

All members of the boost family have continuous input current, discontinuous output current, and an output voltage that is greater than the input voltage. The basic boost regulator, shown in Fig. 15.19, is best analyzed from a conservation of energy standpoint.

During the ON period, τ, of Q_1, current begins to flow from V_{in} to ground, storing energy in the inductor which is equal to $V_{in}^2\ \tau^2/2L$ joules. As Q_1 cuts off, current continues to flow in the inductor through the diode to the load capacitor, dissipating the inductor's stored energy. During one full cycle, T, the load energy is I_oV_oT joules. By equating energy-in to energy-out, neglecting inductor, switch, and capacitor losses, and assuming constant-frequency operation, the switch ON time may be used to control V_o for changes in I_o and V_{in}.

Dc-Dc Transformer

The bridge dc-dc transformer of Fig. 15.20*a* obtains transformer action by chopping and reversing the primary current. When Q_1 is saturated, current is conducted from $V+$ through the primary winding to C_1 and C_2. Transistor Q_1 is cutoff, and after an appropriate delay determined by the controller, Q_2 turns ON. Current is now conducted from the capacitors to ground through Q_2. The secondary voltage may be rectified and filtered. The quasi-square wave converter of Fig. 15.20*b* operates in a similar manner.

Control Circuit

In either the buck, boost, or the dc-dc transformer configurations, the key to regulation is duty cycle control. Duty cycle may be controlled using any of several techniques, among them variable frequency with constant on-time, variable frequency with constant off-time, and constant frequency with variable on-time.

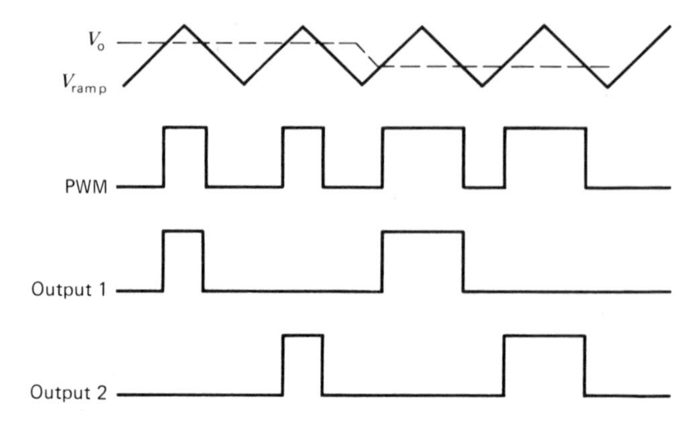

Figure 15.22 Internal waveforms of the MC3520 controller.

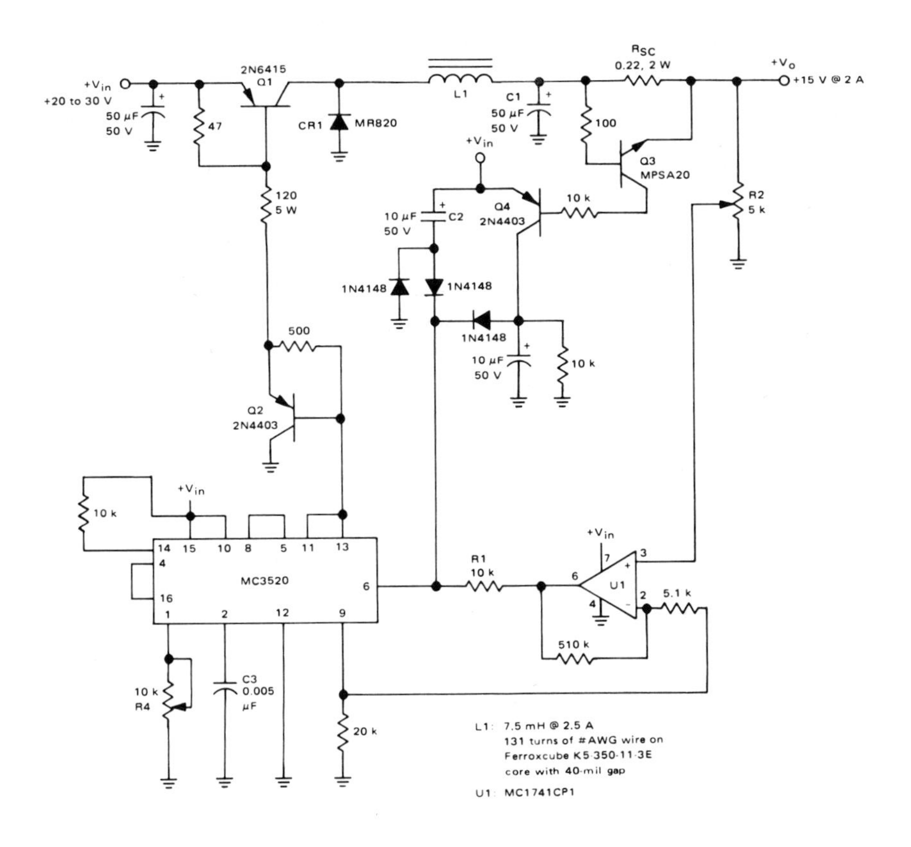

Figure 15.23 A 15 V, 2 A buck regulator employing the MC3520 controller. (Courtesy Motorola, Inc.)

The MC3520 switch-mode controller, typical of those found in the industry (Fig. 15.21), uses the latter method for duty cycle control.

In normal operation, the output of a constant-frequency ramp generator, pin 8, is connected to the noninverting input of the pulse width modulation (PWM) comparator, pin 5. The regulated output voltage, either directly or from a divider, is connected to the inverting input of the PWM comparator. Figure 15.22 illustrates the resultant output for two levels of regulated voltage. The output of the PWM comparator is fed to a phase splitter (flip flop) that alternately drives outputs 1 and 2. These outputs may then drive either the two transistors of a dc-dc transformer or they can be wire-ORed together to drive a buck or boost regulator.

EXAMPLE 15.9 Design a buck regulator using the MC3520 controller to give 15 V out at 2 A.

PROCEDURE

1 The resulting circuit is given in Fig. 15.23. The series switching transistor, Q_1, chops the dc input voltage V_{in} at a frequency of approximately 25 kHz. The resulting waveform is filtered by L_1 and C_1 to provide the dc output voltage.

2 The frequency is set by R_4 and C_3. Output voltage V_o is regulated by comparing its value to the MC3520's reference and amplifying the error with U_1. The output of U_1 is fed into pin 6 to provide PWM to Q_1, thereby controlling its duty cycle and the value of V_o.

3 Capacitor C_2 provides a soft-start feature during power-up to prevent output voltage overshoots and excessive start-up currents through Q_1.

4 Short-circuit protection is provided by R_{sc}, Q_3, and Q_4. When an over current condition occurs, Q_3 is turned ON by the voltage across R_{sc}. Transistor Q_3 drives Q_4 ON, which raises the voltage at pin 6. The duty cycle of Q_1 is thereby reduced and a constant output current of 2 A is maintained.

15.7 REFERENCES

Brokaw, A. P., "Solid-State Regulated Voltage Supply," U.S. Patent 3,887,863, Nov. 28, 1973.

Haver, R. J., "A New Approach to Switching Regulators," Application Note AN-719, Motorola, Inc., Phoenix, AZ, 1975.

Intersil, *Switchmode Converter Topologies—Make Them Work For You,* Intersil, Inc., Cupertino, CA, 1980.

Spencer, J. D. and D. E. Pippenger, *The Voltage Regulator Handbook,* Texas Instruments, Inc., Dallas, TX, 1977.

Vaeches, T., *μA78S40 Switching Voltage Regulator Applications,* Fairchild Camera and Instrument Corp., Mountain View, CA, 1978.

Wurzburg, H., *Voltage Regulator Handbook,* Motorola, Inc., Phoenix, AZ, 1976.

CHAPTER 16

IC Interface Applications

Theodore Vaeches
Digital Signal Processing Product Marketing Manager
American Microsystems, Inc.

16.1 INTRODUCTION

Linear devices, such as operational amplifiers, are designed to operate primarily in a linear mode where the output is an accurate representation of the input waveform. Interface devices, on the other hand, are not concerned with preserving the input waveform in a continuous manner. Interface devices are generally used as information translators—bridging the gap between analog and digital worlds, or level shifters—providing larger voltage swings or current capabilities than the driving circuit is capable of providing.

In this chapter, some of the more universally used interface devices will be discussed. Section 16.2 begins with a review of comparator characteristics. A collection of comparator applications then follows. Section 16.4 covers practical aspects of semiconductor memory interfacing. The relationship between power dissipation and operating speed will be explored. System considerations, such as power supply decoupling, crosstalk, and output waveshaping also will be covered. In Sec. 16.6, the controlling factors in line driver and receiver interfaces will be reviewed. The primary focus of this section will be on the EIA defined standard interfaces (RS-232C, RS-423, and RS-422).

16.2 VOLTAGE COMPARATORS

The dc voltage comparator outwardly appears to be very similar to the operational amplifier (Chap. 9). Both have differential inputs, high gain, and many input and output parameters in common. In fact, the comparator appears to be an operational amplifier functioning in an open loop mode; however, there are fundamental differences.

Voltage comparators are specifically designed around properties of *speed* and *accuracy* for differential comparison purposes. There is no intention of reproducing any part of the original input signal waveform. Therefore, most voltage comparators are essentially uncompensated, high-gain differential amplifiers driving an output stage which is normally saturated high or low. Operational amplifiers, on the other hand, are internally or externally compensated, designed to provide a linear relationship between input and output, and generally have response times too slow for most comparator applications. General-purpose op amps, however, may be used in noncritical applications.

Voltage comparators appear almost universally in digital systems which require that a logic output be made available when an analog voltage or current within the system is greater than, less than, or between some critical threshold(s). This section deals with some of the many applications in which comparators are used.

Level Detector

One of the primary applications of voltage comparators is *level detection* of an input signal relative to a known reference level. When the input signal, V_{in}, exceeds the established reference limit, V_{ref}, the comparator output changes state. Figure 16.1 is an example of a simple level detector.

EXAMPLE 16.1 Design a level detector to act as a zero-crossing detector. In a zero-crossing detector, $V_{ref} = 0$ and the output changes state whenever the input, V_{in}, is equal to zero. The input signal range is ± 12 V. The circuit should be able to be gated by a TTL signal and be capable of driving TTL gates.

PROCEDURE

1 The design is given in Fig. 16.2. A level detector circuit employing the μA311 was chosen because of its wide common mode, input voltage range, offset nulling capability, strobe capability, and ability to drive TTL.
2 Operating the μA311 from a ± 15 V supply ensures a common mode range of ± 14 V, which exceeds the ± 12 V required.
3 Selecting a 2N2222 transistor with a 1 kΩ emitter resistor allows the strobe input to be driven by TTL.
4 The 3 kΩ pot and 3 kΩ series resistor allow adequate nulling of the comparator's offset voltage and provides current limiting at the null inputs.

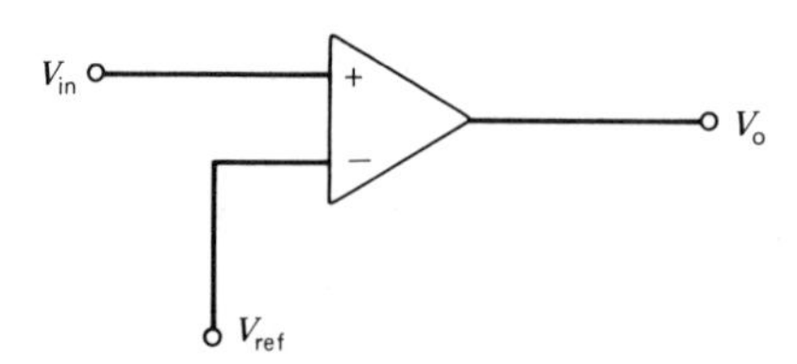

Figure 16.1 An example of a simple level detector.

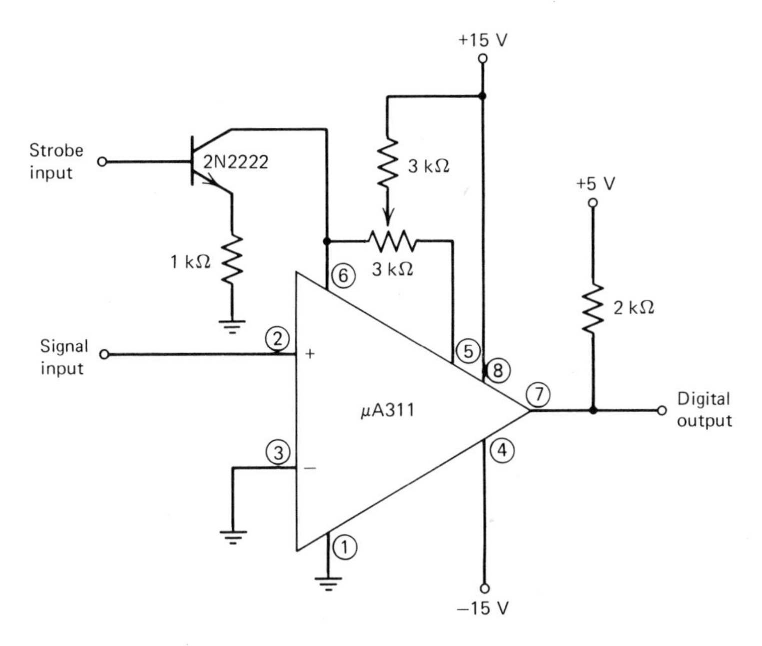

Figure 16.2 A zero-crossing detector with input strobe control and offset null adjustment.

5 A 2 kΩ resistor to +5 V acts as a pullup for the output of the comparator and allows it to drive TTL.

In level detector circuits, especially when procesing slowly varying signals, noise levels riding on these signals can be particularly troublesome at the decision point where the comparator operates in a linear mode. Noise signals are amplified and can cause the comparator output to change state at random. To avoid this difficulty, positive feedback is employed to alter the comparator's transfer characteristic. Figure 16.3 is an example of a level detector employing feedback, which gives rise to *hysteresis,* (Fig. 16.4*a*). The modified transfer characteristic has two threshold levels: V_2 and V_1 (*upper* and *lower threshold levels*). The

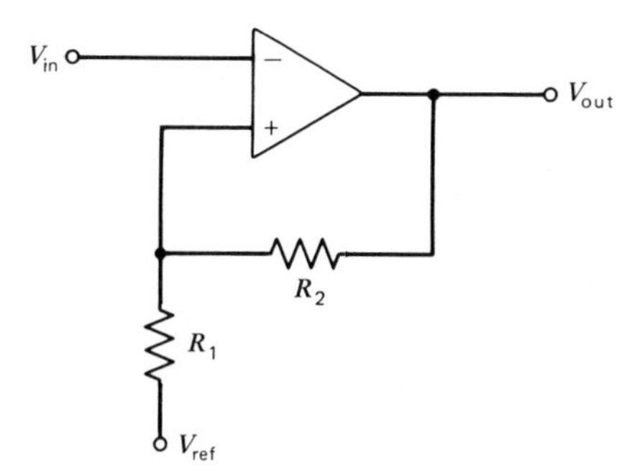

Figure 16.3 A level detector with positive feedback.

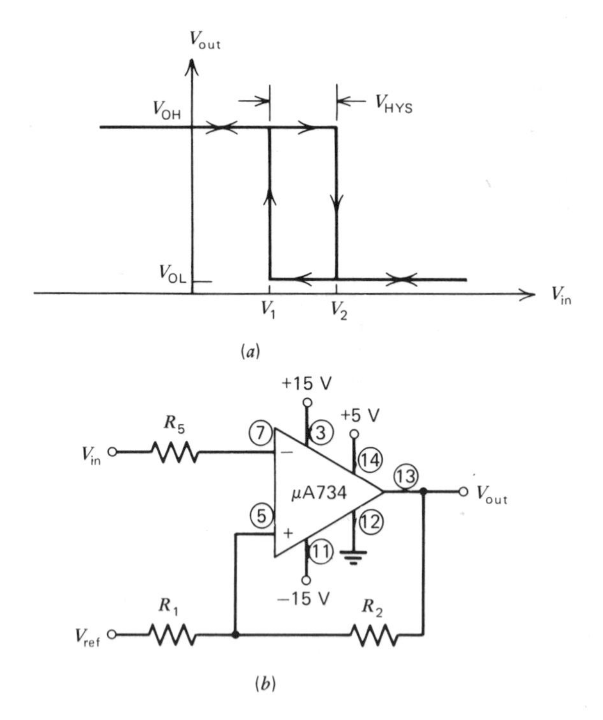

Figure 16.4 A level detector using a μA734 chip. (*a*) Input/output characteristics showing hysteresis. (*b*) Circuit.

difference between V_2 and V_1 is called the *hysteresis voltage,* V_{HYS}. With the hysteresis included, the comparator is less sensitive to noise and responds only to changes in the input signal level.

EXAMPLE 16.2 Design a level detector with a decision threshold V_{TH}, at -2 V for an input signal range of ± 8 V. The expected noise level will be less than 75 mV. The output is required to drive a single TTL load.

PROCEDURE

1 A level detector employing hysteresis (Fig. 16.4*b*) is used in the solution to this problem. The μA734 is chosen because of its high speed, wide common mode voltage range, and ability to drive TTL.

2 To ensure a ± 8 V input range, the μA734 is operated at ± 15 V, which allows a ± 10 V range. Worst case differential voltages are $+6$ and -10 V, as indicated below. The differential input voltage range of the μA734 is ± 10 V, which allows proper operation.

$$V_{\text{diff(max)}} = V_{(+)} - V_{(-)} = V_{\text{TH}} - V_{\text{in(min)}}$$
$$= -2 + 8 = +6 \text{ V}$$
$$V_{\text{diff(min)}} = V_{(+)} - V_{(-)} = V_{\text{TH}} - V_{\text{in(max)}}$$
$$= -2 - 8 = -10 \text{ V}$$

3 The output pullup is connected to +5 V to allow output compatibility with TTL.

4 Resistors R_1, R_2, and a reference, must be chosen to allow adequate noise immunity about the decision point and minimum loading of the output.

a. A 100 mV hysteresis window is chosen around the −2 V decision point to allow for adequate noise immunity.

b.
$$V_{HYS} = \frac{R_1}{R_1 + R_2} \times (V_{OH} - V_{OL}) = 0.1\text{ V} \tag{16.1}$$

For the μA734: $V_{OH} = 2.4$ V and $V_{OL} = 0.4$ V; therefore:

$$\frac{R_1}{R_1 + R_2} = \frac{0.1}{2.4 - 0.4} = 0.05 \text{ and } \frac{R_2}{R_1 + R_2} = 0.95$$

c.
$$V_1 = V_{ref}\frac{R_2}{R_1 + R_2} + V_{OL}\frac{R_1}{R_1 + R_2} = -2.05\text{ V} \tag{16.2}$$

$$V_2 = V_{ref}\frac{R_2}{R_1 + R_2} + V_{OH}\frac{R_1}{R_1 + R_2} = -1.95\text{ V} \tag{16.3}$$

Solving either Eq. 16.2 or 16.3 for V_{ref} gives us:

$$V_{ref} \approx -2.2\text{ V}$$

d. Resistors R_1 and R_2 are selected to minimize the output loading, to ensure that the output can drive one TTL load. A 40 μA hysteresis network current is assumed under worst case conditions (comparator output in the high state):

$$R_1 + R_2 = \frac{V_{OH} - V_{ref}}{I} = \frac{2.4 + 2.2}{40 \cdot 10^{-6}} = 115\text{ k}\Omega \tag{16.4}$$

Therefore,

$$\frac{R_1}{R_1 + R_2} = \frac{R_1}{115\text{ k}\Omega} = 0.05;$$ solving for R_1 we obtain, $R_1 = 5.75\text{ k}\Omega \simeq 6\text{ k}\Omega$.

Hence, $R_2 = 115\text{ k}\Omega - 6\text{ k}\Omega = 109\text{ k}\Omega$.

Window Detector

Voltage comparators are frequently employed in the design of test equipment to detect either the presence of a signal within a specified voltage range, or to indicate when a signal has stepped outside the specified range. A circuit that performs this function is called a *min-max limit,* or *window detector*.

EXAMPLE 16.3 Design a window detector to detect the presence of a signal within the range of +5 to −10 V. The signal input is limited to ±15 V.

PROCEDURE

1 The design is given in Fig. 16.5*a*. An LM193 dual comparator was chosen for this application. It features low offset voltage, low input bias current, wide input common mode and differential voltage range, and TTL compatible outputs.

2 Values for R_1, R_2, and R_3 must be calculated. A reference current of 1 mA is chosen to minimize the power dissipated in the network. The input bias currents of the comparator are negligible compared to this value and do not enter the calculations:

$$R_T = R_1 + R_2 + R_3 = \frac{V_{\text{ref}(+)} - V_{\text{ref}(-)}}{I} = \frac{15 - (-15)}{1} = \frac{30}{1} = 30\ \text{k}\Omega$$

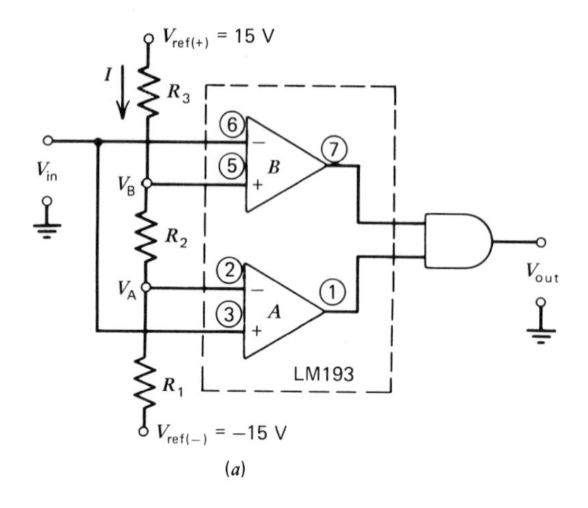

(*a*)

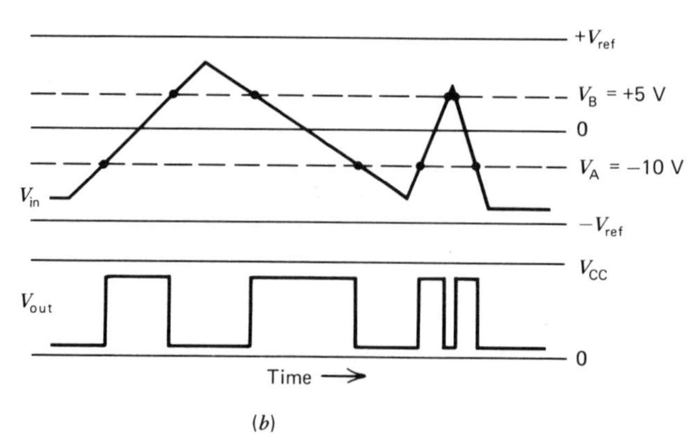

(*b*)

Figure 16.5 A window detector using the LM193 chip. (*a*) Circuit. (*b*) Input/output characteristics.

In Fig. 16.5*b*, $V_A = -10$ V and $V_B = +5$ V.

$$R_1 = \frac{V_A - V_{ref(-)}}{I} = \frac{-10 - (-15)}{1} = \frac{5}{1} = 5 \text{ k}\Omega$$

$$R_2 = \frac{V_B - V_A}{I} = \frac{5 - (-10)}{1} = \frac{15}{1} = 15 \text{ k}\Omega$$

$$R_3 = R_T - R_1 - R_2 = 10 \text{ k}\Omega$$

3 The LM193 has an input voltage range of ± 15 V, a common mode voltage range of -15 and $+13.5$ V, and a differential voltage range of ± 36 V. A calculation must be made to ensure that none of these voltage limitations are exceeded. The input voltage range of ± 15 V is satisfied for both comparators. Solving for comparator (B) gives us:

$$V_{CM(upper)} = \frac{V_{in(max)} + V_B}{2} \tag{16.5}$$

$$= \frac{15 + 5}{2} = 10 \text{ V}$$

$$V_{CM(lower)} = \frac{V_{in(min)} + V_B}{2} \tag{16.6}$$

$$= \frac{-15 + 5}{2} = -5 \text{ V}$$

$$V_{diff(+)} = V_B - V_{in(min)} \tag{16.7}$$

$$= 5 + 15 = 20 \text{ V}$$

$$V_{diff(-)} = V_B - V_{in(max)} \tag{16.8}$$

$$= 5 - 15 = -10 \text{ V}$$

Solving for comparator (A) yields:

$$V_{CM(upper)} = \frac{V_{in(max)} + V_A}{2} = \frac{15 - 10}{2}$$

$$= 2.5 \text{ V}$$

$$V_{CM(lower)} = \frac{V_{in(min)} + V_A}{2} = \frac{-15 - 10}{2}$$

$$= -12.5 \text{ V}$$

$$V_{diff(+)} = V_{in(max)} - V_A = 15 + 10$$

$$= 25 \text{ V}$$

$$V_{diff(-)} = V_{in(min)} - V_A = -15 + 10$$

$$= -5 \text{ V}$$

All voltage limitations are satisfied.

Peak Detector

Peak detection of a signal can be easily accomplished by using the circuit of Fig. 16.6*a*. When a signal is applied to the noninverting input, the output of the comparator switches to a high state. The capacitor at the inverting input is then charged to the peak value of the input signal through the diode. The stored charge on the capacitor is then used to provide the input bias current to the comparator and leakage currents through the diode and capacitor, thereby reducing the voltage on the inverting input to slightly below the peak signal. At the peak of the following cycle, when the input signal exceeds the threshold level, an output pulse is generated and additional charge is supplied to the capacitor. The output of the comparator goes to a low state, ending the pulse. This process results in a train of pulses, where each pulse begins at the peak of the input waveform and has a duration that is determined by the turn-on time of the diode and the RC time constant (Fig. 16.6*b*).

EXAMPLE 16.4 Design a positive peak detector that can respond to a 200 mV pp, 5 MHz sinusoidal input signal.

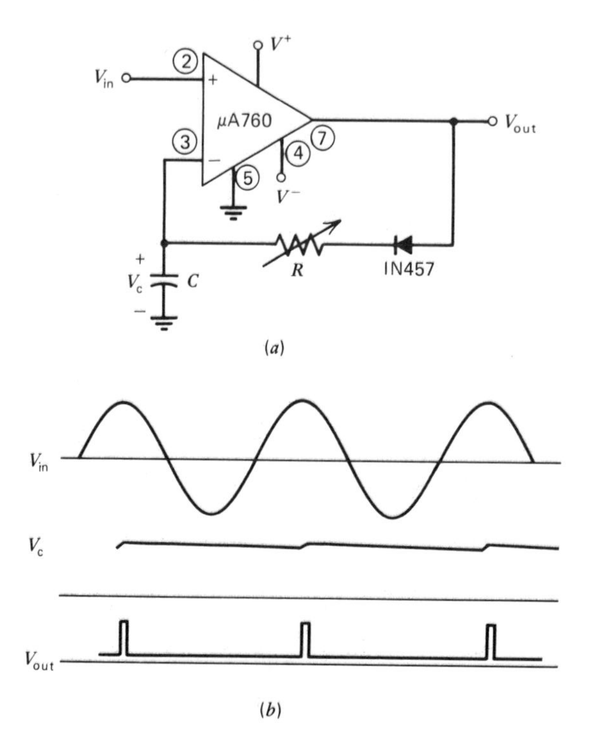

Figure 16.6 A positive peak detector using the μA760. (*a*) Circuit. (*b*) Waveform relationships.

PROCEDURE

1 The circuit shown in Fig. 16.6*a* is used to accomplish the peak detector function. The waveforms V_{in}, V_c, and V_{out} are given in Fig. 16.6*b*. Because the period of the incoming signal is very short (200 ns), a high-speed comparator is required. The μA760, which has response and propagation times of better than 25 ns and 12 ns, respectively, was chosen for this application.

2 In order for an output pulse to be generated at the right time, the capacitor must discharge to a level low enough to allow the input signal to be larger than the threshold level in a time slightly larger than the propagation delay of the comparator. Let this time (τ) equal, to say, 15 ns; then, the change in voltage across capacitor C, ΔV_c, is:

$$\Delta V_c = V_p \left[1 - \cos\left(\frac{4\,\tau}{T} \times 90°\right)\right] \tag{16.9}$$

$$= 100 \times 10^{-3} \left[1 - \cos\left(\frac{4 \times 15 \times 10^{-9}}{200 \times 10^{-9}} \times 90°\right)\right]$$

$$\approx 10 \text{ mV}$$

where V_p is the peak voltage.

3 The size of the capacitor can now be determined:

$$C = \frac{I_B}{\left(\dfrac{\Delta V}{\Delta t}\right)} \tag{16.10}$$

$$= \frac{8 \times 10^{-6}}{\left(\dfrac{10 \times 10^{-3}}{200 \times 10^{-9}}\right)}$$

$$= 160 \text{ pF}$$

where $I_B = 8$ μA is the input bias current to a μA760.

4 A 1N457 diode was selected because of its relatively slow turn-on time, to widen the output pulse width. The output pulse width may be further increased by placing a resistor (R) in series with the diode to increase the RC time constant.

16.3 GENERAL GUIDELINES FOR USING VOLTAGE COMPARATORS

1 Stray capacitance between the output and the offset-adjustment terminals can occasionally cause oscillations. During circuit layout, every attempt should be made to keep these leads separated as much as possible. The same is true of the input and output leads.

2 When driving the input of a comparator from a low-impedance source, a limiting resistance should be placed in series with the input leads to limit the input current. This is especially important if the input voltage can exceed either the supply voltages of the comparator, or if the input and comparator circuits are powered independently.

3 If there is a possibility of power supply reversal, clamping diodes with adequate peak-current handling capability should be installed across the power supply lines.

16.4 MOS/CCD MEMORY DRIVERS

In the past ten years, MOS memories have progressed from medium-performance, p-channel devices to high-performance, n-channel devices. CCD memories also have improved, exhibiting high performance, high density, and cost effectiveness. Advances in both technologies have extended overall system capabilities and have allowed greater flexibility in system configurations; however, inherent in these semiconductor memories are stringent interface requirements.

Today, many MOS (Chap. 6) and CCD (Chap. 5) memory devices must interface with TTL circuits, but they require MOS logic levels. A basic function of memory drivers used in these systems is to translate the TTL signals into signal levels acceptable to the memory devices. Also, because MOS and CCD inputs are primarily capacitive, the driver must be capable of supplying large peak currents rather than large dc currents. Large peak currents are needed to rapidly charge and discharge the total capacitance seen by the driver. The following examples illustrate the use of MOS/CCD memory drivers and some of the practical aspects of semiconductor memory interfacing.

Figure 16.7 shows the 9643 driving the chip enable (CE) inputs for an array of MM5280s. The 9643 is a dual positive logic AND TTL-to-MOS driver with separate address inputs and a common strobe. The MM5280 is a 4096 word by 1-bit dynamic MOS RAM. Only the chip enable input of the MM5280 requires MOS levels; all other inputs are TTL compatible.

EXAMPLE 16.5 Assume a 500 ns period and a 350 ns on-time for the chip enable signal. Transition times should be of the order of 20 ns. Calculate how many MM5280 chip enable (CE) inputs can be driven by a single 9643. The maximum junction temperature, $T_{j(max)}$, of the 9643 is 150°C and the thermal resistance, θ_{jA}, is 90°C/W. The maximum ambient temperature, $T_{A(max)}$, is 70°C.

PROCEDURE

1 Power dissipation capability of the driver package determines the maximum number of memory elements that can be driven at a specific operating frequency (assuming other specifications are met, such as transition time). As the number of memory elements driven, or the operating frequency is increased, the power dissipation in the driver increases because of ac losses. These losses stem from the load capacitance and internal switching. As the power dissipated by the driver increases, the junction temperature of the IC increases as well. For good reliability, the junction temperature should be kept below a critical value—usually specified as 150°C.

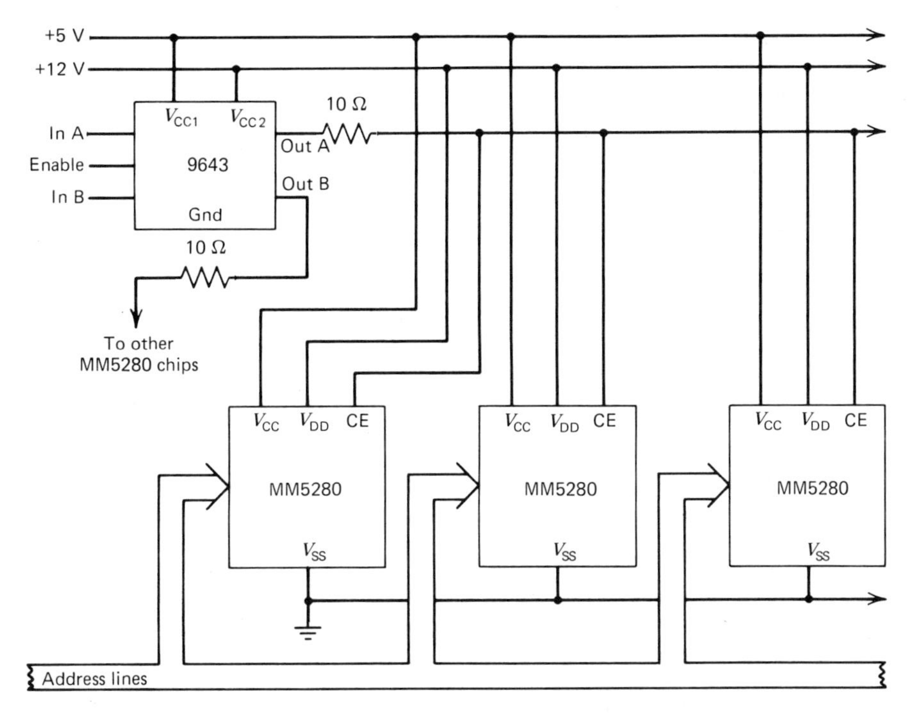

Figure 16.7 A 9643 memory driver used for driving MM5280 4K dynamic RAMs.

In most cases, because of driver loading, this maximum junction temperature is reached before other driver limitations are reached, such as transition time. Temperature, therefore, becomes the limiting factor in determining the maximum load permitted at a given operating frequency.

2 The maximum power handling capability of the package, $P_{d(max)}$, is:

$$P_{d(max)} = [T_{j(max)} - T_{A(max)}]/\theta_{jA} \tag{16.11}$$

Substitution of given values in Eq. 16.11 yields:

$$P_{d(max)} = (150°C - 70°C)/(90°C/W) = 889 \text{ mW}$$

3 The average dc power, P_{dc}, is:

$$P_{dc} = P_{in} + P_Q \tag{16.12}$$

where P_{in} is the input power and P_Q is the quiescent power. The major portion of the average dc power dissipated in the package is the quiescent power, P_Q, used to bias the driver to the proper state. The input power in this case is usually negligible.

The quiescent power is proportional to the supply voltages, supply currents, and the duty cycle, D. The duty cycle is expressed by:

$$D = T_{HIGH}/(T_{HIGH} + T_{LOW}) \tag{16.13}$$

where T_{HIGH} is the time the output is in a high state and T_{LOW} is the time the output is in a low state. Their sum is equal to the period of the switching signal. Hence, substituting the given values in Eq. 16.13 yields:

$$D = 350/(350 + 150) = 0.7$$

4 The quiescent power may be calculated from:

$$P_Q = [V_{CC1} \times I_{CC1(H)} + V_{CC2} \times I_{CC2(H)}] \times D + [V_{CC1} \times I_{CC1(L)} + V_{CC2} \times I_{CC2(L)}] \times (1 - D) \quad (16.14)$$

where I_{CC1} is the supply current for V_{CC1} and I_{CC2} is the supply current for V_{CC2}. Letters L and H refer to the low- and high-output conditions, respectively.

For the 9643, $I_{CC1(L)} = 15$ mA, $I_{CC1(H)} = 9$ mA, $I_{CC2(L)} = 5.5$ mA, and $I_{CC2(H)} = 12$ mA. Substituting the values in Eq. 16.14 gives:

$$P_Q = (5 \times 9 + 12 \times 5.5) \times 0.7 + (5 \times 15 + 12 \times 5.5) \times (1 - 0.7)$$
$$= 120 \text{ mW}$$

5 Switching losses in a driver occur in the output stage because of the charging and discharging of internal capacitors and turn-off delays between switching elements. Switching losses are usually not indicated on data sheets and must be determined empirically for the frequency of interest. The circuit in Fig. 16.8 can be used to estimate the switching losses in the 9643 chip.

The average power dissipation, P_S, from switching losses is estimated by averaging the current spikes during the transistions over their duration, multiplying by the supply voltage, and averaging the result over the entire period (see Fig. 16.9).

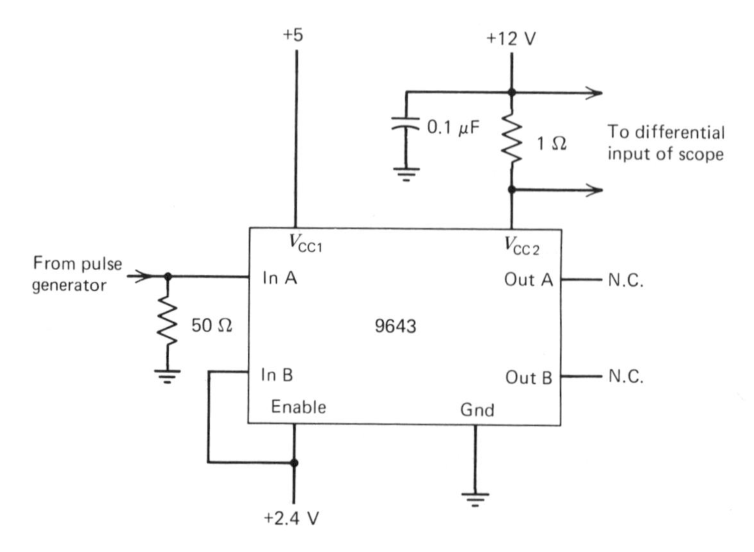

Figure 16.8 Test circuit for determining switching losses in a 9643 driver.

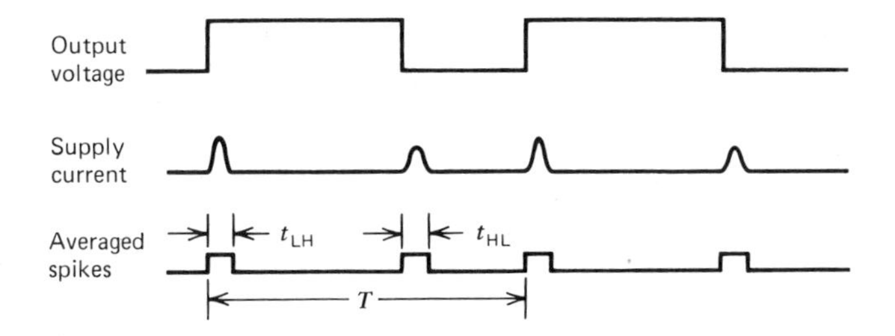

Figure 16.9 Typical waveforms used to determine switching losses.

The result may be expressed by:

$$P_S = \frac{t_{HL}i_{HL}V_{CC2} + t_{LH}i_{LH}V_{CC2}}{T} \tag{16.15}$$

where $t_{HL}i_{HL}$ is the area under the supply current pulse when the output makes a transition from the high to the low state, and $t_{LH}i_{LH}$ is the area under the supply current pulse when the output makes a transition from the low to the high state. For the 9643, $t_{HL}i_{HL} = 7.5 \times 10^{-12}$ and $t_{LH}i_{LH} = 5 \times 10^{-12}$. Substitution of values in Eq. 16.15 yields $P_S = 300$ mW/driver. Because there are two drivers in a package, the total switching loss is 600 mW.

6 The maximum loss from capacitive loading, $P_{C(max)}$, is:

$$\begin{aligned} P_{C(max)} &= P_{d(max)} - P_Q - P_{S(total)} \\ &= 889 - 120 - 600 = 169 \text{ mW} \end{aligned} \tag{16.16}$$

7 The maximum capacitive load, $C_{L(max)}$, is:

$$C_{L(max)} = \frac{P_{C(max)}}{V^2 f} \tag{16.17}$$

where f is the frequency of operation and is equal to one over the period. In this example, $f = 1/(500 \text{ ns}) = 2$ MHz. Substitution of values in Eq. 16.17 gives $C_{L(max)} = 586$ pF. Because there are two drivers in the package, we assume equal loading. The maximum load capacitance for each driver, therefore, is 586/2 or 293 pF.

8 The input capacitance, C_{in}, across the chip enable terminal with respect to ground is 15 pF for the MM5280. The number of memory unit loads, N, per driver is:

$$\begin{aligned} N &= C_{L(max)}/C_{in} \\ &= 293/15 = 19 \end{aligned} \tag{16.18}$$

Therefore, 19 is the maximum number of units that may be driven without exceeding the power dissipation limitations of the 9643.

9 After computing the value of N, we can then check the transition time to see if the driver is fast enough for the application. If it is, then the power dissipation of the package is the limiting factor. If not, N is reduced until the transition time requirement is satisfied. A check of the transition time specification for the 9643 indicates that with 300 pF loads, the transition times will be under 20 ns. This is within the specified requirements.

10 NOTE: In many memory system applications, the output level of the drivers tends to overshoot during switching transitions. Placing a resistor in series with the driver output can control overshoot, but the value of the resistor must be chosen carefully to effectively dampen the output overshoot without increasing the transition times beyond the acceptable limit of the memory system. Typical values for the resistor range between 10 and 50 Ω.

Figure 16.10 shows a 9645 driving the clock inputs of an F464. The 9645 is a high-speed driver intended to drive both dynamic NMOS RAMs and CCD memories. The 9645 features two common enable inputs and a clock control input. Its internal gating structure is organized so that all four drivers may be deactivated for standby operation, a single driver activated during read/write operation, or all four drivers activated during a refresh operation.

The F464 is a 65,534-bit dynamic CCD serial memory organized internally as 16 dynamic shift register blocks of 4096 bits each. Each of the shift register blocks is implemented using a serial-parallel-serial (SPS) interlaced structure (Fig. 16.11).

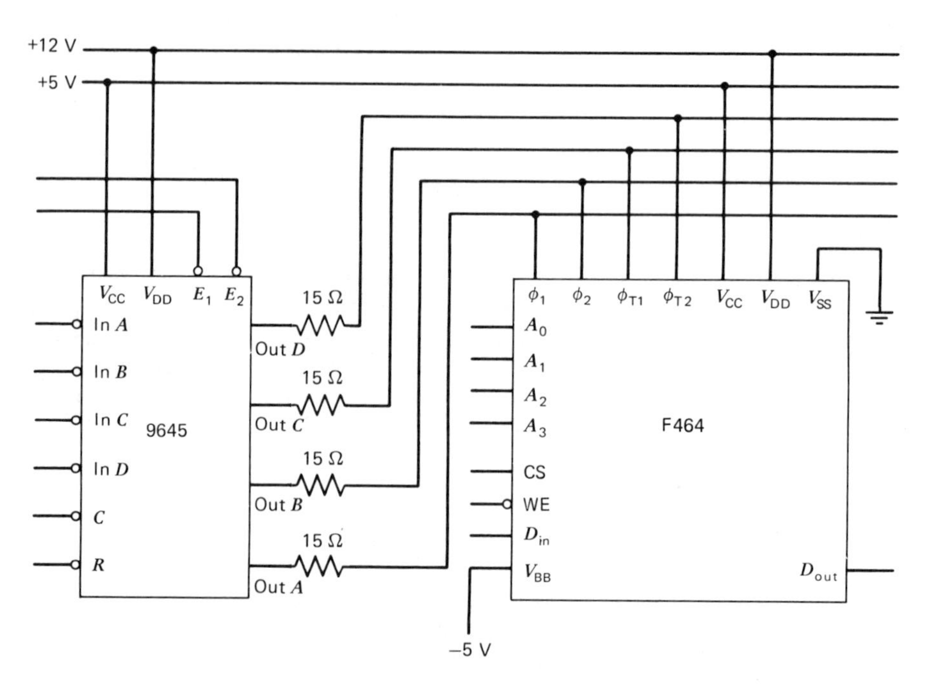

Figure 16.10 A 9645 memory driver used for driving a F464 CCD memory.

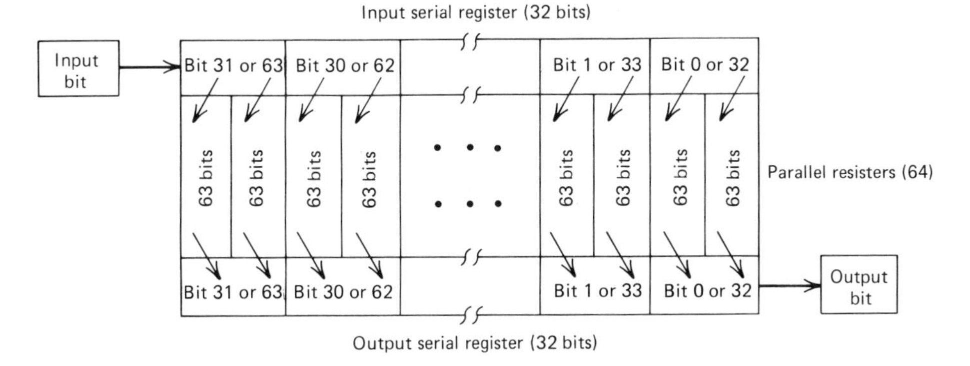

Figure 16.11 Serial-parallel-serial (SPS) interlaced structure of the F464 memory.

Two high-frequency serial clocks (ϕ_1 and ϕ_2) control the movement of data within the input and output serial registers and have a frequency equal to the data rate. Two low-frequency clocks (ϕ_{T1} and ϕ_{T2}) control the data transfers from the input register and to the output register, respectively. These clocks operate at $\frac{1}{32}$ the serial clock frequencies (Fig. 16.12).

EXAMPLE 16.6 Assume a data rate of 2 MHz, 0 to 70°C operation, and the 9645 supplied in a 16 pin plastic package. Clock pulse widths are 100 ns and transition times must be less than 20 ns. How many F464 CCD memories can a single 9645 drive? (Fig. 16.10.) Determine the power dissipation and time limitations of the 9645. $\theta_{jA} = 75°C/W$.

PROCEDURE

1 The power handling capability of the package, by Eq. 16.11, is:

$$P_{d(max)} = \frac{T_{j(max)} - T_{A(max)}}{\theta_{jA}}$$

$$= \frac{150°C - 70°C}{75°C/W} = 1067 \text{ mW}$$

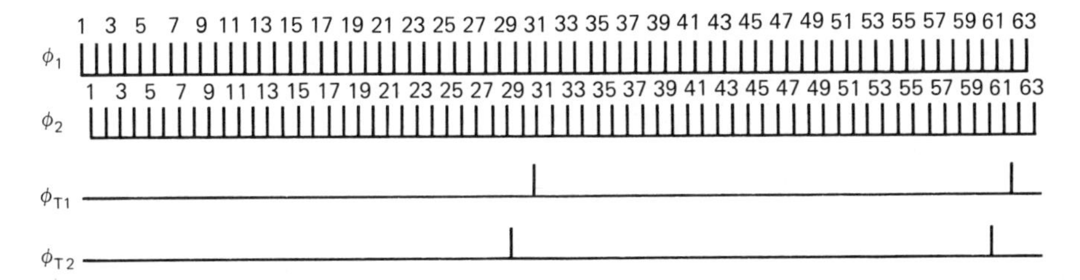

Figure 16.12 Clock relationships for a F464 memory.

2 The input power (P_{in}) once again is negligible. The duty cycle D, by Eq. 16.13 is:

$$D = \frac{T_{HIGH}}{T_{HIGH} + T_{LOW}} = \frac{100}{100 + 400} = 0.2$$

and P_Q, based on Eq. 16.14 is:

$$P_Q = [V_{CC} \times I_{CC(H)} + V_{DD} \times I_{DD(H)}] \times D + [V_{CC} \times I_{CC(L)} + V_{DD} \times I_{DD(L)}] \times (1 - D)$$

For the 9645, $I_{CC(H)} = 13$ mA, $I_{CC(L)} = 27$ mA, $I_{DD(H)} = 14$ mA, and $I_{DD(L)} = 12$ mA; hence,

$$P_Q = [5 \times 13 + 12 \times 14] \times 0.2 + [5 \times 27 + 12 \times 12]\,(1-0.2)$$
$$= 270 \text{ mW}$$

3 Once again the switching losses must be determined empirically (Fig. 16.13). Using similar calculations as in Ex. 16.5, we obtain: $P_S = 300$ mW/driver. Each of the high-speed (2 MHz) outputs exhibits a 300 mW loss. The switching losses of the two low-speed (125 kHz) outputs are negligible.

4 Losses stemming from capacitive loading by Eq. 16.16 are:

$$P_{C(max)} = P_{d(max)} - P_Q - P_{S(total)}$$
$$= 1067 - 270 - 600 = 197 \text{ mW}$$

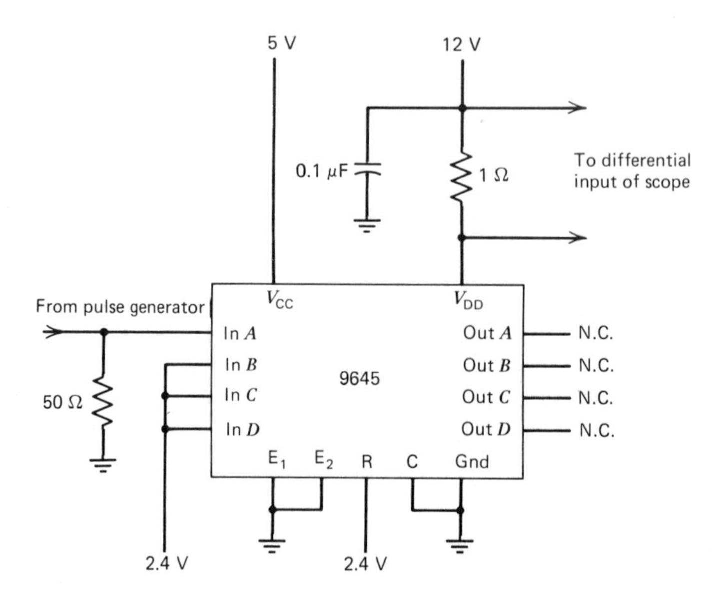

Figure 16.13 Test circuit for determining switching losses in a 9645 driver.

5 A check of the F464 data sheet indicates that the input capacitance for ϕ_1 and ϕ_2 is 100 pF and for ϕ_{T1} and ϕ_{T2} 30 pF. The losses, from Eq. 16.17, are:

$$P_{C1} = C_1 V^2 f_1 = 100 \times 10^{-12} \times (12)^2 \times 2 \times 10^6$$
$$= 28.8 \text{ mW/input}$$
$$P_{C2} = C_2 V^2 f_2 = 30 \times 10^{-12} \times (12)^2 \times 0.125 \times 10^6$$
$$= 0.54 \text{ mW/input}$$

$$P_{C(\text{total})} = 2\,P_{C1} + 2\,P_{C2} \qquad (16.19)$$
$$= 59 \text{ mW/Unit}$$

For each F464 driver, 59 mW is dissipated in the 9645 as a result of capacitive loading.

6 The number of memory unit loads, N, may be expressed by:

$$N = \frac{P_{C(\max)}}{P_{C(\text{per unit})}} \qquad (16.20)$$
$$= \frac{197}{59} \rightarrow 3 \text{ Units}$$

A check of the transition time specifications for the 9645 indicate that for:

$$C_{\text{LOAD}}\ (\phi_1 \text{ and } \phi_2) = 300 \text{ pF each}$$

and

$$C_{\text{LOAD}}\ (\phi_{T1} \text{ and } \phi_{T2}) = 30 \text{ pF each}$$

the transition times will be less than 20 ns.

16.5 GENERAL GUIDELINES FOR USING MEMORY DRIVERS

1 Crosstalk, or cross coupling of signals between lines, is a result of mutual capacitance between lines and between the lines and ground. As a result of this coupling, voltage spikes may be transmitted between lines during transition. The best approach to minimizing this problem is to reduce the coupling effects by keeping the lines as separated and as short as possible.

2 Power supply decoupling is very important in memory system applications. Large peak currents resulting from large capacitive loads, fast switching speeds, and distributed inductance in the supply line can result in voltage spikes feeding back into the supply. These voltage spikes may generate system errors. The problem can be minimized by keeping the length of the supply lines at a minimum and using low-inductance decoupling capacitors at each supply lead.

16.6 LINE DRIVERS AND RECEIVERS

The rapid growth of digital communication systems has increased the need for transmitting digital information over distances varying from only a few feet to several thousand feet. These lines are frequently located in noisy environments that can generate false information if steps are not taken to overcome the effects of the noise on the system. The noise margins of most logic circuits are adequate for the transmission of digital data over distances of a few inches; however, the transmission of error-free data over longer lines in noisy environments requires the use of special line drivers and line receivers. These devices are used to interface between the transmission line and logic circuitry.

It is possible to design a wide variety of interfaces. To increase intersystem compatibility a number of organizations have defined standard interfaces. One such organization is the Electronic Industries Association (EIA). The EIA has formulated a set of specifications governing the electrical characteristics of line interface circuits used for the transmission of serial binary signals between data terminal equipment (DTE) and data communications equipment (DCE). These EIA standards, RS-232-C, RS-423, and RS-422 (Table 16.1) specify the electrical characteristics for both balanced and unbalanced circuits and are the subject of this section.

Interface circuits for data transmissions fall into two major categories: balanced and unbalanced. The mode used depends upon the modulation rate, the distance between the driver and receiver (line length), noise and grounding conditions, and error rates that can be tolerated.

The most widely used standards for interfacing between data terminals and data communications equipment is the EIA RS-232-C, which defines a single-ended, bipolar, unterminated circuit. It is intended for serial data transmission over 150 feet or less at rates under 20 kilobauds. (One baud, Bd, is equal to one transition per second.) This single-ended circuit uses one conductor to carry the signal with the voltage referenced to a single return conductor (Fig. 16.14). This conductor may also be the common return for other signal conductors. It is the simplest way to send data, as it requires only one signal line per circuit. This simplicity, however, is often offset by an inability to discriminate between a valid signal produced by the driver and the sum of the driver signal and externally induced noise signals, commonly called *crosstalk*. Also, operation in multiwire systems tends to produce radiated emissions on neighboring circuits.

Because noise and crosstalk are both directly proportional to transmission line length and bandwidth, the RS-232-C standard restricts both. To control radiated emission, it limits the slew rate of drivers to 30 V/μs, which works with the length limitation (50 feet), to limit reflections on unterminated lines.

EXAMPLE 16.7 Design a transmission interface to exchange data at a rate of 10 kBd over a distance of 40 feet. Noise is not a problem in the system design. Eight data lines are to be provided.

Table 16.1 **Key parameters of EIA specifications.**

Characteristics	*EIA RS-232-C*	*EIA RS-423*	*EIA RS-422*
Form of operation	single-ended	single-ended	differential
Maximum cable length	50 ft	2000 ft	4000 ft
Maximum data rate	20 kilobauds	300 kilobauds	10 megabauds
Driver maximum output voltage, open circuit	±25 V	±6 V	6 V between outputs
Driver minimum output voltage, loaded output	±5 to ±15 V	±3.6 V	2 V between outputs
Driver minimum output resistance, power off	R_o = 300 Ω	100 μA between −6 to +6 V	100 μA between +6 and −0.25 V
Driver maximum output, short-circuit current I_{SC}	±500 mA	±150 mA	±150 mA
Driver output slew rate	30 V/μsec max	slew rate must be controlled based upon cable length and modulation rate	no control necessary
Receiver input resistance R_{IN}	3 to 7 kΩ	⩾4 kΩ	⩾4 kΩ
Receiver maximum input thresholds	−3 to +3 V	−0.2 to +0.2 V	−0.2 to +0.2 V
Receiver maximum input voltage	−25 to +25 V	−12 to +12 V	−12 to +12 V

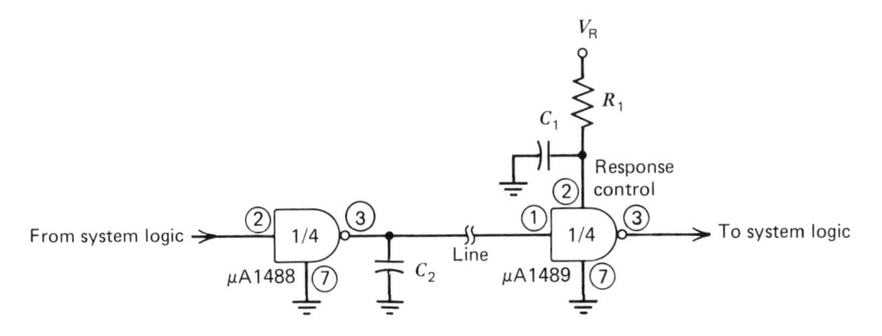

Figure 16.14 An RS-232-C interface using the μA1488 driver and μA1489 receiver.

PROCEDURE

1 The speed and distance limitations, as well as lack of noise within the system, dictate that a single-ended interface may be used, in particular, an RS-232-C interface.

2 Because eight lines are required, the μA1488 quad line driver and a μA1489 line receiver were selected. Both meet the RS-232-C standard (Fig. 16.14).

3 The RS-232-C specification limits the driver output to 30 V/μs. The inherent slew rate of the μA1488 is much too fast for this requirement. The current limited output of the driver, therefore, is used to control the slew rate by connecting a capacitor to each driver output. The value of the required capacitor can be easily determined by Eq. 16.10:

$$C_2 = \frac{I_B}{\frac{\Delta V}{\Delta t}} = \frac{10 \times 10^{-3}}{30 \times 10^6} = 333 \text{ pF}$$

4 The μA1489 receiver has internal feedback from the second stage to the input stage, providing input hysteresis for improved noise immunity. Resistor R_1 and reference voltage V_{ref} can be adjusted, as indicated on the manufacturer's data sheet, to vary the threshold point and the hysteresis for optimal reception, as illustrated in Fig. 16.15.

5 Connecting a capacitor to the response control node allows filtering of high-frequency, high-energy noise pulses. Capacitor values range from 10 to 500 pF.

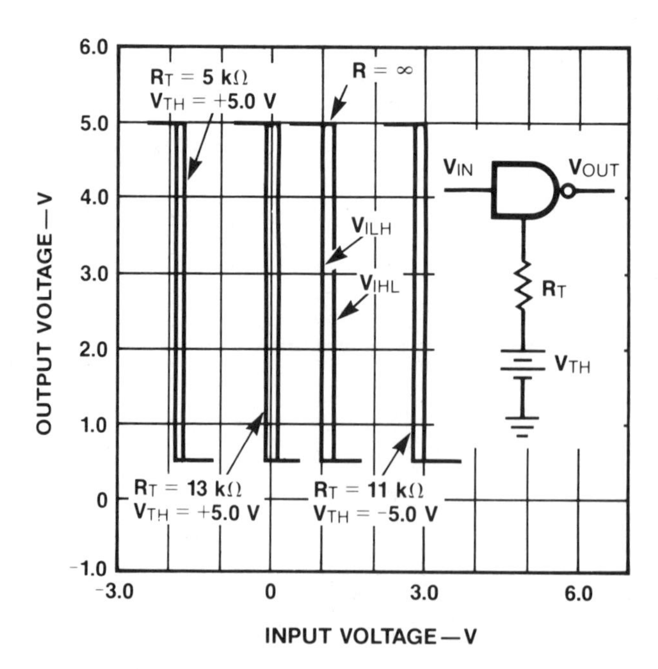

Figure 16.15 Input threshold voltage adjustment for the μA1489 receiver. (Courtesy Fairchild Camera and Instrument Corp.)

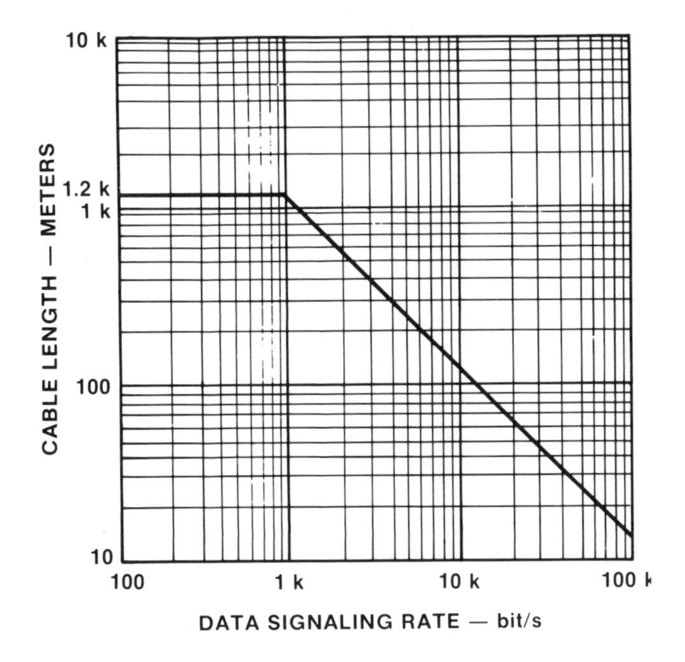

Figure 16.16 Cable length as a function of data signaling rate for an unbalanced interface using 24 AWG twisted-pair cable. (Courtesy Fairchild Camera and Instrument Corp.)

To allow the newer data communication systems to utilize their full performance capabilities, the EIA issued RS-422 and RS-423. They define interface standards for higher data rates and longer distances than possible with RS-232-C (Table 16.1). RS-423 defines a single-ended, bipolar voltage, unterminated circuit. It extends the distance and data rate capabilities to 4000 feet and 3 kBd. Higher rates of up to 300 kBd are permitted at a distance of 40 feet, as indicated in Fig. 16.16.

RS-422 defines a balanced-voltage differential operation capable of significantly higher performance than single-ended configurations. It can accommodate rates of up to 100 kBd at 4000 feet, or up to 10 megabaud (MBd) over shorter distances. These improvements stem from the advantages of a balanced configuration. (Fig. 16.17.)

EXAMPLE 16.8 Design a system to exchange data at a 10 kBd rate. The estimated line length is 100 meters.

PROCEDURE

1 At this line length and data rate either a single-ended or differential system may be employed. If noise is not a critical factor, then the more economical single-ended system can be used.

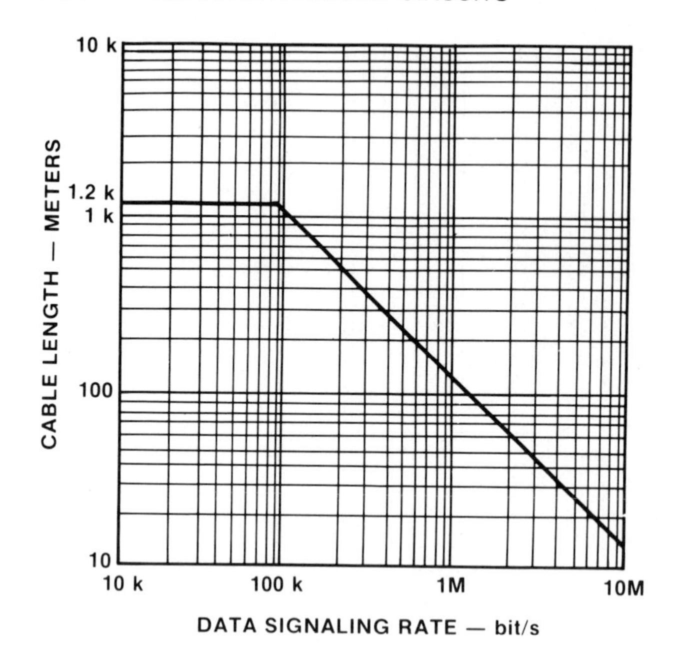

Figure 16.17 Cable length as a function of data signaling rate for a balanced interface using 24 AWG twisted-pair cable. (Courtesy Fairchild Camera and Instrument Corp.)

2 Checking the data signal versus line length curve (Fig. 16.16), it is determined that EIA RS-423 can meet the data rate and line length requirements of this system.
3 The 9636A dual programmable slew rate line driver and 9637A dual differential line receiver are selected for this application (Fig. 16.18).
4 A termination resistor is not required. If employed, however, its value should be made equal to the characteristic impedance of the line.
5 RS-423 requires slew rate limiting of the driver to limit crosstalk. As defined by RS-423, for minimum crosstalk, the rise and fall times (t_r, t_f) $\leq$ 0.3/Baud rate. Hence, the maximum rise or fall time that may be used is:

$$t_r,\ t_f = \frac{0.3}{10{,}000} = 30\ \mu\text{s}$$

The slew rate of the 9636A is controlled by the waveshaping resistor R_{ws}. The value of R_{ws} may be read directly from Fig. 16.19:

$$R_{ws} \approx 300\ \text{k}\Omega$$

6 Diode D_1 is inserted into the negative supply line to satisfy a test requirement specified in RS-423.

EXAMPLE 16.9 Design a system to exchange data at a 10 kBd rate. The estimated line length is 1000 meters.

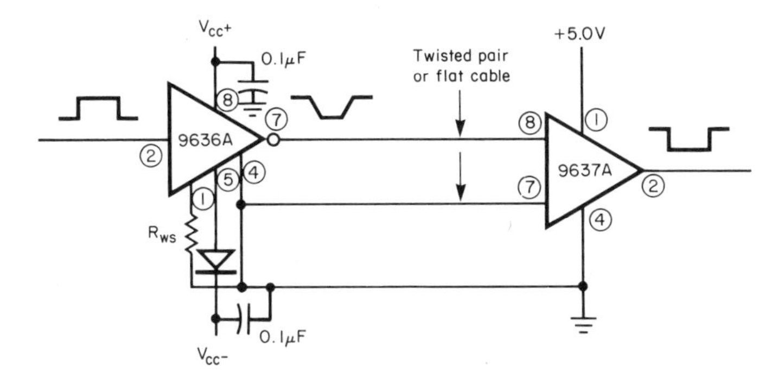

Figure 16.18 An RS-423 system application using the 9636A driver and 9637A receiver. (Courtesy Fairchild Camera and Instrument Corp.)

PROCEDURE

1 Because of the line distance involved, a differential system is preferred. Lines are balanced in a differential system; therefore, externally induced noise appears equally on both inputs to the line receiver. The receiver uses a differential input stage to respond primarily to a differential signal, rejecting the common mode noise signal.
2 Checking the data signal rate versus line length curve (Fig. 16.17), it is determined that EIA RS-422 can meet the requirements of line length and data rate for this application.

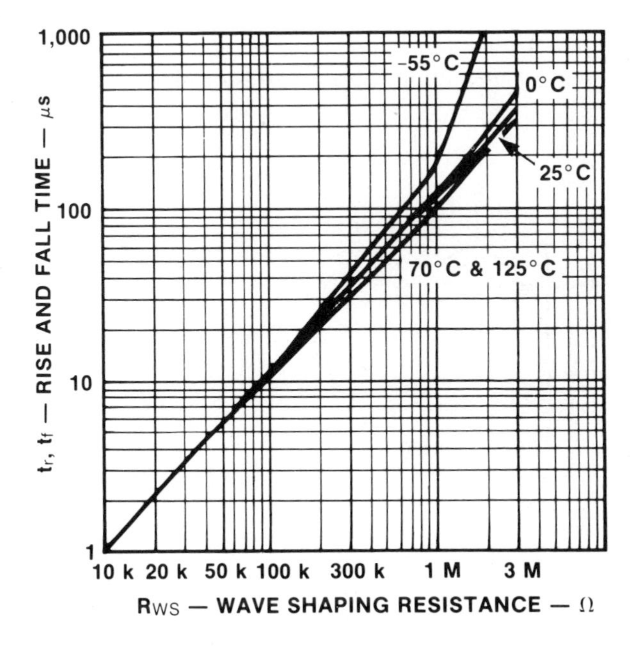

Figure 16.19 Rise and fall times as a function of the waveshaping resistor, R_{WS}, for the 9636A driver. (Courtesy Fairchild Camera and Instrument Corp.)

3 A suitable RS-422 driver and receiver must be selected. The 9638 dual channel line driver and 9637A dual differential line receiver are employed (Fig. 16.20). The 9637A has a wide common mode range (±7 V) and built-in hysteresis.
4 The 35 mV hysteresis of the receiver (Fig. 16.21) provides additional noise immunity to the system during threshold transition periods.
5 It is recommended that a termination resistor be placed across the receiver inputs to reduce, or eliminate, possible reflections occurring from impedance mismatches. The value of the resistor should equal the characteristic impedance of the line.

16.7 GENERAL GUIDELINES FOR USING LINE DRIVERS AND RECEIVERS

When using line drivers and receivers to meet the various EIA interface standards, consideration should be given to some of the problems that may arise because of effects in the interconnecting cable and cable termination.

1 The maximum line length at which a line driver and receiver will operate is governed by the modulation rate of the data. The modulation rate, or baud rate, is defined as the reciprocal of the minimum data unit interval. Characteristics that influence the maximum modulation rate and line length are related to signal quality and include: maximum acceptable signal distortion, amount of noise induced in the line, ground potential differences between the driver and receiver, and imbalances between signal conductors and ground return lines. As the length of the line increases, these characteristics have a greater effect on signal quality deteriorating.
2 RS-422 and RS-423 do not specify maximum cable lengths; however, they do provide conservative guidelines for maximum cable lengths versus data rates.

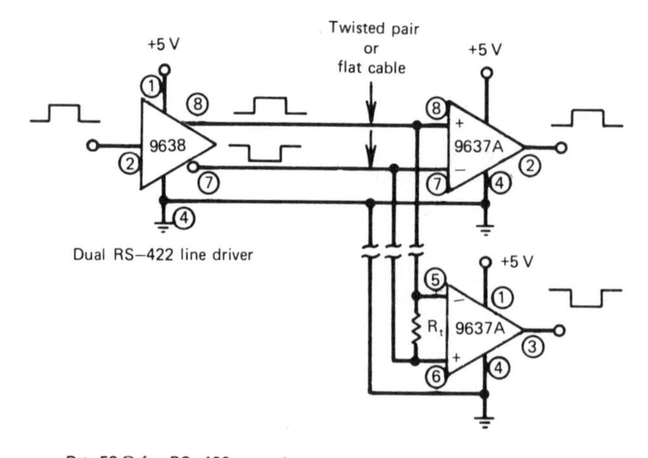

Figure 16.20 An RS-422 system application using the 9638 driver and 9637A receiver. (Courtesy Fairchild Camera and Instrument Corp.)

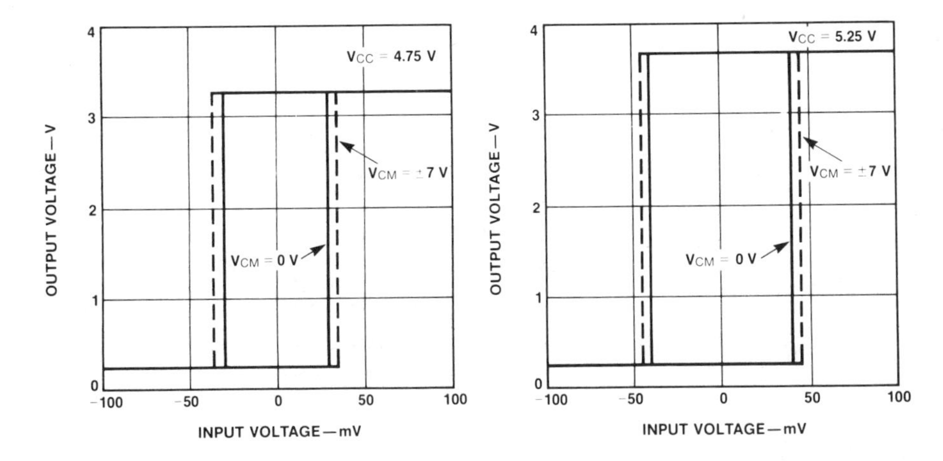

Figure 16.21 Typical input/output characteristics for the 9637A receiver. (Courtesy Fairchild Camera and Instrument Corp.)

This is based on signal rise and fall times, equal to, or less than, one-half unit interval at the applicable modulation rate. This ensures reliable transmission of data where the error rate is minimized. A maximum voltage loss between the driver and receiver of only 6 dB will be realized.

3 The curves of Figs. 16.16 and 16.17 were constructed using empirical data measurements on 24 AWG twisted-pair line. When using wire other than 24 AWG twisted-pair, the following rule of thumb may be used: the unit interval should be greater than twice of 10 to 90% of the rise and fall times of the line. Adherence to this rule will keep the peak-to-peak jitter to less than 5% of the unit interval.

4 Line termination is optional with RS-422 and RS-423, but is recommended to minimize noise and reflections that may appear at the driver and receiver. The object of line termination is to match the characteristic impedance of the line. At lower modulation rates, where threshold crossing ambiguity and signal rise time are not critical, the line need not be terminated.

16.8 REFERENCES

EIA, *Standard RS-232-C,* June 1981.

EIA, *Standard RS-422A,* December 1978.

EIA, *Standard RS-423-A,* December 1978.

Fairchild, *Linear Interface Data Book,* 1978.

Intel, *Memory Design Handbook,* 1977.

Motorola, "Revised Data Interface Standards," AN-781.

NCR, *Data Communication Concepts,* 1971.

National Semiconductor, "Applying Modern Clock Drivers to MOS Memories," AN-76.

National Semiconductor, "Comparing the High Speed Comparators," AN-87.

CHAPTER 17

Phase-Locked Loops for RF Frequency Synthesizers

John D. Hatchett
Principal Staff Engineer
Motorola Inc.
Semiconductor Products Sector

17.1 INTRODUCTION

A frequency synthesizer is a combination of circuit functions that generate, or synthesize, many output signals from only a few input signals. Preferably, only one input signal is required and typically, a single output signal out of several hundred or thousand possibilities is generated at a time. The signal selected is dictated by the user through a program to the synthesizer. The selected signal will generally be much higher in frequency than the signal or signals from which it is derived and, usually, it must also exhibit high stability and excellent spectral purity. The need to generate such signals is common to many applications, including most instrumentation and communications equipment (Chap. 14).

The phase locked loop (PLL) offers an ideal solution to frequency synthesis. Because only one input signal is often required, the number of crystals and circuits can be greatly reduced with a resulting savings in both cost and space. Depending on the number of output signals, their frequencies, and performance requirements, the reduction in crystal count can be dramatic. This is depicted in Fig. 17.1 which compares a non-PLL and a PLL solution for a frequency synthesizer. Such a PLL synthesizer, for example, is used in a CB radio where 40 channel frequencies are derived from a single crystal frequency.

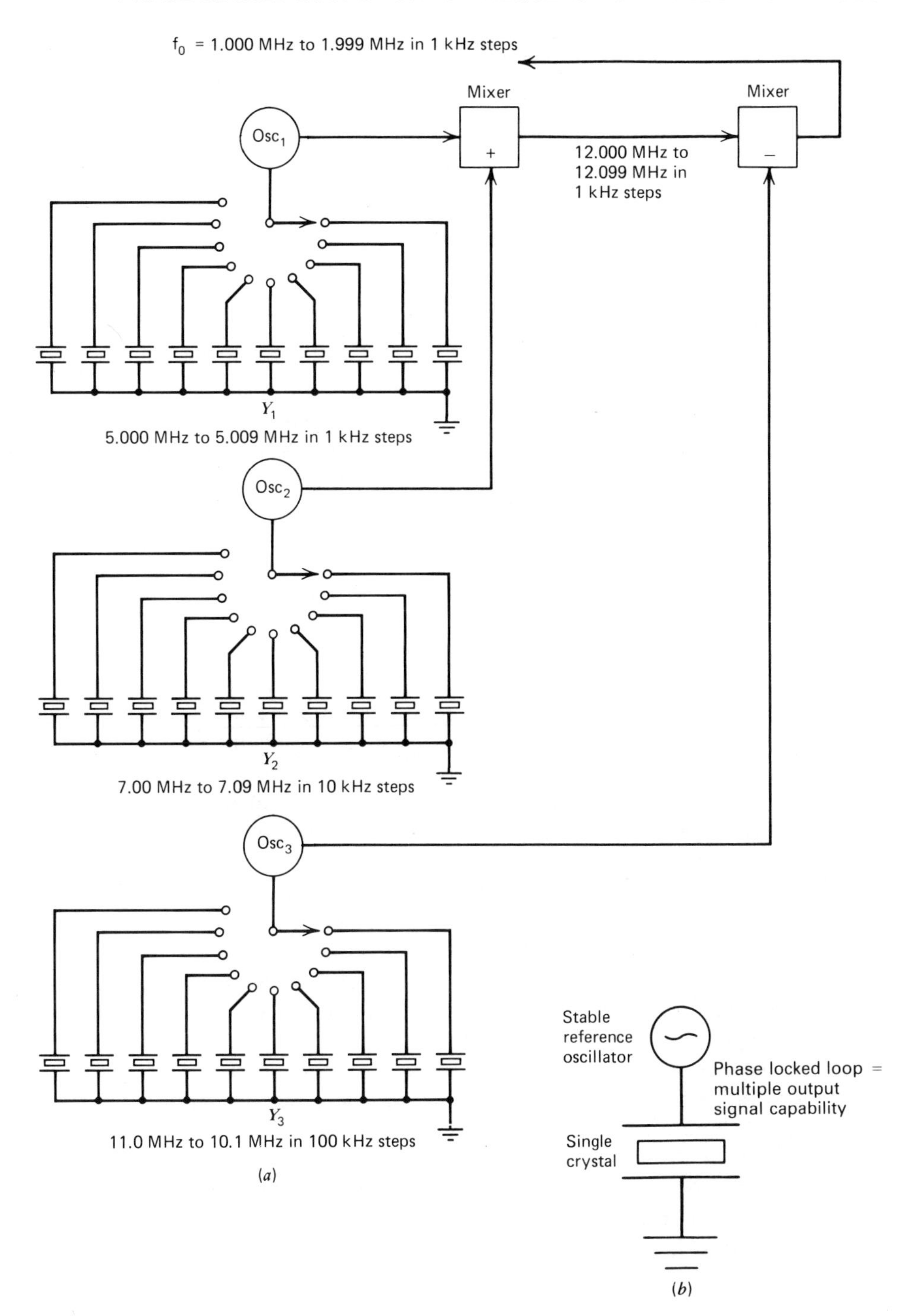

Figure 17.1 Comparison of frequency synthesis approaches. (*a*) Non-PLL solution. (*b*) PLL solution.

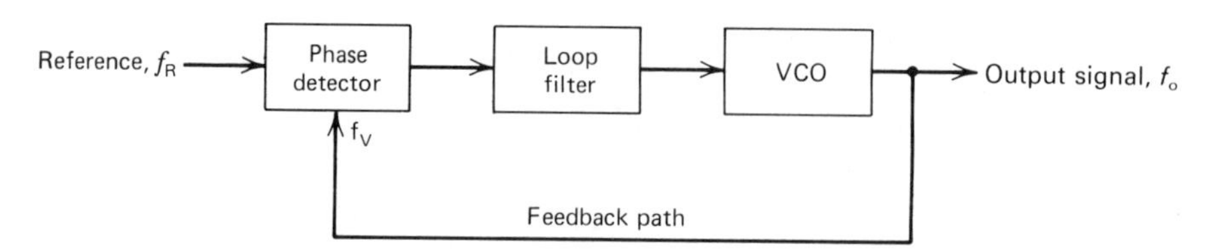

Figure 17.2 Block diagram of a basic phase-locked loop.

17.2 BASIC OPERATION OF THE PLL

The PLL is essentially a closed loop electronic servomechanism whose output is locked onto, and will track with, a reference signal (Fig. 17.2). It can be accomplished using three basic elements: a phase detector, a loop filter, and a voltage controlled oscillator (VCO). A reference signal, f_R, also is required as a second input to the phase detector.

The detector compares the phase between the reference signal, f_R, and the feedback signal f_V, which is returned from the VCO output. An error signal is produced by the phase detector that is proportional to the phase difference between the two signals. The error signal is filtered (and sometimes amplified) by the loop filter to smooth and shape it into a voltage suitable for controlling the VCO. When applied to the VCO, it causes the VCO's phase, and therefore its frequency* to change in a direction to establish a constant phase difference (typically "zero") between the two signals, f_R and f_V applied to the detector.

It also follows that f_V will be maintained at a frequency equal to that of f_R. This results in a "locked" condition. If the VCO's frequency starts to drift, a phase/frequency shift between f_V and f_R results and the detector's output signal increases or decreases just enough to force the VCO back on frequency, thereby preserving the locked condition. A potentially unstable VCO is thus made to approach the stability of a very stable reference signal.

In Fig. 17.2, the output signal is always maintained at the same frequency as that of the reference. The synthesis of various output frequencies, therefore, is not possible. To turn this system into a frequency synthesizer at least one additional function, such as a programmable counter, must be added in the feedback path (Fig. 17.3). The same rules still apply in that f_V and f_R are maintained equal in frequency. Now, however, because the output is divided by N to produce f_V, the output frequency, f_o, is N times f_V, or $f_o = Nf_V = Nf_R$. Because the value of N can be changed by programming the counter, a host of output signals is possible. Each output signal appears individually and is separated in frequency from an adjacent frequency by an amount equal to the reference frequency, f_R. The result is a *frequency synthesizer*.

* Phase and frequency are related because frequency is defined as the rate of change of phase with respect to time.

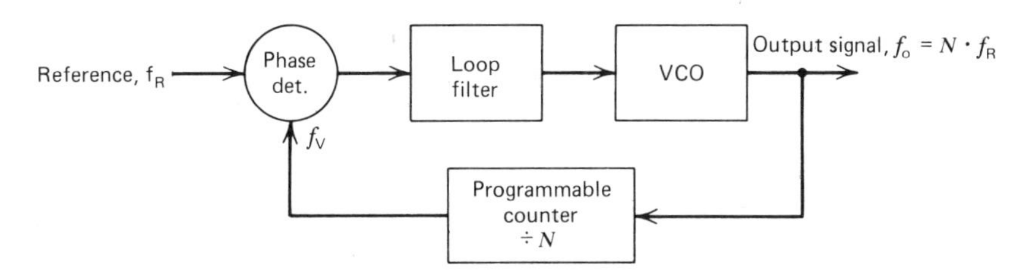

Figure 17.3 Basic PLL employing a divide-by-N counter in the feedback path.

17.3 MAJOR PLL BUILDING BLOCKS—CHARACTERISTICS AND IMPLEMENTATION

The major functions of a PLL RF frequency synthesizer are shown in Fig. 17.4. They can be implemented using individual ICs for each building block, LSI, or discrete components.

Reference Oscillator

The reference oscillator must generate a signal of suitable frequency accuracy and stability (both short and long term) to satisfy the system requirements. Low noise performance with respect to both *noise floor* and *close-in noise* is also important. (Noise floor refers to noise components falling at frequencies far removed from the oscillator frequency. Close-in noise is noise components present at frequencies near the oscillator's frequency.)

For frequency synthesizer designs employing counters in the feedback path, the VCO's frequency accuracy in percent corresponds to that of the reference oscillator. Requirements for ± 0.1 to $\pm 0.00015\%$ accuracies are common. Invariably, this dictates a crystal controlled reference source. Temperature compensation of the crystal and oscillator components may be needed. For more stringent requirements, an oven is used to maintain all frequency determining

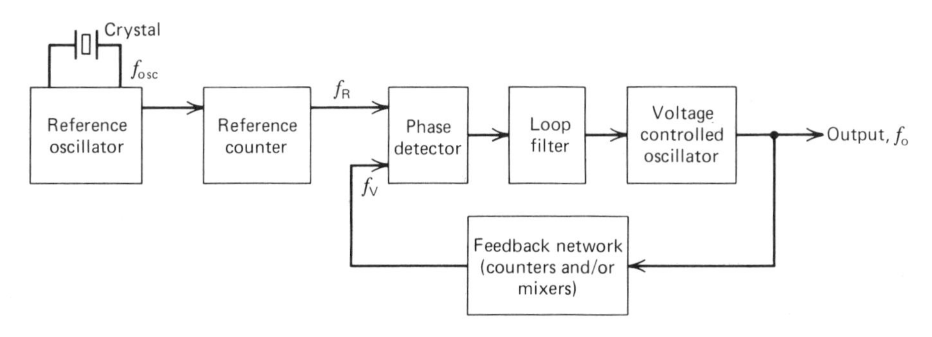

Figure 17.4 Major building blocks of a PLL RF frequency synthesizer.

components at a fixed temperature somewhat above the highest expected ambient temperature.

The reference oscillator frequency, f_{osc}, is normally much higher than the reference comparison frequency, f_R. Typically, f_R is below 100 kHz and sometimes 1 kHz or less while reference oscillator frequencies from 3 to 15 MHz are common. The oscillator frequency must be selected to allow it to be readily divided to form f_R. Otherwise, the choice of frequency is largely a consideration of the crystal parameters and how they relate to the system requirements. Depending on the application, the crystal's temperature sensitivity, aging characteristics, ruggedness, size, and cost will vary in importance.

A crystal controlled oscillator employs a crystal in its feedback path to provide a very high Q (narrow bandwidth) which enhances frequency stability far beyond what is possible using only RC or LC components. Crystal oscillators can be divided into two groups. One group requires the feedback function to be of low impedance and introduces little or no phase shift (Fig. 17.5). These are commonly called *series resonant*, even though the crystal may be pulled away from its actual series resonant frequency because of reactance presented to it by the other oscillator components. The other group requires a phase shift (approximately 180°) to be provided in the feedback path (Fig. 17.6). This type uses the crystal as a reactance (usually inductive) as part of the phase shifting

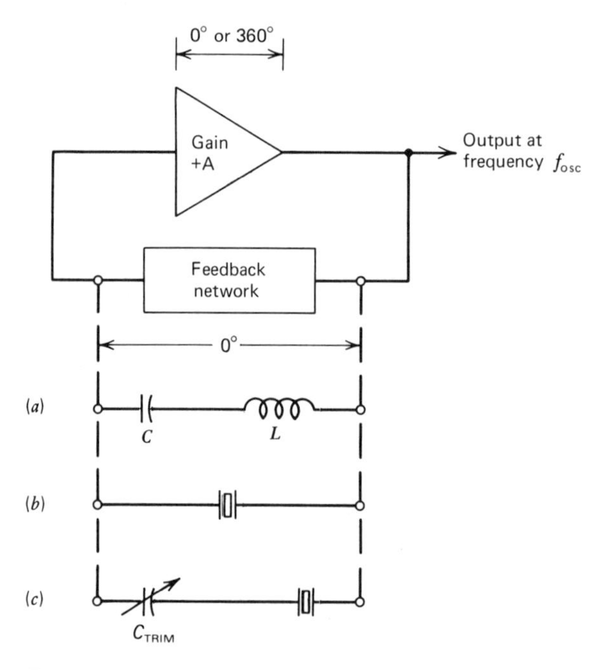

Figure 17.5 Series resonant oscillator with the frequency determining network being a (*a*) series resonant LC circuit, (*b*) crystal operating at its series resonant frequency, and (*c*) crystal operating on the inductive side of its series resonant frequency to offset the capacitive reactance of C_{TRIM}.

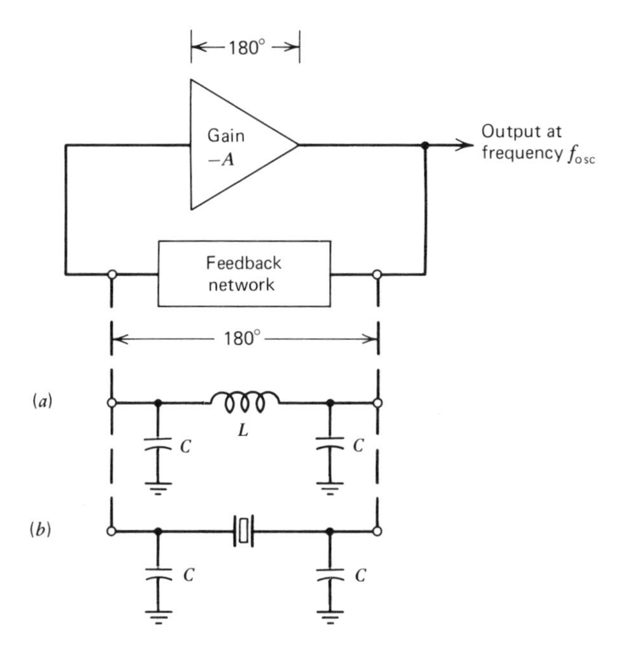

Figure 17.6 Parallel resonant oscillator with the required 180° phase shift in the feedback path provided by (*a*) LC circuit and (*b*) crystal operating as an inductive reactance.

network. These are referred to as *parallel resonant* oscillators because the frequency of oscillation is at, or near, the crystal's parallel resonant frequency.

IC logic gates and inverters have been used with varying degrees of success for the gain element in crystal oscillator designs. Best results, however, can usually be achieved by using ICs developed specifically for crystal oscillator applications. The MC12061/12561 series in Fig. 17.7 are representative examples of such devices. Their frequency of oscillation is established by a fundamental mode crystal placed between pins 5 and 6 and will be very near the crystal's series resonant frequency.

The amount of *frequency pull*, or deviation from the crystal's true series resonant point, will vary somewhat with different crystals. (An idea of what to expect is given in Table 17.1.) The MC12061/12561, which operate from 2 to 20 MHz, exhibit an equivalent inductance, L_{eq}, in series with the crystal. This establishes the frequency of oscillation, f_{osc}, slightly below series resonance for the crystal (Table 17.1). If desired, a capacitor, C_{eq}, can be placed in series with the crystal to resonate out L_{eq} and set the oscillator to the crystal's series resonant frequency. The value of C_{eq} required is:

$$C_{eq} = 1/[L_{eq}(f_{osc})^2(2\pi)^2] \tag{17.1}$$

The nominal value for L_{eq}, and therefore C_{eq}, will vary with frequency, as illustrated in Fig. 17.8. For frequency adjustment, C_{eq} can be made variable.

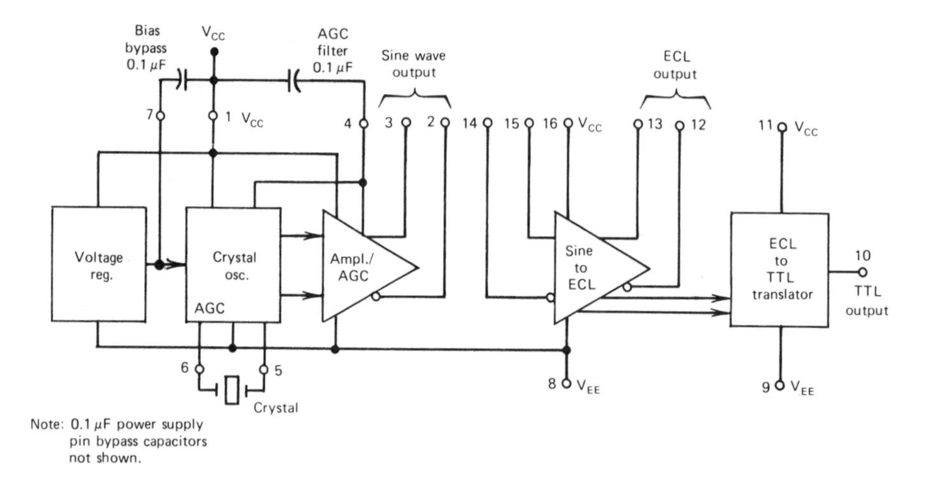

Figure 17.7 Block diagram of MC12060/12560 (100 kHz to 2 MHz) and MC12061/12561 (2 MHz to 20 MHz) crystal oscillator ICs.

EXAMPLE 17.1 The MC12061 is used to generate f_{osc} = 12.0 MHz. A crystal is specified for a series resonant frequency equal to 12.0 MHz. It is anticipated that a trimmer capacitor, C_{TRIM}, will be used in series with the crystal to offset frequency pulling effects of the IC and permit adjustment to the desired 12.0 MHz. Calculate the value needed for C_{TRIM}.

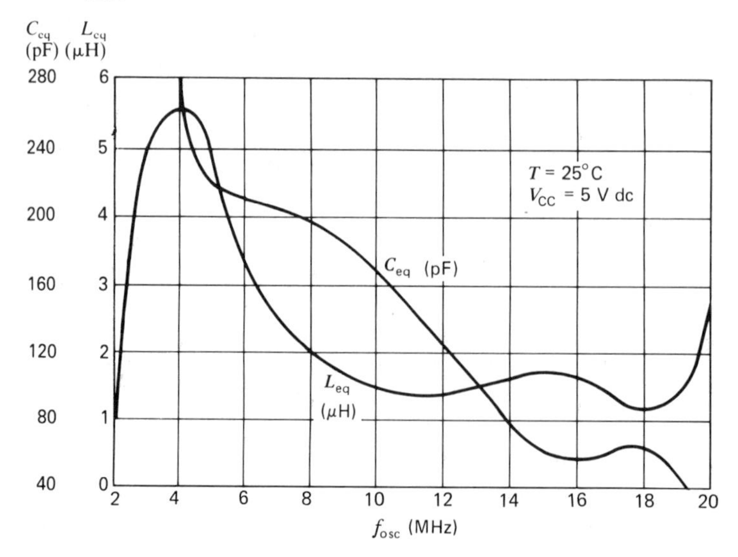

Figure 17.8 Nominal variation in reactance with operating frequency, f_{osc}, for the MC12061/12561 crystal oscillator ICs. (Courtesy Northern Engineering Laboratories.)

Table 17.1 **Crystal frequency pull in percent for MC12060/61 ICs (nominal, 25°C)**

Device	*MC12060*				*MC12061*					
Nominal crystal frequency (MHz)	0.100	0.200	0.500	1.00	2.00	2.50	8.08	13.41	18.75	20.0
Pull in percent from crystal's series resonant frequency	[a]	−0.0005	−0.0012	−0.0040	−0.03	−0.0002	−0.004	−0.01	−0.03	−0.05

[a] Less than 1 Hz, measurement limited by resolution of test equipment.

PROCEDURE

1 Because the crystal has been specified for series resonance equal to f_{osc}, C_{TRIM} must be chosen to just cancel the reactance added in series with the crystal by the 12061 chip.

2 From Fig. 17.8, for $f_{osc} = 12.0$ MHz, $L_{eq} = 1.4$ μH. Hence by Eq. 17.1,

$$C_{TRIM} = 1/[1.4 \times 10^{-6})(12 \times 10^{6})^2(2\pi)^2] = 126 \text{ pF}$$

3 A value of $C_{TRIM} = 126$ pF is too large to be conveniently provided by a trimmer capacitor. This large capacitance value results because there is an inverse relationship between the value of C_{TRIM} and its effect on frequency. Also, the 12061 has affected the crystal's series resonant point, f_{osc}, only slightly, thus resulting in the need for only a small frequency correction.

The solution is to specify the crystal to be series resonant slightly below, rather than at, the desired operating frequency. This can be done so that a realistic value for C_{TRIM} is obtained (Ex. 17.2).

EXAMPLE 17.2 The MC12061 of Ex. 17.1 is to be used with a 10 pF trimmer capacitor in series with the nominal 12 MHz crystal to adjust f_{osc} exactly to 12.0 MHz. Determine how the crystal's series resonant frequency should be specified.

PROCEDURE

1 The crystal should be specified so that the *series combination* of the crystal and a capacitor, C_{EQUIV}, is in series resonance at f_{osc}. The value of C_{EQUIV} is:

$$C_{EQUIV} = 1/(2\pi f_{osc}|X_{EQUIV}|) \tag{17.2}$$

where $X_{EQUIV} = 2\pi f_{osc}L_{eq} - 1/(2\pi f_{osc}C_{TRIM})$.

2 From Fig. 17.8, $L_{eq} = 1.4$ μH; hence,

$$|X_{EQUIV}| = |(2\pi)(12 \times 10^6)(1.4 \times 10^{-6}) - 1/[(2\pi)(12 \times 10^6)(10 \times 10^{-12})]|$$
$$= 1220.73\ \Omega$$

By Eq. 17.2,

$$C_{EQUIV} = 1/[(2\pi)(12 \times 10^6)(1220.73)] = 10.86 \text{ pF}$$

3 C_{EQUIV} deviates only slightly from the 10 pF value desired for C_{TRIM} because the IC's pulling effect is small. A greater influence from the IC, and thus a greater difference between C_{EQUIV} and C_{TRIM}, can be expected for higher frequencies (and a lesser one for lower frequencies).

Reference Counter

The reference counter, or divider, reduces the reference oscillator frequency to the value needed for f_R (Fig. 17.4). The need to divide by a number in the range of one hundred to several thousand is common, but normally only one value is needed for a given application. Programmable counters are, therefore, usually not a requirement, but they may be useful in establishing a particular divide value.

Most phase detectors employed in PLL frequency synthesizer designs are edge sensitive. Depending on the choice of device, either the positive or negative edge of f_R will be the critical one. The reference divider must provide a signal whose edge speed, as well as amplitude, is compatible with the phase detector being used. Sometimes, the reference counter can introduce phase jitter and cause the stability of f_R to be degraded below acceptable levels. The probability of this is greater for larger divide values. This problem can usually be resolved by rephasing the reference divider output with a signal edge derived directly from the reference oscillator or one of the early counter stages.

A variety of ICs is available for the reference counter function (see Table 17.2 for representative examples). They range from standard CMOS and TTL counters to LSI devices offering divide values to over sixteen thousand (see Table 17.8). (If a programmable counter is desired, see Table 17.6.) Reference counter speeds will usually be less than 20 MHz and frequently 10 MHz or lower. On occasion, however, higher frequency devices may be necessary (Table 17.5).

Phase Detector

In a PLL system (Fig. 17.4), the phase detector generates an output error signal proportional to the phase difference between f_R and f_V. Detectors for frequency synthesizer applications typically provide error information in the form of pulses of varying width. The error signal is processed by the loop filter to turn it into a suitably constant level and then applied to the VCO's control input to maintain loop lock. The degree of filtering required depends on the spectral purity needed for the loop output signal and on the characteristics of the error signal from the detector.

Phase detectors for PLL applications are of two kinds: *linear* (or analog) and *digital*. The loops in which they are used are also frequently called *linear* and *digital loops*, respectively.

A linear detector usually is configured as a double-balanced modulator, or mixer (Fig. 17.9). ICs of this type include the MC1496 and 12002 series. A similar form of linear detector is also employed in the NE565 general-purpose PLL and the NE567 tone decoder, as well as several other ICs. Linear phase detectors and linear loops serve well in signal retrieval, signal tracking, and signal demodulation applications. They are *not*, however, well-suited for use in frequency synthesizers because:

1 They can allow the loop to lock at harmonics or subharmonics of its proper frequency and, under some conditions, to "false lock" at other frequencies.

2 Their phase transfer characteristics are a cosine function and, therefore, the linear region is limited to less than approximately $\pm\pi/4$ radians.

3 The error signal generated and, therefore, the loop gain, is a function of the input signal amplitude and duty cycle.

4 Linear loops are limited to the frequency difference between f_R and f_V for which lock or *capture* can be achieved. This capture range is a function of the loop filter characteristics and imposes a constraint on the filter design that can hinder proper filtering of the error signal.

5 Loop lock occurs with the two input signals to the detector separated in phase by 90 degrees. At lock, substantial error signal remains which must be adequately filtered.

When properly configured, digital phase detectors can eliminate the disadvantages noted above. Not all digital detectors accomplish this, however. For example, both single flip flop and exclusive-OR gate (Chap. 1) implementations can also permit loop lock to occur at undesired VCO frequencies and the linear range of their transfer function is limited to approximately $\pm\pi$ and $\pm\pi/2$ radians, respectively.

An example of a digital detector exhibiting many of the properties needed for frequency synthesizer applications is shown in Fig. 17.10. Similar implementations employing two and four flip flops also can provide the characteristics needed. They are available in TTL, ECL, and MOS technologies (Table 17.3). Frequently, the two outputs making up the error information (designated U_1 and D_1 in Fig. 17.10a) are combined on a chip to form a single-ended output error signal. In CMOS designs, the combining circuitry typically results in a three-state output. This provides a relatively low impedance to either the positive supply or to ground to charge or discharge the filter capacitance when error corrections are made. When no correction is required, the output becomes a very high impedance—the third state.

Digital detectors of the type described above are desirable for frequency synthesizer applications because:

1 They fully discriminate *both* phase and frequency differences between their input signals and, therefore, prevent the VCO from locking at incorrect frequencies.

Table 17.2 **A selection of ICs useful in Reference Counter applications**

Function	*Family*	*Maximum frequency (MHz)*	*Device type and temp range: 0 to +70°C*	*−40 to +85°C*	*−55 to +125°C*	*Number of pins*	*Pin equivalent device*
		min at 25°C 5 V dc					
Dual D flip flop (÷2, ÷4)	LSTTL	30	SN74LS74		SN54LS74	14	
Decade counter (÷2, ÷5, ÷10)	LSTTL	32(÷2,10); 16(÷5)	SN74LS90		SN54LS90	14	
Binary counter (÷2, ÷8, ÷16)	LSTTL	32(÷2,16); 16(÷8)	SN74LS93		SN54LS93	14	
Dual decade counter	LSTTL	40(÷2,10); 20(÷5)	SN74LS390		SN54LS390	16	
Dual binary counter	LSTTL	40	SN74LS393		SN54LS393	14	
		typ at 25°C *5 V dc 10 V dc*					
Universal counter (÷2 through 12 except 7 and 11)	MTTL	30	MC4023[a]		MC4323	14	

Dual D flip flop (÷2, ÷4)	CMOS	4.0	10	MC14013BC	MS14013BA	14	CD4013B
Binary counter (division to 2^{14})	CMOS	3.5	9.0	MC14020BC	MC14020BA	16	CD4020B
Binary counter (division to 2^{12})	CMOS	3.5	9.0	MC14040BC	MC14040BA	16	CD4040B
Dual decade counter	CMOS	2.5	6.0	MC14518BC	MC14518BA	16	CD4518B
Dual binary counter	CMOS	2.5	6.0	MC14520BC	MC14520BA	16	CD4520B
Counters D1, D2 and phase detector	CMOS			MC14568BC	MC14568BA	16	
D1 (÷4, 64, 100)		6.0	16				
D1 (÷16)		2.5	6.3				
D2 (÷1 thru 15)		1.8	8.5				
Counter, osc, phase det. (÷32 thru 8176 in steps of 16)	CMOS	25		MC145143		16	

[a] T_A = 0 to +75°C.

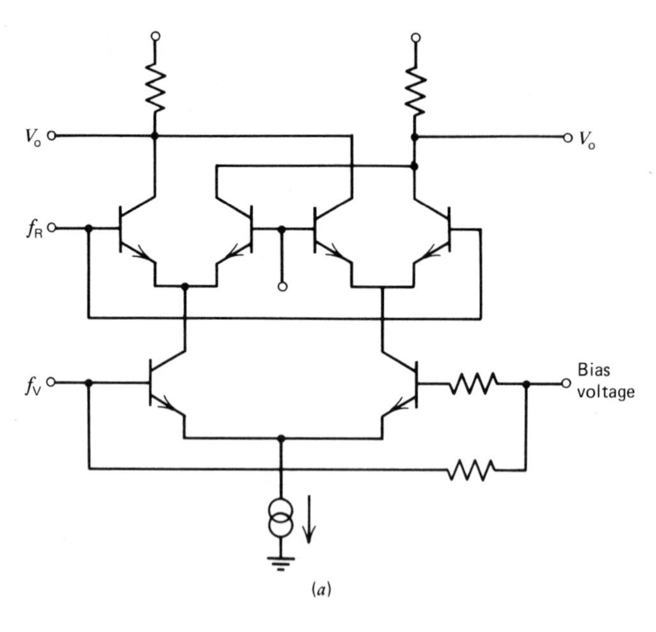

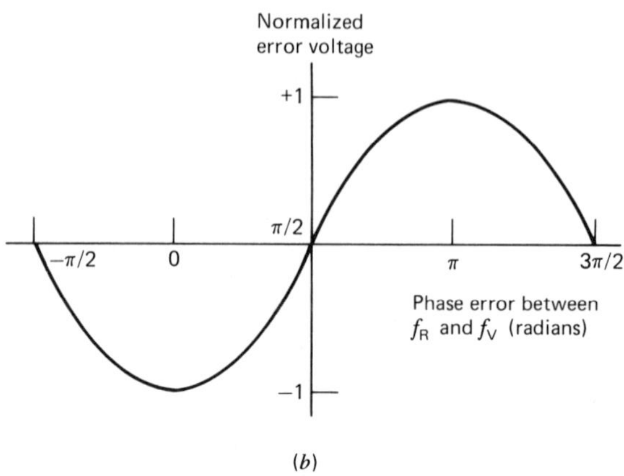

Figure 17.9 Linear phase detector. (*a*) Typical circuit. (*b*) Normalized transfer function.

2 Their phase transfer function, K_ϕ, is linear over approximately $\pm 2\pi$ radians of phase error.

3 They are edge sensitive and, therefore, independent of duty cycle of the two input signals, f_R and f_V. Also, f_R and f_V amplitude variations are not critical as long as the proper logic levels are satisfied.

4 They allow the loop to achieve lock over the full VCO frequency range, regardless of the initial frequency difference between f_R and f_V. The loop capture range is also independent of the loop filter.

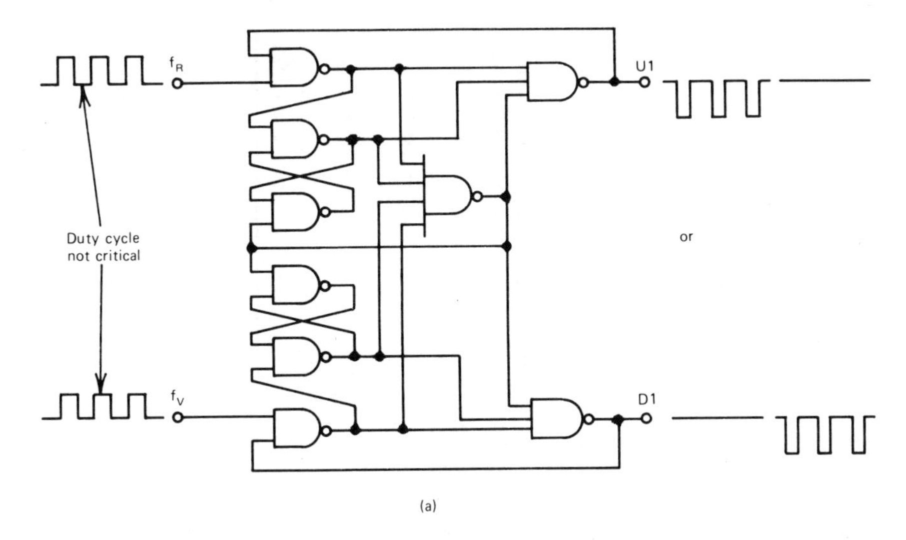

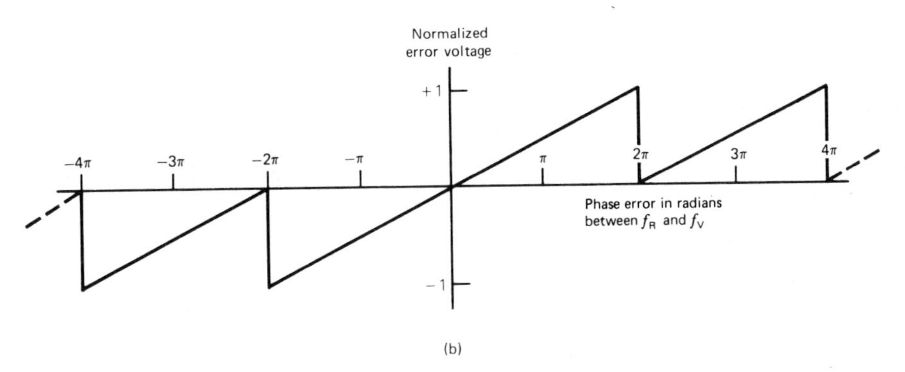

Figure 17.10 Typical digital phase frequency detector. (*a*) Logic diagram. (*b*) Normalized phase transfer function K_ϕ.

5 Loop lock occurs at approximately zero phase error between f_R and f_V and, at this time, the error signal remaining to be filtered can be quite low.

Although these detectors offer many features, two aspects of their transfer functions require consideration:

1 The detector's output typically requires a specific average load voltage to provide symmetrical corrections in either direction. As the average voltage is shifted away from its optimum value, different values of the phase shift transfer function, K_ϕ, are realized depending on the direction of error correction being made. This can be particularly noticeable when passive loop filters are employed. In this case, the average voltage at the detector's output must vary over the full range required by the VCO's control input as the VCO is tuned over its frequency range.

Table 17.3 **Digital Phase/Frequency Detectors for PLL frequency synthesizers**

Device type and temperature range						*Error signal*				
0 to +75°C	*−40 to +85°C*	*−55 to +125°C*	*Family*	*Nominal operating voltage (V dc)*	*Max. frequency MHz (typ)*	*Single ended*	*Double ended*	*Number of pins*	*Pin Equivalent device*	*Comments*
MC12040		MC12540	MECL	5.0	80		Yes	14		
MC4044		MC4344	MTTL	5.0	8	Yes	Yes	14	11C44	Also contains an exclusive-OR gate detector.
	MC14046BC	MC14046BA	CMOS	3 to 15		Yes 3-state		16	CD4046B	Also contains an exclusive-OR gate detector and voltage-controlled multivibrator.
	MC14568BC	MC14568BA	CMOS	3 to 15		Yes 3-state		16		Also contains two counters.

2 In the vicinity of zero phase difference between f_R and f_V, K_ϕ may vary in excess of five to 10 times nominal unless the detector is very high speed or contains special correction circuitry. Although K_ϕ may increase or decrease significantly in the zero phase area, this will not necessarily result in problems for a given loop. Some loop designs can tolerate rather wide fluctuations in K_ϕ and still perform adequately. Also, when a loop is locked, f_R and f_V are actually separated in phase just enough to generate the error signal necessary to offset leakage, filter bias currents, and so on, and maintain the locked condition. For many loops, the phase difference between f_R and f_V will move the operating position at lock to a linear region and allow K_ϕ to assume its nominal value.

Filter

The filter section is best described by considering it to be made up of two types of filters: a *loop* and one or more *sideband* filters. Fundamental loop characteristics, such as loop bandwidth, lock time, and transient response are controlled primarily by the loop filter design. This is done by establishing values for the loop natural frequency, ω_n and damping factor, δ.

The sideband filter elements provide additional filtering of the error signal in order to generate a stable and clean control voltage to the VCO. Disturbances on this line will frequency modulate the VCO and cause undesired frequencies or "sidebands" to be created. The result is a degradation in stability and spectral purity of the loop output signal. Because the error signal from the phase detector tends to be in the form of pulses appearing at the rate of f_R, the frequency f_R and its lower harmonics represent the major spectrum components that must be removed by filtering. The loop filter can also provide attenuation of these frequencies and, depending on design requirements, additional sideband filtering may not be necessary.

Either active (Chap. 11) or passive filters, or a combination of both, can be used for the loop and sideband filters. Each type has its advantages and disadvantages. Active filters provide voltage gain and, therefore, can furnish a wide control voltage range to the VCO while still maintaining an optimum voltage level for the detector to interface with. When used for the loop filter, their capability to provide high dc gain and low input current can result in approximately zero static phase error between f_R and f_V and thus very low error signal energy to be filtered exists when the loop is locked. Unfortunately, active filters also contribute noise components that can appear on the VCO signal. The level of this noise is low enough to be of no problem in many applications. In others, such as narrowband FM radios, the problem can be quite troublesome.

Passive RC filters contribute relatively no additional noise, but can result in a significant variation in the value of K_ϕ for the phase detector as the VCO is tuned over its frequency range. Passive filters also require a somewhat greater static phase error between f_R and f_V to maintain lock than an active filter. This is, however, normally not a severe problem.

Several active and passive filters for use as the loop filter are provided in Fig. 17.11, along with their respective transfer functions and defining loop relations. Two inputs (Fig. 17.11*b*) are required when using a phase detector such as shown in Fig. 17.10*a*. Single-ended phase detector outputs, such as the MC4044 and the 3-state outputs commonly provided by CMOS devices, require a suitable voltage at the filter input to establish equal positive and negative error signal amplitudes. This voltage should be approximately 1.5 V and $V_{DD}/2$ V (where V_{DD} is the supply voltage to a CMOS device), respectively, for the MC4044 and CMOS 3-state outputs. These levels are accomplished by V_{ref} in Fig. 17.11*a* and by the charge stored on capacitor *C* for the passive filters. In the passive case, the voltage level varies from the ideal because it is influenced by the control voltage required to keep the VCO at the correct frequency.

When the error signal is provided by a 3-state output, a current source is frequently most useful. In this case, a current rather than a voltage transfer function is used to describe the detector, thus giving K_ϕ the unit of A/rad. An impedance rather than a voltage function is used for the loop filter so that units of V/rad are still maintained for the detector-filter combination. A suitable loop filter configuration for this approach is described in Fig. 17.11d.

Active loop filters shown in Figs. 17.11*a* and *b* are likely to suffer from saturation or pulse clipping when large error corrections are being made. This arises because a maximum phase detector output can occur simultaneous with a transient overshoot. The result is a poor and uncertain settling time. A small R_2/R_1 ratio helps to minimize this effect; preferably, the ratio should be less than one. This, however, is not always practical because of other design constraints.

A remedy for pulse clipping in many cases is to modify R_1 to be an RC section as illustrated in Fig. 17.12. To ensure stable operation, the cutoff frequency of the filter, ω_c, should be chosen at least five to ten times the value of ω_n. To be effective, however, the upper limit on ω_c must remain significantly below $\omega_R = 2\pi f_R$ because the reference frequency, f_R, is the primary signal that must be attenuated. The R_1C_c network of Fig. 17.12 also helps to reduce VCO sidebands and can be considered as part of the sideband filter. From this standpoint, the approach is also useful with passive loop filters, such as the one in Fig. 17.11*c*.

Other sideband filtering can be incorporated by using additional passive or active filters. The low pass design of Fig. 17.13*a* can be used following the loop filters of Fig. 17.11*a*, *b*, or *c*. A similar sideband filter can be realized for the filter in Fig. 17.11*d* by adding capacitor C' in parallel with *R* to yield a corner frequency $\omega_c = 1/RC'$ rad/s. If more than one sideband section is required, an active filter is suggested to minimize phase shift buildup. Passive *RC* sections tend to accumulate phase shift more rapidly than signal suppression. This excess phase shift subtracts from the PLL's phase margin and can lead to instability. A possible active filter for use in sideband filtering is given in Fig. 17.13*b*. A choice of damping coefficient of about 0.5 provides a good compromise between attenuation properties above cutoff, and excess phase shift below cutoff.

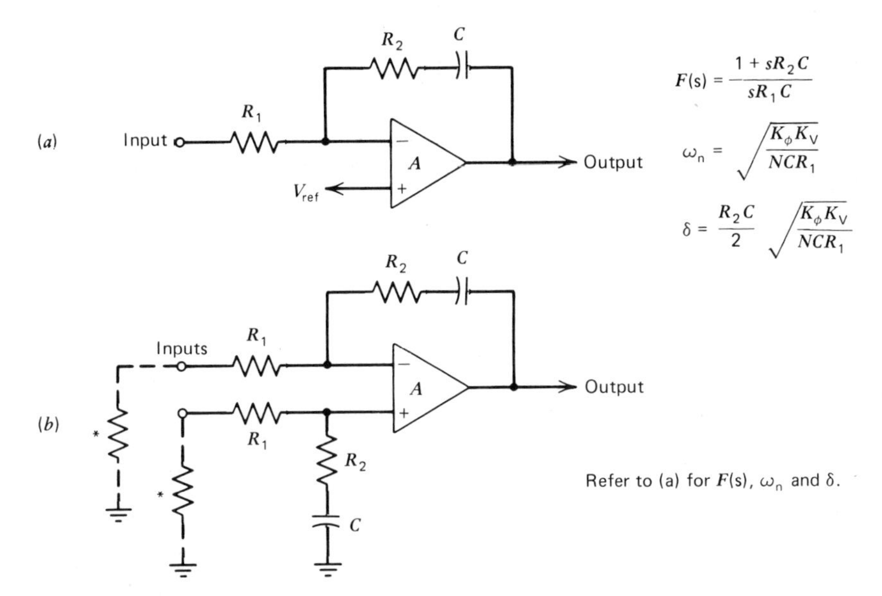

*Emitter pull down resistors of approximately 510 Ω required when operating with the MC12040/12540 phase detector.

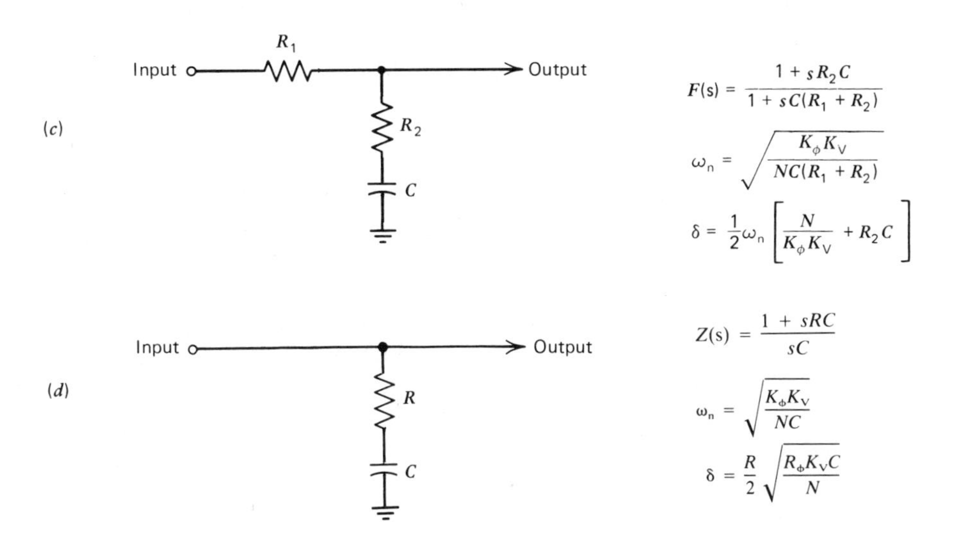

Figure 17.11 Typical loop filters with their voltage, $F(s)$, or impedance, $Z(s)$, transfer functions and relations defining loop natural frequency, ω_n, and damping factor, δ. $F(s)$, ω_n and δ in *a* and *b* assume that open loop gain, A, of the active element is sufficiently large to reduce the complete transfer function $F(s)' = \frac{A(1 + sR_2C)}{1 + s[R_1C(A+1) + R_2C]}$ to $F(s)$. The phase detector transfer function, K_ϕ, is in terms of volts per radian for *a*, *b* and *c* and in amps per radian for *d*.

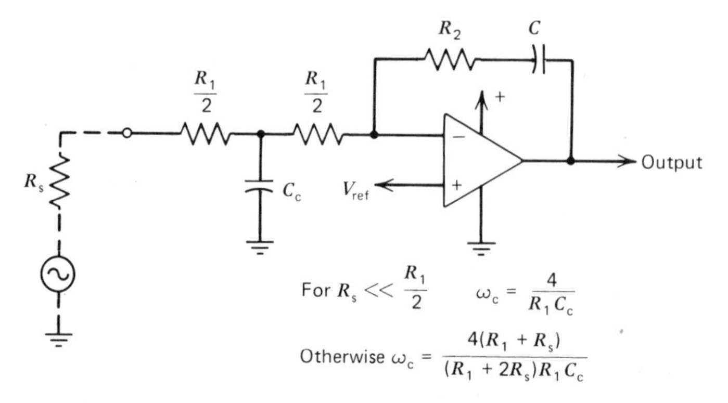

Figure 17.12 Loop filter of Fig. 17.11*a* modified to include additional filtering for minimizing pulse clipping and improving sideband levels.

When using a passive sideband filter with a passive loop filter, it becomes important to consider the loading, or interaction effects, upon the loop filter. Although a given ω_c fixes the RC product for a sideband filter, a degree of control over loading can still be exercised by the choice of values for R and C.

Suppression of a signal at the reference frequency will correspond approximately to the ratio of ω_c to ω_R; expressed in decibels:

$$\text{Signal suppression} \simeq n20\log_{10}(\omega_c/\omega_R) \tag{17.3}$$

where n is the order of the filter (Chap. 11). The amount of suppression calculated by Eq. 17.3 should be added to whatever may already exist owing to the loop filter and the R_1C_c section of Fig. 17.12 if it is employed. Note that the amount of signal suppression is *not* the level of the VCO sidebands. The amount of suppression, combined with knowledge of the error pulse configuration out of the phase detector and the VCO's transfer function will, however, allow an approximation of the sideband levels to be made.

EXAMPLE 17.3 Design the loop and sideband filters for a PLL frequency synthesizer employing the MC12040 phase detector and a VCO having a sensitivity of 1 MHz/V. The loop is to cover 220 to 225 MHz in 25 kHz steps and its natural frequency is 500 radians/second.

PROCEDURE

1 From the problem statement:

$$\omega_n = 500 \text{ rad/s},$$

the VCO's transfer function, K_v, is $2\pi \times 10^6$ rad/(s-V) and $f_R = 25$ kHz (assuming a system approach that allows f_R to be equal to the VCO step size). From the MC12040 data sheet, $K_\phi = 0.16$ V/rad.

2 N, the divide value in the loop's feedback path may be expressed by:

$$N_{min} = f_{VCO(min)}/f_R \tag{17.4a}$$
$$= (220 \text{ MHz})/(25 \text{ kHz}) = 8800$$
$$N_{max} = f_{VCO(max)}/f_R \tag{17.4b}$$
$$= (225 \text{ MHz})/(25 \text{ kHz}) = 9000$$

3 The loop filter must accept double-ended error signal information as provided by the MC12040. The filter of Fig. 17.11*b* is chosen and will be implemented using a 741 op amp for the active element. From Fig. 17.11*b* we obtain:

$$\omega_n = \sqrt{(K_\phi K_v)/(NR_1C)} \tag{17.5}$$

$$\delta = (R_2C/2)\sqrt{(K_\phi K_v)/(NR_1C)} \tag{17.6}$$

Therefore,

$$R_1C = K_\phi K_v/(N\omega_n^2) \tag{17.7}$$

and

$$R_2C = 2\delta\sqrt{(NR_1C/K_\phi K_v)}$$
$$= 2\delta/\omega_n \tag{17.8}$$

4 A practical range of values for the loop damping factor, δ, is 0.3 to 1. When not otherwise restricted, choose $\delta = 0.5$ as a nominal design value. More overshoot occurs as δ decreases; therefore, $\delta = 0.5$ will be established for N_{max}. By Eq. 17.8 we obtain:

$$R_2C = (2 \times 0.5)/500 = 20 \times 10^{-4}$$

and by Eq. 17.7,

$$R_1C = (0.16 \times 2\pi \times 10^6)/(9000 \times 500^2) = 4.47 \times 10^{-4}$$

These two requirements are satisfied by choosing a value for R_1, R_2, or C and then solving for the other two components. General guidelines are: $1\text{ k}\Omega < R_1 < 5\text{ k}\Omega$; $R_2 > 50\ \Omega$; $R_2/R_1 < 10$; C should be a low-leakage capacitor. Choosing $R_1 = 2.2\text{ k}\Omega$ yields:

$$C = (4.47 \times 10^{-4})/(2.2 \times 10^3) = 2.02 \times 10^{-7}$$

Use $C = 0.22\ \mu\text{F}$. The value of R_2 is:

$$R_2 = (20 \times 10^{-4})/(0.22 \times 10^{-6}) = 9.09 \times 10^3$$

Use $R_2 = 9.1\text{ k}\Omega$.

5 The loop filter will exhibit a voltage gain at the 25 kHz reference frequency of approximately $R_2/R_1 = 4.14$. A break frequency given in Fig. 17.12 will be added to

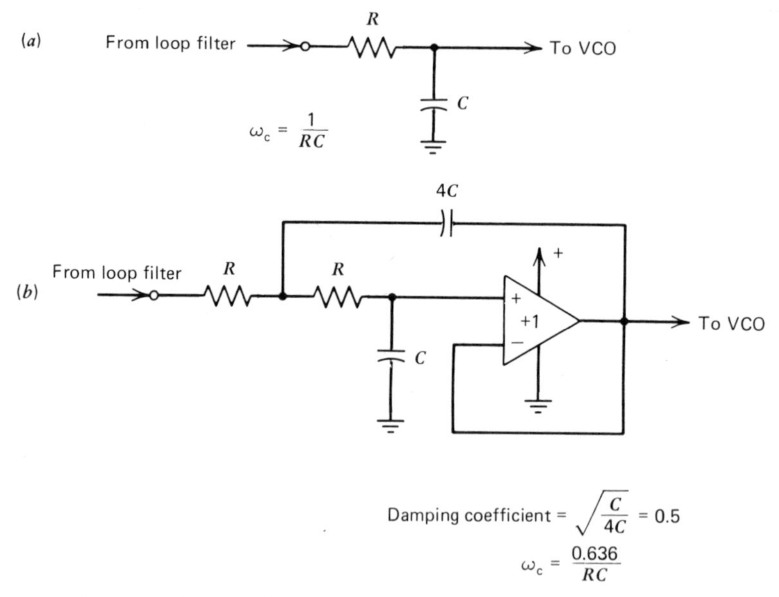

Figure 17.13 Sideband filter examples. (*a*) Passive. (*b*) Two pole, unity gain filter.

minimize pulse clipping and to provide integration of very narrow error pulse information that would otherwise not normally be integrated by the filter. Referring to Fig. 17.12 and choosing $\omega_c = 8\omega_n$ gives:

$$4/(R_1 C_c) = 8 \times 500 = 4 \times 10^3 \text{ rad/s}$$

where $R_1 = 2.2\ \text{k}\Omega$. Then we get:

$$C_c = 4/(4 \times 10^3 \times 2.2 \times 10^3) = 0.45 \times 10^{-6}$$

Use $C_c = 0.47\ \mu\text{F}$ and $R_1/2 = 1.1\ \text{k}\Omega$.

6 Additional sideband filtering can be included by following the loop filter with one of the circuits of Fig. 17.13. Selecting Fig. 17.13*a* and letting $\omega_c = 8\omega_n$ yields $RC = 1/(4 \times 10^3) = 0.25 \times 10^{-3}$. Use $C = 0.22\ \mu\text{F}$. Then,

$$R = (0.25 \times 10^{-3})/(0.22 \times 10^{-6}) = 1.14 \times 10^3$$

Use $R = 1.1\ \text{k}\Omega$.

7 The completed filter design, including component values, is given in Fig. 17.14.

Factors contributing to system phase error and, therefore, the error signal pulse widths at lock include input offset voltage and current of the op amp (see Chap. 9); mismatching of nominally equal resistors; and mismatching of the "high" output states of $\overline{\text{U}}$ and $\overline{\text{D}}$. Phase error can be initially set to "zero" by adjusting either the input offset voltage or one of the $R_1/2$ resistors.

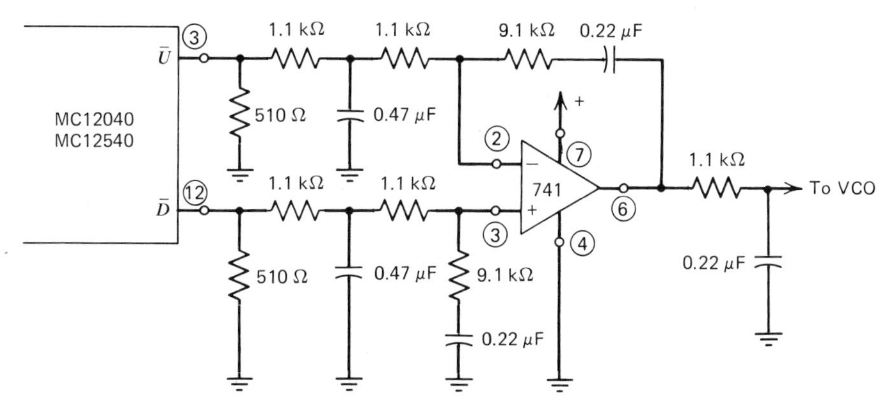

Figure 17.14 Filter design for Ex. 17.3

Voltage Controlled Oscillator (VCO)

The VCO is an oscillator whose frequency of oscillation varies in response to an input voltage. The transfer function, K_v, for such a device has the unit of radians per second per volt [rad/(s-V)] and represents how much frequency shift occurs for a given change in control voltage. Oscillators of this type can usually be placed in one of two categories: multivibrators (VCMs) and LC oscillators (VCOs).

When they can be used, VCMs provide the simplest and most straightforward solution. A single external capacitor (or sometimes a capacitor and resistor) is all that is required to establish their nominal frequency of oscillation. Variations about this value are obtained by simply applying a voltage level to the device's control input. Also, with suitable external components, the nominal operating frequency can be set over a wide range. For example, the MC4024 VCM can oscillate from less than one hertz to 25 MHz with an appropriate value of an external capacitor.

Voltage-controlled multivibrators are available in a variety of IC technologies and operating frequency ranges. Table 17.4 provides a summary of several industry devices. VCMs serve best in such PLL applications as clock signal generation and data recovery. Generally, RF frequency synthesizers demand better oscillator stability, spectral purity, and noise characteristics. To achieve the necessary improved performance for these applications, oscillator designs employing high Q LC resonant elements are normally used. In these cases, voltage control is accomplished by deriving all, or a portion of, the tank capacitance from a voltage capacitor diode, for example, a varactor. A diode of this type changes its capacitance value as a function of a reverse bias voltage, V_R, across it. This in turn shifts the resonant frequency of the LC tank circuit and, thus, the frequency of oscillation.

In a PLL, the value of V_R is controlled by the error signal from the loop filter. VCOs are frequently implemented using low-noise, high-frequency, discrete transistors. Suitable IC oscillators of this type are rare. One type, the

Table 17.4 Voltage-controlled multivibrator (VCM) ICs

Device type and temperature range								
0 to +70°C	*−40 to +85°C*	*−55 to +125°C*	*Family*	*Nominal operating voltage (V dc)*	*Maximum frequency typ at 25°C (MHz)*	*Number of pins*	*Pin equivalent device*	*Comments*
	MC1658[a]		MECL	5.0	150	16	SP1658 11C58	
MC4024[b]		MC4324	MTTL	5.0	30	14	11C24	Contains dual oscillators.
	MC14046BC	MC14046BA	CMOS	3 to 15	1.4 @ 10 V_{dc}	16	CD4046B	Also contains two phase detectors.
SN74LS629		SN54LS629	LSTTL	5.0	20[c]	16		Dual oscillators with range control and enable to start/stop output pulses.
SN74LS624		SN54LS624	LSTTL	5.0	20[c]	14		Provides complementary outputs, enable and range controls.

SN74LS625	SN54LS625	LSTTL	5.0	20[c]	16	Dual oscillators with complementary outputs. Each oscillator has independent supply and ground pins.
SN74LS626	SN54LS626	LSTTL	5.0	20[c]	16	Dual oscillators with complementary outputs and enable controls.
SN74LS627	SN54LS627	LSTTL	5.0	20[c]	14	Dual oscillators with each having independent supply and ground pins.

[a] $T_A = -30$ to $+85°C$.
[b] $T_A = 0$ to $+75°C$.
[c] Maximum frequency specification over temperature range.
The low-power Schottky (LSTTL) devices are manufactured by Texas Instruments.

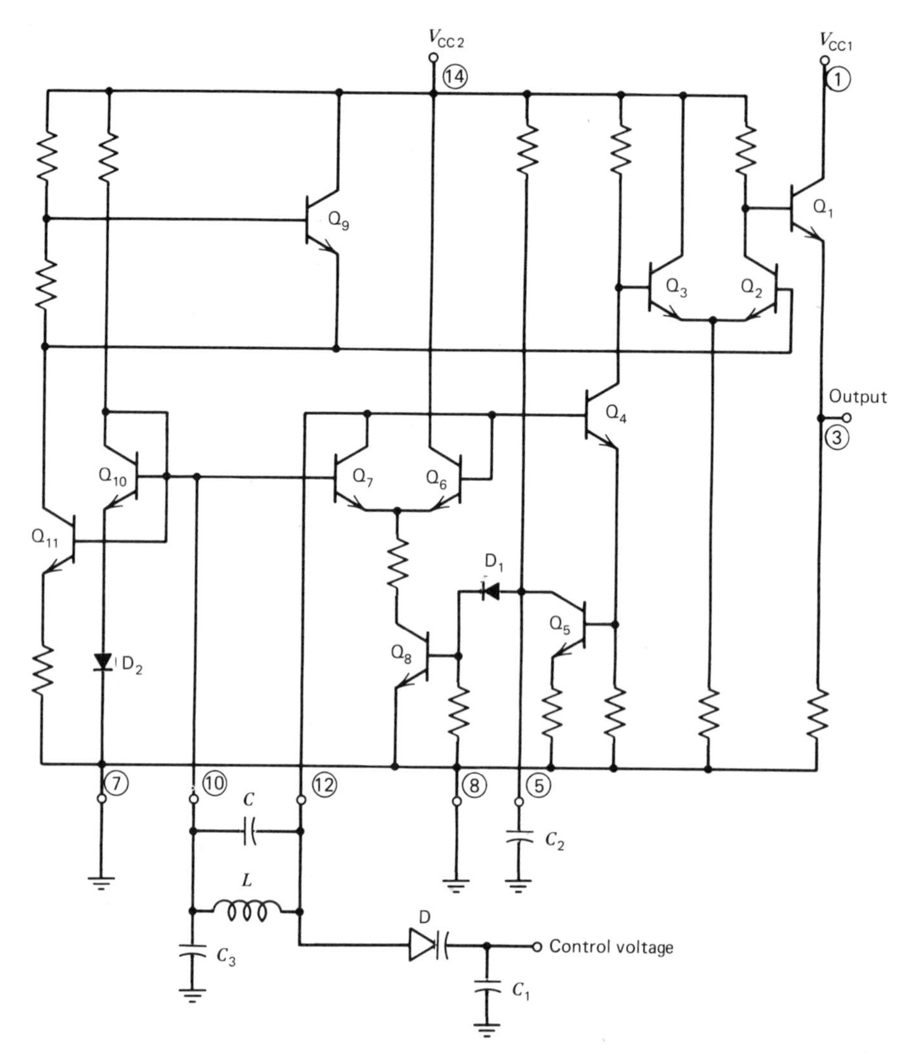

Figure 17.15 Circuit diagram of a VCO. Primary frequency determining components are C, L, and the voltage-variable capacitor, D.

MC1648, may serve the VCO needs for many synthesizer applications. This device is illustrated in Fig. 17.15, along with an example of the external frequency determining components that are required across pins 10 and 12 of the chip. The 1648 will operate from a single 5 V supply connected between pins 1, 14, and ground. It is usable up to 225 MHz. The emitter follower output at pin 3 provides a signal of approximately 800 mV pp at lower frequencies and 500 mV pp at 225 MHz.

Frequency of oscillation, f_o, for the 1648 corresponds to the parallel resonant frequency of the tank circuit connected across pins 10 and 12. Pin 10 is established at ac ground by C_3 and, therefore, tuning diode D appears in parallel with L and capacitor C (assuming the cathode of D is at ac ground via C_1). If this assumption is not true, the impedance from the cathode of D to ac ground must be taken in series with the diode and the combination then treated in parallel with L and C.

Assuming then that L, C, C_D (the capacitance of D which is a function of the control voltage), and C_s (shunt capacitance owing to the IC input capacitance at pin 12 which is typically 6 pF plus stray capacitance) are all in parallel, the frequency of oscillation, f_o, is:

$$f_o = 1/[2\pi\sqrt{L(C_D + C + C_s)}] \text{ Hz} \qquad (17.9)$$

Capacitance C_D varies from $C_{D(min)}$ to $C_{D(max)}$ as the control voltage changes from its most positive to least positive levels, respectively. This establishes the oscillator's maximum, f_{max} and minimum, f_{min} frequency limits:

$$f_{max} = 1/[2\pi\sqrt{L(C_{D(min)} + C + C_s)}] \text{ Hz} \qquad (17.10a)$$

$$f_{min} = 1/[2\pi\sqrt{L(C_{D(max)} + C + C_s)}] \text{ Hz} \qquad (17.10b)$$

The oscillator's transfer function, K_v, is determined by the diode's capacitance versus voltage characteristic and by the value of C. Making C larger relative to C_D tends to swamp the variations in C_D, causing the control voltage to have less influence on f_o. The result is a lower K_v. If C is eliminated, the effect of C_D is maximized and K_v is increased.

EXAMPLE 17.4 Calculate the tank circuit component values to permit the circuit of Fig. 17.15 to be used in a frequency synthesizer that tunes the 144 to 148 MHz amateur radio band. A passive loop filter is used and the phase detector provides correction error pulse swings from 0.5 to 8.5 V about a midpoint voltage of 4.5 V.

PROCEDURE

1 To allow for overshoot, design for a VCO range of 140 to 152 MHz. Based on the error pulse swings and the use of a passive filter, assume the filtered error signal available for tuning the VCO will be 1 to 8 V with a midpoint of 4.5 V. Consistent with the tuning requirements, K_v will be kept low to minimize sensitivity to undesired signals.

2 A low tuning voltage of 2.5 V will be designed to ensure that tuning diode D is always reverse biased when the tank circuit elements are connected as shown in Fig. 17.15. Considering that the 1648 establishes a nominal bias level at pin 10 of 1.5 V, then 1 to 6.5 V is the actual reverse bias voltage available for the diode.

3 Based on the above values,

$$K_v = 2\pi(152 - 140) \times 10^6/(6.5 - 1) = 13.7 \times 10^6 \text{ rad/(s-V)}$$

$$f_{min} = 140 \text{ MHz at a control voltage of 2.5 V } (V_R = 1 \text{ V})$$

$$f_{max} = 152 \text{ MHz at a control voltage of 8.0 V } (V_R = 6.5 \text{ V})$$

Applying Eqs. 17.10*a* and *b*, we obtain:

$$f_{max}/f_{min} = (152 \times 10^6)/(140 \times 10^6) = 1.09$$

or

$$[C_{D(max)} + C + C_s]/[C_{D(min)} + C + C_s] = 1.09^2 = 1.18$$

4 Assume $C_s = 8$ pF (6 pF IC input capacitance at pin 12 and 2 pF stray capacitance). The voltage-variable diode determines $C_{D(max)}$ and $C_{D(min)}$ at the reverse bias levels of 1 and 6.5 V, respectively. This, in turn, fixes the values for C and L. Calculations based on the MV2105 diode will be made; other diodes may be used which will result in different values for L and C. (A change in K_v will also alter the values of L and C.)

5 For reverse bias voltage of 1 and 6.5 V, the MV2105 data sheet specifies C_D values of 23 pF and 13 pF, respectively. Hence,

$$(23 \times 10^{-12} + C + 8 \times 10^{-12})/(13 \times 10^{-12} + C + 8 \times 10^{-12}) = 1.18$$

Solving for C yields $C = 34.5$ pF; use $C = 33$ pF.

6 A midrange frequency of 146 MHz and a reverse voltage of approximately 3.75 V will be used to calculate L. From the MV2105 data sheet, $C_D = 15$ pF (typical) for $V_R = 3.75$ V. By Eq. 17.9,

$$146 \times 10^6 = 1/[2\pi\sqrt{L(15 + 33 + 8) \times 10^{-12}}]$$

Solving gives us the value $L = 21.2$ nH. This inductance value represents about two turns of #20 wire of 0.1 inch diameter using air as the dielectric. A larger value for L is possible by choosing a lower value capacitance diode or allowing a larger value for K_v.

Feedback Path

The only major function of Fig. 17.4 left to be discussed is the feedback path. The feedback loop between the VCO and phase detector must:

1 Provide a proper interface to the VCO.

2 Convert the relatively high VCO frequency (tens or hundreds of megahertz) into a low frequency (kilohertz) which forms the second input signal, f_V, to the phase detector.

3 Ensure that the amplitude, edge speeds, and pulse width of f_V are compatible with the phase detector being used.

4 Be capable of being programmed. This provides the means by which the VCO is stepped over the frequency range of interest.

The most straightforward way of implementing the feedback function is with IC programmable counters. The chief limitation of this type of counter is the maximum operating speeds that can be realized. CMOS and TTL provide programmable counter speeds to about 25 MHz; Schottky TTL goes up to somewhat higher frequencies; ECL 10K operates to about 100 MHz. When the VCO must operate at higher frequencies, which is often the case, the feedback function must be modified to accommodate the higher input speeds.

A method of accomplishing this without sacrificing system performance is to incorporate a dual-modulus prescaler between the VCO and divide-by-N counters. The prescaler divides by either P or $P + 1$ in a time format established by the lower frequency programmable counters. Off-the-shelf dual-modulus chips allow such systems to operate up through 500 to 600 MHz (Sec. 17.4).

Fixed prescaling is straightforward. An IC counter that divides by a single value, P', is inserted just prior to the programmable counter (Fig. 17.3). The value for P' should be kept small—just large enough so that the prescaler's output frequency (the VCO frequency divided by P') is low enough to be compatible with the following divide-by-N counters. Values commonly used for P' include 4, 10, 16, 20, 64, and 256. Any IC counter/divider is a potential candidate for use as a fixed prescaler and available frequency capabilities extend to over 1 GHz. A selection guide to fixed prescalers is given in Table 17.5.

Table 17.6 provides a summary of programmable and presettable counter ICs useful for divide-by-N functions. The basic programmable counter consists primarily of four flip flops and has four programming inputs which the user takes to logic 1 or 0 levels to form a 4-bit code that determines the divide value, N (Fig. 17.16). Programmable *down counters* are common and for these N normally correspond to the binary number formed by the four programming bits $D0$ through $D3$, where $D0$ is the least significant bit (LSB).

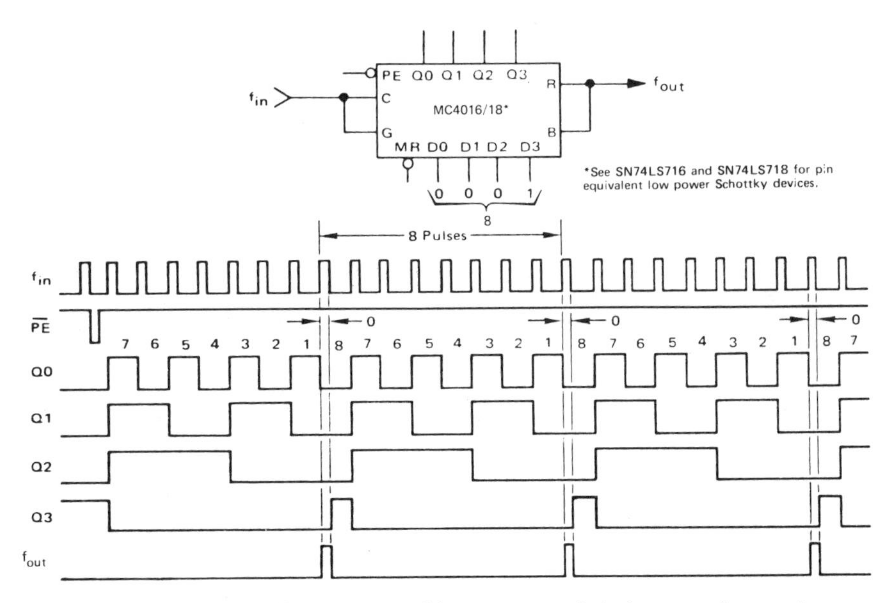

Figure 17.16 Typical 4-bit programmable counter and timing waveforms when programmed for divide-by-8 operation.

Table 17.5 ICs useful as high frequency Fixed Prescalers

Function	Family	Maximum frequency typ at 25°C (MHz)	Device type and temperature range			Number of pins	Pin equivalent device
			0 to +75°C	−30 to +85°C	−55 to +125°C		
÷4	MECL	1100	MC1697			8	
÷4	MECL	1100		MC1699		16	
÷64 ÷256	MECL	350 1200	MC12071[a]			14	CA3179
÷256	MECL	1400	MC12075[a]			14	
÷2	MECL	540		MC1690		16	SP1690
÷2, ÷4	MECL	160		MC10131	MC10531	16	Signetics Fairchild
÷2, ÷5, ÷10	MECL	150		MC10138	MC10538	16	
÷2, ÷4, ÷8, ÷16	MECL	150		MC10178	MC10578	16	Up counters—see MC10154 for pin equivalent down counter.
÷20		200		MC3396P[b]		8	

[a] T_A = 0 to +70°C.
[b] T_A = −40 to +85°C; buffered input, open collector output capable of driving TTL and CMOS.

The terms binary, hexadecimal, decade, or binary coded decimal (BCD) also are used to describe programmable counters. A 4-bit BCD or decade counter accepts input codes which result in N values of zero through nine and has a base, B, of 10. A 4-bit binary or hexadecimal counter accepts input codes resulting in values of zero through 15 for N and has a base of 16. The significance of $N = 0$ and of the counter's base will become apparent when we discuss cascading the individual counter ICs to form multistage counter strings to provide larger maximum divide values of N.

Presettable counters are similar to, but not the same as, programmable divide-by-N counters. To convert presettable counters to fully programmable counter operation usually requires additional decoding logic. This is especially the case when counter stages are cascaded. More importantly, however, is that the counter's frequency capability is normally degraded significantly from the value specified for presettable operation. Also, the amount of speed degradation is proportional to the number of cascaded stages.

When IC programmable counters are cascaded, N, for the complete string, is given by:

$$N = N_0 + B_0N_1 + B_0B_1N_2 + \cdots + B_0 \cdots B_{x-2}B_{x-1}N_X \qquad (17.11)$$

where subscript 0 refers to the least significant stage (signal input stage), subscript 1 to the next significant stage, and so on, up to the most significant stage denoted by subscript X. The subscripted terms N and B represent the programmed divide value and base, respectively, for the stage designated by the subscript value. Note that the number of stages making up the counter is $X + 1$.

Application of Eq. 17.11 is demonstrated in Fig. 17.17 which represents a 3-stage (12-bit) programmable counter formed by three MC4016 (BCD, B = 10) or three MC4018 (binary, B = 16) counters. Observe that two different programming codes are required to obtain the same N value (245), depending on whether BCD or binary counters are used. Divide values include all integers from 1 to 999 when using the BCD string and 1 to 4095 when using all binary counters. In each case 12 bits, or flip flops, are involved but the binary approach provides a much wider range for N. The advantage of BCD lies mainly in the method of programming a particular value of N. With BCD counters, all that is required is to program each stage with the 4-bit binary code representing the digit in N associated with that stage—the least significant digit of N with the least significant counter stage, and so on.

Typical maximum-frequency performance for the counter string of Fig. 17.17 is 10 MHz (the same as for each individual stage). This can be increased to 25 MHz by implementing an "early decode" technique using additional circuitry. For example, the MC12014 is available for this purpose and may be incorporated as shown in Fig. 17.18. To operate in this higher frequency mode, the only sacrifice is that N values less than three for the total cascade are no longer possible. This restriction is practically never a problem in synthesizer divide-by-N applications.

Table 17.6 **Programmable and Presettable Counter ICs useful for synthesizer divide-by-*N* functions**

		Max. frequency typ at 25°C (MHz)		*Device type and temperature range*				
Function	*Family*	*5* V_{dc}	*10* V_{dc}	*0 to +75°C*[b]	*−40 to +85°C*	*−55 to +125°C*	*Number of pins*	*Comments*
Programmable ÷*N* decade (÷0-9)	TTL	10[a]		MC4016		MC4316	16	For pin equivalent LSTTL devices, see Motorola SN74LS716, 18, or SN54LS716, 18 chips.
Programmable ÷*N* hexadecimal (÷0-15)	TTL	10[a]		MC4018		MC4318	16	
Two programmable ÷*N* (÷0-1, ÷0-4)	TTL	10[a]		MC4017		MC4317	16	
Two programmable ÷*N* (÷0-3, ÷0-3)	TTL	10[a]		MC4019		MC4319	16	
Programmable ÷*N* BCD (÷0-9)	CMOS	2.0	5.0		MC14522BC	MC14522BA	16	For pin equivalent see Fairchild, SSS, TI.
Programmable ÷*N* binary (÷0-15)	CMOS	2.0	5.0		MC14526BC	MC14526BA	16	
Dual 4-bit programmable ÷*N*	CMOS	3.5	9.5		MC14569BC	MC14569BA	16	Select pin allows either BCD or binary operation.

Programmable ÷ N binary (÷ 1-15)	CMOS	1.8	8.5		MC14568BC	MC14568BA	16	Also contains a phase detector and ÷ 4, 16, 64 or 100 counter.
Presettable decade	LSTTL	35		SN74LS162		SN54LS162	16	
Presettable binary	LSTTL	35		SN74LS163		SN54LS163	16	
Presettable decade	LSTTL	60		SN74LS196		SN54LS196	14	
Presettable binary	LSTTL	60		SN74LS197		SN54LS197	14	
Presettable up/down decade	LSTTL	40		SN74LS192		SN54LS192	16	
Presettable up/down binary	LSTTL	40		SN74LS193		SN54LS193	16	
Presettable up/down decade	MECL	150			MC10137[c]	MC10537	16	See Motorola application note AN-584 for dsn. information. For pin equivalent see Fairchild and Signetics.
Presettable up/down binary	MECL	150			MC10136[c]	MC10536	16	

[a] Speed can be increased to 25 MHz using early decode function provided by MC12014/12514.
[b] Except for 74LS devices which are 0 to +70°C.
[c] $T_A = -30$ to $+85$°C.

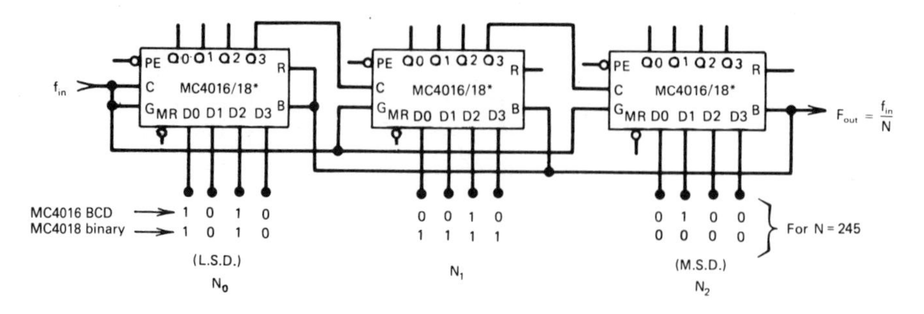

Figure 17.17 Three 4-bit TTL cascaded programmable counter stages and the program codes required for division by 245 when using all BCD or all binary counters.

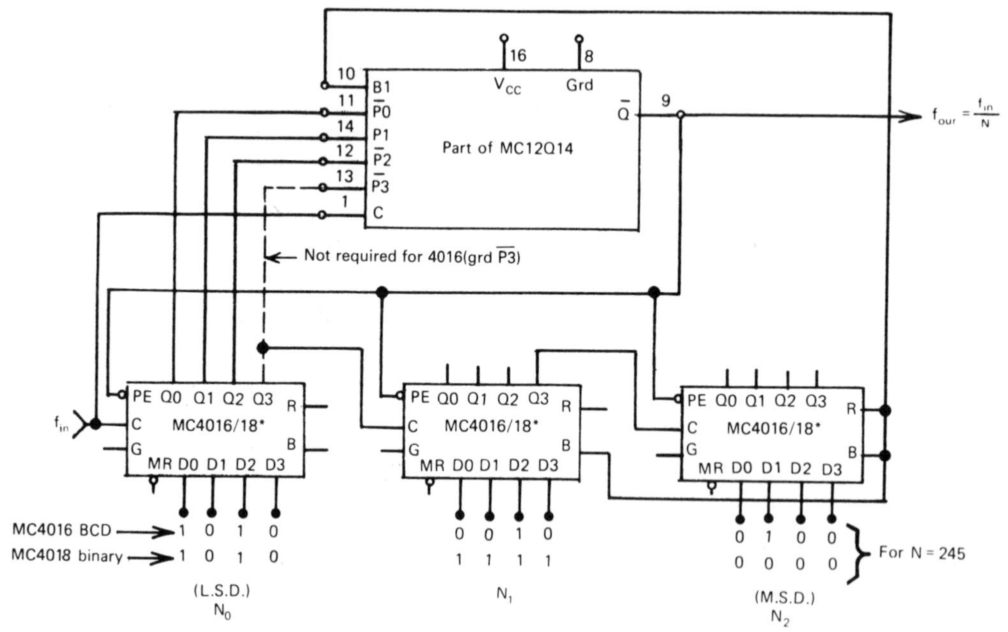

Figure 17.18 The 3-stage counter of Fig. 17.17 modified for increased operating frequency by incorporating "early decoding" provided by the MC12014 chip. The 12014 can also provide the modulus control signal for interfacing the programmable counters with a dual-modulus prescaler for still higher system frequency performance (Sec. 17.4).

For low dc power drain applications, CMOS programmable counters are available. Two examples of such devices are the MC14522B (BCD) and the MC14526B (binary) 4-bit cascadable down counters.

EXAMPLE 17.5 For the cascaded counter of Fig. 17.17, **(a)** determine the divide value obtained using the indicated BCD program codes for binary counters; **(b)** repeat (a) using the given binary program codes to program the BCD counters; **(c)** find the maximum value, N_{max}, possible with the 3-stage cascade using BCD counters for each stage and also when using 4-bit binary counters for each stage.

PROCEDURE

1 The BCD program codes in Fig. 17.17 give $N_0 = 5$, $N_1 = 4$, and $N_2 = 2$. When using 4-bit binary counters their base, B, is 16. Then, by Eq. 17.11, we get:

$$N = 5 + 16 \times 4 + 16 \times 16 \times 2 = 581$$

2 The binary program codes in Fig. 17.17 give $N_0 = 5$, $N_1 = 15$, and $N_2 = 0$. The solution is indeterminate because $N_1 = 15$ and division by a number greater than nine is invalid for a BCD decade programmable counter.

3 For each BCD counter stage the base is 10 and the maximum divide value is nine. By Eq. 17.11,

$$N_{max} = 9 + 10 \times 9 + 10 \times 10 \times 9 = 999$$

This may also be arrived at by inspection because with BCD counters, each stage or decade represents a digit in the total divide number. Hence, each stage at its maximum value, nine, gives $N_{max} = 999$ for the three-decade counter.

For the binary string, the base and maximum divide values per stage are 16 and 15, respectively. This results in:

$$N_{max} = 15 + 16 \times 15 + 16 \times 16 \times 15 = 4095$$

EXAMPLE 17.6 A frequency synthesizer requires a programmable counter in the feedback path to operate from 1.2 to 3.3 MHz while providing a 2.5-kHz output, f_V, to the phase detector. Determine several approaches for the counter and the program code required when the incoming frequency is 2.115 MHz. Consider both BCD and binary solutions.

PROCEDURE

1 The counter divide range must cover:

$$N_{min} = (1.2 \times 10^6)/(2.5 \times 10^3) = 480$$

and

$$N_{max} = (3.3 \times 10^6)/(2.5 \times 10^3) = 1320$$

The divide, N, for an input frequency of 2.115 MHz is:

$$N = (2.115 \times 10^6)/(2.5 \times 10^3) = 846$$

2 Because as many as four digits are needed for N, BCD solutions require four decades with stages N_0, N_1, and N_2 programmed zero through nine and the most significant stage, N_3, programmed for either zero or one (Fig. 17.19). Solutions include the following cascades:

MC4016—MC4016—MC4016—MC4016

SN74LS716—SN74LS716—SN74LS716—SN74LS716

MC14522B—MC14522B—MC14569B

For $N = 846$:

$$N_0 = 6 \quad (0110)$$
$$N_1 = 4 \quad (0100)$$
$$N_2 = 8 \quad (1000)$$
$$N_3 = 0 \quad (0000)$$

where the required program codes for each stage are shown in parentheses with the right-most bit being the least significant.

3 A binary solution requires a minimum of 11 bits to achieve $N_{max} = 1320$. Each of the following cascade examples provide 12 bits and, therefore, satisfy the requirement:

MC4018—MC4018—MC4018

SN74LS718—SN74LS718—SN74LS718

MC14526B—MC14569B

4 The following steps provide a method for determining the program code needed to obtain a specific N value for a binary cascaded counter employing down counters.

a. Consider the total counter as a unit. The counter in step 3 thus becomes a 12-bit binary counter having a least significant bit (LSB) at its input side and a most significant bit (MSB) at its output side.

b. Assign each bit a weight of zero for a logic 0 on the program line for that bit.

c. Assign each bit a weight of 2^Y for a logic "1" on the program line for that bit, where $Y = 0$ for the LSB, $Y = 1$ for the next bit, . . . $Y = b - 1$ for the MSB and b is the number of total bits in the counter.

d. The counter divide value, N is then obtained by summing together all the bits with each bit having a weighted value as described in steps b and c. For a specific value

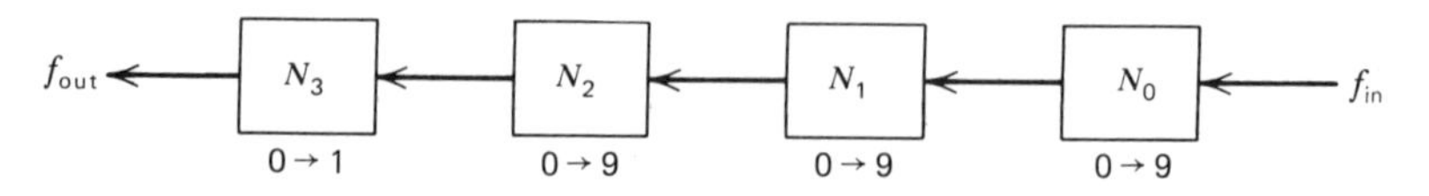

Figure 17.19 A 4-stage BCD counter used in Ex. 17.6.

of N, the counter program line for each bit is made either a logic 0 or a logic 1 so that the sum of the bits results in the desired value for N.

5 Applying steps a through d to the 12-bit binary counter in step 3 for $N = 846$ yields:

Counter bit	*MSB*											*LSB*
Bit	2^{11}	2^{10}	2^9	2^8	2^7	2^6	2^5	2^4	2^3	2^2	2^1	2^0
Weight	2048	1024	512	256	128	64	32	16	8	4	2	1
Program Code for $N = 846$	0	0	1	1	0	1	0	0	1	1	1	0

17.4 DUAL-MODULUS PRESCALING

Dual-modulus prescaling provides a method of achieving high-speed performance in the feedback path without compromising system performance. Frequency tuning resolution and a high value for f_R can be maintained as if only divide-by-N counters were being used. This is a marked advantage over fixed prescaling which requires that f_R be reduced by the prescaling value in order to maintain the same VCO frequency step capability. A reduction in f_R is, of course undesirable because it corresponds to a degradation in potential performance of the system.

Basically, the dual-modulus technique (Fig. 17.20) combines relatively low-frequency programmable counters with a dual-modulus prescaling IC, such that the combination behaves like a programmable counter having a speed capability equal to that of the prescaler. Because the prescaler is required to divide by only two values, P and $P + X$, it can be designed for operating speeds comparable to those of fixed dividers. Proper operation is accomplished by configuring the lower speed counters in two sections and employing special decoding/control logic that makes it possible to select the divide value P or $P + X$ for the prescaler in a specific, timed, format. As shown later, the value of X is normally made equal to one.

The total system divide value, N_{total}, that results with dual-modulus prescaling is a function of P, X, the value (A) programmed into the $\div A$ counter and the value (N) programmed into the $\div N$ counter. A typical system is configured so that at the beginning of a count sequence the modulus control line goes low and causes the prescaler to divide by $P + X$ until the $\div A$ counter has counted down from its programmed value, A. During this time, for every $P + X$ pulse into the prescaler from the VCO, both the $\div A$ counter and the $\div N$ counter are decremented by one. Therefore, after A counts [$(P + X)(A)$ pulses from the VCO] the $\div N$ counter will be at $(N - A)$ where N is the number programmed into the $\div N$ counter and the $\div A$ counter will have reached zero.

This terminal (zero) count is detected and used to trigger a latch circuit causing the modulus control line to be held high and the dual-modulus prescaler to start dividing by P. The prescaler continues to divide by P for the remaining

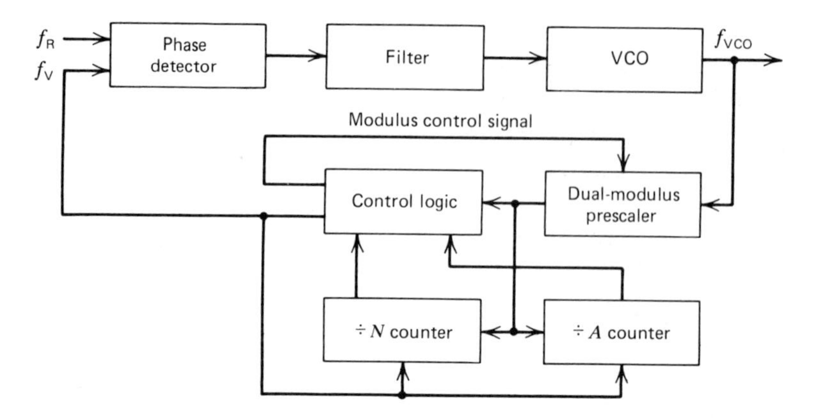

Figure 17.20 Block diagram of a dual-modulus prescaling approach to frequency synthesis.

N count sequence, that is, until $(N - A)$ additional pulses have reached the $\div N$ counter. This occurs after $(N - A)P$ more VCO pulses—corresponding to $N_{total} = (P + X)A + (N - A)P = (N)(P) + XA$ total pulses from the VCO since beginning the count cycle. The $\div N$ counter has now reached zero and both the $\div A$ and $\div N$ counters are preset and the modulus control line taken low, causing the prescaler to once again start dividing by $P + X$. The count cycle now repeats. Note that a necessary condition for dual-modulus prescaling is that the value N must always be equal to or greater than the value A. This is normally not a serious constraint.

The system divide value, $N_{total} = (N)(P) + XA$, that has been accomplished, as noted above, permits one to change the VCO in increments (channel spacings) of X times f_R as A is programmed in unit steps. Therefore, by using a prescaler whose divide values are P and $P + 1$ (for example, making $X = 1$) the VCO step size becomes f_R in value and the defining division equation is:

$$N_{total} = (N)(P) + A \tag{17.12}$$

where $N \geqslant A$.

The value required for N_{total} is dictated by the VCO frequency and step size required, that is, $N_{total} = f_{VCO}/f_R$. To cover a range of N_{total} values in sequence, A is typically required to assume values zero through $P - 1$ in unit steps for a particular N value. N is then incremented to $N + 1$ and the sequence zero through $P - 1$ repeated for A. This procedure is continued until the desired N_{total} range is covered. The $\div A$ counter size is, therefore, dictated by the value used for P and the $\div N$ counter size is dictated by the maximum divide value, $N_{total(max)}$ that must be achieved.

The value chosen for P, the constraint $N \geqslant A$, and the range of the $\div A$ and $\div N$ counters, set a minimum and maximum boundary on N_{total}. If A_{max} =

$P - 1$, then $N_{(\min)} \geqslant P - 1$ gives:

$$N_{\text{total(min)}} = (P - 1)(P) + A_{\min} \tag{17.13}$$

Also,

$$N_{\text{total(max)}} = (N_{\max})P + A_{\max} \tag{17.14}$$

Because a choice of P and $P + 1$ values is usually possible, the $N_{\text{total(min)}}$ and $N_{\text{total(max)}}$ restrictions rarely become a problem in frequency synthesizer designs. Other considerations in choosing the value of P follow.

Rules for Choosing *P*

For the maximum frequency into the prescaler, $f_{\text{VCO(max)}}$, the value used for P must be large enough such that:

1 $f_{\text{VCO(max)}}$ divided by P does not exceed the frequency capability of the $\div A$ and $\div N$ programmable counters.

2 The period of the prescaler's output signal frequency, $f_{\text{VCO(max)}}/P$, must be greater than the sum of the times designated in a, b, and c below:

- **a.** Propagation delay through the dual-modulus prescaler is defined from the prescaler's input to an output edge from the prescaler, which is of the appropriate transition direction to trigger the $\div A$ and $\div N$ counters. This prescaler output occurs once for each group of P or $P + 1$ input cycles to the prescaler.
- **b.** Prescaler setup *or* release time relative to its modulus control signal.
- **c.** Propagation time from the input of the $\div A$ and $\div N$ counters to the modulus control signal output is the time it takes to change the modulus control signal level once an output signal edge corresponding to the completion of $A \div P$ or a $\div(P + 1)$ count sequence is received from the prescaler.

Also, the choice of P and $P + 1$ can sometimes simplify the program code required for programming the $\div A$ and $\div N$ counters. Values such as 10/11, 20/21, and 40/41 usually work best in this regard when using BCD counters for $\div A$ and $\div N$ and values of 8/9, 16/17, 32/33, 64/65, and so on work best when binary down counters are employed for the $\div A$ and $\div N$ functions. For the binary cases, the value needed for N_{total} will result when N_{total} in binary is used as the program code to the $\div N$ and $\div A$ counters in accordance with the following rules.

Rules for Programming Counters

1 Assume the $\div A$ counter contains b bits where $2^b = P$.

2 If the $\div A$ counter contains more than b bits, always program all higher order $\div A$ bits above b to zero.

3 Assume the $\div N$ and $\div A$ counters (with all higher-order bits $\div A$ above b ignored) to be combined into a single binary counter. The MSB of this hypothetical counter is to correspond to the MSB of $\div N$ and the LSB is to correspond to the LSB of $\div A$. The system divide value, N_{total}, now results when the value of N_{total} in binary code is used to program this "new" counter (see Ex. 17.7).

In regard to dual-modulus systems, two additional observations are appropriate. To maximize system frequency capability, the prescaler's output signal (which signifies each group of P or $P + 1$ input signal cycles) must make a low-to-high transition when positive edge sensitive $\div A$ and $\div N$ counters are used. Similarly, for negative edge sensitive $\div A$ and $\div N$ counters, the prescaler's output should make a high-to-low transition to signify each P or $P + 1$ pulse group. The opposite output phasing or a delay in the occurrence of the proper output edge transition from the prescaler subtracts from the amount of time potentially available for the $\div A$ and $\div N$ counters and control logic function to generate the modulus control signal.

The prescaler must divide by the appropriate value as dictated by the modulus control signal level. It is common for this to be P when the modulus control line is high and $P + 1$ when the control line is low.

Several representative prescalers for operation in dual-modulus synthesizers are summarized in Table 17.7. The high-frequency, low current drain of such chips as the MC12015-18 makes dual-modulus synthesizers practical for portable radios.

Using the same PC board layout, several divide values are possible. In addition to being used individually, the $\div 5/6$, $\div 8/9$, and $\div 10/11$ prescalers also can be combined with one additional counter to provide a variety of expanded $P/(P + 1)$ values. Five modulus control signal input ports (three for ECL levels and two for TTL levels) and an on-chip ECL to TTL translator are provided by the prescalers to facilitate this expansion (Fig. 17.21). Several such expansions are summarized in Figs. 17.22 and 17.23.

LSI chips also are available for use in dual-modulus synthesizers. These devices provide the necessary control logic as well as $\div A$ and $\div N$ programmable counters suitable for many synthesizer applications. In addition, they also offer several other features (Sec. 17.5).

EXAMPLE 17.7 A dual-modulus synthesizer is to be designed to provide 118.000 to 135.975 MHz in 25-kHz steps for use in an aircraft communications band transmitter. (a) Determine appropriate $P/(P + 1)$ values. (b) Determine the range of divide values to be provided by the $\div A$ and $\div N$ counters. (c) Calculate the specific counter divide values required for operation on the 124.125 MHz channel and the program code for this channel. Assume binary counters are used.

PROCEDURE

1 Dual-modulus prescaling allows f_R to equal the VCO step size desired. Therefore, $f_R = 25$ kHz.

Table 17.7 **Dual-Modulus Prescaling ICs**

Device type and temp. range				*Maximum input frequency (MHz)*									*Number of*	
				−55°C	*−40°C*		*−30°C*	*25°C*		*85°C*		*125°C*		
−30 to +85°C	*−40 to +85°C*	*−55 to +125°C*	$\div P/(P+1)$	*min.*	*min.*	*typ*	*min.*	*min.*	*typ*	*min.*	*typ*	*min.*	*pins*	*Comments*
MC12009		MC12509	5/6	420			440	480	600	440		420	16	Same pinouts. 5 V dc supply.
MC12011		MC12511	8/9	500			500	550	600	500		500	16	Each contains a nondedicated ECL to TTL translator.
MC12013		MC12513	10/11	500			500	550	600	500		500	16	Input buffer.
	MC12015		32/33		225				300	225			8	4.5 to 9.5 V dc.
	MC12016		40/41		225				300	225			8	Low power: 7mAmax., 6.8 V; 4.5mA, 5.0 V typ.
	MC12017		64/65		225				300	225			8	Same pinouts. Input buffer.
	MC12018		128/129		520					520			8	4.5 to 9.5 V dc. Low power. Input buffer.
	MC3393		15/16			140			180		140		8	5 V dc supply. Input buffer. Low cost.

NOTES: All devices are phased for optimum system speed performance when operating with positive edge triggered $\div N$ and $\div A$ counters. All ICs listed are available from Motorola Semiconductor. Other sources include Fairchild Semiconductor and Plessey Semiconductor.

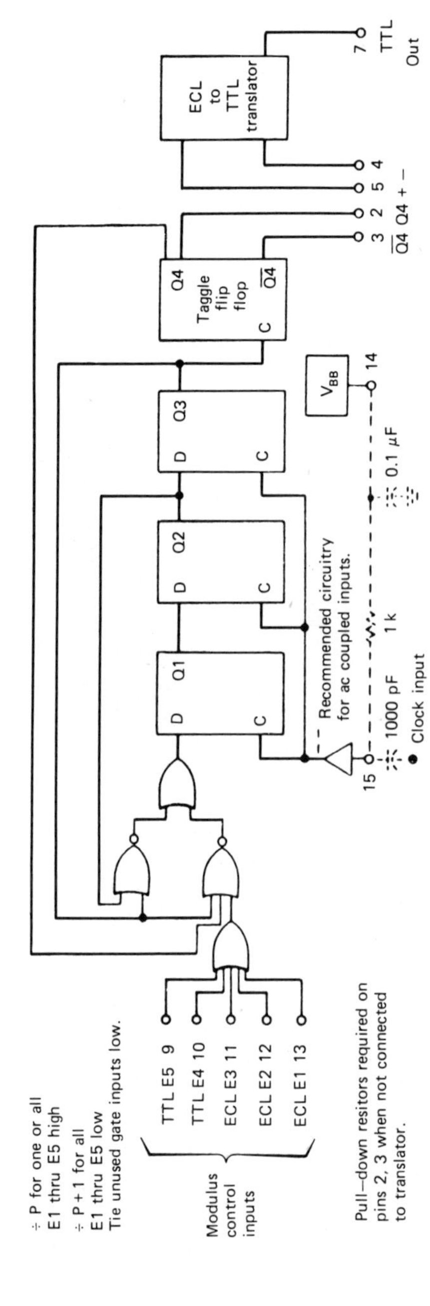

Figure 17.21 High-frequency divide-by 10/11 dual-modulus prescaler logic diagram. On-chip decoding is modified to give divide-by 8/9 and divide-by 5/6 values.

Device B \ Device A	*MC12009* ÷*5/6*	*MC12011* ÷*8/9*	*MC12013* ÷*10/11*
MC10131 dual flip flop	÷20/21	÷32/33	÷40/41
MC10154 binary counter:			
(÷8 section)	÷40/41	÷64/65	÷80/81
(÷16 section)	÷80/81	÷128/129	÷160/161
MC10138 BCD counter:			
(÷5 section)	÷25/26	÷40/41	÷50/51
(÷10 section)	÷50/51	÷80/81	÷100/101

Note: The MC12009, MC12011, and MC12013 are pin-equivalent.

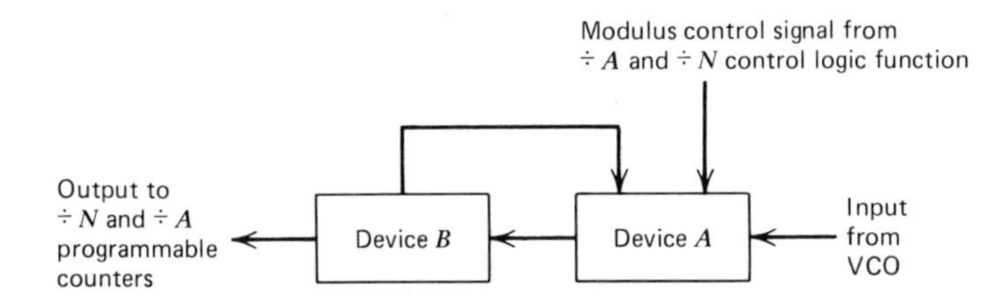

Figure 17.22 Examples of expanded divide-by P/(P + 1) dual-modulus values realized using an MC12009, 12011, or a 12013 with another IC.

2 The minimum divide value needed is $f_{VCO(min)}/f_R = (118.000 \times 10^6)/(25 \times 10^3) = 4720$. By Eq. 17.13, $N_{total(min)} = (P - 1)P + 0 = 4720$; therefore, $P \leq 69$. A value for P equal to 2 raised to an integer power will be chosen to simplify the program code, assuming binary counters are used for the $\div A$ and $\div N$ functions. Also, $f_{VCO(max)}/P$ must be low enough to satisfy the requirements listed under *Rules for Choosing P*. Hence, $P = 32$ or 64 would meet the above objectives. Choose $P = 32$ and, therefore, $P/(P + 1) = 32/33$.

3 For $P = 32$,

a. The $\div A$ counter must be programmed for zero through 31 in integer steps. A counter containing a minimum of five bits is, therefore, required.

b. The maximum divide value needed is $f_{VCO(max)}/f_R = (135.975 \times 10^6)/(25 \times 10^3) = 5439$ and from step 2, the minimum value is 4720.

4 By Eq. 17.12, $(N_{min})(32) + A = 4720$ and $4720/32 = 147.5000$ making $N_{min} = 147$. Also, $(N_{max})(32) + A = 5439$ and $5439/32 = 169.969$ making $N_{max} = 169$. The $\div N$ counter range required is thus 147 to 169 in integer steps. A counter having at least eight bits is required: $2^8 - 1 = 255$.

5 For $f_{VCO} = 124.125$ MHz, $N_{total} = (124.125 \times 10^6)/(25 \times 10^3) = 4965$. Therefore, by Eq. 17.12, $(N)(P) + A = 4965$. Dividing 4965 by $P = 32$ and taking the integer portion for N yields: $4965/32 = 155.15625$ and $N = 155$. The value for A is the remainder multiplied by P or A is equal to $(0.15625)(32) = 5$. By checking we prove,

$$\begin{aligned} N_{\text{total}} &= (N)(P) + A = 4965 \\ &= (155)(32) + 5 = 4965 \\ &= 4960 + 5 = 4965 \\ 4965 &= 4965 \end{aligned}$$

6 By the *Rules for Programming Counters* and steps 4 and 5 of Ex. 17.6, the program code for this example is:

Counter bit	← ÷N → MSB							LSB	← ÷A → MSB				LSB
Bit	2^{12}	2^{11}	2^{10}	2^9	2^8	2^7	2^6	2^5	2^4	2^3	2^2	2^1	2^0
Weight	4096	2048	1024	512	256	128	64	32	16	8	4	2	1
Program code for $N_{\text{total}} =$ 4965	1	0	0	1	1	0	1	1	0	0	1	0	1

If the ÷A counter used contains more than five bits, then all bits over five will always be programmed to zero and ignored when establishing the program code, as outlined in the above table.

EXAMPLE 17.8 Various ÷$P/(P + 1)$ values can be obtained using basic ÷5/6, ÷8/9, and ÷10/11 dual-modulus chips in conjunction with one additional counter as noted in Fig. 17.22. Outline a general procedure to follow for achieving this while specifically determining division values of 100/101. Assume that the ÷A and ÷N are positive edge triggered down counters.

PROCEDURE

1 From Fig. 17.21, one observes that the dual-modulus chips divide by P when any *one* of the E1 through E5 inputs is high and divide by $P + 1$ when *all* E1 through E5 inputs are low. New, expanded, $P/(P + 1)$ values can be obtained by applying logic levels to these E inputs that will cause each divide value to be repeated the correct number of times to form the desired new value before an output pulse is allowed to pass on to the ÷A and ÷N programmable counters.

For example, ÷100 can be achieved by using a chip, such as an MC12013, and causing 10 sequences of ÷10 (requiring one or more of the E inputs to be high) to occur before an output pulse is generated, and ÷101 by causing nine sequences of ÷10 (one or more E inputs high) and one of ÷11 (all the E inputs low) to occur before giving an output pulse. The problem thus becomes one of how to achieve the proper logic levels on the inputs and how to control them when an output pulse is provided to the ÷A and ÷N counters. As illustrated in Fig. 17.22, this may be accomplished by using an appropriate counter in conjunction with a dual-modulus chip.

2 From the truth tables for the Q outputs of basic decade (÷10) and binary (÷16) counters, one observes that all or certain ones of these outputs may be used to keep at least one E input high and to keep all E inputs low for the correct number of times

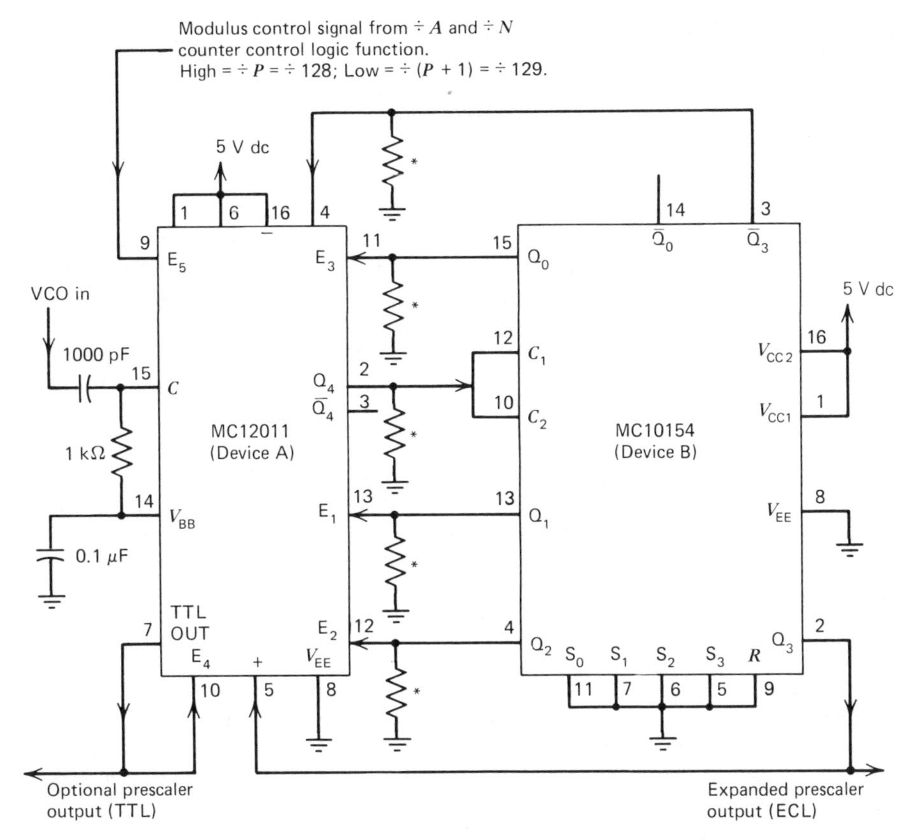

Figure 17.23 Divide-by 128/129 dual-modulus prescaler formed by connecting an MC10154 positive edge triggered binary down counter with an MC12011 divide-by 8/9 dual-modulus IC.

to achieve the desired new $P/(P + 1)$ values. Similarly, the Q outputs of single and dual flip flops also are useful in this regard. In each case, the counter or flip flop is clocked by the dual-modulus prescaler's output and the appropriate Q outputs from the counter or flip flop are coupled to the prescaler's E inputs. Also, one E input always remains under control of the modulus control signal from the $\div A$ and $\div N$ counter control logic function. Any E inputs not being used are tied low. In addition, the $\div A$ and $\div N$ counters no longer receive their input directly from the dual-modulus chip but, instead, from the counter or flip flop output.

3 From step 2, to achieve $\div 101$, it requires a counter that will have one or more of its Q outputs high for nine MC12013 output pulses (giving nine sequences of $\div 10$ operation because at least one E input will be high during this time) and all of its Q outputs low for a single output pulse (giving one cycle in the $\div 11$ mode). Only after this total of 10 outputs shall an output to the $\div A$ and $\div N$ counters be allowed to occur. This combination of events must then be repeated for every ten outputs from the MC12013.

Similarly, ÷100 is accomplished with the modulus control signal derived from the $\div A$ and $\div N$ counters going high to cause the one cycle of ÷11 to become a ÷10 cycle. The truth table (Fig. 17.24) for a decade BCD up counter shows that the counter's four Q outputs display the qualities.

4 The design procedure is not complete without a careful review of system timing. In this case, best performance is *not* achieved by using only Q outputs of the counter of Fig. 17.24. Using $\overline{Q}_0$ rather than Q_0 to drive one of the E inputs reduces the counter's propagation time contribution from approximately four to one flip flop delay. and thereby increases the prescaler's maximum input frequency capability.

In Fig. 17.24, Q_2 rather than Q_3, is more optimum for the output signal to the $\div A$ and $\div N$ counters. This is true because Q_2 provides a low-to-high transition sooner than Q_3 after completion of a divide-by-ten count cycle. Because the $\div A$ and $\div N$ counters are assumed to be positive edge triggered, Q_2 will allow more time for their modulus control signal to be generated. It is appropriate to use Q_2 to drive the counters because it, like Q_3, provides only one positive transition for every 10 clock pulses.

With positive edge triggered down counters for $\div A$ and $\div N$, optimum performance is realized by using a positive edge triggered down counter for Device B. The optimum circuit connections are then also easily defined. In this case, the Q outputs of Device B are always used to drive the E inputs (Fig. 17.23).

EXAMPLE 17.9 A "universal" frequency synthesizer suitable for use in several different equipment lines is to be designed. Five different dual-modulus prescaler divide values are required to achieve the flexibility needed: 40/41, 64/65, 80/81, 128/129, and 160/161. Design the five prescalers in a manner which results in the need for only one PC board to be stocked.

PROCEDURE

1 The prescaler design in Fig. 17.23 will meet the ÷128/129 requirement.

2 The divide values 80/81 and 160/161 can be obtained from the same design used for ÷128/129 by replacing the MC12011 with the MC12009 and MC12013, respectively. No circuit modifications are necessary.

Count	Q_0	Q_1	Q_2	Q_3
0	L	L	L	L
1	H	L	L	L
2	L	H	L	L
3	H	H	L	L
4	L	L	H	L
5	H	L	H	L
6	L	H	H	L
7	H	H	H	L
8	L	L	L	H
9	H	L	L	H

Figure 17.24 MC10138/10538 BCD counter truth table for input clock connected to C1 and Q_0 connected to C2.

3 Applying the technique outlined in Ex. 17.8 shows that the circuit of Fig. 17.23, with only slight modifications, can also be used to yield ÷40/41 and ÷64/65. To accomplish this:

a. Omit the connection from pin 7 to pin 10 of Device A and tie pin 10 low.
b. Take the prescaler's expanded output from Q_2 rather than Q_3 of the MC10154.
c. Use MC12009 for Device A to obtain ÷40/41 and the MC12011 for ÷64/65.

Jumper wires can be used to realize steps a and b. They also may be used to remove the connections from Q_3 and $\overline{Q}_3$ of Device B to Device A if desired. These connections are not required for ÷40/41 and ÷64/65.

EXAMPLE 17.10 A wide bandwidth/fast response frequency synthesizer is required to generate frequencies from 260 to 290 MHz in 250 kHz steps. Dual-modulus prescaling is to be used so that the comparison frequency, f_R, can be made equal to the step size to allow a wide loop bandwidth and enhance response time. A portion of an existing synthesizer will be used for the $\div A$, $\div N$, and control logic functions. These circuits are characterized by: $f_{in(max)} = 10$ MHz; time for generating the modulus control signal = 100 ns. Determine suitable $P/(P + 1)$ values for the prescaler. Assume that the prescaler's propagation time is 30 ns and setup or release time is 5 ns.

PROCEDURE

1 Under *Rules for Choosing P*, by Rule 1,

$$f_{VCO(max)}/P \leqslant f_{in(max)}$$

or

$$(290 \times 10^6)/P \leqslant 10 \times 10^6$$

Hence, $P \geqslant 29$ is a requirement.

By Rule 2,

$$P/f_{VCO(max)} \geqslant (30 + 5 + 100) \text{ ns}$$

Therefore,

$$P \geqslant (290 \times 10^6)(135 \times 10^{-9}) \geqslant 39$$

is also a requirement.

2 From step 1, $P/(P + 1)$ must be made equal to, or greater than, 40/41.

3 The synthesizer's total divide range is $f_{VCO(min)}/f_R$ to $f_{VCO(max)}/f_R$ which gives N_{total} = 1040 to 1160.

4 The minimum divide that can be achieved using $P = 40$ is obtained by Eq. 17.13:

$$N_{total(min)} = (40 - 1)(40) + 0 = 1560$$

5 The value $N_{total(min)} = 1560$ does not satisfy the minimum divide requirement of 1040. Thus, $P \geqslant 40$ cannot be used. The requirements imposed on the synthesizer design, therefore, result in one of the rare instances where dual-modulus prescaling cannot be used. To make it feasible requires one of the following:

a. A lower value for f_R.
b. Less time than 100 ns for generating the modulus control signal.

Table 17.8 **Summary of device features for the MC145100 LSI CMOS synthesizer family**

Device	*Dual-Modulus operation*	*Program-ming format*	*Pkg pins*	*Dual-Modulus Prescaler range*	*÷A Counter range*	*÷N Counter range*	*$\div N_{total(max)}$ (Dual-Modulus operation)*	*Reference Counter divide values*
MC145144	No	4-bit data bus	16	N.A.	N.A.	4 → 4092 in steps of 8	N.A.	3584 → 3839
MC145145	No	4-bit data bus	18	N.A.	N.A.	3 → 16,384	N.A.	3 → 4096
MC145146	Yes	4-bit data bus	20	÷3/4 through ÷128/129	0 → 127	3 → 1024	131,199	3 → 4096
MC145151	No	Parallel	28	N.A.	N.A.	3 → 16,384	N.A.	8, 128, 256, 512, 1024, 2048, 2410, 8192
MC145152	Yes	Parallel	28	÷3/4 through ÷64/65	0 → 63	3 → 1024	65,599	8, 64, 128, 256, 512, 1024, 1160, 2048
MC145155	No	Serial	18	N.A.	N.A.	3 → 16,384	N.A.	16, 512, 1024, 2048, 3668, 4096, 6144, 8192
MC145156	Yes	Serial	20	÷3/4 through ÷128/129	0 → 127	3 → 1024	131,199	8, 64, 128, 256, 640, 1000, 1024, 2048
MC145157	No	Serial	16	N.A.	N.A.	3 → 16,384	N.A.	3 → 16,384
MC145158	Yes	Serial	16	÷3/4 through ÷128/129	0 → 127	3 → 1024	131,199	3 → 16,384
MC145159	Yes	Serial	20	÷3/4 through ÷128/129	0 → 127	3 → 1024	131,199	3 → 16,384

[a] $\div N$ can be increased by 856 without changing program code.
[b] Provides a digital detector to achieve frequency lock and a sample and hold detector to maintain phase lock.
NOTES: N.A. = not applicable. The $\div A$, $\div N$, and reference counters can be programmed in unit steps unless otherwise indicated. Standard package style is dual-in-line plastic or ceramic. Nominal width is 0.3 inch, except for pin 28 which is 0.6 inch.

17.5 CONSOLIDATED FUNCTIONS AND LSI DEVICES

The earlier sections of the chapter defined major functions which make up a typical PLL frequency synthesizer, stressed the important characteristics of each

Table 17.8 **Continued**

Detector error signal		*Phase of 3-state error signal for*	*Reference oscillator*	*Buffered oscillator output*	*Lock detector signal*	*Receive/ Transmit shift*	*Latched switch outputs*	*Other outputs*		
3-state	*double ended*	$f_V < f_R$						f_R	f_V	$f_R/5$
Yes	No	Negative	Yes	No	No	No	None	No	No	Yes
Yes	Yes	Positive	Yes	Yes	Yes	No	None	No	No	No
Yes	Yes	Positive	Yes	No	Yes	No	None	Yes	Yes	No
Yes	Yes	Positive	Yes	No	Yes	Yes[a]	None	No	Yes	No
No	Yes	Positive	Yes	No	Yes	No	None	No	No	No
Yes	Yes	Positive	Yes	Yes	Yes	No	Two	No	No	No
Yes	Yes	Positive	Yes	Yes	Yes	No	Two	No	No	No
Yes	Yes	Positive	Yes	Yes	Yes	No	None	Yes	Yes	No
Yes	Yes	Positive	Yes	Yes	Yes	No	None	Yes	Yes	No
[b]	[b]	[b]	Yes	No	Yes	No	None	No	No	No

function, and designated several chips which are available for implementing a synthesizer system. The ICs covered, for the most part, contained a single loop function, for example, an oscillator, counter, or phase detector. This allows for maximum flexibility and in some instances improved performance at the expense of cost, space, and equipment design time.

Several ICs also are available that contain more than one loop function. They range from the simpler two-function devices such as a phase detector/VCO to complex LSI chips that contain not only the PLL functions, but also a controller capable of programming the synthesizer and providing inputs/outputs to interface with other portions of the system. The MC6195, 96 are example devices. In some cases (e.g., the MC6805T2), the on-chip controller is essentially a microcomputer which can be software programmed to provide customized features. PLL/controller chips are typically optimized for specific applications, such as TV/CATV.

Other LSI PLLs, such as the CMOS family summarized in Table 17.8, do not contain controllers, but do offer most of the needed PLL functions along with low power drain and enough flexibility for a variety of applications. Also, five of these devices provide the modulus control signal required for operation with dual-modulus prescalers.

17.6 REFERENCES

Egan, W. F., *Frequency Synthesis by Phase Lock*, Wiley-Interscience, New York, 1981.

Fadrhons, J., "Design and Analyze PLLs on a Programmable Calculator," *EDN*, pp. 135–142, March 5, 1980.

Gardner, F. M., *Phaselock Techniques*, second edition, Wiley-Interscience, New York, 1979.

Hatchett, J., "Adding to PLL Chip's Functions Speeds RF Synthesizer Design," *Electronics*, pp. 148–154, Oct. 9, 1980.

Hatchett, J. and R. Janikowski, "Predict Frequency Accuracy for MC12060 and MC12061 Oscillator Circuits," Engineering Bulletin EB-59, Motorola Inc., Semiconductor Sector, Phoenix, 1975.

Holmbeck, J. D., "Frequency Tolerance Limitations with Logic Gate Clock Oscillators," 31st Annual Frequency Control Symposium Conference, Record, Atlantic City, June 1977.

Klapper, J. and J. T. Frankle, *Phase-Locked and Frequency-Feedback Systems*, Academic Press, New York, 1972.

Manassewitsch, V., *Frequency Synthesizers Theory and Design*, Wiley-Interscience, New York, 1976.

Motorola, "Phase-Frequency Detector, MC4044, MC4344," data sheet, Motorola Inc., Semiconductor Products Sector, Phoenix.

Motorola, "The Technique of Direct Programming by Using a Two-Modulus Prescaler," Application Note AN-827, Motorola Inc., Semiconductor Products Sector, Phoenix, 1981.

Nash, G., "Phase-Locked Loop Design Fundamentals," Application Note AN-535, Motorola Inc., Semiconductor Products Sector, Phoenix, 1970.

Nichols, J. and C. Shinn, "Pulse Swallowing," *EDN*, pp. 39–42, Oct. 1, 1970.

Sharpe, C. A., "A 3-State Phase Detector Can Improve Your Next PLL Design," *EDN*, pp. 55–59, Sept. 20, 1976.

Signetics, *Linear Phase-Locked Loops Applications Book*, Signetics Corp., Sunnyvale, CA, 1972.

Williams, R. L., "Making It Clear As Crystal," *Communications*, pp. 20–28, March 1978.

PART THREE
FABRICATION

CHAPTER 18

Thick Film Technology

Roydn D. Jones
Manager Instrumentation Systems Research
Tektronix, Inc.

18.1 INTRODUCTION

Thick film is the field of microelectronics in which specially formulated pastes are applied and fired on an insulating, usually ceramic, substrate to produce sets of components such as conductors, resistors, and capacitors. The thick film approach utilizes materials such as glass, noble metals, and metal oxides. The materials are formulated into a paste with appropriate binders, organic solvents, and plasticizers (an additive that provides flexibility), the final paste having a viscosity (flow-resistance) suitable for silk screening operations. The pastes are available as conductors, resistors, or dielectrics and are silk screened onto a substrate and fired to form the final circuit.

Using this approach, high-density circuits containing resistors, conductors, and capacitors can be fabricated at low cost. Active devices such as diodes, transistors, and integrated circuits can be added to form low-cost, fully functional hybrid microcircuits. Such circuits are finding a wide range of applications, particularly in areas where size and/or high-frequency requirements are at a premium. Figure 18.1 shows examples of thick film circuits and indicates the flexibility of this technique.

18.2 THICK FILM MATERIALS

There are three main categories of thick film pastes or inks, namely, conductors, resistors, and dielectrics. A thick film paste has three main constituents:

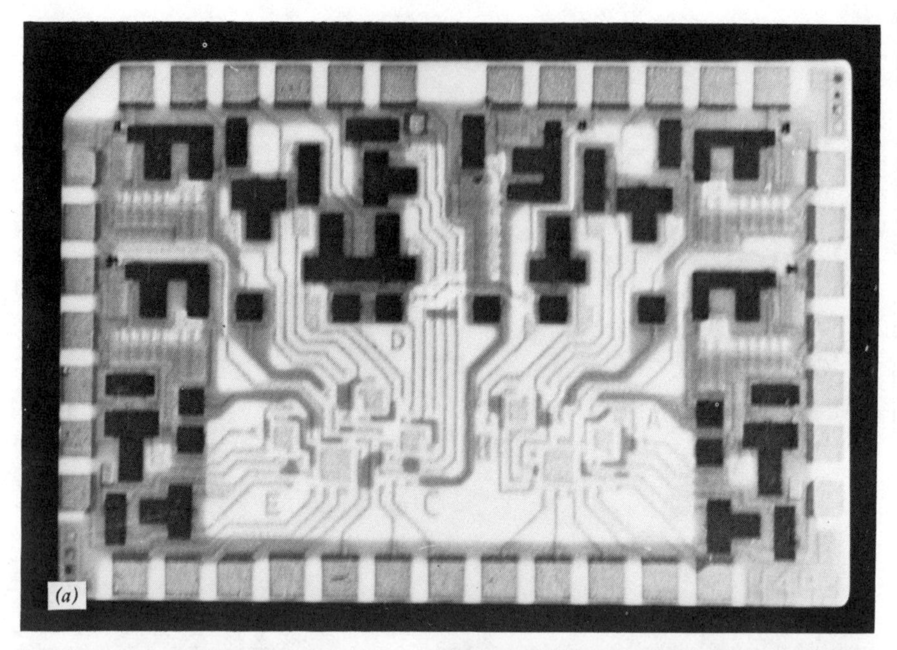

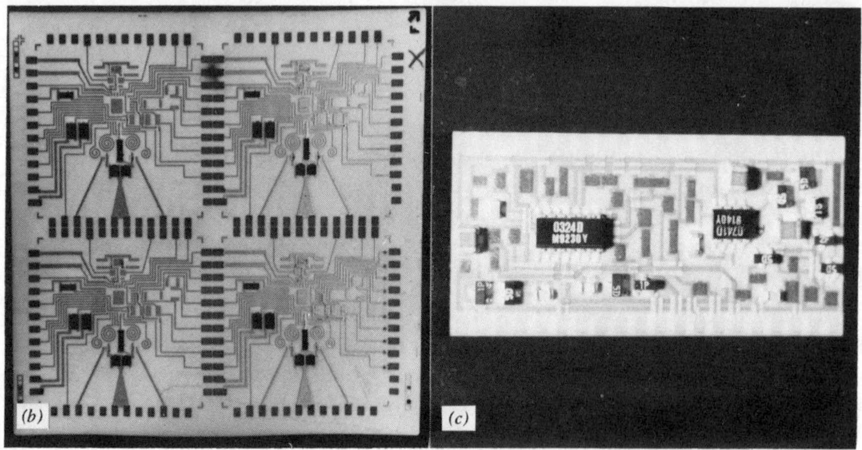

Figure 18.1 Examples of thick film circuits. (*a*) A thick film substrate containing screened resistors (black areas) and conductors (gray areas). (*b*) Multi-image thick film circuit. (*c*) Assembled thick film circuit.

1 The binder—usually a glass frit.

2 The vehicle—organic solvents and plasticizers.

3 The functional elements—metals, alloys, oxides, or ceramic compounds.

The relative proportions of binder and functional elements determine the resistive, conductive, or capacitive properties of the final film.

Thick Film Conductors

Conductors are essential to any thick film circuit and are used in large quantities. Owing to the precious metal content of these conductors, they can contribute considerably to the cost of a thick film hybrid.

Conductors are used to interconnect components, as terminations for resistors, as attachment pads for discrete components, and as electrodes for thick film capacitors. The requirements for a good thick film conductor are:

1. Good conductivity—0.002 to 0.15 Ω/square.
2. Good adhesion—2000 psi for a 100 mil square pad.
3. Provide good wire bondability.
4. Provide good eutectic* die attachment.
5. Solderable with high leach resistance.
6. Good line definition and resolution.
7. Good screening properties.
8. Stable during processing.
9. Long shelf life.

Conductor Formulations

Binder The binder consists of low melting point glasses that hold the metal particles in contact with, and bond the film, to the substrate. Properties of interest are glass temperature/viscosity relationships, surface tension properties, chemical reactivity, and coefficient of thermal expansion.

Vehicle The vehicle helps define the printing characteristic of the paste. The vehicle consists mainly of organic solvents.

Conductive material The conductive materials are small metallic particles less than about 5 μm in size. The amount of metal, particle size, size distribution, and particle shape have a significant influence on the electrical and physical properties of the final thick film conductor. Metals used are gold, silver, palladium, and platinum because these precious metals do not oxidize during the high-temperature firing. Low-cost pastes containing copper (which are fired in an inert atmosphere) are also available.

Conductor Types

Silver (Ag) Silver provides the lowest-cost material for conductors. It exhibits good substrate adhesion, solders easily, and is easily processed. The major problem with silver is its tendency to migrate under the influence of a dc electric

* See Sec. 18.7 for definitions of commonly used terms in thick film technology.

field. This can result in conductive shorts and a reaction of the silver conductor with resistive films. It also suffers from poor solder leach resistance and tends to tarnish—a well-known fact in the jewelry business.

Gold (Au) Gold is used frequently as a conductor material for thick film. It provides a good high-frequency conductor with low resistance, usually about 0.003 to 0.01 Ω/square. Gold has excellent bonding characteristics and is used both for wire bonding and eutectic die attach of semiconductor die.

Palladium Silver (PdAg) Palladium silver accounts for about 90% of the applications for thick film conductors. It has most of the required properties such as good adhesion, good solderability, good solder leach resistance, and low cost. However, because of its silver content, the material has a slight silver migration problem, and care needs to be taken in high humidity and high dc field applications. The material has a higher sheet resistivity than gold, being approximately 0.03 Ω/square. A major disadvantage is that it gives very poor results with wire bonding.

Palladium Gold (PdAu) and Platinum Gold(PtAu) These are fairly popular thick film pastes made by adding palladium or platinum to a gold paste. The addition of these metals improves the solderability characteristics of the pastes while maintaining reasonable wire bonding characteristics. The major disadvantages are the poor resistivity, approximately 0.04 to 0.08 Ω/square, and their high cost which approaches that of gold pastes.

Thick Film Dielectrics

Dielectrics are used as crossover insulators for multilayer conductors, as capacitor dielectrics for thick film capacitors, and as an overall encapsulation glaze. Crossover dielectrics are designed with low dielectric constant materials to minimize coupling capacitance between conductors. Capacitor dielectrics, however, are doped with various amounts of barium titanate to give a wide range of dielectric constants.

The requirements for a good thick film dielectric are:

1 Breakdown strength: 250 V/mil.

2 Insulation resistance greater than 10^{10} Ω.

3 Dielectric constant of 6 to 10 for crossovers.

4 Dielectric constants of 10 to 2000 for capacitors.

5 Minimum tendency to form pin holes.

6 High resistance to thermal shock crazing (cracking).

7 Good screening properties.

8 Long shelf life.

Dielectric Formulations

Binder To achieve a dense pin-hole free film the material needs to flow reasonably well, and a low-viscosity glass binder is required. However a low-viscosity glass tends to allow top conductors to sink or swim in the material during subsequent firings. This could result in shorted crossovers or capacitors. Manufacturers have solved this problem by utilizing crystalline glass formulations which result in a higher viscosity material after the initial firing operation.

Vehicle The vehicle used in dielectric pastes is similar to that for conductor pastes, consisting mainly of organic solvents.

Dielectric materials The characteristics of the final capacitor depend on the nature of the dielectric material incorporated in the paste. There are two major classes of material, those with low permittivity and those with high permittivity.

Dielectric Types

The materials with high dielectric constants, the so-called "high K" types, are largely based on the ferroelectric ceramic, barium titanate. This material has a dielectric constant of about 1200 at room temperature. Addition of strontium, calcium, tin, or zirconium oxides affects the dielectric constant and reduces the temperature coefficient of capacitance to reasonable values. Capacitors with dielectric constants from 1000 to 3000 with temperature coefficients of ±5000 ppm/°C have been formulated.

Several low permittivity, nonferroelectric dielectric materials based on magnesium titanate, zinc titanate, titanium oxide, and calcium titanate are also available. Dielectric constants are in the range 12 to 160 with temperature coefficients of ±200 ppm/°C.

The high K dielectrics are not as stable as the low K dielectrics, exhibiting a gradual decrease in capacitance with time. Changes may be of the order of 5 to 10%. Only low K dielectrics should be used in applications where capacitance drift is a critical parameter.

Glass/ceramic pastes with dielectric constants close to 10 have been developed for crossovers and multilayer circuits. These materials also exhibit the required increase in viscosity after the initial firing operation. After firing at temperatures close to 850°C, these materials can be refired to 1000°C without any loss in dimensional stability.

Glass encapsulation pastes comprise vitreous materials whose melting points are much lower than any of the other thick film pastes. Firing temperatures are normally in the 450 to 500°C range. Metal oxide pigments are often added to these glass pastes to give a colored appearance.

Thick Film Resistors

Resistor pastes are second to conductors as far as usage is concerned. The resistor pastes are available in sheet resistivities from 1 Ω per square to 10 MΩ per square, giving a wide range of possible resistor values. The requirements for a good thick film resistor paste are:

1 Wide range of resistor values.
2 Compatible with substrate and termination conductor materials.
3 Low-temperature coefficient of resistance (TCR).
4 Good tracking between resistors.
5 Low-voltage coefficient of resistance (VCR).
6 Stable at elevated temperatures and during thermal cycling.
7 Stable in high-humidity conditions.
8 Low current noise owing to hot spots and general homogeneity of the material.
9 Capable of being blended to form intermediate sheet resistivities.

Resistor Formulations

Binder For thick film resistor pastes, the glass binder selection is very critical, because the glass system has a significant effect on resistor properties. Particle size and firing requirements can affect both the sheet resistivity and TCR of the film. Compositions vary from manufacturer to manufacturer with lead bismuth borosilicate, lead bismuth, and lead zirconate being used in conjunction with various oxides.

Vehicle This defines the printing characteristic of the paste and typically consists of resin materials and organic solvents.

Conductive material The conductive materials used are usually metals, metal oxides, or a combination of these. Often several different metals or oxides are employed to achieve the desired properties. The conduction mechanisms involved are not fully understood but can be related to semiconductor conduction mechanisms.

Resistor Types

Early thick film resistor pastes were based on compositions of palladium, palladium oxide, and silver. However, these were very sensitive to the firing profile and the newer improved pastes are based on ruthenium, iridium, rhenium, and their oxides. The latter systems exhibit low TCR and are fairly insensitive to the firing profile.

Ruthenium oxide is probably the most commonly used material in thick film resistor pastes. The oxide is very stable and can be heated to 1000°C without undergoing chemical change. Formulations using ruthenium oxide and lead glass provide pastes with a wide range of sheet resistivity, good stability, low TCR, and low electrical noise.

The ruthenium oxide pastes are more expensive than the palladium-palladium, oxide-silver compositions, although the cost is usually insignificant, being maybe 5 to 10 cents per square inch of fired film. There is usually very little cost saving in choosing a low-cost paste over an improved expensive paste.

18.3 THICK FILM PROCESSING

Screening

The process of selectively depositing the thick film materials is almost identical to the silk screening process used extensively in the production of decorative finishes, labeling of containers, and in the production of etched or printed circuit (PC) boards. The process depends on forcing the thick film paste through a stencil pattern comprising a fine mesh, usually of stainless steel. The screens can have mesh counts from 105 to 325 lines per inch. Line widths down to 0.005 inches can be achieved using the finer screens and high quality screening equipment. The screens are evenly coated with a photographic emulsion, dried, and the required pattern defined by exposing the emulsion to an ultraviolet source through a photomask. After exposure the nonexposed areas, corresponding to the pattern to be printed, are washed away with a developer solution.

A typical patterned thick film screen is shown in Fig. 18.2. Note that in this case there is a multiple image of the circuit formed on the screen. This allows multiple substrates to be screened in each screening operation, thus reducing the overall processing cost. The multiple screenings are usually made on a large master substrate which is scribed and broken into individual circuits on completion of all the screening and firing operations.

The screen is mounted on a screen printer, the pattern aligned with the substrate, and the required thick film paste forced through the screen onto the substrate by a squeegee traversing the screen at a controlled rate. The amount of material deposited depends on the following factors:

1 Rheology properties of the materials.

2 Mesh type and tension.

3 Mesh thickness.

4 Squeegee material.

5 Vertical distance between the screen and the substrate (*snap-off* distance).

6 Pattern being printed.

7 Cleanliness of the process.

Firing

Immediately after screening, the solvents are removed by air drying or low-temperature, heat-assisted drying at about 100°C. After this air drying, the substrates can be fairly easily handled as the dry paste is reasonably resistant to scratches.

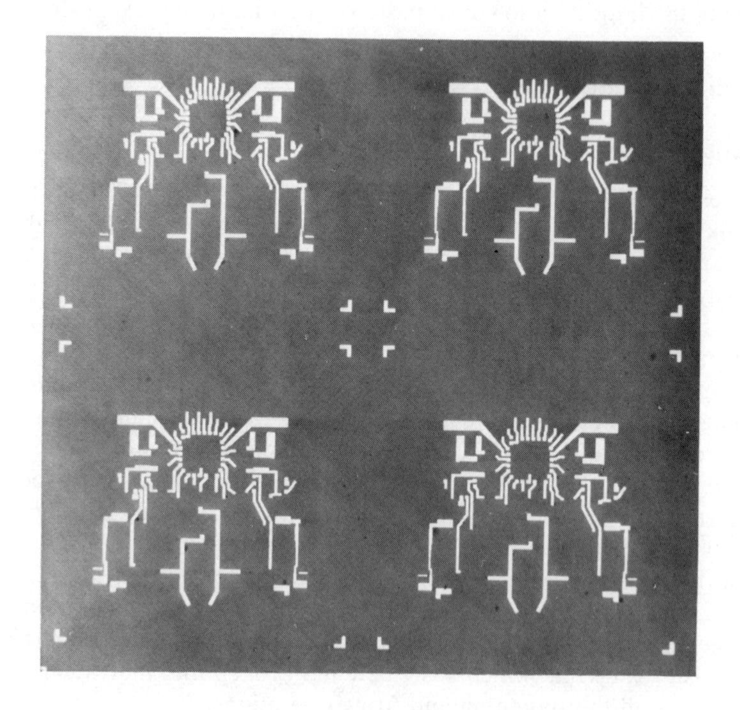

Figure 18.2 A multiple patterned thick film screen.

Following the initial drying, the substrates are fired at temperatures between 500 and 1000°C, depending on the materials being fired. The substrates are fired in conveyor-belt kilns with several zones of controllable heat. This allows an accurate temperature profile to be set up for a particular material. The kiln temperature profile is very critical, particularly with resistive pastes. Temperature variations of the high temperature zone may affect some resistor films as much as 3% per °C.

There are three main zones in the firing operation, as shown in Fig. 18.3. These are:

1 The preheat section where remaining organic constituents are burned off—usually about 250 to 350°C.

2 The hot (firing) zone where the materials are sintered (coherent formed bond) at temperatures from 500 to 1000°C, depending on the material being fired. During this sintering process the glass frit melts and wets the substrate, and the glass/metal matrix forms the functional properties of the final film.

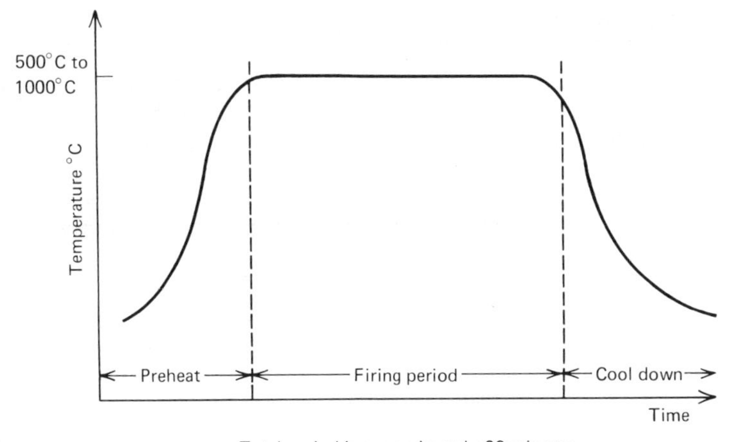

Figure 18.3 Typical temperature profile of a conveyor belt kiln used in firing thick film substrates.

3 The cooling zone allows the substrates to return to room temperature with minimal thermal stressing of the films and substrates.

The usual firing time is 60 minutes, with about 10 minutes at the peak temperatures in the hot zone. Typical firing temperatures for the various thick film materials are given in Table 18.1.

The material sequence is selected so that the highest firing temperature material is applied and fired first, with each succeeding screening and firing at a lower temperature. This is done to minimize interaction between the various materials, and to minimize variations in the properties of the earliest fired materials.

Process Flow

A typical process flow for a thick film hybrid comprising metal conductors on the front side, electrical ground plane metal on the back side, a crossover dielectric on the front side, two different sheet resistivities, and a final encapsulation would be:

Screen metal front side
Air dry
Screen ground plane reverse side
Fire at 950°C
Screen crossover dielectric on front side
Air dry
Screen crossover metal
Air dry
Fire at 900°C

Table 18.1 **Typical firing temperatures**

Material	*Firing temp.* °C
Metallization	900 to 1000
Dielectric capacitors	850 to 1000
Resistors	750 to 900
Dielectric encapsulation	500 to 750

Screen resistor #1
Air dry
Screen resistor #2
Air dry
Fire at 850°C
Screen encapsulation dielectric
Air dry
Fire at 550°C

Although there are many process steps in such a sequence, many parts can be processed simultaneously at each step resulting in a low-cost process.

18.4 THICK FILM SUBSTRATES

The substrate is the mechanical base and electrically insulating material used for hybrid fabrication. The material properties of the substrate affect both the processes used and the final characteristics of the final hybrid circuit.

Engineering Concerns

1. *Dielectric constant* Capacitance effects are directly proportional to the dielectric constant.
2. *Dielectric strength* Determines the voltage breakdown properties of the substrates.
3. *Dissipation factor* Determines the electrical losses in the substrate, and may be particularly important in high-frequency and microwave circuits.
4. *Thermal conductivity* Determines how much power the film components or added discrete components can dissipate without excessive increases in temperature.
5. *Thermal coefficient of expansion (TCE)* A major factor in determining the compatibility between the substrate and the thick film materials. It has a significant effect on the film adhesion and the temperature coefficients of resistance and capacitance.
6. *Volume resistivity* Determines the electrical insulation between the circuit elements on the substrate.

Manufacturing Concerns

1 *Ability to withstand high temperatures* Typically 500 to 1000°C for most thick film materials.

2 *Mechanical strength* Important for ease of handling.

3 *Surface finish* Important for good line definition and printed film uniformity. However, the surface needs to be sufficiently rough to ensure adequate adhesion of the fired thick films.

4 *Camber* Should be a minimum of distortion or bowing of the substrate. Too much camber can result in thick film screening problems.

5 *Visual defects* Small surface defects, such as pits or burrs, can result in circuit defects such as open circuits and pin holes.

6 *Materials compatibility* Substrate should be chemically and physically compatible with the chemicals and materials used for the thick film pastes.

7 *Low cost* Substrate should be low cost in large-quantity production.

8 *Tolerances* Substrate should have reasonable tolerances both in overall dimensions and in the positioning of holes or apertures in the substrate.

Substrate Materials

The majority of substrates used for thick film hybrids are produced from ceramic materials such as alumina, beryllia, magnesia, thoria, zirconia, or combinations of these materials. The industrial standard ceramic substrate is alumina, which offers an excellent combination of the electrical and mechanical properties required. Newer substrate materials, particularly *porcelain steel* substrates similar to the material used in commercial equipment such as washing machines, are gaining in popularity.

Alumina substrates are usually in the 94 to 96% purity range with grain sizes of from 3 to 5 μm. The glassy grain boundaries in the alumina react with the thick film binder glasses during firing to give significantly higher bond strength in comparison with most other substrate materials.

Steatite or fosterite have often been used where lower cost and lower dielectric constant material than alumina is required. However, they are incompatible with most thick film materials and processes and have lower tensile strength and thermal conductivity.

When power dissipation is a problem, beryllia (BeO) with its high thermal conductivity is often used. Although beryllia has a thermal conductivity about eight times that of alumina, it is expensive in substrate form, and slightly weaker. Also, beryllia is highly toxic in either powder or vapor form and requires special handling and safety equipment when machining or firing at high temperatures.

Porcelain steel substrates are very low-cost substrates and are available in large sizes. They can be used in reasonably high-power applications because the steel backing acts as an excellent heat sink. These substrates are finding increased use in high-volume, low-cost applications. Capacitive effects to the steel substrate

Table 18.2 Properties of thick film ceramic substrates

Characteristic	*Units*	*Conditions*	*96% Alumina*	*99.5% Alumina*	*99.5% Beryllia*
Dielectric constant		1 MHz	9.3	9.9	6.9
		1 GHz	9.2	9.8	6.8
Dielectric strength	V/mil		210	220	230
Dissipation factor		1 MHz	0.0003	0.0001	0.0002
		1 GHz	0.0009	0.0004	0.0003
Thermal conductivity	W/(cm-°C)	25°C	0.351	0.367	2.5
		300°C	0.171	0.187	1.21
Thermal coefficient of expansion		25° to 300°C	6.4×10^{-6}	6.6×10^{-6}	7.5×10^{-6}
Bulk resistivity	Ω-cm	25°C	10^{14}	10^{14}	10^{14}
		100°C	2×10^{13}	7.3×10^{13}	10^{14}
Tensile strength	psi		25,000	28,000	23,000
Surface finish	Micro-inches		25 to 40	10	20
Camber	Mils/inch		4	4	4

through the thin porcelain layer, usually about 0.005 inches thick, often preclude their use for high-frequency hybrids. Table 18.2 compares the properties of some commercially available thick film ceramic substrates.

18.5 THICK FILM DESIGN GUIDELINES

General Design Procedures

The transition from an electrical schematic to a thick film hybrid circuit can be a complex process. The designer must keep in mind constraints such as electrical specifications, size, weight, thermal requirements, cost and reliability, as well as the design rules for the various thick film elements.

Given an electrical schematic, the design steps for a successful thick film hybrid design are:

1. Worst case electrical design to ensure the circuit meets specifications over all operating conditions. Temperature effects such as temperature coefficients of resistance and capacitance, temperature drift, and process tolerances all need to be considered. Often computer-aided design tools are used in simulating the circuit under worst case conditions.
2. A preliminary layout is attempted to determine both substrate size and packaging requirements. Process and product flow charts can now be determined and the final layout completed.
3. Photographic masks and screens are processed and prototype hybrids fabricated.
4. These circuits are tested to meet the component specifications and modifications are made to the design if required.
5. Generally one or more subsequent design cycles are required before the component meets all its specifications.

Resistor Design Guidelines

Sheet resistivity The thick film resistor shown in Fig. 18.4 can be considered as a conductive block of uniform thickness T, width W, and length L. If the bulk resistivity (resistance per unit volume) of the thick film paste is ρ Ω-cm, then the resistance R measured across the block in the direction of length L is given by:

$$R = \frac{\text{Bulk resistivity} \times \text{length}}{\text{Cross sectional area}}$$

or,

$$R = \frac{\rho L}{TW} \tag{18.1}$$

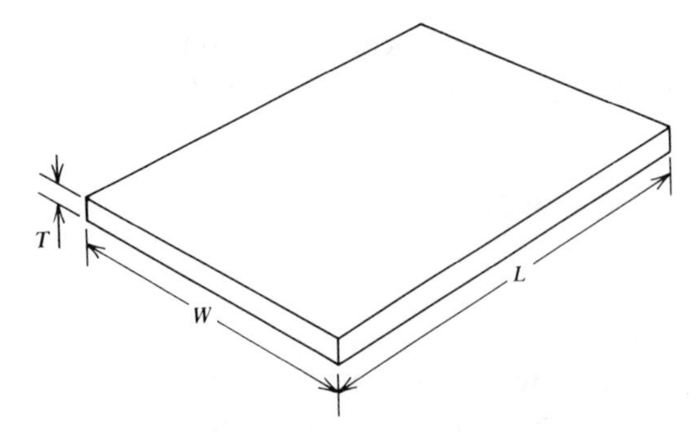

Figure 18.4 Geometry of a thick film resistor.

If ρ/T in this expression is now defined as the sheet resistivity (also called *sheet resistance*) ρ_s, then,

$$R = \rho_s \times \frac{L}{W} \tag{18.2}$$

or,

$$R = \rho_s \times \text{number of squares}$$

where L/W is the number of squares in the direction in which the resistor is being measured. Quantity L/W is called the *aspect ratio*. The sheet resistivity is defined in terms of Ω per square, and is the basic design parameter specified for thick and thin film, and diffused resistors (Chap. 20).

EXAMPLE 18.1 If a resistor is screened with a paste of sheet resistivity 10,000 Ω per square, and the resistor is defined as 0.200 inches long and 0.050 inches wide, calculate the resistance R.

PROCEDURE

By Eq. 18.2,

$$R = 10{,}000 \frac{\Omega}{\text{square}} \times \frac{0.200}{0.050} \text{ squares} = 40{,}000\ \Omega$$

EXAMPLE 18.2 Calculate the sheet resistivity of a square of thick film resistor material with the following properties: bulk resistivity = 10^{-1} Ω-cm and film thickness = 20 μm.

PROCEDURE

1 By definition, the sheet resistivity = the bulk resistivity divided by the film thickness.
2 Substitution of values in the above expression yields:

$$\rho_s = \frac{10^{-1}\ \Omega\text{-cm}}{20 \times 10^{-4}\ \text{cm}} = 50\ \Omega/\text{sq.}$$

It is very important to remember that for any given film of constant thickness, increasing the length and width by equal ratios will not alter the overall resistance value. For example, a 50 mil by 50 mil resistor has the same value as a 200 mil by 200 mil resistor. They both have the dimensions of 1 square.

Normally, sheet resistivity of a thick film paste is specified for a given film thickness *T*, and this value can be affected significantly by the screening and firing parameters. Most resistive pastes are specified at about 0.8 mil to 1.0 mil thickness.

Resistor Design

The aspect ratio should not be greater than about 25:1 or less than about 1:3. Rather than designing with aspect ratios outside these ranges it is usually more convenient and, more importantly, area saving to change to a different sheet resistivity. For example, rather than designing a 100 Ω resistor with 100 squares of 1 Ω per square paste, it is better to design it with 1 square of 100 Ω per square paste. However, it requires a separate piece of artwork, a separate screen, and a separate screening operation for each different sheet resistivity paste used. For low-cost production it is important to select sheet resistivities to minimize the number of screenings required.

EXAMPLE 18.3 Select the minimum number of pastes required to design 50, 110, 350 and 2200 Ω resistors on the same thick film substrate. The available pastes are 10, 100, and 1000 Ω/square, respectively. The aspect ratios must be less than 10:1 and greater than 1:3.

PROCEDURE

1 Construct a table as illustrated in Fig. 18.5.
2 The minimum number of pastes, from the table, is two: the 100 Ω/sq. and the 1000 Ω/sq. There is a choice with the 350 Ω resistor of using either 100 Ω/sq. or 1000 Ω/sq. pastes.

Other requirements such as power dissipation, temperature tracking, voltage stress, and laser trim also are factors in choosing particular resistor geometries. Conservative design rules are power densities less than 50 W per square inch and voltage stress of less than 1 V per mil.

	50 Ω	110 Ω	350 Ω	2200 Ω
10 Ω	(5:1)	11:1	35:1	220:1
100 Ω	(1:2)	(1.1:1)	(3.5:1)	22:1
1000 Ω	1:20	1:9	(1:2.8)	(2.2:1)

Note: Circled quantities are acceptable aspect ratios.

Figure 18.5 Selecting the minimum number of resistor pastes (Ex. 18.3).

Typical resistor geometries are shown in Fig. 18.6, including rectangular, zig-zag (meander) and top hat. The zig-zag geometry is used for high square count resistors, the corners being counted as approximately 0.5 square. Resistors are usually designed to be at least 20 mils wide and 30 mils long.

EXAMPLE 18.4 Calculate the resistance of the 40 mil meander resistor shown in Fig. 18.7. Assume 0.5 squares for the corners. The sheet resistivity is 100 Ω/sq.

PROCEDURE

1. Divide the figure into squares.
2. Count the number of squares, including corners.
3. The result will be 45 squares. Multiplying 45 by 100 Ω/sq. results in a resistance of 4500 Ω.

Resistor Tolerances and Laser Trimming

Owing to variances in materials, screened thickness, screened geometry, and firing parameters, the typical thick film resistor will have a wide tolerance of about ±20%. Resistors requiring this wide tolerance can be designed at layout to nominal value. However, tighter tolerance resistors are designed low at layout and are trimmed into value after firing.

Various trim geometries and trim cuts are available and some of the more popular are shown in Fig. 18.8. Resistors can only be increased in value during trimming and can be trimmed as a passive network (*passive trim*) or trimmed after active devices have been added and the complete circuit is operational (*active* or *functional trim*).

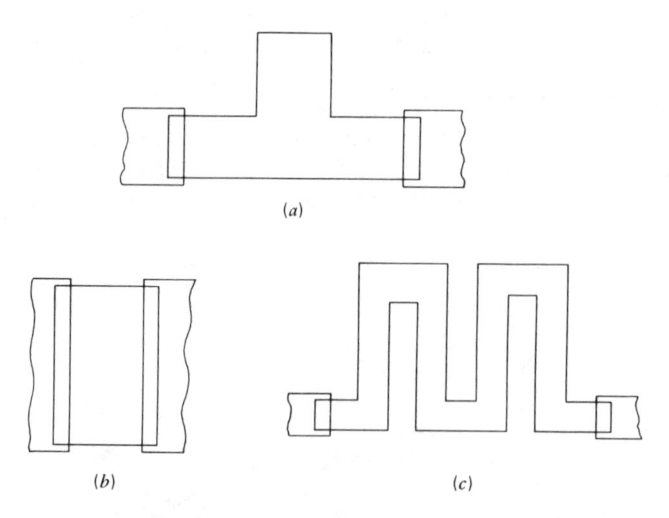

Figure 18.6 Typical thick film resistor geometries. (*a*) Top hat. (*b*) Rectangular. (*c*) Meander (zig-zag).

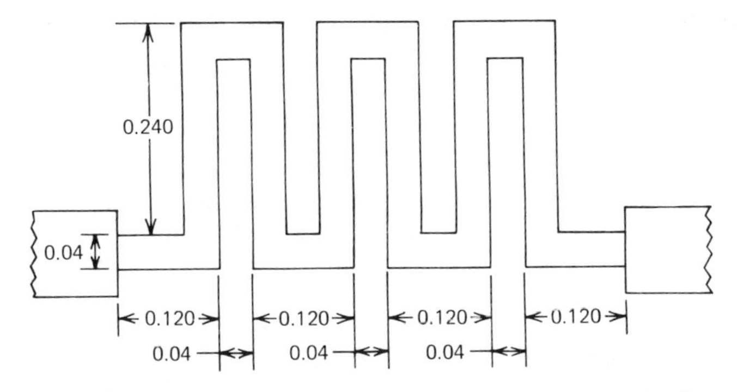

Figure 18.7 Calculating the value of a meander resistor (Ex. 18.4).

The two well-known methods of trimming use a sand, or abrasive trimmer or a laser trimmer. Both processes remove the resistor material down to the ceramic substrate forming a clean cut or *kerf*. The laser trimmer is probably the most popular method giving a very fine kerf of about 1 mil, and is often done under computer control. Using these methods of trimming, very stable resistors with end of life tolerances better than 0.5% can be easily obtained.

Typical Characteristics of Thick Film Resistors

Temperature coefficient of resistance (TCR) The best thick film resistors have TCRs in the range ±100 ppm/°C over the temperature range −55 to 125°C.

TCR tracking Resistors of the same thick film paste (same resistivity) fabricated in the same screening operation will track to ±15 ppm/°C.

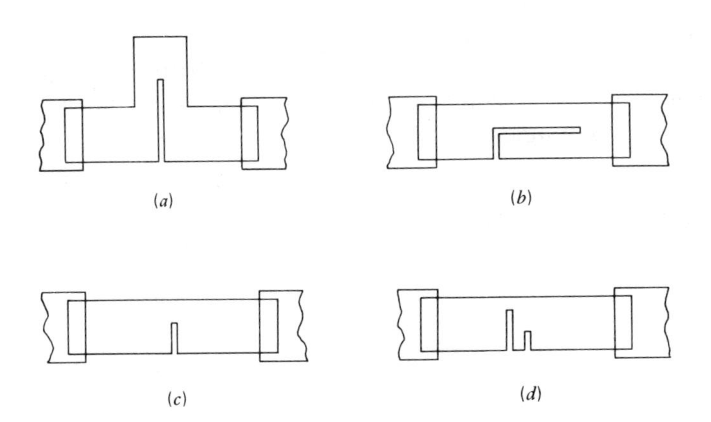

Figure 18.8 Typical resistor trim cuts. (*a*) Top hat plunge. (*b*) L cut. (*c*) Plunge. (*d*) Double plunge.

Long term stability Well-designed resistors will drift less than 0.5% after 5000 hours at 125°C. The tracking between resistors will be better than 0.2% after 5000 hours at 125°C.

Power dissipation Most commercial resistor pastes are specified as capable of dissipating power densities of up to 50 W/in^2. However, this assumes that the total power dissipation on the substrate does not exceed 1 to 2 W/in^2.

Conductor Design Guidelines

There are many thick film conductor compositions each with different properties. The various functions required of a conductor are solderable connections, resistor terminations, crossover connections, capacitance electrodes, wire bonding pads, eutectic die bond pads, and general low-impedance connections or runs on the hybrid.

Materials in general usage are gold for good conductivity and wire bonding, platinum gold for optimal solder leach resistance, and palladium silver for low-cost, solderable conductors. Copper is finding increased usage as a low-cost, general-purpose thick film conductor, and improvements are continually being made in these copper-based pastes. Table 18.3 compares the properties of the most commonly used thick film conductors.

Conductors can be designed as narrow as 8 mil lines and 7 mil spaces, although larger widths and spaces are preferred. Conductors with 5 mil lines and 5 mil spaces can be fabricated with some yield loss. Conductors can be screened on the front and back surfaces of the substrate and can be connected through holes in the substrate (called *vias*) or around the edges.

One advantage of the thick film process is the ability to fabricate conductive crossovers with high yield. A crossover is made by printing a conductor followed by an insulator/dielectric layer, and followed by a second crossover connection. Care needs to be taken in designing such crossovers to minimize the capacitive coupling through the crossover dielectric layer.

Using a succession of conductor and dielectric layers, *multilayer* thick film circuits can be fabricated. Connections are made from one layer to the next through small vias in the dielectric layers. Circuits with six or more conducting layers can be fabricated in this manner.

Table 18.3 **Comparison of thick film conductors**

	Gold	*Platinum gold*	*Palladium silver*
Cost	High	Medium	Low
Sheet resistivity (mΩ per square)	3 to 4	50 to 80	25 to 35
Solderable	No	Yes	Yes
Wire bondable	Good	Poor	Poor
High frequency	Good	Poor	Poor

Capacitor Design Guidelines

Thick film capacitors can be made in several ways, with values from fractions of a picofarad to thousands of picofarads being feasible. Dielectric pastes are available with dielectric constants ranging from 10 to 20,000. The low K dielectrics are very stable, but the higher K dielectrics have poor temperature coefficients of capacitance and poor long term stability.

Substrate Capacitors

A convenient way of fabricating low values of capacitance is to utilize the substrate itself as the dielectric, as shown in Fig. 18.9. The capacitor is across the substrate from an electrode on one side to another electrode or ground plane on the other side. Its value is a function of the substrate dielectric constant and thickness:

$$C = 0.225 \times \frac{\varepsilon_r \times A}{t} \tag{18.3}$$

where C = capacitance (picofarads, pF)
ε_r = dielectric constant
A = electrode area (inches^2)
t = substrate thickness (inches)

For the more common alumina substrates with dielectric constant of 9.6 and 25 mil thick,the capacitance is:

$$C = \frac{0.225 \times 9.6 \times 10^{-6}}{25 \times 10^{-3}} \text{ pF/mil}^2$$
$$= 8.6 \times 10^{-5} \text{ pF/mil}^2 \tag{18.4}$$

Thus a good rule of thumb is that for a 25 mil thick alumina substrate the capacitance from one side of the substrate to the other is approximately 10^{-4}

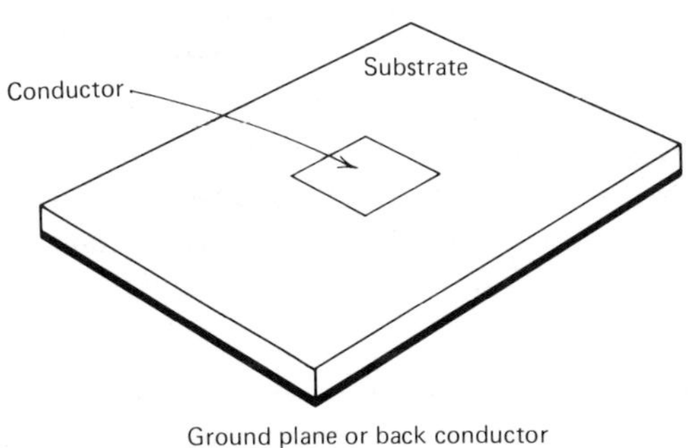

Figure 18.9 An example of a thick film substrate capacitor.

pF/mil^2. Because of this low value, only small values of capacitance up to about 3 pF are fabricated in this manner. Larger capacitance values take up too much substrate area. For example, a 1 pF capacitor requires an area of 100 mil × 100 mil. Although the values achievable are very small, such capacitors are often used in matching networks and compensation circuitry in high-frequency hybrids.

EXAMPLE 18.5 Calculate the capacitance of a 50 mil × 50 mil thick film capacitor if the dielectric constant = 100 and dielectric film thickness = 0.8 mil.

PROCEDURE

1 By Eq. 18.3,

$$C = 0.225 \times \frac{\varepsilon_r \times A}{t}$$

2 Substitution of values in Eq. 18.3 yields:

$$C = \frac{0.225 \times 100 \times 0.05 \times 0.05}{0.0008} = 70.3 \text{ pF}$$

EXAMPLE 18.6 Design a circular 100 pF capacitor with the same thick film dielectric as in Ex. 18.5.

PROCEDURE

1 The area of a circle whose radius is r is πr^2.
2 Substitution of values in Eq. 18.3 gives us:

$$100 = \frac{0.225 \times 100 \times \pi r^2}{0.0008}$$

3 Solving for r, we find that a circle of radius = 34 mils is required.

Interdigital Capacitors

This type of capacitor is fabricated using interdigital conducting fingers screened on one side of the substrate, as shown in Fig. 18.10. The capacitance is approximately 1 pF/inch of run for an alumina dielectric with equal width lines and spaces. Typical values achievable with thick film interdigital capacitors are from 0.5 to 3 pF. Above this value, the capacitors take up too much area because of the minimum 5 mil lines and 5 mil spaces required for thick film screening. This type of capacitor is better suited to the fine line definition of thin films (Chap. 19).

EXAMPLE 18.7 Design a 5 pF interdigital thick film capacitor on an alumina substrate using 5 mil wide runs and spaces. The capacitor should have a final aspect ratio of approximately unity. Use the approximation that C = 1 pF/inch of run.

Figure 18.10 An interdigital capacitor.

PROCEDURE

1 The layout for the interdigital capacitor is drawn in Fig. 18.11.
2 The capacitance = $2x$ pF/digit. For N digits,

$$C = 5 \text{ pF} = 2xN$$

3 For a square topology (to yield an aspect ratio of unity),

$$N = \frac{x}{0.020}$$

4 Substitution of this expression into $C = 2xN$ yields:

$$5 = \frac{2x^2}{0.020}$$

5 Solving for x gives us $x = 0.22$ inch. Hence, $N = 0.22/0.02 \simeq 11$.

Parallel Plate Capacitors

The structure of a parallel plate capacitor is shown in Fig. 18.12. The capacitor consists of a first conductor, a dielectric layer, and a second top conductor. The capacitance is given by Eq. 18.3 where now ε_r = dielectric constant of the dielectric layer and t = dielectric thickness in inches. The range of dielectric thickness t is from 0.5 to 2.5 mils, the nominal value being approximately 1.4

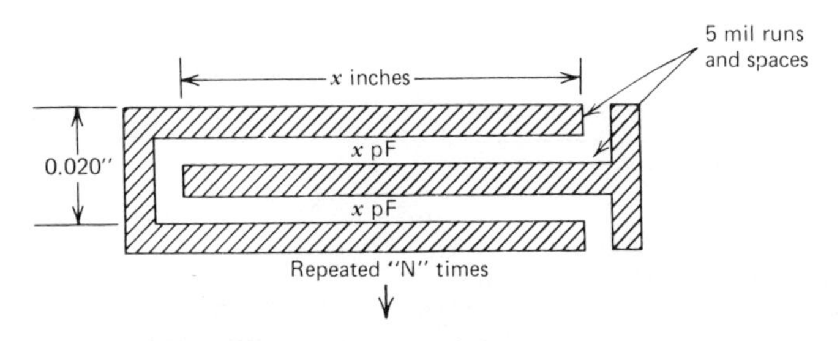

Figure 18.11 Designing an interdigital capacitor (Ex. 18.7).

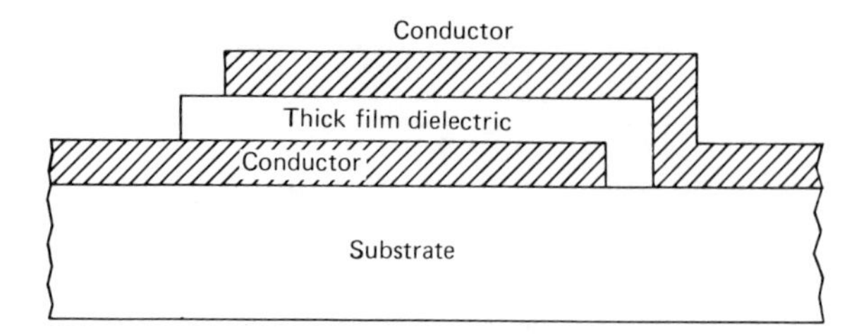

Figure 18.12 An example of a thick film parallel plate capacitor.

mils. For good capacitor design the dielectric should overlap the top and bottom electrodes by 10 mils on all edges. With a dielectric constant of 100 and with a 1.4 mil thick film then a 50 mil × 50 mil capacitor is approximately 40 pF.

Capacitor Tolerances

Substrate capacitors will have about a 10% tolerance because of variances in the substrate thickness and dielectric constant. The interdigital capacitor tolerance is a function of the processing resolution and can be from 10 to 30% depending on the width variations of the interdigital lines. The parallel plate capacitor using a thick film dielectric will have a wide tolerance of about plus or minus 30%, owing to the wide variations in the fired dielectric thickness. Some typical characteristics of thick film capacitors are summarized in Table 18.4.

18.6 THICK FILM APPLICATIONS

The attractiveness of thick film technology stems from its cost effectiveness and high reliability as a hybrid circuit assembly technique. About 75% of all hybrids produced are based on thick film techniques, the remaining 25% being thin film (Chap. 19). The basic advantages of producing thick film hybrids are:

1 Miniaturization, leading to reduced size and weight.

2 Higher frequency performance.

3 Predictability and repeatability in design and manufacturing.

4 High density.

Table 18.4 **Typical thick film capacitor characteristics**

Quality factor Q at 1 MHz = 1000–1500
Dissipation factor = 0.05 to 0.25%
Insulation resistance at 100 V dc > 10^{12} Ω
Breakdown voltage > 400 V ac rms
Temperature coefficient of capacitance < 200 ppm/°C from −55 to 125°C

5 Close tolerance components because of trimming.

6 Functional trimming of components.

7 High thermal conductivity of substrates.

8 Long-term stability and reliability.

9 Relatively simple processing and assembly techniques.

10 Low development costs.

Thick film circuits are finding wide range use in the mass markets of automobile electronics and digital watches. They also are finding increased usage in the fields of telecommunications, computers, business machines, toys, instrumentation, and industrial equipment.

As large scale integrated circuits (LSI) have developed, there has been an increasing need for multilayer thick film hybrids as an interconnect package. This has led to hybrids with up to 10 to 15 conductive layers, and chip densities of 20 to 50 integrated circuits per square inch. This packaging density cannot be achieved by any other assembly technique.

Thick film hybrids also are finding increased use in high-frequency components up to about 1 GHz. Above this frequency, hybrids are generally fabricated using thin film technology (see Chap. 19) because of finer line geometries, higher resolution, and lower loss components available with thin film.

With the development of low-loss substrates, such as porcelain-enameled steel and low-cost, copper-based inks, thick film hybrids will continue to be a major technology in the electronics industry.

18.7 COMMONLY USED TERMS IN THICK FILM TECHNOLOGY

Aspect ratio The ratio between the length and width of a film resistor. Measured in the number of squares.

Eutectic An alloy composition having the lowest melting temperature.

Functional trimming Trimming of circuit elements, such as resistors or capacitors, on an electrically operating circuit in order to achieve a specified response.

Kerf The slit or channel cut in a component during trimming by a laser beam or abrasive jet.

Kiln A high temperature furnace used in firing ceramics.

Leaching In soldering, the dissolving (alloying) of the material to be soldered into the molten solder.

Passivation An insulating layer directly over a circuit element to protect the surface from contaminants, moisture, or other particulate matter.

Profile (firing) A graph of time versus temperature or position in a continuous thick film furnace.

Rheology The science dealing with the deformation and flow of matter.

Scavenging Same as *leaching*.

Sheet resistivity (resistance) The resistivity of material measured in Ω/square.

Vitreous A term used in ceramic technology indicating fired characteristics approaching (but not necessarily totally) glass.

18.8 REFERENCES

Agnew, J., *Thick Film Technology,* Hayden, New Jersey, 1973.

Hamer, D. W. and J. V. Biggers, *Thick Film Hybrid Microcircuit Technology,* McGraw-Hill, New York, 1972.

Harper, A., ed., *Handbook of Thick Film Hybrid Microelectronics,* McGraw-Hill, New York, 1974.

Jones, R.D., *Hybrid Circuit Design and Manufacture,* Marcel Dekker, New York, 1982.

Kaufman, M. and A. H. Seidman, *Handbook of Electronics Calculations,* McGraw-Hill, New York, 1979, Chap. 24.

Topfer, M. L., *Thick-Film Microelectronics,* VNR, New York, 1971.

CHAPTER 19

Thin Film Technology

Roydn D. Jones
Manager Instrumentation Systems Research
Tektronix, Inc.

19.1 INTRODUCTION

Thin film is the field of microelectronics in which conductive, resistive, or dielectric (insulating) films are deposited or sputtered on an insulating substrate or other microelectronic circuit. The films are sometimes deposited in the required pattern; more frequently, however, they are deposited as a planar film, photo processed, and etched to form the required pattern.

Thin film thicknesses are typically from a few hundred angstroms (1 Å = 10^{-10} m) to a few micrometers (1 μm = 10^{-6} m). This compares with the typical 10 to 50 μm of screened thick films. Figure 19.1 illustrates representative thin film dimensions. However, thin film conductors are often electroplated to improve their conductivity. In this case, the plating may be as thick as 10 to 25 μm, making the plated thin film dimensions comparable to that of thick film materials (see Chap. 18).

The characteristics of thin films are not necessarily similar to the characteristics of the same bulk material. Properties such as resistivity, conductivity, dielectric constant, and temperature coefficient vary with the method of deposition, film thickness, substrate material, and substrate surface properties.

Figure 19.2 indicates how sheet resistivity may vary with film thickness. Up to about 50 Å, the film is not continuous because it consists of isolated islands of atoms and the resistivity can be quite high. From 50 to 150 Å these islands begin to touch and the resistivity decreases. The resistivity in this region depends to a great extent on the surface properties of the substrate. From 150 to 400 Å the film becomes smoother and more continuous with a further slight decrease in resistivity. Above about 400 Å the film properties usually approach those of the bulk material.

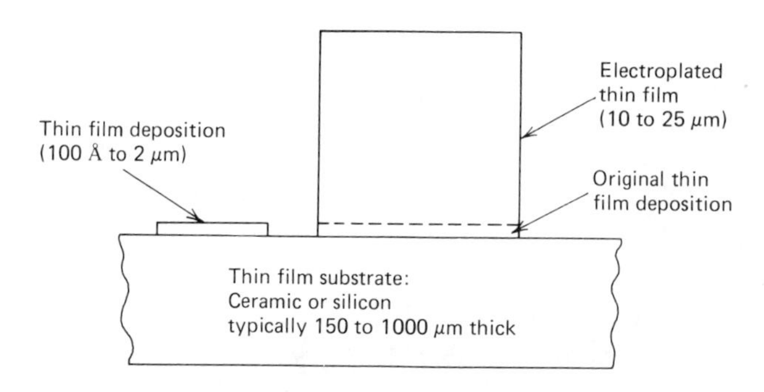

Figure 19.1 Typical thin film dimensions (not drawn to scale).

Thin film technology is used extensively in precision microelectronic circuits. It is used in hybrid microcircuits to produce precision resistors and capacitors and is the dominant technology used in the fabrication of hybrid microwave integrated circuits. Thin film resistors and capacitors are also frequently deposited on semiconductor integrated circuits, having superior properties to equivalent diffused resistors or junction capacitors (see Chap. 20).

Thin film active devices also have been produced, although little commercial success has been achieved in this area. However, recent advances in more exotic thin film materials, such as hydrogenated or fluorinated amorphous silicon, and chalcogenide thin film materials, may increase interest in research and development of active thin film devices. Such active thin film devices may find major applications in the display area, for example, fully addressable liquid crystal displays. Table 19.1 indicates areas where thin film technology is presently being utilized.

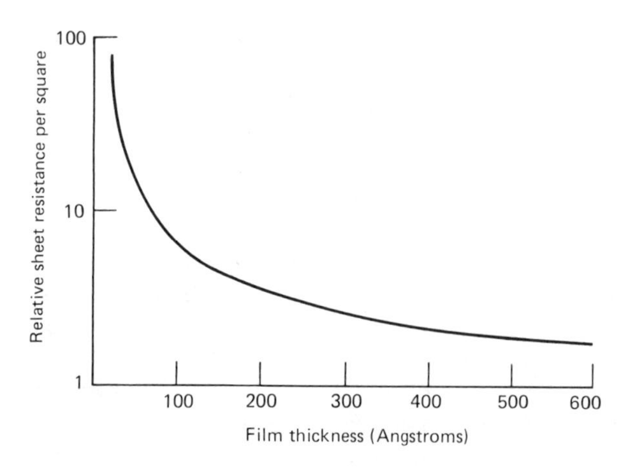

Figure 19.2 Variation of sheet resistivity with film thickness.

Table 19.1 **Applications of thin film technology**

Circuit applications	*Equipment applications*
Integrated circuits	Electronic instruments
Microwave circuits	Communications equipment
Acquisition products	Microwave equipment
Display circuitry	Radar systems
Precision circuitry	Flat panel displays
	Medical instruments
	Satellite equipment

19.2 THIN FILM MATERIALS

There are three main categories of thin film materials: conductors, resistors, and dielectrics.

Thin Film Conductor Materials

The most common thin film metal for conductors on hybrid microcircuits is gold, while the most common thin film metallization for monolithic integrated circuits is aluminum (see Chap. 20). Other metals that are occasionally used are silver, tantalum, nickel, and chromium.

Gold is usually vacuum deposited to a thickness of 2000 to 5000 Å, having a sheet resistivity of from 0.025 to 0.050 Ω/square. On hybrid circuits, particularly in precision or high-frequency hybrids, this fairly high sheet resistivity may pose problems, and the conductor areas are usually electroplated through a photoresistive image to about 5 to 10 μm thickness. This reduces the sheet resistivity to about 0.01 to 0.005 Ω/square.

Aluminum is also vacuum deposited and is an ideal metal for integrated circuits. Aluminum provides nonohmic contacts to devices, pads, or bonding areas for external connections and low resistance interconnections between devices. The aluminum is deposited to a thickness of about 1 μm and at this thickness has a sheet resistivity of about 0.04 Ω/square.

Thin Film Resistor Materials

The most common thin film resistor materials are nickel-chromium, tantalum nitride, and chromium silicide. Nickel-chromium and tantalum nitride films have low sheet resistivities, typically less than 200 Ω/square. Chromium silicide films have higher sheet resistivities, up to about 2000 Ω/square. Tin-oxide film is sometimes used, having sheet resistivities from about 100 to 5000 Ω/square. Because of its transparency, tin-oxide is being used extensively for the X–Y address lines in flat panel display applications, for example, liquid crystal and electroluminescent displays.

Radio frequency (RF) sputtering of nickel-chromium films is preferred over vacuum deposition, because the RF sputtered film retains the composition of the source material. This provides better control over the final film properties. The ratio of nickel to chromium affects both the sheet resistivity and the temperature coefficient of resistance. Composition ratios of 60/40 nickel to chromium, and careful control of the sputtering process, can give low temperature coefficients of resistance in the 50 to 100 ppm/°C region. Although sheet resistivities up to about 200 Ω/square can be fabricated, the upper limit for good control is about 75 to 100 Ω/square. For most applications, 50 Ω/square is used.

Tantalum nitride films also produce very stable resistor properties. Tantalum is sputtered in the presence of nitrogen and this results in reactively sputtered films of tantalum nitride. The nitrogen gas concentration is usually in the range of 0.5 to 5%. The final film exhibits sheet resistivities from 25 to 100 Ω/square, with 50 Ω/square being a typical value. The resistors have a negative temperature coefficient of resistance of about 100 ppm/°C.

For many applications, *ratio tracking of resistance* with temperature in a circuit may be more important than the absolute changes with temperature. The definitions of absolute temperature coefficient of resistance and ratio temperature coefficient of resistance in ppm are given below:

$$\text{Absolute TCR} = \frac{R_{\text{T2}} - R_{\text{T1}}}{R_{\text{T1}}} \times \frac{10^6}{(T_2 - T_1)} \tag{19.1}$$

where R_{T2} = resistance at temperature T_2

R_{T1} = resistance at temperature T_1

$$\text{Ratio TCR} = \frac{\dfrac{R_{\text{T2}}}{R_{\text{BT2}}} - \dfrac{R_{\text{AT1}}}{R_{\text{BT1}}}}{\dfrac{R_{\text{AT1}}}{R_{\text{BT1}}}} \times \frac{10^6}{(T_2 - T_1)} \tag{19.2}$$

where $\dfrac{R_{\text{AT1}}}{R_{\text{BT1}}}$ = ratio of resistors R_{A} and R_{B} at temperature T_1

$\dfrac{R_{\text{AT2}}}{R_{\text{BT2}}}$ = ratio of resistors R_{A} and R_{B} at temperature T_2

Well-designed thin film resistors, both nichrome and tantalum nitride, have ratio TCR's less than 2 to 5 ppm/°C.

EXAMPLE 19.1 A thin film resistor measures 150 Ω at 25°C and 151.5 Ω at 100°C. Calculate its absolute temperature coefficient of resistance in parts per million (ppm) per °C.

PROCEDURE

1 $T_2 - T_1 = 100 - 25 = 75°C$. $R_{T1} = 150\ \Omega$ and $R_{T2} = 151.2\ \Omega$.
2 Substitution of the above values in Eq. 19.1 yields:

$$\text{Absolute TCR} = \frac{151.5 - 150}{150} \times \frac{10^6}{75} = 133 \text{ ppm/°C}$$

EXAMPLE 19.2 Two thin film resistors are measured at 25 and 75°C and are found to have the following values:

Temperature (°C)	R_1 (Ω)	R_2 (Ω)
25	100	200
75	101	202.1

Calculate the ratio TCR in ppm/°C.

PROCEDURE

1 $R_{AT1}/R_{BT1} = 200/100$, $R_{AT2}/R_{BT2} = 202.1/101$, $T_2 - T_1 = 75 - 25 = 50°C$.
2 Substitution of the above values in Eq. 19.2 yields:

$$\text{Ratio TCR} = \frac{202.1/101 - 200/100}{200/100} \times 10^6/50 = 9.9 \text{ ppm/°C}$$

Thin Film Dielectric Materials

Films of silicon monoxide, silicon dioxide, silicon nitride, and tantalum pentoxide are frequently used as thin film dielectrics. Silicon monoxide films have a relative dielectric constant of about five and are usually vacuum evaporated. Silicon dioxide is usually sputtered on hybrid circuitry and produces a fairly good dielectric film with a relative dielectric constant of about five. On silicon wafers, thermally grown silicon dioxide is frequently used as a dielectric. Silicon nitride films are usually sputtered to form good dielectrics with a relative dielectric constant of about nine.

Resistive films of tantalum can be anodized to produce a dielectric film of tantalum pentoxide with a fairly high dielectric constant of about 25. A major advantage of a tantalum-based thin film system is that both resistive and dielectric films can be fabricated using a common material for both.

19.3 THIN FILM PROCESSING

Thin film technology involves a wide variety of processing techniques such as vacuum evaporation, sputtering, vapor-phase deposition, etching, and plating. These processes are described in this section.

Vacuum Evaporation

Vacuum evaporation is probably the most widely used deposition process. Both conducting metals and resistive films can be deposited with this method. The process is usually carried out in a bell jar under high vacuum conditions, typically 10^{-4} to 10^{-6} torr (1 torr = 1 mm Hg). Leak-proof feed-throughs are used to provide controls and power to heating filaments within the bell jar. A typical setup is shown in Fig. 19.3.

The substrate and source material to be evaporated are placed in a bell jar. The system is pumped down to the appropriate pressure and the source material is heated by an electrical element until it vaporizes. Frequently the substrate is also heated to improve film adhesion.

When the source material's vapor pressure exceeds that existing in the bell jar the material vaporizes rapidly. Under a high vacuum, the mean free path of the vaporized atoms or molecules is greater than the distance from the source to the substrate. The vaporized atoms radiate in all directions and condense on all lower temperature surfaces with which they collide, including the substrate, forming a fairly uniform thin film.

The vacuum evaporated films exhibit a fine-grain structure. The uniformity and repeatability of the films are improved if the angle of incidence of the radiating vapor on the substrate is made steeper. This, however, usually limits the number of substrates that may be processed at any one time.

The heating filament is generally made from a refractory metal having a high melting point and low volatility, for example, tungsten or molybdenum. In cases where higher energies are required to vaporize the source material, electron beam bombardment is used rather than thermal heating.

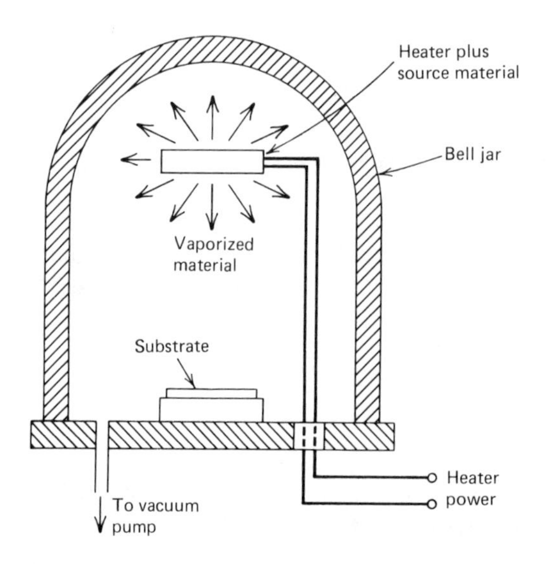

Figure 19.3 Basic vacuum deposition system.

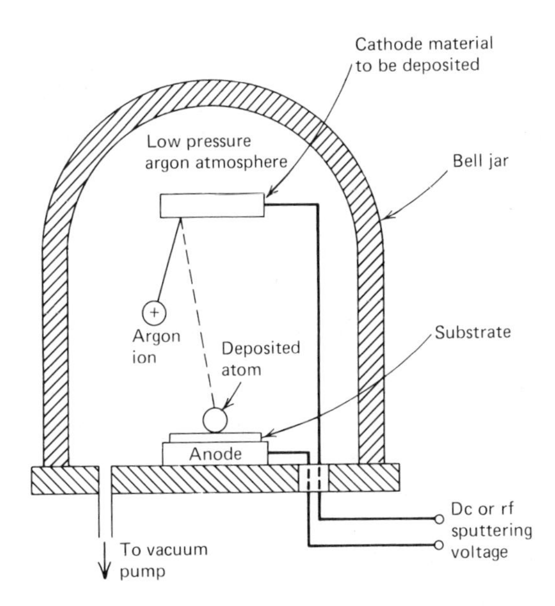

Figure 19.4 Basic sputtering system.

Vacuum deposition can be used to deposit most single element conductors, resistors, and dielectrics. However, alloy depositions are difficult to control because each alloy component has a different evaporation rate at a given temperature.

Dc Cathode Sputtering

In cathode sputtering, metal atoms are released from a cathode constituting the source material as a result of high energy bombardment. Some of the released atoms or molecules are intercepted by the substrate to form a thin film layer.

A typical cathode sputtering system is shown in Fig. 19.4. The bell jar contains a low-pressure, inert gas atmosphere, usually argon, at about 10^{-2} to 10^{-1} torr. A glow discharge is formed by applying a dc potential of 2000 to 5000 V across the anode and cathode. The cathode is heated to release some free electrons that collide with the argon atoms causing them to ionize and form the glow discharge. The positively charged argon atoms accelerate toward the cathode and cause atoms or molecules of the cathode to break away, or "sputter" from the cathode. The cathode atoms are attracted toward the anode and deposit on the substrate that is in close proximity to it.

Owing to the relatively high gas pressure of the argon gas, the mean free path of the sputtered material atoms is shorter than in vacuum evaporation. This results in lower deposition rates for sputtering processes.

The chemical composition of the deposited films can be modified by adding small amounts of reactive gases such as oxygen, nitrogen, or hydrogen to the argon gas. For example, tantalum nitride is sputtered using a tantalum cathode in a nitrogen/argon atmosphere. This process is known as "reactive sputtering."

Thin films of many refractive metals, such as tantalum and molybdenum, which are difficult to evaporate can be obtained by the sputtering process. Also, sputtered films of alloys, such as nickel-chromium, retain the chemical composition of the original sputtering target.

Radio Frequency (RF) Sputtering

Insulators cannot be deposited through dc sputtering techniques, because the cathode potential cannot be applied to the insulator's surface. This problem is overcome by applying a high frequency potential to a metal electrode attached to the insulator material. Power is capacitively coupled to the plasma through the capacitance of the insulator.

The process is called RF sputtering and can be used for sputtering dielectrics, resistors, or conductors. The limiting factors on materials that can be sputtered are the vacuum compatibility of the material and the availability of cathode targets.

Vapor-Phase Deposition

This process is very similar to the epitaxial process used for growing silicon crystal films in semiconductor processes (Chap. 20). Halide compounds of the required metal or metal oxide are chemically reduced in a hydrogen or steam atmosphere at high temperatures. (A halide is a compound that contains fluorine, chlorine, or similar elements.) The reduction of the halide obeys the following equations, depending on the use of steam or hydrogen.

$$2AX \xrightarrow{H_2O} A_2O + 2HX \cdots \text{oxide deposition}$$

$$2\,AX \xrightarrow{H_2} 2A + 2HX \cdots \text{metal deposition}$$

where A is the metal and X may be fluorine, chlorine, or a similar element.

The process can be used to deposit thick layers of material, up to about 20 μm. For example, aluminum chloride ($AlCl_3$) can be reduced to aluminum oxide (Al_2O_3) in a steam atmosphere. Films of aluminum oxide, tin oxide and tantalum pentoxide are frequently deposited using vapor phase deposition.

Plating Techniques

The two types of plating used in thin film processing are electroplating and electroless plating. Electroplating takes place in an electrolytic plating solution with the metal to be plated as the anode and the substrate as the cathode. When a direct current flows through the solution, metal ions migrate from the anode and deposit at the cathode. The metal image is usually plated up through a precision photoresist image.

In electroless plating, simultaneous reduction and oxidation of a chemical agent is used in forming a free metal atom or molecule. The method does not require electrical conduction during plating, and can be used to deposit metals directly on an insulating substrate.

Electroplating is usually used for gold and copper while electroless plating can be used for plating nickel, copper, and gold. In microelectronic applications electroplating is usually used to increase the thickness and, hence, the conductivity of vacuum deposited conductors. It finds wide use in hybrid microcircuitry fabrication.

Patterning of Thin Films

Thin films are usually deposited as a planar sheet over the complete substrate. In the majority of cases in microelectronics the thin film needs to be patterned to form the functional resistors, conductors, or capacitors. This is done by a series of photolithographic and etching techniques; a typical flow of steps is shown in Fig. 19.5.

Photoresist is applied to the substrate by spin or dip coating. The photoresist is dried and exposed to ultraviolet (UV) light through a photographic mask of the required pattern. The photoresist is then developed and, for negative photoresist, the unexposed resist washed away. The opposite occurs for positive photoresists. The substrate is then heated to polymerize the remaining photoresist which acts as a protective layer during the etching step. The photoresist is then removed leaving the required thin film pattern on the substrate. (See Chap. 20 for further details.)

Generally there are two types of photoresists:

1. Materials that are polymerized by illumination yield a negative pattern of the photomask and are called negative photoresists.
2. Materials which become soluble when exposed to light leave a positive image of the mask and are called positive photoresists.

19.4 THIN FILM SUBSTRATES

Thin film technology requires substrates with a smoother surface finish than those used with thick film technology, usually in the range 1–10 microinches, center line average (CLA). Highly polished ceramic, glass, quartz, and sapphire are frequently used for hybrid thin film circuits. Quartz and sapphire are often used for microwave hybrids because of their excellent high frequency properties. High purity alumina substrates (99.9% purity) with polished or glazed surfaces are the most common thin film substrates. Glass substrates are extensively used in lower power, lower frequency applications. Tape cast beryllia substrates are also used in high power thin film hybrids.

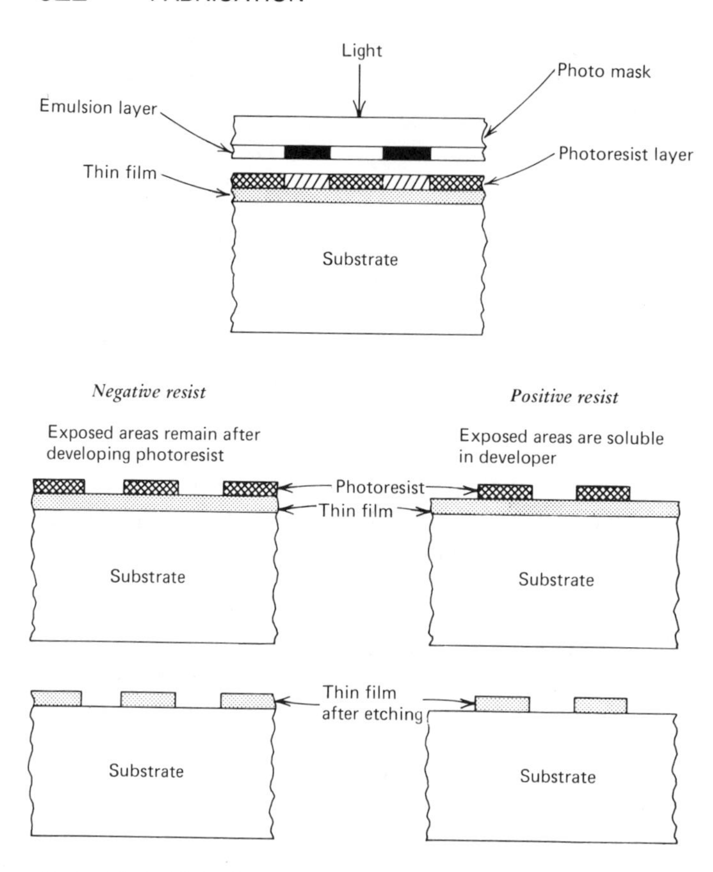

Figure 19.5 Thin film patterning steps.

Table 19.2 lists the properties of many of the commonly used thin film substrates.

19.5 THIN FILM DESIGN GUIDELINES

General Design Procedures

The transition from an electrical schematic to a thin film circuit can be a complex process. The thin film elements may require high precision as in the case of resistor networks for digital to analog (D/A) and analog to digital (A/D) converters (Chap. 13). In such cases the resistor sizes, aspect ratios (length-to-width ratios), and thermal considerations may have a significant impact on the final circuit accuracy. Even the resistance of the thin film conductors connecting the resistors to the active components may affect the final circuit specifications.

In the case of microwave circuitry, the losses in the conductors owing to poor edge definition and surface roughness, as well as dielectric losses in the

***Table 19.2* Properties of thin film substrates**

Characteristics	*Units*	*Alumina*	*Tape cast beryllia*	*Glass*	*Fused silica (quartz)*	*Sapphire*
Relative dielectric constant		10.1	6.9	5.8	3.8	9.4
Dielectric strength	V/mil	750	230		400	200
Dissipation factor	8 GHz	0.0002	0.0003	0.0035	0.00012	0.00005
Thermal conductivity	W/(cm-°C)	0.367	2.5	0.017	0.014	0.417
Thermal coefficient of expansion		6.7×10^{-6}	7.5×10^{-6}	4.6×10^{-6}	0.49×10^{-6}	
Surface finish	Microinches	1.0	15 to 20	1.0	1.0	1.0

substrate, need to be considered. Sapphire and quartz substrates are excellent in this regard, yielding very low-loss transmission lines.

The general steps in the fabrication of a thin film circuit are very similar to those described in Sec. 18.5 on thick film hybrids. However, in the thin film case, screens are not required as the photographic masks are used directly in the fabrication cycle.

Thin Film Resistor Design Guidelines

Sheet resistivity The thin film resistor shown in Fig. 19.6 can be considered as a conductive block of uniform thickness T, width W, and length L. If the bulk resistivity (resistance per unit volume) of the thin film material is ρ in Ω-cm, then the resistance R measured across the block in the direction of the length L is given by:

$$R = \frac{\text{bulk resistivity} \times \text{length}}{\text{Cross sectional area}}$$

$$= \frac{\rho L}{TW} \tag{19.3}$$

If $\frac{\rho}{T}$ in this expression is now defined as the sheet resistivity (resistance) ρ_s, then,

$$R = \rho_s \times \frac{L}{W} \tag{19.4a}$$

$$= \rho_s \times \text{number of squares} \tag{19.4b}$$

where $\frac{L}{W}$, the aspect ratio, is the number of squares in the direction in which the resistor is being measured.

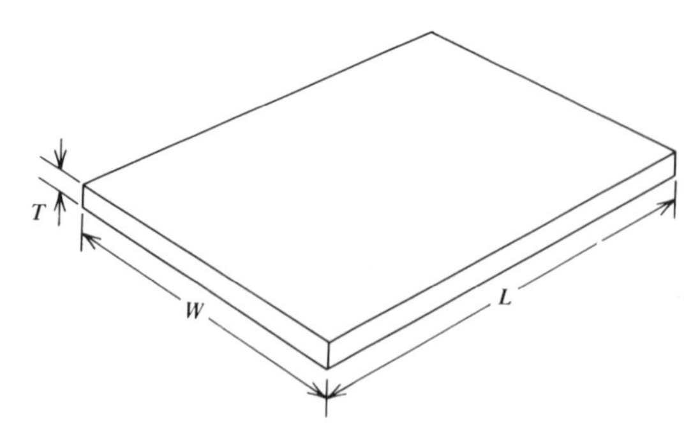

Figure 19.6 Geometry of a thin film resistor.

The sheet resistivity is defined in terms of ohms per square, and is the basic design parameter specified for both thick and thin film resistors. It is very important to remember that for any given film of constant thickness, increasing the length and width by equal ratios will not alter the overall resistance value. For example, a 5 by 5 mil resistor has the same value as a 20 by 20 mil resistor. They both have the dimensions of 1 square.

EXAMPLE 19.3 The bulk resistivity of nichrome is 120 μΩ-cm. Calculate the thickness T in angstroms of a film with sheet resistivity of 50 Ω/square.

PROCEDURE

1 By definition,

$$\rho_s = \frac{\rho}{T} = \frac{120 \times 10^{-6}\ \Omega\text{-cm}}{T\ \text{cm}} = 50\ \Omega/\text{square}$$

2 Hence,

$$T = \frac{120 \times 10^{-6}\ \text{cm}}{50} = \frac{120 \times 10^{-6}}{50} \times 10^{8} = 240\ \text{Å}$$

EXAMPLE 19.4 Calculate the length of a 500 Ω thin film resistor given a sheet resistivity of 100 Ω/square and a resistor width of 200 μm.

PROCEDURE

1 By Eq. 19.4a,

$$500 = \frac{100L}{200}$$

2 Solving for L,

$$L = \frac{(500 \times 200)}{100} = 1000\ \mu\text{m}$$

Resistor design In thin film technology, the substrate is usually coated completely with thin film resistive material and then conductive material. The conductors and resistors are then patterned using conventional photolithographic and etching techniques. Because of the resolution obtained with state of the art photolithography, very fine line widths as low as 20 to 50 μm can be achieved. The film uniformity can be controlled to about ±10% giving as processed tolerances of 10 to 15%. Many thin film resistors can be trimmed to about ±5%

by annealing or oxidizing at elevated temperatures. This final heat treatment also improves the stability of the resistors.

Often a major disadvantage of thin film design is that only one sheet resistivity material is available. This can place restrictions on designing a wide range of resistor values. For example, with a 50 Ω/sq. film it takes 1 square to produce a 50 Ω resistor, but 2000 squares to produce a 100-kΩ resistor. However, 2000 squares can still be achieved fairly easily without consuming large amounts of surface area because of the fine line widths possible.

Other restrictions on resistor size may be power dissipation and voltage stress. Rule of thumb design guidelines are 50 W/inch^2 and 1 V/mil although, under certain circumstances, these figures may be exceeded.

Close tolerance resistors down to better than 0.05% can be achieved using laser trimming. Trim geometries used are similar to those discussed in Chap. 18 on thick films. Laser trimmed resistors are designed to be about 15 to 20% low during the layout stage with about 40% trim capability.

Typical properties of thin film resistors With good quality substrates and close control of deposition parameters, the following characteristics can be obtained:

Temperature coefficient of resistance	±100 ppm
TCR tracking	±5 ppm
Long-term stability	±0.01%

Thin Film Capacitor Design Guidelines

Substrate capacitor The thin film substrate can be used to obtain small values of capacitance from the front substrate surface to the back substrate surface, usually a ground plane. The capacitance of the substrate capacitor of Fig. 19.7 is given by:

$$C = 0.225 \frac{\varepsilon_r A}{t} \tag{19.5}$$

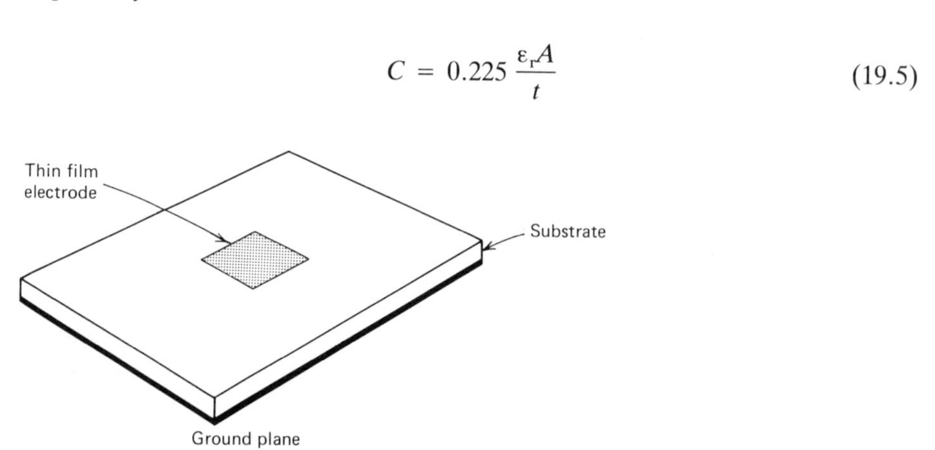

Figure 19.7 Example of a substrate thin film capacitor.

where C = capacitance value in picofarads (10^{-12} farads)
ε_r = dielectric constant of substrate
A = area of conductor, inches^2
t = thickness of substrate, inches

For a 25 mil thick alumina substrate with a dielectric constant of 9.6, the capacitance is approximately 10^{-4}/mil^2. A 100 mil by 100 mil area will yield a capacitance of about 1 pF. Capacitors in the range 0.1 to 3 pF can be obtained in this manner with tolerances of about $\pm 15\%$. Such capacitors can be used in matching networks in very high frequency hybrid circuitry.

EXAMPLE 19.5 Calculate the stray capacitance to a ground plane of a 10 mil × 200 mil gold conductor run on a 15 mil thick alumina substrate; $\varepsilon_r = 9.6$. Neglect fringing effects at the edge of the capacitor.

PROCEDURE

Substitution of the given values in Eq. 19.5 yields:

$$C = \frac{0.225 \times 9.6 \times 0.01 \times 0.2}{0.015} = 0.29 \text{ pF}$$

Interdigital capacitors The planar interdigital capacitor shown in Fig. 19.8 can be used to obtain capacitors in the range 0.5 to about 20 pF. The thin film digits and spaces can be fabricated with 1 mil geometries, producing large total run lengths in relatively small areas. The capacitance for alumina substrates is about 1 pF/inch of conductor run.

EXAMPLE 19.6 Using 1 mil line widths and spaces on an alumina substrate, calculate the approximate dimension of a 5 pF interdigital capacitor with an approximate square aspect ratio.

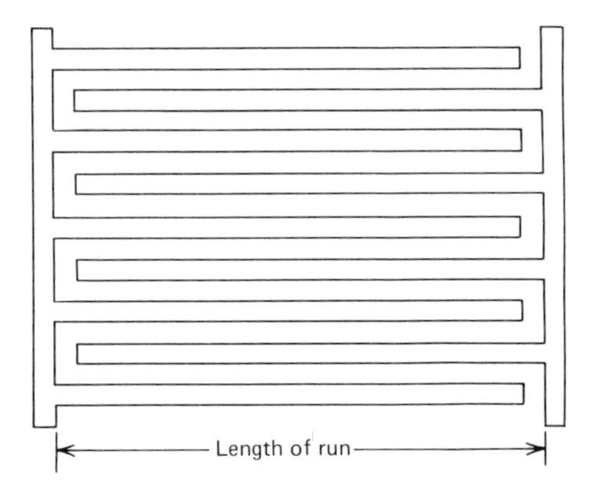

Figure 19.8 An interdigital thin film capacitor.

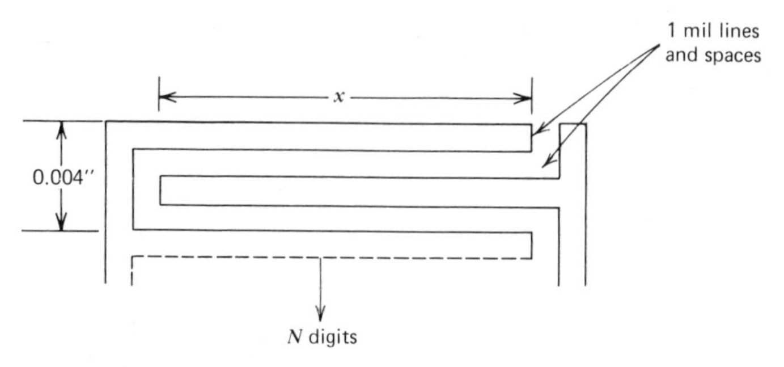

Figure 19.9 Designing an interdigital capacitor (Ex. 19.6).

PROCEDURE

1 The geometry of the capacitor is illustrated in Fig. 19.9.
2 Capacitance = $2x$ pF/digit. For N digits,

$$C = 2xN = 5 \text{ pF} \tag{19.6}$$

3 For an approximate square aspect ratio,

$$N = \frac{x}{0.004} \tag{19.7}$$

4 Substitution of Eq. 19.6 in Eq. 19.7 yields:

$$\frac{2x^2}{0.004} = 5$$

$$x^2 = 0.01$$

$$x = 0.1 \text{ inch} = 100 \text{ mils}$$

Parallel plate capacitors The parallel plate capacitor shown in Fig. 19.10 can be used for both capacitors and crossovers. The capacitor consists of a lower metal electrode, dielectric film, and upper metal electrode. The dielectric film can be very thin, for example, a few hundred to several thousand angstroms, giving fairly high values of capacitance.

Dielectrics that are frequently used for thin film capacitors are silicon monoxide ($\varepsilon_r = 5$), aluminum oxide ($\varepsilon_r = 10$), and tantalum pentoxide ($\varepsilon_r =$

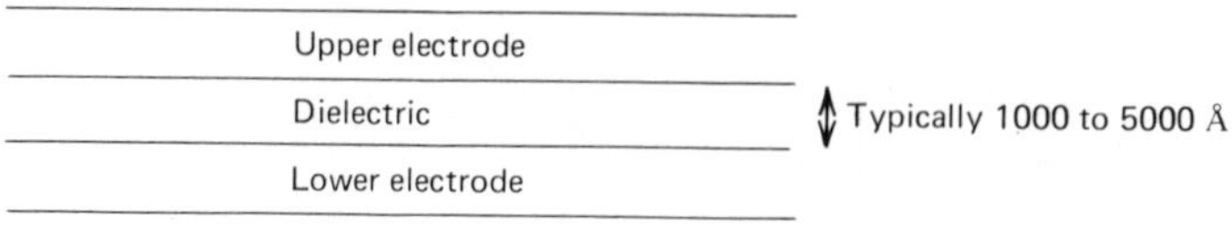

Upper electrode
Dielectric — Typically 1000 to 5000 Å
Lower electrode

Figure 19.10 Cross section view of a parallel plate capacitor.

26). Silicon monoxide and aluminum oxide are usually sputtered. Tantalum pentoxide can be formed by thermal or anodic oxidation of a tantalum film, the original tantalum film forming the bottom electrode.

Because of the thin film thicknesses, the breakdown voltages are usually fairly low and yield loss stemming from pin holes in the dielectric can be a problem. Often dual dielectric layers are used to improve breakdown voltages and to reduce yield losses. The capacitance equation is identical to Eq. 19.5 where t = dielectric thickness.

EXAMPLE 19.7 Calculate the capacitance of a 200 μm square parallel plate capacitor with dual thin film dielectrics consisting of 100 Å of silicon monoxide and 200 Å of tantalum pentoxide (Fig. 19.11).

PROCEDURE

1 Area A and thickness t have dimensions of inches^2 and inches, respectively. Also, 1 μm = 3.937×10^{-5} inches and 1 Å = 3.937×10^{-9} inches.

2 Substituting the given values in Eq. 19.5 obtains:

$$C_1 = \frac{0.225 \times 5 \times 200 \times 200 \times 3.937 \times 3.937 \times 10^{-10}}{100 \times 3.937 \times 10^{-9}} = 177 \text{ pF}$$

$$C_2 = \frac{0.225 \times 26 \times 200 \times 200 \times 3.937 \times 3.937 \times 10^{-10}}{200 \times 3.937 \times 10^{-9}} = 461 \text{ pF}$$

3 The effective capacitance, C_T, is:

$$C_T = \frac{C_1 C_2}{(C_1 + C_2)} = \frac{(177 \times 461)}{(177 + 461)} = 128 \text{ pF}$$

Typical capacitor characteristics Well-designed thin film capacitors have the following typical characteristics:

Quality factor at 1 MHz	500 to 2000
Dissipation factor	0.1 to 0.25%
Insulation resistance	10^{10} Ω
Breakdown voltage	10 to 100 V
Temperature coefficient of capacitance	200 ppm/°C

19.6 THIN FILM APPLICATIONS

The major applications of thin film technology are in metallization schemes for integrated circuits (see Chap. 20) and for thin film hybrid microcircuits. All integrated circuits and discrete semiconductors require thin film metallization to form circuit interconnects and external bonding pads. The most commonly used metals are aluminum and gold, the former being used in the majority of

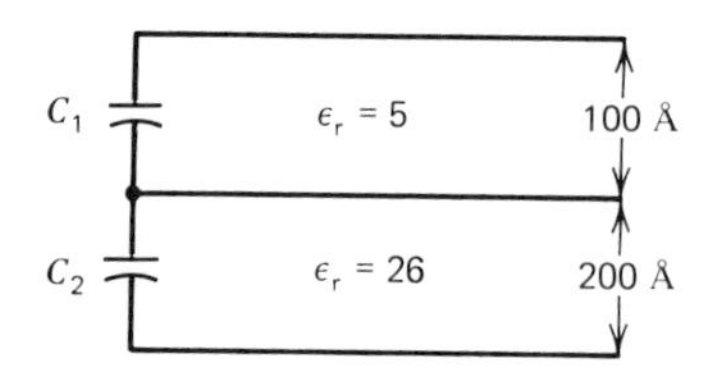

Figure 19.11 Parallel plate capacitor with dual thin film dielectrics (Ex. 19.7).

cases. Thin film resistors are also being used in integrated circuits such as gain setting elements and in ladder-type networks for D/A and A/D convertors. The excellent temperature characteristics of thin film resistors give them an edge over diffused or ion implanted resistors. However, the use of thin film resistors increases the complexity of the process with corresponding increases in cost.

About 10 to 20% of all hybrids produced are based on thin film technology. The major application areas are digital to analog converters, analog to digital convertors, and microwave hybrids. The basic advantages of thin film hybrids are:

1 Higher frequency performance.
2 High density components.
3 Miniaturization leading to reduced size and weight.
4 Precision resistors.
5 Long-term stability and reliability.
6 Excellent thermal characteristics of resistors.
7 Relatively simple processing techniques.

The major disadvantage of thin film hybrids is that they are not suitable for very high-volume, low-cost applications. This has limited the use of thin film to precision and microwave circuits which are generally produced in low quantities.

Thin film technology will dominate the microwave integrated circuits industry for many years to come, and will find increasing usage in both integrated circuit and hybrid precision components.

19.7 REFERENCES

Holland, L., *Thin Film Microelectronics,* Wiley, New York, 1965.

Jones, R.D., *Hybrid Circuit Design and Manufacture*, Marcel Dekker, New York, 1982.

Maissel, L. I. and Glang, R., *Handbook of Thin Film Technology,* McGraw-Hill, New York, 1970.

Vossen, J. L. and Kern, W., *Thin Film Processes,* Academic Press, New York, 1978.

CHAPTER 20

IC Fabrication Technology

Dr. Joseph Stach and Dr. Joseph R. Monkowski
Department of Electrical Engineering
The Pennsylvania State University

20.1 INTRODUCTION

Since the invention of the integrated circuit, the rate of progress in IC fabrication technology has rarely been matched in other nonelectronic technologies. The single fastest growing IC technology is MOS n-channel which has made tremendous strides in the area of RAMs (random access memories), ROMs (read only memories), and the 8- and 16-bit microprocessors. Although the early driving forces in IC technology were space and military, the key factors now are automotive and consumer electronics, particularly the personal home and/or office computer.

This chapter describes briefly various facets of IC fabrication technology. The topics to be covered include crystal growth and wafer preparation, chemical vapor deposition, oxidation, junction formation, lithography, and metallization. Although these areas will be described separately, it should be mentioned that these processes are interactive such that the final yield (number of good chips on a wafer), is determined by how well all the processes, as well as the IC design, have been executed.

20.2 OXIDATION

The function of a layer of silicon dioxide (SiO_2) on a chip is multifaceted. Silicon dioxide's role in silicon IC technology is vital because no other semiconductor material has a native oxide which is able to achieve all the properties of SiO_2. Table 20.1 lists the various properties that SiO_2 has and their role in IC fabri-

Table 20.1 **Properties of SiO_2 in IC fabrication**

Property	*Role or function*
Diffusion mask	Permits selective patterning of junction regions.
Passivation	Electrical neutralization of the transition (interface) of Si to SiO_2.
Insulator	High relative dielectric constant, 3.8, which enables metal line to pass over active silicon regions.
Active electrode	As the gate electrode in an MOS device structure, SiO_2 is the key.

cation. Because of the importance of SiO_2, its growth and electrical properties have been studied extensively.

The most extensive studies of SiO_2 growth on Si for IC use were done in the early 1960's and showed that initially the growth rate of SiO_2 is controlled by the reaction rate at the silicon surface. Once the oxide thickness increases, the growth rate slows down and the controlling or limiting factor is the diffusion of the oxidizing species through the previously grown oxide layer. The growth property of the oxidizing species moving through the oxide differs from that of other materials in which a metal ion moves through the oxide to react at the outer surface of the oxide. Thus, SiO_2 grows from the inside, that is, at the Si–SiO_2 interface.

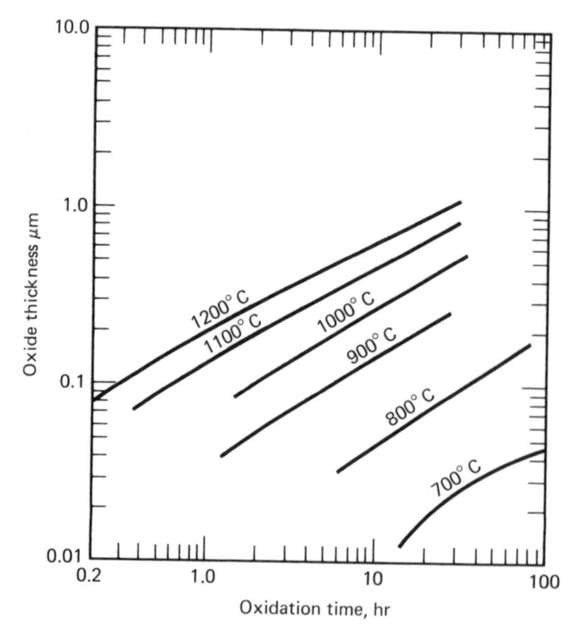

Figure 20.1 Growth rate of SiO_2 in a dry oxygen ambient.

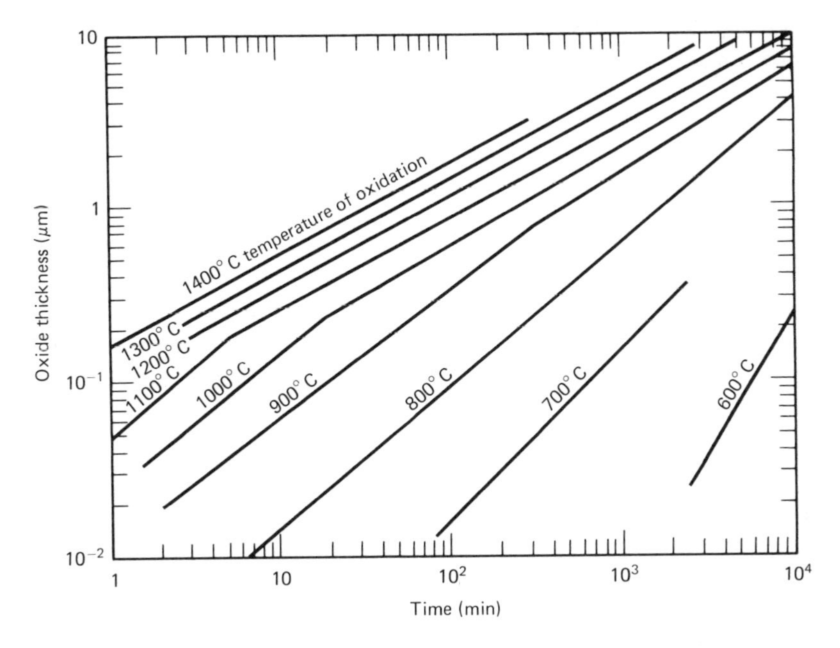

Figure 20.2 Growth rate of SiO_2 in a steam ambient.

The growth rate of silicon dioxide is influenced by the oxidizing ambient and temperature. Figure 20.1 shows the growth of SiO_2 in a dry oxygen ambient for various temperatures. The growth rate of SiO_2 in a steam ambient is given in Fig. 20.2. Notice that the rate of growth in steam is much greater than in dry oxygen. In the past, steam was obtained by boiling ultra high-purity water and passing it into the high-temperature furnace containing the silicon wafers; however, present day technologists generally use hydrogen, H_2, and oxygen, O_2, which are ignited in the furnace tube to form the ultra high-purity water vapor.

The process of silicon oxidation takes place many times during the fabrication of an integrated circuit. It should be noted that once silicon has been oxidized the further growth of oxide is controlled by the thickness of the initial or existing oxide layer. The following examples illustrate the use of the SiO_2 growth rate curves. These process steps are common to several industrial processes.

EXAMPLE 20.1 Determine the thickness of SiO_2 which was grown in a steam ambient at 1100°C for 30 minutes.

PROCEDURE

1 Using Fig. 20.2, find the line representing 1100°C and determine its intersection with the 30 minute line.

2 Next, proceed from the point of intersection to the left of the graph and note the intersection to be approximately 0.45 μm (4500 Å).

EXAMPLE 20.2 Determine the thickness of SiO_2 which was grown in a dry oxygen ambient at 1000°C for 90 minutes.

PROCEDURE

1 Using Fig. 20.1, find the line representing 1000°C and determine its intersection with the 90 minute line.
2 Next, proceed from the point of intersection to the left of the graph and note the intersection to be 0.1 μm (1000 Å).

EXAMPLE 20.3 Determine the final thickness of SiO_2 subjected to the following sequence of oxidations:

Step 1	Dry O_2	1100°C	90 minutes
Step 2	Steam	900°C	50 minutes
Step 3	Steam	1000°C	20 minutes

PROCEDURE

1 Refer to Fig. 20.1 and under the dry O_2 conditions given, the oxide thickness after step 1 is determined to be 0.18 μm (1800 Å).
2 For the second step, refer to Fig. 20.2. Locate the intersection of 0.18 μm on the 900°C line and note the corresponding time. To this specific time add the 50 minutes and note the corresponding total thickness is now 0.3 μm after step 2 is completed.
3 Finally, for the third step, again refer to Fig. 20.2. Locate the intersection of 0.3 μm on the 1000°C line and note the corresponding time. To this specific time add the 20 minutes and note the corresponding total thickness is now 0.4 μm (4000 Å) after the final step 3 is completed.

20.3 SILICON GROWTH

The starting point for single silicon crystel growth is the deposition of high-purity polysilicon. This process is typically done by the Siemens process in which dichlorosilane, SiH_2Cl_2, is reacted with hydrogen to form polycrystalline silicon. Both the dichlorosilane and the hydrogen must be the highest purity available. The polycrystalline silicon is used to grow the single crystal ingots by either the float or Czochralski techniques. The Czochralski technique is more commonly used for IC material while the float zone technique is often used for silicon material employed in power device fabrication.

The Czochralski technique is illustrated in Fig. 20.3. The seed crystal is brought in contact with the molten silicon and then slowly withdrawn. The growth rate of the ingot is controlled by the rate of withdrawal as well as the rate of rotation. Diameter control is an important factor because most silicon wafers presently used in IC processing are 100 mm (4 inches) in diameter with 125 mm (5 inches) just over the horizon. The electrical conductivity, often referred to in terms of resistivity, of the silicon is controlled by the addition of *dopants,* such as boron, phosphorus, or arsenic. Typically the dopants are added in the form of heavily doped silicon to facilitate incorporation into the melt.

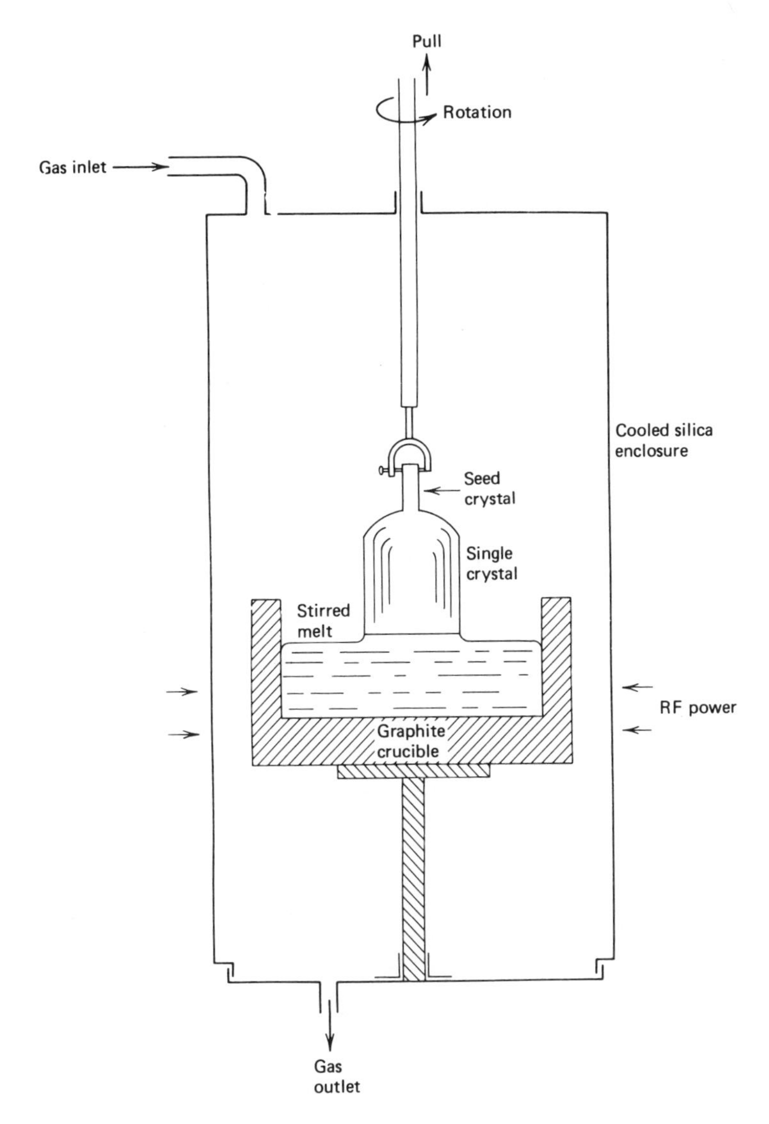

Figure 20.3 The Czochralski technique for growing single-crystal silicon.

Once the ingot has been grown, it is placed in a centerless (cylindrical) grinder and ground to the precise diameter specified. The ingot is then wafered using inside diameter diamond cutting wheels. Then the wafers are mechanically and chemically polished on one side to a specified finish. The backside of the wafer is generally only mechanically polished or ground to a specific texture. The silicon wafers are now ready for IC fabrication.

20.4 JUNCTION FORMATION

The process of junction formation, that is, the transition from p to n type or vice versa, is typically accomplished by the process of *diffusing* the appropriate dopant species during a high temperature furnace process. More recently a technique, *ion implantation,* has become popular and has made significant inroads into diffusion technology in several areas.

Ion implantation doping technology makes use of the ability to accelerate ions, of the desired dopant, by use of a high-energy electric field. The high-energy ions are aimed at the silicon and become imbedded, that is, implanted in the silicon substrate. Doping is controlled by the current of the ion beam and the depth of implantation is controlled by the electric field. The average depth of penetration of the implanted ions is called the *range,* R_p, and is typically expressed in micrometers.

Figure 20.4 depicts an ion implantation system. The ions are generated and repelled from their source in a diverging beam that is focused by an electrical lens before it passes through a magnet that directs only the ions of the desired species through a narrow aperture. A second lens focuses this resolved beam, which then passes through an accelerator that brings the ions to their required energy before they strike the target and become implanted in the exposed areas

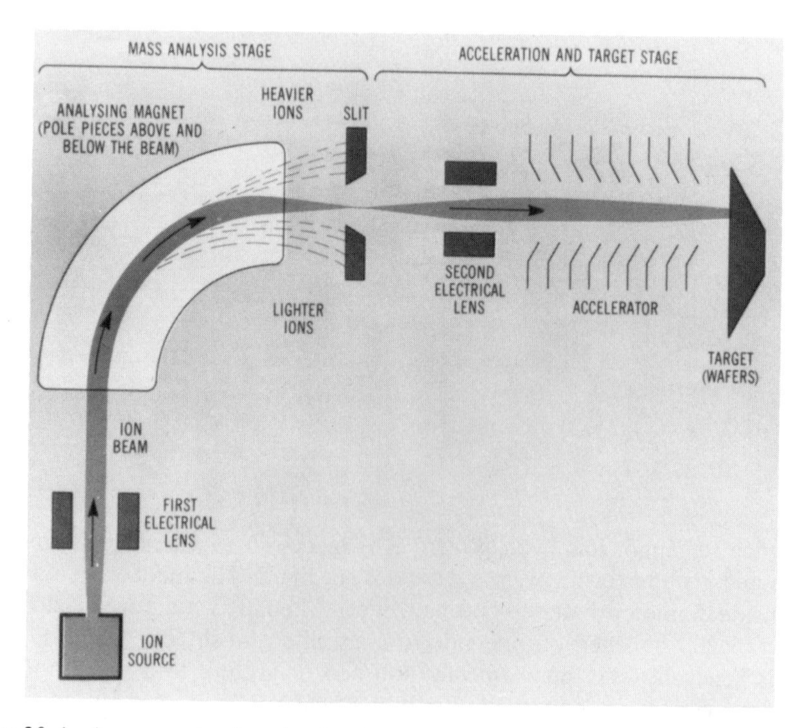

Figure 20.4 An example of an ion-implantation system. (Courtesy Western Electric.)

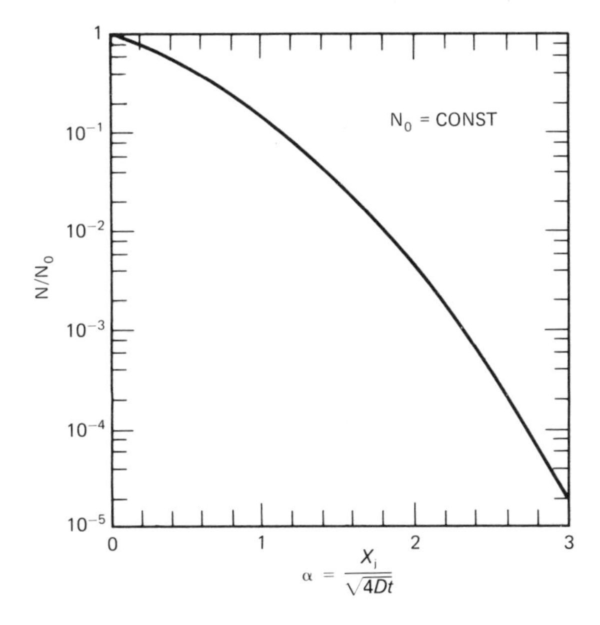

Figure 20.5 Complementary error function (erfc) distribution.

of the silicon wafers. Because the ion beam is small, means must be provided for scanning it uniformly across the wafers.

The basis for diffusion theory is Fick's Law which, when solved for appropriate boundary conditions, gives rise to various dopant distributions, called *profiles,* which are approximated during actual diffusion processes. The two most common diffusion processes yield dopant distributions which are called *Gaussian* or *complementary error function* (erfc). These names are the result of the mathematical description of the distribution function and are often referred to as the two step (Gaussian) and the one step (complementary error function) by the practitioner.

Figure 20.5 shows the typical form of the solution or dopant distribution for the complementary error function. The junction is formed at the intersection of the distribution with the background concentration of the silicon wafer because at this point, the n and p type dopant concentrations are equal.

Figure 20.6 shows how the Gaussian distribution changes with time at a given temperature. Again, the intersection of the dopant distribution with the background concentration of the wafer indicates the location of the junction.

When comparing the Gaussian with the erfc it should be noted that the surface concentration of the Gaussian decreases with time owing to the fixed available dopant concentration, Q. In the case of the erfc, however, the boundary condition states that the surface concentration is a *constant,* typically the maximum solute concentration at that temperature or solid solubility limit (see Fig. 20.7), independent of time. For the case of modeling the depletion layer of a

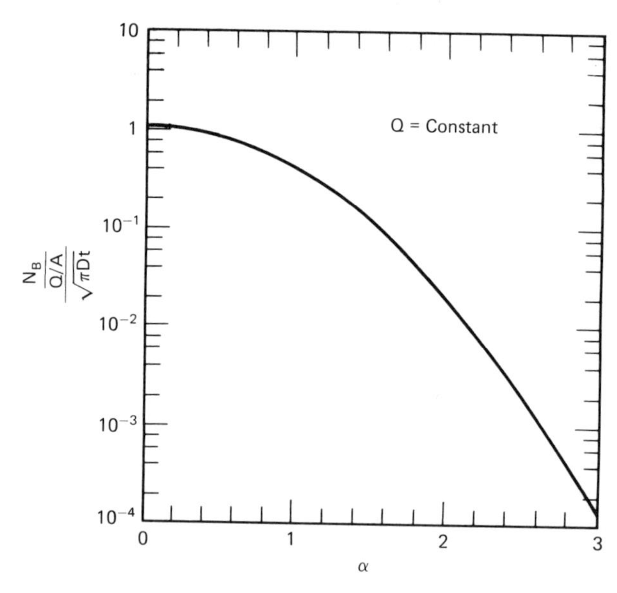

Figure 20.6 Gaussian distribution.

pn junction, the erfc is modeled as a step junction and the Gaussian as a linear graded junction.

The dopant materials used in diffused junction formation are given in Table 20.2. Diffusion processes are generally characterized by measuring, using a *four-point probe,* the electrical resistivity of the diffused layer. Figure 20.8 shows schematically how this measurement is performed. Notice that the measurement is the average resistivity of the sheet of dopant atom in the silicon. This resistivity is called *sheet resistance* or resistivity, ρ_s, and has the unusual dimensions of

Table 20.2 **Dopant source materials for IC diffusions**

Dopant type	*Dopant material*	*Solid*	*Liquid*	*Gas*
p	B_2H_6			X
	BCl_3			X
	BBr_3		X	
	BN	X(wafer)		
n	PBr_3		X	
	$POCl_3$		X	
	PH_3			X
	SiP_2O_7	X(wafer)		
	AsH_3			X
	As_2O_3	X(powder)		
	As_2O_5	X(powder)		

ohms per square (see Chap. 18). This arises from the fact that resistance of a bar of material, is given by,

$$R = \rho\frac{L}{A} \qquad (20.1)$$

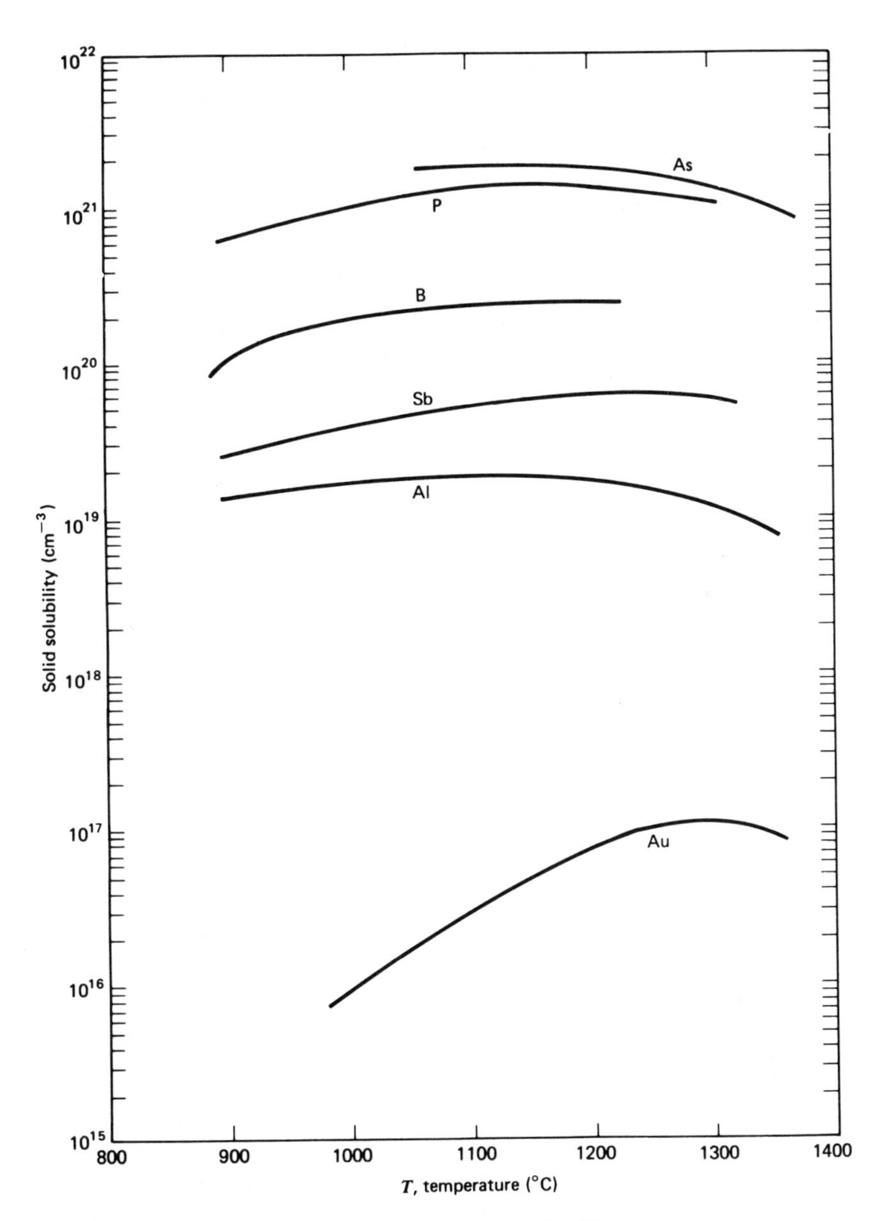

Figure 20.7 Solid solubility of various dopants in silicon.

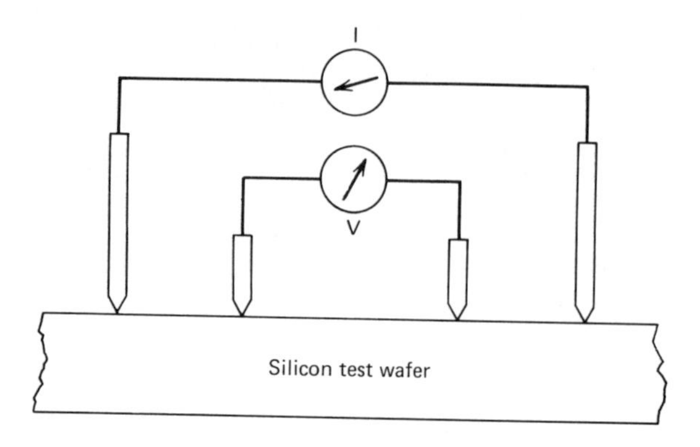

Figure 20.8 Four-point probe used to measure the electrical resistivity of a diffused layer.

where ρ = resistivity in Ω-cm, L = length, A = cross-sectional area, perpendicular to current direction.

In the case of a diffused layer, the cross-sectional area is given by,

$$A = X_j W \tag{20.2}$$

Substituting for A in Eq. 20.1 yields;

$$R = \rho \frac{L}{X_j W} \tag{20.3}$$

If ρ/X_j is separated in Eq. 20.3, then the measure quantity of sheet resistance, ρ_s, is given by,

$$\rho_s = \frac{\rho}{X_j} \ \Omega/\text{sq.} \tag{20.4}$$

and the value of a diffused resistor is given by:

$$R = \rho_s \frac{L}{W} \tag{20.5}$$

where L/W is the number of squares of size W by W along the current path. Examples of diffusion and calculations typical of some IC processing follow.

EXAMPLE 20.4 The base of an npn transistor has a doping of $N_A = 10^{17}/\text{cm}^3$, where N_A is the number of acceptor ions in the base. Phosphorus is to be diffused to form the emitter. How long will the necessary diffusion take at 1100°C if the emitter-base junction is to be 1 μm deep?

PROCEDURE

1 The equation for the one-step diffusion is a complementary error function of the form:

$$N(x) = N_S \operatorname{erfc} \frac{X_j}{2\sqrt{Dt}} \tag{20.6}$$

where N_S is the number of impurity atoms at the surface of Si. The partial pressure of the phosphorus dopant inside the diffusion tube is assumed to be high enough such that the solid solubility limit of phosphorus is obtained. Therefore, from Fig. 20.7,

$$N_S = N_D = \frac{10^{21}}{\text{cm}^3}$$

2 The diffusivity of phosphorus at 1100°C can be obtained from Fig. 20.9 and is found to be:

$$D = 2 \times 10^{-13}\ \text{cm}^2/\text{s}$$

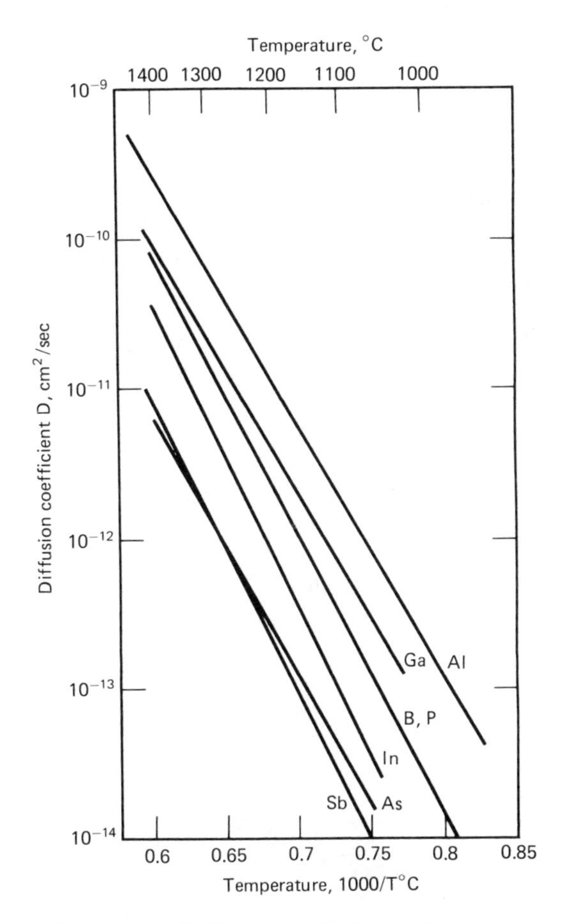

Figure 20.9 Diffusion coefficients of various dopants in silicon.

3 The junction is defined at the point where:

$$N(x) = N_B = N_E = \frac{10^{17}}{cm^3}$$

where N_B is the background concentration which, in this example, equals $10^{17}/cm^3$.

4 Therefore:

$$\frac{N_B}{N_S} = \text{erfc}\,\frac{X_j}{2\sqrt{Dt}} = \frac{10^{17}}{10^{21}} = 10^{-4}$$

5 From Fig. 20.5, this corresponds to:

$$\alpha = 2.75 = \frac{X_j}{2\sqrt{Dt}}$$

6 Using $X_j = 1\ \mu m$ and solving for t yields:

$$t = \frac{1}{D}\left(\frac{X_j}{2\alpha}\right)^2 = \left(\frac{1}{\dfrac{2 \times 10^{-13}cm^2}{s}}\right)\left(\frac{10^{-4}cm}{2(2.75)}\right)^2$$

$$t = 1653\ s \simeq 27.5\ \text{minutes}.$$

EXAMPLE 20.5 A phosphorus ion-implanted dose of $4 \times 10^{14}/cm^2$ is used to form the channel region in an NMOS depletion mode transistor. The mean distance of implant is 0.25 μm from the surface. The ions are to be driven in at 1150°C for 15 minutes. What is the resulting junction depth if the substrate doping, N_B, is $5 \times 10^{15}/cm^3$?

PROCEDURE

1 The equation for the drive-in diffusion is a Gaussian distribution of the form:

$$N(x) = \frac{Q/A}{\sqrt{\pi D't'}}\, e^{-X^2/4D't'} \tag{20.7}$$

2 The dose is equal to Q/A. The diffusivity of phosphorus at 1150°C can be found from Fig. 20.8:

$$D' = 6 \times 10^{-13}\ cm^2/s$$

Therefore:

$$\frac{N_B}{\dfrac{Q/A}{\sqrt{\pi D't'}}} = e^{-\alpha 2} = \frac{5 \times 10^{15}/cm^3}{\dfrac{4 \times 10^{14}}{\sqrt{\pi(6 \times 10^{-13}cm^2/s)\,(900\ s)}}} = 5.15 \times 10^{-4}$$

where $\alpha = X_j/\sqrt{4D't'j}$.

3 From Fig. 20.6, the value of α is found to be:

$$\alpha = 2.75 = \frac{X_j}{2\sqrt{Dt}} \qquad X_j = 2\alpha\sqrt{Dt}$$

$$X_j = 2(2.75)\sqrt{\left(\frac{6 \times 10^{-13}\text{cm}^2}{\text{s}}\right)(900\ \text{s})}$$

$$= 1.278\ \mu\text{m}$$

EXAMPLE 20.6 The base of an npn transistor has a surface concentration of $10^{19}/\text{cm}^3$. If the emitter-base junction is 1.0 μm deep and the base-collector junction 2.5 μm deep, find the average conductivity of the base region given that the collector concentration is $10^{17}/\text{cm}^3$.

PROCEDURE

1 The p-type base is typically a Gaussian distribution. The background concentration is $10^{17}/\text{cm}^3$. Therefore, the Irvin curves given in Fig. 20.10 are used.

2 To find the average conductivity ($\bar{\sigma}$) the following parameter must be calculated:

$$\frac{X}{X_j} = \frac{X_{je-b(\text{emitter-base})}}{X_{jb-c(\text{emitter-collector})}} = \frac{1.0\ \mu\text{m}}{2.5\ \mu\text{m}} = 0.4$$

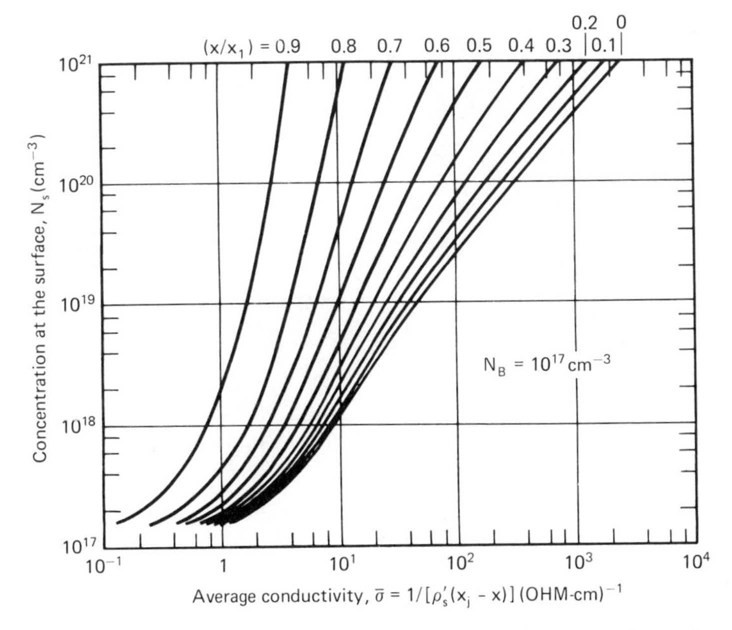

Figure 20.10 Irvin curves for a p-type Gaussian distribution used in determining the average conductivity of a diffused region.

3 Then, by using the curve representing 0.4 in Fig. 20.10, determine its intersection with the horizontal surface concentration line of $10^{19}/cm^3$.
4 Next, proceed downward from the point of intersection to the vertical average conductivity line and note the intersection to be 20 $(\Omega\text{-cm})^{-1}$.

EXAMPLE 20.7 The emitter region of an npn transistor is 0.8 μm deep and from four-point probe measurements is found to have a sheet resistance of 150 Ω/sq. Assuming only a one-step diffusion for emitter processing and a background concentration of $10^{17}/cm^3$, what is the surface concentration?

PROCEDURE

1 The n-type emitter is a complementary error function distribution and the background concentration is $10^{17}/cm^3$. Therefore, the Irvin curves given in Fig. 20.11 are used.
2 From the formula:

$$\bar{\sigma} = \frac{1}{\rho_s(X_j - X)} \tag{20.8}$$

where $X = 0$ (surface) $X_j = 0.8$ μm (emitter-base junction),

$$\rho_s = 150\ \Omega/\text{sq}.$$

3 Solving for the average conductivity yields:

$$\bar{\sigma} = 83.3\ (\Omega\text{-cm})^{-1}$$

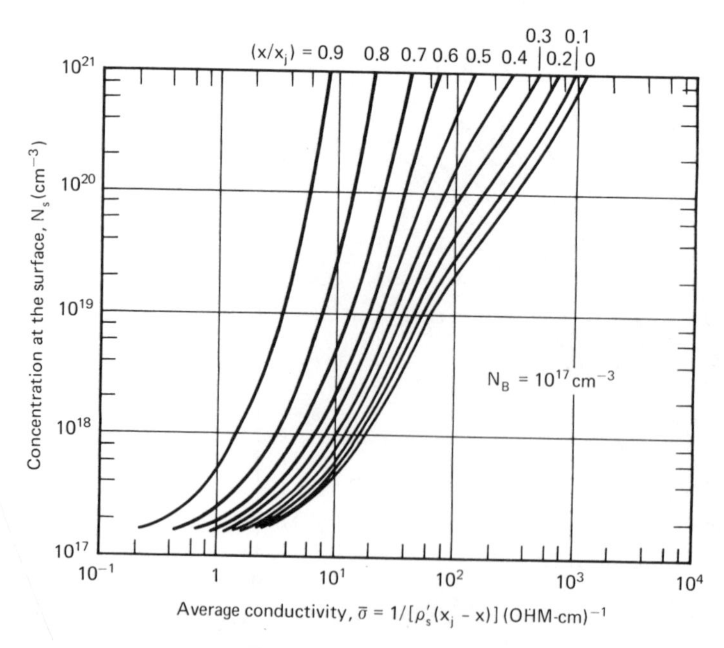

Figure 20.11 Irvin curves for an n-type erfc distribution used in determining the average conductivity of a diffused region.

4 Also,

$$\frac{X}{X_j} = 0$$

Then by using the curve representing 0 in Fig. 20.11, determine its intersection with the vertical average conductivity line of 83.3 $(\Omega\text{-cm})^{-1}$.

5 Next, proceed to the left of the point of intersection to the horizontal surface concentration line and note the intersection to be $2 \times 10^{19}/\text{cm}^3$.

EXAMPLE 20.8 The base-collector depth for a particular npn transistor is 2 μm. The surface concentration of the base is $10^{18}/\text{cm}^3$ and the doping of the collector region is $10^{16}/\text{cm}^3$. Calculate the total depletion capacitance and thicknesses under: (a) 0 V bias, (b) +10-V reverse bias.

PROCEDURE

1 The base of the npn is typically a Gaussian distribution and since:

$$\frac{N_B}{N_0} = \frac{\text{background conc.}}{\text{surface conc.}} = \frac{10^{16}}{10^{18}} = 10^{-2}$$

then Fig. 20.12 is utilized. For (a),

$$V = V_T \ln \frac{N_{\text{base}} N_{\text{collector}}}{n_i^2} \tag{20.9}$$

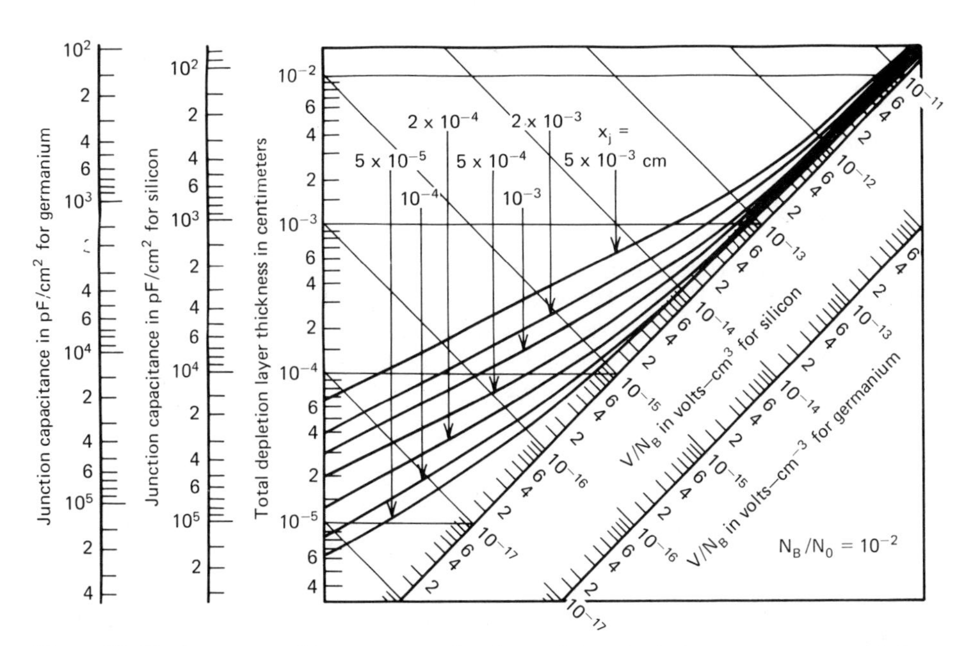

Figure 20.12 Lawrence–Warner curves for determining depletion capacitance and depletion layer thickness.

where $V_T = kT/q$ and n_i is the intrinsic concentration of electrons. At room temperature, $V_T = 0.026$ V and $n_i = 1.5 \times 10^{10}$. Assuming a linear-graded junction: $N_{base} = N_{collector} = 10^{16}/cm^3$; therefore:

$$V = 0.026 \ln \frac{(10^{16})^2}{(1.5 \times 10^{10})^2}$$

$$= 0.70 \text{ V}$$

Hence:

$$\frac{V}{N_B} = 7 \times 10^{-17} \text{ V cm}^3$$

2 Then by using the curve representing 2 μm (2×10^{-4} cm) determine its intersection with the diagonal line representing $V/N_B = 7 \times 10^{-17}$ V-cm³.
3 Next, proceed left to the vertical axis where the total depletion capacitance is found to be 2.0×10^4 pF/cm² and the total depletion thickness is found to be 5×10^{-5} cm.
4 For (b), everything remains the same as in (a) except that:

$$V = 0.7 + 10 = 10.7 \text{ V}$$

Therefore:

$$\frac{V}{N_B} = \frac{10.7 \text{ V}}{\frac{10^{16}}{\text{cm}^3}} = 1.07 \times 10^{-15} \text{ V-cm}^3$$

5 Continuing with the same procedure as in (a), the total depletion capacitance is found to be 7×10^3 pF/cm² and the total depletion thickness is found to be 1.5×10^{-4} cm.

EXAMPLE 20.9 For the data in Ex. 20.8, find the ratios and values of the depletion widths on both the base and collector sides.

PROCEDURE

1 Figure 20.13 is used for the same reasons Fig. 20.12 was used as explained in Ex. 20.8. For (a), $V = .70$ V and $V/N_B = 7 \times 10^{-17}$ V-cm³. Then, by using the curve representing 2 μm, determine its intersection with the vertical line representing $V/N_B = 7 \times 10^{-17}$ V-cm³. Next, proceed left to the vertical axis where the ratio a_1/a_{total} is found to be 0.41 where $a_1 = W_{base\ side} = 0.41a_{total}$ and $a_{total} = 5 \times 10^{-5}$ cm (from Ex. 20.8*a*).

2 Therefore,

$$W_{base\ side} = 2.05 \times 10^{-5} \text{ cm} \simeq 0.2 \text{ μm}$$

$$W_{collector\ side} = 5 \times 10^{-5} \text{ cm} - 2.05 \times 10^{-5} \text{ cm}$$

$$= 2.95 \times 10^{-5} \text{ cm} \simeq 0.3 \text{ μm}$$

3 For (b), everything again remains the same except that:

$$V = 0.7 + 10 = 10.7 \text{ V}$$

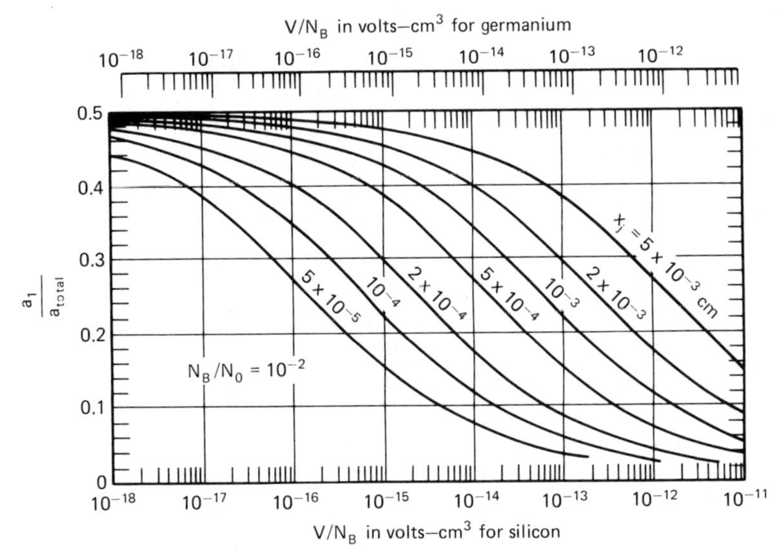

Figure 20.13 Lawrence–Warner curves for finding ratio and depletion widths on both the base and collector sides of a diffused transistor.

Therefore,

$$\frac{V}{N_B} = \frac{10.7 \text{ V}}{\frac{10^{16}}{\text{cm}^3}} = 1.07 \times 10^{-15} \text{ V-cm}^3$$

4 Continuing with the same procedure as above, the following values were obtained:

$$\frac{a_1}{a_T} = 0.30$$

$$a_1 = W_{\text{base side}} = 4.5 \times 10^{-5} \text{ cm} = 0.45 \ \mu\text{m}$$

$$W_{\text{collector side}} = 1.05 \times 10^{-4} \text{ cm} = 1.05 \ \mu\text{m}$$

20.5 PHOTOLITHOGRAPHY

One of the primary steps in putting down the various layers which comprise an IC is the patterning of these layers. This is carried out through a process called photolithography. In general, photolithography can be broken down into two broad categories: fabrication of the mask, and use of the mask to define the patterns on the wafer.

Mask Fabrication

The pattern for the mask is designed from the circuit layout. Many years ago, "breadboarding" of the circuit was typical. This meant that before integration, the circuit was actually built and tested with discrete components. However, at present, when LSI (large-scale integration) and VLSI (very large-scale integration) circuits contain from a thousand to several hundred thousand components, and switching speeds are in the realm where propagation delay times between devices is significant, breadboarding is clearly out of the question. Present-day mask layout is many times done with the assistance of a computer. Important points include high packing density, avoidance of interference between components and subsystems, and a topology where the crossovers do not require more than the desired level of interconnections.

The original pattern from which the mask is produced is usually 100 to 500 times larger on a side than the final IC. Two reductions are required to obtain the final size. The first reduction produces an image which is typically 10 times the final size. This image is then simultaneously reduced to the final size and stepped and repeated many times in order to completely cover a glass plate with identical patterns. This plate is called the "master." From the master is made a "submaster" from which many copies of the "working plates" are made. These working plates are used in the actual transfer of the pattern to the silicon wafers. In the fabrication of any particular IC, approximately one half dozen masks are required. This number will vary depending upon the circuit complexity.

Despite that the entire sequence just described can be done with plates containing a photosensitive emulsion, typically the emulsion is considered too vulnerable to abrasion and tears. For this reason, masks are often made of harder materials such as chrome or iron oxide. These materials are more expensive; however, they last longer and yield better performance.

Photoresist Patterning

In general, the mask just described is used to shield certain portions of a wafer from some form of radiation (Fig. 20.14). The wafer is coated with a layer of "photoresist," which is sensitive to the particular type of radiation. The most common form of radiation is ultraviolet (UV) light, although the present trend is toward radiation with shorter wavelengths, such as electron beams or X rays. Electron beam exposure is based on the principle of focusing electrons from a thermionic source and accelerating these electrons after they are focused to describe the pattern on the photosensitive layer. This technology is being considered in order to overcome the diffraction limitation of optical exposure systems. The shorter wavelength radiation is able to yield a higher resolution, which is very important as the industry moves toward dimensions which are below 1 μm.

The photoresist can be either of the "positive" type or "negative" type. In the former case, the film is originally polymerized (hardened), and the UV light depolymerizes the film. In the negative case (Fig. 20.14) the film is originally unpolymerized, and the UV light polymerizes the film. For the positive film, the photoresist after developing remains in those areas covered by the opaque part of the mask; hence, the photoresist pattern is a "positive" of the mask. For the negative photoresist, a negative is produced.

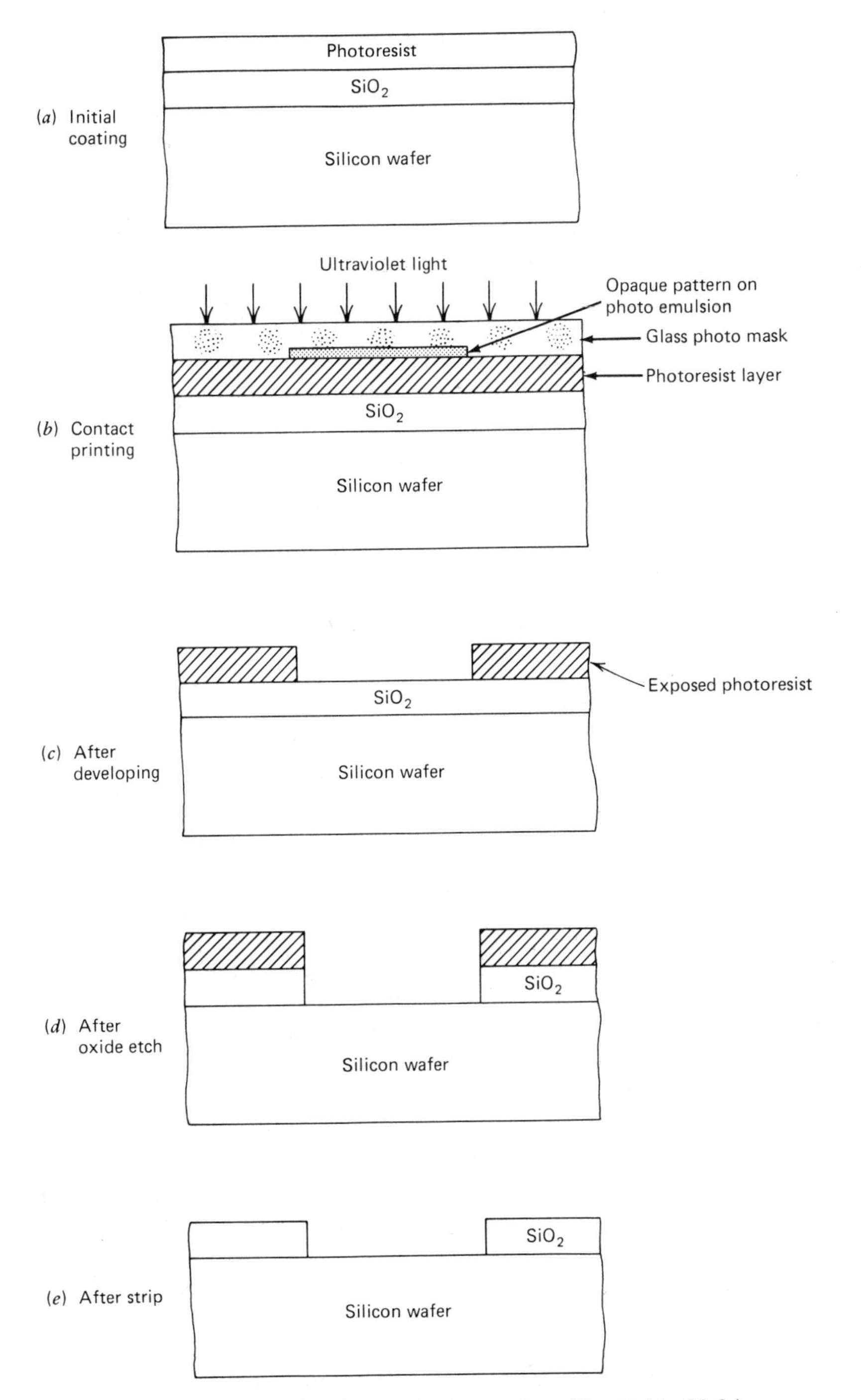

Figure 20.14 The steps in photoresist patterning. (See Table 20.3.)

In principle, either type of photoresist could be used assuming the mask was designed appropriately. In practice, there are differences in the chemistry of the photoresists which makes one type preferable under certain conditions. These differences relate to such properties as resolution and adhesion.

The various steps required for application, exposure, and developing of the photoresist, as well as patterning of the wafer, are listed in Table 20.3. For proper adhesion of the resist, the surface of the wafer must be thoroughly clean and dry. Application of the resist is done either via spraying, or spinning after application of several drops of the resist, which is a viscous liquid, in the center of the wafer. The spin speed determines the resist thickness. The soft bake dries and somewhat hardens the resist. After exposure, the developing defines the desired pattern. The hard bake prepares the resist for the subsequent etching. The etching can be done either with wet chemicals or *plasma etching*. After etching, the photoresist is removed (stripped), and the wafer cleaned.

Table 20.3 **Photoresist steps**

Step	*Results*
Film growth	Oxidation, and so on
Surface preparation	Clean, dry, and so on
Photoresist application	Spin, spray, and so on
Soft bake	Dry resist
Exposure	Align and expose to UV light
Develop	Remove unpolymerized regions
Hard bake	Further drying and resist hardening
Strip resist	Resist removed
Clean	Prepare for next process

Plasma etching makes use of the ability to ionize a gas in an electric field. The ionized gas species can be used to etch the pattern on the surface of the wafer. By controlling the ionized species as well as the neutrally charged gas species, etch rate, etch selectivity, and directionality can be controlled.

Except for the first mask in the IC fabrication process, every mask must be aligned to the pattern produced by the previous masks. This is done with the *mask aligner* (Fig. 20.15). Typical mask aligners are of either the contact type, the proximity type, or projection type. In the contact aligner, the mask is held in contact with the wafer. This produces the best resolution (submicrometer dimensions are possible), but the mask has a short life because of abrasion and wear. The proximity type is much like the contact aligner, except a gap on the order of micrometers exists between the mask and wafer. In this system, the mask lasts longer, but resolution is limited to several micrometers. In the projection aligner, the image is actually projected, via a system of lenses, onto the wafer. This is the costliest of the conventional systems; however, mask life is good, and resolutions of approximately one micrometer can be obtained.

Future trends, aimed at better resolution and better alignment, include

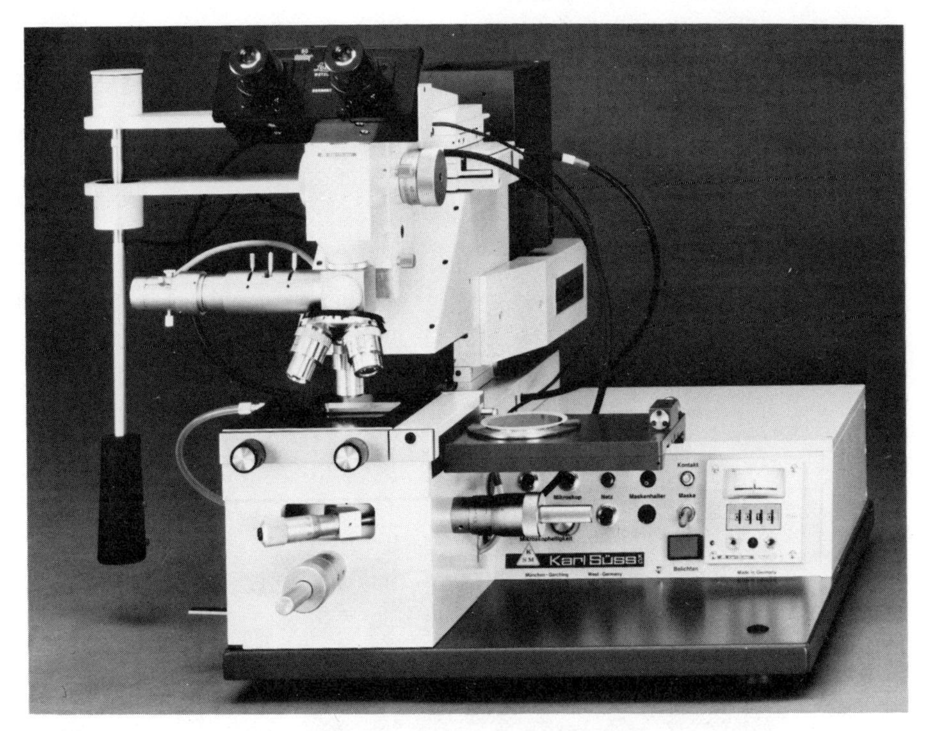

Figure 20.15 An example of a mask aligner. (Courtesy Karl Süss.)

the direct-step-on wafer (DWS) system (Fig. 20.16), where the image is stepped and repeated directly onto the wafer instead of onto an intervening mask, and the direct write on wafer electron-beam (e-beam) system (Fig. 20.17), where an electron beam "writes" the pattern onto the wafer. Presently, e-beam systems are primarily used for mask generation rather than for photoresist exposure. This is expected to change in the near future, however.

20.6 CHEMICAL VAPOR DEPOSITION

Chemical vapor deposition (CVD) is the deposition of a solid material onto a heated substrate via decomposition or chemical reaction of compounds contained in the gas passing over the substrate. Many materials can be deposited via CVD, but those of most interest to the semiconductor industry include silicon nitride, silicon dioxide, polycrystalline silicon (usually called polysilicon), and single-crystal silicon.

The deposition of single-crystal silicon, which can be carried out only on a substrate of single-crystal silicon or single-crystal sapphire, is called *epitaxial deposition*. Epitaxial deposition was the initial form in which CVD was used in IC fabrication, and it continues to play a very important role. A thin epitaxial layer is deposited on a substrate of opposite conductivity type, typically an n

Figure 20.16 An example of a direct-step-wafer (DSW) system. (Courtesy GCA Corp.)

epitaxial layer on a p-type substrate. This process is followed by oxidation, photolithography, and diffusion during which the p-doped regions extend through the epitaxial layer to the p-type substrate. The n regions are now electrically isolated when the n regions are at a more positive potential than the substrate.

The epitaxial process uses either silane, SiH_4, silicon tetrachloride, $SiCl_4$, or trichlorosilane, $SiHCl_3$, as the source chemical with hydrogen as the carrier

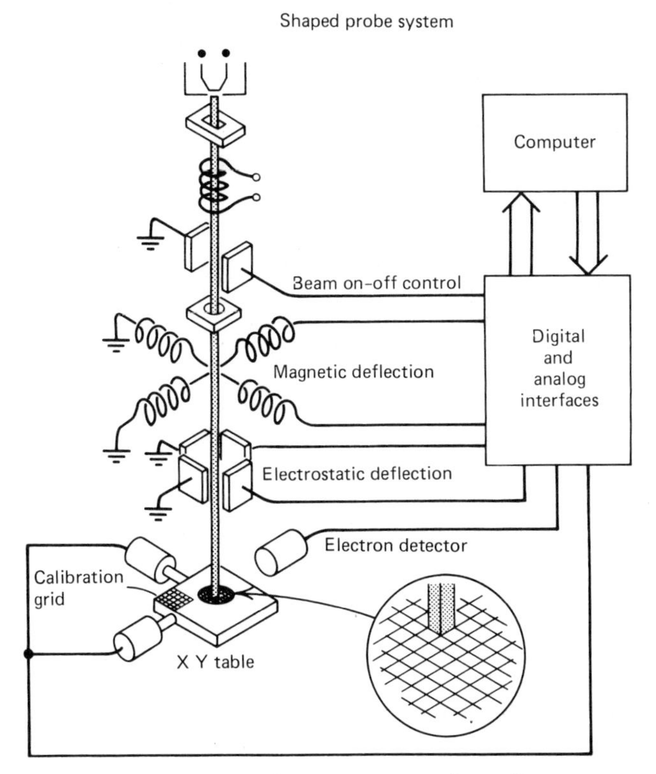

Figure 20.17 Electron-beam exposure of a wafer.

gas. The process is done at atmospheric pressure, although low-presure (subatmospheric) epitaxy is presently under intensive study. Typical reactions are:

$$SiH_4 + H_2 \rightarrow Si + 3H_2$$

$$SiCl_4 + 2H_2 \rightarrow Si + 4HCl$$

$$SiHCl_3 + H_2 \rightarrow Si + 3HCl$$

In the epitaxial process a single extension of the substrate is formed by the deposited silicon atoms. Figure 20.18 shows a typical RF inductively heated system. Radiant heated systems are also currently in use. The deposition rate for an epitaxial process, using an $SiCl_4$ source, is illustrated in Fig. 20.19. High deposition rates are desirable for thick epitaxial layers, although the trend is towards thin, less than 4 μm layers.

Silicon Nitride, Silicon Dioxide, and Polysilicon

The primary difference between the CVD of epitaxial layers and the CVD of other materials, such as oxides and nitrides, is that in the latter case there is no

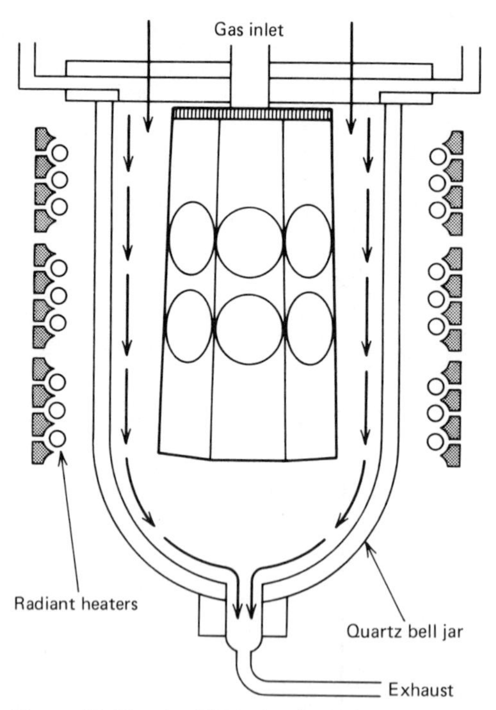

Figure 20.18 An RF inductively-heated epitaxial system.

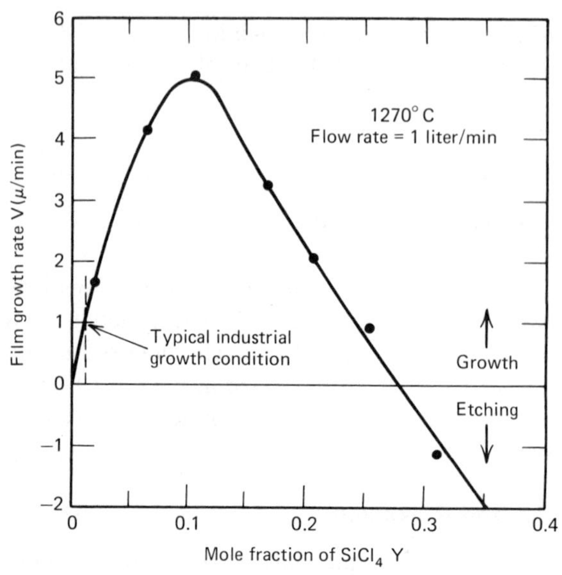

Figure 20.19 Curve of epitaxial silicon deposition rate.

chemical or crystallographic relationship between the substrate and the deposited film. This, in fact, is the advantage of CVD. Whereas an SiO_2 layer can be thermally grown only on silicon, SiO_2 can be deposited via CVD onto silicon, metals, nitrides, polysilicon, and almost any other material that can withstand the CVD growth conditions. Silicon nitride, which plays an important role in oxide masking, cannot even be thermally grown on silicon at the present time. For these reasons, CVD is an integral part of the IC fabrication process where multiple layers are concerned.

Table 20.4 lists some of the sources and reactions most prevalently used to obtain the different films. It should be noted that in most cases either N_2 or H_2 is used as a carrier gas. For the deposition of any particular material, usually a number of sources and possible reactions are available. The primary requirement placed on the source gas is that the reaction or reactions which occur in the deposition chamber yield only one condensed product, and that must be the desired material. All other products must be volatile. Furthermore, the reaction should be predominantly heterogeneous; that is, the reaction should occur primarily on the substrate or growing film, and not homogeneously throughout the gaseous ambient. Reaction in the ambient yields solid particles in the gas phase which can deposit as particulate matter on the growing film.

Table 20.4 **Reactions for obtaining different films**

Reactions for silicon dioxide

$SiH_4 + 2O_2 \rightarrow SiO_2 + 2H_2O$

$SiH_4 + 4CO_2 \rightarrow SiO_2 + 4CO + 2H_2O$

$SiH_2Cl_2 + 2N_2O \rightarrow SiO_2 + 2N_2 + 2HCl$

$SiCl_4 + 2H_2O \rightarrow SiO_2 + 4HCl$

Reactions for silicon nitride

$3SiH_4 + 4NH_3 \rightarrow Si_3N_4 + 12H_2$

$3SiH_2Cl_2 + 10NH_3 \rightarrow Si_3N_4 + 6NH_4Cl + 6H_2$

Reaction for polysilicon

$SiH_4 \rightarrow Si + 2H_2$

In general, the particular source must be chemically appropriate, both in terms of kinetics and thermodynamics, for the desired deposition temperature. For example, SiH_4 reacts with the other compounds at lower temperatures than does SiH_2Cl_2 or $SiCl_4$. For this reason, silane (SiH_4) is used for the deposition of SiO_2 at temperatures of approximately 200–500°C, while silicon tetrachloride ($SiCl_4$) is used in the range of 800 to 1000°C.

The deposition temperature significantly affects the morphology (structure) of the deposited films. Silicon films deposited below approximately 600°C will be amorphous, that is, noncrystalline. The reason for this is that at these low temperatures the silicon atoms do not have enough thermal energy to orient themselves into their crystalline positions. At temperatures slightly above 600°C, the atoms have sufficient energy to form small crystallites or grains; however, the size of these grains is in the range of tens of angstroms.

As the deposition temperature is raised, the grain size increases, until at temperatures above 1000°C, the grain size is in the range of micrometers. For SiO_2 and Si_3N_4, depositions over the entire range of temperatures yields amorphous films; however, films deposited at lower temperatures are usually less dense. Again, this low density stems from low atom mobility. The density of these films can be increased by subsequently annealing the film at a higher temperature. For example, SiO_2 deposited at several hundred degrees, when heated to approximately 900 to 1000°C for 30 minutes, exhibits a density very close to thermally grown SiO_2.

Until several years ago, CVD of oxides, nitrides, and polysilicon was carried out at atmospheric pressure in reactors very similar to those used for epitaxy. The horizontal, vertical, and barrel (or cylindrical) systems are shown in Fig. 20.20. All of these systems are of the cold-wall type; the heating is done via RF energy or radiant energy, where the energy is coupled only into the wafers or wafer holders. Consequently, the chamber remains relatively cool, thus preventing deposition on the chamber walls.

A major problem with these types of reactors is that the molecules taking part in the reaction must diffuse through a stagnant boundary layer which surrounds each wafer. This boundary layer can be made thin by placing the wafers parallel to the gas flow and maintaining a high gas flow rate. However, these measures tend to reduce the throughput and increase gas consumption. Fur-

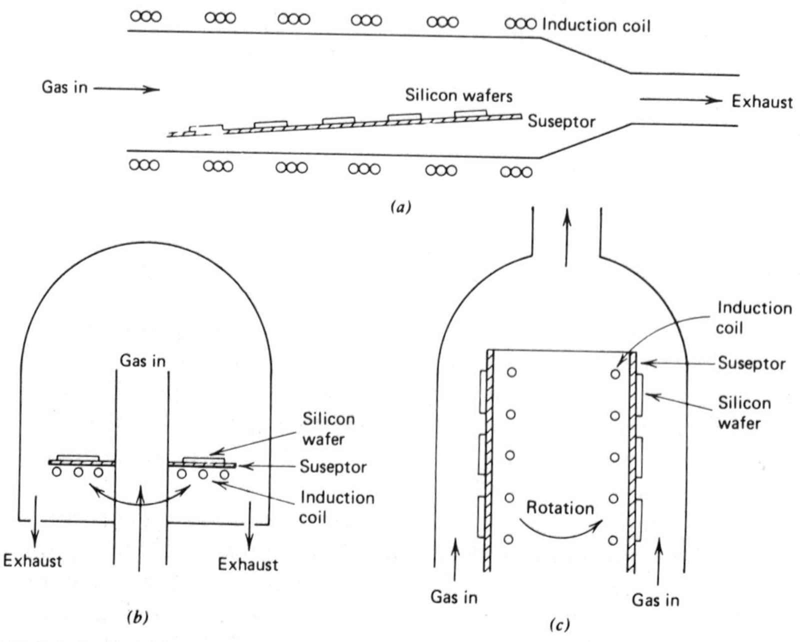

Figure 20.20 Epitaxial reactor systems. (a) Horizontal. (b) Vertical. (c) Barrel.

Figure 20.21 An example of a continuous operation reactor. (Courtesy AMS.)

thermore, if the boundary layer thickness is nonuniform, the deposition rate, which falls in the range of several hundred to approximately 1000 Å/min, will be nonuniform.

One alternative to enhance throughput and uniformity is a continuous operation system shown in Fig. 20.21. By avoiding batch operation, the throughput is increased, and by subjecting all wafers to exactly the same processing, uniformity is increased.

Probably the most significant improvement in throughput and uniformity, however, was brought about through low-pressure CVD (LPCVD). As mentioned earlier, the limiting step in CVD at atmospheric pressure is the transport or diffusion of the molecules through the stagnant boundary layer. If the total pressure of the system is reduced, there are fewer molecules per unit volume and hence less interference between molecules. This has the effect of drastically increasing the transport of the source molecules to the substrate surface.

Because the transport through the ambient is so rapid, now the reaction at the substrate surface is the rate-limiting step. This reaction is primarily temperature dependent, maintaining a uniform temperature across the wafers and

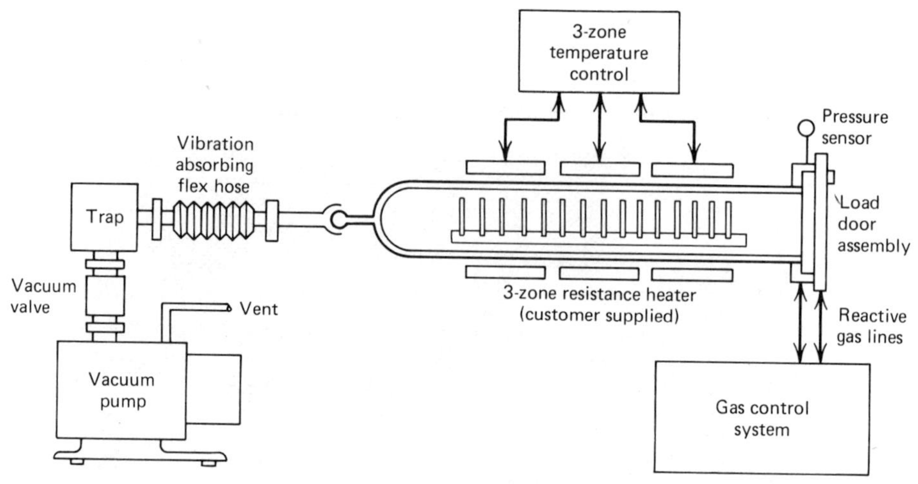

Figure 20.22 A low-pressure CVD (LPCVD) system.

from wafer to wafer, thus allowing a very uniform deposition. Furthermore, since the gas transport is no longer a problem, the wafers can be stacked perpendicular to the gas flow just as in a diffusion process. In fact, because there is no enhanced deposition on the chamber walls, such as occurs in atmospheric systems, hot-wall systems can be used. Many LPCVD systems presently used are in fact diffusion furnaces fitted with the appropriate gas mixing and vacuum pumping systems (Fig. 20.22).

Despite the reduced pressure, where total pressures of less than 1 torr (1 torr = 1/760 atmosphere = 1 mm Hg) are used, owing to the rapid gaseous transport the growth rate can still be maintained at values as high as several hundred angstroms per minute. This is slightly less than some atmospheric systems, but batches of 100 to 200 wafers can be accommodated in LPCVD systems, substantially increasing the throughput.

Typical deposition temperatures for an LPCVD system are in the range of 600 to 900°C. In practice, owing to slight variations in the reactions occurring down the length of the tube, a slight temperature gradient may improve uniformity from wafer to wafer. Other factors which can be used to improve uniformity and deposition rate are the gas pressure, composition, and flow rate. At present, most of the conditions yielding optimum characteristics have been empirically derived.

In an effort to reduce operating temperatures even further, plasma-enhanced CVD and photoexcited CVD are being pursued. The concept behind these processes is that at temperatures insufficient to bring about the appropriate reactions, techniques other than thermal can be used to excite the molecules and stimulate the various reactions. In one case, energetic electrons in a glow discharge plasma are used, and in the other case, ultraviolet light energy is used. The plasma-enhanced CVD is used in the temperature range of 300 to 400°C, while the photoexcited deposition can be done at temperatures as low as 100°C.

The uses of CVD films are listed in Table 20.5. Most of the uses are self-explanatory; however, a few may need further elaboration. When SiO_2 is used as a dopant source, the dopant is incorporated during CVD growth by adding an appropriate dopant source, such as diborane or phosphine. After deposition of the film, the wafer is heated in a furnace, allowing the dopant to diffuse into the silicon.

Table 20.5 **Uses of CVD films**

SiO_2
- Diffusion mask
- Insulation between conducting layers
- MOS gate
- Dopant source (dopant atoms introduced during CVD)
- Passivation of finished devices

Si_3N_4
- Diffusion mask
- Oxide mask for local oxidation
- MIS gate
- Passivation of finished devices
- Gettering of impurities and defects

Polysilicon
- Interconnections
- High-value resistors
- Dopant source
- Gettering of impurities and defects

For the passivation of finished devices, a layer of either SiO_2 or Si_3N_4 is deposited over the entire device. Sometimes, to prevent alkali contaminant motion, phosphorus is added to the SiO_2 film. This is referred to as a phosphosilicate glass (PSG).

The gettering mentioned for the nitride and polysilicon is a result of the stress and the consequent strain in the silicon introduced by these layers. The strained silicon acts as an effective sink for impurities and defects. The films are typically deposited on the back side of the wafers.

The new technique of photoexcited CVD has recently allowed the deposition of a nitride film for passivation of a fully wire-bonded hybrid microcircuit. In general, it can safely be stated that as new CVD technologies emerge, new uses and new materials will be the ultimate outcome.

20.7 BIPOLAR IC PROCESS

To show how the various processes are combined, the fabrication of a standard, junction-isolated bipolar IC will now be briefly described. The major steps in

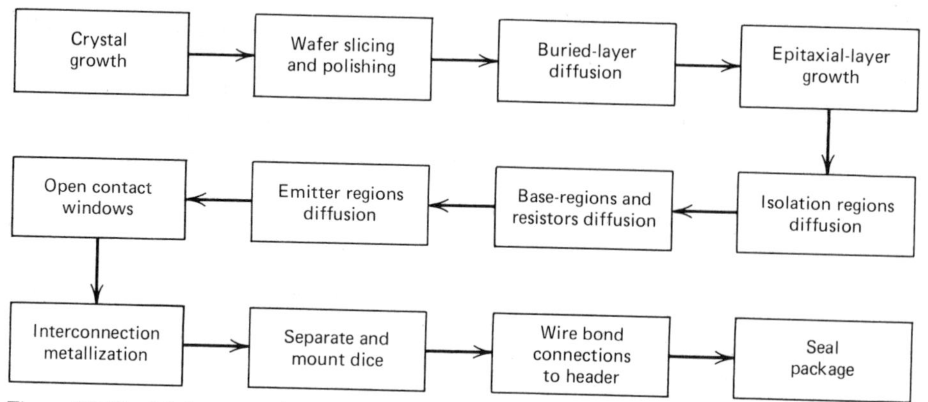

Figure 20.23 Major steps in bipolar IC fabrication process.

the IC bipolar process are listed in Fig. 20.23. The starting material is typically a p-type, single-crystal, silicon wafer, 5 to 20 Ω-cm resistivity, with thickness of approximately several hundred micrometers. The diameter can be 50, 75, 100, 125, or 150 mm; the most standard size at present is 100 mm (4 inches).

By epitaxial deposition, an n-type layer, doped so that its resistivity is in the range of 0.1 to 1 Ω-cm, is grown on top of the silicon wafer. This layer can range in thickness from several micrometers to approximately 10 μm. The crystallographic orientation of this epitaxial layer exactly follows that of the starting wafer.

The wafer with the epitaxial layer is then oxidized at an elevated temperature in an H_2O ambient. This forms a layer of SiO_2, approximately 0.5 μm thick, over the entire surface of the silicon.

A photosensitive material (*photoresist*), is then deposited onto the front of the wafer. This film is approximately 1 μm thick. The film is then exposed to ultraviolet (UV) light through a "mask" which contains a particular "isolation" pattern. For the particular photoresist used in this example, developing has the effect of etching away the exposed regions. Subsequent immersion into an appropriate acid etches the underlying SiO_2 in the region unprotected by the resist, while leaving the resist and protected oxide unaffected. The resist is then removed in a solvent, leaving the pattern in the SiO_2. The process just described is called *photolithography* and is repeated many times during IC fabrication.

The wafer with the patterned SiO_2 is then placed into a furnace containing a p-type dopant, typically boron, in the gas stream. At these high temperatures the boron diffuses into the silicon in the regions unprotected by oxide. If the diffusion time and temperature are appropriately chosen, the dopant will diffuse through the entire epitaxial layer, leaving isolated n-type regions. Into each of these regions will go a particular component, such as a transistor.

The wafer is again thermally oxidized, and via photolithography, the region for the base is defined. Again, in a high temperature furnace, boron is diffused into the silicon, but this time not as deeply.

After another oxidation and photolithographic process, the emitter is dif-

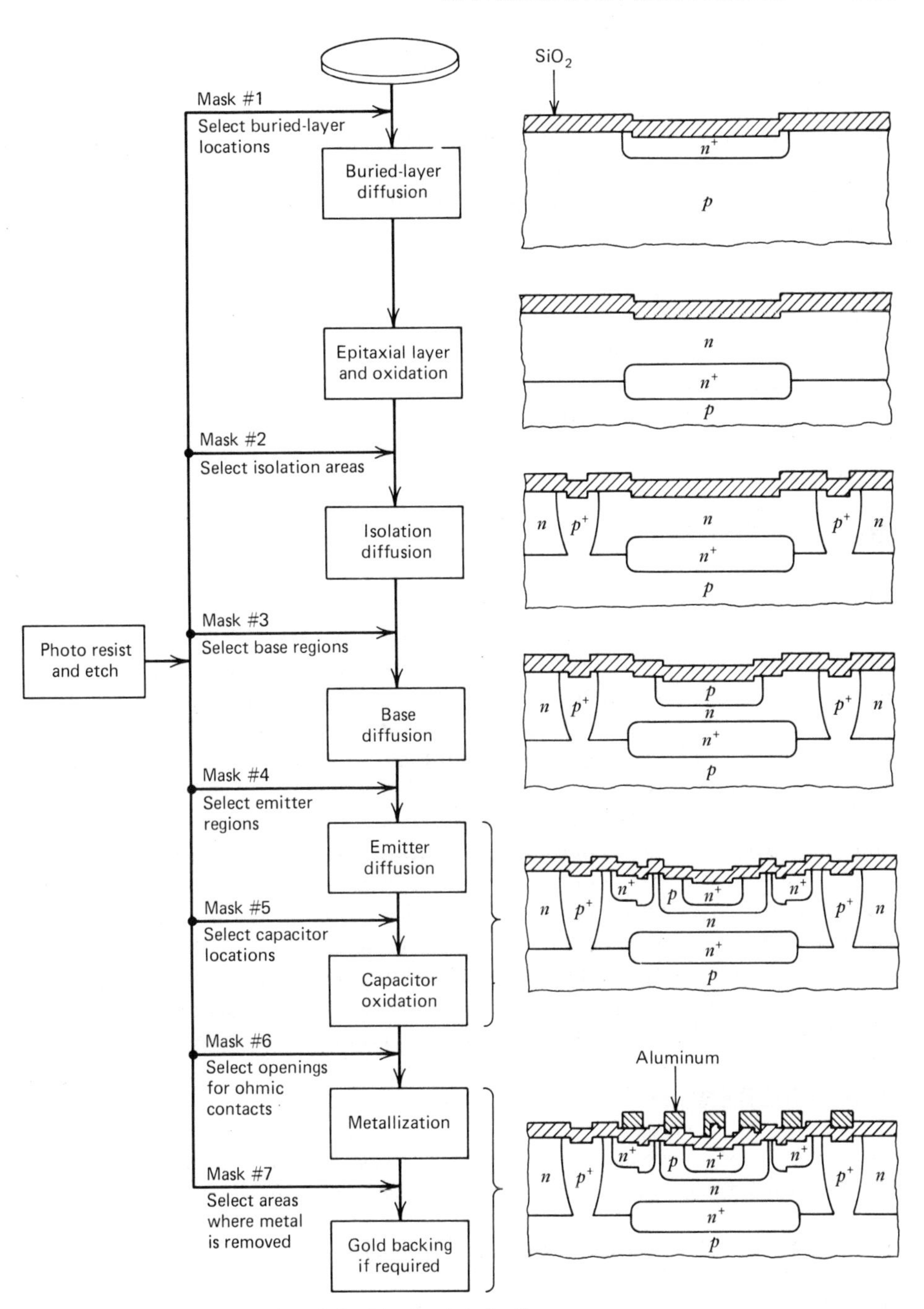

Figure 20.24 An example of the bipolar fabrication process.

fused into the silicon. At the same time, a heavily doped n-type region (n^+) is diffused into the collector region. This is to allow a low resistivity contact.

An oxide is again thermally grown over the entire wafer, and via photolithography those regions where contact is to be made to the silicon are defined. Metal is then deposited by vacuum evaporation. The photolithographic process is then used to define the appropriate metallization interconnect pattern, and the remaining metal is removed. At this point, except for a passivating glass layer which would be deposited over the whole wafer by chemical vapor deposition, the ICs are finished. The IC chips are then cut apart, typically with a diamond saw, tested, and packaged.

In general, each type of IC has its own process, but every IC process will contain the general scheme of film growth, patterning, and selective addition of material, such as a dopant. Moreover, these steps will be repeated a number of times for every IC. An example of the bipolar fabrication process is illustrated in Fig. 20.24.

20.8 THE FUTURE

The impact of IC technology on telecommunications, home computers, office systems, and automobiles will be significant in the next 10 years. The trends toward higher density circuits will require improvements in IC fabrication technology. The need for higher density will drive technological developments in patterning, junction formation, interconnection, and insulating films. However, the key factor will be the ability to control and/or eliminate defects in order to achieve a reasonable yield of good chips. Because the overall yield is the product of the individual yields, the need to eliminate human error also will be important and most probably lead to automated fabrication facilities.

Although the past achievements have been significant, those needed to reach the density of future circuits will require more effort since the increment gains will most probably be smaller. More understanding of the fundamental chemistry of processing will need to be developed in order to be able to predict electrical performance. And, as in the past, increased production will serve to reduce unit cost of a chip which will then open new areas of application.

20.9 REFERENCES

Colclaser, R. A., *Microelectronics: Processing and Device Design,* Wiley, New York, 1980.

Gise, P. E. and R. Blanchard, *Semiconductor and Integrated Circuit Fabrication Techniques,* Reston, Reston, VA, 1979.

Glaser, A. B. and G. E. Subak-Sharpe, *Integrated Circuit Engineering,* Addison-Wesley, Reading, MA, 1977.

Grebene, A. B., *Analog Integrated Circuit Design,* VNR, New York, 1972.

Grove, A. S., *Physics and Technology of Semiconductor Devices,* Wiley, New York, 1967.

Hamilton, D. J. and W. G. Howard, *Basic Integrated Circuit Engineering,* McGraw-Hill, New York, 1975.

Roddy, D., *Introduction to Microelectronics,* second edition, Pergamon, New York, 1978.

INDEX